continued on back endsheets

Seventh Edition

Calculus

An Applied Approach

RON LARSON
The Pennsylvania State University
The Behrend College

BRUCE H. EDWARDS
University of Florida

with the assistance of
DAVID C. FALVO
The Pennsylvania State University
The Behrend College

HOUGHTON MIFFLIN COMPANY
Boston   New York

Publisher: Jack Shira
Associate Sponsoring Editor: Cathy Cantin
Development Manager: Maureen Ross
Development Editor: David George
Editorial Assistant: Elizabeth Kassab
Supervising Editor: Karen Carter
Senior Project Editor: Patty Bergin
Editorial Assistant: Julia Keller
Production Technology Supervisor: Gary Crespo
Senior Marketing Manager: Danielle Potvin Curran
Marketing Coordinator: Nicole Mollica
Senior Manufacturing Coordinator: Marie Barnes

We have included examples and exercises that use real-life data as well as technology output from a variety of software. This would not have been possible without the help of many people and organizations. Our wholehearted thanks goes to all for their time and effort.

Trademark Acknowledgments: TI is a registered trademark of Texas Instruments, Inc. Mathcad is a registered trademark of MathSoft, Inc. Windows, Microsoft, Excel, and MS-DOS are registered trademarks of Microsoft, Inc. Mathematica is a registered trademark of Wolfram Research, Inc. DERIVE is a registered trademark of Soft Warehouse, Inc. IBM is a registered trademark of International Business Machines Corporation. Maple is a registered trademark of the University of Waterloo. Graduate Record Examinations and GRE are registered trademarks of Educational Testing Service. Graduate Management Admission Test and GMAT are registered trademarks of the Graduate Management Admission Council.

Cover credit: © Ryan McVay/Getty Images

Printed in the United States of America

Library of Congress Catalog Number: 2004116465

ISBN 0-618-54718-5

456789-DOW-10 09 08 07 06

Contents

*Available at the text-specific website at *college.hmco.com.*

A Word from the Authors

Welcome to *Calculus: An Applied Approach*, Seventh Edition. In this revision, we have focused on making the text even more student-oriented. To encourage mastery and understanding, we have outlined a straightforward program of study with continual reinforcement and applicability to the real world.

Student-Oriented Approach

Each chapter begins with "What you should learn" and "Why you should learn it." The "What you should learn" is a list of *Objectives* that students will examine in the chapter. The "Why you should learn it" lists sample applications that appear throughout the chapter. Each section begins with a list of learning *Objectives*, enabling students to identify and focus on the key points of the section.

Following every example is a *Try It* exercise. The new problem allows for students to immediately practice the concept learned in the example.

It is crucial for a student to understand an algebraic concept before attempting to master a related calculus concept. To help students in this area, *Algebra Review* tips appear at point of use throughout the text. A two-page *Algebra Review* appears at the end of each chapter, which emphasizes key algebraic concepts discussed in the chapter.

Before students are exposed to selected topics, *Discovery* projects allow them to explore concepts on their own, making them more likely to remember the results. These optional boxed features can be omitted, if the instructor desires, with no loss of continuity in the coverage of the material.

Throughout the text, *Study Tips* address special cases, expand on concepts, and help students avoid common errors. *Side Comments* help explain the steps of a solution. State-of-the-art graphics help students with visualization, especially when working with functions of several variables.

Advances in *Technology* are helping to change the world around us. We have updated and increased technology coverage to be even more readily available at point of use. Students are encouraged to use a graphing utility, computer program, or spreadsheet software as a tool for exploration, discovery, and problem solving. Students are not required to have access to a graphing utility to use this text effectively. In addition to describing the benefits of using technology, the text also pays special attention to its possible misuse or misinterpretation.

Just before each section exercise set, the *Take Another Look* feature asks students to look back at one or more concepts presented in the section, using questions designed to enhance understanding of key ideas.

Each chapter presents many opportunities for students to assess their progress, both at the end of each section (*Prerequisite Review* and *Section Exercises*) and at the end of each chapter (*Chapter Summary*, *Study Strategies*, *Study Tools*, and *Review Exercises*). The test items in *Sample Post-Graduation Exam Questions* show the relevance of calculus. The test questions are representative of types of questions on several common post-graduation exams.

Business Capsules appear at the ends of numerous sections. These capsules and their accompanying exercises deal with business situations that are related to the mathematical concepts covered in the chapter.

Application to the Changing World Around Us

Students studying calculus need to understand how the subject matter relates to the real world. In this edition, we have focused on increasing the variety of applications, especially in the life sciences, economics, and finance. All real-data applications have been revised to use the most current information available. Exercises containing material from textbooks in other disciplines have been included to show the relevance of calculus in other areas. In addition, exercises involving the use of spreadsheets have been incorporated throughout.

We hope you enjoy the Seventh Edition. A readable text with a straightforward approach, it provides effective study tools and direct application to the lives and futures of calculus students.

Ron Larson

Bruce H. Edwards

Supplements

The integrated learning system for *Calculus: An Applied Approach*, Seventh Edition, addresses the changing needs of today's instructors and students, offering dynamic teaching tools for instructors and interactive learning resources for students in print, CD-ROM, and online formats.

Resources

Eduspace®, Houghton Mifflin's Online Learning Tool

Eduspace® is an online learning environment that combines algorithmic tutorials, homework capabilities, and testing. Text-specific content, organized by section, is available to help students understand the mathematics covered in this text.

For the Instructor

Instructor ClassPrep CD-ROM with HM Testing (Windows, Macintosh)

ClassPrep offers complete instructor solutions and other instructor resources. *HM Testing* is a computerized test generator with algorithmically generated test items.

Instructor Website (math.college.hmco.com/instructors)

This website contains pdfs of the *Complete Solutions Guide* and *Test Item File and Instructor's Resource Guide*. Digital Figures and Lessons are available (ppts) for use as handouts or slides.

For the Student

HM mathSpace® Student CD-ROM

HM mathSpace contains a prerequisite algebra review, a link to our online graphing calculator, and graphing calculator programs.

Excel Made Easy: Video Instruction with Activities CD-ROM

Excel Made Easy uses easy-to-follow videos to help students master mathematical concepts introduced in class. The CD-ROM includes electronic spreadsheets and detailed tutorials.

SMARTHINKING™ Online Tutoring

Instructional Video and DVD Series by Dana Mosely

The video and DVD series complement the textbook topic coverage should a student struggle with the calculus concepts or miss a class.

Student Solutions Guide

This printed manual features step-by-step solutions to the odd-numbered exercises. A practice test with full solutions is available for each chapter.

Excel Guide for Finite Math and Applied Calculus

The *Excel Guide* provides useful information, including step-by-step examples and sample exercises.

Student Website (math.college.hmco.com/students)

The website contains self-quizzing content to help students strengthen their calculus skills, a link to our online graphing calculator, graphing calculator programs, and printable formula cards.

Acknowledgments

We would like to thank the many people who have helped us at various stages of this project during the past 24 years. Their encouragement, criticisms, and suggestions have been invaluable to us.

A special note of thanks goes to the instructors who responded to our survey and to all the students who have used the previous editions of the text.

Reviewers of the Seventh Edition

Scott Perkins
Lake Sumter Community College

Jose Gimenez
Temple University

Keng Deng
University of Louisiana at Lafayette

George Anastassiou
University of Memphis

Peggy Luczak
Camden County College

Bernadette Kocyba
J. Sergeant Reynolds Community College

Shane Goodwin
Brigham Young University of Idaho

Harvey Greenwald
California Polytechnic State University

Randall McNiece
San Jacinto College

Reviewers of Previous Editions

Carol Achs, *Mesa Community College*; David Bregenzer, *Utah State University*; Mary Chabot, *Mt. San Antonio College*; Joseph Chance, *University of Texas—Pan American*; John Chuchel, *University of California*; Miriam E. Connellan, *Marquette University*; William Conway, *University of Arizona*; Karabi Datta, *Northern Illinois University*; Roger A. Engle, *Clarion University of Pennsylvania*; Betty Givan, *Eastern Kentucky University*; Mark Greenhalgh, *Fullerton College*; Karen Hay, *Mesa Community College*; Raymond Heitmann, *University of Texas at Austin*; William C. Huffman, *Loyola University of Chicago*; Arlene Jesky, *Rose State College*; Ronnie Khuri, *University of Florida*; Duane Kouba, *University of California—Davis*; James A. Kurre, *The Pennsylvania State University*; Melvin Lax, *California State University—Long Beach*; Norbert Lerner, *State University of New York at Cortland*; Yuhlong Lio, *University of South Dakota*; Peter J. Livorsi, *Oakton Community College*; Samuel A. Lynch, *Southwest Missouri State University*; Kevin McDonald, *Mt. San Antonio College*; Earl H. McKinney, *Ball State University*; Philip R. Montgomery, *University of Kansas*; Mike Nasab, *Long Beach City College*; Karla Neal, *Louisiana State University*; James Osterburg, *University of Cincinnati*; Rita Richards, *Scottsdale Community College*; Stephen B. Rodi, *Austin Community College*; Yvonne Sandoval-Brown, *Pima Community College*; Richard Semmler, *Northern Virginia Community College—Annandale*; Bernard Shapiro, *University of Massachusetts, Lowell*; Jane Y. Smith, *University of Florida*; DeWitt L. Sumners, *Florida State University*; Jonathan Wilkin, *Northern Virginia Community College*; Carol G. Williams, *Pepperdine University*; Melvin R. Woodard, *Indiana University of Pennsylvania*; Carlton Woods, *Auburn University at Montgomery*; Jan E. Wynn, *Brigham Young University*; Robert A. Yawin, *Springfield Technical Community College*; Charles W. Zimmerman, *Robert Morris College*

Our thanks to David Falvo, The Behrend College, The Pennsylvania State University, for his contributions to this project. Our thanks also to Robert Hostetler, The Behrend College, The Pennsylvania State University, for his significant contributions to previous editions of this text.

We would also like to thank the staff at Larson Texts, Inc. who assisted with proofreading the manuscript, preparing and proofreading the art package, and checking and typesetting the supplements.

On a personal level, we are grateful to our spouses, Deanna Gilbert Larson and Consuelo Edwards, for their love, patience, and support. Also, a special thanks goes to R. Scott O'Neil.

If you have suggestions for improving this text, please feel free to write to us. Over the past two decades we have received many useful comments from both instructors and students, and we value these comments very highly.

Ron Larson

Bruce H. Edwards

Features

CHAPTER OPENERS

Each chapter opens with *Strategies for Success*, a checklist that outlines what students should learn and lists several applications of those objectives. Each chapter opener also contains a list of the section topics and a photo referring students to an interesting application in the section exercises.

chapter

2

Differentiation

2.1 The Derivative and the Slope of a Graph

2.2 Some Rules for Differentiation

2.3 Rates of Change: Velocity and Marginals

2.4 The Product and Quotient Rules

2.5 The Chain Rule

2.6 Higher-Order Derivatives

2.7 Implicit Differentiation

2.8 Related Rates

Higher-order derivatives are used to determine the acceleration function of a sports car. The acceleration function shows the changes in the car's velocity. As the car reaches its "cruising" speed, is the acceleration increasing or decreasing?

STRATEGIES FOR SUCCESS

WHAT YOU SHOULD LEARN:

- How to find the slope of a graph and calculate derivatives using the limit definition
- How to use the Constant Rule, Power Rule, Constant Multiple Rule, and Sum and Difference Rules
- How to find rates of change: velocity, marginal profit, marginal revenue, and marginal cost
- How to use the Product, Quotient, Chain, and General Power Rules
- How to calculate higher-order derivatives and derivatives using implicit differentiation
- How to solve related-rate problems and applications

WHY YOU SHOULD LEARN IT:

Derivatives have many applications in real life, as can be seen by the examples below, which represent a small sample of the applications in this chapter.

- Increasing Revenue, Example 10 on page 101
- Psychology: Migraine Prevalence, Exercise 62 on page 104
- Average Velocity, Exercises 15 and 16 on page 117
- Demand Function, Exercises 53 and 54 on page 129
- Quality Control, Exercise 58 on page 129
- Velocity and Acceleration, Exercises 41–44 and 50 on pages 145 and 146

81

2.5   THE CHAIN RULE

- Find derivatives using the Chain Rule.
- Find derivatives using the General Power Rule.
- Write derivatives in simplified form.
- Use derivatives to answer questions about real-life situations.
- Use the differentiation rules to differentiate algebraic functions.

The Chain Rule

In this section, you will study one of the most powerful rules of differential calculus—the **Chain Rule.** This differentiation rule deals with composite functions and adds versatility to the rules presented in Sections 2.2 and 2.4. For example, compare the functions below. Those on the left can be differentiated without the Chain Rule, whereas those on the right are best done with the Chain Rule.

Without the Chain Rule	*With the Chain Rule*
$y = x^2 + 1$	$y = \sqrt{x^2 + 1}$
$y = x + 1$	$y = (x + 1)^{-1/2}$
$y = 3x + 2$	$y = (3x + 2)^5$
$y = \dfrac{x + 5}{x^2 + 2}$	$y = \left(\dfrac{x + 5}{x^2 + 2}\right)^2$

The Chain Rule

If $y = f(u)$ is a differentiable function of u, and $u = g(x)$ is a differentiable function of x, then $y = f(g(x))$ is a differentiable function of x, and

$$\frac{dy}{dx} = \frac{dy}{du} \cdot \frac{du}{dx}$$

or, equivalently,

$$\frac{d}{dx}[f(g(x))] = f'(g(x))g'(x).$$

Basically, the Chain Rule states that if y changes dy/du times as fast as u, and u changes du/dx times as fast as x, then y changes

$$\frac{dy}{du} \cdot \frac{du}{dx}$$

times as fast as x, as illustrated in Figure 2.28. One advantage of the dy/dx notation for derivatives is that it helps you remember differentiation rules, such as the Chain Rule. For instance, in the formula

$$dy/dx = (dy/du)(du/dx)$$

you can imagine that the du's divide out.

x

Input

Function g

Output

Rate of change of *u* with respect to *x* is $\dfrac{du}{dx}$

$u = g(x)$

u

Input

Function f

Output

Rate of change of *y* with respect to *u* is $\dfrac{dy}{du}$

$y = f(u) = f(g(x))$

Rate of change of *y* with respect to *x* is $\dfrac{dy}{dx} = \dfrac{dy}{du}\dfrac{du}{dx}$

FIGURE 2.28

SECTION OBJECTIVES

Each section begins with a list of objectives covered in that section. This outline helps instructors with class planning and students in studying the material in the section.

DEFINITIONS AND THEOREMS

All definitions and theorems are highlighted for emphasis and easy reference.

EXAMPLES

To increase the usefulness of the text as a study tool, the Seventh Edition presents a wide variety of examples, each titled for easy reference. Many of these detailed examples display solutions that are presented graphically, analytically, and/or numerically to provide further insight into mathematical concepts. Side comments clarify the steps of the solution as necessary. Examples using real-life data are identified with a globe icon and are accompanied by the types of illustrations that students are used to seeing in newspapers and magazines.

TRY ITS

Appearing after every example, these new problems help students reinforce concepts right after they are presented.

DISCOVERY

Before students are exposed to selected topics, *Discovery* projects allow them to explore concepts on their own, making them more likely to remember the results. These optional boxed features can be omitted, if the instructor desires, with no loss of continuity in the coverage of material.

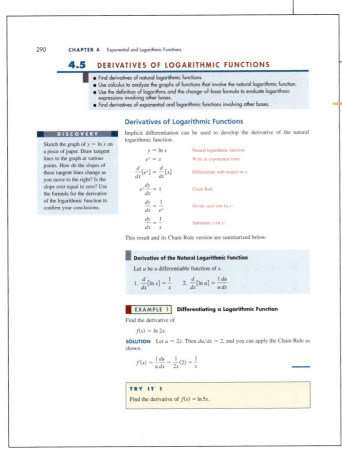

ALGEBRA REVIEWS

Algebra Reviews appear throughout each chapter and offer students algebraic support at point of use. These smaller reviews are then revisited in the Algebra Review at the end of each chapter, where additional details of examples with solutions and explanations are provided.

---

176   **CHAPTER 3**   Applications of the Derivative

Not only is the function in Example 3 continuous on the entire real line, it is also differentiable there. For such functions, the only critical numbers are those for which $f'(x) = 0$. The next example considers a continuous function that has *both* types of critical numbers—those for which $f'(x) = 0$ and those for which f' is undefined.

ALGEBRA REVIEW

For help on the algebra in Example 4, see Example 2(d) in the *Chapter 3 Algebra Review*, on page 249.

EXAMPLE 4   **Finding Increasing and Decreasing Intervals**

Find the open intervals on which the function

$$f(x) = (x^2 - 4)^{2/3}$$

is increasing or decreasing.

SOLUTION   Begin by finding the derivative of the function.

$$f'(x) = \frac{2}{3}(x^2 - 4)^{-1/3}(2x)$$ Differentiate.

$$= \frac{4x}{3(x^2 - 4)^{1/3}}$$ Simplify.

From this, you can see that the derivative is zero when $x = 0$ and the derivative is undefined when $x = \pm 2$. So, the critical numbers are

$$x = -2, \quad x = 0, \quad \text{and} \quad x = 2.$$ Critical numbers

This implies that the test intervals are

$$(-\infty, -2), \quad (-2, 0), \quad (0, 2), \quad \text{and} \quad (2, \infty).$$ Test intervals

The table summarizes the testing of these four intervals, and the graph of the function is shown in Figure 3.6.

Interval	$-\infty < x < -2$	$-2 < x < 0$	$0 < x < 2$	$2 < x < \infty$
Test value	$x = -3$	$x = -1$	$x = 1$	$x = 3$
Sign of $f'(x)$	$f'(-3) < 0$	$f'(-1) > 0$	$f'(1) < 0$	$f'(3) > 0$
Conclusion	Decreasing	Increasing	Decreasing	Increasing

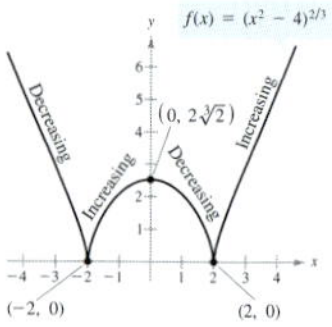

FIGURE 3.6

TRY IT 4

Find the open intervals on which the function $f(x) = x^{2/3}$ is increasing or decreasing.

ALGEBRA REVIEW

To test the intervals in the table, it is not necessary to *evaluate* $f'(x)$ at each test value—you only need to determine its sign. For example, you can determine the sign of $f'(-3)$ as shown.

$$f'(-3) = \frac{4(-3)}{3(9-4)^{1/3}} = \frac{\text{negative}}{\text{positive}} = \text{negative}$$

---

320   **CHAPTER 5**   Integration and Its Applications

Finding Antiderivatives

The inverse relationship between the operations of integration and differentiation can be shown symbolically, as shown.

$$\frac{d}{dx}\left[\int f(x)\,dx\right] = f(x)$$ Differentiation is the inverse of integration.

$$\int f'(x)\,dx = f(x) + C$$ Integration is the inverse of differentiation.

This inverse relationship between integration and differentiation allows you to obtain integration formulas directly from differentiation formulas. The following summary lists the integration formulas that correspond to some of the differentiation formulas you have studied.

Basic Integration Rules

1. $\int k\,dx = kx + C, \quad k$ is a constant.   Constant Rule

2. $\int kf(x)\,dx = k\int f(x)\,dx$ Constant Multiple Rule

3. $\int [f(x) + g(x)]\,dx = \int f(x)\,dx + \int g(x)\,dx$ Sum Rule

4. $\int [f(x) - g(x)]\,dx = \int f(x)\,dx - \int g(x)\,dx$ Difference Rule

5. $\int x^n\,dx = \frac{x^{n+1}}{n+1} + C, \quad n \neq -1$ Simple Power Rule

STUDY TIP

You will study the General Power Rule for integration in Section 5.2 and the Exponential and Log Rules in Section 5.3.

STUDY TIP

In Example 2(b), the integral $\int 1\,dx$ is usually shortened to the form $\int dx$.

TRY IT 2

Find each indefinite integral.

(a) $\int 5\,dx$

(b) $\int -1\,dr$

(c) $\int 2\,dt$

Be sure you see that the Simple Power Rule has the restriction that n cannot be -1. So, you *cannot* use the Simple Power Rule to evaluate the integral

$$\int \frac{1}{x}\,dx.$$

To evaluate this integral, you need the Log Rule, which is described in Section 5.3.

EXAMPLE 2   **Finding Indefinite Integrals**

Find each indefinite integral.

(a) $\int \frac{1}{2}\,dx$ (b) $\int 1\,dx$ (c) $\int -5\,dt$

SOLUTION

(a) $\int \frac{1}{2}\,dx = \frac{1}{2}x + C$ (b) $\int 1\,dx = x + C$ (c) $\int -5\,dt = -5t + C$

STUDY TIPS

Throughout the text, *Study Tips* help students avoid common errors, address special cases, and expand on theoretical concepts.

TAKE ANOTHER LOOK

Starting with Chapter 1, each section in the text closes with a *Take Another Look* problem asking students to look back at one or more concepts presented in the section, using questions designed to enhance understanding of key ideas. These problems can be completed as group projects in class or as homework assignments. Because these problems encourage students to think, reason, and write about calculus, they emphasize the synthesis or the further exploration of the concepts presented in the section.

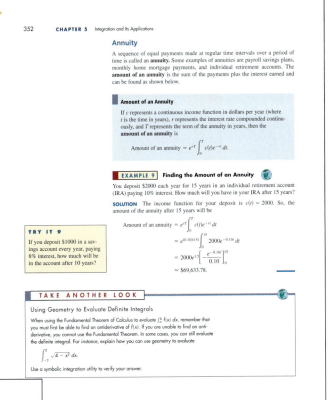

Annuity

A sequence of equal payments made at regular time intervals over a period of time is called an **annuity.** Some examples of annuities are payroll savings plans, monthly home mortgage payments, and individual retirement accounts. The **amount of an annuity** is the sum of the payments plus the interest earned and can be found as shown below.

Amount of an Annuity

If c represents a continuous income function in dollars per year (where t is the time in years), r represents the interest rate compounded continuously, and T represents the term of the annuity in years, then the **amount of an annuity** is

$$\text{Amount of an annuity} = e^{rT} \int_0^T c(t)e^{-rt}\, dt.$$

EXAMPLE 9   **Finding the Amount of an Annuity**

You deposit $2000 each year for 15 years in an individual retirement account (IRA) paying 10% interest. How much will you have in your IRA after 15 years?

SOLUTION   The income function for your deposit is $c(t) = 2000$. So, the amount of the annuity after 15 years will be

$$
\begin{aligned}
\text{Amount of an annuity} &= e^{rT}\int_0^T c(t)e^{-rt}\, dt \\
&= e^{(0.10)(15)}\int_0^{15} 2000 e^{-0.10t}\, dt \\
&= 2000e^{1.5}\left[-\frac{e^{-0.10t}}{0.10}\right]_0^{15} \\
&\approx \$69{,}633.78.
\end{aligned}
$$

TRY IT 9

If you deposit $1000 in a savings account every year, paying 8% interest, how much will be in the account after 10 years?

TAKE ANOTHER LOOK

Using Geometry to Evaluate Definite Integrals

When using the Fundamental Theorem of Calculus to evaluate $\int_a^b f(x)\, dx$, remember that you must first be able to find an antiderivative of $f(x)$. If you are unable to find an antiderivative, you cannot use the Fundamental Theorem. In some cases, you can still evaluate the definite integral. For instance, explain how you can use geometry to evaluate

$$\int_{-2}^{2} \sqrt{4 - x^2}\, dx.$$

Use a symbolic integration utility to verify your answer.

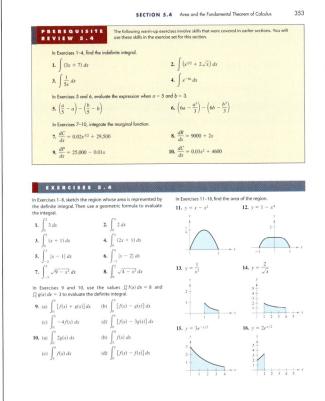

PREREQUISITE REVIEW 5.4   The following warm-up exercises involve skills that were covered in earlier sections. You will use these skills in the exercise set for this section.

In Exercises 1–4, find the indefinite integral.

1. $\int (3x + 7)\, dx$
2. $\int (x^{3/2} + 2\sqrt{x})\, dx$
3. $\int \frac{1}{5x}\, dx$
4. $\int e^{-6x}\, dx$

In Exercises 5 and 6, evaluate the expression when $a = 5$ and $b = 3$.

5. $\left(\frac{a}{5} - a\right) - \left(\frac{b}{5} - b\right)$
6. $\left(6a - \frac{a^3}{3}\right) - \left(6b - \frac{b^3}{3}\right)$

In Exercises 7–10, integrate the marginal function.

7. $\frac{dC}{dx} = 0.02x^{3/2} + 29{,}500$
8. $\frac{dR}{dx} = 9000 + 2x$
9. $\frac{dP}{dx} = 25{,}000 - 0.01x$
10. $\frac{dC}{dx} = 0.03x^2 + 4600$

EXERCISES 5.4

In Exercises 1–8, sketch the region whose area is represented by the definite integral. Then use a geometric formula to evaluate the integral.

1. $\int_0^2 3\, dx$
2. $\int_0^4 2\, dx$
3. $\int_0^5 (x + 1)\, dx$
4. $\int_0^4 (2x + 1)\, dx$
5. $\int_{-2}^2 |x - 1|\, dx$
6. $\int_{-1}^4 |x - 2|\, dx$
7. $\int_{-3}^3 \sqrt{9 - x^2}\, dx$
8. $\int_{-2}^2 \sqrt{4 - x^2}\, dx$

In Exercises 9 and 10, use the values $\int_0^5 f(x)\, dx = 8$ and $\int_0^5 g(x)\, dx = 3$ to evaluate the definite integral.

9. (a) $\int_0^5 [f(x) + g(x)]\, dx$
   (b) $\int_0^5 [f(x) - g(x)]\, dx$
   (c) $\int_0^5 -4f(x)\, dx$
   (d) $\int_0^5 [f(x) - 3g(x)]\, dx$
10. (a) $\int_0^5 2g(x)\, dx$
    (b) $\int_5^0 f(x)\, dx$
    (c) $\int_5^5 f(x)\, dx$
    (d) $\int_0^5 [f(x) - f(x)]\, dx$

In Exercises 11–18, find the area of the region.

11. $y = x - x^2$
12. $y = 1 - x^4$
13. $y = \frac{1}{x^2}$
14. $y = \frac{2}{\sqrt{x}}$
15. $y = 3e^{-x/2}$
16. $y = 2e^{x/2}$

PREREQUISITE REVIEW

Starting with Chapter 1, each text section has a set of *Prerequisite Review* exercises. The exercises enable students to review and practice the previously learned skills necessary to master the new skills presented in the section. Answers to these sections appear in the back of the text.

EXERCISES

The text now contains almost 6000 exercises. Each exercise set is graded, progressing from skill-development problems to more challenging problems, to build confidence, skill, and understanding. The wide variety of types of exercises include many technology-oriented, real, and engaging problems. Answers to all odd-numbered exercises are included in the back of the text. To help instructors make homework assignments, many of the exercises in the text are labeled to indicate the area of application.

GRAPHING UTILITIES

Many exercises in the text can be solved using technology; however, the  symbol identifies all exercises for which students are specifically instructed to use a graphing utility, computer algebra system, or spreadsheet software.

TEXTBOOK EXERCISES

The Seventh Edition includes a number of exercises that contain material from textbooks in other disciplines, such as biology, chemistry, economics, finance, geology, physics, and psychology. These applications make the point to students that they will need to use calculus in future courses outside of the math curriculum. These exercises are identified by the 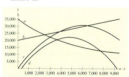 icon and are labeled to indicate the subject area.

SECTION 6.2   Integration by Parts and Present Value   405

(a) Use a graphing utility to decide whether the board of trustees expects the gift income to increase or decrease over the five-year period.

(b) Find the expected total gift income over the five-year period.

(c) Determine the average annual gift income over the five-year period. Compare the result with the income given when $t = 3$.

61. Learning Theory   A model for the ability M of a child to memorize, measured on a scale from 0 to 10, is

$$M = 1 + 1.6t \ln t, \quad 0 < t \le 4$$

where t is the child's age in years. Find the average value of this model between

(a) the child's first and second birthdays.

(b) the child's third and fourth birthdays.

62. Revenue   A company sells a seasonal product. The revenue R (in dollars per year) generated by sales of the product can be modeled by

$$R = 410.5t^2 e^{-t/30} + 25,000, \quad 0 \le t \le 365$$

where t is the time in days.

(a) Find the average daily receipts during the first quarter, which is given by $0 \le t \le 90$.

(b) Find the average daily receipts during the fourth quarter, which is given by $274 \le t \le 365$.

(c) Find the total daily receipts during the year.

Present Value   In Exercises 63–68, find the present value of the income c (measured in dollars) over t_1 years at the given annual inflation rate r.

63. $c = 5000$, $r = 5\%$, $t_1 = 4$ years

64. $c = 450$, $r = 4\%$, $t_1 = 10$ years

65. $c = 150,000 + 2500t$, $r = 4\%$, $t_1 = 10$ years

66. $c = 30,000 + 500t$, $r = 7\%$, $t_1 = 6$ years

67. $c = 1000 + 50e^{t/2}$, $r = 6\%$, $t_1 = 4$ years

68. $c = 5000 + 25te^{t/10}$, $r = 6\%$, $t_1 = 10$ years

69. Present Value   A company expects its income c during the next 4 years to be modeled by

$$c = 150,000 + 75,000t.$$

(a) Find the actual income for the business over the 4 years.

(b) Assuming an annual inflation rate of 4%, what is the present value of this income?

70. Present Value   A professional athlete signs a three-year contract in which the earnings can be modeled by

$$c = 300,000 + 125,000t.$$

(a) Find the actual value of the athlete's contract.

(b) Assuming an annual inflation rate of 5%, what is the present value of the contract?

Future Value   In Exercises 71 and 72, find the future value of the income (in dollars) given by $f(t)$ over t_1 years at the annual interest rate of r. If the function f represents a continuous investment over a period of t_1 years at an annual interest rate of r (compounded continuously), then the future value of the investment is given by

$$\text{Future value} = e^{rt_1} \int_0^{t_1} f(t)e^{-rt}\, dt.$$

71. $f(t) = 3000$, $r = 8\%$, $t_1 = 10$ years

72. $f(t) = 3000e^{0.05t}$, $r = 10\%$, $t_1 = 5$ years

73. Finance: Future Value   Use the equation from Exercises 71 and 72 to calculate the following. *(Source: Adapted from Garman/Forgue, Personal Finance, Fifth Edition)*

(a) The future value of $1200 saved each year for 10 years earning 7% interest.

(b) A person who wishes to invest $1200 each year finds one investment choice that is expected to pay 9% interest per year and another, riskier choice that may pay 10% interest per year. What is the difference in return (future value) if the investment is made for 15 years?

74. Consumer Awareness   In 2004, the total cost to attend Pennsylvania State University for 1 year was estimated to be $19,843. If your grandparents had continuously invested in a college fund according to the model

$$f(t) = 400t$$

for 18 years, at an annual interest rate of 10%, would the fund have grown enough to allow you to cover 4 years of expenses at Pennsylvania State University? *(Source: Pennsylvania State University)*

75. Use a program similar to the Midpoint Rule program on page 366 with $n = 10$ to approximate

$$\int_1^4 \frac{4}{\sqrt{x} + \sqrt[3]{x}}\, dx.$$

76. Use a program similar to the Midpoint Rule program on page 366 with $n = 12$ to approximate the volume of the solid generated by revolving the region bounded by the graphs of

$$y = \frac{10}{\sqrt{xe^x}}, \quad y = 0, \quad x = 1, \quad \text{and} \quad x = 4$$

about the x-axis.

SECTION 3.5   Business and Economics Applications   219

36. Minimum Cost   The ordering and transportation cost C of the components used in manufacturing a product is modeled by

$$C = 100\left(\frac{200}{x^2} + \frac{x}{x + 30}\right), \quad x \ge 1$$

where C is measured in thousands of dollars and x is the order size in hundreds. Find the order size that minimizes the cost. (*Hint:* Use the *root* feature of a graphing utility.)

37. Revenue   The demand for a car wash is

$$x = 600 - 50p$$

where the current price is $5.00. Can revenue be increased by lowering the price and thus attracting more customers? Use price elasticity of demand to determine your answer.

38. Revenue   Repeat Exercise 37 for a demand function of

$$x = 800 - 40p.$$

39. Demand   A demand function is modeled by $x = a/p^m$, where a is a constant and $m > 1$. Show that $\eta = -m$. In other words, show that a 1% increase in price results in an $m\%$ decrease in the quantity demanded.

40. Sales   The sales S (in millions of dollars per year) for Lowe's for the years 1994 through 2003 can be modeled by

$$S = 201.556t^2 - 502.29t + 2622.8 + \frac{9286}{t},$$

$$4 \le t \le 13$$

where $t = 4$ corresponds to 1994. *(Source: Lowe's Companies)*

(a) During which year, from 1994 to 2003, were Lowe's sales increasing most rapidly?

(b) During which year were the sales increasing at the lowest rate?

(c) Find the rate of increase or decrease for each year in parts (a) and (b).

(d) Use a graphing utility to graph the sales function. Then use the *zoom* and *trace* features to confirm the results in parts (a), (b), and (c).

41. Revenue   The revenue R (in millions of dollars per year) for Papa John's for the years 1994 through 2003 can be modeled by

$$R = \frac{-18.0 + 24.74t}{1 - 0.16t + 0.008t^2}, \quad 4 \le t \le 13$$

where $t = 4$ corresponds to 1994. *(Source: Papa John's Int'l.)*

(a) During which year, from 1994 to 2003, was Papa John's revenue the greatest? the least?

(b) During which year was the revenue increasing at the greatest rate? decreasing at the greatest rate?

(c) Use a graphing utility to graph the revenue function, and confirm your results in parts (a) and (b).

42. Match each graph with the function it best represents—a demand function, a revenue function, a cost function, or a profit function. Explain your reasoning. (The graphs are labeled a–d.)

43. Research Project   Choose an innovative product like the one described above. Use your school's library, the Internet, or some other reference source to research the history of the product or service. Collect data about the revenue that the product or service has generated, and find a mathematical model of the data. Summarize your findings.

BUSINESS CAPSULE

While graduate students, Elizabeth Elting and Phil Shawe co-founded TransPerfect Translations in 1992. They used a rented computer and a $5000 credit card cash advance to market their service-oriented translation firm, now one of the largest in the country. Currently, they have a network of 4000 certified language specialists in North America, Europe, and Asia, which translates technical, legal, business, and marketing materials. In 2004, the company estimates its gross sales will be $35 million.

BUSINESS CAPSULES

Business Capsules appear at the ends of numerous sections. These capsules and their accompanying exercises deal with business situations that are related to the mathematical concepts covered in the chapter.

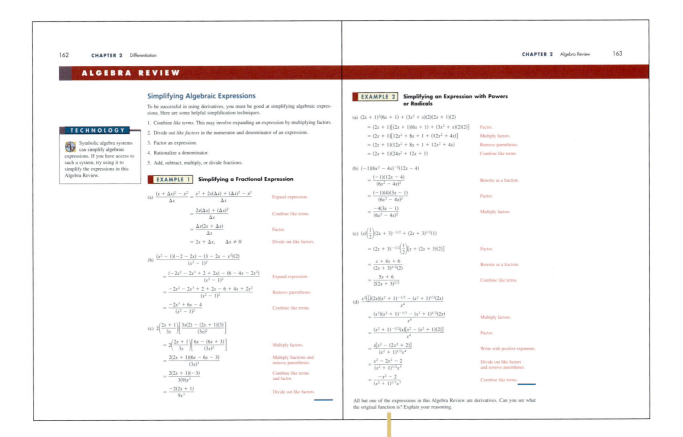

ALGEBRA REVIEW

At the end of each chapter, the *Algebra Review* illustrates the key algebraic concepts used in the chapter. Often, rudimentary steps are provided in detail for selected examples from the chapter. This review offers additional support to those students who have trouble following examples as a result of poor algebra skills.

4   CHAPTER SUMMARY AND STUDY STRATEGIES

After studying this chapter, you should have acquired the following skills. The exercise numbers are keyed to the Review Exercises that begin on page 312. Answers to odd-numbered Review Exercises are given in the back of the text.*

■ Use the properties of exponents to evaluate and simplify exponential expressions. *(Section 4.1 and Section 4.2)* — *Review Exercises 1–16*

$$a^0 = 1, \quad a^x a^y = a^{x+y}, \quad \frac{a^x}{a^y} = a^{x-y}, \quad (a^x)^y = a^{xy}$$

$$(ab)^x = a^x b^x, \quad \left(\frac{a}{b}\right)^x = \frac{a^x}{b^x}, \quad a^{-x} = \frac{1}{a^x}$$

■ Use properties of exponents to answer questions about real life. *(Section 4.1)* — *Review Exercises 17, 18*
■ Sketch the graphs of exponential functions. *(Section 4.1 and Section 4.2)* — *Review Exercises 19–28*
■ Evaluate limits of exponential functions in real life. *(Section 4.2)* — *Review Exercises 29, 30*
■ Evaluate and graph functions involving the natural exponential function. *(Section 4.2)* — *Review Exercises 31–34*
■ Graph logistic growth functions. *(Section 4.2)* — *Review Exercises 35, 36*
■ Solve compound interest problems. *(Section 4.2)* — *Review Exercises 37–40*

$$A = P(1 + r/n)^{nt}, \quad A = Pe^{rt}$$

■ Solve effective rate of interest problems. *(Section 4.2)* — *Review Exercises 41, 42*

$$r_{\text{eff}} = (1 + r/n)^n - 1$$

■ Solve present value problems. *(Section 4.2)* — *Review Exercises 43, 44*

$$P = \frac{A}{(1 + r/n)^{nt}}$$

■ Answer questions involving the natural exponential function as a real-life model. *(Section 4.2)* — *Review Exercises 45, 46*
■ Find the derivatives of natural exponential functions. *(Section 4.3)* — *Review Exercises 47–54*

$$\frac{d}{dx}[e^x] = e^x, \quad \frac{d}{dx}[e^u] = e^u \frac{du}{dx}$$

■ Use calculus to analyze the graphs of functions that involve the natural exponential function. *(Section 4.3)* — *Review Exercises 55–62*
■ Use the definition of the natural logarithmic function to write exponential equations in logarithmic form, and vice versa. *(Section 4.4)* — *Review Exercises 63–66*

$$\ln x = b \quad \text{if and only if} \quad e^b = x.$$

■ Sketch the graphs of natural logarithmic functions. *(Section 4.4)* — *Review Exercises 67–70*
■ Use properties of logarithms to expand and condense logarithmic expressions. *(Section 4.4)* — *Review Exercises 71–76*

$$\ln xy = \ln x + \ln y, \quad \ln \frac{x}{y} = \ln x - \ln y, \quad \ln x^n = n \ln x$$

■ Use inverse properties of exponential and logarithmic functions to solve exponential and logarithmic equations. *(Section 4.4)* — *Review Exercises 77–92*

$$\ln e^x = x, \quad e^{\ln x} = x$$

* Use a wide range of valuable study aids to help you master the material in this chapter. The *Student Solutions Guide* includes step-by-step solutions to all odd-numbered exercises to help you review and prepare. The *HM mathSpace® Student CD-ROM* helps you brush up on your algebra skills. The *Graphing Technology Guide*, available on the Web at *math.college.hmco.com/students*, offers step-by-step commands and instructions for a wide variety of graphing calculators, including the most recent models.

■ Use properties of natural logarithms to answer questions about real life. *(Section 4.4)* — *Review Exercises 93, 94*
■ Find the derivatives of natural logarithmic functions. *(Section 4.5)* — *Review Exercises 95–108*

$$\frac{d}{dx}[\ln x] = \frac{1}{x}, \quad \frac{d}{dx}[\ln u] = \frac{1}{u}\frac{du}{dx}$$

■ Use calculus to analyze the graphs of functions that involve the natural logarithmic function. *(Section 4.5)* — *Review Exercises 109–112*
■ Use the definition of logarithms to evaluate logarithmic expressions involving other bases. *(Section 4.5)* — *Review Exercises 113–116*

$$\log_a x = b \quad \text{if and only if} \quad a^b = x$$

■ Use the change-of-base formula to evaluate logarithmic expressions involving other bases. *(Section 4.5)* — *Review Exercises 117–120*

$$\log_a x = \frac{\ln x}{\ln a}$$

■ Find the derivatives of exponential and logarithmic functions involving other bases. *(Section 4.5)* — *Review Exercises 121–124*

$$\frac{d}{dx}[a^x] = (\ln a)a^x, \quad \frac{d}{dx}[a^u] = (\ln a)a^u \frac{du}{dx}$$

$$\frac{d}{dx}[\log_a x] = \left(\frac{1}{\ln a}\right)\frac{1}{x}, \quad \frac{d}{dx}[\log_a u] = \left(\frac{1}{\ln a}\right)\left(\frac{1}{u}\right)\frac{du}{dx}$$

■ Use calculus to answer questions about real-life rates of change. *(Section 4.5)* — *Review Exercises 125, 126*
■ Use exponential growth and decay to model real-life situations. *(Section 4.6)* — *Review Exercises 127–132*
■ *Classifying Differentiation Rules* Differentiation rules fall into two basic classes: (1) general rules that apply to all differentiable functions; and (2) specific rules that apply to special types of functions. At this point in the course, you have studied six general rules: the Constant Rule, the Constant Multiple Rule, the Sum Rule, the Difference Rule, the Product Rule, and the Quotient Rule. Although these rules were introduced in the context of algebraic functions, remember that they can also be used with exponential and logarithmic functions. You have also studied three specific rules: the Power Rule, the derivative of the natural exponential function, and the derivative of the natural logarithmic function. Each of these rules comes in two forms: the "simple" version, such as $D_x[e^x] = e^x$, and the Chain Rule version, such as $D_x[e^u] = e^u(du/dx)$.
■ *To Memorize or Not to Memorize?* When studying mathematics, you need to memorize some formulas and rules. Much of this will come from practice—the formulas that you use most often will be committed to memory. Some formulas, however, are used only infrequently. With these, it is helpful to be able to *derive* the formula from a *known* formula. For instance, knowing the Log Rule for differentiation and the change-of-base formula, $\log_a x = (\ln x)/(\ln a)$, allows you to derive the formula for the derivative of a logarithmic function to base a.

Study Tools   *Additional resources that accompany this chapter*

■ Algebra Review   (pages 308 and 309)
■ Chapter Summary and Study Strategies (pages 310 and 311)
■ Review Exercises   (pages 312–315)
■ Sample Post-Graduation Exam Questions   (page 316)
■ Web Exercises   (page 289, Exercise 80; page 298, Exercise 83)
■ Student Solutions Guide
■ HM mathSpace® Student CD-ROM
■ Graphing Technology Guide (math.college.hmco.com/students)

CHAPTER SUMMARY AND STUDY STRATEGIES

The *Chapter Summary* reviews the skills covered in the chapter and correlates each skill to the *Review Exercises* that test those skills. Following each *Chapter Summary* is a short list of *Study Strategies* for addressing topics or situations specific to the chapter, and a list of *Study Tools* that accompany each chapter.

REVIEW EXERCISES

The *Review Exercises* offer students opportunities for additional practice as they complete each chapter. Answers to all odd-numbered *Review Exercises* appear at the end of the text.

---

544   **CHAPTER 7**   Functions of Several Variables

7 CHAPTER REVIEW EXERCISES

In Exercises 1 and 2, plot the points.

1. $(2, -1, 4), (-1, 3, -3)$
2. $(1, -2, -3), (-4, -3, 5)$

In Exercises 3 and 4, find the distance between the two points.

3. $(0, 0, 0), (2, 5, 9)$
4. $(-4, 1, 5), (1, 3, 7)$

In Exercises 5 and 6, find the midpoint of the line segment joining the two points.

5. $(2, 6, 4), (-4, 2, 8)$
6. $(5, 0, 7), (-1, -2, 9)$

In Exercises 7–10, find the standard form of the equation of the sphere.

7. Center: $(0, 1, 0)$; radius: 5
8. Center: $(4, -5, 3)$; radius: 10
9. Diameter endpoints: $(3, 4, 0), (5, 8, 2)$
10. Diameter endpoints: $(-2, 5, 1), (4, -3, 3)$

In Exercises 11 and 12, find the center and radius of the sphere.

11. $x^2 + y^2 + z^2 + 4x - 2y - 8z + 5 = 0$
12. $x^2 + y^2 + z^2 + 4y - 10z - 7 = 0$

In Exercises 13 and 14, sketch the xy-trace of the sphere.

13. $(x + 2)^2 + (y - 1)^2 + (z - 3)^2 = 25$
14. $(x - 1)^2 + (y + 3)^2 + (z - 6)^2 = 72$

In Exercises 15–18, find the intercepts and sketch the graph of the plane.

15. $x + 2y + 3z = 6$
16. $2y + z = 4$
17. $6x + 3y - 6z = 12$
18. $4x - y + 2z = 8$

In Exercises 19–26, identify the surface.

19. $x^2 + y^2 + z^2 - 2x + 4y - 6z + 5 = 0$
20. $16x^2 + 16y^2 - 9z^2 = 0$
21. $x^2 + \dfrac{y^2}{16} + \dfrac{z^2}{9} = 1$
22. $-x^2 + \dfrac{y^2}{16} + \dfrac{z^2}{9} = 1$
    $\dfrac{x^2}{9} + y^2$

24. $-4x^2 + y^2 + z^2 = 4$
25. $z = \sqrt{x^2 + y^2}$
26. $z = 9x + 3y - 5$

In Exercises 27 and 28, find the function values.

27. $f(x, y) = xy^2$
    (a) $f(2, 3)$ (b) $f(0, 1)$
    (c) $f(-5, 7)$ (d) $f(-2, -4)$

28. $f(x, y) = \dfrac{x^2}{y}$
    (a) $f(6, 9)$ (b) $f(8, 4)$
    (c) $f(t, 2)$ (d) $f(r, r)$

In Exercises 29 and 30, describe the region R in the xy-plane that corresponds to the domain of the function. Then find the range of the function.

29. $f(x, y) = \sqrt{1 - x^2 - y^2}$
30. $f(x, y) = \dfrac{1}{x + y}$

In Exercises 31–34, describe the level curves of the function. Sketch the level curves for the given c-values.

31. $z = 10 - 2x - 5y$, $c = 0, 2, 4, 5, 10$
32. $z = \sqrt{9 - x^2 - y^2}$, $c = 0, 1, 2, 3$
33. $z = (xy)^2$, $c = 1, 4, 9, 12, 16$
34. $z = 2e^{xy}$, $c = 1, 2, 3, 4, 5$

35. *Meteorology* The contour map shown below represents the average yearly precipitation for Iowa. *(Source: U.S. National Oceanic and Atmospheric Administration)*
    (a) Discuss the use of color to represent the level curves.
    (b) Which part of Iowa receives the most precipitation?
    (c) Which part of Iowa receives the least precipitation?

---

316   **CHAPTER 4**   Exponential and Logarithmic Functions

4 SAMPLE POST-GRADUATION EXAM QUESTIONS

CPA
GMAT
GRE
Actuarial
CLAST

The following questions represent the types of questions that appear on certified public accountant (CPA) exams, Graduate Management Admission Tests (GMAT), Graduate Records Exams (GRE), actuarial exams, and College-Level Academic Skills Tests (CLAST). The answers to the questions are given in the back of the book.

1. 10^x means that 10 is to be used as a factor x times, and 10^{-x} is equal to
    $$\dfrac{1}{10^x}$$
    A very large or very small number, therefore, is frequently written as a decimal multiplied by 10^x, where x is an integer. Which, if any, are false?
    (a) $470{,}000 = 4.7 \times 10^5$
    (b) 450 billion $= 4.5 \times 10^{11}$
    (c) $0.00000000075 = 7.5 \times 10^{-10}$
    (d) 86 hundred-thousandths $= 8.6 \times 10^2$

2. The rate of decay of a radioactive substance is proportional to the amount of the substance present. Three years ago there was 6 grams of substance. Now there is 5 grams. How many grams will there be 3 years from now?
    (a) 4   (b) $\frac{25}{6}$ (c) $\frac{125}{36}$ (d) $\frac{75}{36}$

3. In a certain town, 45% of the people have brown hair, 30% have brown eyes, and 15% have both brown hair and brown eyes. What percent of the people in the town have neither brown hair nor brown eyes?
    (a) 25%   (b) 35%   (c) 40%   (d) 50%

4. You deposit $900 in a savings account that is compounded continuously at 4.76%. After 16 years, the amount in the account will be
    (a) $1927.53   (b) $1077.81   (c) $943.58   (d) $2827.53

5. A bookstore orders 75 books. Each book costs the bookstore $29 and is sold for $42. The bookstore must pay a $4 service charge for each unsold book returned. If the bookstore returns seven books, how much profit will the bookstore make?
    (a) $975   (b) $947   (c) $856   (d) $681

For Questions 6–9, use the data given in the graph.

6. In how many of the years were expenses greater than in the preceding year?
    (a) 2   (b) 4   (c) 1   (d) 3

7. In which year was the profit the greatest?
    (a) 1997   (b) 2000   (c) 1996   (d) 1998

8. In 1999, profits decreased by x percent from 1998 with x equal to
    (a) 60%   (b) 140%   (c) 340%   (d) 40%

9. In 2000, profits increased by y percent from 1999 with y equal to
    (a) 64%   (b) 136%   (c) 178%   (d) 378%

Figure for 6–9

POST-GRADUATION EXAM QUESTIONS

To emphasize the relevance of calculus, every chapter concludes with sample questions representative of the types of questions on certified public accountant (CPA) exams, Graduate Management Admission Tests® (GMAT®), Graduate Record Examinations® (GRE®), actuarial exams, and College-Level Academic Skills Tests (CLAST). The answers to all *Post-Graduation Exam Questions* are given in the back of the text.

A Plan for You as a Student

Study Strategies

Your success in mathematics depends on your active participation both in class and outside of class. Because the material you learn each day builds on the material you have learned previously, it is important that you keep up with your course work every day and develop a clear plan of study. This set of guidelines highlights key study strategies to help you learn how to study mathematics.

Preparing for Class   The syllabus your instructor provides is an invaluable resource that outlines the major topics to be covered in the course. Use it to help you prepare. As a general rule, you should set aside two to four hours of study time for each hour spent in class. Being prepared is the first step toward success. Before class:

- Review your notes from the previous class.
- Read the portion of the text that will be covered in class.
- Use the objectives listed at the beginning of each section to keep you focused on the main ideas of the section.
- Pay special attention to the definitions, rules, and concepts highlighted in boxes. Also, be sure you understand the meanings of mathematical symbols and terms written in boldface type. Keep a vocabulary journal for easy reference.
- Read through the solved examples. Use the side comments given in the solution steps to help you in the solution process. Also, read the *Study Tips* given in the margins.
- Make notes of anything you do not understand as you read through the text. If you still do not understand after your instructor covers the topic in question, ask questions before your instructor moves on to a new topic.
- Try the *Discovery* and *Technology* exercises to get a better grasp of the material before the instructor presents it.

Keeping Up   Another important step toward success in mathematics involves your ability to keep up with the work. It is very easy to fall behind, especially if you miss a class. To keep up with the course work, be sure to:

- Attend every class. Bring your text, a notebook, a pen or pencil, and a calculator (scientific or graphing). If you miss a class, get the notes from a classmate as soon as possible and review them carefully.
- Participate in class. As mentioned above, if there is a topic you do not understand, ask about it before the instructor moves on to a new topic.
- Take notes in class. After class, read through your notes and add explanations so that your notes make sense to *you*. Fill in any gaps and note any questions you might have.
- Reread the portion of the text that was covered in class. This time, work each example *before* reading through the solution.
- Do your homework as soon as possible, while concepts are still fresh in your mind. Allow at least two hours of homework time for each hour spent in class so you do not fall behind. Learning mathematics is a step-by-step process, and you must understand each topic in order to learn the next one.
- When you are working problems for homework assignments, show every step in your solution. Then, if you make an error, it will be easier to find where the error occurred.
- Use your notes from class, the text discussion, the examples, and the *Study Tips* as you do your homework. Many exercises are keyed to specific examples in the text for easy reference.

Getting Extra Help   It can be very frustrating when you do not understand concepts and are unable to complete homework assignments. However, there are many resources available to help you with your studies.

- Your instructor may have office hours. If you are feeling overwhelmed and need help, make an appointment to discuss your difficulties with your instructor.
- Find a study partner or a study group. Sometimes it helps to work through problems with another person.
- Arrange to get regular assistance from a tutor. Many colleges have math resource centers available on campus as well.
- Consult one of the many ancillaries available with this text: the *HM mathSpace® Student CD-ROM*, the *Student Solutions Guide*, videotapes, and additional study resources available at this text's website at *college.hmco.com*.
- Special assistance with algebra appears in the *Algebra Reviews*, which appear throughout each chapter. These short reviews are tied together in the larger *Algebra Review* section at the end of each chapter.

Preparing for an Exam   The last step toward success in mathematics lies in how you prepare for and complete exams. If you have followed the suggestions given above, then you are almost ready for exams. Do not assume that you can cram for the exam the night before—this seldom works. As a final preparation for the exam:

- Read the *Chapter Summary* and *Study Strategies* keyed to each section, and review the concepts and terms.
- Work through the *Review Exercises* if you need extra practice on material from a particular section. You can practice for an exam by first trying to work through the exercises with your book and notebook closed.
- Take practice tests offered online at this text's website at *college.hmco.com*.
- When you study for an exam, first look at all definitions, properties, and formulas until you know them. Review your notes and the portion of the text that will be covered on the exam. Then work as many exercises as you can, especially any kinds of exercises that have given you trouble in the past, reworking homework problems as necessary.
- Start studying for your exam well in advance (at least a week). The first day or two, study only about two hours. Gradually increase your study time each day. Be completely prepared for the exam two days in advance. Spend the final day just building confidence so you can be relaxed during the exam.
- Avoid studying up until the last minute. This will only make you anxious. Allow yourself plenty of time to get to the testing location. When you take the exam, go in with a clear mind and a positive attitude.
- Once the exam begins, read through the directions and the entire exam before beginning. Work the problems that you know how to do first to avoid spending too much time on any one problem. Time management is extremely important when taking an exam.
- If you finish early, use the remaining time to go over your work.
- When you get an exam back, review it carefully and go over your errors. Rework the problems you answered incorrectly. Discovering the mistakes you made will help you improve your test-taking ability. Understanding how to correct your errors will help you build on the knowledge you have gained before you move on to the next topic.

chapter

0

A Precalculus Review

0.1 The Real Number Line and Order

0.2 Absolute Value and Distance on the Real Number Line

0.3 Exponents and Radicals

0.4 Factoring Polynomials

0.5 Fractions and Rationalization

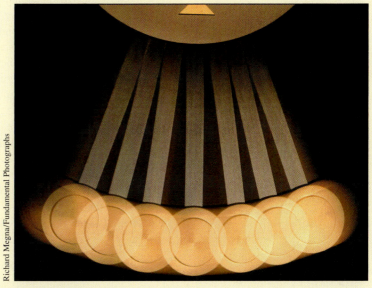

Richard Megna/Fundamental Photographs

The period of a pendulum is dependent only on the length of the pendulum. Changing the length of the pendulum can correct a slow running grandfather clock, regardless of the weight of the bob or the amplitude.

STRATEGIES FOR SUCCESS

WHAT YOU SHOULD LEARN:

This initial chapter, A Precalculus Review, is just that—a review chapter. Make sure that you have a solid understanding of all the material in this chapter before beginning Chapter 1. You can also use it as a reference as you progress through the text, coming back to brush up on an algebra skill that you may have forgotten over the course of the semester. As with all math courses, calculus is a building process—that is, you use what you know to go on to the next topic.

WHY YOU SHOULD LEARN IT:

Precalculus concepts have many applications in real life, as can be seen by the examples below, which represent a small sample of the applications in this chapter.

■ Biology: pH Values, Exercise 29 on page 0-7

■ Budget Variance, Exercises 47-50 on page 0-12

■ Compound Interest, Exercises 57-60 on page 0-18

■ Period of a Pendulum, Exercise 61 on page 0-18

■ Chemistry: Finding Concentrations, Exercise 71 on page 0-24

■ Installment Loan, Exercise 47 on page 0-32

0.1   THE REAL NUMBER LINE AND ORDER

- Represent, classify, and order real numbers.
- Use inequalities to represent sets of real numbers.
- Solve inequalities.
- Use inequalities to model and solve real-life problems.

The Real Number Line

FIGURE 0.1   The Real Number Line

Real numbers can be represented with a coordinate system called the **real number line** (or *x*-axis), as shown in Figure 0.1. The **positive direction** (to the right) is denoted by an arrowhead and indicates the direction of increasing values of *x*. The real number corresponding to a particular point on the real number line is called the **coordinate** of the point. As shown in Figure 0.1, it is customary to label those points whose coordinates are integers.

The point on the real number line corresponding to zero is called the **origin.** Numbers to the right of the origin are **positive,** and numbers to the left of the origin are **negative.** The term **nonnegative** describes a number that is either positive or zero.

The importance of the real number line is that it provides you with a conceptually perfect picture of the real numbers. That is, each point on the real number line corresponds to one and only one real number, and each real number corresponds to one and only one point on the real number line. This type of relationship is called a **one-to-one correspondence** and is illustrated in Figure 0.2.

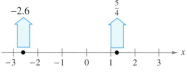

Every point on the real number line corresponds to one and only one real number.

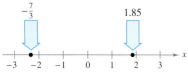

Every real number corresponds to one and only one point on the real number line.

FIGURE 0.2

Each of the four points in Figure 0.2 corresponds to a real number that can be expressed as the ratio of two integers.

$$-2.6 = -\frac{13}{5} \qquad \frac{5}{4}$$
$$-\frac{7}{3} \qquad 1.85 = \frac{37}{20}$$

Such numbers are called **rational.** Rational numbers have either terminating or infinitely repeating decimal representations.

Terminating Decimals	*Infinitely Repeating Decimals*
$\frac{2}{5} = 0.4$	$\frac{1}{3} = 0.333\ldots = 0.\overline{3}*$
$\frac{7}{8} = 0.875$	$\frac{12}{7} = 1.714285714285\ldots = 1.\overline{714285}$

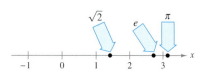

FIGURE 0.3

Real numbers that are not rational are called **irrational,** and they cannot be represented as the ratio of two integers (or as terminating or infinitely repeating decimals). So, a decimal approximation is used to represent an irrational number. Some irrational numbers occur so frequently in applications that mathematicians have invented special symbols to represent them. For example, the symbols $\sqrt{2}$, π, and e represent irrational numbers whose decimal approximations are as shown. (See Figure 0.3.)

$$\sqrt{2} \approx 1.4142135623$$
$$\pi \approx 3.1415926535$$
$$e \approx 2.7182818284$$

*The bar indicates which digit or digits repeat infinitely.

Order and Intervals on the Real Number Line

One important property of the real numbers is that they are **ordered:** 0 is less than 1, -3 is less than -2.5, π is less than $\frac{22}{7}$, and so on. You can visualize this property on the real number line by observing that a is less than b if and only if a lies to the left of b on the real number line. Symbolically, "a is less than b" is denoted by the inequality

   $a < b.$

For example, the inequality $\frac{3}{4} < 1$ follows from the fact that $\frac{3}{4}$ lies to the left of 1 on the real number line, as shown in Figure 0.4.

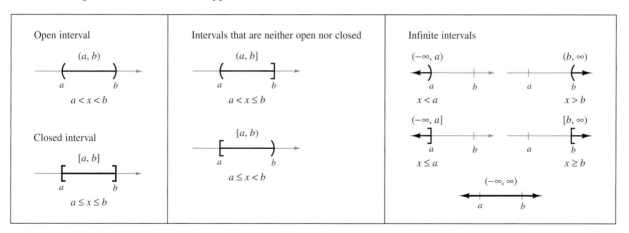

FIGURE 0.4

   When three real numbers a, x, and b are ordered such that $a < x$ and $x < b$, we say that x is **between** a and b and write

   $a < x < b.$ *x is between a and b.*

The set of *all* real numbers between a and b is called the **open interval** between a and b and is denoted by (a, b). An interval of the form (a, b) does not contain the "endpoints" a and b. Intervals that include their endpoints are called **closed** and are denoted by $[a, b]$. Intervals of the form $[a, b)$ and $(a, b]$ are neither open nor closed. Figure 0.5 shows the nine types of intervals on the real number line.

FIGURE 0.5 Intervals on the Real Number Line

STUDY TIP

Note that a square bracket is used to denote "less than or equal to" ($\leq$) or "greater than or equal to" ($\geq$). Furthermore, the symbols ∞ and $-\infty$ denote **positive** and **negative infinity.** These symbols do not denote real numbers; they merely let you describe unbounded conditions more concisely. For instance, the interval $[b, \infty)$ is unbounded to the right because it includes *all* real numbers that are greater than or equal to b.

Solving Inequalities

In calculus, you are frequently required to "solve inequalities" involving variable expressions such as $3x - 4 < 5$. The number a is a **solution** of an inequality if the inequality is true when a is substituted for x. The set of all values of x that satisfy an equality is called the **solution set** of the inequality. The following properties are useful for solving inequalities. (Similar properties are obtained if $<$ is replaced by $\leq$ and $>$ is replaced by $\geq$.)

> **Properties of Inequalities**
>
> Let a, b, c, and d be real numbers.
>
> 1. Transitive property: $a < b$ and $b < c$ ⟹ $a < c$
> 2. Adding inequalities: $a < b$ and $c < d$ ⟹ $a + c < b + d$
> 3. Multiplying by a (positive) constant: $a < b$ ⟹ $ac < bc, \quad c > 0$
> 4. Multiplying by a (negative) constant: $a < b$ ⟹ $ac > bc, \quad c < 0$
> 5. Adding a constant: $a < b$ ⟹ $a + c < b + c$
> 6. Subtracting a constant: $a < b$ ⟹ $a - c < b - c$

Note that you *reverse the inequality* when you multiply by a negative number. For example, if $x < 3$, then $-4x > -12$. This principle also applies to division by a negative number. So, if $-2x > 4$, then $x < -2$.

EXAMPLE 1   **Solving an Inequality**

Find the solution set of the inequality $3x - 4 < 5$.

SOLUTION

$3x - 4 < 5$	Write original inequality.
$3x - 4 + 4 < 5 + 4$	Add 4 to each side.
$3x < 9$	Simplify.
$\frac{1}{3}(3x) < \frac{1}{3}(9)$	Multiply each side by $\frac{1}{3}$.
$x < 3$	Simplify.

So, the solution set is the interval $(-\infty, 3)$, as shown in Figure 0.6. ▬▬▬

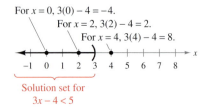

For $x = 0$, $3(0) - 4 = -4$.
For $x = 2$, $3(2) - 4 = 2$.
For $x = 4$, $3(4) - 4 = 8$.

Solution set for
$3x - 4 < 5$

FIGURE 0.6

In Example 1, all five inequalities listed as steps in the solution have the same solution set, and they are called **equivalent inequalities.**

The inequality in Example 1 involves a first-degree polynomial. To solve inequalities involving polynomials of higher degree, you can use the fact that a polynomial can change signs *only* at its real zeros (the real numbers that make the polynomial zero). Between two consecutive real zeros, a polynomial must be entirely positive or entirely negative. This means that when the real zeros of a polynomial are put in order, they divide the real number line into **test intervals** in which the polynomial has no sign changes. That is, if a polynomial has the factored form

$$(x - r_1)(x - r_2), \ldots, (x - r_n), \qquad r_1 < r_2 < r_3 < \cdots < r_n$$

then the test intervals are

$$(-\infty, r_1), \quad (r_1, r_2), \quad \ldots, \quad (r_{n-1}, r_n), \quad \text{and} \quad (r_n, \infty).$$

For example, the polynomial

$$x^2 - x - 6 = (x - 3)(x + 2)$$

can change signs only at $x = -2$ and $x = 3$. To determine the sign of the polynomial in the intervals $(-\infty, -2)$, $(-2, 3)$, and $(3, \infty)$, you need to test only *one value* from each interval.

EXAMPLE 2   **Solving a Polynomial Inequality**

Find the solution set of the inequality $x^2 < x + 6$.

SOLUTION

$$\begin{aligned} x^2 &< x + 6 && \text{Write original inequality.}\\ x^2 - x - 6 &< 0 && \text{Polynomial form}\\ (x - 3)(x + 2) &< 0 && \text{Factor.} \end{aligned}$$

So, the polynomial $x^2 - x - 6$ has $x = -2$ and $x = 3$ as its zeros. You can solve the inequality by testing the sign of the polynomial in each of the following intervals.

$$x < -2, \quad -2 < x < 3, \quad x > 3$$

To test an interval, choose a representative number in the interval and compute the sign of each factor. For example, for any $x < -2$, both of the factors $(x - 3)$ and $(x + 2)$ are negative. Consequently, the product (of two negative numbers) is positive, and the inequality is *not* satisfied in the interval

$$x < -2.$$

A convenient testing format is shown in Figure 0.7. Because the inequality is satisfied only by the center test interval, you can conclude that the solution set is given by the interval

$$-2 < x < 3. \qquad \text{Solution set}$$

Sign of $(x - 3)(x + 2)$

x	Sign	< 0?
-3	$(-)(-)$	No
-2	$(-)(0)$	No
-1	$(-)(+)$	Yes
0	$(-)(+)$	Yes
1	$(-)(+)$	Yes
2	$(-)(+)$	Yes
3	$(0)(+)$	No
4	$(+)(+)$	No

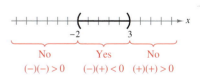

FIGURE 0.7   Is $(x - 3)(x + 2) < 0$?

TRY IT 2

Find the solution set of the inequality $x^2 > 3x + 10$.

Application

Inequalities are frequently used to describe conditions that occur in business and science. For instance, the inequality

$$144 \le W \le 180$$

describes the recommended weight W for a man whose height is 5 feet 10 inches. Example 3 shows how an inequality can be used to describe the production level of a manufacturing plant.

EXAMPLE 3   **Production Levels**

In addition to fixed overhead costs of $500 per day, the cost of producing x units of an item is $2.50 per unit. During the month of August, the total cost of production varied from a high of $1325 to a low of $1200 per day. Find the high and low *production levels* during the month.

SOLUTION   Because it costs $2.50 to produce one unit, it costs $2.5x$ to produce x units. Furthermore, because the fixed cost per day is $500, the total daily cost of producing x units is

$$C = 2.5x + 500.$$

Now, because the cost ranged from $1200 to $1325, you can write the following.

$1200 \le$	$2.5x + 500$	≤ 1325	Write original inequality.
$1200 - 500 \le$	$2.5x + 500 - 500$	$\le 1325 - 500$	Subtract 500 from each side.
$700 \le$	$2.5x$	≤ 825	Simplify.
$\dfrac{700}{2.5} \le$	$\dfrac{2.5x}{2.5}$	$\le \dfrac{825}{2.5}$	Divide each side by 2.5.
$280 \le$	x	≤ 330	Simplify.

So, the daily production levels during the month of August varied from a low of 280 units to a high of 330 units, as shown in Figure 0.8.

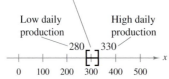

Each day's production during the month fell in this interval.

Low daily production   High daily production

FIGURE 0.8

TRY IT 3

Use the information in Example 3 to find the high and low production levels if, during October, the total cost of production varied from a high of $1500 to a low of $1000 per day.

EXERCISES 0.1

In Exercises 1–10, determine whether the real number is rational or irrational.

*1. 0.7

2. -3678

3. $\dfrac{3\pi}{2}$

4. $3\sqrt{2} - 1$

5. $4.3\overline{451}$

6. $\dfrac{22}{7}$

7. $\sqrt[3]{64}$

8. $0.\overline{8177}$

9. $\sqrt[3]{60}$

10. $2e$

In Exercises 11–14, determine whether each given value of x satisfies the inequality.

11. $5x - 12 > 0$

    (a) $x = 3$ (b) $x = -3$

    (c) $x = \frac{5}{2}$ (d) $x = \frac{3}{2}$

12. $x + 1 < \dfrac{2x}{3}$

    (a) $x = 0$ (b) $x = 4$

    (c) $x = -4$ (d) $x = -3$

13. $0 < \dfrac{x - 2}{4} < 2$

    (a) $x = 4$ (b) $x = 10$

    (c) $x = 0$ (d) $x = \frac{7}{2}$

14. $-1 < \dfrac{3 - x}{2} \leq 1$

    (a) $x = 0$ (b) $x = \sqrt{5}$

    (c) $x = 1$ (d) $x = 5$

In Exercises 15–28, solve the inequality and sketch the graph of the solution on the real number line.

15. $x - 5 \geq 7$

16. $2x > 3$

17. $4x + 1 < 2x$

18. $2x + 7 < 3$

19. $4 - 2x < 3x - 1$

20. $x - 4 \leq 2x + 1$

21. $-4 < 2x - 3 < 4$

22. $0 \leq x + 3 < 5$

23. $\dfrac{3}{4} > x + 1 > \dfrac{1}{4}$

24. $-1 < -\dfrac{x}{3} < 1$

25. $\dfrac{x}{2} + \dfrac{x}{3} > 5$

26. $\dfrac{x}{2} - \dfrac{x}{3} > 5$

27. $2x^2 - x < 6$

28. $2x^2 + 1 < 9x - 3$

---

* The answers to the odd-numbered and selected even exercises are given in the back of the text. Worked-out solutions to the odd-numbered exercises are given in the *Student Solutions Guide*.

29. **Biology: pH Values** The pH scale measures the concentration of hydrogen ions in a solution. Strong acids produce low pH values, while strong bases produce high pH values. Represent the following approximate pH values on a real number line: hydrochloric acid, 0.0; lemon juice, 2.0; oven cleaner, 13.0; baking soda, 9.0; pure water, 7.0; black coffee, 5.0. *(Source: Adapted from Levine/Miller, Biology: Discovering Life, Second Edition)*

30. **Physiology** The maximum heart rate of a person in normal health is related to the person's age by the equation

$$r = 220 - A$$

where r is the maximum heart rate in beats per minute and A is the person's age in years. Some physiologists recommend that during physical activity a person should strive to increase his or her heart rate to at least 60% of the maximum heart rate for sedentary people and at most 90% of the maximum heart rate for highly fit people. Express as an interval the range of the target heart rate for a 20-year-old.

31. **Profit** The revenue for selling x units of a product is

$$R = 115.95x$$

and the cost of producing x units is

$$C = 95x + 750.$$

To obtain a profit, the revenue must be *greater than* the cost. For what values of x will this product return a profit?

32. **Sales** A doughnut shop at a shopping mall sells a dozen doughnuts for $3.50. Beyond the fixed cost (for rent, utilities, and insurance) of $170 per day, it costs $1.75 for enough materials (flour, sugar, etc.) and labor to produce each dozen doughnuts. If the daily profit *varies between* $40 and $250, between what levels (in dozens) do the daily sales vary?

33. **Reimbursement** A pharmaceutical company reimburses their sales representatives $0.35 per mile driven and $100 for meals per week. The company allocates from $200 to $250 per sales representative each week. What are the minimum and maximum numbers of miles the company expects each representative to drive each week?

34. **Area** A square region is to have an area of *at least* 500 square meters. What must the length of the sides of the region be?

In Exercises 35 and 36, determine whether each statement is true or false, given $a < b$.

35. (a) $-2a < -2b$

    (b) $a + 2 < b + 2$

    (c) $6a < 6b$

    (d) $\dfrac{1}{a} < \dfrac{1}{b}$

36. (a) $a - 4 < b - 4$

    (b) $4 - a < 4 - b$

    (c) $-3b < -3a$

    (d) $\dfrac{a}{4} < \dfrac{b}{4}$

0.2   ABSOLUTE VALUE AND DISTANCE ON THE REAL NUMBER LINE

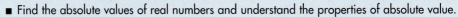

- Find the absolute values of real numbers and understand the properties of absolute value.
- Find the distance between two numbers on the real number line.
- Define intervals on the real number line.
- Find the midpoint of an interval and use intervals to model and solve real-life problems.

Absolute Value of a Real Number

Definition of Absolute Value

The **absolute value** of a real number a is

$$|a| = \begin{cases} a, & \text{if } a \geq 0 \\ -a, & \text{if } a < 0. \end{cases}$$

At first glance, it may appear from this definition that the absolute value of a real number can be negative, but this is not possible. For example, let $a = -3$. Then, because $-3 < 0$, you have

$$|a| = |-3|$$
$$= -(-3)$$
$$= 3.$$

The following properties are useful for working with absolute values.

Properties of Absolute Value

1. Multiplication: $|ab| = |a||b|$

2. Division: $\left|\dfrac{a}{b}\right| = \dfrac{|a|}{|b|}, \quad b \neq 0$

3. Power: $|a^n| = |a|^n$

4. Square root: $\sqrt{a^2} = |a|$

Be sure you understand the fourth property in this list. A common error in algebra is to imagine that by squaring a number and then taking the square root, you come back to the original number. But this is true only if the original number is nonnegative. For instance, if $a = 2$, then

$$\sqrt{2^2} = \sqrt{4} = 2$$

but if $a = -2$, then

$$\sqrt{(-2)^2} = \sqrt{4} = 2.$$

The reason for this is that (by definition) the square root symbol $\sqrt{}$ denotes only the nonnegative root.

Distance on the Real Number Line

Consider two distinct points on the real number line, as shown in Figure 0.9.

1. The **directed distance from a to b** is $b - a$.

2. The **directed distance from b to a** is $a - b$.

3. The **distance between a and b** is $|a - b|$ or $|b - a|$.

In Figure 0.9, note that because b is to the right of a, the directed distance from a to b (moving to the right) is positive. Moreover, because a is to the left of b, the directed distance from b to a (moving to the left) is negative. The distance *between* two points on the real number line can never be negative.

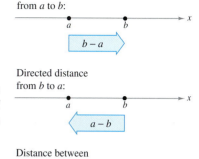

Directed distance from a to b:

Directed distance from b to a:

Distance between a and b:

FIGURE 0.9

Distance Between Two Points on the Real Number Line

The distance d between points x_1 and x_2 on the real number line is given by

$$d = |x_2 - x_1| = \sqrt{(x_2 - x_1)^2}.$$

Note that the order of subtraction with x_1 and x_2 does not matter because

$$|x_2 - x_1| = |x_1 - x_2| \quad \text{and} \quad (x_2 - x_1)^2 = (x_1 - x_2)^2.$$

EXAMPLE 1   Finding Distance on the Real Number Line

Determine the distance between -3 and 4 on the real number line. What is the directed distance from -3 to 4? What is the directed distance from 4 to -3?

SOLUTION   The distance between -3 and 4 is given by

$$|-3 - 4| = |-7| = 7 \quad \text{or} \quad |4 - (-3)| = |7| = 7 \qquad {\scriptstyle |a - b| \text{ or } |b - a|}$$

as shown in Figure 0.10.

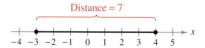

Distance = 7

FIGURE 0.10

The directed distance from -3 to 4 is

$$4 - (-3) = 7. \qquad {\scriptstyle b - a}$$

The directed distance from 4 to -3 is

$$-3 - 4 = -7. \qquad {\scriptstyle a - b}$$

TRY IT 1

Determine the distance between -2 and 6 on the real number line. What is the directed distance from -2 to 6? What is the directed distance from 6 to -2?

Intervals Defined by Absolute Value

| EXAMPLE 2 | **Defining an Interval on the Real Number Line** |

Find the interval on the real number line that contains all numbers that lie no more than two units from 3.

SOLUTION   Let x be any point in this interval. You need to find all x such that the distance between x and 3 is less than or equal to 2. This implies that

$$|x - 3| \leq 2.$$

Requiring the absolute value of $x - 3$ to be less than or equal to 2 means that $x - 3$ must lie between -2 and 2. So, you can write

$$-2 \leq x - 3 \leq 2.$$

Solving this pair of inequalities, you have

$$-2 + 3 \leq x - 3 + 3 \leq 2 + 3$$

$$1 \leq \quad x \quad \leq 5. \qquad \text{Solution set}$$

So, the interval is $[1, 5]$, as shown in Figure 0.11.

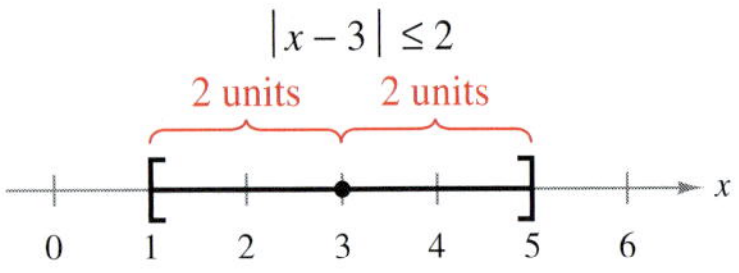

$|x-3| \leq 2$

2 units   2 units

$$\begin{array}{cccccccc} & 0 & 1 & 2 & 3 & 4 & 5 & 6 \end{array} \to x$$

FIGURE 0.11

TRY IT 2

Find the interval on the real number line that contains all numbers that lie no more than four units from 6.

Two Basic Types of Inequalities Involving Absolute Value

Let a and d be real numbers, where $d > 0$.

$|x - a| \leq d$ if and only if $a - d \leq x \leq a + d$.

$|x - a| \geq d$ if and only if $x \leq a - d$ or $a + d \leq x$.

Inequality	*Interpretation*	*Graph*
$|x - a| \leq d$	All numbers x whose distance from a is less than or equal to d.	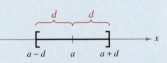
$|x - a| \geq d$	All numbers x whose distance from a is greater than or equal to d.	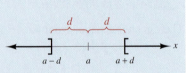

ALGEBRA REVIEW

Be sure you see that inequalities of the form $|x - a| \geq d$ have solution sets consisting of two intervals. To describe the two intervals without using absolute values, you must use *two* separate inequalities, connected by an "or" to indicate union.

Application

EXAMPLE 3 **Quality Control**

A large manufacturer hired a quality control firm to determine the reliability of a product. Using statistical methods, the firm determined that the manufacturer could expect 0.35% ± 0.17% of the units to be defective. If the manufacturer offers a money-back guarantee on this product, how much should be budgeted to cover the refunds on 100,000 units? (Assume that the retail price is $8.95.)

SOLUTION   Let r represent the percent of defective units (written in decimal form). You know that r will differ from 0.0035 by at most 0.0017.

$$0.0035 - 0.0017 \le r \le 0.0035 + 0.0017$$
$$0.0018 \le r \le 0.0052 \qquad \text{Figure 0.12(a)}$$

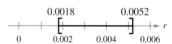

(a) Percent of defective units

Now, letting x be the number of defective units out of 100,000, it follows that $x = 100{,}000r$ and you have

$$0.0018(100{,}000) \le 100{,}000r \le 0.0052(100{,}000)$$
$$180 \le x \le 520. \qquad \text{Figure 0.12(b)}$$

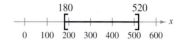

(b) Number of defective units

Finally, letting C be the cost of refunds, you have $C = 8.95x$. So, the total cost of refunds for 100,000 units should fall within the interval given by

$$180(8.95) \le 8.95x \le 520(8.95)$$
$$\$1611 \le C \le \$4654. \qquad \text{Figure 0.12(c)}$$

(c) Cost of refunds

FIGURE 0.12

TRY IT 3

Use the information in Example 3 to determine how much should be budgeted to cover refunds on 250,000 units.

In Example 3, the manufacturer should expect to spend between $1611 and $4654 for refunds. Of course, the safer budget figure for refunds would be the higher of these estimates. However, from a statistical point of view, the most representative estimate would be the average of these two extremes. Graphically, the average of two numbers is the **midpoint** of the interval with the two numbers as endpoints, as shown in Figure 0.13.

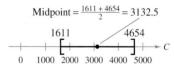

FIGURE 0.13

Midpoint of an Interval

The **midpoint** of the interval with endpoints a and b is found by taking the average of the endpoints.

$$\text{Midpoint} = \frac{a+b}{2}$$

In Exercises 1–6, find (a) the directed distance from a to b, (b) the directed distance from b to a, and (c) the distance between a and b.

1. $a = 126, b = 75$ **2.** $a = -126, b = -75$

3. $a = 9.34, b = -5.65$ **4.** $a = -2.05, b = 4.25$

5. $a = \frac{16}{5}, b = \frac{112}{75}$ **6.** $a = -\frac{18}{5}, b = \frac{61}{15}$

In Exercises 7–18, use absolute values to describe the given interval (or pair of intervals) on the real number line.

7. $[-2, 2]$ **8.** $(-3, 3)$

9. $(-\infty, -2) \cup (2, \infty)$ **10.** $(-\infty, -3] \cup [3, \infty)$

11. $[2, 6]$ **12.** $(-7, -1)$

13. $(-\infty, 0) \cup (4, \infty)$ **14.** $(-\infty, 20) \cup (24, \infty)$

15. All numbers *less than* two units from 4

16. All numbers *more than* six units from 3

17. y is *at most* two units from a.

18. y is *less than* h units from c.

In Exercises 19–34, solve the inequality and sketch the graph of the solution on the real number line.

19. $|x| < 5$ **20.** $|2x| < 6$

21. $\left|\frac{x}{2}\right| > 3$ **22.** $|5x| > 10$

23. $|x + 2| < 5$ **24.** $|3x + 1| \geq 4$

25. $\left|\frac{x-3}{2}\right| \geq 5$ **26.** $|2x + 1| < 5$

27. $|10 - x| > 4$ **28.** $|25 - x| \geq 20$

29. $|9 - 2x| < 1$

30. $\left|1 - \frac{2x}{3}\right| < 1$

31. $|x - a| \leq b, \ b > 0$

32. $|2x - a| \geq b, \ b > 0$

33. $\left|\frac{3x - a}{4}\right| < 2b, \ b > 0$

34. $\left|a - \frac{5x}{2}\right| > b, \ b > 0$

In Exercises 35–40, find the midpoint of the given interval.

35. $[7, 21]$ **36.** $[8.6, 11.4]$

37. $[-6.85, 9.35]$ **38.** $[-4.6, -1.3]$

39. $\left[-\frac{1}{2}, \frac{3}{4}\right]$ **40.** $\left[\frac{5}{6}, \frac{5}{2}\right]$

41. *Chemistry*   Copper has a melting point M within $0.2°C$ of $1083.4°C$. Use absolute values to write the range as an inequality.

42. *Stock Price*   A stock market analyst predicts that over the next year the price p of a stock will not change from its current price of $\$33\frac{1}{8}$ by more than $\$2$. Use absolute values to write this prediction as an inequality.

43. *Statistics*   The heights h of two-thirds of the members of a population satisfy the inequality

$$\left|\frac{h - 68.5}{2.7}\right| \leq 1$$

where h is measured in inches. Determine the interval on the real number line in which these heights lie.

44. *Biology*   The American Kennel Club has developed guidelines for judging the features of various breeds of dogs. For collies, the guidelines specify that the weights for males satisfy the inequality

$$\left|\frac{w - 57.5}{7.5}\right| \leq 1$$

where w is measured in pounds. Determine the interval on the real number line in which these weights lie.

45. *Production*   The estimated daily production x at a refinery is given by

$$|x - 200,000| \leq 25,000$$

where x is measured in barrels of oil. Determine the high and low production levels.

46. *Manufacturing*   The acceptable weights for a 20-ounce cereal box are given by

$$|x - 20| \leq 0.75$$

where x is measured in ounces. Determine the high and low weights for the cereal box.

Budget Variance   In Exercises 47–50, (a) use absolute value notation to represent the two intervals in which expenses must lie if they are to be within $\$500$ and within 5% of the specified budget amount and (b) using the more stringent constraint, determine whether the given expense is at variance with the budget restriction.

Item	Budget	Expense
47. Utilities	$4750.00	$5116.37
48. Insurance	$15,000.00	$14,695.00
49. Maintenance	$20,000.00	$22,718.35
50. Taxes	$7500.00	$8691.00

0.3 EXPONENTS AND RADICALS

- Evaluate expressions involving exponents or radicals.
- Simplify expressions with exponents.
- Find the domains of algebraic expressions.

Expressions Involving Exponents or Radicals

Properties of Exponents

1. Whole-number exponents: $x^n = \underbrace{x \cdot x \cdot x \cdots x}_{n \text{ factors}}$

2. Zero exponent: $x^0 = 1, \quad x \neq 0$

3. Negative exponents: $x^{-n} = \dfrac{1}{x^n}, \quad x \neq 0$

4. Radicals (principal nth root): $\sqrt[n]{x} = a \implies x = a^n$

5. Rational exponents $(1/n)$: $x^{1/n} = \sqrt[n]{x}$

6. Rational exponents (m/n): $x^{m/n} = (x^{1/n})^m = \left(\sqrt[n]{x}\right)^m$

   $x^{m/n} = (x^m)^{1/n} = \sqrt[n]{x^m}$

7. Special convention (square root): $\sqrt[2]{x} = \sqrt{x}$

> **ALGEBRA REVIEW**
>
> If n is even, then the principal nth root is positive. For example, $\sqrt{4} = +2$ and $\sqrt[4]{81} = +3$.

EXAMPLE 1 Evaluating Expressions

Expression	x-Value	Substitution
(a) $y = -2x^2$	$x = 4$	$y = -2(4^2) = -2(16) = -32$
(b) $y = 3x^{-3}$	$x = -1$	$y = 3(-1)^{-3} = \dfrac{3}{(-1)^3} = \dfrac{3}{-1} = -3$
(c) $y = (-x)^2$	$x = \dfrac{1}{2}$	$y = \left(-\dfrac{1}{2}\right)^2 = \dfrac{1}{4}$
(d) $y = \dfrac{2}{x^{-2}}$	$x = 3$	$y = \dfrac{2}{3^{-2}} = 2(3^2) = 18$

> **TRY IT 1**
>
> Evaluate $y = 4x^{-2}$ for $x = 3$.

EXAMPLE 2 Evaluating Expressions

Expression	x-Value	Substitution
(a) $y = 2x^{1/2}$	$x = 4$	$y = 2\sqrt{4} = 2(2) = 4$
(b) $y = \sqrt[3]{x^2}$	$x = 8$	$y = 8^{2/3} = (8^{1/3})^2 = 2^2 = 4$

> **TRY IT 2**
>
> Evaluate $y = 4x^{1/3}$ for $x = 8$.

Operations with Exponents

Operations with Exponents

1. Multiplying like bases: $x^n x^m = x^{n+m}$ Add exponents.

2. Dividing like bases: $\dfrac{x^n}{x^m} = x^{n-m}$ Subtract exponents.

3. Removing parentheses: $(xy)^n = x^n y^n$

   $\left(\dfrac{x}{y}\right)^n = \dfrac{x^n}{y^n}$

   $(x^n)^m = x^{nm}$

4. Special conventions: $-x^n = -(x^n), \quad -x^n \neq (-x)^n$

   $cx^n = c(x^n), \quad cx^n \neq (cx)^n$

   $x^{n^m} = x^{(n^m)}, \quad x^{n^m} \neq (x^n)^m$

EXAMPLE 3   **Simplifying Expressions with Exponents**

Simplify each expression.

(a) $2x^2(x^3)$ (b) $(3x)^2 \sqrt[3]{x}$ (c) $\dfrac{3x^2}{(x^{1/2})^3}$

(d) $\dfrac{5x^4}{(x^2)^3}$ (e) $x^{-1}(2x^2)$ (f) $\dfrac{-\sqrt{x}}{5x^{-1}}$

SOLUTION

(a) $2x^2(x^3) = 2x^{2+3} = 2x^5$ $x^n x^m = x^{n+m}$

(b) $(3x)^2 \sqrt[3]{x} = 9x^2 x^{1/3} = 9x^{2+(1/3)} = 9x^{7/3}$ $x^n x^m = x^{n+m}$

(c) $\dfrac{3x^2}{(x^{1/2})^3} = 3\left(\dfrac{x^2}{x^{3/2}}\right) = 3x^{2-(3/2)} = 3x^{1/2}$ $(x^n)^m = x^{nm}, \ \dfrac{x^n}{x^m} = x^{n-m}$

(d) $\dfrac{5x^4}{(x^2)^3} = \dfrac{5x^4}{x^6} = 5x^{4-6} = 5x^{-2} = \dfrac{5}{x^2}$ $(x^n)^m = x^{nm}, \ \dfrac{x^n}{x^m} = x^{n-m}$

(e) $x^{-1}(2x^2) = 2x^{-1}x^2 = 2x^{2-1} = 2x$ $x^n x^m = x^{n+m}$

(f) $\dfrac{-\sqrt{x}}{5x^{-1}} = -\dfrac{1}{5}\left(\dfrac{x^{1/2}}{x^{-1}}\right) = -\dfrac{1}{5}x^{(1/2)+1} = -\dfrac{1}{5}x^{3/2}$ $\dfrac{x^n}{x^m} = x^{n-m}$

TRY IT 3

Simplify each expression.

(a) $3x^2(x^4)$ (b) $(2x)^3 \sqrt{x}$ (c) $\dfrac{4x^2}{(x^{1/3})^2}$

Note in Example 3 that one characteristic of simplified expressions is the absence of negative exponents. Another characteristic of simplified expressions is that sums and differences are written in *factored form*. To do this, you can use the **Distributive Property.**

$$abx^n + acx^{n+m} = ax^n(b + cx^m)$$

Study the next example carefully to be sure that you understand the concepts involved in the factoring process.

EXAMPLE 4 **Simplifying by Factoring**

Simplify each expression by factoring.

(a) $2x^2 - x^3$ (b) $2x^3 + x^2$ (c) $2x^{1/2} + 4x^{5/2}$ (d) $2x^{-1/2} + 3x^{5/2}$

SOLUTION

(a) $2x^2 - x^3 = x^2(2 - x)$

(b) $2x^3 + x^2 = x^2(2x + 1)$

(c) $2x^{1/2} + 4x^{5/2} = 2x^{1/2}(1 + 2x^2)$

(d) $2x^{-1/2} + 3x^{5/2} = x^{-1/2}(2 + 3x^3) = \dfrac{2 + 3x^3}{\sqrt{x}}$

> **TRY IT 4**
>
> Simplify each expression by factoring.
>
> (a) $x^3 - 2x$
>
> (b) $2x^{1/2} + 8x^{3/2}$

Many algebraic expressions obtained in calculus occur in unsimplified form. For instance, the two expressions shown in the following example are the result of an operation in calculus called **differentiation.** [The first is the derivative of $2(x + 1)^{3/2}(2x - 3)^{5/2}$, and the second is the derivative of $2(x + 1)^{1/2}(2x - 3)^{5/2}$.]

> **STUDY TIP**
>
> To check that the simplified expression is equivalent to the original expression, try substituting values for x into each expression.

EXAMPLE 5 **Simplifying by Factoring**

Simplify each expression by factoring.

(a) $3(x + 1)^{1/2}(2x - 3)^{5/2} + 10(x + 1)^{3/2}(2x - 3)^{3/2}$

(b) $(x + 1)^{-1/2}(2x - 3)^{5/2} + 10(x + 1)^{1/2}(2x - 3)^{3/2}$

SOLUTION

(a) $3(x + 1)^{1/2}(2x - 3)^{5/2} + 10(x + 1)^{3/2}(2x - 3)^{3/2}$

$$= (x + 1)^{1/2}(2x - 3)^{3/2}[3(2x - 3) + 10(x + 1)]$$

$$= (x + 1)^{1/2}(2x - 3)^{3/2}(6x - 9 + 10x + 10)$$

$$= (x + 1)^{1/2}(2x - 3)^{3/2}(16x + 1)$$

(b) $(x + 1)^{-1/2}(2x - 3)^{5/2} + 10(x + 1)^{1/2}(2x - 3)^{3/2}$

$$= (x + 1)^{-1/2}(2x - 3)^{3/2}[(2x - 3) + 10(x + 1)]$$

$$= (x + 1)^{-1/2}(2x - 3)^{3/2}(2x - 3 + 10x + 10)$$

$$= (x + 1)^{-1/2}(2x - 3)^{3/2}(12x + 7)$$

$$= \dfrac{(2x - 3)^{3/2}(12x + 7)}{(x + 1)^{1/2}}$$

> **TRY IT 5**
>
> Simplify the expression by factoring.
>
> $(x + 2)^{1/2}(3x - 1)^{3/2}$
> $+ 4(x + 2)^{-1/2}(3x - 1)^{5/2}$

Example 6 shows some additional types of expressions that can occur in calculus. [The expression in Example 6(d) is an antiderivative of $(x + 1)^{2/3}(2x + 3)$, and the expression in Example 6(e) is the derivative of $(x + 2)^3/(x - 1)^3$.]

EXAMPLE 6   **Factors Involving Quotients**

Simplify each expression by factoring.

(a) $\dfrac{3x^2 + x^4}{2x}$

(b) $\dfrac{\sqrt{x} + x^{3/2}}{x}$

(c) $(9x + 2)^{-1/3} + 18(9x + 2)$

(d) $\dfrac{3}{5}(x + 1)^{5/3} + \dfrac{3}{4}(x + 1)^{8/3}$

(e) $\dfrac{3(x + 2)^2(x - 1)^3 - 3(x + 2)^3(x - 1)^2}{[(x - 1)^3]^2}$

SOLUTION

(a) $\dfrac{3x^2 + x^4}{2x} = \dfrac{x^2(3 + x^2)}{2x} = \dfrac{x^{2-1}(3 + x^2)}{2} = \dfrac{x(3 + x^2)}{2}$

(b) $\dfrac{\sqrt{x} + x^{3/2}}{x} = \dfrac{x^{1/2}(1 + x)}{x} = \dfrac{1 + x}{x^{1-(1/2)}} = \dfrac{1 + x}{\sqrt{x}}$

(c) $(9x + 2)^{-1/3} + 18(9x + 2) = (9x + 2)^{-1/3}[1 + 18(9x + 2)^{4/3}]$

$= \dfrac{1 + 18(9x + 2)^{4/3}}{\sqrt[3]{9x + 2}}$

(d) $\dfrac{3}{5}(x + 1)^{5/3} + \dfrac{3}{4}(x + 1)^{8/3} = \dfrac{12}{20}(x + 1)^{5/3} + \dfrac{15}{20}(x + 1)^{8/3}$

$= \dfrac{3}{20}(x + 1)^{5/3}[4 + 5(x + 1)]$

$= \dfrac{3}{20}(x + 1)^{5/3}(4 + 5x + 5)$

$= \dfrac{3}{20}(x + 1)^{5/3}(5x + 9)$

(e) $\dfrac{3(x + 2)^2(x - 1)^3 - 3(x + 2)^3(x - 1)^2}{[(x - 1)^3]^2}$

$= \dfrac{3(x + 2)^2(x - 1)^2[(x - 1) - (x + 2)]}{(x - 1)^6}$

$= \dfrac{3(x + 2)^2(x - 1 - x - 2)}{(x - 1)^{6-2}}$

$= \dfrac{-9(x + 2)^2}{(x - 1)^4}$

TRY IT 6

Simplify the expression by factoring.

$\dfrac{2(x + 1)^2(x - 4) - 2(x + 1)(x - 4)^2}{[(x + 1)^2]^2}$

Domain of an Algebraic Expression

When working with algebraic expressions involving x, you face the potential difficulty of substituting a value of x for which the expression is not defined (does not produce a real number). For example, the expression $\sqrt{2x + 3}$ is *not defined* when $x = -2$ because $\sqrt{2(-2) + 3}$ is not a real number.

The set of all values for which an expression is defined is called its **domain.** So, the domain of $\sqrt{2x + 3}$ is the set of all values of x such that $\sqrt{2x + 3}$ is a real number. In order for $\sqrt{2x + 3}$ to represent a real number, it is necessary that $2x + 3 \geq 0$. In other words, $\sqrt{2x + 3}$ is defined only for those values of x that lie in the interval $\left[-\frac{3}{2}, \infty\right)$, as shown in Figure 0.14.

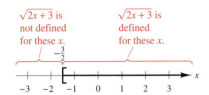

$\sqrt{2x+3}$ is not defined for these x.

$\sqrt{2x+3}$ is defined for these x.

FIGURE 0.14

EXAMPLE 7   **Finding the Domain of an Expression**

Find the domain of each expression.

(a) $\sqrt{3x - 2}$

(b) $\dfrac{1}{\sqrt{3x - 2}}$

(c) $\sqrt[3]{9x + 1}$

SOLUTION

(a) The domain of $\sqrt{3x - 2}$ consists of all x such that

$$3x - 2 \geq 0 \qquad \text{\color{red}{Expression must be nonnegative.}}$$

which implies that $x \geq \frac{2}{3}$. So, the domain is $\left[\frac{2}{3}, \infty\right)$.

(b) The domain of $1/\sqrt{3x - 2}$ is the same as the domain of $\sqrt{3x - 2}$, except that $1/\sqrt{3x - 2}$ is not defined when $3x - 2 = 0$. Because this occurs when $x = \frac{2}{3}$, the domain is $\left(\frac{2}{3}, \infty\right)$.

(c) Because $\sqrt[3]{9x + 1}$ is defined for all real numbers, its domain is $(-\infty, \infty)$.

TRY IT 7

Find the domain of each expression.

(a) $\sqrt{x - 2}$

(b) $\dfrac{1}{\sqrt{x - 2}}$

(c) $\sqrt[3]{x - 2}$

In Exercises 1–20, evaluate the expression for the given value of x.

Expression	x-Value		Expression	x-Value
1. $-3x^3$	$x = 2$		**2.** $\dfrac{x^2}{2}$	$x = 6$
3. $4x^{-3}$	$x = 2$		**4.** $7x^{-2}$	$x = 4$
5. $\dfrac{1 + x^{-1}}{x^{-1}}$	$x = 2$		**6.** $x - 4x^{-2}$	$x = 3$
7. $3x^2 - 4x^3$	$x = -2$		**8.** $5(-x)^3$	$x = 3$
9. $6x^0 - (6x)^0$	$x = 10$		**10.** $\dfrac{1}{(-x)^{-3}}$	$x = 4$
11. $\sqrt[3]{x^2}$	$x = 27$		**12.** $\sqrt{x^3}$	$x = \frac{1}{9}$
13. $x^{-1/2}$	$x = 4$		**14.** $x^{-3/4}$	$x = 16$
15. $x^{-2/5}$	$x = -32$		**16.** $(x^{2/3})^3$	$x = 10$
17. $500x^{60}$	$x = 1.01$		**18.** $\dfrac{10{,}000}{x^{120}}$	$x = 1.075$
19. $\sqrt[3]{x}$	$x = -154$		**20.** $\sqrt[6]{x}$	$x = 325$

In Exercises 21–30, simplify the expression.

21. $6y^{-2}(2y^4)^{-3}$

22. $z^{-3}(3z^4)$

23. $10(x^2)^2$

24. $(4x^3)^2$

25. $\dfrac{7x^2}{x^{-3}}$

26. $\dfrac{x^{-3}}{\sqrt{x}}$

27. $\dfrac{12(x + y)^3}{9(x + y)^{-2}}$

28. $\left(\dfrac{12s^2}{9s}\right)^3$

29. $\dfrac{3x\sqrt{x}}{x^{1/2}}$

30. $\left(\sqrt[3]{x^2}\right)^3$

In Exercises 31–36, simplify by removing all possible factors from the radical.

31. (a) $\sqrt{8}$ (b) $\sqrt{18}$

32. (a) $\sqrt[3]{\frac{16}{27}}$ (b) $\sqrt[3]{\frac{24}{125}}$

33. (a) $\sqrt[3]{16x^5}$ (b) $\sqrt[4]{32x^4z^5}$

34. (a) $\sqrt[4]{(3x^2y^3)^4}$ (b) $\sqrt[3]{54x^7}$

35. (a) $\sqrt[3]{144x^9y^{-4}z^5}$ (b) $\sqrt{12(3x + 5)^7}$

36. (a) $\sqrt[4]{32xy^5z^{-8}}$ (b) $\sqrt{90(2x - 3y)^6}$

In Exercises 37–46, simplify each expression by factoring.

37. $4x^3 - 6x$

38. $8x^4 - 6x^2$

39. $2x^{5/2} + x^{-1/2}$

40. $5x^{3/2} - x^{-3/2}$

41. $3x(x + 1)^{3/2} - 6(x + 1)^{1/2}$

42. $2x(x - 1)^{5/2} - 4(x - 1)^{3/2}$

43. $\dfrac{(x + 1)(x - 1)^2 - (x - 1)^3}{(x + 1)^2}$

44. $\dfrac{(x - 4)(2x - 1)^3 - (2x - 1)^4}{(x - 4)^2}$

45. $(x^2 + 1)^2(x - 1)^{-1/2} + 2x(x - 1)^{1/2}(x^2 + 1)$

46. $(x^4 + 2)^3(x + 3)^{-1/2} + 4x^3(x^4 + 2)^2(x + 3)^{1/2}$

In Exercises 47–56, find the domain of the given expression.

47. $\sqrt{x - 1}$

48. $\sqrt{5 - 2x}$

49. $\sqrt{x^2 + 3}$

50. $\sqrt{4x^2 + 1}$

51. $\dfrac{1}{\sqrt[3]{x - 1}}$

52. $\dfrac{1}{\sqrt[3]{x + 4}}$

53. $\dfrac{\sqrt{x + 2}}{x - 4}$

54. $\dfrac{\sqrt{x - 1}}{x + 1}$

55. $\sqrt{x - 1} + \sqrt{5 - x}$

56. $\dfrac{1}{\sqrt{2x + 3}} + \sqrt{6 - 4x}$

⊕ **Compound Interest** In Exercises 57–60, a certificate of deposit has a principal of P and an annual percentage rate of r (expressed as a decimal) compounded n times per year. Enter the compound interest formula

$$A = P\left(1 + \frac{r}{n}\right)^N$$

into a graphing utility and use it to find the balance after N compoundings.

57. $P = \$10{,}000$, $r = 6.5\%$, $n = 12$, $N = 120$

58. $P = \$7000$, $r = 5\%$, $n = 365$, $N = 1000$

59. $P = \$5000$, $r = 5.5\%$, $n = 4$, $N = 60$

60. $P = \$8000$, $r = 7\%$, $n = 12$, $N = 180$

61. The Period of a Pendulum The period of a pendulum is

$$T = 2\pi\sqrt{\frac{L}{32}}$$

where T is the period in seconds and L is the length of the pendulum in feet. Find the period of a pendulum whose length is 4 feet.

62. Annuity A balance A, after n annual payments of P dollars have been made into an annuity earning an annual percentage rate of r compounded annually, is given by

$$A = P(1 + r) + P(1 + r)^2 + \cdots + P(1 + r)^n.$$

Rewrite this formula by completing the following factorization: $A = P(1 + r)(\quad)$.

The symbol ⊕ indicates an exercise in which you are instructed to use graphing technology or a symbolic computer algebra system. The solutions of other exercises may also be facilitated by use of appropriate technology.

0.4   FACTORING POLYNOMIALS

- Use special products and factorization techniques to factor polynomials.
- Find the domains of radical expressions.
- Use synthetic division to factor polynomials of degree three or more.
- Use the Rational Zero Theorem to find the real zeros of polynomials.

Factorization Techniques

The **Fundamental Theorem of Algebra** states that every nth-degree polynomial

$$a_n x^n + a_{n-1} x^{n-1} + \cdots + a_1 x + a_0, \quad a_n \neq 0$$

has precisely n **zeros**. (The zeros may be repeated or imaginary.) The problem of finding the zeros of a polynomial is equivalent to the problem of factoring the polynomial into linear factors.

Special Products and Factorization Techniques

Quadratic Formula

$$ax^2 + bx + c = 0 \implies x = \frac{-b \pm \sqrt{b^2 - 4ac}}{2a}$$

Example

$$x^2 + 3x - 1 = 0 \implies x = \frac{-3 \pm \sqrt{13}}{2}$$

Special Products

$x^2 - a^2 = (x - a)(x + a)$

$x^3 - a^3 = (x - a)(x^2 + ax + a^2)$

$x^3 + a^3 = (x + a)(x^2 - ax + a^2)$

$x^4 - a^4 = (x - a)(x + a)(x^2 + a^2)$

Examples

$x^2 - 9 = (x - 3)(x + 3)$

$x^3 - 8 = (x - 2)(x^2 + 2x + 4)$

$x^3 + 64 = (x + 4)(x^2 - 4x + 16)$

$x^4 - 16 = (x - 2)(x + 2)(x^2 + 4)$

Binomial Theorem

$(x + a)^2 = x^2 + 2ax + a^2$

$(x - a)^2 = x^2 - 2ax + a^2$

$(x + a)^3 = x^3 + 3ax^2 + 3a^2x + a^3$

$(x - a)^3 = x^3 - 3ax^2 + 3a^2x - a^3$

$(x + a)^4 = x^4 + 4ax^3 + 6a^2x^2 + 4a^3x + a^4$

$(x - a)^4 = x^4 - 4ax^3 + 6a^2x^2 - 4a^3x + a^4$

$(x + a)^n = x^n + nax^{n-1} + \dfrac{n(n-1)}{2!}a^2x^{n-2} + \dfrac{n(n-1)(n-2)}{3!}a^3x^{n-3} + \cdots + na^{n-1}x + a^n{}^*$

$(x - a)^n = x^n - nax^{n-1} + \dfrac{n(n-1)}{2!}a^2x^{n-2} - \dfrac{n(n-1)(n-2)}{3!}a^3x^{n-3} + \cdots \pm na^{n-1}x \mp a^n$

Examples

$(x + 3)^2 = x^2 + 6x + 9$

$(x^2 - 5)^2 = x^4 - 10x^2 + 25$

$(x + 2)^3 = x^3 + 6x^2 + 12x + 8$

$(x - 1)^3 = x^3 - 3x^2 + 3x - 1$

$(x + 2)^4 = x^4 + 8x^3 + 24x^2 + 32x + 16$

$(x - 4)^4 = x^4 - 16x^3 + 96x^2 - 256x + 256$

Factoring by Grouping

$acx^3 + adx^2 + bcx + bd = ax^2(cx + d) + b(cx + d)$

$\qquad\qquad\qquad\qquad\qquad = (ax^2 + b)(cx + d)$

Example

$3x^3 - 2x^2 - 6x + 4 = x^2(3x - 2) - 2(3x - 2)$

$\qquad\qquad\qquad\qquad = (x^2 - 2)(3x - 2)$

* The factorial symbol ! is defined as follows:
   $0! = 1, \; 1! = 1, \; 2! = 2 \cdot 1 = 2, \; 3! = 3 \cdot 2 \cdot 1 = 6, \; 4! = 4 \cdot 3 \cdot 2 \cdot 1 = 24$, and so on.

| **EXAMPLE 1** | **Adding and Subtracting Rational Expressions** |

Perform the indicated operation and simplify.

(a) $x + \dfrac{1}{x}$ (b) $\dfrac{1}{x + 1} - \dfrac{2}{2x - 1}$

SOLUTION

(a) $x + \dfrac{1}{x} = \dfrac{x^2}{x} + \dfrac{1}{x}$ Write with common denominator.

$\qquad = \dfrac{x^2 + 1}{x}$ Add fractions.

(b) $\dfrac{1}{x + 1} - \dfrac{2}{2x - 1} = \dfrac{(2x - 1)}{(x + 1)(2x - 1)} - \dfrac{2(x + 1)}{(x + 1)(2x - 1)}$

$\qquad = \dfrac{2x - 1 - 2x - 2}{2x^2 + x - 1} = \dfrac{-3}{2x^2 + x - 1}$

TRY IT 1

Perform each indicated operation and simplify.

(a) $x + \dfrac{2}{x}$ (b) $\dfrac{2}{x + 1} - \dfrac{1}{2x + 1}$

In adding (or subtracting) fractions whose denominators have no common factors, it is convenient to use the following pattern.

$$\frac{a}{b} + \frac{c}{d} = \frac{a}{b} \times \frac{c}{d} = \frac{ad + bc}{bd}$$

For instance, in Example 1(b), you could have used this pattern as shown.

$$\frac{1}{x + 1} - \frac{2}{2x - 1} = \frac{(2x - 1) - 2(x + 1)}{(x + 1)(2x - 1)}$$

$$= \frac{2x - 1 - 2x - 2}{(x + 1)(2x - 1)} = \frac{-3}{2x^2 + x - 1}$$

In Example 1, the denominators of the rational expressions have no common factors. When the denominators do have common factors, it is best to find the least common denominator before adding or subtracting. For instance, when adding $1/x$ and $2/x^2$, you can recognize that the least common denominator is x^2 and write

$$\frac{1}{x} + \frac{2}{x^2} = \frac{x}{x^2} + \frac{2}{x^2}$$ Write with common denominator.

$$= \frac{x + 2}{x^2}.$$ Add fractions.

This is further demonstrated in Example 2.

| EXAMPLE 2 | **Adding and Subtracting Rational Expressions** |

Perform the indicated operation and simplify.

(a) $\dfrac{x}{x^2 - 1} + \dfrac{3}{x + 1}$ (b) $\dfrac{1}{2(x^2 + 2x)} - \dfrac{1}{4x}$

SOLUTION

(a) Because $x^2 - 1 = (x + 1)(x - 1)$, the least common denominator is $x^2 - 1$.

$$\dfrac{x}{x^2 - 1} + \dfrac{3}{x + 1} = \dfrac{x}{(x - 1)(x + 1)} + \dfrac{3}{x + 1} \qquad \text{Factor.}$$

$$= \dfrac{x}{(x - 1)(x + 1)} + \dfrac{3(x - 1)}{(x - 1)(x + 1)} \qquad \begin{array}{l}\text{Write with} \\ \text{common} \\ \text{denominator.}\end{array}$$

$$= \dfrac{x + 3x - 3}{(x - 1)(x + 1)} \qquad \text{Add fractions.}$$

$$= \dfrac{4x - 3}{x^2 - 1} \qquad \text{Simplify.}$$

(b) In this case, the least common denominator is $4x(x + 2)$.

$$\dfrac{1}{2(x^2 + 2x)} - \dfrac{1}{4x} = \dfrac{1}{2x(x + 2)} - \dfrac{1}{2(2x)} \qquad \text{Factor.}$$

$$= \dfrac{2}{2(2x)(x + 2)} - \dfrac{x + 2}{2(2x)(x + 2)} \qquad \begin{array}{l}\text{Write with} \\ \text{common} \\ \text{denominator.}\end{array}$$

$$= \dfrac{2 - x - 2}{4x(x + 2)} \qquad \begin{array}{l}\text{Subtract} \\ \text{fractions.}\end{array}$$

$$= \dfrac{-\cancel{x}}{4\cancel{x}(x + 2)} \qquad \begin{array}{l}\text{Divide out} \\ \text{like factor.}\end{array}$$

$$= \dfrac{-1}{4(x + 2)}, \quad x \neq 0 \qquad \text{Simplify.}$$

TRY IT 2

Perform each indicated operation and simplify.

(a) $\dfrac{x}{x^2 - 4} + \dfrac{2}{x - 2}$ (b) $\dfrac{1}{3(x^2 + 2x)} - \dfrac{1}{3x}$

ALGEBRA REVIEW

To add more than two fractions, you must find a denominator that is common to all the fractions. For instance, to add $\frac{1}{2}, \frac{1}{3}$, and $\frac{1}{5}$, use a (least) common denominator of 30 and write

$$\dfrac{1}{2} + \dfrac{1}{3} + \dfrac{1}{5} = \dfrac{15}{30} + \dfrac{10}{30} + \dfrac{6}{30} \qquad \text{Write with common denominator.}$$

$$= \dfrac{31}{30}. \qquad \text{Add fractions.}$$

To add more than two rational expressions, use a similar procedure, as shown in Example 3. (Expressions such as those shown in this example are used in calculus to perform an integration technique called integration by partial fractions.)

EXAMPLE 3　Adding More than Two Rational Expressions

Perform the indicated addition of rational expressions.

(a) $\dfrac{A}{x+2} + \dfrac{B}{x-3} + \dfrac{C}{x+4}$

(b) $\dfrac{A}{x+2} + \dfrac{B}{(x+2)^2} + \dfrac{C}{x-1}$

SOLUTION

(a) The least common denominator is $(x+2)(x-3)(x+4)$.

$$\dfrac{A}{x+2} + \dfrac{B}{x-3} + \dfrac{C}{x+4}$$

$$= \dfrac{A(x-3)(x+4) + B(x+2)(x+4) + C(x+2)(x-3)}{(x+2)(x-3)(x+4)}$$

$$= \dfrac{A(x^2+x-12) + B(x^2+6x+8) + C(x^2-x-6)}{(x+2)(x-3)(x+4)}$$

$$= \dfrac{Ax^2 + Bx^2 + Cx^2 + Ax + 6Bx - Cx - 12A + 8B - 6C}{(x+2)(x-3)(x+4)}$$

$$= \dfrac{(A+B+C)x^2 + (A+6B-C)x + (-12A+8B-6C)}{(x+2)(x-3)(x+4)}$$

(b) Here the least common denominator is $(x+2)^2(x-1)$.

$$\dfrac{A}{x+2} + \dfrac{B}{(x+2)^2} + \dfrac{C}{x-1}$$

$$= \dfrac{A(x+2)(x-1) + B(x-1) + C(x+2)^2}{(x+2)^2(x-1)}$$

$$= \dfrac{A(x^2+x-2) + B(x-1) + C(x^2+4x+4)}{(x+2)^2(x-1)}$$

$$= \dfrac{Ax^2 + Cx^2 + Ax + Bx + 4Cx - 2A - B + 4C}{(x+2)^2(x-1)}$$

$$= \dfrac{(A+C)x^2 + (A+B+4C)x + (-2A-B+4C)}{(x+2)^2(x-1)}$$

TRY IT 3

Perform each indicated addition of rational expressions.

(a) $\dfrac{A}{x+1} + \dfrac{B}{x-1} + \dfrac{C}{x+2}$

(b) $\dfrac{A}{x+1} + \dfrac{B}{(x+1)^2} + \dfrac{C}{x-2}$

Expressions Involving Radicals

In calculus, the operation of differentiation tends to produce "messy" expressions when applied to fractional expressions. This is especially true when the fractional expression involves radicals. When differentiation is used, it is important to be able to simplify these expressions so that you can obtain more manageable forms. All of the expressions in Examples 4 and 5 are the results of differentiation. In each case, note how much *simpler* the simplified form is than the original form.

EXAMPLE 4   **Simplifying an Expression with Radicals**

Simplify each expression.

(a) $\dfrac{\sqrt{x+1} - \dfrac{x}{2\sqrt{x+1}}}{x+1}$
(b) $\left(\dfrac{1}{x + \sqrt{x^2+1}}\right)\left(1 + \dfrac{2x}{2\sqrt{x^2+1}}\right)$

SOLUTION

(a) $\dfrac{\sqrt{x+1} - \dfrac{x}{2\sqrt{x+1}}}{x+1} = \dfrac{\dfrac{2(x+1)}{2\sqrt{x+1}} - \dfrac{x}{2\sqrt{x+1}}}{x+1}$ Write with common denominator.

$= \dfrac{\dfrac{2x+2-x}{2\sqrt{x+1}}}{\dfrac{x+1}{1}}$ Subtract fractions.

$= \dfrac{x+2}{2\sqrt{x+1}}\left(\dfrac{1}{x+1}\right)$ To divide, invert and multiply

$= \dfrac{x+2}{2(x+1)^{3/2}}$ Multiply.

(b) $\left(\dfrac{1}{x + \sqrt{x^2+1}}\right)\left(1 + \dfrac{2x}{2\sqrt{x^2+1}}\right)$

$= \left(\dfrac{1}{x + \sqrt{x^2+1}}\right)\left(1 + \dfrac{x}{\sqrt{x^2+1}}\right)$

$= \left(\dfrac{1}{x + \sqrt{x^2+1}}\right)\left(\dfrac{\sqrt{x^2+1}}{\sqrt{x^2+1}} + \dfrac{x}{\sqrt{x^2+1}}\right)$

$= \left(\dfrac{1}{x + \sqrt{x^2+1}}\right)\left(\dfrac{x + \sqrt{x^2+1}}{\sqrt{x^2+1}}\right)$

$= \dfrac{1}{\sqrt{x^2+1}}$

TRY IT 4

Simplify each expression.

(a) $\dfrac{\sqrt{x+2} - \dfrac{x}{4\sqrt{x+2}}}{x+2}$
(b) $\left(\dfrac{1}{x + \sqrt{x^2+4}}\right)\left(1 + \dfrac{x}{\sqrt{x^2+4}}\right)$

| **EXAMPLE 5** | **Simplifying an Expression with Radicals** |

Simplify the expression.

$$\frac{-x\left(\dfrac{2x}{2\sqrt{x^2+1}}\right) + \sqrt{x^2+1}}{x^2} + \left(\frac{1}{x+\sqrt{x^2+1}}\right)\left(1 + \frac{2x}{2\sqrt{x^2+1}}\right)$$

SOLUTION   From Example 4(b), you already know that the second part of this sum simplifies to $1/\sqrt{x^2+1}$. The first part simplifies as shown.

$$\frac{-x\left(\dfrac{2x}{2\sqrt{x^2+1}}\right) + \sqrt{x^2+1}}{x^2} = \frac{-x^2}{x^2\sqrt{x^2+1}} + \frac{\sqrt{x^2+1}}{x^2}$$

$$= \frac{-x^2}{x^2\sqrt{x^2+1}} + \frac{x^2+1}{x^2\sqrt{x^2+1}}$$

$$= \frac{-x^2+x^2+1}{x^2\sqrt{x^2+1}}$$

$$= \frac{1}{x^2\sqrt{x^2+1}}$$

So, the sum is

$$\frac{-x\left(\dfrac{2x}{2\sqrt{x^2+1}}\right) + \sqrt{x^2+1}}{x^2} + \left(\frac{1}{x+\sqrt{x^2+1}}\right)\left(1 + \frac{2x}{2\sqrt{x^2+1}}\right)$$

$$= \frac{1}{x^2\sqrt{x^2+1}} + \frac{1}{\sqrt{x^2+1}}$$

$$= \frac{1}{x^2\sqrt{x^2+1}} + \frac{x^2}{x^2\sqrt{x^2+1}}$$

$$= \frac{x^2+1}{x^2\sqrt{x^2+1}}$$

$$= \frac{\sqrt{x^2+1}}{x^2}.$$

TRY IT 5

Simplify the expression.

$$\frac{-x\left(\dfrac{3x}{3\sqrt{x^2+4}}\right) + \sqrt{x^2+4}}{x^2} + \left(\frac{1}{x+\sqrt{x^2+4}}\right)\left(1 + \frac{3x}{3\sqrt{x^2+4}}\right)$$

ALGEBRA REVIEW

To check that the simplified expression in Example 5 is equivalent to the original expression, try substituting values of x into each expression. For instance, when you substitute $x = 1$ into each expression, you obtain $\sqrt{2}$.

Rationalization Techniques

In working with quotients involving radicals, it is often convenient to move the radical expression from the denominator to the numerator, or vice versa. For example, you can move $\sqrt{2}$ from the denominator to the numerator in the following quotient by multiplying by $\sqrt{2}/\sqrt{2}$.

Radical in Denominator	*Rationalize*	*Radical in Numerator*
$\dfrac{1}{\sqrt{2}}$	$\dfrac{1}{\sqrt{2}}\left(\dfrac{\sqrt{2}}{\sqrt{2}}\right)$	$\dfrac{\sqrt{2}}{2}$

This process is called **rationalizing the denominator.** A similar process is used to **rationalize the numerator.**

Rationalizing Techniques

1. If the denominator is $\sqrt{a}$, multiply by $\dfrac{\sqrt{a}}{\sqrt{a}}$.

2. If the denominator is $\sqrt{a} - \sqrt{b}$, multiply by $\dfrac{\sqrt{a} + \sqrt{b}}{\sqrt{a} + \sqrt{b}}$.

3. If the denominator is $\sqrt{a} + \sqrt{b}$, multiply by $\dfrac{\sqrt{a} - \sqrt{b}}{\sqrt{a} - \sqrt{b}}$.

The same guidelines apply to rationalizing numerators.

ALGEBRA REVIEW

The success of the second and third rationalizing techniques stems from the following.

$$\left(\sqrt{a} - \sqrt{b}\right)\left(\sqrt{a} + \sqrt{b}\right)$$
$$= a - b$$

EXAMPLE 6 **Rationalizing Denominators and Numerators**

Rationalize the denominator or numerator.

(a) $\dfrac{3}{\sqrt{12}}$ (b) $\dfrac{\sqrt{x+1}}{2}$ (c) $\dfrac{1}{\sqrt{5} + \sqrt{2}}$ (d) $\dfrac{1}{\sqrt{x} - \sqrt{x+1}}$

SOLUTION

(a) $\dfrac{3}{\sqrt{12}} = \dfrac{3}{2\sqrt{3}} = \dfrac{3}{2\sqrt{3}}\left(\dfrac{\sqrt{3}}{\sqrt{3}}\right) = \dfrac{3\sqrt{3}}{2(3)} = \dfrac{\sqrt{3}}{2}$

(b) $\dfrac{\sqrt{x+1}}{2} = \dfrac{\sqrt{x+1}}{2}\left(\dfrac{\sqrt{x+1}}{\sqrt{x+1}}\right)$

$\qquad = \dfrac{x+1}{2\sqrt{x+1}}$

(c) $\dfrac{1}{\sqrt{5} + \sqrt{2}} = \dfrac{1}{\sqrt{5} + \sqrt{2}}\left(\dfrac{\sqrt{5} - \sqrt{2}}{\sqrt{5} - \sqrt{2}}\right) = \dfrac{\sqrt{5} - \sqrt{2}}{5 - 2} = \dfrac{\sqrt{5} - \sqrt{2}}{3}$

(d) $\dfrac{1}{\sqrt{x} - \sqrt{x+1}} = \dfrac{1}{\sqrt{x} - \sqrt{x+1}}\left(\dfrac{\sqrt{x} + \sqrt{x+1}}{\sqrt{x} + \sqrt{x+1}}\right)$

$\qquad = \dfrac{\sqrt{x} + \sqrt{x+1}}{x - (x+1)}$

$\qquad = -\sqrt{x} - \sqrt{x+1}$

TRY IT 6

Rationalize the denominator or numerator.

(a) $\dfrac{5}{\sqrt{8}}$

(b) $\dfrac{\sqrt{x+2}}{4}$

(c) $\dfrac{1}{\sqrt{6} - \sqrt{3}}$

(d) $\dfrac{1}{\sqrt{x} + \sqrt{x+2}}$

EXERCISES 0.5

In Exercises 1–16, perform the indicated operations and simplify your answer.

1. $\dfrac{5}{x-1}+\dfrac{x}{x-1}$

2. $\dfrac{2x-1}{x+3}+\dfrac{1-x}{x+3}$

3. $\dfrac{2x}{x^2+2}-\dfrac{1-3x}{x^2+2}$

4. $\dfrac{5x+10}{2x-1}-\dfrac{2x+10}{2x-1}$

5. $\dfrac{2}{x^2-4}-\dfrac{1}{x-2}$

6. $\dfrac{x}{x^2+x-2}-\dfrac{1}{x+2}$

7. $\dfrac{5}{x-3}+\dfrac{3}{3-x}$

8. $\dfrac{x}{2-x}+\dfrac{2}{x-2}$

9. $\dfrac{A}{x+1}+\dfrac{B}{(x+1)^2}+\dfrac{C}{x-2}$

10. $\dfrac{A}{x-5}+\dfrac{B}{x+5}+\dfrac{C}{(x+5)^2}$

11. $\dfrac{A}{x-6}+\dfrac{Bx+C}{x^2+3}$

12. $\dfrac{Ax+B}{x^2+2}+\dfrac{C}{x-4}$

13. $-\dfrac{1}{x}+\dfrac{2}{x^2+1}$

14. $\dfrac{2}{x+1}+\dfrac{1-x}{x^2-2x+3}$

15. $\dfrac{1}{x^2-x-2}-\dfrac{x}{x^2-5x+6}$

16. $\dfrac{x-1}{x^2+5x+4}+\dfrac{2}{x^2-x-2}+\dfrac{10}{x^2+2x-8}$

In Exercises 17–30, simplify each expression.

17. $\dfrac{-x}{(x+1)^{3/2}}+\dfrac{2}{(x+1)^{1/2}}$

18. $2\sqrt{x}(x-2)+\dfrac{(x-2)^2}{2\sqrt{x}}$

19. $\dfrac{2-t}{2\sqrt{1+t}}-\sqrt{1+t}$

20. $-\dfrac{\sqrt{x^2+1}}{x^2}+\dfrac{1}{\sqrt{x^2+1}}$

21. $\left(2x\sqrt{x^2+1}-\dfrac{x^3}{\sqrt{x^2+1}}\right)\div(x^2+1)$

22. $\left(\sqrt{x^3+1}-\dfrac{3x^3}{2\sqrt{x^3+1}}\right)\div(x^3+1)$

23. $\dfrac{(x^2+2)^{1/2}-x^2(x^2+2)^{-1/2}}{x^2}$

24. $\dfrac{x(x+1)^{-1/2}-(x+1)^{1/2}}{x^2}$

25. $\dfrac{\dfrac{\sqrt{x+1}}{\sqrt{x}}-\dfrac{\sqrt{x}}{\sqrt{x+1}}}{2(x+1)}$

26. $\dfrac{\dfrac{2x^2}{3(x^2-1)^{2/3}}-(x^2-1)^{1/3}}{x^2}$

27. $\dfrac{x}{2(x+2)^{1/2}}+(x+2)^{1/2}$

28. $\dfrac{x}{(x-5)^{1/2}}+2(x-5)^{1/2}$

29. $\dfrac{-x^2}{(2x+3)^{3/2}}+\dfrac{2x}{(2x+3)^{1/2}}$

30. $\dfrac{-x}{2(3+x^2)^{3/2}}+\dfrac{3}{(3+x^2)^{1/2}}$

In Exercises 31–44, rationalize the numerator or denominator and simplify.

31. $\dfrac{3}{\sqrt{6}}$

32. $\dfrac{5}{\sqrt{10}}$

33. $\dfrac{x}{\sqrt{x-4}}$

34. $\dfrac{4y}{\sqrt{y+8}}$

35. $\dfrac{49(x-3)}{\sqrt{x^2-9}}$

36. $\dfrac{10(x+2)}{\sqrt{x^2-x-6}}$

37. $\dfrac{5}{\sqrt{14}-2}$

38. $\dfrac{13}{6+\sqrt{10}}$

39. $\dfrac{2x}{5-\sqrt{3}}$

40. $\dfrac{x}{\sqrt{2}+\sqrt{3}}$

41. $\dfrac{1}{\sqrt{6}+\sqrt{5}}$

42. $\dfrac{\sqrt{15}+3}{12}$

43. $\dfrac{2}{\sqrt{x}+\sqrt{x-2}}$

44. $\dfrac{10}{\sqrt{x}+\sqrt{x+5}}$

In Exercises 45 and 46, perform the indicated operations and rationalize as needed.

45. $\dfrac{\dfrac{\sqrt{4-x^2}}{x^4}-\dfrac{2}{x^2\sqrt{4-x^2}}}{4-x^2}$

46. $\dfrac{\dfrac{\sqrt{x^2+1}}{x^2}-\dfrac{1}{x\sqrt{x^2+1}}}{x^2+1}$

47. Installment Loan The monthly payment M for an installment loan is given by the formula

$$M=P\left[\dfrac{r/12}{1-\left(\dfrac{1}{(r/12)+1}\right)^{N}}\right]$$

where P is the amount of the loan, r is the annual percentage rate, and N is the number of monthly payments. Enter the formula into a graphing utility, and use it to find the monthly payment for a loan of \$10,000 at an annual percentage rate of 14% ($r=0.14$) for 5 years ($N=60$ monthly payments).

48. Inventory A retailer has determined that the cost C of ordering and storing x units of a product is

$$C=6x+\dfrac{900,000}{x}.$$

(a) Write the expression for cost as a single fraction.

(b) Determine the cost for ordering and storing $x=240$ units of this product.

1

Functions, Graphs, and Limits

© Jeff Zaruba/CORBIS

A graph showing changes in a company's earnings and other financial indicators can depict the company's general financial trends over time.

STRATEGIES FOR SUCCESS

WHAT YOU SHOULD LEARN:

- How to plot points in the Cartesian plane and find the distance between two points
- How to sketch the graph of an equation, and find the x- and y-intercepts
- How to write equations of lines and sketch the lines
- How to evaluate, simplify, and find inverse functions
- How to find the limit of a function and discuss the continuity of a variety of functions

WHY YOU SHOULD LEARN IT:

Functions have many applications in real life, as can be seen by the examples below, which represent a small sample of the applications in this chapter.

- Dow Jones Industrial Average, Exercises 35 and 36 on page 9
- Break-Even Analysis, Exercises 63–68 on page 22
- Linear Depreciation, Exercises 86 and 87 on page 35
- Demand Function, Exercise 71 on page 47
- Market Equilibrium, Exercise 74 on page 48

1.1   THE CARTESIAN PLANE AND THE DISTANCE FORMULA

- Plot points in a coordinate plane and read data presented graphically.
- Find the distance between two points in a coordinate plane.
- Find the midpoints of line segments connecting two points.
- Translate points in a coordinate plane.

The Cartesian Plane

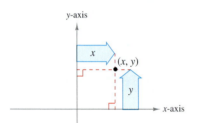

FIGURE 1.1   The Cartesian Plane

Just as you can represent real numbers by points on a real number line, you can represent ordered pairs of real numbers by points in a plane called the **rectangular coordinate system,** or the **Cartesian plane,** after the French mathematician René Descartes (1596–1650).

The Cartesian plane is formed by using two real number lines intersecting at right angles, as shown in Figure 1.1. The horizontal real number line is usually called the **x-axis,** and the vertical real number line is usually called the **y-axis.** The point of intersection of these two axes is the **origin,** and the two axes divide the plane into four parts called **quadrants.**

Each point in the plane corresponds to an **ordered pair** (x, y) of real numbers x and y, called **coordinates** of the point. The **x-coordinate** represents the directed distance from the y-axis to the point, and the **y-coordinate** represents the directed distance from the x-axis to the point, as shown in Figure 1.2.

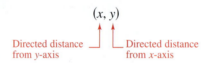

FIGURE 1.2

STUDY TIP

The notation (x, y) denotes both a point in the plane and an open interval on the real number line. The context will tell you which meaning is intended.

EXAMPLE 1   Plotting Points in the Cartesian Plane

Plot the points $(-1, 2)$, $(3, 4)$, $(0, 0)$, $(3, 0)$, and $(-2, -3)$.

SOLUTION   To plot the point

$$(-1, 2)$$

x-coordinate ⎦   ⎣ *y*-coordinate

imagine a vertical line through -1 on the x-axis and a horizontal line through 2 on the y-axis. The intersection of these two lines is the point $(-1, 2)$. The other four points can be plotted in a similar way and are shown in Figure 1.3.

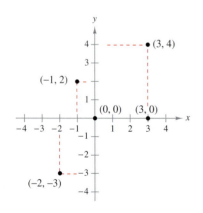

FIGURE 1.3

TRY IT 1

Plot the points $(-3, 2)$, $(4, -2)$, $(3, 1)$, $(0, -2)$, and $(-1, -2)$.

Using a rectangular coordinate system allows you to visualize relationships between two variables. It would be difficult to overestimate the importance of Descartes's introduction of coordinates to the plane. Today his ideas are in common use in virtually every scientific and business-related field. In Example 2, notice how much your intuition is enhanced by the use of a graphical presentation.

EXAMPLE 2 Sketching a Scatter Plot

The amounts A (in millions of dollars) spent on snowmobiles in the United States from 1993 through 2002 are shown in the table, where t represents the year. Sketch a scatter plot of the data. *(Source: National Sporting Goods Association)*

t	1993	1994	1995	1996	1997	1998	1999	2000	2001	2002
A	515	715	910	974	975	883	820	894	784	808

SOLUTION To sketch a *scatter plot* of the data given in the table, you simply represent each pair of values by an ordered pair (t, A), and plot the resulting points, as shown in Figure 1.4. For instance, the first pair of values is represented by the ordered pair $(1993, 515)$. Note that the break in the t-axis indicates that the numbers between 0 and 1992 have been omitted.

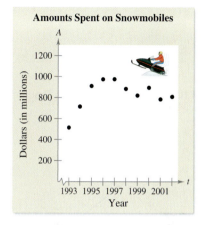

FIGURE 1.4

TRY IT 2

From 1991 to 2000, the enrollments E (in millions) of students in U.S. public colleges are shown, where t represents the year. Sketch a scatter plot of the data. *(Source: U.S. National Center for Education Statistics)*

t	1991	1992	1993	1994	1995	1996	1997	1998	1999	2000
E	11.3	11.4	11.2	11.1	11.1	11.1	11.2	11.1	11.3	11.8

STUDY TIP

In Example 2, you could let $t = 1$ represent the year 1993. In that case, the horizontal axis would not have been broken, and the tick marks would have been labeled 1 through 10 (instead of 1993 through 2002).

TECHNOLOGY

The scatter plot in Example 2 is only one way to represent the given data graphically. Two other techniques are shown at the right. The first is a *bar graph* and the second is a *line graph*. All three graphical representations were created with a computer. If you have access to computer graphing software, try using it to represent graphically the data given in Example 2.

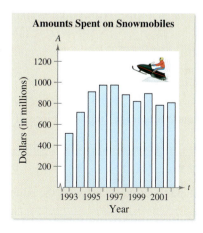

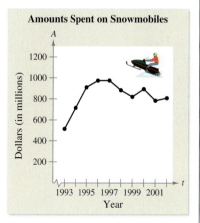

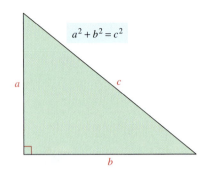

$$a^2 + b^2 = c^2$$

FIGURE 1.5   Pythagorean Theorem

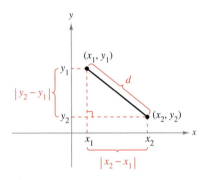

FIGURE 1.6   Distance Between Two Points

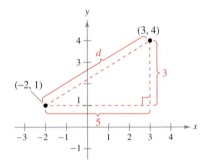

FIGURE 1.7

The Distance Formula

Recall from the Pythagorean Theorem that, for a right triangle with hypotenuse of length c and sides of lengths a and b, you have

$$a^2 + b^2 = c^2 \qquad \text{Pythagorean Theorem}$$

as shown in Figure 1.5. (The converse is also true. That is, if $a^2 + b^2 = c^2$, then the triangle is a right triangle.)

Suppose you want to determine the distance d between two points (x_1, y_1) and (x_2, y_2) in the plane. With these two points, a right triangle can be formed, as shown in Figure 1.6. The length of the vertical side of the triangle is

$$|y_2 - y_1|$$

and the length of the horizontal side is

$$|x_2 - x_1|.$$

By the Pythagorean Theorem, you can write

$$d^2 = |x_2 - x_1|^2 + |y_2 - y_1|^2$$
$$d = \sqrt{|x_2 - x_1|^2 + |y_2 - y_1|^2}$$
$$d = \sqrt{(x_2 - x_1)^2 + (y_2 - y_1)^2}.$$

This result is the **Distance Formula.**

The Distance Formula

The distance d between the points (x_1, y_1) and (x_2, y_2) in the plane is

$$d = \sqrt{(x_2 - x_1)^2 + (y_2 - y_1)^2}.$$

EXAMPLE 3   **Finding a Distance**

Find the distance between the points $(-2, 1)$ and $(3, 4)$.

SOLUTION   Let $(x_1, y_1) = (-2, 1)$ and $(x_2, y_2) = (3, 4)$. Then apply the Distance Formula as shown.

$$d = \sqrt{(x_2 - x_1)^2 + (y_2 - y_1)^2} \qquad \text{Distance Formula}$$
$$= \sqrt{[3 - (-2)]^2 + (4 - 1)^2} \qquad \text{Substitute for } x_1, y_1, x_2, \text{ and } y_2.$$
$$= \sqrt{(5)^2 + (3)^2} \qquad \text{Simplify.}$$
$$= \sqrt{34}$$
$$\approx 5.83 \qquad \text{Use a calculator.}$$

Note in Figure 1.7 that a distance of 5.83 looks about right.

TRY IT 3

Find the distance between the points $(-2, 1)$ and $(2, 4)$.

EXAMPLE 4   **Verifying a Right Triangle**

Use the Distance Formula to show that the points $(2, 1), (4, 0)$, and $(5, 7)$ are vertices of a right triangle.

SOLUTION   The three points are plotted in Figure 1.8. Using the Distance Formula, you can find the lengths of the three sides as shown below.

$$d_1 = \sqrt{(5 - 2)^2 + (7 - 1)^2} = \sqrt{9 + 36} = \sqrt{45}$$
$$d_2 = \sqrt{(4 - 2)^2 + (0 - 1)^2} = \sqrt{4 + 1} = \sqrt{5}$$
$$d_3 = \sqrt{(5 - 4)^2 + (7 - 0)^2} = \sqrt{1 + 49} = \sqrt{50}$$

Because

$$d_1^2 + d_2^2 = 45 + 5 = 50 = d_3^2$$

you can apply the converse of the Pythagorean Theorem to conclude that the triangle must be a right triangle.

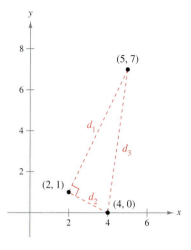

FIGURE 1.8

TRY IT 4

Use the Distance Formula to show that the points $(2, -1), (5, 5)$, and $(6, -3)$ are vertices of a right triangle.

The figures provided with Examples 3 and 4 were not really essential to the solution. *Nevertheless*, we strongly recommend that you develop the habit of including sketches with your solutions—even if they are not required.

EXAMPLE 5   **Finding the Length of a Pass**

In a football game, a quarterback throws a pass from the five-yard line, 20 yards from the sideline. The pass is caught by a wide receiver on the 45-yard line, 50 yards from the same sideline, as shown in Figure 1.9. How long was the pass?

SOLUTION   You can find the length of the pass by finding the distance between the points $(20, 5)$ and $(50, 45)$.

$$d = \sqrt{(50 - 20)^2 + (45 - 5)^2}$$ Distance Formula
$$= \sqrt{900 + 1600}$$
$$= 50$$ Simplify.

So, the pass was 50 yards long.

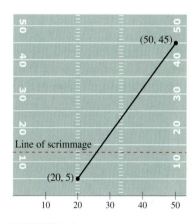

FIGURE 1.9

STUDY TIP

In Example 5, the scale along the goal line showing distance from the sideline does not normally appear on a football field. However, when you use coordinate geometry to solve real-life problems, you are free to place the coordinate system in any way that is convenient to the solution of the problem.

TRY IT 5

A quarterback throws a pass from the 10-yard line, 10 yards from the sideline. The pass is caught by a wide receiver on the 30-yard line, 25 yards from the same sideline. How long was the pass?

The Midpoint Formula

To find the **midpoint** of the line segment that joins two points in a coordinate plane, you can simply find the average values of the respective coordinates of the two endpoints.

> **The Midpoint Formula**
>
> The midpoint of the segment joining the points (x_1, y_1) and (x_2, y_2) is
> $$\text{Midpoint} = \left(\frac{x_1 + x_2}{2}, \frac{y_1 + y_2}{2} \right).$$

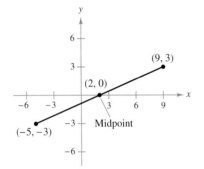

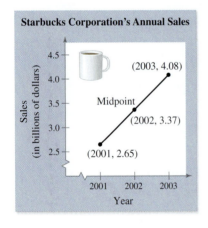

FIGURE 1.10

EXAMPLE 6   **Finding a Segment's Midpoint**

Find the midpoint of the line segment joining the points $(-5, -3)$ and $(9, 3)$, as shown in Figure 1.10.

SOLUTION   Let $(x_1, y_1) = (-5, -3)$ and $(x_2, y_2) = (9, 3)$.
$$\text{Midpoint} = \left(\frac{x_1 + x_2}{2}, \frac{y_1 + y_2}{2} \right) = \left(\frac{-5 + 9}{2}, \frac{-3 + 3}{2} \right) = (2, 0)$$

> **TRY IT 6**
>
> Find the midpoint of the line segment joining $(-6, 2)$ and $(2, 8)$.

EXAMPLE 7   **Estimating Annual Sales**

Starbucks Corporation had annual sales of $2.65 billion in 2001 and $4.08 billion in 2003. Without knowing any additional information, what would you estimate the 2002 sales to have been?   *(Source: Starbucks Corp.)*

SOLUTION   One solution to the problem is to assume that sales followed a linear pattern. With this assumption, you can estimate the 2002 sales by finding the midpoint of the segment connecting the points $(2001, 2.65)$ and $(2003, 4.08)$.
$$\text{Midpoint} = \left(\frac{2001 + 2003}{2}, \frac{2.65 + 4.08}{2} \right) \approx (2002, 3.37)$$

So, you would estimate the 2002 sales to have been about $3.37 billion, as shown in Figure 1.11. (The actual 2002 sales were $3.29 billion.)

FIGURE 1.11

> **TRY IT 7**
>
> Maytag Corporation had annual sales of $4.32 billion in 2001 and $4.79 billion in 2003. What would you estimate the 2002 annual sales to have been? *(Source: Maytag Corp.)*

Translating Points in the Plane

| EXAMPLE 8 | **Translating Points in the Plane** |

Figure 1.12(a) shows the vertices of a parallelogram. Find the vertices of the parallelogram after it has been translated two units down and four units to the right.

SOLUTION   To translate each vertex two units down, subtract 2 from each y-coordinate. To translate each vertex four units to the right, add 4 to each x-coordinate.

Original Point	Translated Point
$(1, 0)$	$(1 + 4, 0 - 2) = (5, -2)$
$(3, 2)$	$(3 + 4, 2 - 2) = (7, 0)$
$(3, 6)$	$(3 + 4, 6 - 2) = (7, 4)$
$(1, 4)$	$(1 + 4, 4 - 2) = (5, 2)$

The translated parallelogram is shown in Figure 1.12(b).

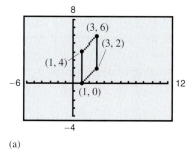

(a)

DREAMWORKS/THE KOBAL COLLECTION

(b)

FIGURE 1.12

Many movies now use extensive computer graphics, much of which consists of transformations of points in two- and three-dimensional space. The photo above shows a scene from Shrek. *The movie's animators used computer graphics to design the scenery, characters, motion, and even the lighting in each scene.*

TRY IT 8

Find the vertices of the parallelogram in Example 8 after it has been translated two units to the left and four units down.

TAKE ANOTHER LOOK

Transforming Points in a Coordinate Plane

Example 8 illustrates points that have been translated (or *slid*) in a coordinate plane. The translated parallelogram is congruent to (has the same size and shape as) the original parallelogram. Try using a graphing utility to graph the transformed parallelogram for each of the following transformations. Describe the transformation. Is it a translation, a reflection, or a rotation? Is the transformed parallelogram congruent to the original parallelogram?

a. (x, y) ⟹ $(-x, y)$

b. (x, y) ⟹ $(x, -y)$

c. (x, y) ⟹ $(-x, -y)$

PREREQUISITE REVIEW 1.1

The following warm-up exercises involve skills that were covered in earlier sections. You will use these skills in the exercise set for this section.

In Exercises 1–6, simplify each expression.

1. $\sqrt{(3-6)^2 + [1-(-5)]^2}$

2. $\sqrt{(-2-0)^2 + [-7-(-3)]^2}$

3. $\dfrac{5 + (-4)}{2}$

4. $\dfrac{-3 + (-1)}{2}$

5. $\sqrt{27} + \sqrt{12}$

6. $\sqrt{8} - \sqrt{18}$

In Exercises 7–10, solve for x or y.

7. $\sqrt{(3-x)^2 + (7-4)^2} = \sqrt{45}$

8. $\sqrt{(6-2)^2 + (-2-y)^2} = \sqrt{52}$

9. $\dfrac{x + (-5)}{2} = 7$

10. $\dfrac{-7 + y}{2} = -3$

EXERCISES 1.1

In Exercises 1–6, (a) find the length of each side of the right triangle and (b) show that these lengths satisfy the Pythagorean Theorem.

*1.
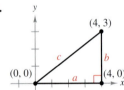

2.
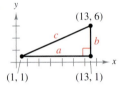

3.
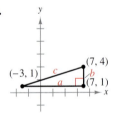

4.
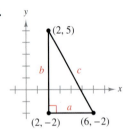

5.
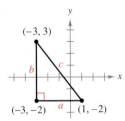

6.
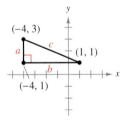

In Exercises 7–14, (a) plot the points, (b) find the distance between the points, and (c) find the midpoint of the line segment joining the points.

7. $(3, 1), (5, 5)$

8. $(-3, 2), (3, -2)$

9. $\left(\frac{1}{2}, 1\right), \left(-\frac{3}{2}, -5\right)$

10. $\left(\frac{2}{3}, -\frac{1}{3}\right), \left(\frac{5}{6}, 1\right)$

11. $(2, 2), (4, 14)$

12. $(-3, 7), (1, -1)$

13. $\left(1, \sqrt{3}\right), (-1, 1)$

14. $(-2, 0), \left(0, \sqrt{2}\right)$

In Exercises 15–18, show that the points form the vertices of the given figure. (A rhombus is a quadrilateral whose sides have the same length.)

Vertices	Figure
15. $(0, 1), (3, 7), (4, -1)$	Right triangle
16. $(1, -3), (3, 2), (-2, 4)$	Isosceles triangle
17. $(0, 0), (1, 2), (2, 1), (3, 3)$	Rhombus
18. $(0, 1), (3, 7), (4, 4), (1, -2)$	Parallelogram

In Exercises 19–22, use the Distance Formula to determine whether the points are collinear (lie on the same line).

19. $(0, -4), (2, 0), (3, 2)$ **20.** $(0, 4), (7, -6), (-5, 11)$

21. $(-2, -6), (1, -3), (5, 2)$ **22.** $(-1, 1), (3, 3), (5, 5)$

In Exercises 23 and 24, find x such that the distance between the points is 5.

23. $(1, 0), (x, -4)$ **24.** $(2, -1), (x, 2)$

In Exercises 25 and 26, find y such that the distance between the points is 8.

25. $(0, 0), (3, y)$ **26.** $(5, 1), (5, y)$

27. Use the Midpoint Formula repeatedly to find the three points that divide the segment joining (x_1, y_1) and (x_2, y_2) into four equal parts.

28. Show that $\left(\frac{1}{3}[2x_1 + x_2], \frac{1}{3}[2y_1 + y_2]\right)$ is one of the points of trisection of the line segment joining (x_1, y_1) and (x_2, y_2). Then, find the second point of trisection by finding the midpoint of the segment joining

$$\left(\frac{1}{3}[2x_1 + x_2], \frac{1}{3}[2y_1 + y_2]\right) \text{ and } (x_2, y_2).$$

29. Use Exercise 27 to find the points that divide the line segment joining the given points into four equal parts.

(a) $(1, -2), (4, -1)$

(b) $(-2, -3), (0, 0)$

30. Use Exercise 28 to find the points of trisection of the line segment joining the given points.

(a) $(1, -2), (4, 1)$

(b) $(-2, -3), (0, 0)$

31. *Building Dimensions* The base and height of the trusses for the roof of a house are 32 feet and 5 feet, respectively (see figure).

(a) Find the distance d from the eaves to the peak of the roof.

(b) The length of the house is 40 feet. Use the result of part (a) to find the number of square feet of roofing.

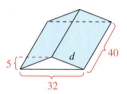

32. *Wire Length* A guy wire is stretched from a broadcasting tower at a point 200 feet above the ground to an anchor 125 feet from the base (see figure). How long is the wire?

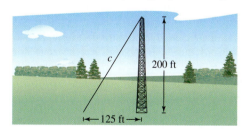

The symbol ⊕ indicates an exercise in which you are instructed to use graphing technology or a symbolic computer algebra system. The solutions of other exercises may also be facilitated by use of appropriate technology.

⊕ In Exercises 33 and 34, use a graphing utility to graph a scatter plot, a bar graph, or a line graph to represent the data. Describe any trends that appear.

33. *Consumer Trends* The numbers (in millions) of cable television subscribers in the United States for 1992–2001 are shown in the table. *(Source: Nielsen Media Research)*

Year	1992	1993	1994	1995	1996
Subscribers	57.2	58.8	60.5	63.0	64.6

Year	1997	1998	1999	2000	2001
Subscribers	65.9	67.0	68.5	69.3	73.0

34. *Consumer Trends* The numbers (in millions) of cellular telephone subscribers in the United States for 1993–2002 are shown in the table. *(Source: Cellular Telecommunications & Internet Association)*

Year	1993	1994	1995	1996	1997
Subscribers	16.0	24.1	33.8	44.0	55.3

Year	1998	1999	2000	2001	2002
Subscribers	69.2	86.0	109.5	128.4	140.8

Dow Jones Industrial Average In Exercises 35 and 36, use the figure below showing the Dow Jones Industrial Average for common stocks. *(Source: Dow Jones, Inc.)*

35. Estimate the Dow Jones Industrial Average for each date.

(a) March 2002   (b) December 2002

(c) May 2003   (d) January 2004

36. Estimate the percent increase or decrease in the Dow Jones Industrial Average (a) from April 2002 to November 2002 and (b) from June 2003 to February 2004.

Figure for 35 and 36

Construction   In Exercises 37 and 38, use the figure, which shows the median sales prices of existing one-family homes sold (in thousands of dollars) in the United States from 1987 to 2002. *(Source: National Association of Realtors)*

37. Estimate the median sales price of existing one-family homes for each year.

    (a) 1987   (b) 1992

    (c) 1997   (d) 2002

38. Estimate the percent increases in the value of existing one-family homes (a) from 1993 to 1994 and (b) from 2001 to 2002.

Figure for 37 and 38

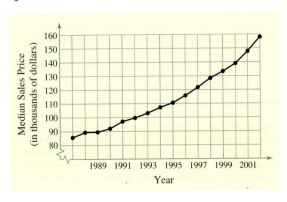

Research Project   In Exercises 39 and 40, (a) use the Midpoint Formula to estimate the revenue and profit of the company in 2001. (b) Then use your school's library, the Internet, or some other reference source to find the actual revenue and profit for 2001. (c) Did the revenue and profit increase in a linear pattern from 1999 to 2003? Explain your reasoning. (d) What were the company's expenses during each of the given years? (e) How would you rate the company's growth from 1999 to 2003? *(Source: Walgreen Company and The Yankee Candle Company)*

39. Walgreen Company

Year	1999	2001	2003
Revenue (millions of $)	17,839		32,505
Profit (millions of $)	624.1		1157.3

40. The Yankee Candle Company

Year	1999	2001	2003
Revenue (millions of $)	256.6		508.6
Profit (millions of $)	34.3		74.8

Computer Graphics   In Exercises 41 and 42, the red figure is translated to a new position in the plane to form the blue figure. (a) Find the vertices of the transformed figure. (b) Then use a graphing utility to draw both figures.

41.

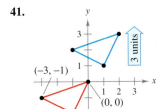

42.

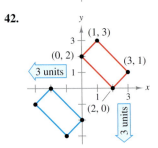

43. ***Economics***   The table shows the numbers of ear infections treated by doctors at HMO clinics of three different sizes: small, medium, and large.

Cases per small clinic	Cases per medium clinic	Cases per large clinic	Number of doctors
0	0	0	0
20	30	35	1
28	42	49	2
35	53	62	3
40	60	70	4

    (a) Show the relationship between doctors and treated ear infections using *three* curves, where the number of doctors is on the horizontal axis and the number of ear infections treated is on the vertical axis.

    (b) Compare the three relationships.

(Source: Adapted from Taylor, Economics, *Fourth Edition)*

The symbol 🌐 indicates an exercise that contains material from textbooks in other disciplines.

1.2   GRAPHS OF EQUATIONS

- Sketch graphs of equations by hand.
- Find the x- and y-intercepts of graphs of equations.
- Write the standard forms of equations of circles.
- Find the points of intersection of two graphs.
- Use mathematical models to model and solve real-life problems.

The Graph of an Equation

In Section 1.1, you used a coordinate system to represent graphically the relationship between two quantities. There, the graphical picture consisted of a collection of points in a coordinate plane (see Example 2 in Section 1.1).

Frequently, a relationship between two quantities is expressed as an equation. For instance, degrees on the Fahrenheit scale are related to degrees on the Celsius scale by the equation

$$F = \tfrac{9}{5}C + 32.$$

In this section, you will study some basic procedures for sketching the graphs of such equations. The **graph** of an equation is the set of all points that are solutions of the equation.

EXAMPLE 1   **Sketching the Graph of an Equation**

Sketch the graph of $y = 7 - 3x$.

SOLUTION   The simplest way to sketch the graph of an equation is the *point-plotting method*. With this method, you construct a table of values that consists of several solution points of the equation, as shown in the table below. For instance, when $x = 0$

$$y = 7 - 3(0) = 7$$

which implies that $(0, 7)$ is a solution point of the graph.

x	0	1	2	3	4
$y = 7 - 3x$	7	4	1	-2	-5

From the table, it follows that

$$(0, 7), (1, 4), (2, 1), (3, -2), \text{ and } (4, -5)$$

are solution points of the equation. After plotting these points, you can see that they appear to lie on a line, as shown in Figure 1.13. The graph of the equation is the line that passes through the five plotted points.

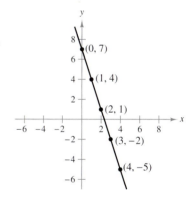

FIGURE 1.13   Solution Points for $y = 7 - 3x$

TRY IT 1

Sketch the graph of $y = 2x + 1$.

STUDY TIP

Even though we refer to the sketch shown in Figure 1.13 as the graph of $y = 7 - 3x$, it actually represents only a *portion* of the graph. The entire graph is a line that would extend off the page.

TECHNOLOGY

Zooming in to Find Intercepts

You can use the *zoom* feature of a graphing utility to approximate the *x*-intercepts of a graph. Suppose you want to approximate the *x*-intercept(s) of the graph of

$$y = 2x^3 - 3x + 2.$$

Begin by graphing the equation, as shown below in part (a). From the viewing window shown, the graph appears to have only one *x*-intercept. This intercept lies between -2 and -1. By zooming in on the intercept, you can improve the approximation, as shown in part (b). To three decimal places, the solution is $x \approx -1.476$.

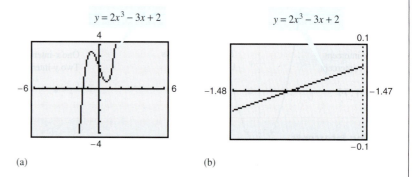

(a)   (b)

STUDY TIP

Some graphing utilities have a built-in program that can find the *x*-intercepts of a graph. If your graphing utility has this feature, try using it to find the *x*-intercept of the graph shown on the left. (Your calculator may call this the *root* or *zero* feature.)

Here are some suggestions for using the *zoom* feature.

1. With each successive zoom-in, adjust the *x*-scale so that the viewing window shows at least one tick mark on each side of the *x*-intercept.

2. The error in your approximation will be less than the distance between two scale marks.

3. The *trace* feature can usually be used to add one more decimal place of accuracy without changing the viewing window.

Part (a) below shows the graph of $y = x^2 - 5x + 3$. Parts (b) and (c) show "zoom-in views" of the two intercepts. From these views, you can approximate the *x*-intercepts to be $x \approx 0.697$ and $x \approx 4.303$.

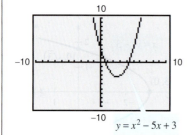

(a)

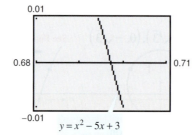

(b)

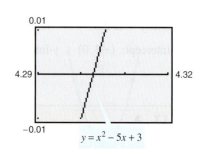

(c)

Circles

Throughout this course, you will learn to recognize several types of graphs from their equations. For instance, you should recognize that the graph of a second-degree equation of the form

$$y = ax^2 + bx + c, \quad a \neq 0$$

is a parabola (see Example 2). Another easily recognized graph is that of a **circle.**

Consider the circle shown in Figure 1.20. A point (x, y) is on the circle if and only if its distance from the center (h, k) is r. By the Distance Formula,

$$\sqrt{(x - h)^2 + (y - k)^2} = r.$$

By squaring both sides of this equation, you obtain the **standard form of the equation of a circle.**

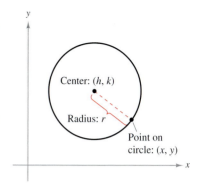

FIGURE 1.20

Standard Form of the Equation of a Circle

The point (x, y) lies on the circle of **radius** r and **center** (h, k) if and only if

$$(x - h)^2 + (y - k)^2 = r^2.$$

From this result, you can see that the standard form of the equation of a circle *with its center at the origin*, $(h, k) = (0, 0)$, is simply

$$x^2 + y^2 = r^2. \qquad \text{Circle with center at origin}$$

EXAMPLE 4 **Finding the Equation of a Circle**

The point $(3, 4)$ lies on a circle whose center is at $(-1, 2)$, as shown in Figure 1.21. Find the standard form of the equation of this circle.

SOLUTION The radius of the circle is the distance between $(-1, 2)$ and $(3, 4)$.

$$r = \sqrt{[3 - (-1)]^2 + (4 - 2)^2} \qquad \text{Distance Formula}$$
$$= \sqrt{16 + 4} \qquad \text{Simplify.}$$
$$= \sqrt{20} \qquad \text{Radius}$$

Using $(h, k) = (-1, 2)$, the standard form of the equation of the circle is

$$(x - h)^2 + (y - k)^2 = r^2$$
$$[x - (-1)]^2 + (y - 2)^2 = \left(\sqrt{20}\right)^2 \qquad \text{Substitute for } h, k, \text{ and } r.$$
$$(x + 1)^2 + (y - 2)^2 = 20. \qquad \text{Write in standard form.}$$

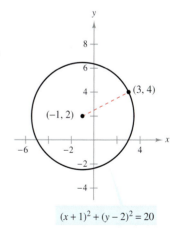

$$(x + 1)^2 + (y - 2)^2 = 20$$

FIGURE 1.21

TRY IT 4

The point $(1, 5)$ lies on a circle whose center is at $(-2, 1)$. Find the standard form of the equation of this circle.

TECHNOLOGY

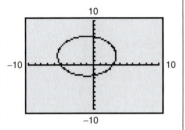 To graph a circle on a graphing utility, you can solve its equation for y and graph the top and bottom halves of the circle separately. For instance, you can graph the circle $(x + 1)^2 + (y - 2)^2 = 20$ by graphing the following equations.

$$y = 2 + \sqrt{20 - (x + 1)^2}$$

$$y = 2 - \sqrt{20 - (x + 1)^2}$$

If you want the result to appear circular, you need to use a square setting, as shown below.

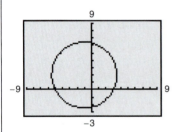

Standard setting

Square setting

General Form of the Equation of a Circle

$$Ax^2 + Ay^2 + Dx + Ey + F = 0, \qquad A \neq 0$$

To change from general form to standard form, you can use a process called **completing the square,** as demonstrated in Example 5.

EXAMPLE 5 **Completing the Square**

Sketch the graph of the circle whose general equation is

$$4x^2 + 4y^2 + 20x - 16y + 37 = 0.$$

SOLUTION   First divide by 4 so that the coefficients of x^2 and y^2 are both 1.

$4x^2 + 4y^2 + 20x - 16y + 37 = 0$ Write original equation.

$x^2 + y^2 + 5x - 4y + \frac{37}{4} = 0$ Divide each side by 4.

$\left(x^2 + 5x + \phantom{\frac{25}{4}}\right) + \left(y^2 - 4y + \right) = -\frac{37}{4}$ Group terms.

$\left(x^2 + 5x + \frac{25}{4}\right) + \left(y^2 - 4y + 4\right) = -\frac{37}{4} + \frac{25}{4} + 4$ Complete the square.

    (Half)²   (Half)²

$\left(x + \frac{5}{2}\right)^2 + (y - 2)^2 = 1$ Write in standard form.

From the standard form, you can see that the circle is centered at $\left(-\frac{5}{2}, 2\right)$ and has a radius of 1, as shown in Figure 1.22.

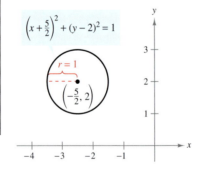

FIGURE 1.22

TRY IT 5

Write the equation of the circle $x^2 + y^2 - 4x + 2y + 1 = 0$ in standard form and sketch its graph.

The general equation $Ax^2 + Ay^2 + Dx + Ey + F = 0$ may not always represent a circle. In fact, such an equation will have no solution points if the procedure of completing the square yields the impossible result

$(x - h)^2 + (y - k)^2 = $ negative number.   No solution points

Further assistance with the calculus and algebra used in this example is available on the CD that accompanies this text.

Points of Intersection

A **point of intersection** of two graphs is an ordered pair that is a solution point of both graphs. For instance, Figure 1.23 shows that the graphs of

$$y = x^2 - 3 \quad \text{and} \quad y = x - 1$$

have two points of intersection: $(2, 1)$ and $(-1, -2)$. To find the points analytically, set the two y-values equal to each other and solve the equation

$$x^2 - 3 = x - 1$$

for x.

A common business application that involves points of intersection is **break-even analysis.** The marketing of a new product typically requires an initial investment. When sufficient units have been sold so that the total revenue has offset the total cost, the sale of the product has reached the **break-even point.** The **total cost** of producing x units of a product is denoted by C, and the **total revenue** from the sale of x units of the product is denoted by R. So, you can find the break-even point by setting the cost C equal to the revenue R, and solving for x.

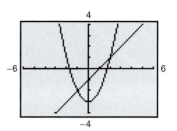

FIGURE 1.23

EXAMPLE 6 **Finding a Break-Even Point**

A business manufactures a product at a cost of $0.65 per unit and sells the product for $1.20 per unit. The company's initial investment to produce the product was $10,000. How many units must the company sell to break even?

SOLUTION The total cost of producing x units of the product is given by

$$C = 0.65x + 10,000. \qquad \textcolor{red}{\text{Cost equation}}$$

The total revenue from the sale of x units is given by

$$R = 1.2x. \qquad \textcolor{red}{\text{Revenue equation}}$$

To find the break-even point, set the cost equal to the revenue and solve for x.

$$R = C \qquad \textcolor{red}{\text{Set revenue equal to cost.}}$$
$$1.2x = 0.65x + 10,000 \qquad \textcolor{red}{\text{Substitute for } R \text{ and } C.}$$
$$0.55x = 10,000 \qquad \textcolor{red}{\text{Subtract } 0.65x \text{ from each side.}}$$
$$x = \frac{10,000}{0.55} \qquad \textcolor{red}{\text{Divide each side by 0.55.}}$$
$$x \approx 18,182 \qquad \textcolor{red}{\text{Use a calculator.}}$$

So, the company must sell 18,182 units before it breaks even. This result is shown graphically in Figure 1.24.

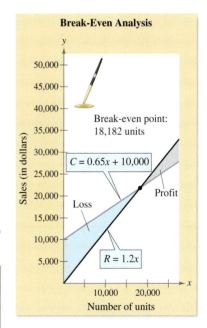

FIGURE 1.24

TRY IT 6

How many units must the company in Example 6 sell to break even if the selling price is $1.45 per unit?

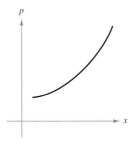

FIGURE 1.25   Supply Curve

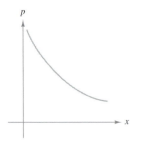

FIGURE 1.26   Demand Curve

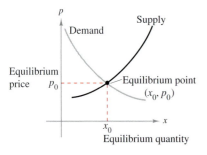

FIGURE 1.27   Equilibrium Point

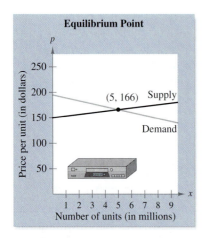

FIGURE 1.28

Two types of applications that economists use to analyze a market are supply and demand equations. A **supply equation** shows the relationship between the unit price p of a product and the quantity supplied x. The graph of a supply equation is called a **supply curve.** (See Figure 1.25.) A typical supply curve rises because producers of a product want to sell more units if the unit price is higher.

A **demand equation** shows the relationship between the unit price p of a product and the quantity demanded x. The graph of a demand equation is called a **demand curve.** (See Figure 1.26.) A typical demand curve tends to show a decrease in the quantity demanded with each increase in price.

In an ideal situation, with no other factors present to influence the market, the production level should stabilize at the point of intersection of the graphs of the supply and demand equations. This point is called the **equilibrium point.** The x-coordinate of the equilibrium point is called the **equilibrium quantity** and the p-coordinate is called the **equilibrium price.** (See Figure 1.27.) You can find the equilibrium point by setting the demand equation equal to the supply equation and solving for x.

EXAMPLE 7   **Finding the Equilibrium Point**

The demand and supply equations for a DVD player are given by

$$p = 195 - 5.8x \qquad \text{Demand equation}$$

$$p = 150 + 3.2x \qquad \text{Supply equation}$$

where p is the price in dollars and x represents the number of units in millions. Find the equilibrium point for this market.

SOLUTION   Begin by setting the demand equation equal to the supply equation.

$$195 - 5.8x = 150 + 3.2x \qquad \text{Set equations equal to each other.}$$

$$45 - 5.8x = 3.2x \qquad \text{Subtract 150 from each side.}$$

$$45 = 9x \qquad \text{Add } 5.8x \text{ to each side.}$$

$$5 = x \qquad \text{Divide each side by 9.}$$

So, the equilibrium point occurs when the demand and supply are each five million units. (See Figure 1.28.) The price that corresponds to this x-value is obtained by substituting $x = 5$ into either of the original equations. For instance, substituting into the demand equation produces

$$p = 195 - 5.8(5) = 195 - 29 = \$166.$$

Substitute $x = 5$ into the supply equation to see that you obtain the same price.

TRY IT 7

The demand and supply equations for a calculator are $p = 136 - 3.5x$ and $p = 112 + 2.5x$, respectively, where p is the price in dollars and x represents the number of units in millions. Find the equilibrium point for this market.

Mathematical Models

In this text, you will see many examples of the use of equations as **mathematical models** of real-life phenomena. In developing a mathematical model to represent actual data, you should strive for two (often conflicting) goals— accuracy and simplicity.

EXAMPLE 8   **Using Mathematical Models**

The table shows the annual sales (in millions of dollars) for Dillard's and Kohl's for 1999 through 2003. In the spring of 2004, the publication *Value Line* listed the projected 2004 sales for the companies as $7740 million and $11,975 million, respectively. How do you think these projections were obtained? *(Source: Dillard's Inc. and Kohl's Corp.)*

Year	1999	2000	2001	2002	2003
t	9	10	11	12	13
Dillard's	8677	8567	8155	7911	7599
Kohl's	4557	6152	7489	9120	10,282

SOLUTION   The projections were obtained by using past sales to predict future sales. The past sales were modeled by equations that were found by a statistical procedure called least squares regression analysis.

$$S = -16.86t^2 + 89.7t + 9269, \quad 9 \le t \le 13 \qquad \text{Dillard's}$$
$$S = -40.86t^2 + 2340.7t - 13{,}202, \quad 9 \le t \le 13 \qquad \text{Kohl's}$$

Using $t = 14$ to represent 2004, you can predict the 2004 sales to be

$$S = -16.86(14)^2 + 89.7(14) + 9269 \approx 7220 \qquad \text{Dillard's}$$
$$S = -40.86(14)^2 + 2340.7(14) - 13{,}202 \approx 11{,}559. \qquad \text{Kohl's}$$

These two projections are close to those projected by *Value Line*. The graphs of the two models are shown in Figure 1.29.

For help in evaluating the expressions in Example 8, see the review of order of operations on page 72.

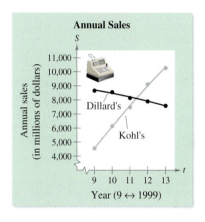

FIGURE 1.29

STUDY TIP

To test the accuracy of a model, you can compare the actual data with the values given by the model. For instance, the table below compares the actual Kohl's sales with those given by the model.

Year	1999	2000	2001
Actual	4557	6152	7489
Model	4554.6	6119	7601.6

Year	2002	2003
Actual	9120	10,282
Model	9002.6	10,322

TRY IT 8

The table shows the annual sales (in millions of dollars) for Dollar General for 1995 through 2002. In the winter of 2004, the publication *Value Line* listed projected 2004 sales for the company as $7800 million. How does this projection compare with the projection obtained using the model below? *(Source: Dollar General Corp.)*

$$S = 32.326t^2 + 78.23t + 530.9, \quad 5 \le t \le 12$$

Year	1995	1996	1997	1998	1999	2000	2001	2002
t	5	6	7	8	9	10	11	12
Sales	1764.2	2134.4	2627.3	3221.0	3888.0	4550.6	5322.9	6100.4

Much of your study of calculus will center around the behavior of the graphs of mathematical models. Figure 1.30 shows the graphs of six basic algebraic equations. Familiarity with these graphs will help you in the creation and use of mathematical models.

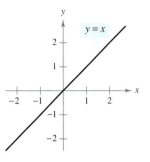

(a) Linear model

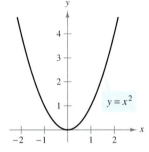

(b) Quadratic model

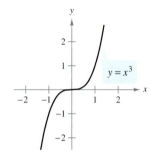

(c) Cubic model

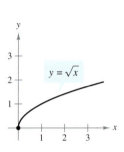

(d) Square root model

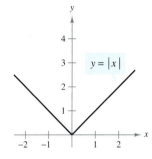

(e) Absolute value model

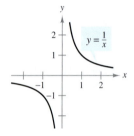

(f) Rational model

FIGURE 1.30

TAKE ANOTHER LOOK

Graphical, Numerical, and Analytic Solutions

Most problems in calculus can be solved in a variety of ways. Often, you can solve a problem graphically, numerically (using a table), and analytically. For instance, Example 6 compares graphical and analytic approaches to finding points of intersection.

In Example 8, suppose you were asked to find the point in time at which Kohl's sales exceeded Dillard's sales. Explain how to use *each* of the three approaches to answer the question. For this question, which approach do you think is best? Explain. Suppose you answered the question and obtained $t = 11.36$. What date does this represent—April 2001 or April 2002? Explain.

The following warm-up exercises involve skills that were covered in earlier sections. You will use these skills in the exercise set for this section.

In Exercises 1–6, solve for y.

1. $5y - 12 = x$

2. $-y = 15 - x$

3. $x^3y + 2y = 1$

4. $x^2 + x - y^2 - 6 = 0$

5. $(x - 2)^2 + (y + 1)^2 = 9$

6. $(x + 6)^2 + (y - 5)^2 = 81$

In Exercises 7–10, complete the square to write the expression as a perfect square trinomial.

7. $x^2 - 4x + $ ▢

8. $x^2 + 6x + $ ▢

9. $x^2 - 5x + $ ▢

10. $x^2 + 3x + $ ▢

In Exercises 11–14, factor the expression.

11. $x^2 - 3x + 2$

12. $x^2 + 5x + 6$

13. $y^2 - 3y + \frac{9}{4}$

14. $y^2 - 7y + \frac{49}{4}$

EXERCISES 1.2

In Exercises 1–6, determine whether the points are solution points of the given equation.

1. $2x - y - 3 = 0$

   (a) $(1, 2)$ (b) $(1, -1)$ (c) $(4, 5)$

2. $7x + 4y - 6 = 0$

   (a) $(6, -9)$ (b) $(-5, 10)$ (c) $\left(\frac{1}{2}, \frac{5}{8}\right)$

3. $x^2 + y^2 = 4$

   (a) $\left(1, -\sqrt{3}\right)$ (b) $\left(\frac{1}{2}, -1\right)$ (c) $\left(\frac{3}{2}, \frac{7}{2}\right)$

4. $x^2y + x^2 - 5y = 0$

   (a) $\left(0, \frac{1}{5}\right)$ (b) $(2, 4)$ (c) $(-2, -4)$

5. $x^2 - xy + 4y = 3$

   (a) $(0, 2)$ (b) $\left(-2, -\frac{1}{6}\right)$ (c) $(3, -6)$

6. $3y + 2xy - x^2 = 5$

   (a) $(-7, -5)$ (b) $(-1, 6)$ (c) $\left(1, \frac{6}{5}\right)$

In Exercises 7–12, match the equation with its graph. Use a graphing utility, set for a square setting, to confirm your result. [The graphs are labeled (a)–(f).]

7. $y = x - 2$

8. $y = -\frac{1}{2}x + 2$

9. $y = x^2 + 2x$

10. $y = \sqrt{9 - x^2}$

11. $y = |x| - 2$

12. $y = x^3 - x$

(a)

(b)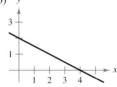

(c)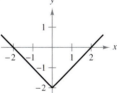

(d)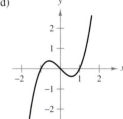

(e)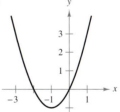

(f)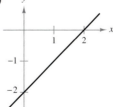

In Exercises 13–22, find the x- and y-intercepts of the graph of the equation.

13. $2x - y - 3 = 0$

14. $4x - 2y - 5 = 0$

15. $y = x^2 + x - 2$

16. $y = x^2 - 4x + 3$

17. $y = x^2\sqrt{9 - x^2}$

18. $y^2 = x^3 - 4x$

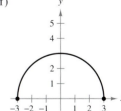

Extended Application: Linear Depreciation

Most business expenses can be deducted the same year they occur. One exception to this is the cost of property that has a useful life of more than 1 year, such as buildings, cars, or equipment. Such costs must be **depreciated** over the useful life of the property. If the *same amount* is depreciated each year, the procedure is called **linear depreciation** or **straight-line depreciation.** The *book value* is the difference between the original value and the total amount of depreciation accumulated to date.

TRY IT 8

Write a linear equation for the machine in Example 8 if the salvage value at the end of 8 years is $1000.

■ **EXAMPLE 8** **Depreciating Equipment**

Your company has purchased a $12,000 machine that has a useful life of 8 years. The salvage value at the end of 8 years is $2000. Write a linear equation that describes the book value of the machine each year.

SOLUTION   Let V represent the value of the machine at the end of year t. You can represent the initial value of the machine by the ordered pair $(0, 12{,}000)$ and the salvage value of the machine by the ordered pair $(8, 2000)$. The slope of the line is

$$m = \frac{2000 - 12{,}000}{8 - 0} = -\$1250 \qquad m = \frac{y_2 - y_1}{t_2 - t_1}$$

which represents the annual depreciation in *dollars per year.* Using the point-slope form, you can write the equation of the line as shown.

$$V - 12{,}000 = -1250(t - 0) \qquad \text{Write in point-slope form.}$$
$$V = -1250t + 12{,}000 \qquad \text{Write in slope-intercept form.}$$

The table shows the book value of the machine at the end of each year.

t	0	1	2	3	4	5	6	7	8
V	12,000	10,750	9500	8250	7000	5750	4500	3250	2000

The graph of this equation is shown in Figure 1.42.

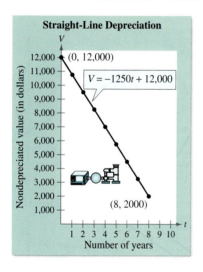

Straight-Line Depreciation

$V = -1250t + 12{,}000$

(0, 12,000)

(8, 2000)

Nondepreciated value (in dollars)

Number of years

FIGURE 1.42

TAKE ANOTHER LOOK

Comparing Different Types of Depreciation

The Internal Revenue Service allows businesses to choose different types of depreciation. Another type is

$$\text{Uniform Declining Balances: } V = 12{,}000\left(\frac{n - 1.605}{n}\right)^t, \quad n = 8.$$

Construct a table that compares this type of depreciation with linear depreciation. What are the advantages of each type?

The following warm-up exercises involve skills that were covered in earlier sections. You will use these skills in the exercise set for this section.

In Exercises 1 and 2, simplify the expression.

1. $\dfrac{5 - (-2)}{-3 - 4}$

2. $\dfrac{-7 - (-0)}{4 - 1}$

3. Evaluate $-\dfrac{1}{m}$ when $m = -3$.

4. Evaluate $-\dfrac{1}{m}$ when $m = \dfrac{6}{7}$.

In Exercises 5–10, solve for y in terms of x.

5. $-4x + y = 7$

6. $3x - y = 7$

7. $y - 2 = 3(x - 4)$

8. $y - (-5) = -1[x - (-2)]$

9. $y - (-3) = \dfrac{4 - (-3)}{2 - 1}(x - 2)$

10. $y - 1 = \dfrac{-3 - 1}{-7 - (-1)}[x - (-1)]$

EXERCISES 1.3

In Exercises 1–4, estimate the slope of the line.

1.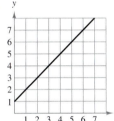

2.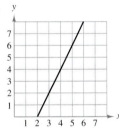

3.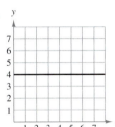

4.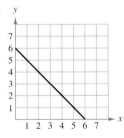

In Exercises 5–16, plot the points and find the slope of the line passing through the pair of points.

5. $(3, -4), (5, 2)$

6. $(1, 2), (-2, 2)$

7. $\left(\frac{1}{2}, 2\right), (6, 2)$

8. $\left(\frac{11}{3}, -2\right), \left(\frac{11}{3}, -10\right)$

9. $(-8, -3), (-8, -5)$

10. $(2, -1), (-2, -5)$

11. $(-2, 1), (4, -3)$

12. $(3, -5), (-2, -5)$

13. $\left(\frac{1}{4}, -2\right), \left(-\frac{3}{8}, 1\right)$

14. $\left(-\frac{3}{2}, -5\right), \left(\frac{5}{6}, 4\right)$

15. $\left(\frac{2}{3}, \frac{5}{2}\right), \left(\frac{1}{4}, -\frac{5}{6}\right)$

16. $\left(\frac{7}{8}, \frac{3}{4}\right), \left(\frac{5}{4}, -\frac{1}{4}\right)$

In Exercises 17–24, use the point on the line and the slope of the line to find three additional points through which the line passes. (There are many correct answers.)

	Point	Slope		Point	Slope
17.	$(2, 1)$	$m = 0$	**18.**	$(-3, -1)$	$m = 0$
19.	$(6, -4)$	$m = \frac{2}{3}$	**20.**	$(-2, -2)$	$m = \frac{5}{2}$
21.	$(1, 7)$	$m = -3$	**22.**	$(10, -6)$	$m = -1$
23.	$(-8, 1)$	m is undefined.			
24.	$(-3, 4)$	m is undefined.			

In Exercises 25–34, find the slope and y-intercept (if possible) of the equation of the line.

25. $x + 5y = 20$

26. $2x + y = 40$

27. $7x - 5y = 15$

28. $6x - 5y = 15$

29. $3x - y = 15$

30. $2x - 3y = 24$

31. $x = 4$

32. $x + 5 = 0$

33. $y - 4 = 0$

34. $y + 1 = 0$

In Exercises 35–46, write an equation of the line that passes through the points. Then use the equation to sketch the line.

35. $(4, 3), (0, -5)$

36. $(-3, -4), (1, 4)$

37. $(0, 0), (-1, 3)$

38. $(-3, 6), (1, 2)$

39. $(2, 3), (2, -2)$

40. $(6, 1), (10, 1)$

41. $(3, -1), (-2, -1)$

42. $(2, 5), (2, -10)$

43. $\left(-\frac{1}{3}, 1\right), \left(-\frac{2}{3}, \frac{5}{6}\right)$

44. $\left(\frac{7}{8}, \frac{3}{4}\right), \left(\frac{5}{4}, -\frac{1}{4}\right)$

45. $\left(-\frac{1}{2}, 4\right), \left(\frac{1}{2}, 8\right)$

46. $(4, -1), \left(\frac{1}{4}, -5\right)$

Inverse Functions

Informally, the inverse function of f is another function g that "undoes" what f has done.

ALGEBRA REVIEW

Don't be confused by the use of the superscript -1 to denote the inverse function f^{-1}. In this text, whenever f^{-1} is written, it *always* refers to the inverse function of f and *not* to the reciprocal of $f(x)$.

Definition of Inverse Function

Let f and g be two functions such that

$$f(g(x)) = x \text{ for each } x \text{ in the domain of } g$$

and

$$g(f(x)) = x \text{ for each } x \text{ in the domain of } f.$$

Under these conditions, the function g is the **inverse function** of f. The function g is denoted by f^{-1}, which is read as "f-inverse." So,

$$f(f^{-1}(x)) = x \text{ and } f^{-1}(f(x)) = x.$$

The domain of f must be equal to the range of f^{-1}, and the range of f must be equal to the domain of f^{-1}.

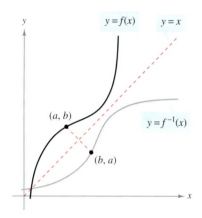

FIGURE 1.48   The graph of f^{-1} is a reflection of the graph of f in the line $y = x$.

EXAMPLE 6   Finding Inverse Functions

Several functions and their inverse functions are shown below. In each case, note that the inverse function "undoes" the original function. For instance, to undo multiplication by 2, you should divide by 2.

(a) $f(x) = 2x$ $f^{-1}(x) = \frac{1}{2}x$

(b) $f(x) = \frac{1}{3}x$ $f^{-1}(x) = 3x$

(c) $f(x) = x + 4$ $f^{-1}(x) = x - 4$

(d) $f(x) = 2x - 5$ $f^{-1}(x) = \frac{1}{2}(x + 5)$

(e) $f(x) = x^3$ $f^{-1}(x) = \sqrt[3]{x}$

(f) $f(x) = \dfrac{1}{x}$ $f^{-1}(x) = \dfrac{1}{x}$

STUDY TIP

You can verify that the functions in Example 6 are inverse functions by substituting specific values of x.

TRY IT 6

Informally find the inverse function of each function.

(a) $f(x) = \frac{1}{5}x$ (b) $f(x) = 3x + 2$

The graphs of f and f^{-1} are mirror images of each other (with respect to the line $y = x$), as shown in Figure 1.48. Try using a graphing utility to confirm this for each of the functions given in Example 6.

The functions in Example 6 are simple enough so that their inverse functions can be found by inspection. The next example demonstrates a strategy for finding the inverse functions of more complicated functions.

EXAMPLE 7   **Finding an Inverse Function**

Find the inverse function of $f(x) = \sqrt{2x - 3}$.

SOLUTION   Begin by replacing $f(x)$ with y. Then, interchange x and y and solve for y.

$f(x) = \sqrt{2x - 3}$	Write original function.
$y = \sqrt{2x - 3}$	Replace $f(x)$ with y.
$x = \sqrt{2y - 3}$	Interchange x and y.
$x^2 = 2y - 3$	Square each side.
$x^2 + 3 = 2y$	Add 3 to each side.
$\dfrac{x^2 + 3}{2} = y$	Divide each side by 2.

So, the inverse function has the form

$$f^{-1}(\;\;\;) = \frac{(\quad)^2 + 3}{2}.$$

Using x as the independent variable, you can write

$$f^{-1}(x) = \frac{x^2 + 3}{2}, \quad x \ge 0.$$

In Figure 1.49, note that the domain of f^{-1} coincides with the range of f.

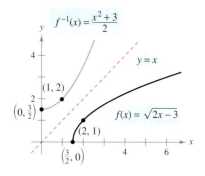

FIGURE 1.49

TRY IT 7

Find the inverse function of $f(x) = x^2 + 2$ for $x \ge 0$.

After you have found an inverse function, you should check your results. You can check your results *graphically* by observing that the graphs of f and f^{-1} are reflections of each other in the line $y = x$. You can check your results *algebraically* by evaluating $f(f^{-1}(x))$ and $f^{-1}(f(x))$—both should be equal to x.

Check that $f(f^{-1}(x)) = x$

$$f(f^{-1}(x)) = f\!\left(\frac{x^2 + 3}{2}\right)$$

$$= \sqrt{2\!\left(\frac{x^2 + 3}{2}\right) - 3}$$

$$= \sqrt{x^2}$$

$$= x, \quad x \ge 0$$

Check that $f^{-1}(f(x)) = x$

$$f^{-1}(f(x)) = f^{-1}\!\left(\sqrt{2x - 3}\right)$$

$$= \frac{\left(\sqrt{2x - 3}\right)^2 + 3}{2}$$

$$= \frac{2x}{2}$$

$$= x, \quad x \ge \frac{3}{2}$$

TECHNOLOGY

A graphing utility can help you check that the graphs of f and f^{-1} are reflections of each other in the line $y = x$. To do this, graph $y = f(x)$, $y = f^{-1}(x)$, and $y = x$ in the same viewing window, using a *square setting*.

The following warm-up exercises involve skills that were covered in earlier sections. You will use these skills in the exercise set for this section.

In Exercises 1–4, evaluate the expression and simplify.

1. $f(x) = x^2 - 3x + 3$

  (a) $f(-1)$ (b) $f(c)$ (c) $f(x + h)$

2. $f(x) = \begin{cases} 2x - 2, & x < 1 \\ 3x + 1, & x \geq 1 \end{cases}$

  (a) $f(-1)$ (b) $f(3)$ (c) $f(t^2 + 1)$

3. $f(x) = x^2 - 2x + 2$ $\dfrac{f(1 + h) - f(1)}{h}$

4. $f(x) = 4x$ $\dfrac{f(2 + h) - f(2)}{h}$

In Exercises 5–8, find the domain and range of the function and sketch its graph.

5. $h(x) = -\dfrac{5}{x}$

6. $g(x) = \sqrt{25 - x^2}$

7. $f(x) = |x - 3|$

8. $f(x) = \dfrac{|x|}{x}$

In Exercises 9 and 10, determine whether y is a function of x.

9. $9x^2 + 4y^2 = 49$

10. $2x^2y + 8x = 7y$

EXERCISES 1.5

In Exercises 1–8, complete the table and use the result to estimate the limit. Use a graphing utility to graph the function to confirm your result.

1. $\lim\limits_{x \to 2} (5x + 4)$

x	1.9	1.99	1.999	2	2.001	2.01	2.1
$f(x)$				?			

2. $\lim\limits_{x \to 2} (x^2 - 3x + 1)$

x	1.9	1.99	1.999	2	2.001	2.01	2.1
$f(x)$				?			

3. $\lim\limits_{x \to 2} \dfrac{x - 2}{x^2 - 4}$

x	1.9	1.99	1.999	2	2.001	2.01	2.1
$f(x)$				?			

4. $\lim\limits_{x \to 2} \dfrac{x^5 - 32}{x - 2}$

x	1.9	1.99	1.999	2	2.001	2.01	2.1
$f(x)$				?			

5. $\lim\limits_{x \to 0} \dfrac{\sqrt{x + 3} - \sqrt{3}}{x}$

x	-0.1	-0.01	-0.001	0	0.001	0.01	0.1
$f(x)$				?			

6. $\lim\limits_{x \to 0} \dfrac{\sqrt{x + 2} - \sqrt{2}}{x}$

x	-0.1	-0.01	-0.001	0	0.001	0.01	0.1
$f(x)$				?			

7. $\lim\limits_{x \to 0^-} \dfrac{\dfrac{1}{x+4} - \dfrac{1}{4}}{x}$

x	-0.5	-0.1	-0.01	-0.001	0
$f(x)$					?

8. $\lim\limits_{x \to 0^+} \dfrac{\dfrac{1}{2+x} - \dfrac{1}{2}}{2x}$

x	0.5	0.1	0.01	0.001	0
$f(x)$					?

In Exercises 9–12, use the graph to find the limit (if it exists).

9.

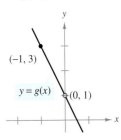

(a) $\lim\limits_{x \to 0} f(x)$

(b) $\lim\limits_{x \to -1} f(x)$

10.

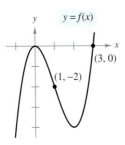

(a) $\lim\limits_{x \to 1} f(x)$

(b) $\lim\limits_{x \to 3} f(x)$

11.

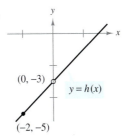

(a) $\lim\limits_{x \to 0} g(x)$

(b) $\lim\limits_{x \to -1} g(x)$

12.

(a) $\lim\limits_{x \to -2} h(x)$

(b) $\lim\limits_{x \to 0} h(x)$

In Exercises 13 and 14, find the limit of (a) $f(x) + g(x)$, (b) $f(x)g(x)$, and (c) $f(x)/g(x)$ as x approaches c.

13. $\lim\limits_{x \to c} f(x) = 3$

$\lim\limits_{x \to c} g(x) = 9$

14. $\lim\limits_{x \to c} f(x) = \frac{3}{2}$

$\lim\limits_{x \to c} g(x) = \frac{1}{2}$

In Exercises 15 and 16, find the limit of (a) $\sqrt{f(x)}$, (b) $[3f(x)]$, and (c) $[f(x)]^2$ as x approaches c.

15. $\lim\limits_{x \to c} f(x) = 16$

16. $\lim\limits_{x \to c} f(x) = 9$

In Exercises 17–22, use the graph to find the limit (if it exists).

(a) $\lim\limits_{x \to c^+} f(x)$

(b) $\lim\limits_{x \to c^-} f(x)$

(c) $\lim\limits_{x \to c} f(x)$

17.

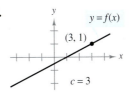

18.

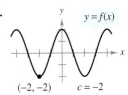

19.

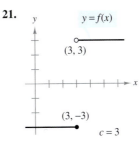

20.

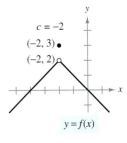

21.

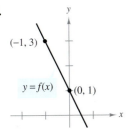

22.

In Exercises 23–40, find the limit.

23. $\lim\limits_{x \to 2} x^4$

24. $\lim\limits_{x \to -2} x^3$

25. $\lim\limits_{x \to -3} (3x + 2)$

26. $\lim\limits_{x \to 0} (2x - 3)$

27. $\lim\limits_{x \to 1} (1 - x^2)$

28. $\lim\limits_{x \to 2} (-x^2 + x - 2)$

29. $\lim\limits_{x \to 3} \sqrt{x + 1}$

30. $\lim\limits_{x \to 4} \sqrt[3]{x + 4}$

31. $\lim\limits_{x \to -3} \dfrac{2}{x + 2}$

32. $\lim\limits_{x \to -2} \dfrac{3x + 1}{2 - x}$

33. $\lim\limits_{x \to -2} \dfrac{x^2 - 1}{2x}$

34. $\lim\limits_{x \to -1} \dfrac{4x - 5}{3 - x}$

35. $\lim\limits_{x \to 7} \dfrac{5x}{x + 2}$

36. $\lim\limits_{x \to 3} \dfrac{\sqrt{x + 1}}{x - 4}$

37. $\lim\limits_{x \to 3} \dfrac{\sqrt{x + 1} - 1}{x}$

38. $\lim\limits_{x \to 5} \dfrac{\sqrt{x + 4} - 2}{x}$

39. $\lim\limits_{x \to 1} \dfrac{\dfrac{1}{x + 4} - \dfrac{1}{4}}{x}$

40. $\lim\limits_{x \to 2} \dfrac{\dfrac{1}{x + 2} - \dfrac{1}{2}}{x}$

In Exercises 41–58, find the limit (if it exists).

41. $\lim\limits_{x \to -1} \dfrac{x^2 - 1}{x + 1}$

42. $\lim\limits_{x \to -1} \dfrac{2x^2 - x - 3}{x + 1}$

43. $\lim\limits_{x \to 2} \dfrac{x - 2}{x^2 - 4x + 4}$

44. $\lim\limits_{x \to 2} \dfrac{2 - x}{x^2 - 4}$

45. $\lim\limits_{t \to 5} \dfrac{t - 5}{t^2 - 25}$

46. $\lim\limits_{t \to 1} \dfrac{t^2 + t - 2}{t^2 - 1}$

47. $\lim\limits_{x \to -2} \dfrac{x^3 + 8}{x + 2}$

48. $\lim\limits_{x \to 1} \dfrac{x^3 - 1}{x - 1}$

49. $\lim\limits_{x \to -2} \dfrac{|x + 2|}{x + 2}$

50. $\lim\limits_{x \to 2} \dfrac{|x - 2|}{x - 2}$

51. $\lim\limits_{x \to 3} f(x)$, where $f(x) = \begin{cases} \frac{1}{3}x - 2, & x \le 3 \\ -2x + 5, & x > 3 \end{cases}$

52. $\lim\limits_{s \to 1} f(s)$, where $f(s) = \begin{cases} s, & s \le 1 \\ 1 - s, & s > 1 \end{cases}$

53. $\lim\limits_{\Delta x \to 0} \dfrac{2(x + \Delta x) - 2x}{\Delta x}$

54. $\lim\limits_{\Delta x \to 0} \dfrac{4(x + \Delta x) - 5 - (4x - 5)}{\Delta x}$

55. $\lim\limits_{\Delta x \to 0} \dfrac{\sqrt{x + 2 + \Delta x} - \sqrt{x + 2}}{\Delta x}$

56. $\lim\limits_{\Delta x \to 0} \dfrac{\sqrt{x + \Delta x} - \sqrt{x}}{\Delta x}$

57. $\lim\limits_{\Delta t \to 0} \dfrac{(t + \Delta t)^2 - 5(t + \Delta t) - (t^2 - 5t)}{\Delta t}$

58. $\lim\limits_{\Delta t \to 0} \dfrac{(t + \Delta t)^2 - 4(t + \Delta t) + 2 - (t^2 - 4t + 2)}{\Delta t}$

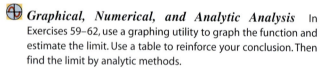 *Graphical, Numerical, and Analytic Analysis* In Exercises 59–62, use a graphing utility to graph the function and estimate the limit. Use a table to reinforce your conclusion. Then find the limit by analytic methods.

59. $\lim\limits_{x \to 1^-} \dfrac{2}{x^2 - 1}$

60. $\lim\limits_{x \to 1^+} \dfrac{5}{1 - x}$

61. $\lim\limits_{x \to -2^-} \dfrac{1}{x + 2}$

62. $\lim\limits_{x \to 0^-} \dfrac{x + 1}{x}$

In Exercises 63–66, use a graphing utility to estimate the limit (if it exists).

63. $\lim\limits_{x \to 2} \dfrac{x^2 - 5x + 6}{x^2 - 4x + 4}$

64. $\lim\limits_{x \to 1} \dfrac{x^2 + 6x - 7}{x^3 - x^2 + 2x - 2}$

65. $\lim\limits_{x \to -4} \dfrac{x^3 + 4x^2 + x + 4}{2x^2 + 7x - 4}$

66. $\lim\limits_{x \to -2} \dfrac{4x^3 + 7x^2 + x + 6}{3x^2 - x - 14}$

67. The limit of

$$f(x) = (1 + x)^{1/x}$$

is a natural base for many business applications, as you will see in Section 4.2.

$$\lim\limits_{x \to 0} (1 + x)^{1/x} = e \approx 2.718$$

(a) Show the reasonableness of this limit by completing the table.

x	-0.01	-0.001	-0.0001	0	0.0001	0.001	0.01
$f(x)$							

 (b) Use a graphing utility to graph f and to confirm the answer in part (a).

(c) Find the domain and range of the function.

68. Find $\lim\limits_{x \to 0} f(x)$, given

$$4 - x^2 \le f(x) \le 4 + x^2, \text{ for all } x.$$

69. *Environment* The cost (in dollars) of removing $p\%$ of the pollutants from the water in a small lake is given by

$$C = \dfrac{25{,}000p}{100 - p}, \quad 0 \le p < 100$$

where C is the cost and p is the percent of pollutants.

(a) Find the cost of removing 50% of the pollutants.

(b) What percent of the pollutants can be removed for $100,000?

(c) Evaluate $\lim\limits_{p \to 100^-} C$. Explain your results.

70. *Compound Interest* You deposit $1000 in an account that is compounded quarterly at an annual rate of r (in decimal form). The balance A after 10 years is

$$A = 1000\left(1 + \dfrac{r}{4}\right)^{40}.$$

Does the limit of A exist as the interest rate approaches 6%? If so, what is the limit?

71. *Compound Interest* Consider a certificate of deposit that pays 10% (annual percentage rate) on an initial deposit of $500. The balance A after 10 years is

$$A = 500(1 + 0.1x)^{10/x}$$

where x is the length of the compounding period (in years).

(a) Use a graphing utility to graph A, where $0 \le x \le 1$.

(b) Use the *zoom* and *trace* features to estimate the balance for quarterly compounding and daily compounding.

(c) Use the *zoom* and *trace* features to estimate

$$\lim\limits_{x \to 0^+} A.$$

What do you think this limit represents? Explain your reasoning.

1.6   CONTINUITY

- Determine the continuity of functions.
- Determine the continuity of functions on a closed interval.
- Use the greatest integer function to model and solve real-life problems.
- Use compound interest models to solve real-life problems.

Continuity

In mathematics, the term "continuous" has much the same meaning as it does in everyday use. To say that a function is continuous at $x = c$ means that there is no interruption in the graph of f at c. The graph of f is unbroken at c, and there are no holes, jumps, or gaps. As simple as this concept may seem, its precise definition eluded mathematicians for many years. In fact, it was not until the early 1800s that a precise definition was finally developed.

Before looking at this definition, consider the function whose graph is shown in Figure 1.60. This figure identifies three values of x at which the function f is not continuous.

1. At $x = c_1$, $f(c_1)$ is not defined.

2. At $x = c_2$, $\lim\limits_{x \to c_2} f(x)$ does not exist.

3. At $x = c_3$, $f(c_3) \neq \lim\limits_{x \to c_3} f(x)$.

At all other points in the interval (a, b), the graph of f is uninterrupted, which implies that the function f is continuous at all other points in the interval (a, b).

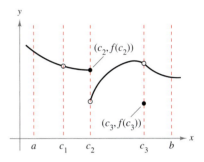

FIGURE 1.60 f is not continuous when $x = c_1, c_2, c_3$.

Definition of Continuity

Let c be a number in the interval (a, b), and let f be a function whose domain contains the interval (a, b). The function f is **continuous at the point c** if the following conditions are true.

1. $f(c)$ is defined.

2. $\lim\limits_{x \to c} f(x)$ exists.

3. $\lim\limits_{x \to c} f(x) = f(c)$.

If f is continuous at every point in the interval (a, b), then it is **continuous on an open interval (a, b).**

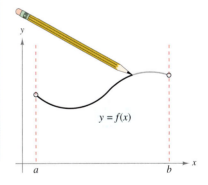

FIGURE 1.61   On the interval (a, b), the graph of f can be traced with a pencil.

Roughly, you can say that a function is continuous on an interval if its graph on the interval can be traced using a pencil and paper without lifting the pencil from the paper, as shown in Figure 1.61.

ALGEBRA REVIEW

Order of Operations

Much of the algebra in this chapter involves evaluation of algebraic expressions. When you evaluate an algebraic expression, you need to know the priorities assigned to different operations. These priorities are called the *order of operations*.

1. Perform operations inside *symbols of grouping or absolute value symbols*, starting with the innermost symbol.

2. Evaluate all *exponential* expressions.

3. Perform all *multiplications* and *divisions* from left to right.

4. Perform all *additions* and *subtractions* from left to right.

EXAMPLE 1   Using Order of Operations

Evaluate each expression.

(a) $7 - [(5 \cdot 3) + 2^3]$

(b) $[36 \div (3^2 \cdot 2)] + 6$

(c) $36 - [3^2 \cdot (2 \div 6)]$

(d) $10 - 2(8 + |5 - 7|)$

SOLUTION

(a) $7 - [(5 \cdot 3) + 2^3] = 7 - [15 + 2^3]$ Multiply inside parentheses.

$= 7 - [15 + 8]$ Evaluate exponential expression.

$= 7 - 23$ Add inside brackets.

$= -16$ Subtract.

(b) $[36 \div (3^2 \cdot 2)] + 6 = [36 \div (9 \cdot 2)] + 6$ Evaluate exponential expression inside parentheses.

$= [36 \div 18] + 6$ Multiply inside parentheses.

$= 2 + 6$ Divide inside brackets.

$= 8$ Add.

(c) $36 - [3^2 \cdot (2 \div 6)] = 36 - \left[3^2 \cdot \frac{1}{3}\right]$ Divide inside parentheses.

$= 36 - \left[9 \cdot \frac{1}{3}\right]$ Evaluate exponential expression.

$= 36 - 3$ Multiply inside brackets.

$= 33$ Subtract.

(d) $10 - 2(8 + |5 - 7|) = 10 - 2(8 + |-2|)$ Subtract inside absolute value symbols.

$= 10 - 2(8 + 2)$ Evaluate absolute value.

$= 10 - 2(10)$ Add inside parentheses.

$= 10 - 20$ Multiply.

$= -10$ Subtract.

TECHNOLOGY

Most scientific and graphing calculators use the same order of operations listed above. Try entering the expressions in Example 1 into your calculator. Do you get the same results?

Solving Equations

A second algebraic skill in this chapter is solving an equation in one variable.

1. To solve a *linear equation*, you can add or subtract the same quantity from each side of the equation. You can also multiply or divide each side of the equation by the same *nonzero* quantity.

2. To solve a *quadratic equation*, you can take the square root of each side, use factoring, or use the Quadratic Formula.

3. To solve a *radical equation*, isolate the radical on one side of the equation and square each side of the equation.

4. To solve an *absolute value equation*, use the definition of absolute value to rewrite the equation as two equations.

EXAMPLE 2   Solving Equations

Solve each equation.

(a) $3x - 3 = 5x - 7$

(b) $2x^2 = 10$

(c) $2x^2 + 5x - 6 = 6$

(d) $\sqrt{2x - 7} = 5$

SOLUTION

(a)
$$3x - 3 = 5x - 7$$ Write original (linear) equation.
$$-3 = 2x - 7$$ Subtract $3x$ from each side.
$$4 = 2x$$ Add 7 to each side.
$$2 = x$$ Divide each side by 2.

(b)
$$2x^2 = 10$$ Write original (quadratic) equation.
$$x^2 = 5$$ Divide each side by 2.
$$x = \pm\sqrt{5}$$ Take the square root of each side.

(c)
$$2x^2 + 5x - 6 = 6$$ Write original (quadratic) equation.
$$2x^2 + 5x - 12 = 0$$ Write in general form.
$$(2x - 3)(x + 4) = 0$$ Factor.
$$2x - 3 = 0 \implies x = \tfrac{3}{2}$$ Set first factor equal to zero.
$$x + 4 = 0 \implies x = -4$$ Set second factor equal to zero.

(d)
$$\sqrt{2x - 7} = 5$$ Write original (radical) equation.
$$2x - 7 = 25$$ Square each side.
$$2x = 32$$ Add 7 to each side.
$$x = 16$$ Divide each side by 2.

ALGEBRA REVIEW

You should be aware that solving radical equations can sometimes lead to *extraneous solutions* (those that do not satisfy the original equation). For example, squaring both sides of the following equation yields two possible solutions, one of which is extraneous.

$$\sqrt{x} = x - 2$$
$$x = x^2 - 4x + 4$$
$$0 = x^2 - 5x + 4$$
$$= (x - 4)(x - 1)$$
$$x - 4 = 0 \implies x = 4 \text{ (solution)}$$
$$x - 1 = 0 \implies x = 1 \text{ (extraneous)}$$

1   CHAPTER SUMMARY AND STUDY STRATEGIES

*After studying this chapter, you should have acquired the following skills. The exercise numbers are keyed to the view Exercises that begin on page 76. Answers to odd-numbered Review Exercises are given in the back of the text.**

■ Plot points in a coordinate plane and read data presented graphically. *(Section 1.1)*   *Review Exercises 1–4*

■ Find the distance between two points in a coordinate plane.   *(Section 1.1)*   *Review Exercises 5–8*

$$d = \sqrt{(x_2 - x_1)^2 + (y_2 - y_1)^2}$$

■ Find the midpoints of line segments connecting two points.   *(Section 1.1)*   *Review Exercises 9–12*

$$\text{Midpoint} = \left(\frac{x_1 + x_2}{2}, \frac{y_1 + y_2}{2}\right)$$

■ Interpret real-life data that is presented graphically.   *(Section 1.1)*   *Review Exercises 13, 14*

■ Translate points in a coordinate plane.   *(Section 1.1)*   *Review Exercises 15, 16*

■ Construct a bar graph from real-life data.   *(Section 1.1)*   *Review Exercise 17*

■ Sketch graphs of equations by hand.   *(Section 1.2)*   *Review Exercises 18–27*

■ Find the x- and y-intercepts of graphs of equations algebraically *and* graphically using a graphing utility.   *(Section 1.2)*   *Review Exercises 28, 29*

■ Write the standard forms of equations of circles, given the center and a point on the circle.   *(Section 1.2)*   *Review Exercises 30, 31*

$$(x - h)^2 + (y - k)^2 = r^2$$

■ Convert equations of circles from general form to standard form by completing the square, and sketch the circles.   *(Section 1.2)*   *Review Exercises 32, 33*

■ Find the points of intersection of two graphs algebraically *and* graphically using a graphing utility.   *(Section 1.2)*   *Review Exercises 34–37*

■ Find the break-even point for a business.   *(Section 1.2)*   *Review Exercises 38, 39*

The break-even point occurs when the revenue R is equal to the cost C.

■ Find the equilibrium points of supply equations and demand equations. *(Section 1.2)*   *Review Exercise 40*

The equilibrium point is the point of intersection of the graphs of the supply and demand equations.

■ Use the slope-intercept form of a linear equation to sketch graphs of lines. *(Section 1.3)*   *Review Exercises 41–46*

$$y = mx + b$$

■ Find slopes of lines passing through two points.   *(Section 1.3)*   *Review Exercises 47–50*

$$m = \frac{y_2 - y_1}{x_2 - x_1}$$

■ Use the point-slope form to write equations of lines and graph equations using a graphing utility.   *(Section 1.3)*   *Review Exercises 51, 52*

$$y - y_1 = m(x - x_1)$$

---

* Use a wide range of valuable study aids to help you master the material in this chapter. The *Student Solutions Guide* includes step-by-step solutions to all odd-numbered exercises to help you review and prepare. The *HM mathSpace® Student CD-ROM* helps you brush up on your algebra skills. The *Graphing Technology Guide*, available on the Web at *math.college.hmco.com/students*, offers step-by-step commands and instructions for a wide variety of graphing calculators, including the most recent models.

■ Find equations of parallel and perpendicular lines. *(Section 1.3)*   *Review Exercises 53, 54*

$$\text{Parallel lines: } m_1 = m_2 \qquad \text{Perpendicular lines: } m_1 = -\frac{1}{m_2}$$

■ Use linear equations to solve real-life problems such as predicting future sales   *Review Exercises 55, 56*
  or creating a linear depreciation schedule. *(Section 1.3)*

■ Use the vertical line test to decide whether equations define functions.   *Review Exercises 57–60*
  (Section 1.4)

■ Use function notation to evaluate functions. *(Section 1.4)*   *Review Exercises 61, 62*

■ Use a graphing utility to graph functions and find the domains and ranges   *Review Exercises 63–68*
  of functions. *(Section 1.4)*

■ Combine functions to create other functions. *(Section 1.4)*   *Review Exercises 69, 70*

■ Use the horizontal line test to determine whether functions have inverse functions.   *Review Exercises 71–74*
  If they do, find the inverse functions. *(Section 1.4)*

■ Determine whether limits exist. If they do, find the limits. *(Section 1.5)*   *Review Exercises 75–92*

■ Use a table to estimate one-sided limits. *(Section 1.5)*   *Review Exercises 93, 94*

■ Determine whether statements about limits are true or false. *(Section 1.5)*   *Review Exercises 95–100*

■ Determine whether functions are continuous at a point, on an open interval, and   *Review Exercises 101–108*
  on a closed interval. *(Section 1.6)*

■ Determine the constant such that *f* is continuous. *(Section 1.6)*   *Review Exercises 109, 110*

■ Use analytic and graphical models of real-life data to solve real-life problems.   *Review Exercises 111–114*
  (Section 1.6)

On pages xxv–xxviii of the preface, we included a feature called A Plan for You as a Student. *If you have not already read this feature, we encourage you to do so now. Here are some other strategies that can help you succeed in this course.*

■ ***Use a Graphing Utility***   A graphing calculator or graphing software for a computer can help you in this course in two important ways. As an *exploratory device*, a graphing utility allows you to learn concepts by allowing you to compare graphs of equations. For instance, sketching the graphs of $y = x^2$, $y = x^2 + 1$, and $y = x^2 - 1$ helps confirm that adding (or subtracting) a constant to (or from) a function shifts the graph of the function vertically. As a *problem-solving tool*, a graphing utility frees you of some of the drudgery of sketching complicated graphs by hand. The time that you save can be spent using mathematics to solve real-life problems.

■ ***Use the Warm-Up Exercises***   Each exercise set in this text begins with a set of warm-up exercises. We urge you to begin each homework session by quickly working all of the warm-up exercises (all are answered in the back of the text). The "old" skills covered in the warm-up exercises are needed to master the "new" skills in the section exercise set. The warm-up exercises remind you that mathematics is cumulative—to be successful in this course, you must retain "old" skills.

■ ***Use the Additional Study Aids***   The additional study aids were prepared specifically to help you master the concepts discussed in the text. They are the *Student Solutions Guide*, the *Calculus: An Applied Approach Learning Tools Student CD-ROM*, *The Algebra of Calculus*, and the *Graphing Technology Guide*.

Study Tools   *Additional resources that accompany this chapter*

- **Algebra Review**   (pages 72 and 73)
- **Chapter Summary and Study Strategies**   (pages 74 and 75)
- **Review Exercises**   (pages 76–79)
- **Sample Post-Graduation Exam Questions**   (page 80)
- **Web Exercises**   (page 10, Exercises 39 and 40; page 23, Exercise 74; page 34, Exercise 79; page 35, Exercise 91; page 48, Exercise 87)

- **Student Solutions Guide**
- **HM mathSpace® Student CD-ROM**
- **Graphing Technology Guide**   (math.college.hmco.com/students)

1   CHAPTER REVIEW EXERCISES

In Exercises 1–4, match the data with the real-life situation that it represents. [The graphs are labeled (a)–(d).]

1. Population of Texas

2. Population of California

3. Number of unemployed workers in the United States

4. Best Buy sales

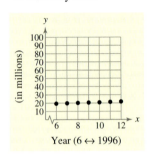

(a)

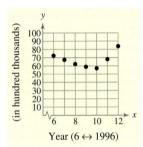

(b)

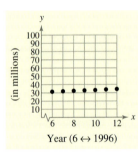

(c)

(d)

In Exercises 5–8, find the distance between the two points.

5. $(0, 0)$, $(5, 2)$

6. $(1, 2)$, $(4, 3)$

7. $(-1, 3)$, $(-4, 6)$

8. $(6, 8)$, $(-3, 7)$

In Exercises 9–12, find the midpoint of the line segment connecting the two points.

9. $(5, 6)$, $(9, 2)$

10. $(0, 0)$, $(-4, 8)$

11. $(-10, 4)$, $(-6, 8)$

12. $(7, -9)$, $(-3, 5)$

In Exercises 13 and 14, use the graph below, which gives the revenues, costs, and profits for Pixar from 1999 through 2003. (Pixar develops and produces animated feature films.) *(Source: Pixar)*

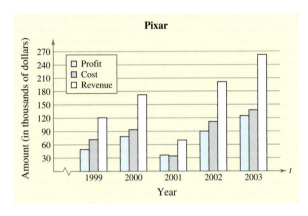

13. Write an equation that relates the revenue R, cost C, and profit P. Explain the relationship between the heights of the bars and the equation.

14. Estimate the revenue, cost, and profit for Pixar for each year.

15. Translate the triangle whose vertices are $(1, 3)$, $(2, 4)$, and $(5, 6)$ three units to the right and four units up. Find the coordinates of the translated vertices.

16. Translate the rectangle whose vertices are $(-2, 1)$, $(-1, 2)$, $(1, 0)$, and $(0, -1)$ four units to the right and one unit down.

17. *Biology*   The following data represent six intertidal invertebrate species collected from four stations along the Maine coast.

Mytilus	107	*Gammarus*	78
Littorina	65	*Arbacia*	6
Nassarius	112	*Mya*	18

Use a graphing utility to construct a bar graph that represents the data. *(Source: Adapted from Haefner,* Exploring Marine Biology: Laboratory and Field Exercises*)*

In Exercises 18–27, sketch the graph of the equation.

18. $y = 4x - 12$

19. $y = 4 - 3x$

20. $y = x^2 + 5$

21. $y = 1 - x^2$

22. $y = |4 - x|$

23. $y = |2x - 3|$

24. $y = x^3 + 4$

25. $y = x^3 + 2x^2 - x + 2$

26. $y = \sqrt{4x + 1}$

27. $y = \sqrt{2x}$

In Exercises 28 and 29, find the *x*- and *y*-intercepts of the graph of the equation algebraically. Use a graphing utility to verify your results.

28. $4x + y + 3 = 0$

29. $y = (x - 1)^3 + 2(x - 1)^2$

In Exercises 30 and 31, write the standard form of the equation of the circle.

30. Center: $(0, 0)$

Solution point: $(2, \sqrt{5})$

31. Center: $(2, -1)$

Solution point: $(-1, 7)$

In Exercises 32 and 33, complete the square to write the equation of the circle in standard form. Determine the radius and center of the circle. Then sketch the circle.

32. $x^2 + y^2 - 6x + 8y = 0$

33. $x^2 + y^2 + 10x + 4y - 7 = 0$

In Exercises 34–37, find the point(s) of intersection of the graphs algebraically. Then use a graphing utility to verify your results.

34. $x + y = 2, \ 2x - y = 1$

35. $x^2 + y^2 = 5, \ x - y = 1$

36. $y = x^3, \ y = x$ **37.** $y = \sqrt{x}, \ y = x$

38. *Break-Even Analysis* The student government association wants to raise money by having a T-shirt sale. Each shirt costs $8. The silk screening costs $200 for the design, plus $2 per shirt. Each shirt will sell for $14.

(a) Find equations for the total cost C and the total revenue R for selling x shirts.

(b) Find the break-even point.

39. *Break-Even Analysis* You are starting a part-time business. You make an initial investment of $6000. The unit cost of the product is $6.50, and the selling price is $13.90.

(a) Find equations for the total cost C and the total revenue R for selling x units of the product.

(b) Find the break-even point.

40. *Supply and Demand* The demand and supply equations for a cordless screwdriver are given by

$p = 91.4 - 0.009x$ Demand equation

$p = 6.4 + 0.008x$ Supply equation

where p is the price in dollars and x represents the number of units. Find the equilibrium point for this market.

In Exercises 41–46, find the slope and *y*-intercept (if possible) of the linear equation. Then sketch the graph of the equation.

41. $3x + y = -2$ **42.** $-\frac{1}{3}x + \frac{5}{6}y = 1$

43. $y = -\frac{5}{3}$ **44.** $x = -3$

45. $-2x - 5y - 5 = 0$ **46.** $3.2x - 0.8y + 5.6 = 0$

In Exercises 47–50, find the slope of the line passing through the two points.

47. $(0, 0), (7, 6)$ **48.** $(-1, 5), (-5, 7)$

49. $(10, 17), (-11, -3)$ **50.** $(-11, -3), (-1, -3)$

 In Exercises 51 and 52, find an equation of the line that passes through the point and has the given slope. Then use a graphing utility to graph the line.

51. Point: $(3, -1)$; slope: $m = -2$

52. Point: $(-3, -3)$; slope: $m = \frac{1}{2}$

In Exercises 53 and 54, find the general form of the equation of the line passing through the point and satisfying the given condition.

53. Point: $(-3, 6)$

(a) Slope is $\frac{7}{8}$.

(b) Parallel to the line $4x + 2y = 7$.

(c) Passes through the origin.

(d) Perpendicular to the line $3x - 2y = 2$.

54. Point: $(1, -3)$

(a) Parallel to the *x*-axis.

(b) Perpendicular to the *x*-axis.

(c) Parallel to the line $-4x + 5y = -3$.

(d) Perpendicular to the line $5x - 2y = 3$.

55. *Demand* When a wholesaler sold a product at $32 per unit, sales were 750 units per week. After a price increase of $5 per unit, however, the sales dropped to 700 units per week.

(a) Write the quantity demanded x as a linear function of the price p.

(b) *Linear Interpolation* Predict the number of units sold at a price of $34.50 per unit.

(c) *Linear Extrapolation* Predict the number of units sold at a price of $42.00 per unit.

 56. *Linear Depreciation* A small business purchases a typesetting system for $117,000. After 9 years, the system will be obsolete and have no value.

(a) Write a linear equation giving the value v of the system in terms of the time t.

(b) Use a graphing utility to graph the function.

(c) Use a graphing utility to estimate the value of the system after 4 years.

(d) Use a graphing utility to estimate the time when the system's value will be $84,000.

In Exercises 57–60, use the vertical line test to determine whether y is a function of x.

57. $y = -x^2 + 2$

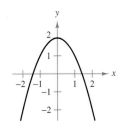

58. $x^2 + y^2 = 4$

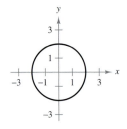

59. $y^2 - \frac{1}{4}x^2 = 4$

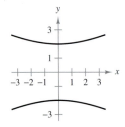

60. $y = |x + 4|$

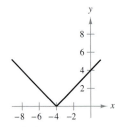

In Exercises 61 and 62, evaluate the function at the specified values of the independent variable. Simplify the result.

61. $f(x) = 3x + 4$

   (a) $f(1)$ (b) $f(x + 1)$ (c) $f(2 + \Delta x)$

62. $f(x) = x^2 + 4x + 3$

   (a) $f(0)$ (b) $f(x - 1)$ (c) $f(x + \Delta x) - f(x)$

 In Exercises 63–68, use a graphing utility to graph the function. Then find the domain and range of the function.

63. $f(x) = x^2 + 3x + 2$

64. $f(x) = 2$

65. $f(x) = \sqrt{x + 1}$

66. $f(x) = \dfrac{x - 3}{x^2 + x - 12}$

67. $f(x) = -|x| + 3$

68. $f(x) = -\frac{12}{13}x - \frac{7}{8}$

In Exercises 69 and 70, use f and g to find the combinations of the functions.

(a) $f(x) + g(x)$ (b) $f(x) - g(x)$ (c) $f(x)g(x)$

(d) $\dfrac{f(x)}{g(x)}$ (e) $f(g(x))$ (f) $g(f(x))$

69. $f(x) = 1 + x^2$, $g(x) = 2x - 1$

70. $f(x) = 2x - 3$, $g(x) = \sqrt{x + 1}$

In Exercises 71–74, find the inverse function of f (if it exists).

71. $f(x) = \frac{3}{2}x$

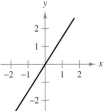

72. $f(x) = |x + 1|$

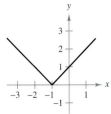

73. $f(x) = -x^2 + \frac{1}{2}$

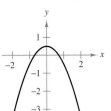

74. $f(x) = x^3 - 1$

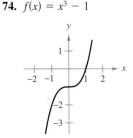

In Exercises 75–92, find the limit (if it exists).

75. $\displaystyle\lim_{x \to 2} (5x - 3)$

76. $\displaystyle\lim_{x \to 2} (2x + 9)$

77. $\displaystyle\lim_{x \to 2} (5x - 3)(2x + 3)$

78. $\displaystyle\lim_{x \to 2} \dfrac{5x - 3}{2x + 9}$

79. $\displaystyle\lim_{t \to 3} \dfrac{t^2 + 1}{t}$

80. $\displaystyle\lim_{t \to 0} \dfrac{t^2 + 1}{t}$

81. $\displaystyle\lim_{t \to 1} \dfrac{t + 1}{t - 2}$

82. $\displaystyle\lim_{t \to 2} \dfrac{t + 1}{t - 2}$

83. $\displaystyle\lim_{x \to -2} \dfrac{x + 2}{x^2 - 4}$

84. $\displaystyle\lim_{x \to 3^-} \dfrac{x^2 - 9}{x - 3}$

85. $\displaystyle\lim_{x \to 0^+} \left(x - \dfrac{1}{x}\right)$

86. $\displaystyle\lim_{x \to 1/2} \dfrac{2x - 1}{6x - 3}$

87. $\displaystyle\lim_{x \to 0} \dfrac{[1/(x - 2)] - 1}{x}$

88. $\displaystyle\lim_{x \to 0} \dfrac{[1/(x - 4)] - (1/4)}{x}$

89. $\displaystyle\lim_{t \to 0} \dfrac{(1/\sqrt{t + 4}) - (1/2)}{t}$

90. $\displaystyle\lim_{s \to 0} \dfrac{(1/\sqrt{1 + s}) - 1}{s}$

91. $\displaystyle\lim_{\Delta x \to 0} \dfrac{(x + \Delta x)^3 - (x + \Delta x) - (x^3 - x)}{\Delta x}$

92. $\displaystyle\lim_{\Delta x \to 0} \dfrac{1 - (x + \Delta x)^2 - (1 - x^2)}{\Delta x}$

In Exercises 93 and 94, use a table to estimate the limit.

93. $\displaystyle\lim_{x \to 1^+} \dfrac{\sqrt{2x + 1} - \sqrt{3}}{x - 1}$

94. $\displaystyle\lim_{x \to 1^+} \dfrac{1 - \sqrt[3]{x}}{x - 1}$

True or False? In Exercises 95–100, determine whether the statement is true or false. If it is false, explain why or give an example that shows it is false.

95. $\lim\limits_{x \to 0} \dfrac{|x|}{x} = 1$

96. $\lim\limits_{x \to 0} x^3 = 0$

97. $\lim\limits_{x \to 0} \sqrt{x} = 0$

98. $\lim\limits_{x \to 0} \sqrt[3]{x} = 0$

99. $\lim\limits_{x \to 2} f(x) = 3, \quad f(x) = \begin{cases} 3, & x \le 2 \\ 0, & x > 2 \end{cases}$

100. $\lim\limits_{x \to 3} f(x) = 1, \quad f(x) = \begin{cases} x - 2, & x \le 3 \\ -x^2 + 8x - 14, & x > 3 \end{cases}$

In Exercises 101–108, describe the interval(s) on which the function is continuous.

101. $f(x) = \dfrac{1}{(x+4)^2}$

102. $f(x) = \dfrac{x+2}{x}$

103. $f(x) = \dfrac{3}{x+1}$

104. $f(x) = \dfrac{x+1}{2x+2}$

105. $f(x) = [\![x + 3]\!]$

106. $f(x) = [\![x]\!] - 2$

107. $f(x) = \begin{cases} x, & x \le 0 \\ x + 1, & x > 0 \end{cases}$

108. $f(x) = \begin{cases} x, & x \le 0 \\ x^2, & x > 0 \end{cases}$

In Exercises 109 and 110, find the constant a such that f is continuous on the entire real line.

109. $f(x) = \begin{cases} -x + 1, & x \le 3 \\ ax - 8, & x > 3 \end{cases}$

110. $f(x) = \begin{cases} x + 1, & x < 1 \\ 2x + a, & x \ge 1 \end{cases}$

 111. ***National Debt*** The table lists the national debt D (in billions of dollars) for selected years. A mathematical model for the national debt is

$$D = 2.7502t^3 - 61.061t^2 + 598.79t + 3103.6,$$

$$0 \le t \le 13$$

where $t = 0$ represents 1990. *(Source: U.S. Department of the Treasury)*

t	0	1	2	3	4
D	3206.3	3598.2	4001.8	4351.0	4643.3

t	5	6	7	8	9
D	4920.6	5181.5	5369.2	5478.2	5605.5

t	10	11	12	13
D	5628.7	5769.9	6198.4	6752.0

(a) Use a graphing utility to graph the model.

(b) Create a table that compares the values given by the model with the actual data.

(c) Use the model to estimate the national debt in 2008.

 112. ***Consumer Awareness*** A cellular phone company charges \$2 for the first minute and \$0.10 for each additional minute or fraction thereof. Use the greatest integer function to create a model for the cost C of a phone call lasting t minutes. Use a graphing utility to graph the function, and discuss its continuity.

113. ***Recycling*** A recycling center pays \$0.25 for each pound of aluminum cans. Twenty-four aluminum cans weigh one pound. A mathematical model for the amount A paid by the recycling center is

$$A = \frac{1}{4}\left[\!\!\left[\frac{x}{24}\right]\!\!\right]$$

where x is the number of cans.

(a) Use a graphing utility to graph the function and then discuss its continuity.

(b) How much does a recycling center pay out for 1500 cans?

114. ***Biology*** A researcher experimenting with strains of corn produced the results in the figure below. From the figure, visually estimate the x- and y-intercepts, and use these points to write an equation for each line.

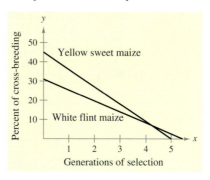

(Source: Adapted from Levine/Miller, Biology: Discovering Life, Second Edition)

1 SAMPLE POST-GRADUATION EXAM QUESTIONS

The following questions represent the types of questions that appear on certified public accountant (CPA) exams, Graduate Management Admission Tests (GMAT), Graduate Records Exams (GRE), actuarial exams, and College-Level Academic Skills Tests (CLAST). The answers to the questions are given in the back of the book.

In Questions 1–5, use the data given in the graphs. *(Source: U.S. Bureau of Labor Statistics)*

Figure for 1–5

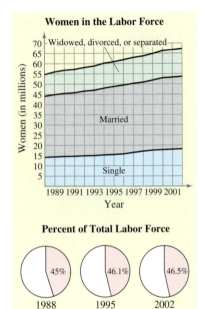

Women in the Labor Force

Percent of Total Labor Force

1988   45%
1995   46.1%
2002   46.5%

1. The total labor force in 2002 was about y million with y equal to

(a) 100   (b) 118   (c) 129   (d) 145   (e) 154

2. In 1995, the percent of married women in the labor force was about

(a) 19   (b) 32   (c) 45   (d) 55   (e) 82

3. What was the first year when more than 60 million women were in the labor force?

(a) 1992   (b) 1994   (c) 1996   (d) 1998   (e) 2000

4. Between 1990 and 2000, the number of women in the labor force

(a) increased by about 17%   (b) increased by about 25%

(c) increased by about 50%   (d) increased by about 100%

(e) increased by about 125%

5. Which of the statements about the labor force can be inferred from the graphs?

 I. Between 1988 and 2002, there were no years when more than 20 million widowed, divorced, or separated women were in the labor force.

 II. In every year between 1988 and 2002, the number of married women in the labor force increased.

 III. In every year between 1988 and 2002, women made up at least $\frac{2}{5}$ of the total labor force.

(a) I only   (b) II only   (c) I and II only

(d) II and III only   (e) I, II, and III

6. What is the length of the line segment connecting $(1, 3)$ and $(-1, 5)$?

(a) $\sqrt{3}$ (b) 2   (c) $2\sqrt{2}$ (d) 4   (e) 8

7. The interest charged on a loan is p dollars per $1000 for the first month and q dollars per $1000 for each month after the first month. How much interest will be charged during the first 3 months on a loan of $10,000?

(a) $30p$ (b) $30q$ (c) $p + 2q$

(d) $20p + 10q$ (e) $10p + 20q$

8. If $x + y > 5$ and $x - y > 3$, which of the following describes the x solutions?

(a) $x > 3$ (b) $x > 4$ (c) $x > 5$ (d) $x < 5$ (e) $x < 3$

9. In the figure at the left, in order for line A to be parallel to line B, the coordinates of C must be $(5, y)$ with y equal to which of the following?

(a) -4 (b) $-\frac{4}{3}$ (c) 0   (d) $\frac{1}{3}$ (e) 5

Figure for 9

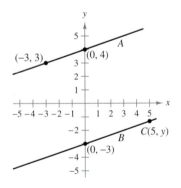

Differentiation

© Martyn Goddard/CORBIS

Higher-order derivatives are used to determine the acceleration function of a sports car. The acceleration function shows the changes in the car's velocity. As the car reaches its "cruising" speed, is the acceleration increasing or decreasing?

STRATEGIES FOR SUCCESS

WHAT YOU SHOULD LEARN:

- How to find the slope of a graph and calculate derivatives using the limit definition
- How to use the Constant Rule, Power Rule, Constant Multiple Rule, and Sum and Difference Rules
- How to find rates of change: velocity, marginal profit, marginal revenue, and marginal cost
- How to use the Product, Quotient, Chain, and General Power Rules
- How to calculate higher-order derivatives and derivatives using implicit differentiation
- How to solve related-rate problems and applications

WHY YOU SHOULD LEARN IT:

Derivatives have many applications in real life, as can be seen by the examples below, which represent a small sample of the applications in this chapter.

- Increasing Revenue, Example 10 on page 101
- Psychology: Migraine Prevalence, Exercise 62 on page 104
- Average Velocity, Exercises 15 and 16 on page 117
- Demand Function, Exercises 53 and 54 on page 129
- Quality Control, Exercise 58 on page 129
- Velocity and Acceleration, Exercises 41–44 and 50 on pages 145 and 146

2.1   THE DERIVATIVE AND THE SLOPE OF A GRAPH

■ Identify tangent lines to a graph at a point.
■ Approximate the slopes of tangent lines to graphs at points.
■ Use the limit definition to find the slopes of graphs at points.
■ Use the limit definition to find the derivatives of functions.
■ Describe the relationship between differentiability and continuity.

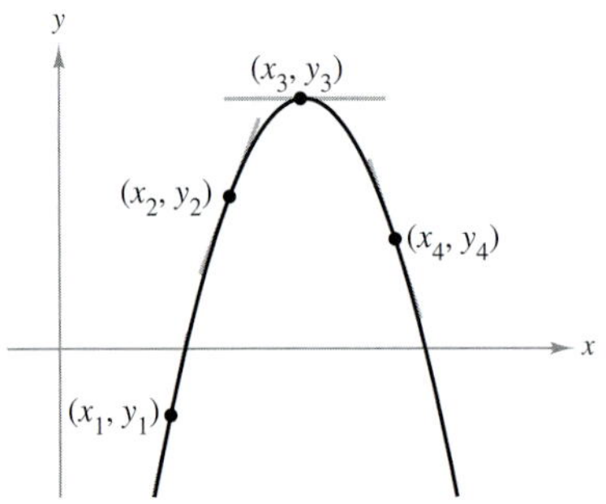

Tangent Line to a Graph

Calculus is a branch of mathematics that studies rates of change of functions. In this course, you will learn that rates of change have many applications in real life. In Section 1.3, you learned how the slope of a line indicates the rate at which the line rises or falls. For a line, this rate (or slope) is the same at every point on the line. For graphs other than lines, the rate at which the graph rises or falls changes from point to point. For instance, in Figure 2.1, the parabola is rising more quickly at the point (x_1, y_1) than it is at the point (x_2, y_2). At the vertex (x_3, y_3), the graph levels off, and at the point (x_4, y_4), the graph is falling.

To determine the rate at which a graph rises or falls at a *single point*, you can find the slope of the **tangent line** at the point. In simple terms, the tangent line to the graph of a function f at a point $P(x_1, y_1)$ is the line that best approximates the graph at that point, as shown in Figure 2.1. Figure 2.2 shows other examples of tangent lines.

FIGURE 2.1   The slope of a non-linear graph changes from one point to another.

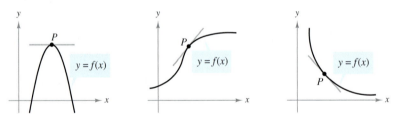

FIGURE 2.2   Tangent Line to a Graph at a Point

When Isaac Newton (1642–1727) was working on the "tangent line problem," he realized that it is difficult to define precisely what is meant by a tangent to a general curve. From geometry, you know that a line is tangent to a circle if the line intersects the circle at only one point, as shown in Figure 2.3. Tangent lines to a noncircular graph, however, can intersect the graph at more than one point. For instance, in the second graph in Figure 2.2, if the tangent line were extended, it would intersect the graph at a point other than the point of tangency. In this section, you will see how the notion of a limit can be used to define a general tangent line.

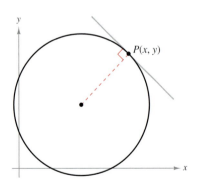

FIGURE 2.3   Tangent Line to a Circle

DISCOVERY

Use a graphing utility to sketch the graph of $f(x) = 2x^3 - 4x^2 + 3x - 5$. On the same screen, sketch the graphs of $y = x - 5$, $y = 2x - 5$, and $y = 3x - 5$. Which of these lines, if any, appears to be tangent to the graph of f at the point $(0, -5)$? Explain your reasoning.

Slope of a Graph

Because a tangent line approximates the graph at a point, the problem of finding the slope of a graph at a point becomes one of finding the slope of the tangent line at the point.

EXAMPLE 1 Approximating the Slope of a Graph

Use the graph in Figure 2.4 to approximate the slope of the graph of $f(x) = x^2$ at the point $(1, 1)$.

SOLUTION From the graph of $f(x) = x^2$, you can see that the tangent line at $(1, 1)$ rises approximately two units for each unit change in x. So, the slope of the tangent line at $(1, 1)$ is given by

$$\text{Slope} = \frac{\text{change in } y}{\text{change in } x} \approx \frac{2}{1} = 2.$$

Because the tangent line at the point $(1, 1)$ has a slope of about 2, you can conclude that the graph has a slope of about 2 at the point $(1, 1)$. ━━━━━

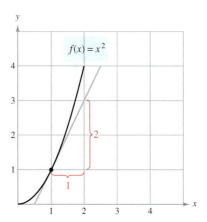

FIGURE 2.4

STUDY TIP

When visually approximating the slope of a graph, note that the scales on the horizontal and vertical axes may differ. When this happens (as it frequently does in applications), the slope of the tangent line is distorted, and you must be careful to account for the difference in scales.

EXAMPLE 2 Interpreting Slope

Figure 2.5 graphically depicts the average daily temperature (in degrees Fahrenheit) in Duluth, Minnesota. Estimate the slope of this graph at the indicated point and give a physical interpretation of the result. *(Source: National Oceanic and Atmospheric Administration)*

SOLUTION From the graph, you can see that the tangent line at the given point falls approximately 27 units for each two-unit change in x. So, you can estimate the slope at the given point to be

$$\text{Slope} = \frac{\text{change in } y}{\text{change in } x} \approx \frac{-27}{2}$$

$$= -13.5 \text{ degrees per month.}$$

This means that you can expect the average daily temperatures in November to be about 13.5 degrees *lower* than the corresponding temperatures in October.

TRY IT 1

Use the graph to approximate the slope of the graph of $f(x) = x^3$ at the point $(1, 1)$.

Average Temperature in Duluth

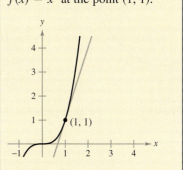

FIGURE 2.5

TRY IT 2

For which months do the slopes of the tangent lines appear to be positive? Negative? Interpret these slopes in the context of the problem.

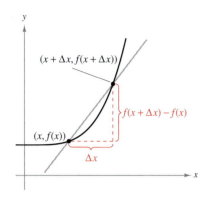

FIGURE 2.6　The Secant Line Through the Two Points $(x, f(x))$ and $(x + \Delta x, f(x + \Delta x))$

Slope and the Limit Process

In Examples 1 and 2, you approximated the slope of a graph at a point by making a careful graph and then "eyeballing" the tangent line at the point of tangency. A more precise method of approximating tangent lines makes use of a **secant line** through the point of tangency and a second point on the graph, as shown in Figure 2.6. If $(x, f(x))$ is the point of tangency and $(x + \Delta x, f(x + \Delta x))$ is a second point on the graph of f, then the slope of the secant line through the two points is

$$m_{\text{sec}} = \frac{f(x + \Delta x) - f(x)}{\Delta x}. \qquad \text{Slope of secant line}$$

The right side of this equation is called the **difference quotient.** The denominator Δx is the **change in x,** and the numerator is the **change in y.** The beauty of this procedure is that you obtain better and better approximations of the slope of the tangent line by choosing the second point closer and closer to the point of tangency, as shown in Figure 2.7.

Using the limit process, you can find the *exact* slope of the tangent line at $(x, f(x))$.

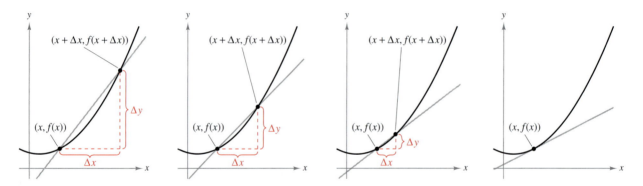

FIGURE 2.7　As Δx approaches 0, the secant lines approach the tangent line.

Definition of the Slope of a Graph

The **slope** m of the graph of f at the point $(x, f(x))$ is equal to the slope of its tangent line at $(x, f(x))$, and is given by

$$m = \lim_{\Delta x \to 0} m_{\text{sec}} = \lim_{\Delta x \to 0} \frac{f(x + \Delta x) - f(x)}{\Delta x}$$

provided this limit exists.

STUDY TIP

Δx is used as a variable to represent the change in x in the definition of the slope of a graph. Other variables may also be used. For instance, this definition is sometimes written as

$$m = \lim_{h \to 0} \frac{f(x + h) - f(x)}{h}.$$

EXAMPLE 3   **Finding Slope by the Limit Process**

Find the slope of the graph of $f(x) = x^2$ at the point $(-2, 4)$.

SOLUTION   Begin by finding an expression that represents the slope of a secant line at the point $(-2, 4)$.

$$m_{sec} = \frac{f(-2 + \Delta x) - f(-2)}{\Delta x}$$ Set up difference quotient.

$$= \frac{(-2 + \Delta x)^2 - (-2)^2}{\Delta x}$$ Use $f(x) = x^2$.

$$= \frac{4 - 4\,\Delta x + (\Delta x)^2 - 4}{\Delta x}$$ Expand terms.

$$= \frac{-4\,\Delta x + (\Delta x)^2}{\Delta x}$$ Simplify.

$$= \frac{\Delta x(-4 + \Delta x)}{\Delta x}$$ Factor and divide out.

$$= -4 + \Delta x, \quad \Delta x \neq 0$$ Simplify.

Next, take the limit of m_{sec} as $\Delta x \to 0$.

$$m = \lim_{\Delta x \to 0} m_{sec} = \lim_{\Delta x \to 0} (-4 + \Delta x) = -4$$

So, the graph of f has a slope of -4 at the point $(-2, 4)$, as shown in Figure 2.8.

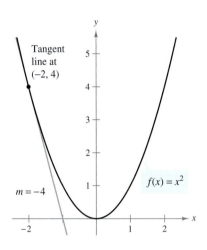

FIGURE 2.8

TRY IT 3

Find the slope of the graph of $f(x) = x^2$ at the point $(2, 4)$.

EXAMPLE 4   **Finding the Slope of a Graph**

Find the slope of $f(x) = -2x + 4$.

SOLUTION   You know from your study of linear functions that the line given by $f(x) = -2x + 4$ has a slope of -2, as shown in Figure 2.9. This conclusion is consistent with the limit definition of slope.

$$m = \lim_{\Delta x \to 0} \frac{f(x + \Delta x) - f(x)}{\Delta x}$$

$$= \lim_{\Delta x \to 0} \frac{[-2(x + \Delta x) + 4] - [-2x + 4]}{\Delta x}$$

$$= \lim_{\Delta x \to 0} \frac{-2x - 2\,\Delta x + 4 + 2x - 4}{\Delta x}$$

$$= \lim_{\Delta x \to 0} \frac{-2\,\Delta x}{\Delta x} = -2$$

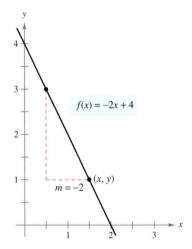

FIGURE 2.9

TRY IT 4

Find the slope of the graph of $f(x) = 2x + 5$.

It is important that you see the distinction between the ways the difference quotients were set up in Examples 3 and 4. In Example 3, you were finding the slope of a graph at a specific point $(c, f(c))$. To find the slope, you can use the following form of a difference quotient.

$$m = \lim_{\Delta x \to 0} \frac{f(c + \Delta x) - f(c)}{\Delta x} \qquad \text{Slope at specific point}$$

In Example 4, however, you were finding a formula for the slope at *any* point on the graph. In such cases, you should use x, rather than c, in the difference quotient.

$$m = \lim_{\Delta x \to 0} \frac{f(x + \Delta x) - f(x)}{\Delta x} \qquad \text{Formula for slope}$$

Except for linear functions, this form will always produce a function of x, which can then be evaluated to find the slope at any desired point.

EXAMPLE 5   Finding a Formula for the Slope of a Graph

Find a formula for the slope of the graph of $f(x) = x^2 + 1$. What are the slopes at the points $(-1, 2)$ and $(2, 5)$?

SOLUTION

$$
\begin{aligned}
m_{sec} &= \frac{f(x + \Delta x) - f(x)}{\Delta x} & \text{Set up difference quotient.}\\[2mm]
&= \frac{[(x + \Delta x)^2 + 1] - (x^2 + 1)}{\Delta x} & \text{Use } f(x) = x^2 + 1.\\[2mm]
&= \frac{x^2 + 2x\,\Delta x + (\Delta x)^2 + 1 - x^2 - 1}{\Delta x} & \text{Expand terms.}\\[2mm]
&= \frac{2x\,\Delta x + (\Delta x)^2}{\Delta x} & \text{Simplify.}\\[2mm]
&= \frac{\Delta x(2x + \Delta x)}{\Delta x} & \text{Factor and divide out.}\\[2mm]
&= 2x + \Delta x, \quad \Delta x \neq 0 & \text{Simplify.}
\end{aligned}
$$

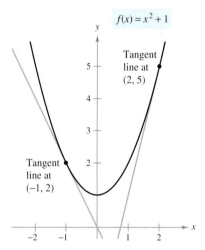

$f(x) = x^2 + 1$

Tangent line at $(2, 5)$

Tangent line at $(-1, 2)$

FIGURE 2.10

Next, take the limit of m_{sec} as $\Delta x \to 0$.

$$
\begin{aligned}
m &= \lim_{\Delta x \to 0} m_{sec}\\[2mm]
&= \lim_{\Delta x \to 0} (2x + \Delta x)\\[2mm]
&= 2x
\end{aligned}
$$

Using the formula $m = 2x$, you can find the slopes at the specified points. At $(-1, 2)$ the slope is $m = 2(-1) = -2$, and at $(2, 5)$ the slope is $m = 2(2) = 4$. The graph of f is shown in Figure 2.10.

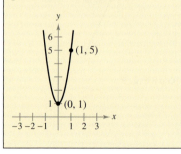
STUDY TIP

The slope of the graph of $f(x) = x^2 + 1$ varies for different values of x. For what value of x is the slope equal to 0?

The Derivative of a Function

In Example 5, you started with the function $f(x) = x^2 + 1$ and used the limit process to derive another function, $m = 2x$, that represents the slope of the graph of f at the point $(x, f(x))$. This derived function is called the **derivative** of f at x. It is denoted by $f'(x)$, which is read as "f prime of x."

Definition of the Derivative

The **derivative of f at x** is given by

$$f'(x) = \lim_{\Delta x \to 0} \frac{f(x + \Delta x) - f(x)}{\Delta x}$$

provided this limit exists. A function is **differentiable** at x if its derivative exists at x. The process of finding derivatives is called **differentiation.**

In addition to $f'(x)$, other notations can be used to denote the derivative of $y = f(x)$. The most common are

$$\frac{dy}{dx}, \quad y', \quad \frac{d}{dx}[f(x)], \quad \text{and} \quad D_x[y].$$

EXAMPLE 6 **Finding a Derivative**

Find the derivative of $f(x) = 3x^2 - 2x$.

SOLUTION

$$f'(x) = \lim_{\Delta x \to 0} \frac{f(x + \Delta x) - f(x)}{\Delta x}$$

$$= \lim_{\Delta x \to 0} \frac{[3(x + \Delta x)^2 - 2(x + \Delta x)] - (3x^2 - 2x)}{\Delta x}$$

$$= \lim_{\Delta x \to 0} \frac{3x^2 + 6x \, \Delta x + 3(\Delta x)^2 - 2x - 2 \, \Delta x - 3x^2 + 2x}{\Delta x}$$

$$= \lim_{\Delta x \to 0} \frac{6x \, \Delta x + 3(\Delta x)^2 - 2 \, \Delta x}{\Delta x}$$

$$= \lim_{\Delta x \to 0} \frac{\cancel{\Delta x}(6x + 3 \, \Delta x - 2)}{\cancel{\Delta x}}$$

$$= \lim_{\Delta x \to 0} (6x + 3 \, \Delta x - 2)$$

$$= 6x - 2$$

So, the derivative of $f(x) = 3x^2 - 2x$ is $f'(x) = 6x - 2$.

TRY IT 6

Find the derivative of $f(x) = x^2 - 5x$.

PREREQUISITE REVIEW 2.1

The following warm-up exercises involve skills that were covered in earlier sections. You will use these skills in the exercise set for this section.

In Exercises 1 and 2, find an equation of the line containing P and Q.

1. $P(2, 1)$, $Q(2, 4)$

2. $P(2, 2)$, $Q(-5, 2)$

In Exercises 3–6, find the limit.

3. $\lim\limits_{\Delta x \to 0} \dfrac{2x\Delta x + (\Delta x)^2}{\Delta x}$

4. $\lim\limits_{\Delta x \to 0} \dfrac{3x^2\Delta x + 3x(\Delta x)^2 + (\Delta x)^3}{\Delta x}$

5. $\lim\limits_{\Delta x \to 0} \dfrac{1}{x(x + \Delta x)}$

6. $\lim\limits_{\Delta x \to 0} \dfrac{(x + \Delta x)^2 - x^2}{\Delta x}$

In Exercises 7–10, find the domain of the function.

7. $f(x) = \dfrac{1}{x - 1}$

8. $f(x) = \dfrac{1}{5}x^3 - 2x^2 + \dfrac{1}{3}x - 1$

9. $f(x) = \dfrac{6x}{x^3 + x}$

10. $f(x) = \dfrac{x^2 - 2x - 24}{x^2 + x - 12}$

EXERCISES 2.1

In Exercises 1–4, trace the graph and sketch the tangent lines at (x_1, y_1) and (x_2, y_2).

1.
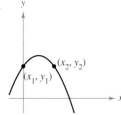

2.
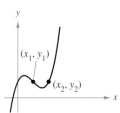

3.
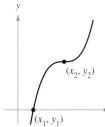

4.
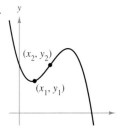

In Exercises 5–10, estimate the slope of the graph at the point (x, y). (Each square on the grid is 1 unit by 1 unit.)

5.
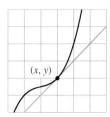

6.
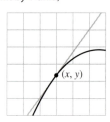

7.
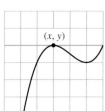

8.
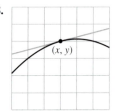

9.
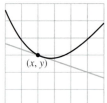

10.
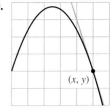

11. *Revenue* The graph (on page 91) represents the revenue R (in millions of dollars per year) for Polo Ralph Lauren from 1996 through 2002, where $t = 6$ corresponds to 1996. Estimate the slopes of the graph for the years 1997, 2000, and 2002. *(Source: Polo Ralph Lauren Corp.)*

Figure for 11

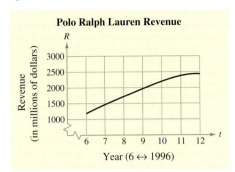

Polo Ralph Lauren Revenue

Year (6 ↔ 1996)

12. Sales The graph represents the sales S (in millions of dollars per year) for Scotts Company from 1997 through 2003, where $t = 7$ corresponds to 1997. Estimate the slopes of the graph for the years 1998, 2001, and 2003. *(Source: Scotts Company)*

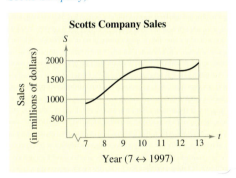

Scotts Company Sales

Year (7 ↔ 1997)

13. Consumer Trends The graph shows the number of visitors V to a national park in hundreds of thousands during a one-year period, where $t = 1$ corresponds to January. Estimate the slopes of the graph at $t = 1$, 8, and 12.

Visitors to a National Park

Month (1 ↔ January)

14. Athletics Two long distance runners starting out side by side begin a 10,000-meter run. Their distances are given by $s = f(t)$ and $s = g(t)$, respectively, where s is measured in thousands of meters and t is measured in minutes.

Figure for 14

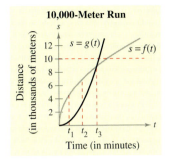

10,000-Meter Run

Time (in minutes)

(a) Which runner is running faster at t_1?

(b) What conclusion can you make regarding their rates at t_2?

(c) What conclusion can you make regarding their rates at t_3?

(d) Which runner finishes the race first? Explain.

In Exercises 15–26, use the limit definition to find the derivative of the function.

15. $f(x) = 3$ **16.** $f(x) = -4$

17. $f(x) = -5x + 3$ **18.** $f(x) = \frac{1}{2}x + 5$

19. $f(x) = x^2 - 4$ **20.** $f(x) = 1 - x^2$

21. $h(t) = \sqrt{t - 1}$ **22.** $f(x) = \sqrt{x + 2}$

23. $f(t) = t^3 - 12t$ **24.** $f(t) = t^3 + t^2$

25. $f(x) = \dfrac{1}{x + 2}$ **26.** $g(s) = \dfrac{1}{s - 1}$

In Exercises 27–36, find the slope of the tangent line to the graph of f at the given point.

27. $f(x) = 6 - 2x; (2, 2)$ **28.** $f(x) = 2x + 4; (1, 6)$

29. $f(x) = -1; (0, -1)$ **30.** $f(x) = 6; (-2, 6)$

31. $f(x) = x^2 - 2; (2, 2)$

32. $f(x) = x^2 + 2x + 1; (-3, 4)$

33. $f(x) = x^3 - x; (2, 6)$ **34.** $f(x) = x^3 + 2x; (1, 3)$

35. $f(x) = \sqrt{1 - 2x}; (-4, 3)$

36. $f(x) = \sqrt{2x - 2}; (9, 4)$

In Exercises 37–44, find an equation of the tangent line to the graph of f at the given point. Then verify your result by sketching the graph of f and the tangent line.

37. $f(x) = \frac{1}{2}x^2; (2, 2)$ **38.** $f(x) = -x^2; (-1, -1)$

39. $f(x) = (x - 1)^2; (-2, 9)$ **40.** $f(x) = 2x^2 - 1; (0, -1)$

41. $f(x) = \sqrt{x} + 1; (4, 3)$ **42.** $f(x) = \sqrt{x + 2}; (7, 3)$

43. $f(x) = \dfrac{1}{x}; (1, 1)$ **44.** $f(x) = \dfrac{1}{x - 1}; (2, 1)$

In Exercises 45–48, find an equation of the line that is tangent to the graph of f and parallel to the given line.

Function	Line
45. $f(x) = -\frac{1}{4}x^2$	$x + y = 0$
46. $f(x) = x^2 + 1$	$2x + y = 0$
47. $f(x) = -\frac{1}{2}x^3$	$6x + y + 4 = 0$
48. $f(x) = x^2 - x$	$x + 2y - 6 = 0$

In Exercises 49–56, describe the x-values at which the function is differentiable. Explain your reasoning.

49. $y = |x + 3|$

50. $y = |x^2 - 9|$

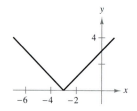

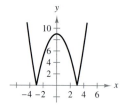

51. $y = (x - 3)^{2/3}$

52. $y = x^{2/5}$

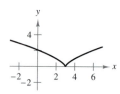

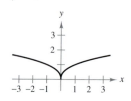

53. $y = \sqrt{x - 1}$

54. $y = \dfrac{x^2}{x^2 - 4}$

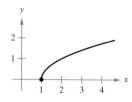

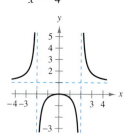

55. $y = \begin{cases} x^3 + 3, & x < 0 \\ x^3 - 3, & x \geq 0 \end{cases}$

56. $y = \begin{cases} x^2, & x \leq 1 \\ -x^2, & x > 1 \end{cases}$

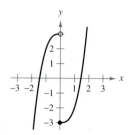

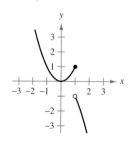

 Graphical, Numerical, and Analytic Analysis   In Exercises 57–60, use a graphing utility to graph f on the interval $[-2, 2]$. Complete the table by graphically estimating the slopes of the graph at the given points. Then evaluate the slopes analytically and compare your results with those obtained graphically.

x	-2	$-\frac{3}{2}$	-1	$-\frac{1}{2}$	0	$\frac{1}{2}$	1	$\frac{3}{2}$	2
$f(x)$									
$f'(x)$									

57. $f(x) = \frac{1}{4}x^3$

58. $f(x) = \frac{1}{2}x^2$

59. $f(x) = -\frac{1}{2}x^3$

60. $f(x) = -\frac{3}{2}x^2$

 In Exercises 61–64, find the derivative of the given function f. Then use a graphing utility to graph f and its derivative in the same viewing window. What does the x-intercept of the derivative indicate about the graph of f?

61. $f(x) = x^2 - 4x$ **62.** $f(x) = 2 + 6x - x^2$

63. $f(x) = x^3 - 3x$ **64.** $f(x) = x^3 - 6x^2$

65. *Think About It* Sketch a graph of a function whose derivative is always negative.

66. *Think About It* Sketch a graph of a function whose derivative is always positive.

 67. *Writing* Use a graphing utility to graph the two functions $f(x) = x^2 + 1$ and $g(x) = |x| + 1$ in the same viewing window. Use the *zoom* and *trace* features to analyze the graphs near the point $(0, 1)$. What do you observe? Which function is differentiable at this point? Write a short paragraph describing the geometric significance of differentiability at a point.

True or False?   In Exercises 68–71, determine whether the statement is true or false. If it is false, explain why or give an example that shows it is false.

68. The slope of the graph of $y = x^2$ is different at every point on the graph of f.

69. If a function is continuous at a point, then it is differentiable at that point.

70. If a function is differentiable at a point, then it is continuous at that point.

71. A tangent line to a graph can intersect the graph at more than one point.

2.2   SOME RULES FOR DIFFERENTIATION

- Find the derivatives of functions using the Constant Rule.
- Find the derivatives of functions using the Power Rule.
- Find the derivatives of functions using the Constant Multiple Rule.
- Find the derivatives of functions using the Sum and Difference Rules.
- Use derivatives to answer questions about real-life situations.

The Constant Rule

In Section 2.1, you found derivatives by the limit process. This process is tedious, even for simple functions, but fortunately there are rules that greatly simplify differentiation. These rules allow you to calculate derivatives without the *direct* use of limits.

The Constant Rule

The derivative of a constant function is zero. That is,

$$\frac{d}{dx}[c] = 0, \qquad c \text{ is a constant.}$$

PROOF   Let $f(x) = c$. Then, by the limit definition of the derivative, you can write

$$f'(x) = \lim_{\Delta x \to 0} \frac{f(x + \Delta x) - f(x)}{\Delta x} = \lim_{\Delta x \to 0} \frac{c - c}{\Delta x} = \lim_{\Delta x \to 0} 0 = 0.$$

So,

$$\frac{d}{dx}[c] = 0.$$

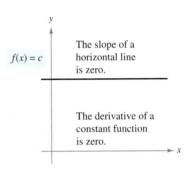

The slope of a horizontal line is zero.

The derivative of a constant function is zero.

FIGURE 2.12

STUDY TIP

Note in Figure 2.12 that the Constant Rule is equivalent to saying that the slope of a horizontal line is zero.

EXAMPLE 1   **Finding Derivatives of Constant Functions**

(a) $\dfrac{d}{dx}[7] = 0$

(b) If $f(x) = 0$, then $f'(x) = 0$.

(c) If $y = 2$, then $\dfrac{dy}{dx} = 0$.

(d) If $g(t) = -\dfrac{3}{2}$, then $g'(t) = 0$.

STUDY TIP

An interpretation of the Constant Rule says that the tangent line to a constant function is the function itself. Find an equation of the tangent line to $f(x) = -4$ at $x = 3$.

TRY IT 1

Find the derivative of each function.

(a) $f(x) = -2$ (b) $y = \pi$ (c) $g(w) = \sqrt{5}$ (d) $s(t) = 320.5$

The Power Rule

The binomial expansion process is used to prove the Power Rule.

$$(x + \Delta x)^2 = x^2 + 2x\,\Delta x + (\Delta x)^2$$

$$(x + \Delta x)^3 = x^3 + 3x^2\,\Delta x + 3x(\Delta x)^2 + (\Delta x)^3$$

$$(x + \Delta x)^n = x^n + nx^{n-1}\,\Delta x + \underbrace{\frac{n(n-1)x^{n-2}}{2}(\Delta x)^2 + \cdots + (\Delta x)^n}_{(\Delta x)^2 \text{ is a factor of these terms.}}$$

The (Simple) Power Rule

$$\frac{d}{dx}[x^n] = nx^{n-1}, \qquad n \text{ is any real number.}$$

PROOF   We prove only the case in which n is a positive integer. Let $f(x) = x^n$. Using the binomial expansion, you can write

$$f'(x) = \lim_{\Delta x \to 0} \frac{f(x + \Delta x) - f(x)}{\Delta x} \qquad \text{Definition of derivative}$$

$$= \lim_{\Delta x \to 0} \frac{(x + \Delta x)^n - x^n}{\Delta x}$$

$$= \lim_{\Delta x \to 0} \frac{x^n + nx^{n-1}\,\Delta x + \dfrac{n(n-1)x^{n-2}}{2}(\Delta x)^2 + \cdots + (\Delta x)^n - x^n}{\Delta x}$$

$$= \lim_{\Delta x \to 0}\left[nx^{n-1} + \frac{n(n-1)x^{n-2}}{2}(\Delta x) + \cdots + (\Delta x)^{n-1} \right]$$

$$= nx^{n-1} + 0 + \cdots + 0 = nx^{n-1}.$$

For the Power Rule, the case in which $n = 1$ is worth remembering as a separate differentiation rule. That is,

$$\frac{d}{dx}[x] = 1. \qquad \text{The derivative of } x \text{ is } 1.$$

This rule is consistent with the fact that the slope of the line given by $y = x$ is 1. (See Figure 2.13.)

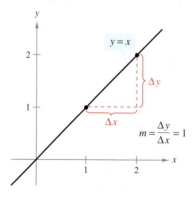

FIGURE 2.13   The slope of the line $y = x$ is 1.

EXAMPLE 2 **Applying the Power Rule**

Find the derivative of each function.

Function	Derivative

(a) $f(x) = x^3$ $f'(x) = 3x^2$

(b) $y = \dfrac{1}{x^2} = x^{-2}$ $\dfrac{dy}{dx} = (-2)x^{-3} = -\dfrac{2}{x^3}$

(c) $g(t) = t$ $g'(t) = 1$

(d) $R = x^4$ $\dfrac{dR}{dx} = 4x^3$

> **TRY IT 2**
>
> Find the derivative of each function.
>
> (a) $f(x) = x^4$ (b) $y = \dfrac{1}{x^3}$
>
> (c) $g(w) = w^2$ (d) $s(t) = \dfrac{1}{t}$

In Example 2(b), note that *before* differentiating, you should rewrite $1/x^2$ as x^{-2}. Rewriting is the first step in *many* differentiation problems.

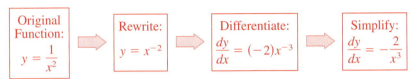

Remember that the derivative of a function f is another function that gives the slope of the graph of f at any point at which f is differentiable. So, you can use the derivative to find slopes, as shown in Example 3.

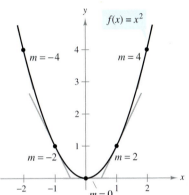

FIGURE 2.14

EXAMPLE 3 **Finding the Slope of a Graph**

Find the slopes of the graph of

$$f(x) = x^2 \qquad \text{Original function}$$

when $x = -2, -1, 0, 1,$ and 2.

SOLUTION Begin by using the Power Rule to find the derivative of f.

$$f'(x) = 2x \qquad \text{Derivative}$$

You can use the derivative to find the slopes of the graph of f, as shown.

x-Value	Slope of Graph of f
$x = -2$	$m = f'(-2) = 2(-2) = -4$
$x = -1$	$m = f'(-1) = 2(-1) = -2$
$x = 0$	$m = f'(0) = 2(0) = 0$
$x = 1$	$m = f'(1) = 2(1) = 2$
$x = 2$	$m = f'(2) = 2(2) = 4$

The graph of f is shown in Figure 2.14.

> **TRY IT 3**
>
> Find the slopes of the graph of $f(x) = x^3$ when $x = -1, 0,$ and 1.
>
>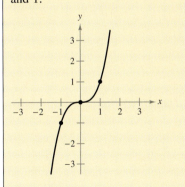

The Constant Multiple Rule

To prove the Constant Multiple Rule, the following property of limits is used.

$$\lim_{x \to a} cg(x) = c\left[\lim_{x \to a} g(x)\right]$$

The Constant Multiple Rule

If f is a differentiable function of x, and c is a real number, then

$$\frac{d}{dx}[cf(x)] = cf'(x), \qquad c \text{ is a constant.}$$

PROOF   Apply the definition of the derivative to produce

$$\frac{d}{dx}[cf(x)] = \lim_{\Delta x \to 0} \frac{cf(x + \Delta x) - cf(x)}{\Delta x} \qquad \text{Definition of derivative}$$

$$= \lim_{\Delta x \to 0} c\left[\frac{f(x + \Delta x) - f(x)}{\Delta x}\right]$$

$$= c\left[\lim_{\Delta x \to 0} \frac{f(x + \Delta x) - f(x)}{\Delta x}\right] = cf'(x).$$

Informally, the Constant Multiple Rule states that constants can be factored out of the differentiation process.

$$\frac{d}{dx}[cf(x)] = c\frac{d}{dx}[\, f(x)] = cf'(x)$$

The usefulness of this rule is often overlooked, especially when the constant appears in the denominator, as shown below.

$$\frac{d}{dx}\left[\frac{f(x)}{c}\right] = \frac{d}{dx}\left[\frac{1}{c}f(x)\right] = \frac{1}{c}\left(\frac{d}{dx}[\, f(x)]\right) = \frac{1}{c}f'(x)$$

To use the Constant Multiple Rule efficiently, look for constants that can be factored out *before* differentiating. For example,

$$\frac{d}{dx}[5x^2] = 5\frac{d}{dx}[x^2] \qquad \text{Factor out 5.}$$

$$= 5(2x) \qquad \text{Differentiate.}$$

$$= 10x \qquad \text{Simplify.}$$

and

$$\frac{d}{dx}\left[\frac{x^2}{5}\right] = \frac{1}{5}\left(\frac{d}{dx}[x^2]\right) \qquad \text{Factor out } \tfrac{1}{5}.$$

$$= \frac{1}{5}(2x) \qquad \text{Differentiate.}$$

$$= \frac{2}{5}x. \qquad \text{Simplify.}$$

EXAMPLE 4   **Using the Power and Constant Multiple Rules**

Differentiate each function.

(a) $y = 2x^{1/2}$ (b) $f(t) = \dfrac{4t^2}{5}$

TECHNOLOGY

If you have access to a symbolic differentiation utility, try using it to confirm the derivatives shown in this section.

SOLUTION

(a) Using the Constant Multiple Rule and the Power Rule, you can write

$$\frac{dy}{dx} = \frac{d}{dx}[2x^{1/2}] = \underbrace{2\frac{d}{dx}[x^{1/2}]}_{\text{Constant Multiple Rule}} = \underbrace{2\left(\frac{1}{2}x^{-1/2}\right)}_{\text{Power Rule}} = x^{-1/2} = \frac{1}{\sqrt{x}}.$$

(b) Begin by rewriting $f(t)$ as

$$f(t) = \frac{4t^2}{5} = \frac{4}{5}t^2.$$

Then, use the Constant Multiple Rule and the Power Rule to obtain

$$f'(t) = \frac{d}{dt}\left[\frac{4}{5}t^2\right] = \frac{4}{5}\left[\frac{d}{dt}(t^2)\right] = \frac{4}{5}(2t) = \frac{8}{5}t.$$

TRY IT 4

Differentiate each function.

(a) $y = 4x^2$

(b) $f(x) = 16x^{1/2}$

You may find it helpful to combine the Constant Multiple Rule and the Power Rule into one combined rule.

$$\frac{d}{dx}[cx^n] = cnx^{n-1}, \qquad n \text{ is a real number, } c \text{ is a constant.}$$

For instance, in Example 4(b), you can apply this combined rule to obtain

$$\frac{d}{dt}\left[\frac{4}{5}t^2\right] = \left(\frac{4}{5}\right)(2)(t) = \frac{8}{5}t.$$

The three functions in the next example are simple, yet errors are frequently made in differentiating functions involving constant multiples of the first power of x. Keep in mind that

$$\frac{d}{dx}[cx] = c, \qquad c \text{ is a constant.}$$

EXAMPLE 5   **Applying the Constant Multiple Rule**

Find the derivative of each function.

Original Function	*Derivative*
(a) $y = -\dfrac{3x}{2}$	$y' = -\dfrac{3}{2}$
(b) $y = 3\pi x$	$y' = 3\pi$
(c) $y = -\dfrac{x}{2}$	$y' = -\dfrac{1}{2}$

TRY IT 5

Find the derivative of each function.

(a) $y = \dfrac{t}{4}$

(b) $y = -\dfrac{2x}{5}$

Parentheses can play an important role in the use of the Constant Multiple Rule and the Power Rule. In Example 6, be sure you understand the mathematical conventions involving the use of parentheses.

EXAMPLE 6 Using Parentheses When Differentiating

Find the derivative of each function.

(a) $y = \dfrac{5}{2x^3}$ (b) $y = \dfrac{5}{(2x)^3}$ (c) $y = \dfrac{7}{3x^{-2}}$ (d) $y = \dfrac{7}{(3x)^{-2}}$

SOLUTION

Function	Rewrite	Differentiate	Simplify
(a) $y = \dfrac{5}{2x^3}$	$y = \dfrac{5}{2}(x^{-3})$	$y' = \dfrac{5}{2}(-3x^{-4})$	$y' = -\dfrac{15}{2x^4}$
(b) $y = \dfrac{5}{(2x)^3}$	$y = \dfrac{5}{8}(x^{-3})$	$y' = \dfrac{5}{8}(-3x^{-4})$	$y' = -\dfrac{15}{8x^4}$
(c) $y = \dfrac{7}{3x^{-2}}$	$y = \dfrac{7}{3}(x^2)$	$y' = \dfrac{7}{3}(2x)$	$y' = \dfrac{14x}{3}$
(d) $y = \dfrac{7}{(3x)^{-2}}$	$y = 63(x^2)$	$y' = 63(2x)$	$y' = 126x$

TRY IT 6

Find the derivative of each function.

(a) $y = \dfrac{9}{4x^2}$ (b) $y = \dfrac{9}{(4x)^2}$

EXAMPLE 7 Differentiating Radical Functions

Find the derivative of each function.

(a) $y = \sqrt{x}$ (b) $y = \dfrac{1}{2\sqrt[3]{x^2}}$ (c) $y = \sqrt{2x}$

SOLUTION

Function	Rewrite	Differentiate	Simplify
(a) $y = \sqrt{x}$	$y = x^{1/2}$	$y' = \left(\dfrac{1}{2}\right)x^{-1/2}$	$y' = \dfrac{1}{2\sqrt{x}}$
(b) $y = \dfrac{1}{2\sqrt[3]{x^2}}$	$y = \dfrac{1}{2}x^{-2/3}$	$y' = \dfrac{1}{2}\left(-\dfrac{2}{3}\right)x^{-5/3}$	$y' = -\dfrac{1}{3x^{5/3}}$
(c) $y = \sqrt{2x}$	$y = \sqrt{2}(x^{1/2})$	$y' = \sqrt{2}\left(\dfrac{1}{2}\right)x^{-1/2}$	$y' = \dfrac{1}{\sqrt{2x}}$

TRY IT 7

Find the derivative of each function.

(a) $y = \sqrt{5x}$

(b) $y = \sqrt[3]{x}$

The Sum and Difference Rules

The next two rules are ones that you might expect to be true, and you may have used them without thinking about it. For instance, if you were asked to differentiate $y = 3x + 2x^3$, you would probably write

$$y' = 3 + 6x^2$$

without questioning your answer. The validity of differentiating a sum of functions term by term is given by the Sum and Difference Rules.

The Sum and Difference Rules

The derivative of the sum (or difference) of two differentiable functions is the sum (or difference) of their derivatives.

$$\frac{d}{dx}[f(x) + g(x)] = f'(x) + g'(x) \qquad \text{Sum Rule}$$

$$\frac{d}{dx}[f(x) - g(x)] = f'(x) - g'(x) \qquad \text{Difference Rule}$$

PROOF   Let $h(x) = f(x) + g(x)$. Then, you can prove the Sum Rule as shown.

$$h'(x) = \lim_{\Delta x \to 0} \frac{h(x + \Delta x) - h(x)}{\Delta x} \qquad \text{Definition of derivative}$$

$$= \lim_{\Delta x \to 0} \frac{f(x + \Delta x) + g(x + \Delta x) - f(x) - g(x)}{\Delta x}$$

$$= \lim_{\Delta x \to 0} \frac{f(x + \Delta x) - f(x) + g(x + \Delta x) - g(x)}{\Delta x}$$

$$= \lim_{\Delta x \to 0} \left[\frac{f(x + \Delta x) - f(x)}{\Delta x} + \frac{g(x + \Delta x) - g(x)}{\Delta x} \right]$$

$$= \lim_{\Delta x \to 0} \frac{f(x + \Delta x) - f(x)}{\Delta x} + \lim_{\Delta x \to 0} \frac{g(x + \Delta x) - g(x)}{\Delta x}$$

$$= f'(x) + g'(x)$$

So,

$$\frac{d}{dx}[f(x) + g(x)] = f'(x) + g'(x).$$

The Difference Rule can be proved in a similar manner.   ━━━━

The Sum and Difference Rules can be extended to the sum or difference of any finite number of functions. For instance, if $y = f(x) + g(x) + h(x)$, then $y' = f'(x) + g'(x) + h'(x)$.

STUDY TIP

Look back at Example 6 on page 87. Notice that the example asks for the derivative of the difference of two functions. Verify this result by using the Difference Rule.

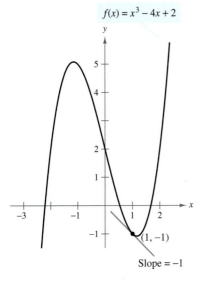

$f(x) = x^3 - 4x + 2$

FIGURE 2.15

With the four differentiation rules listed in this section, you can differentiate *any* polynomial function.

EXAMPLE 8   **Using the Sum and Difference Rules**

Find the slope of the graph of $f(x) = x^3 - 4x + 2$ at the point $(1, -1)$.

SOLUTION   The derivative of $f(x)$ is

$$f'(x) = 3x^2 - 4.$$

So, the slope of the graph of f at $(1, -1)$ is

$$\text{Slope} = f'(1) = 3(1)^2 - 4 = -1$$

as shown in Figure 2.15.

TRY IT 8

Find the slope of the graph of $f(x) = x^2 - 5x + 1$ at the point $(2, -5)$.

Example 8 illustrates the use of the derivative for determining the shape of a graph. A rough sketch of the graph of $f(x) = x^3 - 4x + 2$ might lead you to think that the point $(1, -1)$ is a minimum point of the graph. After finding the slope at this point to be -1, however, you can conclude that the minimum point (where the slope is 0) is farther to the right. (You will study techniques for finding minimum and maximum points in Section 3.2.)

EXAMPLE 9   **Using the Sum and Difference Rules**

Find an equation of the tangent line to the graph of

$$g(x) = -\frac{1}{2}x^4 + 3x^3 - 2x$$

at the point $\left(-1, -\frac{3}{2}\right)$.

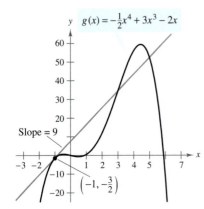

$g(x) = -\frac{1}{2}x^4 + 3x^3 - 2x$

FIGURE 2.16

SOLUTION   The derivative of $g(x)$ is $g'(x) = -2x^3 + 9x^2 - 2$, which implies that the slope of the graph at the point $\left(-1, -\frac{3}{2}\right)$ is

$$\begin{aligned}
\text{Slope} = g'(-1) &= -2(-1)^3 + 9(-1)^2 - 2 \\
&= 2 + 9 - 2 \\
&= 9
\end{aligned}$$

as shown in Figure 2.16. Using the point-slope form, you can write the equation of the tangent line at $\left(-1, -\frac{3}{2}\right)$ as shown.

$$y - \left(-\frac{3}{2}\right) = 9[x - (-1)] \qquad \text{Point-slope form}$$

$$y = 9x + \frac{15}{2} \qquad \text{Equation of tangent line}$$

TRY IT 9

Find an equation of the tangent line to the graph of $f(x) = -x^2 + 3x - 2$ at the point $(2, 0)$.

Application

EXAMPLE 10   **Modeling Revenue**

From 1998 through 2003, the revenue R (in millions of dollars per year) for Microsoft Corporation can be modeled by

$$R = 174.343t^3 - 5630.45t^2 + 63,029.8t - 218,635, \quad 8 \le t \le 13$$

where $t = 8$ represents 1998. At what rate was Microsoft's revenue changing in 1999?   *(Source: Microsoft Corporation)*

SOLUTION   One way to answer this question is to find the derivative of the revenue model with respect to time.

$$\frac{dR}{dt} = 523.029t^2 - 11,260.90t + 63,029.8, \quad 8 \le t \le 13$$

In 1999 (when $t = 9$), the rate of change of the revenue with respect to time is given by

$$523.029(9)^2 - 11,260.90(9) + 63,029.8 \approx 4047.$$

Because R is measured in millions of dollars and t is measured in years, it follows that the derivative dR/dt is measured in millions of dollars per year. So, at the end of 1999, Microsoft's revenues were increasing at a rate of about $4047 million per year, as shown in Figure 2.17.

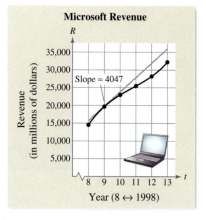

Microsoft Revenue

FIGURE 2.17

TRY IT 10

From 1995 through 2002, the revenue per share R (in dollars) for McDonald's Corporation can be modeled by

$$R = 0.0043t^2 + 0.689t + 3.39, \quad 5 \le t \le 12$$

where $t = 5$ represents 1995. At what rate was McDonald's revenue per share changing in 1998?   *(Source: McDonald's Corporation)*

TAKE ANOTHER LOOK

Units for Rates of Change

In Example 10, the units for R are millions of dollars and the units for dR/dt are millions of dollars per year. State the units for the derivatives of each model.

a. A population model that gives the population P (in millions of people) of the United States in terms of the year t. What are the units for dP/dt?

b. A position model that gives the height s (in feet) of an object in terms of the time t (in seconds). What are the units for ds/dt?

c. A demand model that gives the price per unit p (in dollars) of a product in terms of the number of units x sold. What are the units for dp/dx?

PREREQUISITE REVIEW 2.2

The following warm-up exercises involve skills that were covered in earlier sections. You will use these skills in the exercise set for this section.

In Exercises 1 and 2, evaluate each expression when $x = 2$.

1. (a) $2x^2$ (b) $(2x)^2$ (c) $2x^{-2}$

2. (a) $\dfrac{1}{(3x)^2}$ (b) $\dfrac{1}{4x^3}$ (c) $\dfrac{(2x)^{-3}}{4x^{-2}}$

In Exercises 3–6, simplify the expression.

3. $4(3)x^3 + 2(2)x$

4. $\frac{1}{2}(3)x^2 - \frac{3}{2}x^{1/2}$

5. $\left(\frac{1}{4}\right)x^{-3/4}$

6. $\frac{1}{3}(3)x^2 - 2\left(\frac{1}{2}\right)x^{-1/2} + \frac{1}{3}x^{-2/3}$

In Exercises 7–10, solve the equation.

7. $3x^2 + 2x = 0$

8. $x^3 - x = 0$

9. $x^2 + 8x - 20 = 0$

10. $x^2 - 10x - 24 = 0$

EXERCISES 2.2

In Exercises 1–4, find the slope of the tangent line to $y = x^n$ at the point $(1, 1)$.

1. (a) $y = x^2$

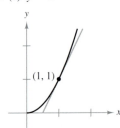

(b) $y = x^{1/2}$

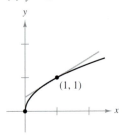

2. (a) $y = x^{3/2}$

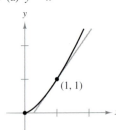

(b) $y = x^3$

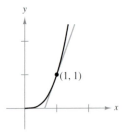

3. (a) $y = x^{-1}$

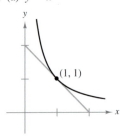

(b) $y = x^{-1/3}$

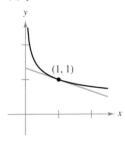

4. (a) $y = x^{-1/2}$

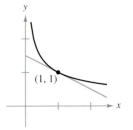

(b) $y = x^{-2}$

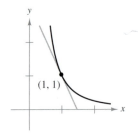

In Exercises 5–20, find the derivative of the function.

5. $y = 3$

6. $f(x) = -2$

7. $f(x) = 4x + 1$

8. $g(x) = 3x - 1$

9. $g(x) = x^2 + 4x - 1$

10. $y = t^2 + 2t - 3$

11. $f(t) = -3t^2 + 2t - 4$

12. $y = x^3 - 9x^2 + 2$

13. $s(t) = t^3 - 2t + 4$

14. $y = 2x^3 - x^2 + 3x - 1$

15. $y = 4t^{4/3}$

16. $h(x) = x^{5/2}$

17. $f(x) = 4\sqrt{x}$

18. $g(x) = 4\sqrt[3]{x} + 2$

19. $y = 4x^{-2} + 2x^2$

20. $s(t) = 4t^{-1} + 1$

In Exercises 21–26, use Example 6 as a model to find the derivative.

Function	Rewrite	Differentiate	Simplify
21. $y = \dfrac{1}{4x^3}$			
22. $y = \dfrac{2}{3x^2}$			
23. $y = \dfrac{1}{(4x)^3}$			
24. $y = \dfrac{\pi}{(3x)^2}$			
25. $y = \dfrac{\sqrt{x}}{x}$			
26. $y = \dfrac{4x}{x^{-3}}$			

In Exercises 27–32, find the value of the derivative of the function at the given point.

Function	Point
27. $f(x) = \dfrac{1}{x}$	$(1, 1)$
28. $f(t) = 4 - \dfrac{4}{3t}$	$\left(\dfrac{1}{2}, \dfrac{4}{3}\right)$
29. $f(x) = -\frac{1}{2}x(1 + x^2)$	$(1, -1)$
30. $y = 3x\left(x^2 - \dfrac{2}{x}\right)$	$(2, 18)$
31. $y = (2x + 1)^2$	$(0, 1)$
32. $f(x) = 3(5 - x)^2$	$(5, 0)$

In Exercises 33–46, find $f'(x)$.

33. $f(x) = x^2 - \dfrac{4}{x} - 3x^{-2}$

34. $f(x) = x^2 - 3x - 3x^{-2} + 5x^{-3}$

35. $f(x) = x^2 - 2x - \dfrac{2}{x^4}$ **36.** $f(x) = x^2 + 4x + \dfrac{1}{x}$

37. $f(x) = x(x^2 + 1)$ **38.** $f(x) = (x^2 + 2x)(x + 1)$

39. $f(x) = (x + 4)(2x^2 - 1)$

40. $f(x) = (3x^2 - 5x)(x^2 + 2)$

41. $f(x) = \dfrac{2x^3 - 4x^2 + 3}{x^2}$ **42.** $f(x) = \dfrac{2x^2 - 3x + 1}{x}$

43. $f(x) = \dfrac{4x^3 - 3x^2 + 2x + 5}{x^2}$

44. $f(x) = \dfrac{-6x^3 + 3x^2 - 2x + 1}{x}$

45. $f(x) = x^{4/5} + x$ **46.** $f(x) = x^{1/3} - 1$

In Exercises 47–50, find an equation of the tangent line to the graph of the function at the given point.

Function	Point
47. $y = -2x^4 + 5x^2 - 3$	$(1, 0)$
48. $y = x^3 + x$	$(-1, -2)$
49. $f(x) = \sqrt[3]{x} + \sqrt[5]{x}$	$(1, 2)$
50. $f(x) = \dfrac{1}{\sqrt[3]{x^2}} - x$	$(-1, 2)$

In Exercises 51–54, determine the point(s), if any, at which the graph of the function has a horizontal tangent line.

51. $y = -x^4 + 3x^2 - 1$

52. $y = x^3 + 3x^2$

53. $y = \frac{1}{2}x^2 + 5x$

54. $y = x^2 + 2x$

In Exercises 55 and 56,

(a) Sketch the graphs of $f, g,$ and h on the same set of coordinate axes.

(b) Find $f'(1)$, $g'(1)$, and $h'(1)$.

(c) Sketch the graph of the tangent line to each graph when $x = 1$.

55. $f(x) = x^3$ **56.** $f(x) = \sqrt{x}$
$\quad\ g(x) = x^3 + 3$ $\qquad\qquad g(x) = \sqrt{x} + 4$
$\quad\ h(x) = x^3 - 2$ $\qquad\qquad h(x) = \sqrt{x} - 2$

57. Use the Constant Rule, the Constant Multiple Rule, and the Sum Rule to find $h'(1)$ given that $f'(1) = 3$.

(a) $h(x) = f(x) - 2$ (b) $h(x) = 2f(x)$

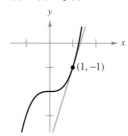

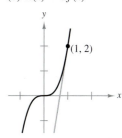

(c) $h(x) = -f(x)$ (d) $h(x) = -1 + 2f(x)$

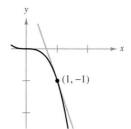

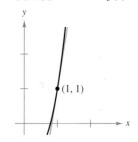

58. *Revenue* The revenue R (in millions of dollars per year) for Polo Ralph Lauren from 1996 through 2002 can be modeled by

$$R = -1.17879t^4 + 38.3641t^3 - 469.994t^2 + 2820.22t - 5577.7$$

where $t = 6$ corresponds to 1996. *(Source: Polo Ralph Lauren Corp.)*

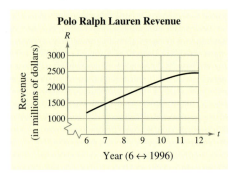

Polo Ralph Lauren Revenue

(a) Find the slopes of the graph for the years 1997, 2000, and 2002.

(b) Compare your results with those obtained in Exercise 11 in Section 2.1.

(c) What are the units for the slope of the graph? Interpret the slope of the graph in the context of the problem.

59. *Sales* The sales S (in millions of dollars per year) for Scotts Company from 1997 through 2003 can be modeled by

$$S = 8.70947t^4 - 341.0927t^3 + 4885.752t^2 - 30,118.17t + 68,395.3$$

where $t = 7$ corresponds to 1997. *(Source: Scotts Company)*

Scotts Company Sales

(a) Find the slopes of the graph for the years 1998, 2001, and 2003.

(b) Compare your results with those obtained in Exercise 12 in Section 2.1.

(c) What are the units for the slope of the graph? Interpret the slope of the graph in the context of the problem.

60. *Cost* The variable cost for manufacturing an electrical component is $7.75 per unit, and the fixed cost is $500. Write the cost C as a function of x, the number of units produced. Show that the derivative of this cost function is a constant and is equal to the variable cost.

61. *Profit* A college club raises funds by selling candy bars for $1.00 each. The club pays $0.60 for each candy bar and has annual fixed costs of $250. Write the profit P as a function of x, the number of candy bars sold. Show that the derivative of the profit function is a constant and is equal to the profit on each candy bar sold.

62. *Psychology: Migraine Prevalence* The graph illustrates the prevalence of migraine headaches in males and females in selected income groups. *(Source: Adapted from Sue/Sue/Sue, Understanding Abnormal Behavior, Seventh Edition)*

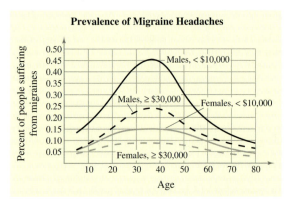

Prevalence of Migraine Headaches

(a) Write a short paragraph describing your general observations about the prevalence of migraines in females and males with respect to age group and income bracket.

(b) Describe the graph of the derivative of each curve, and explain the significance of each derivative. Include an explanation of the units of the derivatives, and indicate the time intervals in which the derivatives would be positive and negative.

In Exercises 63 and 64, use a graphing utility to graph f and f' over the given interval. Determine any points at which the graph of f has horizontal tangents.

Function	Interval
63. $f(x) = 4.1x^3 - 12x^2 + 2.5x$	$[0, 3]$
64. $f(x) = x^3 - 1.4x^2 - 0.96x + 1.44$	$[-2, 2]$

True or False? In Exercises 65 and 66, determine whether the statement is true or false. If it is false, explain why or give an example that shows it is false.

65. If $f'(x) = g'(x)$, then $f(x) = g(x)$.

66. If $f(x) = g(x) + c$, then $f'(x) = g'(x)$.

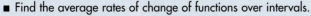

2.3   RATES OF CHANGE: VELOCITY AND MARGINALS

- Find the average rates of change of functions over intervals.
- Find the instantaneous rates of change of functions at points.
- Find the marginal revenues, marginal costs, and marginal profits for products.

Average Rate of Change

In Sections 2.1 and 2.2, you studied the two primary applications of derivatives.

1. **Slope** The derivative of f is a function that gives the slope of the graph of f at a point $(x, f(x))$.

2. **Rate of Change** The derivative of f is a function that gives the rate of change of $f(x)$ with respect to x at the point $(x, f(x))$.

In this section, you will see that there are many real-life applications of rates of change. A few are velocity, acceleration, population growth rates, unemployment rates, production rates, and water flow rates. Although rates of change often involve change with respect to time, you can investigate the rate of change of one variable with respect to any other related variable.

When determining the rate of change of one variable with respect to another, you must be careful to distinguish between *average* and *instantaneous* rates of change. The distinction between these two rates of change is comparable to the distinction between the slope of the secant line through two points on a graph and the slope of the tangent line at one point on the graph.

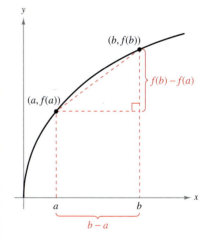

FIGURE 2.18

Definition of Average Rate of Change

If $y = f(x)$, then the **average rate of change** of y with respect to x on the interval $[a, b]$ is

$$\text{Average rate of change} = \frac{f(b) - f(a)}{b - a}$$

$$= \frac{\Delta y}{\Delta x}.$$

Note that $f(a)$ is the value of the function at the *left* endpoint of the interval, $f(b)$ is the value of the function at the *right* endpoint of the interval, and $b - a$ is the width of the interval, as shown in Figure 2.18.

STUDY TIP

In real-life problems, it is important to list the units of measure for a rate of change. The units for $\Delta y/\Delta x$ are "y-units" per "x-units." For example, if y is measured in miles and x is measured in hours, then $\Delta y/\Delta x$ is measured in *miles per hour.*

EXAMPLE 1   **Medicine**

The concentration C (in milligrams per milliliter) of a drug in a patient's bloodstream is monitored over 10-minute intervals for 2 hours, where t is measured in minutes, as shown in the table. Find the average rate of change over each interval.

(a) $[0, 10]$ (b) $[0, 20]$ (c) $[100, 110]$

t	0	10	20	30	40	50	60	70	80	90	100	110	120
C	0	2	17	37	55	73	89	103	111	113	113	103	68

SOLUTION

(a) For the interval $[0, 10]$, the average rate of change is

$$\frac{\Delta C}{\Delta t} = \frac{2 - 0}{10 - 0} = \frac{2}{10} = 0.2 \text{ milligram per milliliter per minute.}$$

Value of C at right endpoint

Value of C at left endpoint

Width of interval

(b) For the interval $[0, 20]$, the average rate of change is

$$\frac{\Delta C}{\Delta t} = \frac{17 - 0}{20 - 0} = \frac{17}{20} = 0.85 \text{ milligram per milliliter per minute.}$$

(c) For the interval $[100, 110]$, the average rate of change is

$$\frac{\Delta C}{\Delta t} = \frac{103 - 113}{110 - 100} = \frac{-10}{10} = -1 \text{ milligram per milliliter per minute.}$$

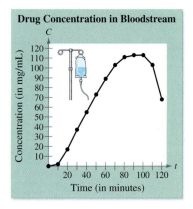

Drug Concentration in Bloodstream

FIGURE 2.19

TRY IT 1

Use the table in Example 1 to find the average rates of change over each interval.

(a) $[0, 120]$ (b) $[90, 100]$ (c) $[90, 120]$

The rates of change in Example 1 are in milligrams per milliliter per minute because the concentration is measured in milligrams per milliliter and the time is measured in minutes.

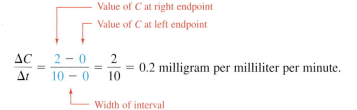

Concentration is measured in milligrams per milliliter.

Rate of change is measured in milligrams per milliliter per minute.

$$\frac{\Delta C}{\Delta t} = \frac{2 - 0}{10 - 0} = \frac{2}{10} = 0.2 \text{ milligram per milliliter per minute}$$

Time is measured in minutes.

A common application of an average rate of change is to find the **average velocity** of an object that is moving in a straight line. That is,

$$\text{Average velocity} = \frac{\text{change in distance}}{\text{change in time}}.$$

This formula is demonstrated in Example 2.

EXAMPLE 2 **Finding an Average Velocity**

If a free-falling object is dropped from a height of 100 feet, and *air resistance is neglected*, the height h (in feet) of the object at time t (in seconds) is given by

$$h = -16t^2 + 100. \qquad \text{(See Figure 2.20.)}$$

Find the average velocity of the object over each interval.

(a) $[1, 2]$ (b) $[1, 1.5]$ (c) $[1, 1.1]$

SOLUTION You can use the position equation $h = -16t^2 + 100$ to determine the heights at $t = 1$, $t = 1.1$, $t = 1.5$, and $t = 2$, as shown in the table.

t (in seconds)	0	1	1.1	1.5	2
h (in feet)	100	84	80.64	64	36

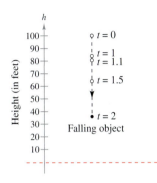

FIGURE 2.20 Some falling objects have considerable air resistance. Other falling objects have negligible air resistance. When modeling a falling-body problem, you must decide whether to account for air resistance or neglect it.

(a) For the interval $[1, 2]$, the object falls from a height of 84 feet to a height of 36 feet. So, the average velocity is

$$\frac{\Delta h}{\Delta t} = \frac{36 - 84}{2 - 1} = \frac{-48}{1} = -48 \text{ feet per second.}$$

(b) For the interval $[1, 1.5]$, the average velocity is

$$\frac{\Delta h}{\Delta t} = \frac{64 - 84}{1.5 - 1} = \frac{-20}{0.5} = -40 \text{ feet per second.}$$

(c) For the interval $[1, 1.1]$, the average velocity is

$$\frac{\Delta h}{\Delta t} = \frac{80.64 - 84}{1.1 - 1} = \frac{-3.36}{0.1} = -33.6 \text{ feet per second.}$$

TRY IT 2

The height h (in feet) of a free-falling object at time t (in seconds) is given by $h = -16t^2 + 180$. Find the average velocity of the object over each interval.

(a) $[0, 1]$ (b) $[1, 2]$ (c) $[2, 3]$

STUDY TIP

In Example 2, the average velocities are negative because the object is moving downward.

Instantaneous Rate of Change and Velocity

Suppose in Example 2 you wanted to find the rate of change of h at the instant $t = 1$ second. Such a rate is called an **instantaneous rate of change.** You can approximate the instantaneous rate of change at $t = 1$ by calculating the average rate of change over smaller and smaller intervals of the form $[1, 1 + \Delta t]$, as shown in the table. From the table, it seems reasonable to conclude that the instantaneous rate of change of the height when $t = 1$ is -32 feet per second.

Δt approaches 0.

Δt	1	0.5	0.1	0.01	0.001	0.0001	0
$\dfrac{\Delta h}{\Delta t}$	-48	-40	-33.6	-32.16	-32.016	-32.0016	-32

$\dfrac{\Delta h}{\Delta t}$ approaches -32.

Definition of Instantaneous Rate of Change

The **instantaneous rate of change** (or simply **rate of change**) of $y = f(x)$ at x is the limit of the average rate of change on the interval $[x, x + \Delta x]$, as Δx approaches 0.

$$\lim_{\Delta x \to 0} \frac{\Delta y}{\Delta x} = \lim_{\Delta x \to 0} \frac{f(x + \Delta x) - f(x)}{\Delta x}$$

If y is a distance and x is time, then the rate of change is a **velocity.**

EXAMPLE 3   **Finding an Instantaneous Rate of Change**

Find the velocity of the object in Example 2 when $t = 1$.

SOLUTION   From Example 2, you know that the height of the falling object is given by

$h = -16t^2 + 100.$ Position function

By taking the derivative of this position function, you obtain the velocity function.

$h'(t) = -32t$ Velocity function

The velocity function gives the velocity at *any* time. So, when $t = 1$, the velocity is

$h'(1) = -32(1)$

$\qquad = -32$ feet per second.

TRY IT 3

Find the velocity of the object in Try It 2 when $t = 1.75$ and $t = 2$.

The general **position function** for a free-falling object, neglecting air resistance, is

$$h = -16t^2 + v_0 t + h_0 \qquad \text{Position function}$$

where h is the height (in feet), t is the time (in seconds), v_0 is the initial velocity (in feet per second), and h_0 is the initial height (in feet). Remember that the model assumes that positive velocities indicate upward motion and negative velocities indicate downward motion. The derivative $h' = -32t + v_0$ is the **velocity function.** The absolute value of the velocity is the **speed** of the object.

EXAMPLE 4 Finding the Velocity of a Diver

At time $t = 0$, a diver jumps from a diving board that is 32 feet high, as shown in Figure 2.21. Because the diver's initial velocity is 16 feet per second, his position function is

$$h = -16t^2 + 16t + 32. \qquad \text{Position function}$$

(a) When does the diver hit the water?

(b) What is the diver's velocity at impact?

SOLUTION

(a) To find the time at which the diver hits the water, let $h = 0$ and solve for t.

$$-16t^2 + 16t + 32 = 0 \qquad \text{Set } h \text{ equal to 0.}$$
$$-16(t^2 - t - 2) = 0 \qquad \text{Factor out common factor.}$$
$$-16(t + 1)(t - 2) = 0 \qquad \text{Factor.}$$
$$t = -1 \ \text{ or } \ t = 2 \qquad \text{Solve for } t.$$

The solution $t = -1$ does not make sense in the problem because it would mean the diver hits the water 1 second before he jumps. So, you can conclude that the diver hits the water when $t = 2$ seconds.

(b) The velocity at time t is given by the derivative

$$h' = -32t + 16. \qquad \text{Velocity function}$$

The velocity at time $t = 2$ is $-32(2) + 16 = -48$ feet per second.

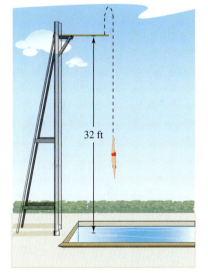

32 ft

FIGURE 2.21

TRY IT 4

Give the position function of a diver who jumps from a board 12 feet high with initial velocity 16 feet per second. Then find the diver's velocity function.

In Example 4, note that the diver's initial velocity is $v_0 = 16$ feet per second (upward) and his initial height is $h_0 = 32$ feet.

Initial velocity is 16 feet per second.

Initial height is 32 feet.

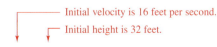

$$h = -16t^2 + 16t + 32$$

Rates of Change in Economics: Marginals

Another important use of rates of change is in the field of economics. Economists refer to *marginal profit*, *marginal revenue*, and *marginal cost* as the rates of change of the profit, revenue, and cost with respect to the number x of units produced or sold. An equation that relates these three quantities is

$$P = R - C$$

where P, R, and C represent the following quantities.

P = total profit

R = total revenue

and

C = total cost

The derivatives of these quantities are called the **marginal profit, marginal revenue,** and **marginal cost,** respectively.

$$\frac{dP}{dx} = \text{marginal profit}$$

$$\frac{dR}{dx} = \text{marginal revenue}$$

$$\frac{dC}{dx} = \text{marginal cost}$$

In many business and economics problems, the number of units produced or sold is restricted to positive integer values, as indicated in Figure 2.22(a). (Of course, it could happen that a sale involves half or quarter units, but it is hard to conceive of a sale involving $\sqrt{2}$ units.) The variable that denotes such units is called a **discrete variable.** To analyze a function of a discrete variable x, you can temporarily assume that x is a **continuous variable** and is able to take on any real value in a given interval, as indicated in Figure 2.22(b). Then, you can use the methods of calculus to find the x-value that corresponds to the marginal revenue, maximum profit, minimum cost, or whatever is called for. Finally, you should round the solution to the nearest sensible x-value—cents, dollars, units, or days, depending on the context of the problem.

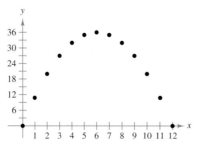

(a) Function of a Discrete Variable

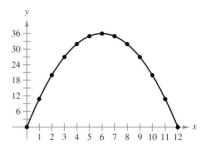

(b) Function of a Continuous Variable

FIGURE 2.22

EXAMPLE 5   **Finding the Marginal Profit**

The profit derived from selling x units of an alarm clock is given by

$$P = 0.0002x^3 + 10x.$$

(a) Find the marginal profit for a production level of 50 units.

(b) Compare this with the actual gain in profit obtained by increasing the production level from 50 to 51 units.

SOLUTION

(a) Because the profit is $P = 0.0002x^3 + 10x$, the marginal profit is given by the derivative

$$dP/dx = 0.0006x^2 + 10.$$

When $x = 50$, the marginal profit is

$$0.0006(50)^2 + 10 = 1.5 + 10$$
$$= \$11.50 \text{ per unit.} \qquad \textcolor{red}{\text{Marginal profit for } x = 50}$$

(b) For $x = 50$, the actual profit is

$$P = (0.0002)(50)^3 + 10(50) \qquad \textcolor{red}{\text{Substitute 50 for } x.}$$
$$= 25 + 500$$
$$= \$525.00 \qquad \textcolor{red}{\text{Actual profit for } x = 50}$$

and for $x = 51$, the actual profit is

$$P = (0.0002)(51)^3 + 10(51) \qquad \textcolor{red}{\text{Substitute 51 for } x.}$$
$$\approx 26.53 + 510$$
$$= \$536.53. \qquad \textcolor{red}{\text{Actual profit for } x = 51}$$

So, the additional profit obtained by increasing the production level from 50 to 51 units is

$$536.53 - 525.00 = \$11.53. \qquad \textcolor{red}{\text{Extra profit for one unit}}$$

Note that the actual profit increase of \$11.53 (when x increases from 50 to 51 units) can be approximated by the marginal profit of \$11.50 per unit (when $x = 50$), as shown in Figure 2.23.

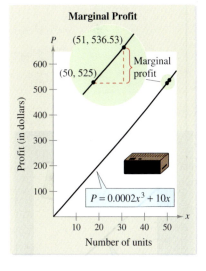

FIGURE 2.23

TRY IT 5

Use the profit function in Example 5 to find the marginal profit for a production level of 100 units. Compare this with the actual gain in profit by increasing production from 100 to 101 units.

STUDY TIP

The reason the marginal profit gives a good approximation of the actual change in profit is that the graph of P is nearly straight over the interval $50 \le x \le 51$. You will study more about the use of marginals to approximate actual changes in Section 3.8.

The following warm-up exercises involve skills that were covered in earlier sections. You will
use these skills in the exercise set for this section.

In Exercises 1 and 2, evaluate the expression.

1. $\dfrac{-63 - (-105)}{21 - 7}$

2. $\dfrac{-37 - 54}{16 - 3}$

In Exercises 3–10, find the derivative of the function.

3. $y = 4x^2 - 2x + 7$

4. $y = -3t^3 + 2t^2 - 8$

5. $s = -16t^2 + 24t + 30$

6. $y = -16x^2 + 54x + 70$

7. $A = \frac{1}{10}(-2r^3 + 3r^2 + 5r)$

8. $y = \frac{1}{9}(6x^3 - 18x^2 + 63x - 15)$

9. $y = 12x - \dfrac{x^2}{5000}$

10. $y = 138 + 74x - \dfrac{x^3}{10,000}$

EXERCISES 2.3

1. Research and Development The graph shows the amounts A (in billions of dollars per year) spent on R&D in the United States from 1980 through 2002. Approximate the average rate of change of A during each period. *(Source: U.S. National Science Foundation)*

(a) 1980–1985   (b) 1985–1990   (c) 1990–1995

(d) 1995–2000   (e) 1980–2002   (f) 1990–2002

Research and Development

2. Trade Deficit The graph shows the values I (in billions of dollars per year) of goods imported to the United States and the value E (in billions of dollars per year) of goods exported from the United States from 1980 through 2002. Approximate each indicated average rate of change. *(Source: U.S. International Trade Administration)*

(a) Imports: 1980–1990   (b) Exports: 1980–1990

(c) Imports: 1990–2000   (d) Exports: 1990–2000

(e) Imports: 1980–2002   (f) Exports: 1980–2002

Figure for 2

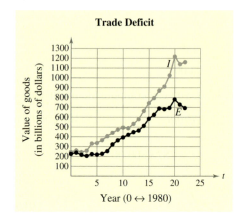

Trade Deficit

In Exercises 3–8, sketch the graph of the function and find its average rate of change on the interval. Compare this rate with the instantaneous rates of change at the endpoints of the interval.

3. $f(t) = 2t + 7; [1, 2]$

4. $h(x) = 1 - x; [0, 1]$

5. $h(x) = x^2 - 4x + 2; [-2, 2]$

6. $f(x) = x^2 - 6x - 1; [-1, 3]$

7. $f(x) = \dfrac{1}{x}; [1, 4]$

8. $f(x) = \dfrac{1}{\sqrt{x}}; [1, 4]$

In Exercises 9 and 10, use a graphing utility to graph the function and find its average rate of change on the interval. Compare this rate with the instantaneous rates of change at the endpoints of the interval.

9. $g(x) = x^4 - x^2 + 2; [1, 3]$

10. $g(x) = x^3 - 1; [-1, 1]$

11. Consumer Trends The graph shows the number of visitors V to a national park in hundreds of thousands during a one-year period, where $t = 1$ represents January.

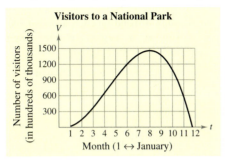

Visitors to a National Park

(a) Estimate the rate of change of V over the interval $[9, 12]$ and explain your results.

(b) Over what interval is the average rate of change approximately equal to the rate of change at $t = 8$? Explain your reasoning.

12. Medicine The graph shows the estimated number of milligrams of a pain medication M in the bloodstream t hours after a 1000-milligram dose of the drug has been given.

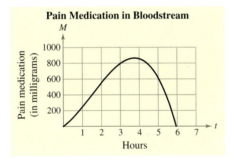

Pain Medication in Bloodstream

(a) Estimate the one-hour interval over which the average rate of change is the greatest.

(b) Over what interval is the average rate of change approximately equal to the rate of change at $t = 4$? Explain your reasoning.

13. Medicine The effectiveness E (on a scale from 0 to 1) of a pain-killing drug t hours after entering the bloodstream is given by

$$E = \frac{1}{27}(9t + 3t^2 - t^3), \quad 0 \le t \le 4.5.$$

Find the average rate of change of E on each indicated interval and compare this rate with the instantaneous rates of change at the endpoints of the interval.

(a) $[0, 1]$ (b) $[1, 2]$ (c) $[2, 3]$ (d) $[3, 4]$

14. Chemistry: Wind Chill At $0°$ Celsius, the heat loss H (in kilocalories per square meter per hour) from a person's body can be modeled by

$$H = 33(10\sqrt{v} - v + 10.45)$$

where v is the wind speed (in meters per second).

(a) Find $\dfrac{dH}{dv}$ and interpret its meaning in this situation.

(b) Find the rates of change of H when $v = 2$ and when $v = 5$.

15. Velocity The height s (in feet) at time t (in seconds) of a silver dollar dropped from the top of the Washington Monument is given by

$$s = -16t^2 + 555.$$

(a) Find the average velocity on the interval $[2, 3]$.

(b) Find the instantaneous velocities when $t = 2$ and when $t = 3$.

(c) How long will it take the dollar to hit the ground?

(d) Find the velocity of the dollar when it hits the ground.

 16. Physics: Velocity A racecar travels northward on a straight, level track at a constant speed, traveling 0.750 kilometer in 20.0 seconds. The return trip over the same track is made in 25.0 seconds.

(a) What is the average velocity of the car in meters per second for the first leg of the run?

(b) What is the average velocity for the total trip?

(Source: Shipman/Wilson/Todd, An Introduction to Physical Science, Tenth Edition)

Marginal Cost In Exercises 17–20, find the marginal cost for producing x units. (The cost is measured in dollars.)

17. $C = 4500 + 1.47x$ **18.** $C = 104,000 + 7200x$

19. $C = 55,000 + 470x - 0.25x^2, \quad 0 \le x \le 940$

20. $C = 100(9 + 3\sqrt{x})$

Marginal Revenue In Exercises 21–24, find the marginal revenue for producing x units. (The revenue is measured in dollars.)

21. $R = 50x - 0.5x^2$ **22.** $R = 30x - x^2$

23. $R = -6x^3 + 8x^2 + 200x$ **24.** $R = 50(20x - x^{3/2})$

Marginal Profit In Exercises 25–28, find the marginal profit for producing x units. (The profit is measured in dollars.)

25. $P = -2x^2 + 72x - 145$

26. $P = -0.25x^2 + 2000x - 1,250,000$

27. $P = -0.00025x^2 + 12.2x - 25,000$

28. $P = -0.5x^3 + 30x^2 - 164.25x - 1000$

PREREQUISITE REVIEW 2.4

The following warm-up exercises involve skills that were covered in earlier sections. You will use these skills in the exercise set for this section.

In Exercises 1–10, simplify the expression.

1. $(x^2 + 1)(2) + (2x + 7)(2x)$

2. $(2x - x^3)(8x) + (4x^2)(2 - 3x^2)$

3. $x(4)(x^2 + 2)^3(2x) + (x^2 + 4)(1)$

4. $x^2(2)(2x + 1)(2) + (2x + 1)^4(2x)$

5. $\dfrac{(2x + 7)(5) - (5x + 6)(2)}{(2x + 7)^2}$

6. $\dfrac{(x^2 - 4)(2x + 1) - (x^2 + x)(2x)}{(x^2 - 4)^2}$

7. $\dfrac{(x^2 + 1)(2) - (2x + 1)(2x)}{(x^2 + 1)^2}$

8. $\dfrac{(1 - x^4)(4) - (4x - 1)(-4x^3)}{(1 - x^4)^2}$

9. $(x^{-1} + x)(2) + (2x - 3)(-x^{-2} + 1)$

10. $\dfrac{(1 - x^{-1})(1) - (x - 4)(x^{-2})}{(1 - x^{-1})^2}$

In Exercises 11–14, find $f'(2)$.

11. $f(x) = 3x^2 - x + 4$

12. $f(x) = -x^3 + x^2 + 8x$

13. $f(x) = \dfrac{1}{x}$

14. $f(x) = x^2 - \dfrac{1}{x^2}$

EXERCISES 2.4

In Exercises 1–14, find the value of the derivative of the function at the given point.

Function	Point
1. $f(x) = x^2(3x^3 - 1)$	$(1, 2)$
2. $f(x) = (x^2 + 1)(2x + 5)$	$(-1, 6)$
3. $f(x) = \frac{1}{3}(2x^3 - 4)$	$\left(0, -\frac{4}{3}\right)$
4. $f(x) = \frac{1}{7}(5 - 6x^2)$	$\left(1, -\frac{1}{7}\right)$
5. $g(x) = (x^2 - 4x + 3)(x - 2)$	$(4, 6)$
6. $g(x) = (x^2 - 2x + 1)(x^3 - 1)$	$(1, 0)$
7. $h(x) = \dfrac{x}{x - 5}$	$(6, 6)$
8. $h(x) = \dfrac{x^2}{x + 3}$	$\left(-1, \frac{1}{2}\right)$
9. $f(t) = \dfrac{2t^2 - 3}{3t + 1}$	$\left(3, \frac{3}{2}\right)$
10. $f(x) = \dfrac{3x}{x^2 + 4}$	$\left(-1, -\frac{3}{5}\right)$
11. $g(x) = \dfrac{2x + 1}{x - 5}$	$(6, 13)$
12. $f(x) = \dfrac{x + 1}{x - 1}$	$(2, 3)$
13. $f(t) = \dfrac{t^2 - 1}{t + 4}$	$(1, 0)$
14. $g(x) = \dfrac{4x - 5}{x^2 - 1}$	$(0, 5)$

In Exercises 15–22, find the derivative of the function. Use Example 7 as a model.

Function	Rewrite	Differentiate	Simplify
15. $y = \dfrac{x^2 + 2x}{x}$			
16. $y = \dfrac{4x^{3/2}}{x}$			
17. $y = \dfrac{7}{3x^3}$			
18. $y = \dfrac{4}{5x^2}$			
19. $y = \dfrac{4x^2 - 3x}{8\sqrt{x}}$			
20. $y = \dfrac{3x^2 - 4x}{6x}$			
21. $y = \dfrac{x^2 - 4x + 3}{x - 1}$			
22. $y = \dfrac{x^2 - 4}{x + 2}$			

In Exercises 23–38, find the derivative of the function.

23. $f(x) = (x^3 - 3x)(2x^2 + 3x + 5)$

24. $h(t) = (t^5 - 1)(4t^2 - 7t - 3)$

25. $g(t) = (2t^3 - 1)^2$

26. $h(p) = (p^3 - 2)^2$

27. $f(x) = \sqrt[3]{x}(\sqrt{x} + 3)$

28. $f(x) = \sqrt[3]{x}(x + 1)$

29. $f(x) = \dfrac{3x - 2}{2x - 3}$

30. $f(x) = \dfrac{x^3 + 3x + 2}{x^2 - 1}$

31. $f(x) = \dfrac{3 - 2x - x^2}{x^2 - 1}$

32. $f(x) = (x^5 - 3x)\left(\dfrac{1}{x^2}\right)$

33. $f(x) = x\left(1 - \dfrac{2}{x + 1}\right)$

34. $h(t) = \dfrac{t + 2}{t^2 + 5t + 6}$

35. $g(s) = \dfrac{s^2 - 2s + 5}{\sqrt{s}}$

36. $f(x) = \dfrac{x + 1}{\sqrt{x}}$

37. $g(x) = \left(\dfrac{x - 3}{x + 4}\right)(x^2 + 2x + 1)$

38. $f(x) = (3x^3 + 4x)(x - 5)(x + 1)$

 In Exercises 39–44, find an equation of the tangent line to the graph of the function at the given point. Then use a graphing utility to graph the function and the tangent line in the same viewing window.

Function	Point
39. $f(x) = (x - 1)^2(x - 2)$	$(0, -2)$
40. $h(x) = (x^2 - 1)^2$	$(-2, 9)$
41. $f(x) = \dfrac{x - 2}{x + 1}$	$\left(1, -\frac{1}{2}\right)$
42. $f(x) = \dfrac{2x + 1}{x - 1}$	$(2, 5)$
43. $f(x) = \left(\dfrac{x + 5}{x - 1}\right)(2x + 1)$	$(0, -5)$
44. $g(x) = (x + 2)\left(\dfrac{x - 5}{x + 1}\right)$	$(0, -10)$

In Exercises 45–48, find the point(s), if any, at which the graph of f has a horizontal tangent.

45. $f(x) = \dfrac{x^2}{x - 1}$

46. $f(x) = \dfrac{x^2}{x^2 + 1}$

47. $f(x) = \dfrac{x^4}{x^3 + 1}$

48. $f(x) = \dfrac{x^4 + 3}{x^2 + 1}$

In Exercises 49–52, use a graphing utility to graph f and f' on the interval $[-2, 2]$.

49. $f(x) = x(x + 1)$

50. $f(x) = x^2(x + 1)$

51. $f(x) = x(x + 1)(x - 1)$

52. $f(x) = x^2(x + 1)(x - 1)$

Demand In Exercises 53 and 54, use the demand function to find the rate of change in the demand x for the given price p.

53. $x = 275\left(1 - \dfrac{3p}{5p + 1}\right)$, $p = \$4$

54. $x = 300 - p - \dfrac{2p}{p + 1}$, $p = \$3$

55. **Environment** The model

$$f(t) = \dfrac{t^2 - t + 1}{t^2 + 1}$$

measures the percent of the normal level of oxygen in a pond, where t is the time (in weeks) after organic waste is dumped into the pond. Find the rates of change of f with respect to t when (a) $t = 0.5$, (b) $t = 2$, and (c) $t = 8$.

56. **Physical Science** The temperature T of food placed in a refrigerator is modeled by

$$T = 10\left(\dfrac{4t^2 + 16t + 75}{t^2 + 4t + 10}\right)$$

where t is the time (in hours). What is the initial temperature of the food? Find the rates of change of T with respect to t when (a) $t = 1$, (b) $t = 3$, (c) $t = 5$, and (d) $t = 10$.

57. **Population Growth** A population of bacteria is introduced into a culture. The number of bacteria P can be modeled by

$$P = 500\left(1 + \dfrac{4t}{50 + t^2}\right)$$

where t is the time (in hours). Find the rate of change of the population when $t = 2$.

58. **Quality Control** The percent P of defective parts produced by a new employee t days after the employee starts work can be modeled by

$$P = \dfrac{t + 1750}{50(t + 2)}.$$

Find the rates of change of P when (a) $t = 1$ and (b) $t = 10$.

59. **Profit** You decide to form a partnership with another business. Your business determines that the demand x for your product is inversely proportional to the square of the price for $x \geq 5$.

(a) The price is $1000 and the demand is 16 units. Find the demand function.

(b) Your partner determines that the product costs $250 per unit and the fixed cost is $10,000. Find the cost function.

 (c) Find the profit function and use a graphing utility to graph it. From the graph, what price would you negotiate with your partner for this product? Explain your reasoning.

 60. Profit   You are managing a store and have been adjusting the price of an item. You have found that you make a profit of $50 when 10 units are sold, $60 when 12 units are sold, and $65 when 14 units are sold.

(a) Fit these data to the model $P = ax^2 + bx + c$.

(b) Use a graphing utility to graph P.

(c) Find the point on the graph at which the marginal profit is zero. Interpret this point in the context of the problem.

 61. Demand Function   Given $f(x) = x^2 + 1$, which function would most likely represent a demand function?

(a) $p = f(x)$

(b) $p = xf(x)$

(c) $p = 1/f(x)$

Explain your reasoning. Use a graphing utility to graph each function, and use each graph as part of your explanation.

 62. Cost   The cost of producing x units of a product is given by

$$C = x^3 - 15x^2 + 87x - 73, \qquad 4 \le x \le 9.$$

(a) Use a graphing utility to graph the marginal cost function and the average cost function, C/x, in the same viewing window.

(b) Find the point of intersection of the graphs of dC/dx and C/x. Does this point have any significance?

63. Inventory Replenishment   The ordering and transportation cost C (in thousands of dollars) of the components used in manufacturing a product is given by

$$C = 100\left(\frac{200}{x^2} + \frac{x}{x + 30}\right), \qquad 1 \le x$$

where x is the order size (in hundreds). Find the rate of change of C with respect to x for each order size.

(a) $x = 10$

(b) $x = 15$

(c) $x = 20$

What do these rates of change imply about increasing the size of an order?

64. Sales Analysis   The monthly sales of memberships M at a newly built fitness center are modeled by

$$M(t) = \frac{300t}{t^2 + 1} + 8$$

where t is the number of months since the center opened.

(a) Find $M'(t)$.

(b) Find $M(3)$ and $M'(3)$ and interpret the results.

(c) Find $M(24)$ and $M'(24)$ and interpret the results.

65. Consumer Awareness   The prices of 1 pound of 100% ground beef in the United States from 1995 to 2002 can be modeled by

$$P = \frac{1.47 - 0.311t + 0.0173t^2}{1 - 0.206t + 0.0112t^2}$$

where t is the year, with $t = 5$ corresponding to 1995. Find dP/dt and evaluate it for $t = 5, 7, 9$, and 11. Interpret the meaning of these values.   *(Source: U.S. Bureau of Labor Statistics)*

BUSINESS CAPSULE

The Blackwood Centre for Adolescent Development, a state-sponsored secondary school for young people at risk in Victoria, Australia, joined forces in 2000 with the Centre for Executive Development (CED). With the CED providing fundraising and logistical support, the Blackwood Centre has been able to transform its program into a model of success, offering their students training and educational opportunities for entry into the business world. The CED has gained team-building skills, improved workplace assessment methods, and a stronger connection to the community.

66. Research Project   Use your school's library, the Internet, or some other reference source to find information about partnerships between companies and federal, state, or local government that have benefited their communities. (One such partnership is described above.) Write a short paper about the partnership.

2.5   THE CHAIN RULE

■ Find derivatives using the Chain Rule.
■ Find derivatives using the General Power Rule.
■ Write derivatives in simplified form.
■ Use derivatives to answer questions about real-life situations.
■ Use the differentiation rules to differentiate algebraic functions.

The Chain Rule

In this section, you will study one of the most powerful rules of differential calculus—the **Chain Rule.** This differentiation rule deals with composite functions and adds versatility to the rules presented in Sections 2.2 and 2.4. For example, compare the functions below. Those on the left can be differentiated without the Chain Rule, whereas those on the right are best done with the Chain Rule.

Without the Chain Rule	*With the Chain Rule*
$y = x^2 + 1$	$y = \sqrt{x^2 + 1}$
$y = x + 1$	$y = (x + 1)^{-1/2}$
$y = 3x + 2$	$y = (3x + 2)^5$
$y = \dfrac{x + 5}{x^2 + 2}$	$y = \left(\dfrac{x + 5}{x^2 + 2}\right)^2$

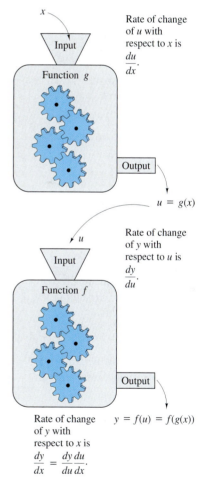

Rate of change of u with respect to x is $\dfrac{du}{dx}$.

Rate of change of y with respect to u is $\dfrac{dy}{du}$.

Rate of change of y with respect to x is $\dfrac{dy}{dx} = \dfrac{dy}{du}\dfrac{du}{dx}$.

FIGURE 2.28

The Chain Rule

If $y = f(u)$ is a differentiable function of u, and $u = g(x)$ is a differentiable function of x, then $y = f(g(x))$ is a differentiable function of x, and

$$\frac{dy}{dx} = \frac{dy}{du} \cdot \frac{du}{dx}$$

or, equivalently,

$$\frac{d}{dx}[f(g(x))] = f'(g(x))g'(x).$$

Basically, the Chain Rule states that if y changes dy/du times as fast as u, and u changes du/dx times as fast as x, then y changes

$$\frac{dy}{du} \cdot \frac{du}{dx}$$

times as fast as x, as illustrated in Figure 2.28. One advantage of the dy/dx notation for derivatives is that it helps you remember differentiation rules, such as the Chain Rule. For instance, in the formula

$$dy/dx = (dy/du)(du/dx)$$

you can imagine that the du's divide out.

When applying the Chain Rule, it helps to think of the composite function $y = f(g(x))$ or $y = f(u)$ as having two parts—an *inside* and an *outside*—as illustrated below.

$$
\begin{array}{c}
\text{Inside} \\
\downarrow \quad \downarrow \\
y = f(g(x)) \ = \ f(u) \\
\uparrow \qquad \uparrow \\
\text{Outside}
\end{array}
$$

The Chain Rule tells you that the derivative of $y = f(u)$ is the derivative of the outer function (at the inner function u) *times* the derivative of the inner function. That is,

$$y' = f'(u) \cdot u'.$$

TRY IT 1

Write each function as the composition of two functions, where $y = f(g(x))$.

(a) $y = \dfrac{1}{\sqrt{x+1}}$

(b) $y = (x^2 + 2x + 5)^3$

EXAMPLE 1   Decomposing Composite Functions

Write each function as the composition of two functions.

(a) $y = \dfrac{1}{x+1}$ (b) $y = \sqrt{3x^2 - x + 1}$

SOLUTION   There is more than one correct way to decompose each function. One way for each is shown below.

$$
\begin{array}{lll}
y = f(g(x)) & u = g(x) \ (inside) & y = f(u) \ (outside) \\
\\
\text{(a)} \ y = \dfrac{1}{x+1} & u = x + 1 & y = \dfrac{1}{u} \\
\\
\text{(b)} \ y = \sqrt{3x^2 - x + 1} & u = 3x^2 - x + 1 & y = \sqrt{u}
\end{array}
$$

EXAMPLE 2   Using the Chain Rule

Find the derivative of $y = (x^2 + 1)^3$.

SOLUTION   To apply the Chain Rule, you need to identify the inside function u.

$$y = \overbrace{(x^2 + 1)}^{u}{}^3 = u^3$$

By the Chain Rule, you can write the derivative as shown.

$$\dfrac{dy}{dx} = \overbrace{3(x^2 + 1)^2}^{\frac{dy}{du}}\overbrace{(2x)}^{\frac{du}{dx}} = 6x(x^2 + 1)^2$$

STUDY TIP

Try checking the result of Example 2 by expanding the function to obtain

$$y = x^6 + 3x^4 + 3x^2 + 1$$

and finding the derivative. Do you obtain the same answer?

TRY IT 2

Find the derivative of $y = (x^3 + 1)^2$.

The General Power Rule

The function in Example 2 illustrates one of the most common types of composite functions—a power function of the form

$$y = [u(x)]^n.$$

The rule for differentiating such functions is called the **General Power Rule,** and it is a special case of the Chain Rule.

The General Power Rule

If $y = [u(x)]^n$, where u is a differentiable function of x and n is a real number, then

$$\frac{dy}{dx} = n[u(x)]^{n-1}\frac{du}{dx}$$

or, equivalently,

$$\frac{d}{dx}[u^n] = nu^{n-1}u'.$$

PROOF   Apply the Chain Rule and the Simple Power Rule as shown.

$$\frac{dy}{dx} = \frac{dy}{du} \cdot \frac{du}{dx}$$

$$= \frac{d}{du}[u^n]\frac{du}{dx}$$

$$= nu^{n-1}\frac{du}{dx}$$

EXAMPLE 3   **Using the General Power Rule**

Find the derivative of

$$f(x) = (3x - 2x^2)^3.$$

SOLUTION   The inside function is $u = 3x - 2x^2$. So, by the General Power Rule,

$$f'(x) = \overset{n}{3}(\underbrace{3x - 2x^2)^2}_{u^{n-1}} \underbrace{\frac{d}{dx}[3x - 2x^2]}_{u'}$$

$$= 3(3x - 2x^2)^2(3 - 4x)$$

$$= (9 - 12x)(3x - 2x^2)^2.$$

TECHNOLOGY

If you have access to a symbolic differentiation utility, try using it to confirm the result of Example 3.

TRY IT 3

Find the derivative of $y = (x^2 + 3x)^4$.

Application

U.S. Cellular is in the process of converting its wireless network to an advanced wireless digital technology. The benefits of wireless digital technology are fewer blocked calls, reduced background noise and interference, improved security and privacy, greater capacity, and expanded coverage.

EXAMPLE 8   Finding Rates of Change

From 1993 through 2002, the revenue per share R (in dollars) for U.S. Cellular can be modeled by $R = (-0.003t^2 + 0.42t + 0.5)^2$ for $3 \leq t \leq 12$, where $t = 3$ corresponds to 1993. Use the model to approximate the rates of change in the revenue per share in 1994, 1996, and 2000. If you had been a U.S. Cellular stockholder from 1993 through 2002, would you have been satisfied with the performance of this stock? *(Source: U.S. Cellular)*

SOLUTION   The rate of change in R is given by the derivative dR/dt. You can use the General Power Rule to find the derivative.

$$\frac{dR}{dt} = 2(-0.003t^2 + 0.42t + 0.5)^1(-0.006t + 0.42)$$

$$= (-0.012t + 0.84)(-0.003t^2 + 0.42t + 0.5)$$

In 1994, the revenue per share was changing at a rate of

$$[-0.012(4) + 0.84][-0.003(4)^2 + 0.42(4) + 0.5] \approx \$1.69 \text{ per year.}$$

In 1996, the revenue per share was changing at a rate of

$$[-0.012(6) + 0.84][-0.003(6)^2 + 0.42(6) + 0.5] \approx \$2.24 \text{ per year.}$$

In 2000, the revenue per share was changing at a rate of

$$[-0.012(10) + 0.84][-0.003(10)^2 + 0.42(10) + 0.5] \approx \$3.17 \text{ per year.}$$

The graph of the revenue per share function is shown in Figure 2.30. For most investors, the performance of U.S. Cellular stock would be considered to be very good.

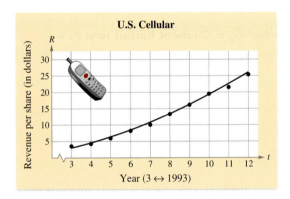

FIGURE 2.30

TRY IT 8

From 1994 through 2002, the sales per share (in dollars) for Dollar Tree can be modeled by $S = (-0.009t^2 + 0.52t - 0.4)^2$ for $4 \leq t \leq 12$, where $t = 4$ corresponds to 1994. Use the model to approximate the rate of change in sales per share in 2000. *(Source: Dollar Tree Stores, Inc.)*

Summary of Differentiation Rules

You now have all the rules you need to differentiate *any* algebraic function. For your convenience, they are summarized below.

Summary of Differentiation Rules

1. Constant Rule $\dfrac{d}{dx}[c] = 0,$ c is a constant.

2. Constant Multiple Rule $\dfrac{d}{dx}[cu] = c\dfrac{du}{dx},$ c is a constant.

3. Sum and Difference Rules $\dfrac{d}{dx}[u \pm v] = \dfrac{du}{dx} \pm \dfrac{dv}{dx}$

4. Product Rule $\dfrac{d}{dx}[uv] = u\dfrac{dv}{dx} + v\dfrac{du}{dx}$

5. Quotient Rule $\dfrac{d}{dx}\left[\dfrac{u}{v}\right] = \dfrac{v\dfrac{du}{dx} - u\dfrac{dv}{dx}}{v^2}$

6. Power Rules $\dfrac{d}{dx}[x^n] = nx^{n-1}$

$\dfrac{d}{dx}[u^n] = nu^{n-1}\dfrac{du}{dx}$

7. Chain Rule $\dfrac{dy}{dx} = \dfrac{dy}{du} \cdot \dfrac{du}{dx}$

TAKE ANOTHER LOOK

Comparing Power Rules

You now have two power rules for differentiation.

$\dfrac{d}{dx}[x^n] = nx^{n-1}$ Simple Power Rule

$\dfrac{d}{dx}[u^n] = nu^{n-1}\dfrac{du}{dx}$ General Power Rule

Explain how you can tell which rule to use. Then state whether you would use the simple or general rule for each function.

a. $y = x^4$ b. $y = (x^2 + 1)$

c. $y = (x^2 + 1)^2$ d. $y = \dfrac{1}{x^4}$

e. $y = \dfrac{1}{x - 1}$ f. $y = \dfrac{1}{\sqrt{2x - 1}}$

138   **CHAPTER 2**   Differentiation

PREREQUISITE REVIEW 2.5

The following warm-up exercises involve skills that were covered in earlier sections. You will use these skills in the exercise set for this section.

In Exercises 1–6, rewrite the expression with rational exponents.

1. $\sqrt[5]{(1-5x)^2}$

2. $\sqrt[4]{(2x-1)^3}$

3. $\dfrac{1}{\sqrt{4x^2+1}}$

4. $\dfrac{1}{\sqrt[3]{x-6}}$

5. $\dfrac{\sqrt{x}}{\sqrt[3]{1-2x}}$

6. $\dfrac{\sqrt{(3-7x)^3}}{2x}$

In Exercises 7–10, factor the expression.

7. $3x^3-6x^2+5x-10$

8. $5x\sqrt{x}-x-5\sqrt{x}+1$

9. $4(x^2+1)^2-x(x^2+1)^3$

10. $-x^5+3x^3+x^2-3$

EXERCISES 2.5

In Exercises 1–8, identify the inside function, $u=g(x)$, and the outside function, $y=f(u)$.

$$y=f(g(x)) \qquad u=g(x) \qquad y=f(u)$$

1. $y=(6x-5)^4$

2. $y=(x^2-2x+3)^3$

3. $y=(4-x^2)^{-1}$

4. $y=(x^2+1)^{4/3}$

5. $y=\sqrt{5x-2}$

6. $y=\sqrt{9-x^2}$

7. $y=(3x+1)^{-1}$

8. $y=(x+1)^{-1/2}$

In Exercises 9–16, match the function with the rule that you would use to find the derivative *most efficiently*.

(a) Simple Power Rule

(b) Constant Rule

(c) General Power Rule

(d) Quotient Rule

9. $f(x)=\dfrac{2}{1-x^3}$

10. $f(x)=\dfrac{2x}{1-x^3}$

11. $f(x)=\sqrt[3]{8^2}$

12. $f(x)=\sqrt[3]{x^2}$

13. $f(x)=\dfrac{x^2+2}{x}$

14. $f(x)=\dfrac{x^4-2x+1}{\sqrt{x}}$

15. $f(x)=\dfrac{2}{x-2}$

16. $f(x)=\dfrac{5}{x^2+1}$

In Exercises 17–34, use the General Power Rule to find the derivative of the function.

17. $y=(2x-7)^3$

18. $y=(3x^2+1)^4$

19. $g(x)=(4-2x)^3$

20. $h(t)=(1-t^2)^4$

21. $h(x)=(6x-x^3)^2$

22. $f(x)=(4x-x^2)^3$

23. $f(x)=(x^2-9)^{2/3}$

24. $f(t)=(9t+2)^{2/3}$

25. $f(t)=\sqrt{t+1}$

26. $g(x)=\sqrt{2x+3}$

27. $s(t)=\sqrt{2t^2+5t+2}$

28. $y=\sqrt[3]{3x^3+4x}$

29. $y=\sqrt[3]{9x^2+4}$

30. $y=2\sqrt{4-x^2}$

31. $f(x)=-3\sqrt[4]{2-9x}$

32. $f(x)=(25+x^2)^{-1/2}$

33. $h(x)=(4-x^3)^{-4/3}$

34. $f(x)=(4-3x)^{-5/2}$

In Exercises 35–40, find an equation of the tangent line to the graph of f at the point $(2,f(2))$. Use a graphing utility to check your result by graphing the original function and the tangent line in the same viewing window.

35. $f(x)=2(x^2-1)^3$

36. $f(x)=3(9x-4)^4$

37. $f(x)=\sqrt{4x^2-7}$

38. $f(x)=x\sqrt{x^2+5}$

39. $f(x)=\sqrt{x^2-2x+1}$

40. $f(x)=(4-3x^2)^{-2/3}$

In Exercises 41–44, use a symbolic differentiation utility to find the derivative of the function. Graph the function and its derivative in the same viewing window. Describe the behavior of the function when the derivative is zero.

41. $f(x)=\dfrac{\sqrt{x}+1}{x^2+1}$

42. $f(x)=\sqrt{\dfrac{2x}{x+1}}$

43. $f(x)=\sqrt{\dfrac{x+1}{x}}$

44. $f(x)=\sqrt{x}(2-x^2)$

In Exercises 45–64, find the derivative of the function.

45. $y=\dfrac{1}{x-2}$

46. $s(t)=\dfrac{1}{t^2+3t-1}$

47. $y=-\dfrac{4}{(t+2)^2}$

48. $f(x)=\dfrac{3}{(x^3-4)^2}$

49. $f(x)=\dfrac{1}{(x^2-3x)^2}$

50. $y=\dfrac{1}{\sqrt{x+2}}$

51. $g(t) = \dfrac{1}{t^2 - 2}$

52. $g(x) = \dfrac{3}{\sqrt[3]{x^3 - 1}}$

53. $f(x) = x(3x - 9)^3$

54. $f(x) = x^3(x - 4)^2$

55. $y = x\sqrt{2x + 3}$

56. $y = t\sqrt{t + 1}$

57. $y = t^2\sqrt{t - 2}$

58. $y = \sqrt{x}(x - 2)^2$

59. $f(x) = \sqrt{\dfrac{3 - 2x}{4x}}$

60. $g(t) = \dfrac{3t^2}{\sqrt{t^2 + 2t - 1}}$

61. $f(x) = \sqrt{x^2 + 1} - \sqrt{x^2 - 1}$

62. $y = \sqrt{x - 1} + \sqrt{x + 1}$

63. $y = \left(\dfrac{6 - 5x}{x^2 - 1}\right)^2$

64. $y = \left(\dfrac{4x^2}{3 - x}\right)^3$

 In Exercises 65–70, find an equation of the tangent line to the graph of the function at the given point. Then use a graphing utility to graph the function and the tangent line in the same viewing window.

Function	Point

65. $f(t) = \dfrac{36}{(3 - t)^2}$ $(0, 4)$

66. $s(x) = \dfrac{1}{\sqrt{x^2 - 3x + 4}}$ $(3, \tfrac{1}{2})$

67. $f(t) = (t^2 - 9)\sqrt{t + 2}$ $(-1, -8)$

68. $y = \dfrac{2x}{\sqrt{x + 1}}$ $(3, 3)$

69. $f(x) = \dfrac{x + 1}{\sqrt{2x - 3}}$ $(2, 3)$

70. $y = \dfrac{x}{\sqrt{25 + x^2}}$ $(0, 0)$

71. *Compound Interest* You deposit $1000 in an account with an annual interest rate of r (in decimal form) compounded monthly. At the end of 5 years, the balance is

$$A = 1000\left(1 + \frac{r}{12}\right)^{60}.$$

Find the rates of change of A with respect to r when (a) $r = 0.08$, (b) $r = 0.10$, and (c) $r = 0.12$.

72. *Environment* An environmental study indicates that the average daily level P of a certain pollutant in the air in parts per million can be modeled by the equation

$$P = 0.25\sqrt{0.5n^2 + 5n + 25}$$

where n is the number of residents of the community in thousands. Find the rate at which the level of pollutant is increasing when the population of the community is 12,000.

73. *Biology* The number N of bacteria in a culture after t days is modeled by

$$N = 400\left[1 - \frac{3}{(t^2 + 2)^2}\right].$$

Complete the table. What can you conclude?

t	0	1	2	3	4
dN/dt					

74. *Depreciation* The value V of a machine t years after it is purchased is inversely proportional to the square root of $t + 1$. The initial value of the machine is $10,000.

(a) Write V as a function of t.

(b) Find the rate of depreciation when $t = 1$.

(c) Find the rate of depreciation when $t = 3$.

75. *Depreciation* Repeat Exercise 74 given that the value of the machine t years after it is purchased is inversely proportional to the cube root of $t + 1$.

 76. *Credit Card Rate* The average annual rate r (in percent form) for commercial bank credit cards from 1994 through 2002 can be modeled by

$$r = \sqrt{-0.14239t^4 + 3.939t^3 - 39.0835t^2 + 161.037t + 22.13}$$

where $t = 4$ corresponds to 1994. *(Source: Federal Reserve Bulletin)*

(a) Find the derivative of this model. Which differentiation rule(s) did you use?

(b) Use a graphing utility to graph the derivative. Use the interval $4 \le t \le 12$.

(c) Use the *trace* feature to find the years during which the finance rate was changing the most.

(d) Use the *trace* feature to find the years during which the finance rate was changing the least.

True or False? In Exercises 77 and 78, determine whether the statement is true or false. If it is false, explain why or give an example that shows it is false.

77. If $y = (1 - x)^{1/2}$, then $y' = \tfrac{1}{2}(1 - x)^{-1/2}$.

78. If y is a differentiable function of u, u is a differentiable function of v, and v is a differentiable function of x, then

$$\frac{dy}{dx} = \frac{dy}{du} \cdot \frac{du}{dv} \cdot \frac{dv}{dx}.$$

2.6   HIGHER-ORDER DERIVATIVES

■ Find higher-order derivatives.
■ Find and use the position functions to determine the velocity and acceleration of moving objects.

Second, Third, and Higher-Order Derivatives

The derivative of f' is the **second derivative** of f and is denoted by f''.

$$\frac{d}{dx}[f'(x)] = f''(x) \qquad \text{Second derivative}$$

The derivative of f'' is the **third derivative** of f and is denoted by f'''.

$$\frac{d}{dx}[f''(x)] = f'''(x) \qquad \text{Third derivative}$$

By continuing this process, you obtain **higher-order derivatives** of f. Higher-order derivatives are denoted as follows.

DISCOVERY

For each function, find the indicated higher-order derivative.

(a) $y = x^2$ (b) $y = x^3$
    y'' y'''

(c) $y = x^4$ (d) $y = x^n$
    $y^{(4)}$ $y^{(n)}$

Notation for Higher-Order Derivatives

1. 1st derivative: y', $f'(x)$, $\dfrac{dy}{dx}$, $\dfrac{d}{dx}[f(x)]$, $D_x[y]$

2. 2nd derivative: y'', $f''(x)$, $\dfrac{d^2y}{dx^2}$, $\dfrac{d^2}{dx^2}[f(x)]$, $D_x^2[y]$

3. 3rd derivative: y''', $f'''(x)$, $\dfrac{d^3y}{dx^3}$, $\dfrac{d^3}{dx^3}[f(x)]$, $D_x^3[y]$

4. 4th derivative: $y^{(4)}$, $f^{(4)}(x)$, $\dfrac{d^4y}{dx^4}$, $\dfrac{d^4}{dx^4}[f(x)]$, $D_x^4[y]$

5. nth derivative: $y^{(n)}$, $f^{(n)}(x)$, $\dfrac{d^ny}{dx^n}$, $\dfrac{d^n}{dx^n}[f(x)]$, $D_x^n[y]$

EXAMPLE 1   **Finding Higher-Order Derivatives**

Find the first five derivatives of $f(x) = 2x^4 - 3x^2$.

$$f(x) = 2x^4 - 3x^2 \qquad \text{Write original function.}$$
$$f'(x) = 8x^3 - 6x \qquad \text{First derivative}$$
$$f''(x) = 24x^2 - 6 \qquad \text{Second derivative}$$
$$f'''(x) = 48x \qquad \text{Third derivative}$$
$$f^{(4)}(x) = 48 \qquad \text{Fourth derivative}$$
$$f^{(5)}(x) = 0 \qquad \text{Fifth derivative}$$

TRY IT 1

Find the first four derivatives of
$$f(x) = 6x^3 - 2x^2 + 1.$$

EXAMPLE 2   **Finding Higher-Order Derivatives**

Find the value of $g'''(2)$ for the function

$g(t) = -t^4 + 2t^3 + t + 4.$ Original function

SOLUTION   Begin by differentiating three times.

$g'(t) = -4t^3 + 6t^2 + 1$ First derivative

$g''(t) = -12t^2 + 12t$ Second derivative

$g'''(t) = -24t + 12$ Third derivative

Then, evaluate the third derivative of g at $t = 2$.

$g'''(2) = -24(2) + 12$

$= -36$ Value of third derivative   _____

TRY IT 2

Find the value of $g'''(1)$ for $g(x) = x^4 - x^3 + 2x$.

 Higher-order derivatives of nonpolynomial functions can be difficult to find by hand. If you have access to a symbolic differentiation utility, try using it to find higher-order derivatives.

Examples 1 and 2 show how to find higher-order derivatives of *polynomial* functions. Note that with each successive differentiation, the degree of the polynomial drops by one. Eventually, higher-order derivatives of polynomial functions degenerate to a constant function. Specifically, the nth-order derivative of an nth-degree polynomial function

$f(x) = a_n x^n + a_{n-1} x^{n-1} + \cdots + a_1 x + a_0$

is the constant function

$f^{(n)}(x) = n! a_n$

where $n! = 1 \cdot 2 \cdot 3 \cdots n$. Each derivative of order higher than n is the zero function. Polynomial functions are the *only* functions with this characteristic. For other functions, successive differentiation never produces a constant function.

EXAMPLE 3   **Finding Higher-Order Derivatives**

Find the first four derivatives of $y = x^{-1}$.

$y = x^{-1} = \dfrac{1}{x}$ Write original function.

$y' = (-1)x^{-2} = -\dfrac{1}{x^2}$ First derivative

$y'' = (-1)(-2)x^{-3} = \dfrac{2}{x^3}$ Second derivative

$y''' = (-1)(-2)(-3)x^{-4} = -\dfrac{6}{x^4}$ Third derivative

$y^{(4)} = (-1)(-2)(-3)(-4)x^{-5} = \dfrac{24}{x^5}$ Fourth derivative

---

TRY IT 3

Find the fourth derivative of

$y = \dfrac{1}{x^2}.$

Acceleration

In Section 2.3, you saw that the velocity of an object moving in a straight path (neglecting air resistance) is given by the derivative of its position function. In other words, the rate of change of the position with respect to time is defined to be the velocity. In a similar way, the rate of change of the velocity with respect to time is defined to be the **acceleration** of the object.

$$s = f(t) \qquad \text{Position function}$$

$$\frac{ds}{dt} = f'(t) \qquad \text{Velocity function}$$

$$\frac{d^2s}{dt^2} = f''(t) \qquad \text{Acceleration function}$$

To find the position, velocity, or acceleration at a particular time t, substitute the given value of t into the appropriate function, as illustrated in Example 4.

EXAMPLE 4   Finding Acceleration

A ball is thrown into the air from the top of a 160-foot cliff, as shown in Figure 2.31. The initial velocity of the ball is 48 feet per second, which implies that the position function is

$$s = -16t^2 + 48t + 160$$

where the time t is measured in seconds. Find the height, the velocity, and the acceleration of the ball when $t = 3$.

SOLUTION   Begin by differentiating to find the velocity and acceleration functions.

$$s = -16t^2 + 48t + 160 \qquad \text{Position function}$$

$$\frac{ds}{dt} = -32t + 48 \qquad \text{Velocity function}$$

$$\frac{d^2s}{dt^2} = -32 \qquad \text{Acceleration function}$$

To find the height, velocity, and acceleration when $t = 3$, substitute $t = 3$ into each of the functions above.

Height $= -16(3)^2 + 48(3) + 160 = 160$ feet

Velocity $= -32(3) + 48 = -48$ feet per second

Acceleration $= -32$ feet per second squared

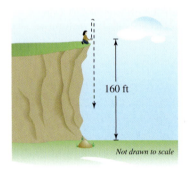

160 ft

Not drawn to scale

FIGURE 2.31

TRY IT 4

A ball is thrown upward from the top of an 80-foot cliff with an initial velocity of 64 feet per second. Give the position function. Then find the velocity and acceleration functions.

In Example 4, notice that the acceleration of the ball is -32 feet per second squared at any time t. This constant acceleration is due to the gravitational force of Earth and is called the **acceleration due to gravity.** Note that the negative value indicates that the ball is being pulled *down*—toward Earth.

Although the acceleration exerted on a falling object is relatively constant near Earth's surface, it varies greatly throughout our solar system. Large planets exert a much greater gravitational pull than do small planets or moons. The next example describes the motion of a free-falling object on the moon.

EXAMPLE 5 **Finding Acceleration on the Moon**

An astronaut standing on the surface of the moon throws a rock into the air. The height s (in feet) of the rock is given by

$$s = -\frac{27}{10}t^2 + 27t + 6$$

where t is measured in seconds. How does the acceleration due to gravity on the moon compare with that on Earth?

SOLUTION

$$s = -\frac{27}{10}t^2 + 27t + 6 \qquad \text{Position function}$$

$$\frac{ds}{dt} = -\frac{27}{5}t + 27 \qquad \text{Velocity function}$$

$$\frac{d^2s}{dt^2} = -\frac{27}{5} \qquad \text{Acceleration function}$$

So, the acceleration at any time is

$$-\frac{27}{5} = -5.4 \text{ feet per second squared}$$

—about one-sixth of the acceleration due to gravity on Earth.

The position function described in Example 5 neglects air resistance, which is appropriate because the moon has no atmosphere—and *no air resistance.* This means that the position function for any free-falling object on the moon is given by

$$s = -\frac{27}{10}t^2 + v_0 t + h_0$$

where s is the height (in feet), t is the time (in seconds), v_0 is the initial velocity, and h_0 is the initial height. For instance, the rock in Example 5 was thrown upward with an initial velocity of 27 feet per second and had an initial height of 6 feet. This position function is valid for all objects, whether heavy ones such as hammers or light ones such as feathers.

In 1971, astronaut David R. Scott demonstrated the lack of atmosphere on the moon by dropping a hammer and a feather from the same height. Both took exactly the same time to fall to the ground. If they were dropped from a height of 6 feet, how long did each take to hit the ground?

NASA

The acceleration due to gravity on the surface of the moon is only about one-sixth that exerted by Earth. So, if you were on the moon and threw an object into the air, it would rise to a greater height than it would on Earth's surface.

TRY IT 5

The position function on Earth, where s is measured in meters, t is measured in seconds, v_0 is the initial velocity in meters per second, and h_0 is the initial height in meters, is

$$s = -4.9t^2 + v_0 t + h_0.$$

If the initial velocity is 2.2 and the initial height is 3.6, what is the acceleration due to gravity on Earth in meters per second per second?

EXAMPLE 6   **Finding Velocity and Acceleration**

The velocity v (in feet per second) of a certain automobile starting from rest is

$$v = \frac{80t}{t + 5}$$ Velocity function

where t is the time (in seconds). The positions of the automobile at 10-second intervals are shown in Figure 2.32. Find the velocity and acceleration of the automobile at 10-second intervals from $t = 0$ to $t = 60$.

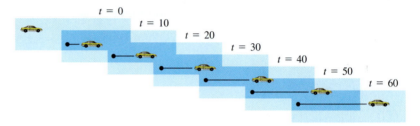

FIGURE 2.32

SOLUTION   To find the acceleration function, differentiate the velocity function.

$$\frac{dv}{dt} = \frac{(t + 5)(80) - (80t)(1)}{(t + 5)^2}$$

$$= \frac{400}{(t + 5)^2}$$ Acceleration function

TRY IT 6

Use a graphing utility to graph the velocity function and acceleration function in Example 6 in the same viewing window. Compare the graphs with the table at the right. As the velocity levels off, what does the acceleration approach?

t (seconds)	0	10	20	30	40	50	60
v (ft/sec)	0	53.3	64.0	68.6	71.1	72.7	73.8
$\frac{dv}{dt}$ (ft/sec²)	16	1.78	0.64	0.33	0.20	0.13	0.09

In the table, note that the acceleration approaches zero as the velocity levels off. This observation should agree with your experience—when riding in an accelerating automobile, you do not feel the velocity, but you do feel the acceleration. In other words, you feel changes in velocity.

TAKE ANOTHER LOOK

Acceleration Due to Gravity

Newton's Law of Universal Gravitation states that the gravitational attraction of two objects is directly proportional to their masses and inversely proportional to the square of the distance between their centers. The Earth has a mass of 5.979×10^{24} kilograms and a radius of 6371 kilometers. The moon has a mass of 7.354×10^{22} kilograms and a radius of 1738 kilometers. Discuss how you could find the ratio of the Earth's gravity to the moon's gravity. What is this ratio?

The following warm-up exercises involve skills that were covered in earlier sections. You will use these skills in the exercise set for this section.

In Exercises 1–4, solve the equation.

1. $-16t^2 + 24t = 0$

2. $-16t^2 + 80t + 224 = 0$

3. $-16t^2 + 128t + 320 = 0$

4. $-16t^2 + 9t + 1440 = 0$

In Exercises 5–8, find dy/dx.

5. $y = x^2(2x + 7)$

6. $y = (x^2 + 3x)(2x^2 - 5)$

7. $y = \dfrac{x^2}{2x + 7}$

8. $y = \dfrac{x^2 + 3x}{2x^2 - 5}$

In Exercises 9 and 10, find the domain and range of f.

9. $f(x) = x^2 - 4$

10. $f(x) = \sqrt{x - 7}$

EXERCISES 2.6

In Exercises 1–14, find the second derivative of the function.

1. $f(x) = 5 - 4x$

2. $f(x) = 3x - 1$

3. $f(x) = x^2 + 7x - 4$

4. $f(x) = 3x^2 + 4x$

5. $g(t) = \frac{1}{3}t^3 - 4t^2 + 2t$

6. $f(x) = 4(x^2 - 1)^2$

7. $f(t) = \dfrac{3}{4t^2}$

8. $g(t) = t^{-1/3}$

9. $f(x) = 3(2 - x^2)^3$

10. $f(x) = x\sqrt[3]{x}$

11. $f(x) = \dfrac{x + 1}{x - 1}$

12. $g(t) = -\dfrac{4}{(t + 2)^2}$

13. $y = x^2(x^2 + 4x + 8)$

14. $h(s) = s^3(s^2 - 2s + 1)$

In Exercises 15–20, find the third derivative of the function.

15. $f(x) = x^5 - 3x^4$

16. $f(x) = x^4 - 2x^3$

17. $f(x) = 5x(x + 4)^3$

18. $f(x) = (x - 1)^2$

19. $f(x) = \dfrac{3}{16x^2}$

20. $f(x) = \dfrac{1}{x}$

In Exercises 21–26, find the given value.

Function	Value
21. $g(t) = 5t^4 + 10t^2 + 3$	$g''(2)$
22. $f(x) = 9 - x^2$	$f''\!\left(-\sqrt{5}\right)$
23. $f(x) = \sqrt{4 - x}$	$f'''(-5)$
24. $f(t) = \sqrt{2t + 3}$	$f''\!\left(\tfrac{1}{2}\right)$
25. $f(x) = x^2(3x^2 + 3x - 4)$	$f'''(-2)$
26. $g(x) = 2x^3(x^2 - 5x + 4)$	$g'''(0)$

In Exercises 27–32, find the higher-order derivative.

Given	Derivative
27. $f'(x) = 2x^2$	$f''(x)$
28. $f''(x) = 20x^3 - 36x^2$	$f'''(x)$
29. $f''(x) = (2x - 2)/x$	$f'''(x)$
30. $f'''(x) = 2\sqrt{x - 1}$	$f^{(4)}(x)$
31. $f^{(4)}(x) = (x + 1)^2$	$f^{(6)}(x)$
32. $f(x) = x^3 - 2x$	$f''(x)$

In Exercises 33–40, find the second derivative and solve the equation $f''(x) = 0$.

33. $f(x) = x^3 - 9x^2 + 27x - 27$

34. $f(x) = 3x^3 - 9x + 1$

35. $f(x) = (x + 3)(x - 4)(x + 5)$

36. $f(x) = (x + 2)(x - 2)(x + 3)(x - 3)$

37. $f(x) = x\sqrt{x^2 - 1}$

38. $f(x) = x\sqrt{4 - x^2}$

39. $f(x) = \dfrac{x}{x^2 + 3}$

40. $f(x) = \dfrac{x}{x^2 + 1}$

41. *Velocity and Acceleration* A ball is propelled straight upward from ground level with an initial velocity of 144 feet per second.

(a) Write the position function of the ball.

(b) Write the velocity and acceleration functions.

(c) When is the ball at its highest point? How high is this point?

(d) How fast is the ball traveling when it hits the ground? How is this speed related to the initial velocity?

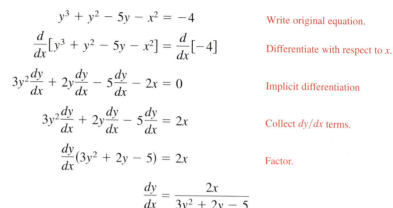

EXAMPLE 4   **Using Implicit Differentiation**

Find dy/dx for the equation $y^3 + y^2 - 5y - x^2 = -4$.

SOLUTION

$$y^3 + y^2 - 5y - x^2 = -4 \qquad \text{Write original equation.}$$

$$\frac{d}{dx}[y^3 + y^2 - 5y - x^2] = \frac{d}{dx}[-4] \qquad \text{Differentiate with respect to } x.$$

$$3y^2\frac{dy}{dx} + 2y\frac{dy}{dx} - 5\frac{dy}{dx} - 2x = 0 \qquad \text{Implicit differentiation}$$

$$3y^2\frac{dy}{dx} + 2y\frac{dy}{dx} - 5\frac{dy}{dx} = 2x \qquad \text{Collect } dy/dx \text{ terms.}$$

$$\frac{dy}{dx}(3y^2 + 2y - 5) = 2x \qquad \text{Factor.}$$

$$\frac{dy}{dx} = \frac{2x}{3y^2 + 2y - 5}$$

The graph of the original equation is shown in Figure 2.34. What are the slopes of the graph at the points $(1, -3)$, $(2, 0)$, and $(1, 1)$?

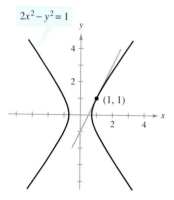

$$y^3 + y^2 - 5y - x^2 = -4$$

FIGURE 2.34

TRY IT 4

Find dy/dx for the equation $y^2 + x^2 - 2y - 4x = 4$.

EXAMPLE 5   **Finding the Slope of a Graph Implicitly**

Find the slope of the graph of $2x^2 - y^2 = 1$ at the point $(1, 1)$.

SOLUTION   Begin by finding dy/dx implicitly.

$$2x^2 - y^2 = 1 \qquad \text{Write original equation.}$$

$$4x - 2y\left(\frac{dy}{dx}\right) = 0 \qquad \text{Differentiate with respect to } x.$$

$$-2y\left(\frac{dy}{dx}\right) = -4x \qquad \text{Subtract } 4x \text{ from each side.}$$

$$\frac{dy}{dx} = \frac{2x}{y} \qquad \text{Divide each side by } -2y.$$

At the point $(1, 1)$, the slope of the graph is

$$\frac{2(1)}{1} = 2$$

as shown in Figure 2.35. The graph is called a **hyperbola.**

$2x^2 - y^2 = 1$

FIGURE 2.35   Hyperbola

TRY IT 5

Find the slope of the graph of $x^2 - 9y^2 = 16$ at the point $(5, 1)$.

Application

EXAMPLE 6   **Using a Demand Function**

The demand function for a product is modeled by

$$p = \frac{3}{0.000001x^3 + 0.01x + 1}$$

where p is measured in dollars and x is measured in thousands of units, as shown in Figure 2.36. Find the rate of change of the demand x with respect to the price p when $x = 100$.

SOLUTION   To simplify the differentiation, begin by rewriting the function. Then, differentiate *with respect to p*.

$$p = \frac{3}{0.000001x^3 + 0.01x + 1}$$

$$0.000001x^3 + 0.01x + 1 = \frac{3}{p}$$

$$0.000003x^2\frac{dx}{dp} + 0.01\frac{dx}{dp} = -\frac{3}{p^2}$$

$$(0.000003x^2 + 0.01)\frac{dx}{dp} = -\frac{3}{p^2}$$

$$\frac{dx}{dp} = -\frac{3}{p^2(0.000003x^2 + 0.01)}$$

When $x = 100$, the price is

$$p = \frac{3}{0.000001(100)^3 + 0.01(100) + 1} = \$1.$$

So, when $x = 100$ and $p = 1$, the rate of change of the demand with respect to the price is

$$-\frac{3}{(1)^2[0.000003(100)^2 + 0.01]} = -75.$$

This means that when $x = 100$, the demand is dropping at the rate of 75 thousand units for each dollar increase in price.

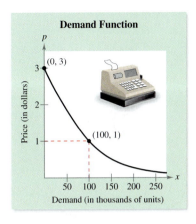

Demand Function

FIGURE 2.36

TRY IT 6

The demand function for a product is given by

$$p = \frac{2}{0.001x^2 + 1}.$$

Find dx/dp implicitly.

TAKE ANOTHER LOOK

Comparing Derivatives

In Example 6, the derivative dx/dp does not represent the slope of the graph of the demand function. Because the demand function is given by $p = f(x)$, the slope of the graph is given by dp/dx. Find dp/dx. Show that the two derivatives are related by $dx/dp = 1/(dp/dx)$. What does dp/dx represent?

In Exercises 1–6, solve the equation for y.

1. $x - \dfrac{y}{x} = 2$

2. $\dfrac{4}{x-3} = \dfrac{1}{y}$

3. $xy - x + 6y = 6$

4. $12 + 3y = 4x^2 + x^2y$

5. $x^2 + y^2 = 5$

6. $x = \pm\sqrt{6 - y^2}$

In Exercises 7–10, evaluate the expression at the given point.

7. $\dfrac{3x^2 - 4}{3y^2}$, $(2, 1)$

8. $\dfrac{x^2 - 2}{1 - y}$, $(0, -3)$

9. $\dfrac{5x}{3y^2 - 12y + 5}$, $(-1, 2)$

10. $\dfrac{1}{y^2 - 2xy + x^2}$, $(4, 3)$

EXERCISES 2.7

In Exercises 1–12, find dy/dx.

1. $5xy = 1$

2. $\frac{1}{2}x^2 - y = 6x$

3. $y^2 = 1 - x^2,\ 0 \le x \le 1$

4. $4x^2y - \dfrac{3}{y} = 0$

5. $x^2y^2 - 4y = 1$

6. $xy^2 + 4xy = 10$

7. $4y^2 - xy = 2$

8. $2xy^3 - x^2y = 2$

9. $\dfrac{2y - x}{y^2 - 3} = 5$

10. $\dfrac{xy - y^2}{y - x} = 1$

11. $\dfrac{x + y}{2x - y} = 1$

12. $\dfrac{2x + y}{x - 5y} = 1$

In Exercises 13–24, find dy/dx by implicit differentiation and evaluate the derivative at the given point.

Equation	Point
13. $x^2 + y^2 = 49$	$(0, 7)$
14. $x^2 - y^2 = 16$	$(4, 0)$
15. $y + xy = 4$	$(-5, -1)$
16. $x^2 - y^3 = 3$	$(2, 1)$
17. $x^3 - xy + y^2 = 4$	$(0, -2)$
18. $x^2y + y^2x = -2$	$(2, -1)$
19. $x^3y^3 - y = x$	$(0, 0)$
20. $x^3 + y^3 = 2xy$	$(1, 1)$
21. $x^{1/2} + y^{1/2} = 9$	$(16, 25)$
22. $\sqrt{xy} = x - 2y$	$(4, 1)$
23. $x^{2/3} + y^{2/3} = 5$	$(8, 1)$
24. $(x + y)^3 = x^3 + y^3$	$(-1, 1)$

In Exercises 25–30, find the slope of the graph at the given point.

25. $3x^2 - 2y + 5 = 0$

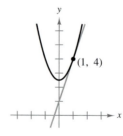

26. $4x^2 + 2y - 1 = 0$

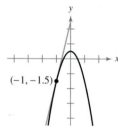

27. $x^2 + y^2 = 4$

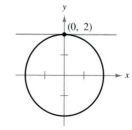

28. $4x^2 + y^2 = 4$

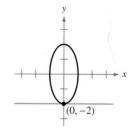

29. $4x^2 + 9y^2 = 36$

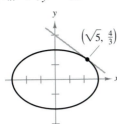

$\left(\sqrt{5}, \frac{4}{3}\right)$

30. $x^2 - y^3 = 0$

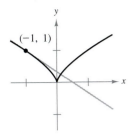

$(-1, 1)$

In Exercises 31–34, find dy/dx implicitly and explicitly (the explicit functions are shown on the graph) and show that the results are equivalent. Use the graph to estimate the slope of the tangent line at the labeled point. Then verify your result analytically by evaluating dy/dx at the point.

31. $x^2 + y^2 = 25$

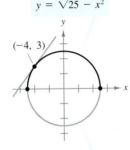

$y = \sqrt{25 - x^2}$

$(-4, 3)$

$y = -\sqrt{25 - x^2}$

32. $9x^2 + 16y^2 = 144$

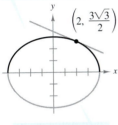

$y = \dfrac{\sqrt{144 - 9x^2}}{4}$

$\left(2, \frac{3\sqrt{3}}{2}\right)$

$y = -\dfrac{\sqrt{144 - 9x^2}}{4}$

33. $x - y^2 - 1 = 0$

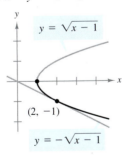

$y = \sqrt{x - 1}$

$(2, -1)$

$y = -\sqrt{x - 1}$

34. $4y^2 - x^2 = 7$

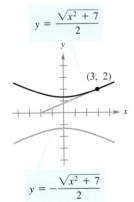

$y = \dfrac{\sqrt{x^2 + 7}}{2}$

$(3, 2)$

$y = -\dfrac{\sqrt{x^2 + 7}}{2}$

In Exercises 35–40, find equations of the tangent lines to the graph at the given points. Use a graphing utility to graph the equation and the tangent lines in the same viewing window.

Equation	Points
35. $x^2 + y^2 = 169$	$(5, 12)$ and $(-12, 5)$

Equation	Points
36. $x^2 + y^2 = 9$	$(0, 3)$ and $\left(2, \sqrt{5}\right)$
37. $y^2 = 5x^3$	$\left(1, \sqrt{5}\right)$ and $\left(1, -\sqrt{5}\right)$
38. $4xy + x^2 = 5$	$(1, 1)$ and $(5, -1)$
39. $x^3 + y^3 = 8$	$(0, 2)$ and $(2, 0)$
40. $y^2 = \dfrac{x^3}{4 - x}$	$(2, 2)$ and $(2, -2)$

Demand In Exercises 41–44, find the rate of change of x with respect to p.

41. $p = 0.006x^4 + 0.02x^2 + 10, \quad x \geq 0$

42. $p = 0.002x^4 + 0.01x^2 + 5, \quad x \geq 0$

43. $p = \sqrt{\dfrac{200 - x}{2x}}, \quad 0 < x \leq 200$

44. $p = \sqrt{\dfrac{500 - x}{2x}}, \quad 0 < x \leq 500$

45. ***Production*** Let x represent the units of labor and y the capital invested in a manufacturing process. When 135,540 units are produced, the relationship between labor and capital can be modeled by $100x^{0.75}y^{0.25} = 135{,}540$.

(a) Find the rate of change of y with respect to x when $x = 1500$ and $y = 1000$.

(b) The model used in the problem is called the *Cobb-Douglas production function*. Graph the model on a graphing utility and describe the relationship between labor and capital.

46. ***Health: U.S. AIDS Epidemic*** The numbers (in millions) of cases y of AIDS reported in the years 1994 to 2001 can be modeled by

$$y^2 + 4436 = -4.2460t^4 + 146.821t^3 - 1728.00t^2$$
$$+ 7456.6t$$

where $t = 4$ corresponds to 1994. *(Source: U.S. Centers for Disease Control and Prevention)*

(a) Use a graphing utility to graph the model and describe the results.

(b) Use the graph to determine the year during which the number of reported cases decreasing most rapidly.

(c) Complete the table to confirm your estimate.

t	4	5	6	7	8	9	10	11
y								
y'								

2.8   RELATED RATES

- Examine related variables.
- Solve related-rate problems.

Related Variables

In this section, you will study problems involving variables that are changing with respect to time. If two or more such variables are related to each other, then their rates of change with respect to time are also related.

For instance, suppose that x and y are related by the equation $y = 2x$. If both variables are changing with respect to time, then their rates of change will also be related.

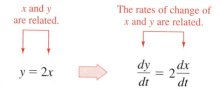

In this simple example, you can see that because y always has twice the value of x, it follows that the rate of change of y with respect to time is always twice the rate of change of x with respect to time.

EXAMPLE 1   Examining Two Rates That Are Related

The variables x and y are differentiable functions of t and are related by the equation

$$y = x^2 + 3.$$

When $x = 1$, $dx/dt = 2$. Find dy/dt when $x = 1$.

SOLUTION   Use the Chain Rule to differentiate both sides of the equation with respect to t.

$$y = x^2 + 3 \qquad \text{Write original equation.}$$

$$\frac{d}{dt}[y] = \frac{d}{dt}[x^2 + 3] \qquad \text{Differentiate with respect to } t.$$

$$\frac{dy}{dt} = 2x\frac{dx}{dt} \qquad \text{Apply Chain Rule.}$$

When $x = 1$ and $dx/dt = 2$, you have

$$2x\frac{dx}{dt} = 2(1)(2)$$

$$= 4.$$

TRY IT 1

When $x = 1$, $dx/dt = 3$. Find dy/dt when $x = 1$ if $y = x^3 + 2$.

Solving Related-Rate Problems

In Example 1, you are *given* the mathematical model.

Given equation: $y = x^2 + 3$

Given rate: $\dfrac{dx}{dt} = 2$ when $x = 1$

Find: $\dfrac{dy}{dt}$ when $x = 1$

In the next example, you are asked to *create* a similar mathematical model.

EXAMPLE 2 **Changing Area**

A pebble is dropped into a calm pool of water, causing ripples in the form of concentric circles, as shown in Figure 2.37. The radius r of the outer ripple is increasing at a constant rate of 1 foot per second. When the radius is 4 feet, at what rate is the total area A of the disturbed water changing?

SOLUTION The variables r and A are related by the equation for the area of a circle, $A = \pi r^2$. To solve this problem, use the fact that the rate of change of the radius is given by dr/dt.

Equation: $A = \pi r^2$

Given rate: $\dfrac{dr}{dt} = 1$ when $r = 4$

Find: $\dfrac{dA}{dt}$ when $r = 4$

FIGURE 2.37

Using this model, you can proceed as in Example 1.

$A = \pi r^2$ Write original equation.

$\dfrac{d}{dt}[A] = \dfrac{d}{dt}[\pi r^2]$ Differentiate with respect to t.

$\dfrac{dA}{dt} = 2\pi r \dfrac{dr}{dt}$ Apply Chain Rule.

When $r = 4$ and $dr/dt = 1$, you have

$2\pi r \dfrac{dr}{dt} = 2\pi(4)(1)$

$= 8\pi$ feet squared per second

> **TRY IT 2**
>
> If the radius r of the outer ripple in Example 2 is increasing at a rate of 2 feet per second, at what rate is the total area changing when the radius is 3 feet?

STUDY TIP

In Example 2, note that the radius changes at a *constant* rate ($dr/dt = 1$ for all t), but the area changes at a *nonconstant* rate.

When $r = 1$ ft	When $r = 2$ ft	When $r = 3$ ft	When $r = 4$ ft
$\dfrac{dA}{dt} = 2\pi$ ft²/sec	$\dfrac{dA}{dt} = 4\pi$ ft²/sec	$\dfrac{dA}{dt} = 6\pi$ ft²/sec	$\dfrac{dA}{dt} = 8\pi$ ft²/sec

ALGEBRA REVIEW

Simplifying Algebraic Expressions

To be successful in using derivatives, you must be good at simplifying algebraic expressions. Here are some helpful simplification techniques.

1. Combine *like terms*. This may involve expanding an expression by multiplying factors.

2. Divide out *like factors* in the numerator and denominator of an expression.

3. Factor an expression.

4. Rationalize a denominator.

5. Add, subtract, multiply, or divide fractions.

EXAMPLE 1   **Simplifying a Fractional Expression**

(a)
$$\frac{(x + \Delta x)^2 - x^2}{\Delta x} = \frac{x^2 + 2x(\Delta x) + (\Delta x)^2 - x^2}{\Delta x} \qquad \text{Expand expression.}$$

$$= \frac{2x(\Delta x) + (\Delta x)^2}{\Delta x} \qquad \text{Combine like terms.}$$

$$= \frac{\Delta x(2x + \Delta x)}{\Delta x} \qquad \text{Factor.}$$

$$= 2x + \Delta x, \qquad \Delta x \neq 0 \qquad \text{Divide out like factors.}$$

(b)
$$\frac{(x^2 - 1)(-2 - 2x) - (3 - 2x - x^2)(2)}{(x^2 - 1)^2}$$

$$= \frac{(-2x^2 - 2x^3 + 2 + 2x) - (6 - 4x - 2x^2)}{(x^2 - 1)^2} \qquad \text{Expand expression.}$$

$$= \frac{-2x^2 - 2x^3 + 2 + 2x - 6 + 4x + 2x^2}{(x^2 - 1)^2} \qquad \text{Remove parentheses.}$$

$$= \frac{-2x^3 + 6x - 4}{(x^2 - 1)^2} \qquad \text{Combine like terms.}$$

(c)
$$2\left(\frac{2x + 1}{3x}\right)\left[\frac{3x(2) - (2x + 1)(3)}{(3x)^2}\right]$$

$$= 2\left(\frac{2x + 1}{3x}\right)\left[\frac{6x - (6x + 3)}{(3x)^2}\right] \qquad \text{Multiply factors.}$$

$$= \frac{2(2x + 1)(6x - 6x - 3)}{(3x)^3} \qquad \text{Multiply fractions and remove parentheses.}$$

$$= \frac{2(2x + 1)(-3)}{3(9)x^3} \qquad \text{Combine like terms and factor.}$$

$$= \frac{-2(2x + 1)}{9x^3} \qquad \text{Divide out like factors.}$$

| EXAMPLE 2 | **Simplifying an Expression with Powers or Radicals** |

(a) $(2x + 1)^2(6x + 1) + (3x^2 + x)(2)(2x + 1)(2)$

$= (2x + 1)[(2x + 1)(6x + 1) + (3x^2 + x)(2)(2)]$ Factor.

$= (2x + 1)[12x^2 + 8x + 1 + (12x^2 + 4x)]$ Multiply factors.

$= (2x + 1)(12x^2 + 8x + 1 + 12x^2 + 4x)$ Remove parentheses.

$= (2x + 1)(24x^2 + 12x + 1)$ Combine like terms.

(b) $(-1)(6x^2 - 4x)^{-2}(12x - 4)$

$= \dfrac{(-1)(12x - 4)}{(6x^2 - 4x)^2}$ Rewrite as a fraction.

$= \dfrac{(-1)(4)(3x - 1)}{(6x^2 - 4x)^2}$ Factor.

$= \dfrac{-4(3x - 1)}{(6x^2 - 4x)^2}$ Multiply factors.

(c) $(x)\left(\dfrac{1}{2}\right)(2x + 3)^{-1/2} + (2x + 3)^{1/2}(1)$

$= (2x + 3)^{-1/2}\left(\dfrac{1}{2}\right)[x + (2x + 3)(2)]$ Factor.

$= \dfrac{x + 4x + 6}{(2x + 3)^{1/2}(2)}$ Rewrite as a fraction.

$= \dfrac{5x + 6}{2(2x + 3)^{1/2}}$ Combine like terms.

(d) $\dfrac{x^2\left(\frac{1}{2}\right)(2x)(x^2 + 1)^{-1/2} - (x^2 + 1)^{1/2}(2x)}{x^4}$

$= \dfrac{(x^3)(x^2 + 1)^{-1/2} - (x^2 + 1)^{1/2}(2x)}{x^4}$ Multiply factors.

$= \dfrac{(x^2 + 1)^{-1/2}(x)[x^2 - (x^2 + 1)(2)]}{x^4}$ Factor.

$= \dfrac{x[x^2 - (2x^2 + 2)]}{(x^2 + 1)^{1/2}x^4}$ Write with positive exponents.

$= \dfrac{x^2 - 2x^2 - 2}{(x^2 + 1)^{1/2}x^3}$ Divide out like factors and remove parentheses.

$= \dfrac{-x^2 - 2}{(x^2 + 1)^{1/2}x^3}$ Combine like terms.

All but one of the expressions in this Algebra Review are derivatives. Can you see what the original function is? Explain your reasoning.

2 CHAPTER SUMMARY AND STUDY STRATEGIES

After studying this chapter, you should have acquired the following skills. The exercise numbers are keyed to the Review Exercises that begin on page 166. Answers to odd-numbered Review Exercises are given in the back of the text.*

Skills

- Approximate the slope of the tangent line to a graph at a point. *(Section 2.1)*   *Review Exercises 1–4*
- Interpret the slope of a graph in a real-life setting. *(Section 2.1)*   *Review Exercises 5–8*
- Use the limit definition to find the derivative of a function and the slope of a graph at a point. *(Section 2.1)*   *Review Exercises 9–16*

$$f'(x) = \lim_{\Delta x \to 0} \frac{f(x + \Delta x) - f(x)}{\Delta x}$$

- Use the derivative to find the slope of a graph at a point. *(Section 2.1)*   *Review Exercises 17–24*
- Use the graph of a function to recognize points at which the function is not differentiable. *(Section 2.1)*   *Review Exercises 25–28*
- Use the Constant Multiple Rule for differentiation. *(Section 2.2)*   *Review Exercises 29, 30*

$$\frac{d}{dx}[cf(x)] = cf'(x)$$

- Use the Sum and Difference Rules for differentiation. *(Section 2.2)*   *Review Exercises 31–38*

$$\frac{d}{dx}[f(x) \pm g(x)] = f'(x) \pm g'(x)$$

- Find the average rate of change of a function over an interval and the instantaneous rate of change at a point. *(Section 2.3)*   *Review Exercises 39, 40*

$$\text{Average rate of change} = \frac{f(b) - f(a)}{b - a}$$

$$\text{Instantaneous rate of change} = \lim_{\Delta x \to 0} \frac{f(x + \Delta x) - f(x)}{\Delta x}$$

- Find the average and instantaneous rates of change of a quantity in a real-life problem. *(Section 2.3)*   *Review Exercises 41–44*

* Use a wide range of valuable study aids to help you master the material in this chapter. The *Student Solutions Guide* includes step-by-step solutions to all odd-numbered exercises to help you review and prepare. The *HM mathSpace® Student CD-ROM* helps you brush up on your algebra skills. The *Graphing Technology Guide*, available on the Web at *math.college.hmco.com/students*, offers step-by-step commands and instructions for a wide variety of graphing calculators, including the most recent models.

- Find the velocity of an object that is moving in a straight line. *(Section 2.3)*   *Review Exercises 45, 46*
- Create mathematical models for the revenue, cost, and profit for a product. *(Section 2.3)*   *Review Exercises 47, 48*

  $$P = R - C, \qquad R = xp$$

- Find the marginal revenue, marginal cost, and marginal profit for a product. *(Section 2.3)*   *Review Exercises 49–58*
- Use the Product Rule for differentiation. *(Section 2.4)*   *Review Exercises 59–62*

  $$\frac{d}{dx}[f(x)g(x)] = f(x)g'(x) + g(x)f'(x)$$

- Use the Quotient Rule for differentiation. *(Section 2.4)*   *Review Exercises 63, 64*

  $$\frac{d}{dx}\left[\frac{f(x)}{g(x)}\right] = \frac{g(x)f'(x) - f(x)g'(x)}{[g(x)]^2}$$

- Use the General Power Rule for differentiation. *(Section 2.5)*   *Review Exercises 65–68*

  $$\frac{d}{dx}[u^n] = nu^{n-1}u'$$

- Use differentiation rules efficiently to find the derivative of any algebraic function, then simplify the result. *(Section 2.5)*   *Review Exercises 69–78*
- Use derivatives to answer questions about real-life situations. *(Sections 2.1–2.5)*   *Review Exercises 79, 80*
- Find higher-order derivatives. *(Section 2.6)*   *Review Exercises 81–88*
- Find and use the position function to determine the velocity and acceleration of a moving object. *(Section 2.6)*   *Review Exercises 89, 90*
- Find derivatives implicitly. *(Section 2.7)*   *Review Exercises 91–98*
- Solve related-rate problems. *(Section 2.8)*   *Review Exercises 99, 100*
- ***Simplify Your Derivatives*** Often our students ask if they have to simplify their derivatives. Our answer is "Yes, if you expect to use them." In the next chapter, you will see that almost all applications of derivatives require that the derivatives be written in simplified form. It is not difficult to see the advantage of a derivative in simplified form. Consider, for instance, the derivative of $f(x) = x/\sqrt{x^2 + 1}$. The "raw form" produced by the Quotient and Chain Rules

  $$f'(x) = \frac{(x^2 + 1)^{1/2}(1) - (x)\left(\frac{1}{2}\right)(x^2 + 1)^{-1/2}(2x)}{\left(\sqrt{x^2 + 1}\right)^2}$$

  is obviously much more difficult to use than the simplified form

  $$f'(x) = \frac{1}{(x^2 + 1)^{3/2}}.$$

- ***List Units of Measure in Applied Problems*** When using derivatives in real-life applications, be sure to list the units of measure for each variable. For instance, if R is measured in dollars and t is measured in years, then the derivative dR/dt is measured in dollars per year.

Study Tools   *Additional resources that accompany this chapter*

- **Algebra Review** (pages 162 and 163)
- **Chapter Summary and Study Strategies** (pages 164 and 165)
- **Review Exercises** (pages 166–169)
- **Sample Post-Graduation Exam Questions**   (page 170)
- **Web Exercises**   (page 130, Exercise 66)
- **Student Solutions Guide**
- **HM mathSpace® Student CD-ROM**
- **Graphing Technology Guide** (*math.college.hmco.com/students*)

2 CHAPTER REVIEW EXERCISES

In Exercises 1–4, approximate the slope of the tangent line to the graph at (x, y).

1.
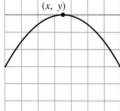

2.
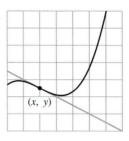

3.
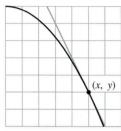

4.
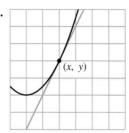

5. *Sales* The graph approximates the annual sales S (in millions of dollars per year) of Home Depot for the years 1996 to 2002, with $t = 6$ corresponding to 1996. Estimate the slopes of the graph when $t = 7$, $t = 10$, and $t = 12$. Interpret each slope in the context of the problem. *(Source: The Home Depot, Inc.)*

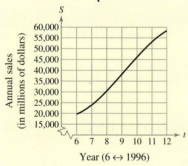

Home Depot Sales

Year $(6 \leftrightarrow 1996)$

6. *Consumer Trends* The graph approximates the number of subscribers S (in thousands per year) of cellular telephones for 1993 to 2002, with $t = 3$ corresponding to 1993. Estimate the slopes of the graph when $t = 4$, $t = 8$, and $t = 12$. Interpret each slope in the context of the problem. *(Source: Cellular Telecommunications & Internet Association)*

Figure for 6

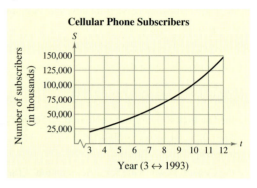

Cellular Phone Subscribers

Year $(3 \leftrightarrow 1993)$

7. *Medicine* The graph shows the estimated number of milligrams of a pain medication M in the bloodstream t hours after a 1000-milligram dose of the drug has been given. Estimate the slopes of the graph at $t = 0$, 4, and 6.

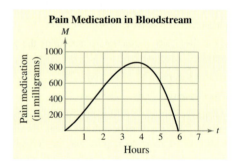

Pain Medication in Bloodstream

Hours

8. *Athletics* Two white-water rafters leave a campsite simultaneously and start downstream. Their distances from the campsite are given by $s = f(t)$ and $s = g(t)$, where s is measured in miles and t is measured in hours.

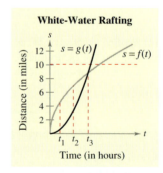

White-Water Rafting

Time (in hours)

(a) Which rafter is traveling at a greater rate at t_1?

(b) What can you conclude about their rates at t_2?

(c) What can you conclude about their rates at t_3?

(d) Which rafter finishes the trip first? Explain your reasoning.

In Exercises 9–16, use the limit definition to find the derivative of the function. Then use the limit definition to find the slope of the tangent line to the graph of f at the given point.

9. $f(x) = -3x - 5$; $(-2, 1)$ **10.** $f(x) = 7x + 3$; $(-1, 4)$

11. $f(x) = x^2 - 4x$; $(1, -3)$ **12.** $f(x) = x^2 + 10$; $(2, 14)$

13. $f(x) = \sqrt{x + 9}$; $(-5, 2)$ **14.** $f(x) = \sqrt{x - 1}$; $(10, 3)$

15. $f(x) = \dfrac{1}{x - 5}$; $(6, 1)$ **16.** $f(x) = \dfrac{1}{x + 4}$; $(-3, 1)$

In Exercises 17–24, find the slope of the graph of f at the given point.

17. $f(x) = 8 - 5x$; $(3, -7)$ **18.** $f(x) = 2 - 3x$; $(1, -1)$

19. $f(x) = -\frac{1}{2}x^2 + 2x$; $(2, 2)$ **20.** $f(x) = 4 - x^2$; $(-1, 3)$

21. $f(x) = \sqrt{x} + 2$; $(9, 5)$ **22.** $f(x) = 2\sqrt{x} + 1$; $(4, 5)$

23. $f(x) = \dfrac{5}{x}$; $(1, 5)$ **24.** $f(x) = \dfrac{2}{x} - 1$; $\left(\frac{1}{2}, 3\right)$

In Exercises 25–28, determine the x-value at which the function is not differentiable.

25. $y = \dfrac{x + 1}{x - 1}$

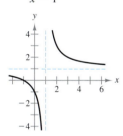

26. $y = -|x| + 3$

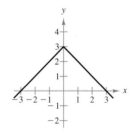

27. $y = \begin{cases} -x - 2, & x \le 0 \\ x^3 + 2, & x > 0 \end{cases}$

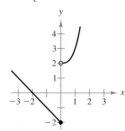

28. $y = (x + 1)^{2/3}$

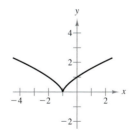

In Exercises 29–38, find the equation of the tangent line at the given point. Then use a graphing utility to graph the function and the equation of the tangent line in the same viewing window.

Function	Point
29. $g(t) = \dfrac{2}{3t^2}$	$\left(1, \dfrac{2}{3}\right)$
30. $h(x) = \dfrac{2}{(3x)^2}$	$\left(2, \dfrac{1}{18}\right)$

Function	Point
31. $f(x) = x^2 + 3$	$(1, 4)$
32. $f(x) = 2x^2 - 3x + 1$	$(2, 3)$
33. $y = 11x^4 - 5x^2 + 1$	$(-1, 7)$
34. $y = x^3 - 5 + \dfrac{3}{x^3}$	$(-1, -9)$
35. $f(x) = \sqrt{x} - \dfrac{1}{\sqrt{x}}$	$(1, 0)$
36. $f(x) = 2x^{-3} + 4 - \sqrt{x}$	$(1, 5)$
37. $f(x) = \dfrac{x^2 + 3}{x}$	$(1, 4)$
38. $f(x) = -x^2 - 4x - 4$	$(-4, -4)$

In Exercises 39 and 40, find the average rate of change of the function over the indicated interval. Then compare the average rate of change with the instantaneous rates of change at the endpoints of the interval.

39. $f(x) = x^2 + 3x - 4$; $[0, 1]$

40. $f(x) = x^3 + x$; $[-2, 2]$

41. ***Sales*** The annual sales S (in millions of dollars per year) of Home Depot for the years 1996 to 2002 can be modeled by

$$S = -172.361t^3 + 4740.75t^2 - 35{,}441.6t + 98{,}831$$

where $t = 6$ corresponds to 1996. A graph of this model appears in Exercise 5. *(Source: The Home Depot, Inc.)*

(a) Find the average rate of change for the interval from 1998 to 2002.

(b) Find the instantaneous rates of change of the model for 1998 and 2002.

(c) Interpret the results of parts (a) and (b) in the context of the problem.

42. ***Consumer Trends*** The numbers of subscribers S (in thousands per year) of cellular telephones for the years 1993 to 2002 can be modeled by

$$S = \dfrac{1168.2751 + 5366.2569t}{1 - 0.046307t}$$

where $t = 3$ corresponds to 1993. A graph of this model appears in Exercise 6. *(Source: Cellular Telecommunications & Internet Association)*

(a) Find the average rate of change for the interval from 1997 to 2002.

(b) Find the instantaneous rates of change of the model for 1997 and 2002.

(c) Interpret the results of parts (a) and (b) in the context of the problem.

43. Retail Price The average retail price P (in dollars) of 1 pound of 100% ground beef from 1996 to 2002 can be modeled by the equation

$$P = -0.001059t^4 + 0.03015t^3 - 0.2850t^2$$
$$+ 1.007t + 0.50$$

where t is the year, with $t = 6$ corresponding to 1996. *(Source: U.S. Bureau of Labor Statistics)*

(a) Find the rate of change of the price with respect to the year.

(b) At what rate was the price of 100% ground beef changing in 1997? in 2000? in 2002?

 (c) Use a graphing utility to graph the function for $6 \leq t \leq 12$. During which years was the price increasing? decreasing?

(d) For what years do the slopes of the tangent lines appear to be positive? negative?

(e) Compare your answers for parts (c) and (d).

44. Recycling The amount T of recycled paper products in millions of tons from 1993 to 2001 can be modeled by the equation

$$T = \sqrt{2.4890t^3 - 62.062t^2 + 553.16t - 509.4}$$

where t is the year, with $t = 3$ corresponding to 1993. *(Source: Franklin Associates, Ltd.)*

 (a) Use a graphing utility to graph the equation. Be sure to choose an appropriate window.

(b) Determine dT/dt. Evaluate dT/dt for 1993, 1998, and 2001.

(c) Is dT/dt positive for $t \geq 3$? Does this agree with the graph of the function? What does this tell you about this situation? Explain your reasoning.

45. Velocity A rock is dropped from a tower on the Brooklyn Bridge, 276 feet above the East River. Let t represent the time in seconds.

(a) Write a model for the position function (assume that air resistance is negligible).

(b) Find the average velocity during the first 2 seconds.

(c) Find the instantaneous velocities when $t = 2$ and $t = 3$.

(d) How long will it take for the rock to hit the water?

(e) When it hits the water, what is the rock's speed?

46. Velocity The straight-line distance s (in feet) traveled by an accelerating bicyclist can be modeled by

$$s = 2t^{3/2}, \qquad 0 \leq t \leq 8$$

where t is the time (in seconds). Complete the table, showing the velocity of the bicyclist at two-second intervals.

Table for 46

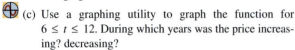

Time, t	0	2	4	6	8
Velocity					

47. Cost, Revenue, and Profit The fixed cost of operating a small flower shop is $2500 per month. The average cost of a floral arrangement is $15 and the average price is $27.50. Write the monthly revenue, cost, and profit functions for the floral shop in terms of x, the number of arrangements sold.

48. Profit The weekly demand and cost functions for a product are given by

$$p = 1.89 - 0.0083x \quad \text{and} \quad C = 21 + 0.65x.$$

Write the profit function for this product.

Marginal Cost In Exercises 49–52, find the marginal cost function.

49. $C = 2500 + 320x$ **50.** $C = 225x + 4500$

51. $C = 370 + 2.55\sqrt{x}$ **52.** $C = 475 + 5.25x^{2/3}$

Marginal Revenue In Exercises 53–56, find the marginal revenue function.

53. $R = 200x - \dfrac{1}{5}x^2$ **54.** $R = 150x - \dfrac{3}{4}x^2$

55. $R = \dfrac{35x}{\sqrt{x-2}}, \quad x \geq 6$ **56.** $R = x\left(5 + \dfrac{10}{\sqrt{x}}\right)$

Marginal Profit In Exercises 57 and 58, find the marginal profit function.

57. $P = -0.0002x^3 + 6x^2 - x - 2000$

58. $P = -\dfrac{1}{15}x^3 + 4000x^2 - 120x - 144,000$

In Exercises 59–78, find the derivative of the function. Simplify your result.

59. $f(x) = x^3(5 - 3x^2)$ **60.** $y = (3x^2 + 7)(x^2 - 2x)$

61. $y = (4x - 3)(x^3 - 2x^2)$ **62.** $s = \left(4 - \dfrac{1}{t^2}\right)(t^2 - 3t)$

63. $f(x) = \dfrac{6x - 5}{x^2 + 1}$ **64.** $f(x) = \dfrac{x^2 + x - 1}{x^2 - 1}$

65. $f(x) = (5x^2 + 2)^3$ **66.** $f(x) = \sqrt[3]{x^2 - 1}$

67. $h(x) = \dfrac{2}{\sqrt{x+1}}$ **68.** $g(x) = \sqrt{x^6 - 12x^3 + 9}$

69. $g(x) = x\sqrt{x^2 + 1}$ **70.** $g(t) = \dfrac{t}{(1-t)^3}$

71. $f(x) = x(1 - 4x^2)^2$ **72.** $f(x) = \left(x^2 + \dfrac{1}{x}\right)^5$

73. $h(x) = [x^2(2x + 3)]^3$

74. $f(x) = [(x - 2)(x + 4)]^2$

75. $f(x) = x^2(x - 1)^5$

76. $f(s) = s^3(s^2 - 1)^{5/2}$

77. $h(t) = \dfrac{\sqrt{3t + 1}}{(1 - 3t)^2}$

78. $g(x) = \dfrac{(3x + 1)^2}{(x^2 + 1)^2}$

79. *Physical Science* The temperature T (in degrees Fahrenheit) of food placed in a freezer can be modeled by

$$T = \frac{1300}{t^2 + 2t + 25}$$

where t is the time (in hours).

(a) Find the rates of change of T when $t = 1$, $t = 3$, $t = 5$, and $t = 10$.

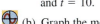 (b) Graph the model on a graphing utility and describe the rate at which the temperature is changing.

80. *Forestry* According to the *Doyle Log Rule*, the volume V (in board-feet) of a log of length L (feet) and diameter D (inches) at the small end is

$$V = \left(\frac{D - 4}{4}\right)^2 L.$$

Find the rates at which the volume is changing with respect to D for a 12-foot-long log whose smallest diameter is (a) 8 inches, (b) 16 inches, (c) 24 inches, and (d) 36 inches.

In Exercises 81–88, find the given derivative.

81. Given $f(x) = 3x^2 + 7x + 1$, find $f''(x)$.

82. Given $f'(x) = 5x^4 - 6x^2 + 2x$, find $f'''(x)$.

83. Given $f'''(x) = -\dfrac{6}{x^4}$, find $f^{(5)}(x)$.

84. Given $f(x) = \sqrt{x}$, find $f^{(4)}(x)$.

85. Given $f'(x) = 7x^{5/2}$, find $f''(x)$.

86. Given $f(x) = x^2 + \dfrac{3}{x}$, find $f''(x)$.

87. Given $f''(x) = 6\sqrt[3]{x}$, find $f'''(x)$.

88. Given $f'''(x) = 20x^4 - \dfrac{2}{x^3}$, find $f^{(5)}(x)$.

89. *Athletics* A person dives from a 30-foot platform with an initial velocity of 5 feet per second (upward).

(a) Find the position function of the diver.

(b) How long will it take for the diver to hit the water?

(c) What is the diver's velocity at impact?

(d) What is the diver's acceleration at impact?

90. *Velocity and Acceleration* The position function of a particle is given by

$$s = \frac{1}{t^2 + 2t + 1}$$

where s is the height (in feet) and t is the time (in seconds). Find the velocity and acceleration functions.

In Exercises 91–94, use implicit differentiation to find dy/dx.

91. $x^2 + 3xy + y^3 = 10$

92. $x^2 + 9xy + y^2 = 0$

93. $y^2 - x^2 + 8x - 9y - 1 = 0$

94. $y^2 + x^2 - 6y - 2x - 5 = 0$

In Exercises 95–98, use implicit differentiation to find an equation of the tangent line at the given point.

Equation	*Point*
95. $y^2 = x - y$	$(2, 1)$
96. $2\sqrt[3]{x} + 3\sqrt{y} = 10$	$(8, 4)$
97. $y^2 - 2x = xy$	$(1, 2)$
98. $y^3 - 2x^2y + 3xy^2 = -1$	$(0, -1)$

99. *Water Level* A swimming pool is 40 feet long, 20 feet wide, 4 feet deep at the shallow end, and 9 feet deep at the deep end (see figure). Water is being pumped into the pool at the rate of 10 cubic feet per minute. How fast is the water level rising when there is 4 feet of water in the deep end?

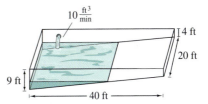

100. *Profit* The demand and cost functions for a product can be modeled by

$$p = 211 - 0.002x$$

and

$$C = 30x + 1{,}500{,}000$$

where x is the number of units produced.

(a) Write the profit function for this product.

(b) Find the marginal profit when 80,000 units are produced.

 (c) Graph the profit function on a graphing utility and use the graph to determine the price you would charge for the product. Explain your reasoning.

2 SAMPLE POST-GRADUATION EXAM QUESTIONS

CPA
GMAT
GRE
Actuarial
CLAST

The following questions represent the types of questions that appear on certified public accountant (CPA) exams, Graduate Management Admission Tests (GMAT), Graduate Records Exams (GRE), actuarial exams, and College-Level Academic Skills Tests (CLAST). The answers to the questions are given in the back of the book.

1. What is the length of the line segment that connects A to B (see graph)?

    (a) 2   (b) 4   (c) $2\sqrt{2}$ (d) 6   (e) $\sqrt{3}$

For Questions 2–4, refer to the following table.   *(Source: U.S. Census Bureau)*

Figure for 1

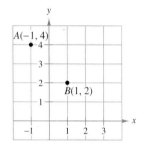

Participation in National Elections (millions of persons)

Characteristic	1992		1996		2000	
	Persons of voting age	Percent voted	Persons of voting age	Percent voted	Persons of voting age	Percent voted
Total	185.7	61.3	193.7	54.2	202.6	54.7
Male	88.6	60.2	92.6	52.8	97.1	53.1
Female	97.1	62.3	101.0	55.5	105.5	56.2
Age 18 to 20	9.7	38.5	10.8	31.2	11.9	28.4
21 to 24	14.6	33.4	13.9	33.4	14.9	35.4
25 to 34	41.6	53.2	40.1	43.1	37.3	43.7
35 to 44	39.7	63.6	43.3	54.9	44.5	55.0
45 to 64	49.1	70.0	53.7	64.4	61.4	64.1
65 years and over	30.8	70.1	31.9	67.0	32.8	67.6

2. Which of the following groups had the highest percent of voters in 1996?

    (a) Male   (b) Age 35 to 44   (c) Age 25 to 34

    (d) Age 18 to 20   (e) Female

3. In 2000, what percent (to the nearest percent) of persons of voting age were female?

    (a) 56   (b) 48   (c) 53   (d) 57   (e) 52

4. In 1992, how many males of voting age voted?

    (a) 60,493,300   (b) 111,791,400   (c) 53,337,200

    (d) 35,262,800   (e) 48,892,800

For Questions 5 and 6, refer to the following example.
The position s at time t of an object in motion is given by $s(t) = t(t^2 - 2)^2$.

5. The acceleration function is

    (a) $5t^4 - 12t^2 + 4$ (b) $5t^4 - 8t$ (c) $20t^3 - 24t$

    (d) $20t^3 - 8$ (e) $t^5 - 4t^2 + 4$

6. The velocity of the object at time $t = 3$ is

    (a) 301   (b) 381   (c) 147   (d) 532   (e) 468

chapter

3

Applications of the Derivative

© Paul Barton/CORBIS

Economists use the derivative to measure the increase or decrease in demand for products, such as food, when the price is lowered or raised.

STRATEGIES FOR SUCCESS

WHAT YOU SHOULD LEARN:

- How to find the open intervals on which a function is increasing or decreasing
- How to determine relative and absolute extrema of a function
- How to determine the concavity and points of inflection of a graph
- How to solve real-life optimization problems
- How to determine vertical and horizontal asymptotes of a graph
- How to use calculus to analyze the shape of the graph of a function

WHY YOU SHOULD LEARN IT:

Derivatives have many applications in real life, as can be seen by the examples below, which represent a small sample of the applications in this chapter.

- Chemistry: Molecular Velocity, Exercise 36 on page 180
- Diminishing Returns, Exercises 51 and 52 on page 199
- Price Elasticity, Exercises 33-35 on page 218
- Learning Curve, Exercises 61 and 62 on page 230
- Marginal Analysis, Exercises 27-36 on page 247

3.1   INCREASING AND DECREASING FUNCTIONS

■ Test for increasing and decreasing functions.
■ Find the critical numbers of functions and find the open intervals on which functions are increasing or decreasing.
■ Use increasing and decreasing functions to model and solve real-life problems.

Increasing and Decreasing Functions

A function is **increasing** if its graph moves up as x moves to the right and **decreasing** if its graph moves down as x moves to the right. The following definition states this more formally.

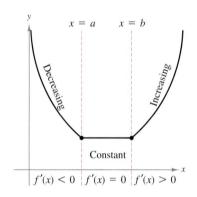

FIGURE 3.1

Definition of Increasing and Decreasing Functions

A function f is **increasing** on an interval if for any x_1 and x_2 in the interval

$$x_2 > x_1 \quad \text{implies} \quad f(x_2) > f(x_1).$$

A function f is **decreasing** on an interval if for any x_1 and x_2 in the interval

$$x_2 > x_1 \quad \text{implies} \quad f(x_2) < f(x_1).$$

The function in Figure 3.1 is decreasing on the interval $(-\infty, a)$, constant on the interval (a, b), and increasing on the interval (b, ∞). Actually, from the definition of increasing and decreasing functions, the function shown in Figure 3.1 is decreasing on the interval $(-\infty, a]$ and increasing on the interval $[b, \infty)$. This text restricts the discussion to finding *open* intervals on which a function is increasing or decreasing.

The derivative of a function can be used to determine whether the function is increasing or decreasing on an interval.

Test for Increasing and Decreasing Functions

Let f be differentiable on the interval (a, b).

1. If $f'(x) > 0$ for all x in (a, b), then f is increasing on (a, b).

2. If $f'(x) < 0$ for all x in (a, b), then f is decreasing on (a, b).

3. If $f'(x) = 0$ for all x in (a, b), then f is constant on (a, b).

STUDY TIP

The conclusions in the first two cases of testing for increasing and decreasing functions are valid even if $f'(x) = 0$ at a finite number of x-values in (a, b).

| EXAMPLE 1 | **Testing for Increasing and Decreasing Functions** |

Show that the function

$$f(x) = x^2$$

is decreasing on the open interval $(-\infty, 0)$ and increasing on the open interval $(0, \infty)$.

SOLUTION The derivative of f is

$$f'(x) = 2x.$$

On the open interval $(-\infty, 0)$, the fact that x is negative implies that $f'(x) = 2x$ is also negative. So, by the test for a decreasing function, you can conclude that f is *decreasing* on this interval. Similarly, on the open interval $(0, \infty)$, the fact that x is positive implies that $f'(x) = 2x$ is also positive. So, it follows that f is *increasing* on this interval, as shown in Figure 3.2.

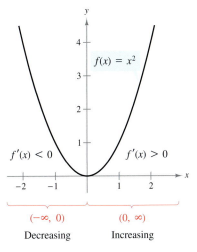

FIGURE 3.2

TRY IT 1

Show that the function $f(x) = x^4$ is decreasing on the open interval $(-\infty, 0)$ and increasing on the open interval $(0, \infty)$.

| EXAMPLE 2 | **Modeling Consumption** | |

From 1993 through 2001, the consumption M of mozzarella cheese (in pounds per person per year) can be modeled by

$$M = 0.007t^2 + 0.15t + 7.1, \qquad 3 \le t \le 11$$

where $t = 3$ corresponds to 1993 (see Figure 3.3). Show that the consumption of mozzarella cheese was increasing from 1993 through 2001. *(Source: U.S. Department of Agriculture)*

SOLUTION The derivative of this model is $dM/dt = 0.014t + 0.15$. As long as t is positive, the derivative is also positive. So, the function is increasing, which implies that the consumption of mozzarella cheese was increasing from 1993 through 2001.

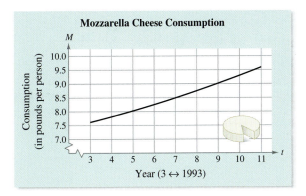

FIGURE 3.3

DISCOVERY

Use a graphing utility to graph $f(x) = 2 - x^2$ and $f'(x) = -2x$ in the same viewing window. On what interval is f increasing? On what interval is f' positive? Describe how the first derivative can be used to determine where a function is increasing and decreasing. Repeat this analysis for $g(x) = x^3 - x$ and $g'(x) = 3x^2 - 1$.

TRY IT 2

From 1990 through 1999, the consumption of bottled water (in gallons per person per year) can be modeled by

$$W = 0.099t^2 + 0.17t + 7.9, \qquad 0 \le t \le 9$$

where $t = 0$ corresponds to 1990. Show that the consumption of bottled water was increasing from 1990 to 1999. *(Source: U.S. Department of Agriculture)*

Critical Numbers and Their Use

In Example 1, you were given two intervals: one on which the function was decreasing and one on which it was increasing. Suppose you had been asked to determine these intervals. To do this, you could have used the fact that for a continuous function, $f'(x)$ can change signs only at x-values where $f'(x) = 0$ or at x-values where $f'(x)$ is undefined, as shown in Figure 3.4. These two types of numbers are called the **critical numbers** of f.

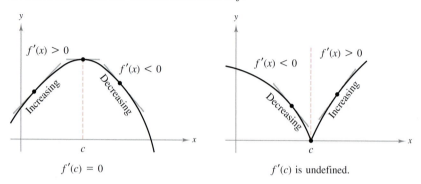

FIGURE 3.4

Definition of Critical Number

If f is defined at c, then c is a critical number of f if $f'(c) = 0$ or if f' is undefined at c.

STUDY TIP

This definition requires that a critical number be in the domain of the function. For example, $x = 0$ is not a critical number of the function $f(x) = 1/x$.

To determine the intervals on which a continuous function is increasing or decreasing, you can use the guidelines below.

Guidelines for Applying Increasing/Decreasing Test

1. Find the derivative of f.

2. Locate the critical numbers of f and use these numbers to determine test intervals. That is, find all x for which $f'(x) = 0$ or $f'(x)$ is undefined.

3. Test the sign of $f'(x)$ at an arbitrary number in each of the test intervals.

4. Use the test for increasing and decreasing functions to decide whether f is increasing or decreasing on each interval.

| **EXAMPLE 3** | **Finding Increasing and Decreasing Intervals** |

Find the open intervals on which the function is increasing or decreasing.

$$f(x) = x^3 - \frac{3}{2}x^2$$

SOLUTION   Begin by finding the derivative of f. Then set the derivative equal to zero and solve for the critical numbers.

$f'(x) = 3x^2 - 3x$	Differentiate original function.
$3x^2 - 3x = 0$	Set derivative equal to 0.
$3(x)(x - 1) = 0$	Factor.
$x = 0, x = 1$	Critical numbers

Because there are no x-values for which f' is undefined, it follows that $x = 0$ and $x = 1$ are the *only* critical numbers. So, the intervals that need to be tested are $(-\infty, 0)$, $(0, 1)$, and $(1, \infty)$. The table summarizes the testing of these three intervals.

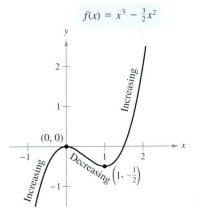

$f(x) = x^3 - \frac{3}{2}x^2$

Interval	$-\infty < x < 0$	$0 < x < 1$	$1 < x < \infty$
Test value	$x = -1$	$x = \frac{1}{2}$	$x = 2$
Sign of $f'(x)$	$f'(-1) = 6 > 0$	$f'(\frac{1}{2}) = -\frac{3}{4} < 0$	$f'(2) = 6 > 0$
Conclusion	Increasing	Decreasing	Increasing

The graph of f is shown in Figure 3.5. Note that the test values in the intervals were chosen for convenience—other x-values could have been used. ━━━ **FIGURE 3.5**

TRY IT 3

Find the open intervals on which the function $f(x) = x^3 - 12x$ is increasing or decreasing.

TECHNOLOGY

You can use the *trace* feature of a graphing utility to confirm the result of Example 3. Begin by graphing the function, as shown at the right. Then activate the *trace* feature and move the cursor from left to right. In intervals on which the function is increasing, note that the y-values increase as the x-values increase, whereas in intervals on which the function is decreasing, the y-values decrease as the x-values increase.

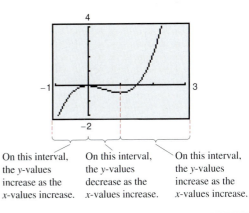

On this interval, the y-values increase as the x-values increase.   On this interval, the y-values decrease as the x-values increase.   On this interval, the y-values increase as the x-values increase.

Not only is the function in Example 3 continuous on the entire real line, it is also differentiable there. For such functions, the only critical numbers are those for which $f'(x) = 0$. The next example considers a continuous function that has *both* types of critical numbers—those for which $f'(x) = 0$ and those for which f' is undefined.

ALGEBRA REVIEW

For help on the algebra in Example 4, see Example 2(d) in the *Chapter 3 Algebra Review*, on page 249.

EXAMPLE 4 **Finding Increasing and Decreasing Intervals**

Find the open intervals on which the function

$$f(x) = (x^2 - 4)^{2/3}$$

is increasing or decreasing.

SOLUTION Begin by finding the derivative of the function.

$$f'(x) = \frac{2}{3}(x^2 - 4)^{-1/3}(2x) \qquad \text{Differentiate.}$$

$$= \frac{4x}{3(x^2 - 4)^{1/3}} \qquad \text{Simplify.}$$

From this, you can see that the derivative is zero when $x = 0$ and the derivative is undefined when $x = \pm 2$. So, the critical numbers are

$$x = -2, \quad x = 0, \quad \text{and} \quad x = 2. \qquad \text{Critical numbers}$$

This implies that the test intervals are

$$(-\infty, -2), \quad (-2, 0), \quad (0, 2), \quad \text{and} \quad (2, \infty). \qquad \text{Test intervals}$$

The table summarizes the testing of these four intervals, and the graph of the function is shown in Figure 3.6.

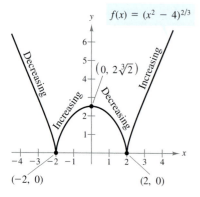

$f(x) = (x^2 - 4)^{2/3}$

$\left(0, 2\sqrt[3]{2}\right)$

$(-2, 0)$ $(2, 0)$

FIGURE 3.6

Interval	$-\infty < x < -2$	$-2 < x < 0$	$0 < x < 2$	$2 < x < \infty$
Test value	$x = -3$	$x = -1$	$x = 1$	$x = 3$
Sign of $f'(x)$	$f'(-3) < 0$	$f'(-1) > 0$	$f'(1) < 0$	$f'(3) > 0$
Conclusion	Decreasing	Increasing	Decreasing	Increasing

TRY IT 4

Find the open intervals on which the function $f(x) = x^{2/3}$ is increasing or decreasing.

ALGEBRA REVIEW

To test the intervals in the table, it is not necessary to *evaluate* $f'(x)$ at each test value—you only need to determine its sign. For example, you can determine the sign of $f'(-3)$ as shown.

$$f'(-3) = \frac{4(-3)}{3(9 - 4)^{1/3}} = \frac{\text{negative}}{\text{positive}} = \text{negative}$$

The functions in Examples 1 through 4 are continuous on the entire real line. If there are isolated x-values at which a function is not continuous, then these x-values should be used along with the critical numbers to determine the test intervals. For example, the function

$$f(x) = \frac{x^4 + 1}{x^2}$$

is not continuous when $x = 0$. Because the derivative of f

$$f'(x) = \frac{2(x^4 - 1)}{x^3}$$

is zero when $x = \pm 1$, you should use the following numbers to determine the test intervals.

$x = -1, x = 1$ Critical numbers

$x = 0$ Discontinuity

After testing $f'(x)$, you can determine that the function is decreasing on the intervals $(-\infty, -1)$ and $(0, 1)$, and increasing on the intervals $(-1, 0)$ and $(1, \infty)$, as shown in Figure 3.7.

The converse of the test for increasing and decreasing functions is *not* true. For instance, it is possible for a function to be increasing on an interval even though its derivative is not positive at every point in the interval.

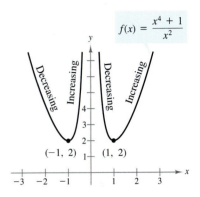

FIGURE 3.7

EXAMPLE 5   **Testing an Increasing Function**

Show that

$$f(x) = x^3 - 3x^2 + 3x$$

is increasing on the entire real line.

SOLUTION From the derivative of f

$$f'(x) = 3x^2 - 6x + 3 = 3(x - 1)^2$$

you can see that the only critical number is $x = 1$. So, the test intervals are $(-\infty, 1)$ and $(1, \infty)$. The table summarizes the testing of these two intervals. From Figure 3.8, you can see that f is increasing on the entire real line, even though $f'(1) = 0$. To convince yourself of this, look back at the definition of an increasing function.

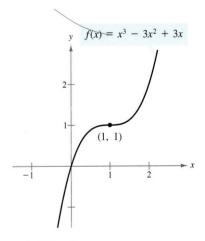

Interval	$-\infty < x < 1$	$1 < x < \infty$
Test value	$x = 0$	$x = 2$
Sign of $f'(x)$	$f'(0) = 3(-1)^2 > 0$	$f'(2) = 3(1)^2 > 0$
Conclusion	Increasing	Increasing

FIGURE 3.8

TRY IT 5

Show that $f(x) = -x^3 + 2$ is decreasing on the entire real line.

Application

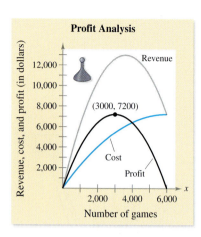

Profit Analysis

Revenue, cost, and profit (in dollars)

12,000
10,000
8,000
6,000
4,000
2,000

Revenue

(3000, 7200)

Cost

Profit

2,000   4,000   6,000 x

Number of games

FIGURE 3.9

EXAMPLE 6   Profit Analysis

A national toy distributor determines the cost and revenue models for one of its games.

$$C = 2.4x - 0.0002x^2, \quad 0 \le x \le 6000$$
$$R = 7.2x - 0.001x^2, \quad 0 \le x \le 6000$$

Determine the interval on which the profit function is increasing.

SOLUTION   The profit for producing x games is

$$P = R - C$$
$$= (7.2x - 0.001x^2) - (2.4x - 0.0002x^2)$$
$$= 4.8x - 0.0008x^2.$$

To find the interval on which the profit is increasing, set the marginal profit P' equal to zero and solve for x.

$$P' = 4.8 - 0.0016x \qquad \text{Differentiate profit function.}$$
$$4.8 - 0.0016x = 0 \qquad \text{Set } P' \text{ equal to 0.}$$
$$-0.0016x = -4.8 \qquad \text{Subtract 4.8 from each side.}$$
$$x = \frac{-4.8}{-0.0016} \qquad \text{Divide each side by } -0.0016.$$
$$x = 3000 \text{ games} \qquad \text{Simplify.}$$

On the interval $(0, 3000)$, P' is positive and the profit is *increasing*. On the interval $(3000, 6000)$, P' is negative and the profit is *decreasing*. The graphs of the cost, revenue, and profit functions are shown in Figure 3.9.

TAKE ANOTHER LOOK

Comparing Cost, Revenue, and Profit

Use the models from Example 6 to answer the questions.

a. What is the demand function for the product described in the example?

b. What price would you set to obtain a maximum profit?

c. What price would you set to obtain a maximum revenue?

d. Why doesn't the maximum revenue occur at the same x-value as the maximum profit?

In Exercises 1–4, solve the equation.

1. $x^2 = 8x$

2. $15x = \dfrac{5}{8}x^2$

3. $\dfrac{x^2 - 25}{x^3} = 0$

4. $\dfrac{2x}{\sqrt{1 - x^2}} = 0$

In Exercises 5–8, find the domain of the expression.

5. $\dfrac{x + 3}{x - 3}$

6. $\dfrac{2}{\sqrt{1 - x}}$

7. $\dfrac{2x + 1}{x^2 - 3x - 10}$

8. $\dfrac{3x}{\sqrt{9 - 3x^2}}$

In Exercises 9–12, evaluate the expression when $x = -2, 0$, and 2.

9. $-2(x + 1)(x - 1)$

10. $4(2x + 1)(2x - 1)$

11. $\dfrac{2x + 1}{(x - 1)^2}$

12. $\dfrac{-2(x + 1)}{(x - 4)^2}$

EXERCISES 3.1

In Exercises 1–4, evaluate the derivative of the function at the indicated points on the graph.

1. $f(x) = \dfrac{x^2}{x^2 + 4}$

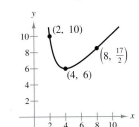

2. $f(x) = x + \dfrac{32}{x^2}$

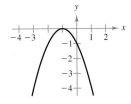

3. $f(x) = (x + 2)^{2/3}$

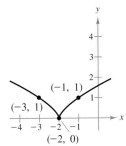

4. $f(x) = -3x\sqrt{x + 1}$

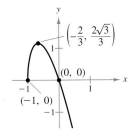

In Exercises 5–8, use the derivative to identify the open intervals on which the function is increasing or decreasing. Verify your result with the graph of the function.

5. $f(x) = -(x + 1)^2$

6. $f(x) = \dfrac{x^3}{4} - 3x$

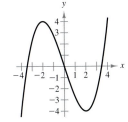

7. $f(x) = x^4 - 2x^2$

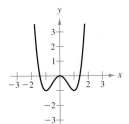

8. $f(x) = \dfrac{x^2}{x + 1}$

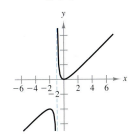

In Exercises 9–18, find the critical numbers and the open intervals on which the function is increasing or decreasing. Sketch the graph of the function.

9. $f(x) = 2x - 3$

10. $f(x) = 5 - 3x$

11. $g(x) = -(x - 1)^2$

12. $g(x) = (x + 2)^2$

13. $y = x^2 - 5x$

14. $y = -x^2 + 2x$

15. $y = x^3 - 6x^2$

16. $y = (x - 2)^3$

17. $f(x) = \sqrt{x^2 - 1}$

18. $f(x) = \sqrt{4 - x^2}$

 In Exercises 19–28, find the critical numbers and the open intervals on which the function is increasing or decreasing. Then use a graphing utility to graph the function.

19. $f(x) = -2x^2 + 4x + 3$

20. $f(x) = x^2 + 8x + 10$

21. $y = 3x^3 + 12x^2 + 15x$

22. $y = x^3 - 3x + 2$

23. $f(x) = x\sqrt{x + 1}$

24. $h(x) = x\sqrt[3]{x - 1}$

25. $f(x) = x^4 - 2x^3$

26. $f(x) = \frac{1}{4}x^4 - 2x^2$

27. $f(x) = \dfrac{x}{x^2 + 4}$

28. $f(x) = \dfrac{x^2}{x^2 + 4}$

In Exercises 29–34, find the critical numbers and the open intervals on which the function is increasing or decreasing. (*Hint:* Check for discontinuities.) Sketch the graph of the function.

29. $f(x) = \dfrac{2x}{16 - x^2}$

30. $f(x) = \dfrac{x}{x + 1}$

31. $y = \begin{cases} 4 - x^2, & x \le 0 \\ -2x, & x > 0 \end{cases}$

32. $y = \begin{cases} 2x + 1, & x \le -1 \\ x^2 - 2, & x > -1 \end{cases}$

33. $y = \begin{cases} 3x + 1, & x \le 1 \\ 5 - x^2, & x > 1 \end{cases}$

34. $y = \begin{cases} -x^3 + 1, & x \le 0 \\ -x^2 + 2x, & x > 0 \end{cases}$

 35. *Cost*   The ordering and transportation cost C (in hundreds of dollars) for an automobile dealership is modeled by

$$C = 10\left(\frac{1}{x} + \frac{x}{x + 3}\right), \qquad 1 \le x$$

where x is the number of automobiles ordered.

(a) Find the intervals on which C is increasing or decreasing.

(b) Use a graphing utility to graph the cost function.

(c) Use the *trace* feature to determine the order sizes for which the cost is $900. Assuming that the revenue function is increasing for $x \ge 0$, which order size would you use? Explain your reasoning.

 36. *Chemistry: Molecular Velocity*   Plots of the relative numbers of N_2 (nitrogen) molecules that have a given velocity at each of three temperatures (in degrees Kelvin) are shown in the figure. Identify the differences in the average velocities (indicated by the peaks of the curves) for the three temperatures, and describe the intervals on which the velocity is increasing and decreasing for each of the three temperatures. (*Source: Adapted from Zumdahl, Chemistry, Sixth Edition*)

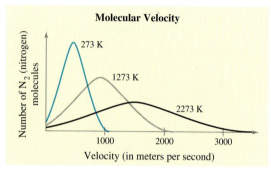

Molecular Velocity

Position Function   In Exercises 37 and 38, the position function gives the height s (in feet) of a ball, where the time t is measured in seconds. Find the time interval on which the ball is rising and the interval on which it is falling.

37. $s = 96t - 16t^2, \qquad 0 \le t \le 6$

38. $s = -16t^2 + 64t, \qquad 0 \le t \le 4$

39. *Law Degrees*   The number y of law degrees conferred in the United States from 1970 to 2000 can be modeled by

$$y = 2.743t^3 - 171.55t^2 + 3462.3t + 15,265,$$

$$0 \le t \le 30$$

where t is the time in years, with $t = 0$ corresponding to 1970. (*Source: U.S. National Center for Education Statistics*)

 (a) Use a graphing utility to graph the model. Then graphically estimate the years during which the model is increasing and the years during which it is decreasing.

(b) Use the test for increasing and decreasing functions to verify the result of part (a).

40. *Profit*   The profit P made by a cinema from selling x bags of popcorn can be modeled by

$$P = 2.36x - \frac{x^2}{25,000} - 3500, \qquad 0 \le x \le 50,000.$$

(a) Find the intervals on which P is increasing and decreasing.

(b) If you owned the cinema, what price would you charge to obtain a maximum profit for popcorn? Explain your reasoning.

3.2 EXTREMA AND THE FIRST-DERIVATIVE TEST

- Recognize the occurrence of relative extrema of functions.
- Use the First-Derivative Test to find the relative extrema of functions.
- Find absolute extrema of continuous functions on a closed interval.
- Find minimum and maximum values of real-life models and interpret the results in context.

Relative Extrema

You have used the derivative to determine the intervals on which a function is increasing or decreasing. In this section, you will examine the points at which a function changes from increasing to decreasing, or vice versa. At such a point, the function has a **relative extremum.** (The plural of extremum is *extrema.*) The **relative extrema** of a function include the **relative minima** and **relative maxima** of the function. For instance, the function shown in Figure 3.10 has two relative extrema—the left point is a relative maximum and the right point is a relative minimum.

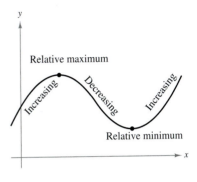

FIGURE 3.10

Definition of Relative Extrema

Let f be a function defined at c.

1. $f(c)$ is a **relative maximum** of f if there exists an interval (a, b) containing c such that $f(x) \leq f(c)$ for all x in (a, b).

2. $f(c)$ is a **relative minimum** of f if there exists an interval (a, b) containing c such that $f(x) \geq f(c)$ for all x in (a, b).

If $f(c)$ is a relative extremum of f, then the relative extremum is said to *occur* at $x = c$.

For a continuous function, the relative extrema must occur at critical numbers of the function, as shown in Figure 3.11.

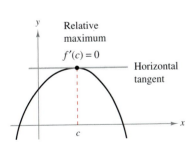

FIGURE 3.11

Occurrence of Relative Extrema

If f has a relative minimum or relative maximum when $x = c$, then c is a critical number of f. That is, either $f'(c) = 0$ or $f'(c)$ is undefined.

The First-Derivative Test

The discussion on the preceding page implies that in your search for relative extrema of a continuous function, you only need to test the critical numbers of the function. Once you have determined that c is a critical number of a function f, the **First-Derivative Test** for relative extrema enables you to classify $f(c)$ as a relative minimum, a relative maximum, or neither.

First-Derivative Test for Relative Extrema

Let f be continuous on the interval (a, b) in which c is the only critical number. If f is differentiable on the interval (except possibly at c), then $f(c)$ can be classified as a relative minimum, a relative maximum, or neither, as shown.

1. On the interval (a, b), if $f'(x)$ is negative to the left of $x = c$ and positive to the right of $x = c$, then $f(c)$ is a relative minimum.

2. On the interval (a, b), if $f'(x)$ is positive to the left of $x = c$ and negative to the right of $x = c$, then $f(c)$ is a relative maximum.

3. On the interval (a, b), if $f'(x)$ has the same sign to the left and right of $x = c$, then $f(c)$ is not a relative extremum of f.

A graphical interpretation of the First-Derivative Test is shown in Figure 3.12.

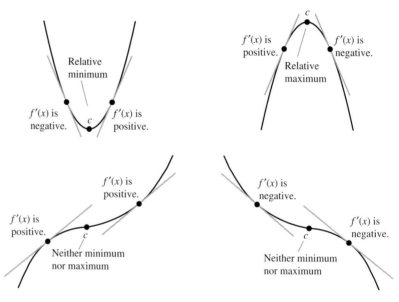

FIGURE 3.12

EXAMPLE 1 **Finding Relative Extrema**

Find all relative extrema of the function

$$f(x) = 2x^3 - 3x^2 - 36x + 14.$$

SOLUTION   Begin by finding the critical numbers of f.

$f'(x) = 6x^2 - 6x - 36$	Find derivative of f.
$6x^2 - 6x - 36 = 0$	Set derivative equal to 0.
$6(x^2 - x - 6) = 0$	Factor out common factor.
$6(x - 3)(x + 2) = 0$	Factor.
$x = -2, x = 3$	Critical numbers

Because $f'(x)$ is defined for all x, the only critical numbers of f are $x = -2$ and $x = 3$. Using these numbers, you can form the three test intervals $(-\infty, -2)$, $(-2, 3)$, and $(3, \infty)$. The testing of the three intervals is shown in the table.

Interval	$-\infty < x < -2$	$-2 < x < 3$	$3 < x < \infty$
Test value	$x = -3$	$x = 0$	$x = 4$
Sign of $f'(x)$	$f'(-3) = 36 > 0$	$f'(0) = -36 < 0$	$f'(4) = 36 > 0$
Conclusion	Increasing	Decreasing	Increasing

Using the First-Derivative Test, you can conclude that the critical number -2 yields a relative maximum [$f'(x)$ changes sign from positive to negative], and the critical number 3 yields a relative minimum [$f'(x)$ changes sign from negative to positive].

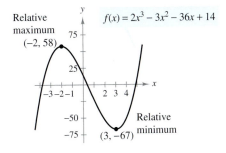

FIGURE 3.13

The graph of f is shown in Figure 3.13. To find the y-coordinates of the relative extrema, substitute the x-coordinates into the function. For instance, the relative maximum is $f(-2) = 58$ and the relative minimum is $f(3) = -67$. ———

STUDY TIP

In Section 2.2, Example 8, you examined the graph of the function $f(x) = x^3 - 4x + 2$ and discovered that it does *not* have a relative minimum at the point $(1, -1)$. Try using the First-Derivative Test to find the point at which the graph *does* have a relative minimum.

TRY IT 1

Find all relative extrema of $f(x) = 2x^3 - 6x + 1$. Sketch a graph of the function and label the relative extrema.

In Example 1, both critical numbers yielded relative extrema. In the next example, only one of the two critical numbers yields a relative extremum.

ALGEBRA REVIEW

For help on the algebra in Example 2, see Example 2(c) in the *Chapter 3 Algebra Review*, on page 249.

EXAMPLE 2   **Finding Relative Extrema**

Find all relative extrema of the function $f(x) = x^4 - x^3$.

SOLUTION   From the derivative of the function

$$f'(x) = 4x^3 - 3x^2 = x^2(4x - 3)$$

you can see that the function has only two critical numbers: $x = 0$ and $x = \frac{3}{4}$. These numbers produce the test intervals $(-\infty, 0)$, $\left(0, \frac{3}{4}\right)$, and $\left(\frac{3}{4}, \infty\right)$, which are tested in the table.

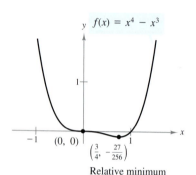

$f(x) = x^4 - x^3$

$(0, 0)$

$\left(\frac{3}{4}, -\frac{27}{256}\right)$

Relative minimum

FIGURE 3.14

Interval	$-\infty < x < 0$	$0 < x < \frac{3}{4}$	$\frac{3}{4} < x < \infty$
Test value	$x = -1$	$x = \frac{1}{2}$	$x = 1$
Sign of $f'(x)$	$f'(-1) = -7 < 0$	$f'\left(\frac{1}{2}\right) = -\frac{1}{4} < 0$	$f'(1) = 1 > 0$
Conclusion	Decreasing	Decreasing	Increasing

By the First-Derivative Test, it follows that f has a relative minimum when $x = \frac{3}{4}$, as shown in Figure 3.14. Note that the critical number $x = 0$ does not yield a relative extremum.

TRY IT 2

Find all relative extrema of $f(x) = x^4 - 4x^3$.

EXAMPLE 3   **Finding Relative Extrema**

Find all relative extrema of the function

$$f(x) = 2x - 3x^{2/3}.$$

SOLUTION   From the derivative of the function

$$f'(x) = 2 - \frac{2}{x^{1/3}} = \frac{2(x^{1/3} - 1)}{x^{1/3}}$$

you can see that $f'(1) = 0$ and f' is undefined at $x = 0$. So, the function has two critical numbers: $x = 1$ and $x = 0$. These numbers produce the test intervals $(-\infty, 0)$, $(0, 1)$, and $(1, \infty)$. By testing these intervals, you can conclude that f has a relative maximum at $(0, 0)$ and a relative minimum at $(1, -1)$, as shown in Figure 3.15.

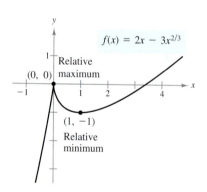

$f(x) = 2x - 3x^{2/3}$

$(0, 0)$ Relative maximum

$(1, -1)$ Relative minimum

FIGURE 3.15

TRY IT 3

Find all relative extrema of $f(x) = 3x^{2/3} - 2x$.

TECHNOLOGY

Finding Relative Extrema

There are several ways to use technology to find relative extrema of a function. One way is to use a graphing utility to graph the function, and then use the *zoom* and *trace* features to find the relative minimum and relative maximum points. For instance, consider the graph of

$$f(x) = 3.1x^3 - 7.3x^2 + 1.2x + 2.5,$$

as shown below.

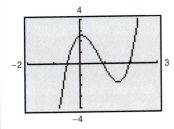

From the graph, you can see that the function has one relative maximum and one relative minimum. You can approximate the coordinates of these points by zooming in and using the *trace* feature, as shown below.

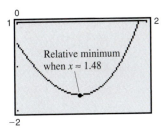

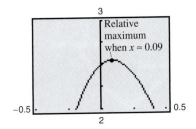

A second way to use technology to find relative extrema is to perform the First-Derivative Test with a symbolic differentiation utility. You can use the utility to differentiate the function, set the derivative equal to zero, and then solve the resulting equation. After obtaining the critical numbers, 1.48287 and 0.0870148, you can graph the function and observe that the first yields a relative minimum and the second yields a relative maximum.

$$f(x) = 3.1x^3 - 7.3x^2 + 1.2x + 2.5 \qquad \text{Write original function.}$$

$$f'(x) = \frac{d}{dx}[3.1x^3 - 7.3x^2 + 1.2x + 2.5] \qquad \text{Differentiate with respect to } x.$$

$$f'(x) = 9.3x^2 - 14.6x + 1.2 \qquad \text{First derivative}$$

$$9.3x^2 - 14.6x + 1.2 = 0 \qquad \text{Set derivative equal to 0.}$$

$$x = \frac{73 \pm \sqrt{4213}}{93} \qquad \text{Solve for } x.$$

$$x \approx 1.48288, x \approx 0.0870148 \qquad \text{Approximate.}$$

STUDY TIP

Some graphing calculators have a special feature that allows you to find the minimum or maximum of a function on an interval. Consult the user's manual for information on the *minimum value* and *maximum value* features of your graphing utility.

Absolute Extrema

The terms *relative minimum* and *relative maximum* describe the *local* behavior of a function. To describe the *global* behavior of the function on an entire interval, you can use the terms **absolute maximum** and **absolute minimum.**

Definition of Absolute Extrema

Let f be defined on an interval I containing c.

1. $f(c)$ is an **absolute minimum of f** on I if $f(c) \leq f(x)$ for every x in I.

2. $f(c)$ is an **absolute maximum of f** on I if $f(c) \geq f(x)$ for every x in I.

The absolute minimum and absolute maximum values of a function on an interval are sometimes simply called the **minimum** and **maximum** of f on I.

Be sure that you understand the distinction between relative extrema and absolute extrema. For instance, in Figure 3.16, the function has a relative minimum that also happens to be an absolute minimum on the interval $[a, b]$. The relative maximum of f, however, is not the absolute maximum on the interval $[a, b]$. The next theorem points out that if a continuous function has a closed interval as its domain, then it *must* have both an absolute minimum and an absolute maximum on the interval. From Figure 3.16, note that these extrema can occur at endpoints of the interval.

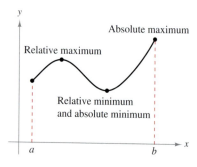

FIGURE 3.16

Extreme Value Theorem

If f is continuous on $[a, b]$, then f has both a minimum value and a maximum value on $[a, b]$.

Although a continuous function has just one minimum and one maximum value on a closed interval, either of these values can occur for more than one x-value. For instance, on the interval $[-3, 3]$, the function $f(x) = 9 - x^2$ has a minimum value of zero when $x = -3$ *and* when $x = 3$, as shown in Figure 3.17.

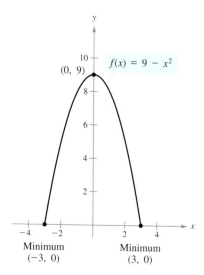

$f(x) = 9 - x^2$

Minimum
$(-3, 0)$

Minimum
$(3, 0)$

FIGURE 3.17

When looking for the extreme value of a function on a *closed* interval, remember that you must consider the values of the function at the endpoints as well as at the critical numbers of the function. You can use the guidelines below to find extrema on a closed interval.

Guidelines for Finding Extrema on a Closed Interval

To find the extrema of a continuous function f on a closed interval $[a, b]$, use the steps below.

1. Evaluate f at each of its critical numbers in (a, b).

2. Evaluate f at each endpoint, a and b.

3. The least of these values is the minimum, and the greatest is the maximum.

TECHNOLOGY

A graphing utility can help you locate the extrema of a function on a closed interval. For instance, try using a graphing utility to confirm the results of Example 4. (Set the viewing window to $-1 \le x \le 6$ and $-8 \le y \le 4$.) Use the *trace* feature to check that the minimum y-value occurs when $x = 3$ and the maximum y-value occurs when $x = 0$.

EXAMPLE 4   **Finding Extrema on a Closed Interval**

Find the minimum and maximum values of

$$f(x) = x^2 - 6x + 2$$

on the interval $[0, 5]$.

SOLUTION   Begin by finding the critical numbers of the function.

$f'(x) = 2x - 6$	Find derivative of f.
$2x - 6 = 0$	Set derivative equal to 0.
$2x = 6$	Add 6 to each side.
$x = 3$	Solve for x.

From this, you can see that the only critical number of f is $x = 3$. Because this number lies in the interval under question, you should test the values of $f(x)$ at this number *and* at the endpoints of the interval, as shown in the table.

x-value	Endpoint: $x = 0$	Critical number: $x = 3$	Endpoint: $x = 5$
$f(x)$	$f(0) = 2$	$f(3) = -7$	$f(5) = -3$
Conclusion	Maximum	Minimum	Neither maximum nor minimum

From the table, you can see that the minimum of f on the interval $[0, 5]$ is $f(3) = -7$. Moreover, the maximum of f on the interval $[0, 5]$ is $f(0) = 2$. This is confirmed by the graph of f, as shown in Figure 3.18.

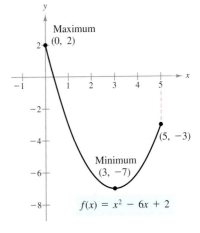

FIGURE 3.18

TRY IT 4

Find the minimum and maximum values of $f(x) = x^2 - 8x + 10$ on the interval $[0, 7]$. Sketch the graph of $f(x)$ and label the minimum and maximum values.

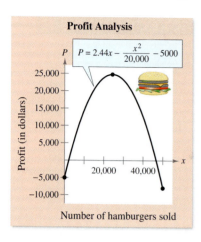

Profit Analysis

$$P = 2.44x - \frac{x^2}{20,000} - 5000$$

FIGURE 3.19

Number of hamburgers sold

TRY IT 5

Verify the results of Example 5 by completing the table.

x (units)	24,000	24,200	24,300
P (profit)			

x (units)	24,400	24,500	24,600
P (profit)			

x (units)	24,800	25,000
P (profit)		

Applications of Extrema

Finding the minimum and maximum values of a function is one of the most common applications of calculus.

EXAMPLE 5   **Finding the Maximum Profit**

Recall the fast-food restaurant in Examples 7 and 8 in Section 2.3. The restaurant's profit function for hamburgers is given by

$$P = 2.44x - \frac{x^2}{20,000} - 5000, \quad 0 \le x \le 50,000.$$

Find the production level that produces a maximum profit.

SOLUTION   To begin, find an equation for marginal profit. Then set the marginal profit equal to zero and solve for x.

$$P' = 2.44 - \frac{x}{10,000} \qquad \text{Find marginal profit.}$$

$$2.44 - \frac{x}{10,000} = 0 \qquad \text{Set marginal profit equal to 0.}$$

$$-\frac{x}{10,000} = -2.44 \qquad \text{Subtract 2.44 from each side.}$$

$$x = 24,400 \text{ hamburgers} \qquad \text{Critical number}$$

From Figure 3.19, you can see that the critical number $x = 24,400$ corresponds to the production level that produces a maximum profit. To find the maximum profit, substitute $x = 24,400$ into the profit function.

$$P = 2.44x - \frac{x^2}{20,000} - 5000$$

$$= 2.44(24,400) - \frac{(24,400)^2}{20,000} - 5000$$

$$= \$24,768$$

TAKE ANOTHER LOOK

Setting the Price of a Product

In Example 5, you discovered that a production level of 24,400 hamburgers corresponds to a maximum profit. Remember that this model assumes that the quantity demanded and the price per unit are related by a demand function. So, the only way to sell more hamburgers is to lower the price, and consequently lower the profit. What is the demand function for Example 5? What is the price per unit that produces a maximum profit?

PREREQUISITE REVIEW 3.2

The following warm-up exercises involve skills that were covered in earlier sections. You will use these skills in the exercise set for this section.

In Exercises 1–6, solve the equation $f'(x) = 0$.

1. $f(x) = 4x^4 - 2x^2 + 1$

2. $f(x) = \frac{1}{3}x^3 - \frac{3}{2}x^2 - 10x$

3. $f(x) = 5x^{4/5} - 4x$

4. $f(x) = \frac{1}{2}x^2 - 3x^{5/3}$

5. $f(x) = \dfrac{x+4}{x^2+1}$

6. $f(x) = \dfrac{x-1}{x^2+4}$

In Exercises 7–10, use $g(x) = -x^5 - 2x^4 + 4x^3 + 2x - 1$ to determine the sign of the derivative.

7. $g'(-4)$

8. $g'(0)$

9. $g'(1)$

10. $g'(3)$

In Exercises 11 and 12, decide whether the function is increasing or decreasing on the given interval.

11. $f(x) = 2x^2 - 11x - 6$, $(3, 6)$

12. $f(x) = x^3 + 2x^2 - 4x - 8$, $(-2, 0)$

EXERCISES 3.2

In Exercises 1–4, use a table similar to that in Example 1 to find all relative extrema of the function.

1. $f(x) = -2x^2 + 4x + 3$

2. $f(x) = x^2 + 8x + 10$

3. $f(x) = x^2 - 6x$

4. $f(x) = -4x^2 + 4x + 1$

In Exercises 5–12, find all relative extrema of the function.

5. $g(x) = 6x^3 - 15x^2 + 12x$

6. $g(x) = \frac{1}{5}x^5 - x$

7. $h(x) = -(x + 4)^3$

8. $h(x) = 2(x - 3)^3$

9. $f(x) = x^3 - 6x^2 + 15$

10. $f(x) = x^4 - 32x + 4$

11. $f(x) = x^4 - 2x^3$

12. $f(x) = x^4 - 12x^3$

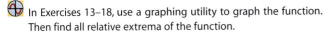 In Exercises 13–18, use a graphing utility to graph the function. Then find all relative extrema of the function.

13. $f(x) = (x - 1)^{2/3}$

14. $f(t) = (t - 1)^{1/3}$

15. $g(t) = t - \dfrac{1}{2t^2}$

16. $f(x) = x + \dfrac{1}{x}$

17. $f(x) = \dfrac{x}{x + 1}$

18. $h(x) = \dfrac{4}{x^2 + 1}$

In Exercises 19–26, find the absolute extrema of the function on the closed interval.

Function	Interval
19. $f(x) = 2(3 - x)$	$[-1, 2]$
20. $f(x) = \frac{1}{3}(2x + 5)$	$[0, 5]$
21. $f(x) = 5 - 2x^2$	$[0, 3]$
22. $f(x) = x^2 + 2x - 4$	$[-1, 1]$
23. $f(x) = x^3 - 3x^2$	$[-1, 3]$
24. $f(x) = x^3 - 12x$	$[0, 4]$
25. $h(s) = \dfrac{1}{3 - s}$	$[0, 2]$
26. $h(t) = \dfrac{t}{t - 2}$	$[3, 5]$

In Exercises 27–30, find the absolute extrema of the function on the closed interval. Use a graphing utility to verify your results.

Function	Interval
27. $f(x) = 3x^{2/3} - 2x$	$[-1, 2]$
28. $g(t) = \dfrac{t^2}{t^2 + 3}$	$[-1, 1]$
29. $h(t) = (t - 1)^{2/3}$	$[-7, 2]$
30. $g(x) = 4\left(1 + \dfrac{1}{x} + \dfrac{1}{x^2}\right)$	$[-4, 5]$

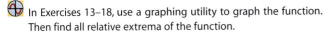 In Exercises 31–34, use a graphing utility to find graphically the absolute extrema of the function on the closed interval.

Function	Interval
31. $f(x) = 0.4x^3 - 1.8x^2 + x - 3$	$[0, 5]$
32. $f(x) = 3.2x^5 + 5x^3 - 3.5x$	$[0, 1]$
33. $f(x) = \frac{4}{3}x\sqrt{3 - x}$	$[0, 3]$
34. $f(x) = 4\sqrt{x} - 2x + 1$	$[0, 6]$

For a *continuous* function f, you can find the intervals on which the graph of f is concave upward and concave downward as follows. [For a function that is not continuous, the test intervals should be formed using points of discontinuity, along with the points at which $f''(x)$ is zero or undefined.]

Guidelines for Applying Concavity Test

1. Locate the x-values at which $f''(x) = 0$ or $f''(x)$ is undefined.

2. Use these x-values to determine the test intervals.

3. Test the sign of $f''(x)$ in each test interval.

EXAMPLE 1 **Applying the Test for Concavity**

(a) The graph of the function

$$f(x) = x^2 \qquad \text{Original function}$$

is concave upward on the entire real line because its second derivative

$$f''(x) = 2 \qquad \text{Second derivative}$$

is positive for all x. (See Figure 3.21.)

(b) The graph of the function

$$f(x) = \sqrt{x} \qquad \text{Original function}$$

is concave downward for $x > 0$ because its second derivative

$$f''(x) = -\frac{1}{4}x^{-3/2} \qquad \text{Second derivative}$$

is negative for all $x > 0$. (See Figure 3.22.)

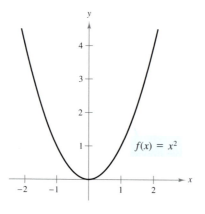

FIGURE 3.21 Concave Upward

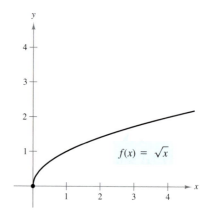

FIGURE 3.22 Concave Downward

EXAMPLE 2 | **Determining Concavity**

Determine the intervals on which the graph of the function is concave upward or concave downward.

$$f(x) = \frac{6}{x^2 + 3}$$

SOLUTION Begin by finding the second derivative of f.

$$f(x) = 6(x^2 + 3)^{-1}$$ Rewrite original function.

$$f'(x) = (-6)(2x)(x^2 + 3)^{-2}$$ Chain Rule

$$= \frac{-12x}{(x^2 + 3)^2}$$ Simplify.

$$f''(x) = \frac{(x^2 + 3)^2(-12) - (-12x)(2)(2x)(x^2 + 3)}{(x^2 + 3)^4}$$ Quotient Rule

$$= \frac{-12(x^2 + 3) + (48x^2)}{(x^2 + 3)^3}$$ Simplify.

$$= \frac{36(x^2 - 1)}{(x^2 + 3)^3}$$ Simplify.

From this, you can see that $f''(x)$ is defined for all real numbers and $f''(x) = 0$ when $x = \pm 1$. So, you can test the concavity of f by testing the intervals $(-\infty, -1)$, $(-1, 1)$, and $(1, \infty)$, as shown in the table. The graph of f is shown in Figure 3.23.

Interval	$-\infty < x < -1$	$-1 < x < 1$	$1 < x < \infty$
Test value	$x = -2$	$x = 0$	$x = 2$
Sign of $f''(x)$	$f''(-2) > 0$	$f''(0) < 0$	$f''(2) > 0$
Conclusion	Concave upward	Concave downward	Concave upward

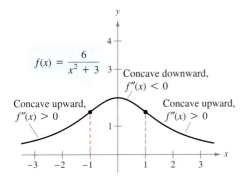

$$f(x) = \frac{6}{x^2 + 3}$$

Concave downward, $f''(x) < 0$

Concave upward, $f''(x) > 0$

Concave upward, $f''(x) > 0$

FIGURE 3.23

ALGEBRA REVIEW

For help on the algebra in Example 2, see Example 1(a) in the *Chapter 3 Algebra Review*, on page 248.

STUDY TIP

In Example 2, f' is increasing on the interval $(1, \infty)$ even though f is decreasing there. Be sure you see that the increasing or decreasing of f' does not necessarily correspond to the increasing or decreasing of f.

TRY IT 2

Determine the intervals on which the graph of the function is concave upward and concave downward.

$$f(x) = \frac{24}{x^2 + 12}$$

The following warm-up exercises involve skills that were covered in earlier sections. You will use these skills in the exercise set for this section.

In Exercises 1–6, find the second derivative of the function.

1. $f(x) = 4x^4 - 9x^3 + 5x - 1$

2. $g(s) = (s^2 - 1)(s^2 - 3s + 2)$

3. $g(x) = (x^2 + 1)^4$

4. $f(x) = (x - 3)^{4/3}$

5. $h(x) = \dfrac{4x + 3}{5x - 1}$

6. $f(x) = \dfrac{2x - 1}{3x + 2}$

In Exercises 7–10, find the critical numbers of the function.

7. $f(x) = 5x^3 - 5x + 11$

8. $f(x) = x^4 - 4x^3 - 10$

9. $g(t) = \dfrac{16 + t^2}{t}$

10. $h(x) = \dfrac{x^4 - 50x^2}{8}$

EXERCISES 3.3

In Exercises 1–8, analytically find the intervals on which the graph is concave upward and those on which it is concave downward. Verify your results using the graph of the function.

1. $y = x^2 - x - 2$

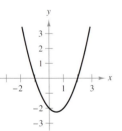

2. $y = -x^3 + 3x^2 - 2$

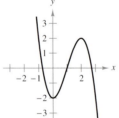

3. $f(x) = \dfrac{x^2 - 1}{2x + 1}$

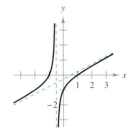

4. $f(x) = \dfrac{x^2 + 4}{4 - x^2}$

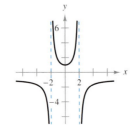

5. $f(x) = \dfrac{24}{x^2 + 12}$

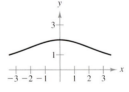

6. $f(x) = \dfrac{x^2}{x^2 + 1}$

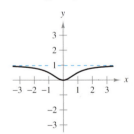

7. $y = -x^3 + 6x^2 - 9x - 1$

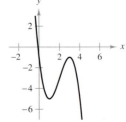

8. $y = x^5 + 5x^4 - 40x^2$

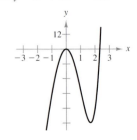

In Exercises 9–18, find all relative extrema of the function. Use the Second-Derivative Test when applicable.

9. $f(x) = 6x - x^2$

10. $f(x) = (x - 5)^2$

11. $f(x) = x^3 - 5x^2 + 7x$

12. $f(x) = x^4 - 4x^3 + 2$

13. $f(x) = x^{2/3} - 3$

14. $f(x) = x + \dfrac{4}{x}$

15. $f(x) = \sqrt{x^2 + 1}$

16. $f(x) = \sqrt{4 - x^2}$

17. $f(x) = \dfrac{x}{x - 1}$

18. $f(x) = \dfrac{x}{x^2 - 1}$

 In Exercises 19–22, use a graphing utility to estimate graphically all relative extrema of the function.

19. $f(x) = \frac{1}{2}x^4 - \frac{1}{3}x^3 - \frac{1}{2}x^2$

20. $f(x) = -\frac{1}{3}x^5 - \frac{1}{2}x^4 + x$

21. $f(x) = 5 + 3x^2 - x^3$

22. $f(x) = 3x^3 + 5x^2 - 2$

In Exercises 23–26, state the signs of $f'(x)$ and $f''(x)$ on the interval $(0, 2)$.

23.
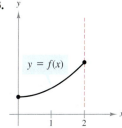

24.
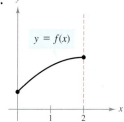

25.
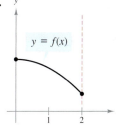

26.
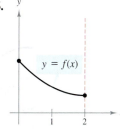

In Exercises 27–34, find the point(s) of inflection of the graph of the function.

27. $f(x) = x^3 - 9x^2 + 24x - 18$

28. $f(x) = x(6 - x)^2$

29. $f(x) = (x - 1)^3(x - 5)$

30. $f(x) = x^4 - 18x^2 + 5$

31. $g(x) = 2x^4 - 8x^3 + 12x^2 + 12x$

32. $f(x) = -4x^3 - 8x^2 + 32$

33. $h(x) = (x - 2)^3(x - 1)$

34. $f(t) = (1 - t)(t - 4)(t^2 - 4)$

 In Exercises 35–46, use a graphing utility to graph the function and identify all relative extrema and points of inflection.

35. $f(x) = x^3 - 12x$

36. $f(x) = x^3 - 3x$

37. $f(x) = x^3 - 6x^2 + 12x$

38. $f(x) = x^3 - \frac{3}{2}x^2 - 6x$

39. $f(x) = \frac{1}{4}x^4 - 2x^2$

40. $f(x) = 2x^4 - 8x + 3$

41. $g(x) = (x - 2)(x + 1)^2$

42. $g(x) = (x - 6)(x + 2)^3$

43. $g(x) = x\sqrt{x + 3}$

44. $g(x) = x\sqrt{9 - x}$

45. $f(x) = \dfrac{4}{1 + x^2}$

46. $f(x) = \dfrac{2}{x^2 - 1}$

In Exercises 47 and 48, sketch a graph of a function f having the given characteristics.

Function	First Derivative	Second Derivative
47. $f(2) = 0$	$f'(x) < 0, \ x < 3$	$f''(x) > 0$
$f(4) = 0$	$f'(3) = 0$	
	$f'(x) > 0, \ x > 3$	
48. $f(2) = 0$	$f'(x) > 0, \ x < 3$	$f''(x) > 0, \ x \neq 3$
$f(4) = 0$	$f'(3)$ is undefined.	
	$f'(x) < 0, \ x > 3$	

In Exercises 49 and 50, use the graph to sketch the graph of f'. Find the intervals on which (a) $f'(x)$ is positive, (b) $f'(x)$ is negative, (c) f' is increasing, and (d) f' is decreasing. For each of these intervals, describe the corresponding behavior of f.

49.
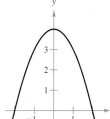

50.
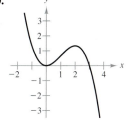

Point of Diminishing Returns In Exercises 51 and 52, identify the point of diminishing returns for the input-output function. For each function, R is the revenue and x is the amount spent on advertising. Use a graphing utility to verify your results.

51. $R = \dfrac{1}{50,000}(600x^2 - x^3), \quad 0 \leq x \leq 400$

52. $R = -\frac{4}{9}(x^3 - 9x^2 - 27), \quad 0 \leq x \leq 5$

Average Cost In Exercises 53 and 54, you are given the total cost of producing x units. Find the production level that minimizes the average cost per unit. Use a graphing utility to verify your results.

53. $C = 0.5x^2 + 15x + 5000$

54. $C = 0.002x^3 + 20x + 500$

Productivity   In Exercises 55 and 56, consider a college student who works from 7 P.M. to 11 P.M. assembling mechanical components. The number N of components assembled after t hours is given by the function. At what time is the student assembling components at the greatest rate?

55. $N = -0.12t^3 + 0.54t^2 + 8.22t,$ $0 \leq t \leq 4$

56. $N = \dfrac{20t^2}{4 + t^2},$ $0 \leq t \leq 4$

Sales Growth   In Exercises 57 and 58, find the time t in years when the annual sales x of a new product are increasing at the greatest rate. Use a graphing utility to verify your results.

57. $x = \dfrac{10,000t^2}{9 + t^2}$ **58.** $x = \dfrac{500,000t^2}{36 + t^2}$

In Exercises 59–62, use a graphing utility to graph f, f', and f'' in the same viewing window. Graphically locate the relative extrema and points of inflection of the graph.

Function	Interval
59. $f(x) = \frac{1}{2}x^3 - x^2 + 3x - 5$	$[0, 3]$
60. $f(x) = -\frac{1}{20}x^5 - \frac{1}{12}x^2 - \frac{1}{3}x + 1$	$[-2, 2]$
61. $f(x) = \dfrac{2}{x^2 + 1}$	$[-3, 3]$
62. $f(x) = \dfrac{x^2}{x^2 + 1}$	$[-3, 3]$

63. *Dow Jones Industrial Average*   The graph shows the Dow Jones Industrial Average y on Black Monday, October 19, 1987, where $t = 0$ corresponds to 8:30 A.M., when the market opens, and $t = 7.5$ corresponds to 4 P.M., the closing time. *(Source: Wall Street Journal)*

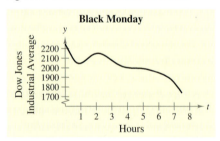

Black Monday

(a) Estimate the relative extrema and absolute extrema of the graph. Interpret your results in the context of the problem.

(b) Estimate the point of inflection of the graph on the interval $[3, 6]$. Interpret your result in the context of the problem.

64. *Think About It*   Let S represent monthly sales of a new model of MP3 player. Write a statement describing S' and S'' for each of the following.

(a) The rate of change of sales is increasing.

(b) Sales are increasing, but at a greater rate.

(c) The rate of change of sales is steady.

(d) Sales are steady.

(e) Sales are declining, but at a lower rate.

(f) Sales have bottomed out and have begun to rise.

65. *Medicine*   The spread of a virus can be modeled by

$$N = -t^3 + 12t^2,   0 \leq t \leq 12$$

where N is the number of people infected in hundreds, and t is the time in weeks.

(a) What is the maximum number of people projected to be infected?

(b) When will the virus be spreading most rapidly?

(c) Use a graphing utility to verify your results.

BUSINESS CAPSULE

Gordon Weinberger won a baking contest two years in a row and decided to start his own pie making company. He raised money and opened Gordon's Top of the Tree Baking Company in Londonderry, NH in 1994. His pies sold well in small stores, but not in the larger markets. He was in debt, so he painted a schoolbus in psychedelic patterns and drove 1500 miles a week to visit as many supermarkets as he could. He eventually found a buyer who ordered 40 tractor-trailers full, which meant $1 million in sales. Gordon's yearly sales hit $5 million. In 2002, he sold the company to Mrs. Smith's Bakeries.

66. *Research Project*   Use your school's library, the Internet, or some other reference source to research the financial history of a small company like the one above. Gather the data on the company's costs and revenues over a period of time, and use a graphing utility to graph a scatter plot of the data. Fit models to the data. Do the models appear to be concave upward or downward? Do they appear to be increasing or decreasing? Discuss the implications of your answers.

3.4   OPTIMIZATION PROBLEMS

■ Solve real-life optimization problems.

Solving Optimization Problems

One of the most common applications of calculus is the determination of optimum (minimum or maximum) values. Before learning a general method for solving optimization problems, consider the next example.

EXAMPLE 1   **Finding the Maximum Volume**

A manufacturer wants to design an open box that has a square base and a surface area of 108 square inches, as shown in Figure 3.31. What dimensions will produce a box with a maximum volume?

SOLUTION   Because the base of the box is square, the volume is

$$V = x^2h. \qquad \text{\color{red}Primary equation}$$

This equation is called the **primary equation** because it gives a formula for the quantity to be optimized. The surface area of the box is

$$S = (\text{area of base}) + (\text{area of four sides})$$
$$108 = x^2 + 4xh. \qquad \text{\color{red}Secondary equation}$$

Because V is to be optimized, it helps to express V as a function of just one variable. To do this, solve the secondary equation for h in terms of x to obtain

$$h = \frac{108 - x^2}{4x}$$

and substitute into the primary equation.

$$V = x^2h = x^2\left(\frac{108 - x^2}{4x}\right) = 27x - \frac{1}{4}x^3 \qquad \text{\color{red}Function of one variable}$$

Before finding which x-value yields a maximum value of V, you need to determine the **feasible domain** of the function. That is, what values of x make sense in the problem? Because x must be nonnegative and the area of the base ($A = x^2$) is at most 108, you can conclude that the feasible domain is

$$0 \le x \le \sqrt{108}. \qquad \text{\color{red}Feasible domain}$$

Using the techniques described in the first three sections of this chapter, you can determine that $\left(\text{on the interval } 0 \le x \le \sqrt{108}\right)$ this function has an absolute maximum when $x = 6$ inches and $h = 3$ inches. ▬▬▬

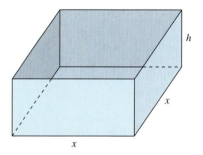

FIGURE 3.31   Open Box with Square Base: $S = x^2 + 4xh = 108$

TRY IT 1

Use a graphing utility to graph the volume function $V = 27x - \frac{1}{4}x^3$ on $0 \le x \le \sqrt{108}$ from Example 1. Verify that the function has an absolute maximum when $x = 6$. What is the maximum volume?

In studying Example 1, be sure that you understand the basic question that it asks. Some students have trouble with optimization problems because they are too eager to start solving the problem by using a standard formula. For instance, in Example 1, you should realize that there are infinitely many open boxes having 108 square inches of surface area. You might begin to solve this problem by asking yourself which basic shape would seem to yield a maximum volume. Should the box be tall, cubical, or squat? You might even try calculating a few volumes, as shown in Figure 3.32, to see if you can get a good feeling for what the optimum dimensions should be.

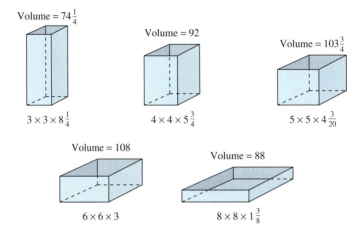

Volume $= 74\frac{1}{4}$

$3 \times 3 \times 8\frac{1}{4}$

Volume $= 92$

$4 \times 4 \times 5\frac{3}{4}$

Volume $= 103\frac{3}{4}$

$5 \times 5 \times 4\frac{3}{20}$

Volume $= 108$

$6 \times 6 \times 3$

Volume $= 88$

$8 \times 8 \times 1\frac{3}{8}$

FIGURE 3.32   Which box has the maximum volume?

STUDY TIP

Remember that you are not ready to begin solving an optimization problem until you have clearly identified what the problem is. Once you are sure you understand what is being asked, you are ready to begin considering a method for solving the problem.

There are several steps in the solution of Example 1. The first step is to sketch a diagram and identify all *known* quantities and all quantities *to be determined*. The second step is to write a primary equation for the quantity to be optimized. Then, a secondary equation is used to rewrite the primary equation as a function of one variable. Finally, calculus is used to determine the optimum value. These steps are summarized below.

STUDY TIP

When performing Step 5, remember that to determine the maximum or minimum value of a continuous function f on a closed interval, you need to compare the values of f at its critical numbers with the values of f at the endpoints of the interval. The largest of these values is the desired maximum and the smallest is the desired minimum.

Guidelines for Solving Optimization Problems

1. Identify all given quantities and all quantities to be determined. When feasible, sketch a diagram.

2. Write a **primary equation** for the quantity that is to be maximized or minimized. (A summary of several common formulas is given in Appendix D.)

3. Reduce the primary equation to one having a single independent variable. This may involve the use of a **secondary equation** that relates the independent variables of the primary equation.

4. Determine the **feasible domain** of the primary equation. That is, determine the values for which the stated problem makes sense.

5. Determine the desired maximum or minimum value by the calculus techniques discussed in Sections 3.1 through 3.3.

EXAMPLE 2   **Finding a Minimum Sum**

The product of two positive numbers is 288. Minimize the sum of the second number and twice the first number.

SOLUTION

1. Let x be the first number, y the second, and S the sum to be minimized.

2. Because you want to minimize S, the primary equation is

$$S = 2x + y.$$ Primary equation

3. Because the product of the two numbers is 288, you can write the secondary equation as

$$xy = 288$$ Secondary equation

$$y = \frac{288}{x}.$$

Using this result, you can rewrite the primary equation as a function of one variable.

$$S = 2x + \frac{288}{x}$$ Function of one variable

4. Because the numbers are positive, the feasible domain is

$$0 < x.$$ Feasible domain

5. To find the minimum value of S, begin by finding its critical numbers.

$$\frac{dS}{dx} = 2 - \frac{288}{x^2}$$ Find derivative of S.

$$0 = 2 - \frac{288}{x^2}$$ Set derivative equal to 0.

$$x^2 = 144$$ Simplify.

$$x = \pm 12$$ Critical numbers

 don't understand step

Choosing the positive x-value, you can use the First-Derivative Test to conclude that S is decreasing on the interval $(0, 12)$ and increasing on the interval $(12, \infty)$, as shown in the table. So, $x = 12$ yields a minimum, and the two numbers are

$$x = 12 \quad \text{and} \quad y = \frac{288}{12} = 24.$$

Interval	$0 < x < 12$	$12 < x < \infty$
Test value	$x = 1$	$x = 13$
Sign of $\dfrac{dS}{dx}$	$\dfrac{dS}{dx} < 0$	$\dfrac{dS}{dx} > 0$
Conclusion	S is decreasing.	S is increasing.

ALGEBRA REVIEW

For help on the algebra in Example 2, see Example 1(b) in the *Chapter 3 Algebra Review,* on page 248.

TECHNOLOGY

After you have written the primary equation as a function of a single variable, you can estimate the optimum value by graphing the function. For instance, the graph of

$$S = 2x + \frac{288}{x}$$

shown below indicates that the minimum value of S occurs when x is about 12.

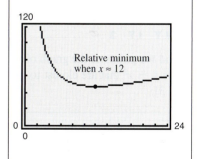

TRY IT 2

The product of two numbers is 72. Minimize the sum of the second number and twice the first number.

As applications go, the four examples described in this section are fairly simple, and yet the resulting primary equations are quite complicated. Real-life applications often involve equations that are at least as complex as these four. Remember that one of the main goals of this course is to enable you to use the power of calculus to analyze equations that at first glance seem formidable.

Also remember that once you have found the primary equation, you can use the graph of the equation to help solve the problem. For instance, the graphs of the primary equations in Examples 1 through 4 are shown in Figure 3.35.

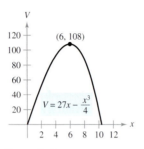

Example 1

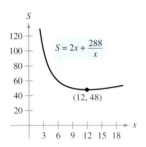

Example 2

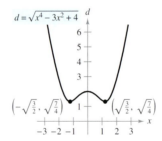

Example 3

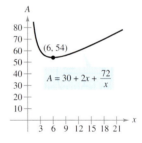

Example 4

FIGURE 3.35

TAKE ANOTHER LOOK

Comparing Graphical, Numerical, and Analytic Approaches

In Examples 1, 2, 3, and 4, an analytic approach was used to find the minimum or maximum value. To compare this type of analysis with other approaches, divide into at most eight groups and solve the examples employing a numerical approach (using a table) or a graphical approach (using a graphing utility).

Example 1 – *numerical* approach   Example 3 – *numerical* approach
Example 1 – *graphical* approach   Example 3 – *graphical* approach
Example 2 – *numerical* approach   Example 4 – *numerical* approach
Example 2 – *graphical* approach   Example 4 – *graphical* approach

When finished, compare all three approaches for each example. What are the advantages of each approach? What are the disadvantages?

PREREQUISITE REVIEW 3.4

The following warm-up exercises involve skills that were covered in earlier sections. You will use these skills in the exercise set for this section.

In Exercises 1–4, write a formula for the written statement.

1. The sum of one number and half a second number is 12.

2. The product of one number and twice another is 24.

3. The area of a rectangle is 24 square units.

4. The distance between two points is 10 units.

In Exercises 5–10, find the critical numbers of the function.

5. $y = x^2 + 6x - 9$
6. $y = 2x^3 - x^2 - 4x$
7. $y = 5x + \dfrac{125}{x}$

8. $y = 3x + \dfrac{96}{x^2}$
9. $y = \dfrac{x^2 + 1}{x}$
10. $y = \dfrac{x}{x^2 + 9}$

EXERCISES 3.4

In Exercises 1–6, find two positive numbers satisfying the given requirements.

1. The sum is 110 and the product is a maximum.

2. The sum is S and the product is a maximum.

3. The sum of the first and twice the second is 36 and the product is a maximum.

4. The sum of the first and twice the second is 100 and the product is a maximum.

5. The product is 192 and the sum is a minimum.

6. The product is 192 and the sum of the first plus three times the second is a minimum.

7. What positive number x minimizes the sum of x and its reciprocal?

8. The difference of two numbers is 50. Find the two numbers such that their product is a minimum.

In Exercises 9 and 10, find the length and width of a rectangle that has the given perimeter and a maximum area.

9. Perimeter: 100 meters   10. Perimeter: P units

In Exercises 11 and 12, find the length and width of the rectangle that has the given area and a minimum perimeter.

11. Area: 64 square feet   12. Area: A square centimeters

13. **Maximum Area** A rancher has 200 feet of fencing to enclose two adjacent rectangular corrals (see figure). What dimensions should be used so that the enclosed area will be a maximum?

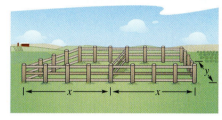

Figure for 13

14. **Area** A dairy farmer plans to enclose a rectangular pasture adjacent to a river. To provide enough grass for the herd, the pasture must contain 180,000 square meters. No fencing is required along the river. What dimensions will use the smallest amount of fencing?

15. **Maximum Volume**
    (a) Verify that each of the rectangular solids shown in the figure has a surface area of 150 square inches.
    (b) Find the volume of each solid.
    (c) Determine the dimensions of a rectangular solid (with a square base) of maximum volume if its surface area is 150 square inches.

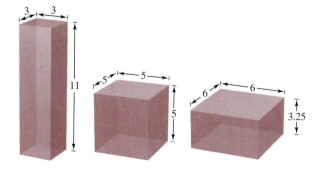

16. *Maximum Volume* Determine the dimensions of a rectangular solid (with a square base) with maximum volume if its surface area is 337.5 square centimeters.

17. *Maximum Area* A Norman window is constructed by adjoining a semicircle to the top of an ordinary rectangular window (see figure). Find the dimensions of a Norman window of maximum area if the total perimeter is 16 feet.

18. *Volume* An open box is to be made from a six-inch by six-inch square piece of material by cutting equal squares from the corners and turning up the sides (see figure). Find the volume of the largest box that can be made.

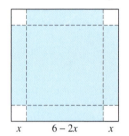

19. *Volume* An open box is to be made from a two-foot by three-foot rectangular piece of material by cutting equal squares from the corners and turning up the sides. Find the volume of the largest box that can be made in this manner.

20. *Minimum Surface Area* A net enclosure for golf practice is open at one end (see figure). The volume of the enclosure is $83\frac{1}{3}$ cubic meters. Find the dimensions that require the smallest amount of netting.

21. *Gardening* A home gardener estimates that if she plants 16 apple trees, the average yield will be 80 apples per tree. But because of the size of the garden, for each additional tree planted the yield will decrease by four apples per tree. How many trees should be planted to maximize the total yield of apples? What is the maximum yield?

22. *Area* A rectangular page is to contain 36 square inches of print. The margins at the top and bottom and on each side are to be $1\frac{1}{2}$ inches. Find the dimensions of the page that will minimize the amount of paper used.

23. *Area* A rectangular page is to contain 30 square inches of print. The margins at the top and bottom of the page are to be 2 inches wide. The margins on each side are to be 1 inch wide. Find the dimensions of the page such that the least amount of paper is used.

24. *Maximum Area* A rectangle is bounded by the *x*- and *y*-axes and the graph of

$$y = \frac{6 - x}{2}$$

(see figure). What length and width should the rectangle have so that its area is a maximum?

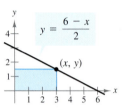   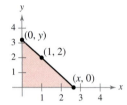

Figure for 24   Figure for 25

25. *Minimum Length* A right triangle is formed in the first quadrant by the *x*- and *y*-axes and a line through the point (1, 2) (see figure).

(a) Write the length *L* of the hypotenuse as a function of *x*.

⊕ (b) Use a graphing utility to approximate *x* graphically such that the length of the hypotenuse is a minimum.

(c) Find the vertices of the triangle such that its area is a minimum.

26. *Maximum Area* A rectangle is bounded by the *x*-axis and the semicircle

$$y = \sqrt{25 - x^2}$$

(see figure). What length and width should the rectangle have so that its area is a maximum?

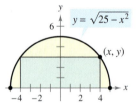

27. *Area* Find the dimensions of the largest rectangle that can be inscribed in a semicircle of radius *r*. (See Exercise 26.)

28. *Volume* You are designing a soft drink container that has the shape of a right circular cylinder. The container is supposed to hold 12 fluid ounces (1 fluid ounce is approximately 1.80469 cubic inches). Find the dimensions that will use a minimum amount of construction material.

29. *Volume* Find the volume of the largest right circular cylinder that can be inscribed in a sphere of radius *r* (see figure on next page).

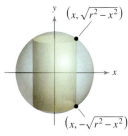
$\left(x, \sqrt{r^2 - x^2}\right)$
$\left(x, -\sqrt{r^2 - x^2}\right)$

Figure for 29

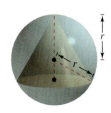

Figure for 30

30. Maximum Volume Find the volume of the largest right circular cone that can be inscribed in a sphere of radius r.

In Exercises 31 and 32, find the points on the graph of the function that are closest to the given point.

Function	Point
31. $f(x) = x^2 + 1$	$(0, 4)$
32. $f(x) = x^2$	$\left(2, \frac{1}{2}\right)$

33. Maximum Volume A rectangular package to be sent by a postal service can have a maximum combined length and girth of 108 inches. Find the dimensions of the package with maximum volume. Assume that the package's dimensions are x by x by y (see figure).

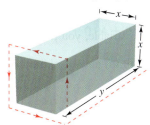

34. Minimum Surface Area A solid is formed by adjoining two hemispheres to the ends of a right circular cylinder. The total volume of the solid is 12 cubic inches. Find the radius of the cylinder that produces the minimum surface area.

35. Minimum Area The combined perimeter of a circle and a square is 16. Find the dimensions of the circle and square that produce a minimum total area.

36. Minimum Area The combined perimeter of an equilateral triangle and a square is 10. Find the dimensions of the triangle and square that produce a minimum total area.

37. Minimum Time You are in a boat 2 miles from the nearest point on the coast. You are to go to point Q, located 3 miles down the coast and 1 mile inland (see figure). You can row at a rate of 2 miles per hour and you can walk at a rate of 4 miles per hour. Toward what point on the coast should you row in order to reach point Q in the least time?

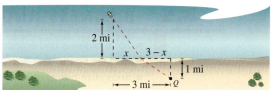

Figure for 37

38. Maximum Area An indoor physical fitness room consists of a rectangular region with a semicircle on each end. The perimeter of the room is to be a 200-meter running track. Find the dimensions that will make the area of the rectangular region as large as possible.

39. Farming A strawberry farmer will receive $4 per bushel of strawberries during the first week of harvesting. Each week after that, the value will drop $0.10 per bushel. The farmer estimates that there are approximately 120 bushels of strawberries in the fields, and that the crop is increasing at a rate of four bushels per week. When should the farmer harvest the strawberries to maximize their value? How many bushels of strawberries will yield the maximum value? What is the maximum value of the strawberries?

40. Beam Strength A wooden beam has a rectangular cross section of height h and width w (see figure). The strength S of the beam is directly proportional to its width and the square of its height. What are the dimensions of the strongest beam that can be cut from a round log of diameter 24 inches? (*Hint:* $S = kh^2w$, where k is the proportionality constant.)

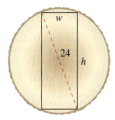

41. Maximum Area Use a graphing utility to graph the primary equation and its first derivative to find the dimensions of the rectangle of maximum area that can be inscribed in a semicircle of radius 10.

42. Area Four feet of wire is to be used to form a square and a circle.

(a) Express the sum of the areas of the square and the circle as a function A of the side of the square x.

(b) What is the domain of A?

(c) Use a graphing utility to graph A on its domain.

(d) How much wire should be used for the square and how much for the circle in order to enclose the smallest total area? the greatest total area?

Business Terms and Formulas

This section concludes with a summary of the basic business terms and formulas used in this section. A summary of the graphs of the demand, revenue, cost, and profit functions is shown in Figure 3.43.

Summary of Business Terms and Formulas

x = number of units produced (or sold)	η = price elasticity of demand
p = price per unit	$= (p/x)/(dp/dx)$
R = total revenue from selling x units $= xp$	dR/dx = marginal revenue
C = total cost of producing x units	dC/dx = marginal cost
P = total profit from selling x units $= R - C$	dP/dx = marginal profit
$\overline{C}$ = average cost per unit $= \dfrac{C}{x}$	

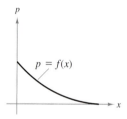

Demand function

Quantity demanded increases as price decreases.

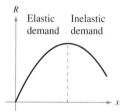

Revenue function

The low prices required to sell more units eventually result in a decreasing revenue.

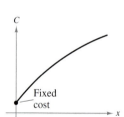

Cost function

The total cost to produce x units includes the fixed cost.

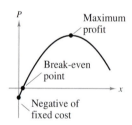

Profit function

The break-even point occurs when $R = C$.

FIGURE 3.43

TAKE ANOTHER LOOK

Demand Function

Throughout this text, it is assumed that demand functions are decreasing. Can you think of a product that has an increasing demand function? That is, can you think of a product that becomes more in demand as its price increases? Explain your reasoning, and sketch a graph of the function.

The following warm-up exercises involve skills that were covered in earlier sections. You will use these skills in the exercise set for this section.

In Exercises 1–4, evaluate the expression for $x = 150$.

1. $\left| -\dfrac{300}{x} + 3 \right|$

2. $\left| -\dfrac{600}{5x} + 2 \right|$

3. $\left| \dfrac{(20x^{-1/2})/x}{-10x^{-3/2}} \right|$

4. $\left| \dfrac{(4000/x^2)/x}{-8000x^{-3}} \right|$

In Exercises 5–10, find the marginal revenue, marginal cost, or marginal profit.

5. $C = 650 + 1.2x + 0.003x^2$

6. $P = 0.01x^2 + 11x$

7. $R = 14x - \dfrac{x^2}{2000}$

8. $R = 3.4x - \dfrac{x^2}{1500}$

9. $P = -0.7x^2 + 7x - 50$

10. $C = 1700 + 4.2x + 0.001x^3$

EXERCISES 3.5

In Exercises 1–4, find the number of units x that produces a maximum revenue R.

1. $R = 800x - 0.2x^2$

2. $R = 48x^2 - 0.02x^3$

3. $R = 400x - x^2$

4. $R = 30x^{2/3} - 2x$

In Exercises 5–8, find the number of units x that produces the minimum average cost per unit $\bar{C}$.

5. $C = 1.25x^2 + 25x + 8000$

6. $C = 0.001x^3 + 5x + 250$

7. $C = 2x^2 + 255x + 5000$

8. $C = 0.02x^3 + 55x^2 + 1250$

In Exercises 9–12, find the price per unit p that produces the maximum profit P.

Cost Function	Demand Function
9. $C = 100 + 30x$	$p = 90 - x$
10. $C = 0.5x + 600$	$p = \dfrac{60}{\sqrt{x}}$
11. $C = 8000 + 50x + 0.03x^2$	$p = 70 - 0.001x$
12. $C = 35x + 500$	$p = 50 - 0.1\sqrt{x}$

Average Cost In Exercises 13 and 14, use the cost function to find the production level for which the average cost is a minimum. For this production level, show that the marginal cost and average cost are equal. Use a graphing utility to graph the average cost function and verify your results.

13. $C = 2x^2 + 5x + 18$

14. $C = x^3 - 6x^2 + 13x$

15. *Maximum Profit* A commodity has a demand function modeled by

$$p = 100 - 0.5x^2$$

and a total cost function modeled by $C = 40x + 37.5$.

(a) What price yields a maximum profit?

(b) When the profit is maximized, what is the average cost per unit?

16. *Maximum Profit* How would the answer to Exercise 15 change if the marginal cost rose from $40 per unit to $50 per unit? In other words, rework Exercise 15 using the cost function $C = 50x + 37.5$.

Maximum Profit In Exercises 17 and 18, find the amount s of advertising that maximizes the profit P. (s and P are measured in thousands of dollars.) Find the point of diminishing returns.

17. $P = -2s^3 + 35s^2 - 100s + 200$

18. $P = -0.1s^3 + 6s^2 + 400$

19. *Maximum Profit* The cost per unit of producing a type of radio is $60. The manufacturer charges $90 per unit for orders of 100 or less. To encourage large orders, however, the manufacturer reduces the charge by $0.10 per radio for each order in excess of 100 units. For instance, an order of 101 radios would be $89.90 per radio, an order of 102 radios would be $89.80 per radio, and so on. Find the largest order the manufacturer should allow to obtain a maximum profit.

20. Maximum Profit A real estate office handles a 50-unit apartment complex. When the rent is $580 per month, all units are occupied. For each $40 increase in rent, however, an average of one unit becomes vacant. Each occupied unit requires an average of $45 per month for service and repairs. What rent should be charged to obtain a maximum profit?

21. Maximum Revenue When a wholesaler sold a product at $40 per unit, sales were 300 units per week. After a price increase of $5, however, the average number of units sold dropped to 275 per week. Assuming that the demand function is linear, what price per unit will yield a maximum total revenue?

22. Maximum Profit Assume that the amount of money deposited in a bank is proportional to the square of the interest rate the bank pays on the money. Furthermore, the bank can reinvest the money at 12% simple interest. Find the interest rate the bank should pay to maximize its profit.

⊕ **23. Minimum Cost** A power station is on one side of a river that is 0.5 mile wide, and a factory is 6 miles downstream on the other side of the river (see figure). It costs $6 per foot to run overland power lines and $8 per foot to run underwater power lines. Write a cost function for running the power lines from the power station to the factory. Use a graphing utility to graph your function. Estimate the value of x that minimizes the cost. Explain your results.

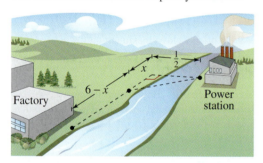

24. Minimum Cost An offshore oil well is 1 mile off the coast. The oil refinery is 2 miles down the coast. Laying pipe in the ocean is twice as expensive as laying it on land. Find the most economical path for the pipe from the well to the oil refinery.

25. Minimum Cost A small business uses a minivan to make deliveries. The cost per hour for fuel is $C = v^2/600$, where v is the speed of the minivan (in miles per hour). The driver is paid $10 per hour. Find the speed that minimizes the cost of a 110-mile trip. (Assume there are no costs other than fuel and wages.)

26. Minimum Cost Repeat Exercise 25 for a fuel cost per hour of

$$C = \frac{v^2 + 360}{720}$$

and a wage of $8 per hour.

⊕ **Elasticity** In Exercises 27–32, find the price elasticity of demand for the demand function at the indicated x-value. Is the demand elastic, inelastic, or of unit elasticity at the indicated x-value? Use a graphing utility to graph the revenue function, and identify the intervals of elasticity and inelasticity.

Demand Function	Quantity Demanded
27. $p = 400 - 3x$	$x = 20$
28. $p = 5 - 0.03x$	$x = 100$
29. $p = 20 - 0.0002x$	$x = 30$
30. $p = \dfrac{500}{x + 2}$	$x = 23$
31. $p = \dfrac{100}{x^2} + 2$	$x = 10$
32. $p = 100 - \sqrt{0.2x}$	$x = 125$

33. Elasticity The demand function for a product is given by

$$x = p^2 - 20p + 100.$$

(a) Consider a price of $2. If the price increases by 5%, what is the corresponding percent change in the quantity demanded?

(b) Average elasticity of demand is defined to be the percent change in quantity divided by the percent change in price. Use the percent in part (a) to find the average elasticity over the interval $[2, 2.1]$.

(c) Find the elasticity for a price of $2 and compare the result with that in part (b).

(d) Find an expression for the total revenue and find the values of x and p that maximize the total revenue.

34. Elasticity The demand function for a product is given by

$$p^3 + x^3 = 9.$$

(a) Find the price elasticity of demand when $x = 2$.

(b) Find the values of x and p that maximize the total revenue.

(c) For the value of x found in part (b), show that the price elasticity of demand has unit elasticity.

35. Elasticity The demand function for a product is given by

$$p = 20 - 0.02x, \quad 0 < x < 1000.$$

(a) Find the price elasticity of demand when $x = 560$.

(b) Find the values of x and p that maximize the total revenue.

(c) For the value of x found in part (b), show that the price elasticity of demand has unit elasticity.

36. Minimum Cost The ordering and transportation cost C of the components used in manufacturing a product is modeled by

$$C = 100\left(\frac{200}{x^2} + \frac{x}{x + 30}\right), \quad x \geq 1$$

where C is measured in thousands of dollars and x is the order size in hundreds. Find the order size that minimizes the cost. (*Hint:* Use the *root* feature of a graphing utility.)

37. Revenue The demand for a car wash is

$$x = 600 - 50p$$

where the current price is $5.00. Can revenue be increased by lowering the price and thus attracting more customers? Use price elasticity of demand to determine your answer.

38. Revenue Repeat Exercise 37 for a demand function of

$$x = 800 - 40p.$$

39. Demand A demand function is modeled by $x = a/p^m$, where a is a constant and $m > 1$. Show that $\eta = -m$. In other words, show that a 1% increase in price results in an $m\%$ decrease in the quantity demanded.

40. Sales The sales S (in millions of dollars per year) for Lowe's for the years 1994 through 2003 can be modeled by

$$S = 201.556t^2 - 502.29t + 2622.8 + \frac{9286}{t},$$

$$4 \leq t \leq 13$$

where $t = 4$ corresponds to 1994. *(Source: Lowe's Companies)*

(a) During which year, from 1994 to 2003, were Lowe's sales increasing most rapidly?

(b) During which year were the sales increasing at the lowest rate?

(c) Find the rate of increase or decrease for each year in parts (a) and (b).

(d) Use a graphing utility to graph the sales function. Then use the *zoom* and *trace* features to confirm the results in parts (a), (b), and (c).

41. Revenue The revenue R (in millions of dollars per year) for Papa John's for the years 1994 through 2003 can be modeled by

$$R = \frac{-18.0 + 24.74t}{1 - 0.16t + 0.008t^2}, \quad 4 \leq t \leq 13$$

where $t = 4$ corresponds to 1994. *(Source: Papa John's Int'l.)*

(a) During which year, from 1994 to 2003, was Papa John's revenue the greatest? the least?

(b) During which year was the revenue increasing at the greatest rate? decreasing at the greatest rate?

(c) Use a graphing utility to graph the revenue function, and confirm your results in parts (a) and (b).

42. Match each graph with the function it best represents— a demand function, a revenue function, a cost function, or a profit function. Explain your reasoning. (The graphs are labeled a–d.)

BUSINESS CAPSULE

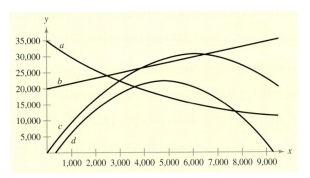

Courtesy of Transperfect Translations

While graduate students, Elizabeth Elting and Phil Shawe co-founded TransPerfect Translations in 1992. They used a rented computer and a $5000 credit card cash advance to market their service-oriented translation firm, now one of the largest in the country. Currently, they have a network of 4000 certified language specialists in North America, Europe, and Asia, which translates technical, legal, business, and marketing materials. In 2004, the company estimates its gross sales will be $35 million.

43. Research Project Choose an innovative product like the one described above. Use your school's library, the Internet, or some other reference source to research the history of the product or service. Collect data about the revenue that the product or service has generated, and find a mathematical model of the data. Summarize your findings.

3.6   ASYMPTOTES

- Find the vertical asymptotes of functions and find infinite limits.
- Find the horizontal asymptotes of functions and find limits at infinity.
- Use asymptotes to answer questions about real-life situations.

Vertical Asymptotes and Infinite Limits

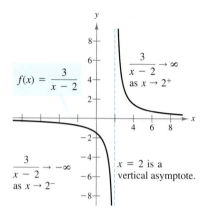

$$f(x) = \frac{3}{x-2}$$

$$\frac{3}{x-2} \to \infty$$ as $x \to 2^+$

$$\frac{3}{x-2} \to -\infty$$ as $x \to 2^-$

$x = 2$ is a vertical asymptote.

FIGURE 3.44

In the first three sections of this chapter, you studied ways in which you can use calculus to help analyze the graph of a function. In this section, you will study another valuable aid to curve sketching: the determination of vertical and horizontal asymptotes.

Recall from Section 1.5, Example 10, that the function

$$f(x) = \frac{3}{x-2}$$

is unbounded as x approaches 2 (see Figure 3.44). This type of behavior is described by saying that the line $x = 2$ is a **vertical asymptote** of the graph of f. The type of limit in which $f(x)$ approaches infinity (or negative infinity) as x approaches c from the left or from the right is an **infinite limit.** The infinite limits for the function $f(x) = 3/(x-2)$ can be written as

$$\lim_{x \to 2^-} \frac{3}{x-2} = -\infty$$

and

$$\lim_{x \to 2^+} \frac{3}{x-2} = \infty.$$

> ### Definition of Vertical Asymptote
>
> If $f(x)$ approaches infinity (or negative infinity) as x approaches c from the right or from the left, then the line $x = c$ is a **vertical asymptote** of the graph of f.

TECHNOLOGY

When you use a graphing utility to graph a function that has a vertical asymptote, the utility may try to connect separate branches of the graph. For instance, the figure at the right shows the graph of

$$f(x) = \frac{3}{x-2}$$

on a graphing calculator.

This line is not part of the graph of the function.

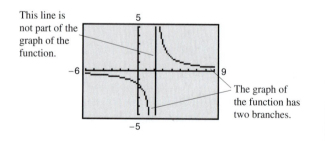

The graph of the function has two branches.

One of the most common instances of a vertical asymptote is the graph of a *rational function*—that is, a function of the form $f(x) = p(x)/q(x)$, where $p(x)$ and $q(x)$ are polynomials. If c is a real number such that $q(c) = 0$ and $p(c) \neq 0$, the graph of f has a vertical asymptote at $x = c$. Example 1 shows four cases.

EXAMPLE 1 Finding Infinite Limits

Find each limit.

Limit from the left *Limit from the right*

(a) $\displaystyle\lim_{x \to 1^-} \frac{1}{x - 1} = -\infty$ $\displaystyle\lim_{x \to 1^+} \frac{1}{x - 1} = \infty$ See Figure 3.45(a).

(b) $\displaystyle\lim_{x \to 1^-} \frac{-1}{x - 1} = \infty$ $\displaystyle\lim_{x \to 1^+} \frac{-1}{x - 1} = -\infty$ See Figure 3.45(b).

(c) $\displaystyle\lim_{x \to 1^-} \frac{-1}{(x - 1)^2} = -\infty$ $\displaystyle\lim_{x \to 1^+} \frac{-1}{(x - 1)^2} = -\infty$ See Figure 3.45(c).

(d) $\displaystyle\lim_{x \to 1^-} \frac{1}{(x - 1)^2} = \infty$ $\displaystyle\lim_{x \to 1^+} \frac{1}{(x - 1)^2} = \infty$ See Figure 3.45(d).

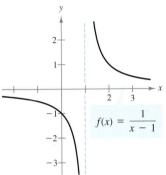

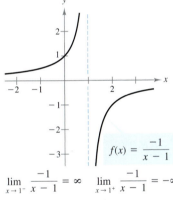

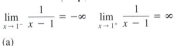
$\displaystyle\lim_{x \to 1^-} \frac{1}{x - 1} = -\infty$ $\displaystyle\lim_{x \to 1^+} \frac{1}{x - 1} = \infty$

(a)

$\displaystyle\lim_{x \to 1^-} \frac{-1}{x - 1} = \infty$ $\displaystyle\lim_{x \to 1^+} \frac{-1}{x - 1} = -\infty$

(b)

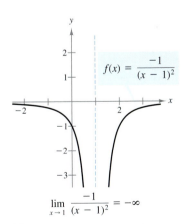
$\displaystyle\lim_{x \to 1} \frac{-1}{(x - 1)^2} = -\infty$

(c)

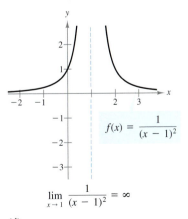
$\displaystyle\lim_{x \to 1} \frac{1}{(x - 1)^2} = \infty$

(d)

FIGURE 3.45

TECHNOLOGY

Use a spreadsheet or table to verify the results shown in Example 1. (Consult the user's manual of a spreadsheet software program for specific instructions on how to create a table.) For instance, in Example 1(a), notice that the values of $f(x) = 1/(x - 1)$ decrease and increase without bound as x gets closer and closer to 1 from the left and the right.

x Approaches 1 from the Left	
x	$f(x) = 1/(x - 1)$
0	-1
0.9	-10
0.99	-100
0.999	-1000
0.9999	$-10,000$

x Approaches 1 from the Right	
x	$f(x) = 1/(x - 1)$
2	1
1.1	10
1.01	100
1.001	1000
1.0001	10,000

TRY IT 1

Find each limit.

(a) *Limit from the left*

$$\lim_{x \to 2^-} \frac{1}{x - 2}$$

Limit from the right

$$\lim_{x \to 2^+} \frac{1}{x - 2}$$

(b) *Limit from the left*

$$\lim_{x \to -3^-} \frac{-1}{x + 3}$$

Limit from the right

$$\lim_{x \to -3^+} \frac{-1}{x + 3}$$

$$f(x) = \frac{x + 2}{x^2 - 2x}$$

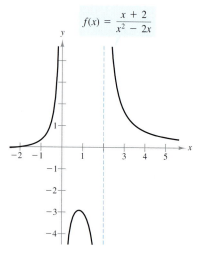

FIGURE 3.46   Vertical Asymptotes at $x = 0$ and $x = 2$

Each of the graphs in Example 1 has only one vertical asymptote. As shown in the next example, the graph of a rational function can have more than one vertical asymptote.

EXAMPLE 2   **Finding Vertical Asymptotes**

Find the vertical asymptotes of the graph of

$$f(x) = \frac{x + 2}{x^2 - 2x}.$$

SOLUTION   The possible vertical asymptotes correspond to the x-values for which the denominator is zero.

$$x^2 - 2x = 0 \qquad \text{Set denominator equal to 0.}$$
$$x(x - 2) = 0 \qquad \text{Factor.}$$
$$x = 0, x = 2 \qquad \text{Zeros of denominator}$$

Because the numerator of f is not zero at either of these x-values, you can conclude that the graph of f has two vertical asymptotes—one at $x = 0$ and one at $x = 2$, as shown in Figure 3.46.

TRY IT 2

Find the vertical asymptote(s) of the graph of

$$f(x) = \frac{x + 4}{x^2 - 4x}.$$

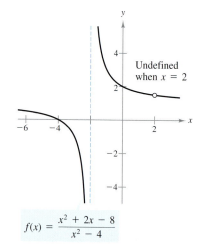

$$f(x) = \frac{x^2 + 2x - 8}{x^2 - 4}$$

FIGURE 3.47   Vertical Asymptote at $x = -2$

TRY IT 3

Find the vertical asymptotes of the graph of

$$f(x) = \frac{x^2 + 4x + 3}{x^2 - 9}.$$

EXAMPLE 3   **Finding Vertical Asymptotes**

Find the vertical asymptotes of the graph of

$$f(x) = \frac{x^2 + 2x - 8}{x^2 - 4}.$$

SOLUTION   First factor the numerator and denominator. Then divide out like factors.

$$f(x) = \frac{x^2 + 2x - 8}{x^2 - 4} \qquad \text{Write original function.}$$
$$= \frac{(x + 4)(x - 2)}{(x + 2)(x - 2)} \qquad \text{Factor numerator and denominator.}$$
$$= \frac{(x + 4)(x - 2)}{(x + 2)(x - 2)} \qquad \text{Divide out like factors.}$$
$$= \frac{x + 4}{x + 2}, \quad x \neq 2 \qquad \text{Simplify.}$$

For all values of x other than $x = 2$, the graph of this simplified function is the same as the graph of f. So, you can conclude that the graph of f has only one vertical asymptote. This occurs at $x = -2$, as shown in Figure 3.47.

From Example 3, you know that the graph of

$$f(x) = \frac{x^2 + 2x - 8}{x^2 - 4}$$

has a vertical asymptote at $x = -2$. This implies that the limit of $f(x)$ as $x \to -2$ from the right (or from the left) is either ∞ or $-\infty$. But without looking at the graph, how can you determine that the limit from the left is *negative* infinity and the limit from the right is *positive* infinity? That is, why is the limit from the left

$$\lim_{x \to -2^-} \frac{x^2 + 2x - 8}{x^2 - 4} = -\infty \qquad \text{Limit from the left}$$

and why is the limit from the right

$$\lim_{x \to -2^+} \frac{x^2 + 2x - 8}{x^2 - 4} = \infty? \qquad \text{Limit from the right}$$

It is cumbersome to determine these limits analytically, and you may find the graphical method shown in Example 4 to be more efficient.

EXAMPLE 4 **Determining Infinite Limits**

Find the limits.

$$\lim_{x \to 1^-} \frac{x^2 - 3x}{x - 1} \quad \text{and} \quad \lim_{x \to 1^+} \frac{x^2 - 3x}{x - 1}$$

SOLUTION Begin by considering the function

$$f(x) = \frac{x^2 - 3x}{x - 1}.$$

Because the denominator is zero when $x = 1$ and the numerator is not zero when $x = 1$, it follows that the graph of the function has a vertical asymptote at $x = 1$. This implies that each of the given limits is either ∞ or $-\infty$. To determine which, use a graphing utility to graph the function, as shown in Figure 3.48. From the graph, you can see that the limit from the left is positive infinity and the limit from the right is negative infinity. That is,

$$\lim_{x \to 1^-} \frac{x^2 - 3x}{x - 1} = \infty \qquad \text{Limit from the left}$$

and

$$\lim_{x \to 1^+} \frac{x^2 - 3x}{x - 1} = -\infty. \qquad \text{Limit from the right}$$

From the left, $f(x)$ approaches positive infinity.

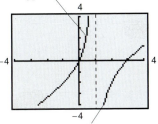

From the right, $f(x)$ approaches negative infinity.

FIGURE 3.48

STUDY TIP

In Example 4, try evaluating $f(x)$ at x-values that are just barely to the left of 1. You will find that you can make the values of $f(x)$ arbitrarily large by choosing x sufficiently close to 1. For instance, $f(0.99999) = 199,999$.

TRY IT 4

Find the limits.

$$\lim_{x \to 2^-} \frac{x^2 - 4x}{x - 2} \quad \text{and} \quad \lim_{x \to 2^+} \frac{x^2 - 4x}{x - 2}$$

Then verify your solution by graphing the function.

Horizontal Asymptotes and Limits at Infinity

Another type of limit, called a **limit at infinity,** specifies a finite value approached by a function as x increases (or decreases) without bound.

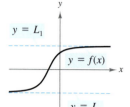

Definition of Horizontal Asymptote

If f is a function and L_1 and L_2 are real numbers, the statements

$$\lim_{x \to \infty} f(x) = L_1 \quad \text{and} \quad \lim_{x \to -\infty} f(x) = L_2$$

denote **limits at infinity.** The lines $y = L_1$ and $y = L_2$ are **horizontal asymptotes** of the graph of f.

Figure 3.49 shows two ways in which the graph of a function can approach one or more horizontal asymptotes. Note that it is possible for the graph of a function to cross its horizontal asymptote.

Limits at infinity share many of the properties of limits discussed in Section 1.5. When finding horizontal asymptotes, you can use the property that

$$\lim_{x \to \infty} \frac{1}{x^r} = 0, \quad r > 0 \quad \text{and} \quad \lim_{x \to -\infty} \frac{1}{x^r} = 0, \quad r > 0.$$

(The second limit assumes that x^r is defined when $x < 0$.)

FIGURE 3.49

EXAMPLE 5   Finding Limits at Infinity

Find the limit: $\displaystyle \lim_{x \to \infty} \left(5 - \frac{2}{x^2} \right).$

SOLUTION

$$\lim_{x \to \infty} \left(5 - \frac{2}{x^2} \right) = \lim_{x \to \infty} 5 - \lim_{x \to \infty} \frac{2}{x^2} \qquad \lim_{x \to \infty} [f(x) - g(x)] = \lim_{x \to \infty} f(x) - \lim_{x \to \infty} g(x)$$

$$= \lim_{x \to \infty} 5 - 2 \left(\lim_{x \to \infty} \frac{1}{x^2} \right) \qquad \lim_{x \to \infty} cf(x) = c \lim_{x \to \infty} f(x)$$

$$= 5 - 2(0)$$

$$= 5$$

You can verify this limit by sketching the graph of

$$f(x) = 5 - \frac{2}{x^2}$$

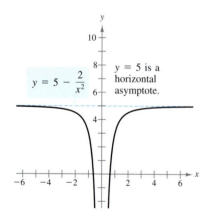

$y = 5 - \dfrac{2}{x^2}$

$y = 5$ is a horizontal asymptote.

FIGURE 3.50

as shown in Figure 3.50. Note that the graph has $y = 5$ as a horizontal asymptote to the right. By evaluating the limit of $f(x)$ as $x \to -\infty$, you can show that this line is also a horizontal asymptote to the left.

TRY IT 5

Find the limit: $\displaystyle \lim_{x \to \infty} \left(2 + \frac{5}{x^2} \right).$

There is an easy way to determine whether the graph of a *rational* function has a horizontal asymptote. This shortcut is based on a comparison of the degrees of the numerator and denominator of the rational function.

Horizontal Asymptotes of Rational Functions

Let $f(x) = p(x)/q(x)$ be a rational function.

1. If the degree of the numerator is less than the degree of the denominator, then $y = 0$ is a horizontal asymptote of the graph of f (to the left and to the right).

2. If the degree of the numerator is equal to the degree of the denominator, then $y = a/b$ is a horizontal asymptote of the graph of f (to the left and to the right), where a and b are the leading coefficients of $p(x)$ and $q(x)$, respectively.

3. If the degree of the numerator is greater than the degree of the denominator, then the graph of f has no horizontal asymptote.

TECHNOLOGY

Some functions have two horizontal asymptotes: one to the right and one to the left. For instance, try sketching the graph of

$$f(x) = \frac{x}{\sqrt{x^2 + 1}}.$$

What horizontal asymptotes does the function appear to have?

EXAMPLE 6 **Finding Horizontal Asymptotes**

Find the horizontal asymptote of the graph of each function.

(a) $y = \dfrac{-2x + 3}{3x^2 + 1}$ (b) $y = \dfrac{-2x^2 + 3}{3x^2 + 1}$ (c) $y = \dfrac{-2x^3 + 3}{3x^2 + 1}$

SOLUTION

(a) Because the degree of the numerator is less than the degree of the denominator, $y = 0$ is a horizontal asymptote. [See Figure 3.51(a).]

(b) Because the degree of the numerator is equal to the degree of the denominator, the line $y = -\frac{2}{3}$ is a horizontal asymptote. [See Figure 3.51(b).]

(c) Because the degree of the numerator is greater than the degree of the denominator, the graph has no horizontal asymptote. [See Figure 3.51(c).]

TRY IT 6

Find the horizontal asymptote of the graph of each function.

(a) $y = \dfrac{2x + 1}{4x^2 + 5}$

(b) $y = \dfrac{2x^2 + 1}{4x^2 + 5}$

(c) $y = \dfrac{2x^3 + 1}{4x^2 + 5}$

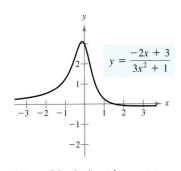

(a) $y = 0$ is a horizontal asymptote.

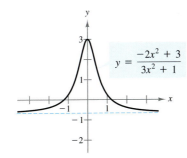

(b) $y = -\frac{2}{3}$ is a horizontal asymptote.

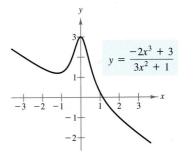

(c) No horizontal asymptote

FIGURE 3.51

Applications of Asymptotes

There are many examples of asymptotic behavior in real life. For instance, Example 7 describes the asymptotic behavior of an average cost function.

EXAMPLE 7   **Modeling Average Cost**

A small business invests $5000 in a new product. In addition to this initial investment, the product will cost $0.50 per unit to produce. Find the average cost per unit if 1000 units are produced, if 10,000 units are produced, and if 100,000 units are produced. What is the limit of the average cost as the number of units produced increases?

SOLUTION   From the given information, you can model the total cost C (in dollars) by

$$C = 0.5x + 5000 \qquad \text{Total cost function}$$

where x is the number of units produced. This implies that the average cost function is

$$\overline{C} = \frac{C}{x} = 0.5 + \frac{5000}{x}. \qquad \text{Average cost function}$$

If only 1000 units are produced, then the average cost per unit is

$$\overline{C} = 0.5 + \frac{5000}{1000} = \$5.50. \qquad \text{Average cost for 1000 units}$$

If 10,000 units are produced, then the average cost per unit is

$$\overline{C} = 0.5 + \frac{5000}{10,000} = \$1.00. \qquad \text{Average cost for 10,000 units}$$

If 100,000 units are produced, then the average cost per unit is

$$\overline{C} = 0.5 + \frac{5000}{100,000} = \$0.55. \qquad \text{Average cost for 100,000 units}$$

As x approaches infinity, the limiting average cost per unit is

$$\lim_{x \to \infty} \left(0.5 + \frac{5000}{x} \right) = \$0.50.$$

As shown in Figure 3.52, this example points out one of the major problems of small businesses. That is, it is difficult to have competitively low prices when the production level is low.

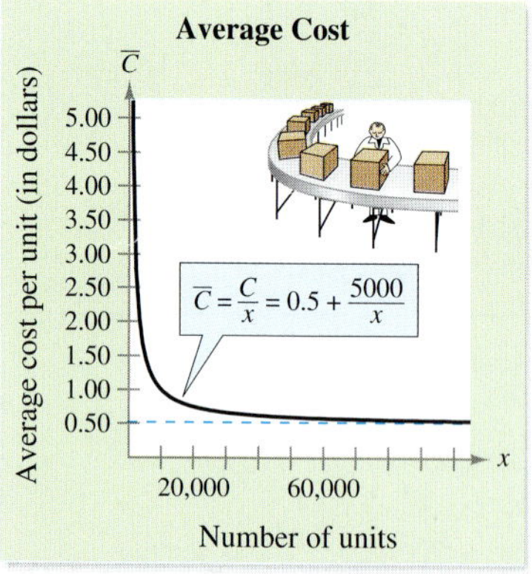

Average Cost

Average cost per unit (in dollars)

$\overline{C}$

5.00
4.50
4.00
3.50
3.00
2.50
2.00
1.50
1.00
0.50

$\overline{C} = \dfrac{C}{x} = 0.5 + \dfrac{5000}{x}$

20,000   60,000 x

Number of units

FIGURE 3.52   As $x \to \infty$, the average cost per unit approaches $0.50.

TRY IT 7

A small business invests $25,000 in a new product. In addition, the product will cost $0.75 per unit to produce. Find the cost function and the average cost function. What is the limit of the average cost function as production increases?

EXAMPLE 8   **Modeling Smokestack Emission**

A manufacturing plant has determined that the cost C (in dollars) of removing $p\%$ of the smokestack pollutants of its main smokestack is modeled by

$$C = \frac{80,000p}{100 - p}, \quad 0 \le p < 100.$$

What is the vertical asymptote of this function? What does the vertical asymptote mean to the plant owners?

SOLUTION   The graph of the cost function is shown in Figure 3.53. From the graph, you can see that $p = 100$ is the vertical asymptote. This means that as the plant attempts to remove higher and higher percents of the pollutants, the cost increases dramatically. For instance, the cost of removing 85% of the pollutants is

$$C = \frac{80,000(85)}{100 - 85} \approx \$453,333 \qquad \text{Cost for 85\% removal}$$

but the cost of removing 90% is

$$C = \frac{80,000(90)}{100 - 90} = \$720,000. \qquad \text{Cost for 90\% removal}$$

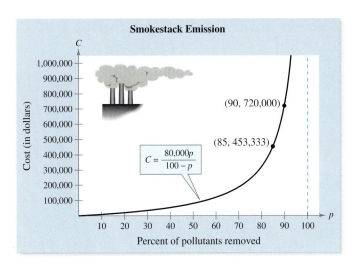

Smokestack Emission

$C = \dfrac{80,000p}{100 - p}$

(90, 720,000)

(85, 453,333)

Cost (in dollars)

Percent of pollutants removed

FIGURE 3.53

Rob Crandall/Rainbow

Since the 1980s, industries in the United States have spent billions of dollars to reduce air pollution.

TRY IT 8

According to the cost function in Example 8, is it possible to remove 100% of the smokestack pollutants? Why or why not?

TAKE ANOTHER LOOK

Indirect Costs

In Example 8, the given cost model considers only the direct cost for the manufacturing plant. Describe the possible indirect costs incurred by society as the plant attempts to remove more and more of the pollutants from its smokestack emission.

PREREQUISITE REVIEW 3.6

The following warm-up exercises involve skills that were covered in earlier sections. You will use these skills in the exercise set for this section.

In Exercises 1–8, find the limit.

1. $\lim\limits_{x \to 2} (x + 1)$

2. $\lim\limits_{x \to -1} (3x + 4)$

3. $\lim\limits_{x \to -3} \dfrac{2x^2 + x - 15}{x + 3}$

4. $\lim\limits_{x \to 2} \dfrac{3x^2 - 8x + 4}{x - 2}$

5. $\lim\limits_{x \to 2^+} \dfrac{x^2 - 5x + 6}{x^2 - 4}$

6. $\lim\limits_{x \to 1^-} \dfrac{x^2 - 6x + 5}{x^2 - 1}$

7. $\lim\limits_{x \to 0^+} \sqrt{x}$

8. $\lim\limits_{x \to 1^+} \left(x + \sqrt{x - 1}\right)$

In Exercises 9–12, find the average cost and the marginal cost.

9. $C = 150 + 3x$

10. $C = 1900 + 1.7x + 0.002x^2$

11. $C = 0.005x^2 + 0.5x + 1375$

12. $C = 760 + 0.05x$

EXERCISES 3.6

In Exercises 1–8, find the vertical and horizontal asymptotes. Write the asymptotes as equations of lines.

1. $f(x) = \dfrac{x^2 + 1}{x^2}$

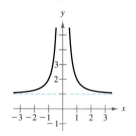

2. $f(x) = \dfrac{4}{(x - 2)^3}$

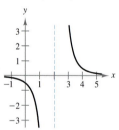

3. $f(x) = \dfrac{x^2 - 2}{x^2 - x - 2}$

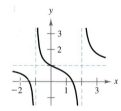

4. $f(x) = \dfrac{2 + x}{1 - x}$

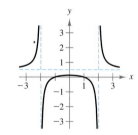

5. $f(x) = \dfrac{3x^2}{2(x^2 + 1)}$

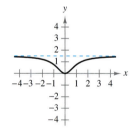

6. $f(x) = \dfrac{-4x}{x^2 + 4}$

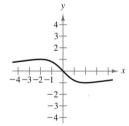

7. $f(x) = \dfrac{x^2 - 1}{2x^2 - 8}$

8. $f(x) = \dfrac{x^2 + 1}{x^3 - 8}$

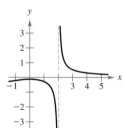

In Exercises 9–14, match the function with its graph. Use horizontal asymptotes as an aid. [The graphs are labeled (a)–(f).]

(a)
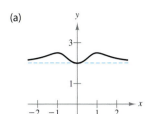

(b)
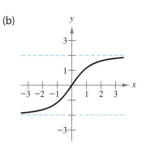

(c)
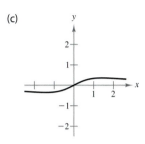

(d)
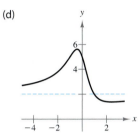

(e)
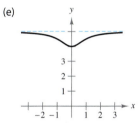

(f)
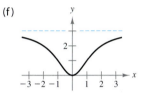

9. $f(x) = \dfrac{3x^2}{x^2 + 2}$

10. $f(x) = \dfrac{2x}{\sqrt{x^2 + 2}}$

11. $f(x) = \dfrac{x}{x^2 + 2}$

12. $f(x) = 2 + \dfrac{x^2}{x^4 + 1}$

13. $f(x) = 5 - \dfrac{1}{x^2 + 1}$

14. $f(x) = \dfrac{2x^2 - 3x + 5}{x^2 + 1}$

In Exercises 15–22, find the limit.

15. $\displaystyle\lim_{x \to -2^-} \dfrac{1}{(x + 2)^2}$

16. $\displaystyle\lim_{x \to -2^-} \dfrac{1}{x + 2}$

17. $\displaystyle\lim_{x \to 3^+} \dfrac{x - 4}{x - 3}$

18. $\displaystyle\lim_{x \to 1^+} \dfrac{2 + x}{1 - x}$

19. $\displaystyle\lim_{x \to 4^-} \dfrac{x^2}{x^2 - 16}$

20. $\displaystyle\lim_{x \to 4} \dfrac{x^2}{x^2 + 16}$

21. $\displaystyle\lim_{x \to 0^-} \left(1 + \dfrac{1}{x}\right)$

22. $\displaystyle\lim_{x \to 0^-} \left(x^2 - \dfrac{1}{x}\right)$

In Exercises 23–32, find the limit.

23. $\displaystyle\lim_{x \to \infty} \dfrac{2x - 1}{3x + 2}$

24. $\displaystyle\lim_{x \to \infty} \dfrac{5x^3 + 1}{10x^3 - 3x^2 + 7}$

25. $\displaystyle\lim_{x \to \infty} \dfrac{3x}{4x^2 - 1}$

26. $\displaystyle\lim_{x \to \infty} \dfrac{2x^{10} - 1}{10x^{11} - 3}$

27. $\displaystyle\lim_{x \to -\infty} \dfrac{5x^2}{x + 3}$

28. $\displaystyle\lim_{x \to \infty} \dfrac{x^3 - 2x^2 + 3x + 1}{x^2 - 3x + 2}$

29. $\displaystyle\lim_{x \to \infty} (2x - x^{-2})$

30. $\displaystyle\lim_{x \to \infty} (2 - x^{-3})$

31. $\displaystyle\lim_{x \to -\infty} \left(\dfrac{2x}{x - 1} + \dfrac{3x}{x + 1}\right)$

32. $\displaystyle\lim_{x \to \infty} \left(\dfrac{2x^2}{x - 1} + \dfrac{3x}{x + 1}\right)$

In Exercises 33 and 34, complete the table. Then use the result to estimate the limit of $f(x)$ as x approaches infinity.

33. $f(x) = \dfrac{x + 1}{x\sqrt{x}}$

x	10^0	10^1	10^2	10^3	10^4	10^5	10^6
$f(x)$							

34. $f(x) = x^2 - x\sqrt{x(x - 1)}$

x	10^0	10^1	10^2	10^3	10^4	10^5	10^6
$f(x)$							

In Exercises 35 and 36, use a spreadsheet software program to complete the table and use the result to estimate the limit of $f(x)$ as x approaches infinity.

35. $f(x) = \dfrac{x^2 - 1}{0.02x^2}$

x	10^0	10^1	10^2	10^3	10^4	10^5	10^6
$f(x)$							

36. $f(x) = \dfrac{3x^2}{0.1x^2 + 1}$

x	10^0	10^1	10^2	10^3	10^4	10^5	10^6
$f(x)$							

In Exercises 37 and 38, use a graphing utility to complete the table and use the result to estimate the limit of $f(x)$ as x approaches infinity and as x approaches negative infinity.

37. $f(x) = \dfrac{2x}{\sqrt{x^2 + 4}}$

x	-10^6	-10^4	-10^2	10^0	10^2	10^4	10^6
$f(x)$							

38. $f(x) = x - \sqrt{x(x-1)}$

x	-10^6	-10^4	-10^2	10^0	10^2	10^4	10^6
$f(x)$							

In Exercises 39–56, sketch the graph of the equation. Use intercepts, extrema, and asymptotes as sketching aids.

39. $y = \dfrac{2+x}{1-x}$

40. $y = \dfrac{x-3}{x-2}$

41. $f(x) = \dfrac{x^2}{x^2 + 9}$

42. $f(x) = \dfrac{x}{x^2 + 4}$

43. $g(x) = \dfrac{x^2}{x^2 - 16}$

44. $g(x) = \dfrac{x}{x^2 - 4}$

45. $xy^2 = 4$

46. $x^2 y = 4$

47. $y = \dfrac{2x}{1-x}$

48. $y = \dfrac{2x}{1-x^2}$

49. $y = 3(1 - x^{-2})$

50. $y = 1 + x^{-1}$

51. $f(x) = \dfrac{1}{x^2 - x - 2}$

52. $f(x) = \dfrac{x-2}{x^2 - 4x + 3}$

53. $g(x) = \dfrac{x^2 - x - 2}{x - 2}$

54. $g(x) = \dfrac{x^2 - 9}{x + 3}$

55. $y = \dfrac{2x^2 - 6}{(x-1)^2}$

56. $y = \dfrac{x}{(x+1)^2}$

57. Cost   The cost C (in dollars) of producing x units of a product is $C = 1.35x + 4570$.

(a) Find the average cost function $\overline{C}$.

(b) Find $\overline{C}$ when $x = 100$ and when $x = 1000$.

(c) What is the limit of $\overline{C}$ as x approaches infinity?

58. Average Cost   A business has a cost (in dollars) of $C = 0.5x + 500$ for producing x units.

(a) Find the average cost function $\overline{C}$.

(b) Find $\overline{C}$ when $x = 250$ and when $x = 1250$.

(c) What is the limit of $\overline{C}$ as x approaches infinity?

59. Cost   The cost C (in millions of dollars) for the federal government to seize $p\%$ of a type of illegal drug as it enters the country is modeled by

$$C = 528p/(100 - p), \qquad 0 \le p < 100.$$

(a) Find the cost of seizing 25%, 50%, and 75%.

(b) Find the limit of C as $p \to 100^-$.

(c) Use a graphing utility to verify the result of part (b).

60. Cost   The cost C (in dollars) of removing $p\%$ of the air pollutants in the stack emission of a utility company that burns coal is modeled by

$$C = 80{,}000p/(100 - p), \qquad 0 \le p < 100.$$

(a) Find the cost of removing 15%, 50%, and 90%.

(b) Find the limit of C as $p \to 100^-$.

(c) Use a graphing utility to verify the result of part (b).

61. Learning Curve   Psychologists have developed mathematical models to predict performance P (the percent of correct responses) as a function of n, the number of times a task is performed. One such model is

$$P = \frac{b + \theta a(n - 1)}{1 + \theta(n - 1)}$$

where a, b, and θ are constants that depend on the actual learning situation. Find the limit of P as n approaches infinity.

62. Learning Curve   Consider the learning curve given by

$$P = \frac{0.5 + 0.9(n - 1)}{1 + 0.9(n - 1)}, \qquad 0 < n.$$

(a) Complete the table for the model.

n	1	2	3	4	5	6	7	8	9	10
P										

(b) Find the limit as n approaches infinity.

(c) Use a graphing utility to graph this learning curve, and interpret the graph in the context of the problem.

63. Biology: Wildlife Management   The state game commission introduces 30 elk into a new state park. The population N of the herd is modeled by

$$N = [10(3 + 4t)]/(1 + 0.1t)$$

where t is the time in years.

(a) Find the size of the herd after 5, 10, and 25 years.

(b) According to this model, what is the limiting size of the herd as time progresses?

64. Average Profit   The cost and revenue functions for a product are $C = 34.5x + 15{,}000$ and $R = 69.9x$.

(a) Find the average profit function

$$\overline{P} = (R - C)/x.$$

(b) Find the average profit when x is 1000, 10,000, and 100,000.

(c) What is the limit of the average profit function as x approaches infinity? Explain your reasoning.

65. Average Profit   The cost and revenue functions for a product are $C = 25.5x + 1000$ and $R = 75.5x$.

(a) Find the average profit function $\overline{P} = \dfrac{R - C}{x}$.

(b) Find the average profit when x is 100, 500, and 1000.

(c) What is the limit of the average profit function as x approaches infinity? Explain your reasoning.

3.7 CURVE SKETCHING: A SUMMARY

- Analyze the graphs of functions.
- Recognize the graphs of simple polynomial functions.

Summary of Curve-Sketching Techniques

It would be difficult to overstate the importance of using graphs in mathematics. Descartes's introduction of analytic geometry contributed significantly to the rapid advances in calculus that began during the mid-seventeenth century.

So far, you have studied several concepts that are useful in analyzing the graph of a function.

- x-intercepts and y-intercepts   (Section 1.2)
- Domain and range   (Section 1.4)
- Continuity   (Section 1.6)
- Differentiability   (Section 2.1)
- Relative extrema   (Section 3.2)
- Concavity   (Section 3.3)
- Points of inflection   (Section 3.3)
- Vertical asymptotes   (Section 3.6)
- Horizontal asymptotes   (Section 3.6)

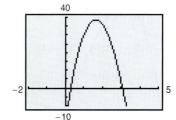

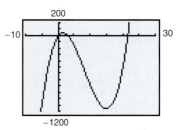

FIGURE 3.54

When you are sketching the graph of a function, either by hand or with a graphing utility, remember that you cannot normally show the *entire* graph. The decision as to which part of the graph to show is crucial. For instance, which of the viewing windows in Figure 3.54 better represents the graph of

$$f(x) = x^3 - 25x^2 + 74x - 20?$$

The lower viewing window gives a more complete view of the graph, but the context of the problem might indicate that the upper view is better. Here are some guidelines for analyzing the graph of a function.

Guidelines for Analyzing the Graph of a Function

1. Determine the domain and range of the function. If the function models a real-life situation, consider the context.

2. Determine the intercepts and asymptotes of the graph.

3. Locate the x-values where $f'(x)$ and $f''(x)$ are zero or undefined. Use the results to determine the relative extrema and points of inflection.

TECHNOLOGY

Which of the viewing windows best represents the graph of the function

$$f(x) = \frac{x^3 + 8x^2 - 33x}{5}?$$

(a) Xmin $= -15$, Xmax $= 1$,
    Ymin $= -10$, Ymax $= 60$

(b) Xmin $= -10$, Xmax $= 10$,
    Ymin $= -10$, Ymax $= 10$

(c) Xmin $= -13$, Xmax $= 5$,
    Ymin $= -10$, Ymax $= 60$

In these guidelines, note the importance of *algebra* (as well as calculus) for solving the equations $f(x) = 0$, $f'(x) = 0$, and $f''(x) = 0$.

Relative
maximum
$(-3, 32)$

$(-1, 16)$
Point of
inflection

$(-5, 0)$ $(0, 5)$

$(1, 0)$
Relative
minimum

$f(x) = x^3 + 3x^2 - 9x + 5$

FIGURE 3.55

EXAMPLE 1   **Analyzing a Graph**

Analyze the graph of

$$f(x) = x^3 + 3x^2 - 9x + 5.$$ Original function

SOLUTION   Begin by finding the intercepts of the graph. This function factors as

$$f(x) = (x - 1)^2(x + 5).$$ Factored form

So, the x-intercepts occur when $x = 1$ and $x = -5$. The derivative is

$$f'(x) = 3x^2 + 6x - 9$$ First derivative

$$= 3(x - 1)(x + 3).$$ Factored form

So, the critical numbers of f are $x = 1$ and $x = -3$. The second derivative of f is

$$f''(x) = 6x + 6$$ Second derivative

$$= 6(x + 1).$$ Factored form

which implies that the second derivative is zero when $x = -1$. By testing the values of $f'(x)$ and $f''(x)$, as shown in the table, you can see that f has one relative minimum, one relative maximum, and one point of inflection. The graph of f is shown in Figure 3.55.

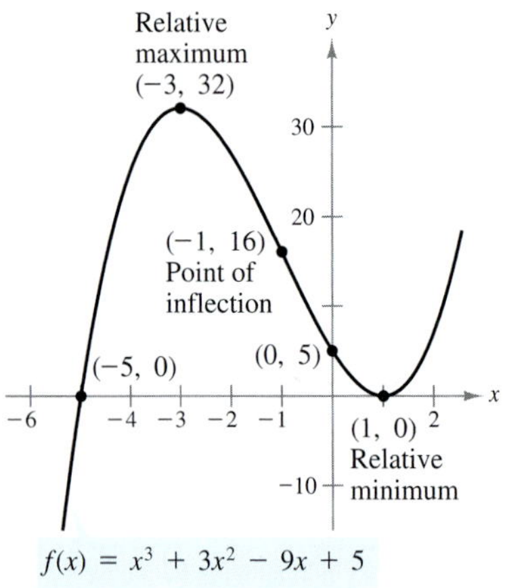

	$f(x)$	$f'(x)$	$f''(x)$	Characteristics of graph
x in $(-\infty, -3)$		$+$	$-$	Increasing, concave downward
$x = -3$	32	0	$-$	Relative maximum
x in $(-3, -1)$		$-$	$-$	Decreasing, concave downward
$x = -1$	16	$-$	0	Point of inflection
x in $(-1, 1)$		$-$	$+$	Decreasing, concave upward
$x = 1$	0	0	$+$	Relative minimum
x in $(1, \infty)$		$+$	$+$	Increasing, concave upward

TRY IT 1

Analyze the graph of $f(x) = -x^3 + 3x^2 + 9x - 27$.

TECHNOLOGY

In Example 1, you are able to find the zeros of f, f', and f'' algebraically (by factoring). When this is not feasible, you can use a graphing utility to find the zeros. For instance, the function

$$g(x) = x^3 + 3x^2 - 9x + 6$$

is similar to the function in the example, but it does not factor with integer coefficients. Using a graphing utility, you can determine that the function has only one x-intercept, $x \approx -5.0275$.

EXAMPLE 2 **Analyzing a Graph**

Analyze the graph of

$$f(x) = x^4 - 12x^3 + 48x^2 - 64x. \qquad \text{Original function}$$

SOLUTION Begin by finding the intercepts of the graph. This function factors as

$$f(x) = x(x^3 - 12x^2 + 48x - 64)$$

$$= x(x - 4)^3. \qquad \text{Factored form}$$

So, the x-intercepts occur when $x = 0$ and $x = 4$. The derivative is

$$f'(x) = 4x^3 - 36x^2 + 96x - 64 \qquad \text{First derivative}$$

$$= 4(x - 1)(x - 4)^2. \qquad \text{Factored form}$$

So, the critical numbers of f are $x = 1$ and $x = 4$. The second derivative of f is

$$f''(x) = 12x^2 - 72x + 96 \qquad \text{Second derivative}$$

$$= 12(x - 4)(x - 2) \qquad \text{Factored form}$$

which implies that the second derivative is zero when $x = 2$ and $x = 4$. By testing the values of $f'(x)$ and $f''(x)$, as shown in the table, you can see that f has one relative minimum and two points of inflection. The graph is shown in Figure 3.56.

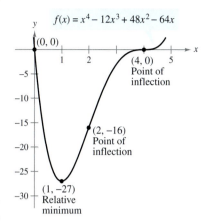

FIGURE 3.56

	$f(x)$	$f'(x)$	$f''(x)$	Characteristics of graph
x in $(-\infty, 1)$		$-$	$+$	Decreasing, concave upward
$x = 1$	-27	0	$+$	Relative minimum
x in $(1, 2)$		$+$	$+$	Increasing, concave upward
$x = 2$	-16	$+$	0	Point of inflection
x in $(2, 4)$		$+$	$-$	Increasing, concave downward
$x = 4$	0	0	0	Point of inflection
x in $(4, \infty)$		$+$	$+$	Increasing, concave upward

TRY IT 2

Analyze the graph of $f(x) = x^4 - 4x^3 + 5$.

DISCOVERY

A polynomial function of degree n can have at most $n - 1$ relative extrema and at most $n - 2$ points of inflection. For instance, the third-degree polynomial in Example 1 has two relative extrema and one point of inflection. Similarly, the fourth-degree polynomial function in Example 2 has one relative extremum and two points of inflection. Is it possible for a third-degree function to have no relative extrema? Is it possible for a fourth-degree function to have no relative extrema?

Show that the function in Example 3 can be rewritten as

$$f(x) = \frac{x^2 - 2x + 4}{x - 2}$$

$$= x + \frac{4}{x - 2}.$$

Use a graphing utility to graph f together with the line $y = x$. How do the two graphs compare as you zoom out? Describe what is meant by a "slant asymptote." Find the slant asymptote of the function $g(x) = \dfrac{x^2 - x - 1}{x - 1}$.

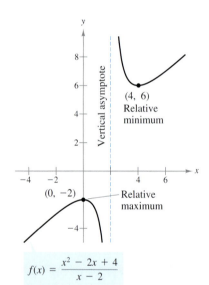

$$f(x) = \frac{x^2 - 2x + 4}{x - 2}$$

FIGURE 3.57

EXAMPLE 3 Analyzing a Graph

Analyze the graph of

$$f(x) = \frac{x^2 - 2x + 4}{x - 2}. \qquad \text{Original function}$$

SOLUTION   The y-intercept occurs at $(0, -2)$. Using the Quadratic Formula on the numerator, you can see that there are no x-intercepts. Because the denominator is zero when $x = 2$ (and the numerator is not zero when $x = 2$), it follows that $x = 2$ is a vertical asymptote of the graph. There are no horizontal asymptotes because the degree of the numerator is greater than the degree of the denominator. The derivative is

$$f'(x) = \frac{(x - 2)(2x - 2) - (x^2 - 2x + 4)}{(x - 2)^2} \qquad \text{First derivative}$$

$$= \frac{x(x - 4)}{(x - 2)^2}. \qquad \text{Factored form}$$

So, the critical numbers of f are $x = 0$ and $x = 4$. The second derivative is

$$f''(x) = \frac{(x - 2)^2(2x - 4) - (x^2 - 4x)(2)(x - 2)}{(x - 2)^4} \qquad \text{Second derivative}$$

$$= \frac{(x - 2)(2x^2 - 8x + 8 - 2x^2 + 8x)}{(x - 2)^4}$$

$$= \frac{8}{(x - 2)^3}. \qquad \text{Factored form}$$

Because the second derivative has no zeros and because $x = 2$ is not in the domain of the function, you can conclude that the graph has no points of inflection. By testing the values of $f'(x)$ and $f''(x)$, as shown in the table, you can see that f has one relative minimum and one relative maximum. The graph of f is shown in Figure 3.57.

	$f(x)$	$f'(x)$	$f''(x)$	Characteristics of graph
x in $(-\infty, 0)$		$+$	$-$	Increasing, concave downward
$x = 0$	-2	0	$-$	Relative maximum
x in $(0, 2)$		$-$	$-$	Decreasing, concave downward
$x = 2$	Undef.	Undef.	Undef.	Vertical asymptote
x in $(2, 4)$		$-$	$+$	Decreasing, concave upward
$x = 4$	6	0	$+$	Relative minimum
x in $(4, \infty)$		$+$	$+$	Increasing, concave upward

TRY IT 3

Analyze the graph of $f(x) = \dfrac{x^2}{x - 1}$.

■ **EXAMPLE 4** **Analyzing a Graph**

Analyze the graph of

$$f(x) = \frac{2(x^2 - 9)}{x^2 - 4}.$$ Original function

SOLUTION Begin by writing the function in factored form.

$$f(x) = \frac{2(x - 3)(x + 3)}{(x - 2)(x + 2)}$$ Factored form

The y-intercept is $\left(0, \frac{9}{2}\right)$, and the x-intercepts are $(-3, 0)$ and $(3, 0)$. There are vertical asymptotes at $x = \pm 2$ and a horizontal asymptote at $y = 2$. The first derivative is

$$f'(x) = \frac{2[(x^2 - 4)(2x) - (x^2 - 9)(2x)]}{(x^2 - 4)^2}$$ First derivative

$$= \frac{20x}{(x^2 - 4)^2}.$$ Factored form

So, the critical number of f is $x = 0$. The second derivative of f is

$$f''(x) = \frac{(x^2 - 4)^2(20) - (20x)(2)(2x)(x^2 - 4)}{(x^2 - 4)^4}$$ Second derivative

$$= \frac{20(x^2 - 4)(x^2 - 4 - 4x^2)}{(x^2 - 4)^4}$$

$$= -\frac{20(3x^2 + 4)}{(x^2 - 4)^3}.$$ Factored form

Because the second derivative has no zeros and $x = \pm 2$ are not in the domain of the function, you can conclude that the graph has no points of inflection. By testing the values of $f'(x)$ and $f''(x)$, as shown in the table, you can see that f has one relative minimum. The graph of f is shown in Figure 3.58.

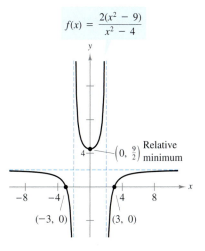

$f(x) = \dfrac{2(x^2 - 9)}{x^2 - 4}$

FIGURE 3.58

	$f(x)$	$f'(x)$	$f''(x)$	Characteristics of graph
x in $(-\infty, -2)$		$-$	$-$	Decreasing, concave downward
$x = -2$	Undef.	Undef.	Undef.	Vertical asymptote
x in $(-2, 0)$		$-$	$+$	Decreasing, concave upward
$x = 0$	$\frac{9}{2}$	0	$+$	Relative minimum
x in $(0, 2)$		$+$	$+$	Increasing, concave upward
$x = 2$	Undef.	Undef.	Undef.	Vertical asymptote
x in $(2, \infty)$		$+$	$-$	Increasing, concave downward

TRY IT 4

Analyze the graph of $f(x) = \dfrac{x^2 + 1}{x^2 - 1}$.

Some graphing utilities will not graph the function in Example 5 properly if the function is entered as

$$f(x) = 2x\wedge(5/3) - 5x\wedge(4/3).$$

To correct for this, you can enter the function as

$$f(x) = 2(\sqrt[3]{x})\wedge 5 - 5(\sqrt[3]{x})\wedge 4.$$

Try entering both functions into a graphing utility to see whether both functions produce correct graphs.

ALGEBRA REVIEW

For help on the algebra in Example 5, see Example 2(a) in the *Chapter 3 Algebra Review*, on page 249.

EXAMPLE 5 **Analyzing a Graph**

Analyze the graph of

$$f(x) = 2x^{5/3} - 5x^{4/3}. \qquad \text{Original function}$$

SOLUTION Begin by writing the function in factored form.

$$f(x) = x^{4/3}(2x^{1/3} - 5) \qquad \text{Factored form}$$

One of the intercepts is $(0, 0)$. A second x-intercept occurs when $2x^{1/3} = 5$.

$$2x^{1/3} = 5$$
$$x^{1/3} = \tfrac{5}{2}$$
$$x = \left(\tfrac{5}{2}\right)^3$$
$$x = \tfrac{125}{8}$$

The first derivative is

$$f'(x) = \tfrac{10}{3}x^{2/3} - \tfrac{20}{3}x^{1/3} \qquad \text{First derivative}$$
$$= \tfrac{10}{3}x^{1/3}(x^{1/3} - 2). \qquad \text{Factored form}$$

So, the critical numbers of f are $x = 0$ and $x = 8$. The second derivative is

$$f''(x) = \tfrac{20}{9}x^{-1/3} - \tfrac{20}{9}x^{-2/3} \qquad \text{Second derivative}$$
$$= \tfrac{20}{9}x^{-2/3}(x^{1/3} - 1)$$
$$= \frac{20(x^{1/3} - 1)}{9x^{2/3}}. \qquad \text{Factored form}$$

So, possible points of inflection occur when $x = 1$ and when $x = 0$. By testing the values of $f'(x)$ and $f''(x)$, as shown in the table, you can see that f has one relative maximum, one relative minimum, and one point of inflection. The graph of f is shown in Figure 3.59.

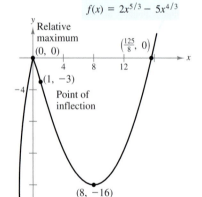

$$f(x) = 2x^{5/3} - 5x^{4/3}$$

FIGURE 3.59

	$f(x)$	$f'(x)$	$f''(x)$	Characteristics of graph
x in $(-\infty, 0)$		$+$	$-$	Increasing, concave downward
$x = 0$	0	0	Undef.	Relative maximum
x in $(0, 1)$		$-$	$-$	Decreasing, concave downward
$x = 1$	-3	$-$	0	Point of inflection
x in $(1, 8)$		$-$	$+$	Decreasing, concave upward
$x = 8$	-16	0	$+$	Relative minimum
x in $(8, \infty)$		$+$	$+$	Increasing, concave upward

TRY IT 5

Analyze the graph of

$$f(x) = 2x^{3/2} - 6x^{1/2}.$$

Summary of Simple Polynomial Graphs

A summary of the graphs of polynomial functions of degrees 0, 1, 2, and 3 is shown in Figure 3.60. Because of their simplicity, lower-degree polynomial functions are commonly used as mathematical models.

Constant function (degree 0):

$$y = a$$

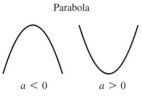

Horizontal line

Linear function (degree 1):

$$y = ax + b$$

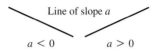

Line of slope a

$a < 0$ $a > 0$

Quadratic function (degree 2):

$$y = ax^2 + bx + c$$

Parabola

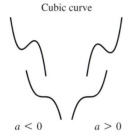

$a < 0$ $a > 0$

Cubic function (degree 3):

$$y = ax^3 + bx^2 + cx + d$$

Cubic curve

$a < 0$ $a > 0$

FIGURE 3.60

STUDY TIP

The graph of any cubic polynomial has one point of inflection. The slope of the graph at the point of inflection may be zero or nonzero.

TAKE ANOTHER LOOK

Graphs of Fourth-Degree Polynomial Functions

In the summary presented above, the graphs of cubic functions are classified into four basic types. How many basic types of graphs are possible with fourth-degree polynomial functions? Make a rough sketch of each type. Then use a graphing utility to classify each of the following. (In each case, use a viewing window that shows all the basic characteristics of the graph.)

a. $y = x^4$

b. $y = -x^4 + 5x^2$

c. $y = x^4 - x^3 + x$

d. $y = -x^4 - 4x^3 - 3x^2 + x$

e. $y = -x^4 + 2x^2$

f. $y = -x^4 + x^3$

g. $y = x^4 - 5x$

h. $y = x^4 - 8x^3 + 8x^2$

The following warm-up exercises involve skills that were covered in earlier sections. You will use these skills in the exercise set for this section.

In Exercises 1–4, find the vertical and horizontal asymptotes of the graph.

1. $f(x) = \dfrac{1}{x^2}$

2. $f(x) = \dfrac{8}{(x-2)^2}$

3. $f(x) = \dfrac{40x}{x+3}$

4. $f(x) = \dfrac{x^2 - 3}{x^2 - 4x + 3}$

In Exercises 5–10, determine the open intervals on which the function is increasing or decreasing.

5. $f(x) = x^2 + 4x + 2$

6. $f(x) = -x^2 - 8x + 1$

7. $f(x) = x^3 - 3x + 1$

8. $f(x) = \dfrac{-x^3 + x^2 - 1}{x^2}$

9. $f(x) = \dfrac{x-2}{x-1}$

10. $f(x) = -x^3 - 4x^2 + 3x + 2$

EXERCISES 3.7

In Exercises 1–20, sketch the graph of the function. Choose a scale that allows all relative extrema and points of inflection to be identified on the graph.

1. $y = -x^2 - 2x + 3$

2. $y = 2x^2 - 4x + 1$

3. $y = x^3 - 4x^2 + 6$

4. $y = -\frac{1}{3}(x^3 - 3x + 2)$

5. $y = 2 - x - x^3$

6. $y = x^3 + 3x^2 + 3x + 2$

7. $y = 3x^3 - 9x + 1$

8. $y = -4x^3 + 6x^2$

9. $y = 3x^4 + 4x^3$

10. $y = 3x^4 - 6x^2$

11. $y = x^3 - 6x^2 + 3x + 10$

12. $y = -x^3 + 3x^2 + 9x - 2$

13. $y = x^4 - 8x^3 + 18x^2 - 16x + 5$

14. $y = x^4 - 4x^3 + 16x - 16$

15. $y = x^4 - 4x^3 + 16x$

16. $y = x^5 + 1$

17. $y = x^5 - 5x$

18. $y = (x - 1)^5$

19. $y = \begin{cases} x^2 + 1, & x \le 0 \\ 1 - 2x, & x > 0 \end{cases}$

20. $y = \begin{cases} x^2 + 4, & x < 0 \\ 4 - x, & x \ge 0 \end{cases}$

 In Exercises 21–32, use a graphing utility to graph the function. Choose a window that allows all relative extrema and points of inflection to be identified on the graph.

21. $y = \dfrac{x^2 + 2}{x^2 + 1}$

22. $y = \dfrac{x}{x^2 + 1}$

23. $y = 3x^{2/3} - 2x$

24. $y = 3x^{2/3} - x^2$

25. $y = 1 - x^{2/3}$

26. $y = (1 - x)^{2/3}$

27. $y = x^{1/3} + 1$

28. $y = x^{-1/3}$

29. $y = x^{5/3} - 5x^{2/3}$

30. $y = x^{4/3}$

31. $y = x\sqrt{x^2 - 9}$

32. $y = \dfrac{x}{\sqrt{x^2 - 4}}$

In Exercises 33–42, sketch the graph of the function. Label the intercepts, relative extrema, points of inflection, and asymptotes. Then state the domain of the function.

33. $y = \dfrac{5 - 3x}{x - 2}$

34. $y = \dfrac{x^2 + 1}{x^2 - 2}$

35. $y = \dfrac{2x}{x^2 - 1}$

36. $y = \dfrac{x^2 - 6x + 12}{x - 4}$

37. $y = x\sqrt{4 - x}$

38. $y = x\sqrt{4 - x^2}$

39. $y = \dfrac{x - 3}{x}$

40. $y = x + \dfrac{32}{x^2}$

41. $y = \dfrac{x^3}{x^3 - 1}$

42. $y = \dfrac{x^4}{x^4 - 1}$

In Exercises 43–46, find values of *a*, *b*, *c*, and *d* such that the graph of $f(x) = ax^3 + bx^2 + cx + d$ will resemble the given graph. Then use a graphing utility to verify your result. (There are many correct answers.)

43.

44.

45.

46.

In Exercises 47–50, use the graph of *f′* or *f″* to sketch the graph of *f*. (There are many correct answers.)

47.

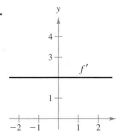

48.

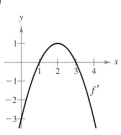

49.

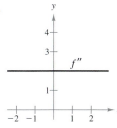

50.

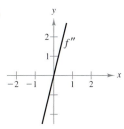

In Exercises 51 and 52, sketch a graph of a function *f* having the given characteristics. (There are many correct answers.)

51. $f(-2) = 0$

$f(0) = 0$

$f'(x) > 0, \quad -\infty < x < -1$

$f'(-1) = 0$

$f'(x) < 0, \quad -1 < x < 0$

$f'(0) = 0$

$f'(x) > 0, \quad 0 < x < \infty$

52. $f(-1) = 0$

$f(3) = 0$

$f'(1)$ is undefined.

$f'(x) < 0, \quad -\infty < x < 1$

$f'(x) > 0, \quad 1 < x < \infty$

$f''(x) < 0, \quad x \neq 1$

$\lim_{x \to \infty} f(x) = 4$

53. *Cost* An employee of a delivery company earns $9 per hour driving a delivery van in an area where gasoline costs $1.80 per gallon. When the van is driven at a constant speed *s* (in miles per hour, with $40 \leq s \leq 65$), the van gets $500/s$ miles per gallon.

(a) Find the cost *C* as a function of *s* for a 100-mile trip on an interstate highway.

 (b) Use a graphing utility to graph the function found in part (a) and determine the most economical speed.

54. *Profit* The management of a company is considering three possible models for predicting the company's profits from 2001 through 2006. Model I gives the expected annual profits if the current trends continue. Models II and III give the expected annual profits for various combinations of increased labor and energy costs. In each model, *p* is the profit (in billions of dollars) and $t = 0$ corresponds to 2001.

Model I: $p = 0.03t^2 - 0.01t + 3.39$

Model II: $p = 0.08t + 3.36$

Model III: $p = -0.07t^2 + 0.05t + 3.38$

(a) Use a graphing utility to graph all three models in the same viewing window.

(b) For which models are profits increasing during the interval from 2001 through 2006?

(c) Which model is the most optimistic? Which is the most pessimistic?

55. *Meteorology* The monthly normal temperature *T* (in degrees Fahrenheit) for Pittsburgh, Pennsylvania can be modeled by

$$T = \frac{23.011 - 1.0t + 0.048t^2}{1 - 0.204t + 0.014t^2}, \quad 1 \leq t \leq 12$$

where *t* is the month, with $t = 1$ corresponding to January. Use a graphing utility to graph the model and find all absolute extrema. Explain the meaning of those values. *(Source: National Climatic Data Center)*

Writing In Exercises 56 and 57, use a graphing utility to graph the function. Explain why there is no vertical asymptote when a superficial examination of the function may indicate that there should be one.

56. $h(x) = \dfrac{6 - 2x}{3 - x}$

57. $g(x) = \dfrac{x^2 + x - 2}{x - 1}$

3.8   DIFFERENTIALS AND MARGINAL ANALYSIS

- Find the differentials of functions.
- Use differentials to approximate changes in functions.
- Use differentials to approximate changes in real-life models.

Differentials

When the derivative was defined in Section 2.1 as the limit of the ratio $\Delta y/\Delta x$, it seemed natural to retain the quotient symbolism for the limit itself. So, the derivative of y with respect to x was denoted by

$$\frac{dy}{dx} = \lim_{\Delta x \to 0} \frac{\Delta y}{\Delta x}$$

even though we did not interpret dy/dx as the quotient of two separate quantities. In this section, you will see that the quantities dy and dx can be assigned meanings in such a way that their quotient, when $dx \neq 0$, is equal to the derivative of y with respect to x.

Definition of Differentials

Let $y = f(x)$ represent a differentiable function. The **differential of x** (denoted by dx) is any nonzero real number. The **differential of y** (denoted by dy) is

$$dy = f'(x)\, dx.$$

One use of differentials is in approximating the change in $f(x)$ that corresponds to a change in x, as shown in Figure 3.61. This change is denoted by

$$\Delta y = f(x + \Delta x) - f(x). \qquad \text{Change in } y$$

In Figure 3.61, notice that as Δx gets smaller and smaller, the values of dy and Δy get closer and closer. That is, when Δx is small, $dy \approx \Delta y$.

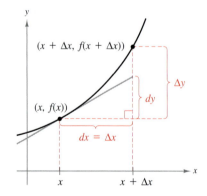

FIGURE 3.61

This **tangent line approximation** is the basis for most applications of differentials.

EXAMPLE 1   Interpreting Differentials Graphically

Consider the function given by

$$f(x) = x^2. \qquad \text{Original function}$$

Find the value of dy when $x = 1$ and $dx = 0.01$. Compare this with the value of Δy when $x = 1$ and $\Delta x = 0.01$. Interpret the results graphically.

SOLUTION   Begin by finding the derivative of f.

$$f'(x) = 2x \qquad \text{Derivative of } f$$

When $x = 1$ and $dx = 0.01$, the value of the differential dy is

$$
\begin{aligned}
dy &= f'(x)\,dx & \text{Differential of } y \\
&= f'(1)(0.01) & \text{Substitute 1 for } x \text{ and } 0.01 \text{ for } dx. \\
&= 2(1)(0.01) & \text{Use } f'(x) = 2x. \\
&= 0.02. & \text{Simplify.}
\end{aligned}
$$

When $x = 1$ and $\Delta x = 0.01$, the value of Δy is

$$
\begin{aligned}
\Delta y &= f(x + \Delta x) - f(x) & \text{Change in } y \\
&= f(1.01) - f(1) & \text{Substitute 1 for } x \text{ and } 0.01 \text{ for } \Delta x. \\
&= (1.01)^2 - (1)^2 \\
&= 0.0201. & \text{Simplify.}
\end{aligned}
$$

Note that $dy \approx \Delta y$, as shown in Figure 3.62.

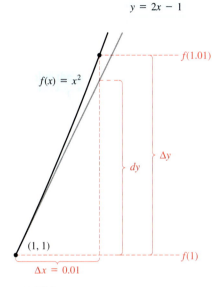

FIGURE 3.62

TRY IT 1

Find the value of dy when $x = 2$ and $dx = 0.01$ for $f(x) = x^4$. Compare this with the value of Δy when $x = 2$ and $\Delta x = 0.01$.

The validity of the approximation

$$dy \approx \Delta y, \quad dx \neq 0$$

stems from the definition of the derivative. That is, the existence of the limit

$$f'(x) = \lim_{\Delta x \to 0} \frac{f(x + \Delta x) - f(x)}{\Delta x}$$

implies that when Δx is close to zero, then $f'(x)$ is close to the difference quotient. So, you can write

$$
\begin{aligned}
\frac{f(x + \Delta x) - f(x)}{\Delta x} &\approx f'(x) \\
f(x + \Delta x) - f(x) &\approx f'(x)\,\Delta x \\
\Delta y &\approx f'(x)\,\Delta x.
\end{aligned}
$$

Substituting dx for Δx and dy for $f'(x)\,dx$ produces

$$\Delta y \approx dy.$$

STUDY TIP

Find an equation of the tangent line $y = g(x)$ to the graph of $f(x) = x^2$ at the point $x = 1$. Evaluate $g(1.01)$ and $f(1.01)$.

3 CHAPTER SUMMARY AND STUDY STRATEGIES

After studying this chapter, you should have acquired the following skills. The exercise numbers are keyed to the Review Exercises that begin on page 252. Answers to odd-numbered Review Exercises are given in the back of the text.*

Skills

■ Find the critical numbers of a function. *(Section 3.1)*   *Review Exercises 1–4*

   c is a critical number of f if $f'(c) = 0$ or $f'(c)$ is undefined.

■ Find the open intervals on which a function is increasing or decreasing. *(Section 3.1)*   *Review Exercises 5–8*

   Increasing if $f'(x) > 0$

   Decreasing if $f'(x) < 0$

■ Find intervals on which a real-life model is increasing or decreasing, and interpret   *Review Exercises 9, 10, 95*
   the results in context. *(Section 3.1)*

■ Use the First-Derivative Test to find the relative extrema of a function. *(Section 3.2)*   *Review Exercises 11–20*

■ Find the absolute extrema of a continuous function on a closed interval. *(Section 3.2)*   *Review Exercises 21–30*

■ Find minimum and maximum values of a real-life model and interpret the results   *Review Exercises 31, 32*
   in context. *(Section 3.2)*

■ Find the open intervals on which the graph of a function is concave upward or   *Review Exercises 33–36*
   concave downward. *(Section 3.3)*

   Concave upward if $f''(x) > 0$

   Concave downward if $f''(x) < 0$

■ Find the points of inflection of the graph of a function. *(Section 3.3)*   *Review Exercises 37–40*

■ Use the Second-Derivative Test to find the relative extrema of a function.   *Review Exercises 41–44*
   (Section 3.3)

■ Find the point of diminishing returns of an input-output model. *(Section 3.3)*   *Review Exercises 45, 46*

■ Solve real-life optimization problems. *(Section 3.4)*   *Review Exercises 47–53, 96*

■ Solve business and economics optimization problems. *(Section 3.5)*   *Review Exercises 54–58, 99*

■ Find the price elasticity of demand for a demand function. *(Section 3.5)*   *Review Exercises 59–62*

■ Find the vertical and horizontal asymptotes of a function and sketch its graph.   *Review Exercises 63–68*
   (Section 3.6)

■ Find infinite limits and limits at infinity. *(Section 3.6)*   *Review Exercises 69–76*

■ Use asymptotes to answer questions about real life. *(Section 3.6)*   *Review Exercises 77, 78*

■ Analyze the graph of a function. *(Section 3.7)*   *Review Exercises 79–86*

■ Find the differential of a function. *(Section 3.8)*   *Review Exercises 87–90*

■ Use differentials to approximate changes in a function. *(Section 3.8)*   *Review Exercises 91–94*

■ Use differentials to approximate changes in real-life models. *(Section 3.8)*   *Review Exercises 97, 98*

---

* Use a wide range of valuable study aids to help you master the material in this chapter. The *Student Solutions Guide* includes step-by-step solutions to all odd-numbered exercises to help you review and prepare. The *HM mathSpace® Student CD-ROM* helps you brush up on your algebra skills. The *Graphing Technology Guide*, available on the Web at *math.college.hmco.com/students*, offers step-by-step commands and instructions for a wide variety of graphing calculators, including the most recent models.

■ **Solve Problems Graphically, Analytically, and Numerically**   When analyzing the graph of a function, use a variety of problem-solving strategies. For instance, if you were asked to analyze the graph of

$$f(x) = x^3 - 4x^2 + 5x - 4$$

you could begin *graphically*. That is, you could use a graphing utility to find a viewing window that appears to show the important characteristics of the graph. From the graph shown below, the function appears to have one relative minimum, one relative maximum, and one point of inflection.

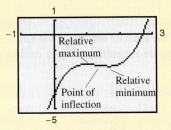

Next, you could use calculus to *analyze* the graph. Because the derivative of f is

$$f'(x) = 3x^2 - 8x + 5 = (3x - 5)(x - 1)$$

the critical numbers of f are $x = \frac{5}{3}$ and $x = 1$. By the First-Derivative Test, you can conclude that $x = \frac{5}{3}$ yields a relative minimum and $x = 1$ yields a relative maximum. Because

$$f''(x) = 6x - 8$$

you can conclude that $x = \frac{4}{3}$ yields a point of inflection. Finally, you could analyze the graph *numerically*. For instance, you could construct a table of values and observe that f is increasing on the interval $(-\infty, 1)$, decreasing on the interval $\left(1, \frac{5}{3}\right)$, and increasing on the interval $\left(\frac{5}{3}, \infty\right)$.

■ **Problem-Solving Strategies**   If you get stuck when trying to solve an optimization problem, consider the strategies below.

1. *Draw a Diagram.*   If feasible, draw a diagram that represents the problem. Label all known values and unknown values on the diagram.

2. *Solve a Simpler Problem.*   Simplify the problem, or write several simple examples of the problem. For instance, if you are asked to find the dimensions that will produce a maximum area, try calculating the areas of several examples.

3. *Rewrite the Problem in Your Own Words.*   Rewriting a problem can help you understand it better.

4. *Guess and Check.*   Try guessing the answer, then check your guess in the statement of the original problem. By refining your guesses, you may be able to think of a general strategy for solving the problem.

Study Tools   *Additional resources that accompany this chapter*

■ **Algebra Review**   (pages 248 and 249)

■ **Chapter Summary and Study Strategies** (pages 250 and 251)

■ **Review Exercises**   (pages 252–255)

■ **Sample Post-Graduation Exam Questions**   (page 256)

■ **Web Exercises**   (page 200, Exercise 66; page 219, Exercise 43)

■ **Student Solutions Guide**

■ **HM mathSpace® Student CD-ROM**

■ **Graphing Technology Guide** (*math.college.hmco.com/students*)

57. *Inventory Cost*   The cost C of inventory modeled by

$$C = \left(\frac{Q}{x}\right)s + \left(\frac{x}{2}\right)r$$

depends on ordering and storage costs, where Q is the number of units sold per year, r is the cost of storing one unit for 1 year, s is the cost of placing an order, and x is the number of units in the order. Determine the order size that will minimize the cost when $Q = 10,000$, $s = 4.5$, and $r = 5.76$.

58. *Profit*   The demand and cost functions for a product are given by

$$p = 600 - 3x$$

and

$$C = 0.3x^2 + 6x + 600$$

where p is the price per unit, x is the number of units, and C is the total cost. The profit for producing x units is given by

$$P = xp - C - xt$$

where t is the excise tax per unit. Find the maximum profits for excise taxes of $t = \$5$, $t = \$10$, and $t = \$20$.

In Exercises 59–62, find the intervals on which the demand is elastic, inelastic, and of unit elasticity.

59. $p = 30 - 0.2x$, $0 \le x \le 150$

60. $p = 60 - 0.04x$, $0 \le x \le 1500$

61. $p = \sqrt{300 - x}$, $0 \le x \le 300$

62. $p = \sqrt{960 - x}$, $0 \le x \le 960$

In Exercises 63–68, find the vertical and horizontal asymptotes of the graph. Then use a graphing utility to graph the function.

63. $h(x) = \dfrac{2x + 3}{x - 4}$

64. $g(x) = \dfrac{5x^2}{x^2 + 2}$

65. $f(x) = \dfrac{\sqrt{9x^2 + 1}}{x}$

66. $h(x) = \dfrac{3x}{\sqrt{x^2 + 2}}$

67. $f(x) = \dfrac{3}{x^2 - 5x + 4}$

68. $h(x) = \dfrac{2x^2 + 3x - 5}{x - 1}$

In Exercises 69–76, find the limit, if it exists.

69. $\displaystyle\lim_{x \to 0^+} \left(x - \frac{1}{x^3}\right)$

70. $\displaystyle\lim_{x \to 0^-} \left(3 + \frac{1}{x}\right)$

71. $\displaystyle\lim_{x \to -1^+} \frac{x^2 - 2x + 1}{x + 1}$

72. $\displaystyle\lim_{x \to 3^-} \frac{3x^2 + 1}{x^2 - 9}$

73. $\displaystyle\lim_{x \to \infty} \frac{5x^2 + 3}{2x^2 - x + 1}$

74. $\displaystyle\lim_{x \to \infty} \frac{3x^2 - 2x + 3}{x + 1}$

75. $\displaystyle\lim_{x \to -\infty} \frac{3x^2}{x + 2}$

76. $\displaystyle\lim_{x \to -\infty} \left(\frac{x}{x - 2} + \frac{2x}{x + 2}\right)$

77. *Health*   For a person with sensitive skin, the maximum amount T (in hours) of exposure to the sun that can be tolerated before skin damage occurs can be modeled by

$$T = \frac{0.37s + 23.8}{s}, \quad 0 < s \le 120$$

where s is the Sunsor Scale reading.   *(Source: Sunsor, Inc.)*

(a) Use a graphing utility to graph the model. Compare your result with the graph below.

(b) Describe the value of T as s increases.

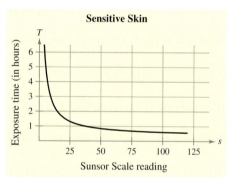

Sensitive Skin

(vertical axis) Exposure time (in hours)

(horizontal axis) Sunsor Scale reading

78. *Average Cost and Profit*   The cost and revenue functions for a product are given by

$$C = 10,000 + 48.9x$$

and

$$R = 68.5x.$$

(a) Find the average cost function.

(b) What is the limit of the average cost as x approaches infinity?

(c) Find the average profits when x is 1 million, 2 million, and 10 million.

(d) What is the limit of the average profit as x increases without bound?

In Exercises 79–86, sketch the graph of the function. Label the intercepts, relative extrema, points of inflection, and asymptotes. State the domain of the function.

79. $f(x) = 4x - x^2$

80. $f(x) = 4x^3 - x^4$

81. $f(x) = x\sqrt{16 - x^2}$

82. $f(x) = x^2\sqrt{9 - x^2}$

83. $f(x) = \dfrac{x + 1}{x - 1}$

84. $f(x) = \dfrac{2x}{1 + x^2}$

85. $f(x) = x^2 + \dfrac{2}{x}$

86. $f(x) = x^{4/5}$

In Exercises 87–90, find the differential dy.

87. $y = 6x^2 - 5$

88. $y = (3x^2 - 2)^3$

89. $y = -\dfrac{5}{\sqrt[3]{x}}$

90. $y = \dfrac{2 - x}{x + 5}$

In Exercises 91–94, use differentials to approximate the change in cost, revenue, or profit corresponding to an increase in sales of one unit.

91. $C = 40x^2 + 1225, \quad x = 10$

92. $C = 1.5\sqrt[3]{x} + 500, \quad x = 125$

93. $R = 6.25x + 0.4x^{3/2}, \quad x = 225$

94. $P = 0.003x^2 + 0.019x - 1200, \quad x = 750$

 95. Revenue Per Share   The revenues per share R (in dollars) for the Walt Disney Company for the years 1992 through 2003 are shown in the table. *(Source: The Walt Disney Company)*

Year, t	2	3	4	5	6	7
Revenue per share, R	4.77	5.31	6.40	7.70	10.50	11.10

Year, t	8	9	10	11	12	13
Revenue per share, R	11.21	11.34	12.09	12.52	12.40	13.23

(a) Use a graphing utility to create a scatter plot of the data, where t is the time in years, with $t = 2$ corresponding to 1992.

(b) Describe any trends and/or patterns of the data.

(c) A model for the data is

$$R = \dfrac{4.72 - 1.605t + 0.1741t^2}{1 - 0.356t + 0.0420t^2 - 0.00112t^3},$$

$$2 \le t \le 13.$$

Graph the model and the data in the same viewing window.

(d) Find the years when the revenue per share was increasing and decreasing.

(e) Find the years when the rate of change of the revenue per share was increasing and decreasing.

(f) Briefly explain your results for parts (d) and (e).

 96. Medicine   The effectiveness E of a pain-killing drug t hours after entering the bloodstream is modeled by

$$E = 22.5t + 7.5t^2 - 2.5t^3, \quad 0 \le t \le 4.5.$$

(a) Use a graphing utility to graph the equation. Choose an appropriate window.

(b) Find the maximum effectiveness the pain-killing drug attains over the interval $[0, 4.5]$.

97. Surface Area and Volume   The diameter of a sphere is measured to be 18 inches with a possible error of 0.05 inch. Use differentials to approximate the possible error in the surface area and the volume of the sphere.

98. Demand   A company finds that the demand for its product is modeled by

$$p = 85 - 0.125x.$$

If x changes from 7 to 8, what is the corresponding change in p? Compare the values of Δp and dp.

 99. Economics: Revenue   Consider the following cost and demand information for a monopoly (in dollars). Complete the table, and then use the information to answer the questions. *(Source: Adapted from Taylor, Economics, Fourth Edition)*

Quantity of output	Price	Total revenue	Marginal revenue
1	14.00		
2	12.00		
3	10.00		
4	8.50		
5	7.00		
6	5.50		

(a) Use the *regression* feature of a graphing utility to find a quadratic model for the total revenue data.

(b) From the total revenue model you found in part (a), use derivatives to find an equation for the marginal revenue. Now use the values for output in the table and compare the results with the values in the marginal revenue column of the table. How close was your model?

(c) What quantity maximizes total revenue for the monopoly?

3 SAMPLE POST-GRADUATION EXAM QUESTIONS

The following questions represent the types of questions that appear on certified public accountant (CPA) exams, Graduate Management Admission Tests (GMAT), Graduate Records Exams (GRE), actuarial exams, and College-Level Academic Skills Tests (CLAST). The answers to the questions are given in the back of the book.

For Questions 1–3, use the data shown in the graph.
(Source: U.S. National Center for Health Statistics)

Figure for 1–3

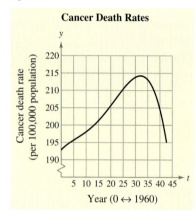

1. The percent decrease in the cancer death rate between 1990 and 2000 was about

(a) 3%   (b) 6.5%   (c) 4%   (d) 16%

2. Which of the following statements about the cancer death rate can be inferred from the graph?

I. The cancer death rate reached a maximum around 1990.

II. The cancer death rate was increasing at a maximum rate around 1980.

III. Between 1960 and 1990, the average rate of change of the cancer death rate was about 0.75 per year.

(a) I and II   (b) I and III   (c) II and III   (d) I, II, and III

3. Let f represent the cancer death rate function. In 1990, $f'(x) =$

(a) 208   (b) 0   (c) 1   (d) undefined

4. In 2000, a company issued 75,000 shares of stock. Each share of stock was worth $85.75. Five years later, each share of stock was worth $72.21. How much less were the shares worth in 2005 than in 2000?

(a) $1,115,500   (b) $1,051,500   (c) $1,155,000   (d) $1,015,500

5. Which of the figures below most resembles the graph of $f(x) = \dfrac{x}{x^2 - 4}$?

(a)
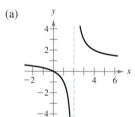

(b)
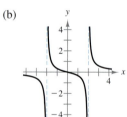

(c)
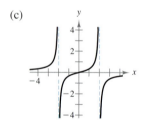

(d)
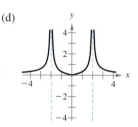

chapter

4

Exponential and Logarithmic Functions

© Ketut Suardana/Stringer/Reuters/CORBIS

On January 2, 2004, the Indonesian islands of Bali and Lombok experienced an earthquake measuring 6.1 on the Richter scale, a logarithmic function that serves as one way to calculate an earthquake's magnitude.

STRATEGIES FOR SUCCESS

WHAT YOU SHOULD LEARN:

- How to graph the exponential function $f(x) = a^x$ and how to graph the natural exponential function $f(x) = e^x$

- How to calculate derivatives of exponential functions

- How to graph the logarithmic function $f(x) = \ln x$ and use it to solve exponential and logarithmic equations

- How to calculate the derivatives of logarithmic functions

- How to solve exponential growth and decay applications

WHY YOU SHOULD LEARN IT:

Exponential and logarithmic functions have many applications in real life, as can be seen by the examples below, which represent a small sample of the applications in this chapter.

- Property Value, Exercise 37 on page 263
- Present Value, Exercises 41 and 42 on page 271
- Normal Probability Density Function, Exercises 45–48 on page 280
- Human Memory Model, Exercise 79 on page 289
- Effective Yield, Exercises 31–34 on page 306
- Earthquake Intensity, Exercise 48 on page 307

257

4.1   EXPONENTIAL FUNCTIONS

■ Use the properties of exponents to evaluate and simplify exponential expressions.
■ Sketch the graphs of exponential functions.

Exponential Functions

You are already familiar with the behavior of algebraic functions such as

$$f(x) = x^2, \quad g(x) = \sqrt{x} = x^{1/2}, \quad \text{and} \quad h(x) = \frac{1}{x} = x^{-1}$$

each of which involves a variable raised to a constant power. By interchanging roles and raising a constant to a variable power, you obtain another important class of functions called **exponential functions.** Some simple examples are

$$f(x) = 2^x, \quad g(x) = \left(\frac{1}{10}\right)^x = \frac{1}{10^x}, \quad \text{and} \quad h(x) = 3^{2x} = 9^x.$$

In general, you can use any positive base $a \neq 1$ as the base of an exponential function.

Definition of Exponential Function

If $a > 0$ and $a \neq 1$, then the **exponential function** with base a is given by

$$f(x) = a^x.$$

STUDY TIP

In the definition above, the base $a = 1$ is excluded because it yields

$$f(x) = 1^x = 1.$$

This is a constant function, not an exponential function.

When working with exponential functions, the properties of exponents, shown below, are useful.

Properties of Exponents

Let a and b be positive numbers.

1. $a^0 = 1$ 2. $a^x a^y = a^{x+y}$ 3. $\dfrac{a^x}{a^y} = a^{x-y}$

4. $(a^x)^y = a^{xy}$ 5. $(ab)^x = a^x b^x$ 6. $\left(\dfrac{a}{b}\right)^x = \dfrac{a^x}{b^x}$

7. $a^{-x} = \dfrac{1}{a^x}$

EXAMPLE 1   **Applying Properties of Exponents**

Simplify each expression using the properties of exponents.

(a) $(2^2)(2^3)$ (b) $(2^2)(2^{-3})$ (c) $(3^2)^3$

(d) $\left(\dfrac{1}{3}\right)^{-2}$ (e) $\dfrac{3^2}{3^3}$ (f) $(2^{1/2})(3^{1/2})$

SOLUTION

(a) $(2^2)(2^3) = 2^{2+3} = 2^5 = 32$ Apply Property 2.

(b) $(2^2)(2^{-3}) = 2^{2-3} = 2^{-1} = \frac{1}{2}$ Apply Properties 2 and 7.

(c) $(3^2)^3 = 3^{2(3)} = 3^6 = 729$ Apply Property 4.

(d) $\left(\dfrac{1}{3}\right)^{-2} = \dfrac{1}{(1/3)^2} = \left(\dfrac{1}{1/3}\right)^2 = 3^2 = 9$ Apply Properties 6 and 7.

(e) $\dfrac{3^2}{3^3} = 3^{2-3} = 3^{-1} = \dfrac{1}{3}$ Apply Properties 3 and 7.

(f) $(2^{1/2})(3^{1/2}) = [(2)(3)]^{1/2} = 6^{1/2} = \sqrt{6}$ Apply Property 5.

> **TRY IT 1**
>
> Simplify each expression using the properties of exponents.
>
> (a) $(3^2)(3^3)$ (b) $(3^2)(3^{-1})$
>
> (c) $(2^3)^2$ (d) $\left(\dfrac{1}{2}\right)^{-3}$
>
> (e) $\dfrac{2^2}{2^3}$ (f) $(2^{1/2})(5^{1/2})$

Although Example 1 demonstrates the properties of exponents with integer and rational exponents, it is important to realize that the properties hold for *all* real exponents. With a calculator, you can obtain approximations of a^x for any base a and any real exponent x. Here are some examples.

$$2^{-0.6} \approx 0.660, \qquad \pi^{0.75} \approx 2.360, \qquad (1.56)^{\sqrt{2}} \approx 1.876$$

EXAMPLE 2   **Dating Organic Material**

In living organic material, the ratio of radioactive carbon isotopes to the total number of carbon atoms is about 1 to 10^{12}. When organic material dies, its radioactive carbon isotopes begin to decay, with a half-life of about 5715 years. This means that after 5715 years, the ratio of isotopes to atoms will have decreased to one-half the original ratio, after a second 5715 years the ratio will have decreased to one-fourth of the original, and so on. Figure 4.1 shows this decreasing ratio. The formula for the ratio R of carbon isotopes to carbon atoms is

$$R = \left(\frac{1}{10^{12}}\right)\left(\frac{1}{2}\right)^{t/5715}$$

where t is the time in years. Find the value of R for each period of time.

(a) 10,000 years   (b) 20,000 years   (c) 25,000 years

SOLUTION

(a) $R = \left(\dfrac{1}{10^{12}}\right)\left(\dfrac{1}{2}\right)^{10,000/5715} \approx 2.973 \times 10^{-13}$ Ratio for 10,000 years

(b) $R = \left(\dfrac{1}{10^{12}}\right)\left(\dfrac{1}{2}\right)^{20,000/5715} \approx 8.842 \times 10^{-14}$ Ratio for 20,000 years

(c) $R = \left(\dfrac{1}{10^{12}}\right)\left(\dfrac{1}{2}\right)^{25,000/5715} \approx 4.821 \times 10^{-14}$ Ratio for 25,000 years

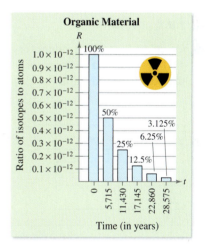

Organic Material

FIGURE 4.1

> **TRY IT 2**
>
> Use the formula for the ratio of carbon isotopes to carbon atoms in Example 2 to find the value of R for each period of time.
>
> (a) 5,000 years
>
> (b) 15,000 years
>
> (c) 30,000 years

Graphs of Exponential Functions

The basic nature of the graph of an exponential function can be determined by the point-plotting method or by using a graphing utility.

> **EXAMPLE 3**   **Graphing Exponential Functions**

Sketch the graph of each exponential function.

(a) $f(x) = 2^x$ (b) $g(x) = \left(\frac{1}{2}\right)^x = 2^{-x}$ (c) $h(x) = 3^x$

SOLUTION   To sketch these functions by hand, you can begin by constructing a table of values, as shown below.

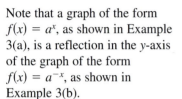

STUDY TIP

Note that a graph of the form $f(x) = a^x$, as shown in Example 3(a), is a reflection in the *y*-axis of the graph of the form $f(x) = a^{-x}$, as shown in Example 3(b).

x	-3	-2	-1	0	1	2	3	4
$f(x) = 2^x$	$\frac{1}{8}$	$\frac{1}{4}$	$\frac{1}{2}$	1	2	4	8	16
$g(x) = 2^{-x}$	8	4	2	1	$\frac{1}{2}$	$\frac{1}{4}$	$\frac{1}{8}$	$\frac{1}{16}$
$h(x) = 3^x$	$\frac{1}{27}$	$\frac{1}{9}$	$\frac{1}{3}$	1	3	9	27	81

The graphs of the three functions are shown in Figure 4.2. Note that the graphs of $f(x) = 2^x$ and $h(x) = 3^x$ are increasing, whereas the graph of $g(x) = 2^{-x}$ is decreasing.

TRY IT 3

Complete the table of values for $f(x) = 5^x$. Sketch the graph of the exponential function.

x	-3	-2	-1	0
$f(x)$				

x	1	2	3
$f(x)$			

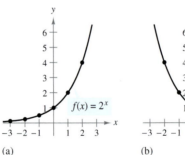

(a)

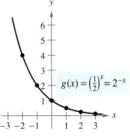

(b)

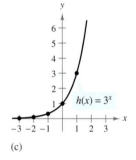

(c)

FIGURE 4.2

> **TECHNOLOGY**
>
> Try graphing the functions $f(x) = 2^x$ and $h(x) = 3^x$ in the same viewing window, as shown at the right. From the display, you can see that the graph of h is increasing more rapidly than the graph of f.

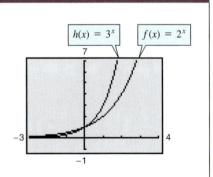

The forms of the graphs in Figure 4.2 are typical of the graphs of the exponential functions $y = a^{-x}$ and $y = a^x$, where $a > 1$. The basic characteristics of such graphs are summarized in Figure 4.3.

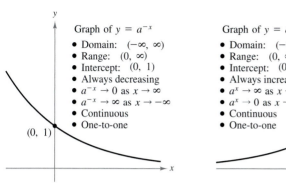

Graph of $y = a^{-x}$
- Domain: $(-\infty, \infty)$
- Range: $(0, \infty)$
- Intercept: $(0, 1)$
- Always decreasing
- $a^{-x} \to 0$ as $x \to \infty$
- $a^{-x} \to \infty$ as $x \to -\infty$
- Continuous
- One-to-one

Graph of $y = a^x$
- Domain: $(-\infty, \infty)$
- Range: $(0, \infty)$
- Intercept: $(0, 1)$
- Always increasing
- $a^x \to \infty$ as $x \to \infty$
- $a^x \to 0$ as $x \to -\infty$
- Continuous
- One-to-one

FIGURE 4.3 Characteristics of the Exponential Functions $y = a^{-x}$ and $y = a^x \, (a > 1)$

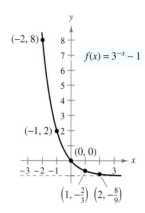

FIGURE 4.4

EXAMPLE 4 **Graphing an Exponential Function**

Sketch the graph of

$$f(x) = 3^{-x} - 1.$$

SOLUTION Begin by creating a table of values, as shown below.

x	-2	-1	0	1	2
$f(x)$	$3^2 - 1 = 8$	$3^1 - 1 = 2$	$3^0 - 1 = 0$	$3^{-1} - 1 = -\frac{2}{3}$	$3^{-2} - 1 = -\frac{8}{9}$

From the limit

$$\lim_{x \to \infty} (3^{-x} - 1) = \lim_{x \to \infty} 3^{-x} - \lim_{x \to \infty} 1$$

$$= \lim_{x \to \infty} \frac{1}{3^x} - \lim_{x \to \infty} 1$$

$$= 0 - 1$$

$$= -1$$

you can see that $y = -1$ is a horizontal asymptote of the graph. The graph is shown in Figure 4.4.

TRY IT 4

Complete the table of values for $f(x) = 2^{-x} + 1$. Sketch the graph of the function. Determine the horizontal asymptote of the graph.

x	-3	-2	-1	0
$f(x)$				

x	1	2	3
$f(x)$			

TAKE ANOTHER LOOK

Finding a Pattern

Use a graphing utility to investigate the function $f(x) = a^x$ for $0 < a < 1$, $a = 1$, and $a > 1$. Discuss the effect that a has on the shape of the graph.

The following warm-up exercises involve skills that were covered in earlier sections. You will use these skills in the exercise set for this section.

In Exercises 1–6, describe how the graph of g is related to the graph of f.

1. $g(x) = f(x + 2)$

2. $g(x) = -f(x)$

3. $g(x) = -1 + f(x)$

4. $g(x) = f(-x)$

5. $g(x) = f(x - 1)$

6. $g(x) = f(x) + 2$

In Exercises 7–10, discuss the continuity of the function.

7. $f(x) = \dfrac{x^2 + 2x - 1}{x + 4}$

8. $f(x) = \dfrac{x^2 - 3x + 1}{x^2 + 2}$

9. $f(x) = \dfrac{x^2 - 3x - 4}{x^2 - 1}$

10. $f(x) = \dfrac{x^2 - 5x + 4}{x^2 + 1}$

In Exercises 11–16, solve for x.

11. $2x - 6 = 4$

12. $3x + 1 = 5$

13. $(x + 4)^2 = 25$

14. $(x - 2)^2 = 8$

15. $x^2 + 4x - 5 = 0$

16. $2x^2 - 3x + 1 = 0$

EXERCISES 4.1

In Exercises 1 and 2, evaluate each expression.

1. (a) $5(5^3)$

(b) $27^{2/3}$

(c) $64^{3/4}$

(d) $81^{1/2}$

(e) $25^{3/2}$

(f) $32^{2/5}$

2. (a) $\left(\frac{1}{5}\right)^3$

(b) $\left(\frac{1}{8}\right)^{1/3}$

(c) $64^{2/3}$

(d) $\left(\frac{5}{8}\right)^2$

(e) $100^{3/2}$

(f) $4^{5/2}$

In Exercises 3–6, use the properties of exponents to simplify the expression.

3. (a) $(5^2)(5^3)$

(b) $(5^2)(5^{-3})$

(c) $(5^2)^2$

(d) 5^{-3}

4. (a) $\dfrac{5^3}{5^6}$

(b) $\left(\dfrac{1}{5}\right)^{-2}$

(c) $(8^{1/2})(2^{1/2})$

(d) $(32^{3/2})\left(\frac{1}{2}\right)^{3/2}$

5. (a) $\dfrac{5^3}{25^2}$

(b) $(9^{2/3})(3)(3^{2/3})$

(c) $[(25^{1/2})(5^2)]^{1/3}$

(d) $(8^2)(4^3)$

6. (a) $(4^3)(4^2)$

(b) $\left(\frac{1}{4}\right)^2(4^2)$

(c) $(4^6)^{1/2}$

(d) $[(8^{-1})(8^{2/3})]^3$

In Exercises 7–10, evaluate the function. If necessary, use a graphing utility, rounding your answers to three decimal places.

7. $f(x) = 2^{x-1}$

(a) $f(3)$

(b) $f\left(\frac{1}{2}\right)$

(c) $f(-2)$

(d) $f\left(-\frac{3}{2}\right)$

8. $f(x) = 3^{x+2}$

(a) $f(-4)$

(b) $f\left(-\frac{1}{2}\right)$

(c) $f(2)$

(d) $f\left(-\frac{5}{2}\right)$

9. $g(x) = 1.05^x$

(a) $g(-2)$

(b) $g(120)$

(c) $g(12)$

(d) $g(5.5)$

10. $g(x) = 1.075^x$

(a) $g(1.2)$

(b) $g(180)$

(c) $g(60)$

(d) $g(12.5)$

In Exercises 11–18, solve the equation for x.

11. $3^x = 81$

12. $5^{x+1} = 125$

13. $\left(\frac{1}{3}\right)^{x-1} = 27$

14. $\left(\frac{1}{5}\right)^{2x} = 625$

15. $4^3 = (x + 2)^3$

16. $4^2 = (x + 2)^2$

17. $x^{3/4} = 8$

18. $(x + 3)^{4/3} = 16$

In Exercises 19–24, match the function with its graph. [The graphs are labeled (a)–(f).]

(a)
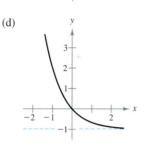

(b)

(c)
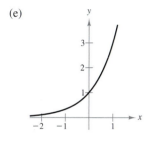

(d)
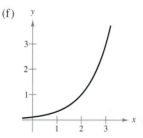

(e)

(f)

19. $f(x) = 3^x$

20. $f(x) = 3^{-x/2}$

21. $f(x) = -3^x$

22. $f(x) = 3^{x-2}$

23. $f(x) = 3^{-x} - 1$

24. $f(x) = 3^x + 2$

In Exercises 25–34, sketch the graph of the function.

25. $f(x) = 6^x$

26. $f(x) = 4^x$

27. $f(x) = \left(\frac{1}{5}\right)^x = 5^{-x}$

28. $f(x) = \left(\frac{1}{4}\right)^x = 4^{-x}$

29. $y = 3^{-x^2}$

30. $y = 2^{-x^2}$

31. $y = 3^{-|x|}$

32. $y = 3^{|x|}$

33. $s(t) = \frac{1}{4}(3^{-t})$

34. $s(t) = 2^{-t} + 3$

35. Population Growth   The population P (in millions) of the United States from 1992 through 2002 can be modeled by the exponential function

$$P(t) = 251.27(1.0118)^t$$

where t is the time in years, with $t = 2$ corresponding to 1992. Use the model to estimate the population in the years (a) 2006 and (b) 2012. *(Source: U.S. Census Bureau)*

36. Sales   The sales S (in millions of dollars) for Starbucks from 1994 through 2003 can be modeled by the exponential function

$$S(t) = 116.59(1.3295)^t$$

where t is the time in years, with $t = 4$ corresponding to 1994. Use the model to estimate the sales in the years (a) 2006 and (b) 2012. *(Source: Starbucks Corp.)*

37. Property Value   Suppose that the value of a piece of property doubles every 15 years. If you buy the property for $64,000, its value t years after the date of purchase should be

$$V(t) = 64,000(2)^{t/15}.$$

Use the model to approximate the value of the property (a) 5 years and (b) 20 years after it is purchased.

38. Inflation Rate   Suppose that the annual rate of inflation averages 5% over the next 10 years. With this rate of inflation, the approximate cost C of goods or services during any year in that decade will be given by

$$C(t) = P(1.05)^t, \quad 0 \leq t \leq 10$$

where t is time in years and P is the present cost. If the price of a movie theater ticket is presently $6.95, estimate the price 10 years from now.

39. Depreciation   After t years, the value of a car that originally cost $16,000 depreciates so that each year it is worth $\frac{3}{4}$ of its value for the previous year. Find a model for $V(t)$, the value of the car after t years. Sketch a graph of the model and determine the value of the car 4 years after it was purchased.

40. Radioactive Decay   After t years, the initial mass of 16 grams of a radioactive element whose half-life is 30 years is given by

$$y = 16\left(\frac{1}{2}\right)^{t/30}, \quad t \geq 0.$$

(a) Use a graphing utility to graph the function.

(b) How much of the initial mass remains after 50 years?

(c) Use the *zoom* and *trace* features of a graphing utility to find the time required for the mass to decay to an amount of 1 gram.

41. Radioactive Decay   After t years, the initial mass of 23 grams of a radioactive element whose half-life is 45 years is given by

$$y = 23\left(\frac{1}{2}\right)^{t/45}, \quad t \geq 0.$$

(a) Use a graphing utility to graph the function.

(b) How much of the initial mass remains after 75 years?

(c) Use the *zoom* and *trace* features of a graphing utility to find the time required for the mass to decay to an amount of 1 gram.

4.2   NATURAL EXPONENTIAL FUNCTIONS

■ Evaluate and graph functions involving the natural exponential function.
■ Solve compound interest problems.
■ Solve present value problems.

Natural Exponential Functions

In Section 4.1, exponential functions were introduced using an unspecified base a. In calculus, the most convenient (or natural) choice for a base is the irrational number e, whose decimal approximation is

$$e \approx 2.71828182846.$$

Although this choice of base may seem unusual, its convenience will become apparent as the rules for differentiating exponential functions are developed in Section 4.3. In that development, you will encounter the limit used in the definition of e.

> **Limit Definition of e**
>
> The irrational number e is defined to be the limit of $(1 + x)^{1/x}$ as $x \to 0$. That is,
>
> $$\lim_{x \to 0} (1 + x)^{1/x} = e.$$

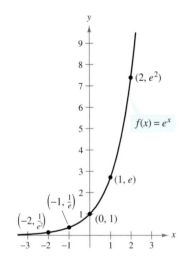

FIGURE 4.5

EXAMPLE 1   Graphing the Natural Exponential Function

Sketch the graph of $f(x) = e^x$.

SOLUTION   Begin by evaluating the function for several values of x, as shown in the table.

x	-2	-1	0	1	2
$f(x)$	$e^{-2} \approx 0.135$	$e^{-1} \approx 0.368$	$e^0 = 1$	$e^1 \approx 2.718$	$e^2 \approx 7.389$

The graph of $f(x) = e^x$ is shown in Figure 4.5. Note that e^x is positive for all values of x. Moreover, the graph has the x-axis as a horizontal asymptote to the left. That is,

$$\lim_{x \to -\infty} e^x = 0.$$

> **TRY IT 1**
>
> Complete the table of values for $f(x) = e^{-x}$. Sketch the graph of the function.
>
> | x | -2 | -1 | 0 | 1 | 2 |
> |---|---|---|---|---|---|
> | $f(x)$ | | | | | |

Exponential functions are often used to model the growth of a quantity or a population. When the quantity's growth *is not* restricted, an exponential model is often used. When the quantity's growth *is* restricted, the best model is often a **logistic growth function** of the form

$$f(t) = \frac{a}{1 + be^{-kt}}.$$

Graphs of both types of population growth models are shown in Figure 4.6.

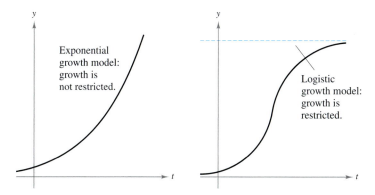

Exponential growth model: growth is not restricted.

Logistic growth model: growth is restricted.

FIGURE 4.6

When a culture is grown in a dish, the size of the dish and the available food limit the culture's growth.

EXAMPLE 2 **Modeling a Population**

A bacterial culture is growing according to the *logistic growth model*

$$y = \frac{1.25}{1 + 0.25e^{-0.4t}}, \quad t \geq 0$$

where y is the culture weight (in grams) and t is the time (in hours). Find the weight of the culture after 0 hours, 1 hour, and 10 hours. What is the limit of the model as t increases without bound?

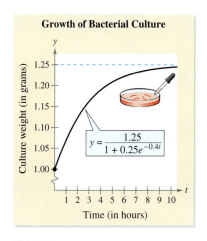

Growth of Bacterial Culture

$$y = \frac{1.25}{1 + 0.25e^{-0.4t}}$$

FIGURE 4.7

SOLUTION

$$y = \frac{1.25}{1 + 0.25e^{-0.4(0)}} = 1 \text{ gram} \qquad \text{Weight when } t = 0$$

$$y = \frac{1.25}{1 + 0.25e^{-0.4(1)}} \approx 1.071 \text{ grams} \qquad \text{Weight when } t = 1$$

$$y = \frac{1.25}{1 + 0.25e^{-0.4(10)}} \approx 1.244 \text{ grams} \qquad \text{Weight when } t = 10$$

As t approaches infinity, the limit of y is

$$\lim_{t \to \infty} \frac{1.25}{1 + 0.25e^{-0.4t}} = \lim_{t \to \infty} \frac{1.25}{1 + (0.25/e^{0.4t})}$$

$$= \frac{1.25}{1 + 0}$$

$$= 1.25.$$

So, as t increases without bound, the weight of the culture approaches 1.25 grams. The graph of the model is shown in Figure 4.7.

TRY IT 2

A bacterial culture is growing according to the model

$$y = \frac{1.50}{1 + 0.2e^{-0.5t}}, \quad t \geq 0$$

where y is the culture weight (in grams) and t is the time (in hours). Find the weight of the culture after 0 hours, 1 hour, and 10 hours. What is the limit of the model as t increases without bound?

Extended Application: Compound Interest

If P dollars is deposited in an account at an annual interest rate of r (in decimal form), what is the balance after 1 year? The answer depends on the number of times the interest is compounded, according to the formula

$$A = P\left(1 + \frac{r}{n}\right)^n$$

where n is the number of compoundings per year. The balances for a deposit of $1000 at 8%, at various compounding periods, are shown in the table.

Number of times compounded per year, n	Balance (in dollars), A
Annually, $n = 1$	$A = 1000\left(1 + \frac{0.08}{1}\right)^1 = \$1080.00$
Semiannually, $n = 2$	$A = 1000\left(1 + \frac{0.08}{2}\right)^2 = \$1081.60$
Quarterly, $n = 4$	$A = 1000\left(1 + \frac{0.08}{4}\right)^4 \approx \$1082.43$
Monthly, $n = 12$	$A = 1000\left(1 + \frac{0.08}{12}\right)^{12} \approx \$1083.00$
Daily, $n = 365$	$A = 1000\left(1 + \frac{0.08}{365}\right)^{365} \approx \$1083.28$

You may be surprised to discover that as n increases, the balance A approaches a limit, as indicated in the following development. In this development, let $x = r/n$. Then $x \to 0$ as $n \to \infty$, and you have

$$A = \lim_{n \to \infty} P\left(1 + \frac{r}{n}\right)^n$$

$$= P \lim_{n \to \infty} \left[\left(1 + \frac{r}{n}\right)^{n/r}\right]^r$$

$$= P\left[\lim_{x \to 0} (1 + x)^{1/x}\right]^r \qquad \text{Substitute } x \text{ for } r/n.$$

$$= Pe^r.$$

This limit is the balance after 1 year of **continuous compounding.** So, for a deposit of $1000 at 8%, compounded continuously, the balance at the end of the year would be

$$A = 1000e^{0.08}$$

$$\approx \$1083.29.$$

Summary of Compound Interest Formulas

Let P be the amount deposited, t the number of years, A the balance, and r the annual interest rate (in decimal form).

1. Compounded n times per year: $A = P\left(1 + \frac{r}{n}\right)^{nt}$

2. Compounded continuously: $A = Pe^{rt}$

The average interest rates paid by banks on savings accounts have varied greatly during the past 30 years. At times, savings accounts have earned as much as 12% annual interest and at times they have earned as little as 3%. The next example shows how the annual interest rate can affect the balance of an account.

EXAMPLE 3 **Finding Account Balances**

You are creating a trust fund for your newborn nephew. You deposit $12,000 in an account, with instructions that the account be turned over to your nephew on his 25th birthday. Compare the balances in the account for each situation.

(a) 7%, compounded continuously

(b) 7%, compounded quarterly

(c) 11%, compounded continuously

(d) 11%, compounded quarterly

SOLUTION

(a) $12,000e^{0.07(25)} \approx 69,055.23$ 7%, compounded continuously

(b) $12,000\left(1 + \dfrac{0.07}{4}\right)^{4(25)} \approx 68,017.87$ 7%, compounded quarterly

(c) $12,000e^{0.11(25)} \approx 187,711.58$ 11%, compounded continuously

(d) $12,000\left(1 + \dfrac{0.11}{4}\right)^{4(25)} \approx 180,869.07$ 11%, compounded quarterly

The growth of the account for parts (a) and (c) is shown in Figure 4.8. Notice the dramatic difference between the balances at 7% and 11%.

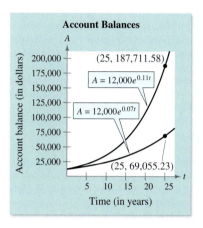

FIGURE 4.8

TRY IT 3

Find the balance in an account if $2000 is deposited for 10 years at an interest of 9%, compounded as follows. Compare the results and make a general statement about compounding.

(a) quarterly   (b) monthly

(c) daily   (d) continuously

In Example 3, note that the interest earned depends on the frequency with which the interest is compounded. The annual percentage rate is called the **stated rate** or **nominal rate.** However, the nominal rate does not reflect the actual rate at which interest is earned, which means that the compounding produced an **effective rate** that is larger than the normal rate. In general, the effective rate corresponding to a nominal rate of r that is compounded n times per year is

$$\text{Effective rate} = r_{eff} = \left(1 + \frac{r}{n}\right)^n - 1.$$

EXAMPLE 4   **Finding the Effective Rate of Interest**

Find the effective rate of interest corresponding to a nominal rate of 6% per year compounded (a) annually, (b) quarterly, and (c) monthly.

SOLUTION

(a) $r_{eff} = \left(1 + \dfrac{r}{n}\right)^n - 1$ Formula for effective rate of interest

$\phantom{r_{eff}} = \left(1 + \dfrac{0.06}{1}\right)^1 - 1$ Substitute for r and n.

$\phantom{r_{eff}} = 1.06 - 1$ Simplify.

$\phantom{r_{eff}} = 0.06$

So, the effective rate is 6% per year.

(b) $r_{eff} = \left(1 + \dfrac{r}{n}\right)^n - 1$ Formula for effective rate of interest

$\phantom{r_{eff}} = \left(1 + \dfrac{0.06}{4}\right)^4 - 1$ Substitute for r and n.

$\phantom{r_{eff}} = (1.015)^4 - 1$ Simplify.

$\phantom{r_{eff}} \approx 0.0614$

So, the effective rate is about 6.14% per year.

(c) $r_{eff} = \left(1 + \dfrac{r}{n}\right)^n - 1$ Formula for effective rate of interest

$\phantom{r_{eff}} = \left(1 + \dfrac{0.06}{12}\right)^{12} - 1$ Substitute for r and n.

$\phantom{r_{eff}} \approx (1.005)^{12} - 1$ Simplify.

$\phantom{r_{eff}} \approx 0.0618$

So, the effective rate is about 6.18% per year.

TRY IT 4

Find the effective rate of interest corresponding to a nominal rate of 7% per year compounded (a) semiannually and (b) daily.

Present Value

In planning for the future, this problem often arises: "How much money P should be deposited now, at a fixed rate of interest r, in order to have a balance of A, t years from now?" The answer to this question is given by the **present value** of A.

To find the present value of a future investment, use the formula for compound interest as shown.

$$A = P\left(1 + \dfrac{r}{n}\right)^{nt}$$ Formula for compound interest

Solving for P gives a present value of

$$P = \frac{A}{\left(1 + \dfrac{r}{n}\right)^{nt}} \quad \text{or} \quad P = \frac{A}{(1 + i)^N}$$

where $i = r/n$ is the interest rate per compounding period and $N = nt$ is the total number of compounding periods. You will learn another way to find the present value of a future investment in Section 6.2.

EXAMPLE 5 **Finding Present Value**

An investor is purchasing a 12-year certificate of deposit that pays an annual percentage rate of 8%, compounded monthly. How much should the person invest in order to obtain a balance of $15,000 at maturity?

SOLUTION Here, $A = 15,000$, $r = 0.08$, $n = 12$, and $t = 12$. Using the formula for present value, you obtain

$$P = \frac{15,000}{\left(1 + \dfrac{0.08}{12}\right)^{12(12)}} \qquad \textcolor{red}{\text{Substitute for } A,\ r,\ n,\ \text{and } t.}$$

$$\approx 5761.72. \qquad \textcolor{red}{\text{Simplify.}}$$

So, the person should invest $5761.72 in the certificate of deposit. ───────

TRY IT 5

How much money should be deposited in an account paying 6% interest compounded monthly in order to have a balance of $20,000 after 3 years?

TAKE ANOTHER LOOK

Compound Interest

You want to invest $5000 in a certificate of deposit for 10 years. You are given the options below. Which would you choose?

a. You can buy a 10-year certificate of deposit that earns 7%, compounded continuously. You are guaranteed a 7% rate for the entire 10 years, but you cannot withdraw the money early without paying a substantial penalty.

b. You can buy a five-year certificate of deposit that earns 6%, compounded continuously. After 5 years, you can reinvest your money at whatever the current interest rate is at that time.

c. You can buy a two-year certificate of deposit that earns 5%, compounded continuously. After 2 years, you can reinvest your money at whatever the current interest rate is at that time.

PREREQUISITE REVIEW 4.2

The following warm-up exercises involve skills that were covered in earlier sections. You will use these skills in the exercise set for this section.

In Exercises 1–4, discuss the continuity of the function.

1. $f(x) = \dfrac{3x^2 + 2x + 1}{x^2 + 1}$

2. $f(x) = \dfrac{x + 1}{x^2 - 4}$

3. $f(x) = \dfrac{x^2 - 6x + 5}{x^2 - 3}$

4. $g(x) = \dfrac{x^2 - 9x + 20}{x - 4}$

In Exercises 5–12, find the limit.

5. $\displaystyle\lim_{x\to\infty} \dfrac{25}{1 + 4x}$

6. $\displaystyle\lim_{x\to\infty} \dfrac{16x}{3 + x^2}$

7. $\displaystyle\lim_{x\to\infty} \dfrac{8x^3 + 2}{2x^3 + x}$

8. $\displaystyle\lim_{x\to\infty} \dfrac{x}{2x}$

9. $\displaystyle\lim_{x\to\infty} \dfrac{3}{2 + (1/x)}$

10. $\displaystyle\lim_{x\to\infty} \dfrac{6}{1 + x^{-2}}$

11. $\displaystyle\lim_{x\to\infty} 2^{-x}$

12. $\displaystyle\lim_{x\to\infty} \dfrac{7}{1 + 5x}$

EXERCISES 4.2

In Exercises 1–4, use the properties of exponents to simplify the expression.

1. (a) $(e^3)(e^4)$
   (b) $(e^3)^4$
   (c) $(e^3)^{-2}$
   (d) e^0

2. (a) $\left(\dfrac{1}{e}\right)^{-2}$
   (b) $\left(\dfrac{e^5}{e^2}\right)^{-1}$
   (c) $\dfrac{e^5}{e^3}$
   (d) $\dfrac{1}{e^{-3}}$

3. (a) $(e^2)^{5/2}$
   (b) $(e^2)(e^{1/2})$
   (c) $(e^{-2})^{-3}$
   (d) $\dfrac{e^5}{e^{-2}}$

4. (a) $(e^{-3})^{2/3}$
   (b) $\dfrac{e^4}{e^{-1/2}}$
   (c) $(e^{-2})^{-4}$
   (d) $(e^{-4})(e^{-3/2})$

In Exercises 5–12, solve the equation for x.

5. $e^{-3x} = e$

6. $e^x = 1$

7. $e^{\sqrt{x}} = e^3$

8. $e^{-1/x} = \sqrt{e}$

9. $x^{2/3} = \sqrt[3]{e^2}$

10. $\dfrac{x^2}{2} = e^2$

11. $3x^3 = 9e^3$

12. $x^{-2} = \dfrac{2}{e^2}$

In Exercises 13–18, match the function with its graph. [The graphs are labeled (a)–(f).]

13. $f(x) = e^{2x+1}$

14. $f(x) = e^{-x/2}$

15. $f(x) = e^{x^2}$

16. $f(x) = e^{-1/x}$

17. $f(x) = e^{\sqrt{x}}$

18. $f(x) = -e^x + 1$

(a)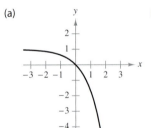

(b)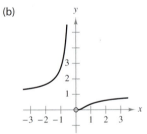

(c)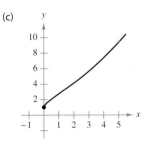

(d)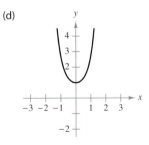

(e) 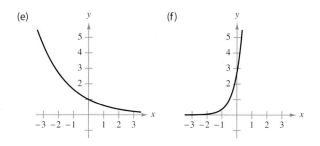   (f)

In Exercises 19–22, sketch the graph of the function.

19. $h(x) = e^{x-2}$

20. $f(x) = e^{2x}$

21. $g(x) = e^{1-x}$

22. $j(x) = e^{-x+2}$

 In Exercises 23–26, use a graphing utility to graph the function. Be sure to choose an appropriate viewing window.

23. $N(t) = 500e^{-0.2t}$

24. $A(t) = 500e^{0.15t}$

25. $g(x) = \dfrac{2}{1 + e^{x^2}}$

26. $g(x) = \dfrac{10}{1 + e^{-x}}$

 In Exercises 27–30, use a graphing utility to graph the function. Determine whether the function has any horizontal asymptotes and discuss the continuity of the function.

27. $f(x) = \dfrac{e^x + e^{-x}}{2}$

28. $f(x) = \dfrac{e^x - e^{-x}}{2}$

29. $f(x) = \dfrac{2}{1 + e^{1/x}}$

30. $f(x) = \dfrac{2}{1 + 2e^{-0.2x}}$

Compound Interest   In Exercises 31–34, complete the table to determine the balance A for P dollars invested at rate r for t years, compounded n times per year.

n	1	2	4	12	365	Continuous compounding
A						

31. $P = \$1000$, $r = 3\%$, $t = 10$ years

32. $P = \$2500$, $r = 5\%$, $t = 20$ years

33. $P = \$1000$, $r = 3\%$, $t = 40$ years

34. $P = \$2500$, $r = 5\%$, $t = 40$ years

Compound Interest   In Exercises 35–38, complete the table to determine the amount of money P that should be invested at rate r to produce a final balance of $\$100,000$ in t years.

t	1	10	20	30	40	50
P						

35. $r = 4\%$, compounded continuously

36. $r = 3\%$, compounded continuously

37. $r = 5\%$, compounded monthly

38. $r = 6\%$, compounded daily

39. *Effective Rate*   Find the effective rate of interest corresponding to a nominal rate of 9% per year compounded (a) annually, (b) semiannually, (c) quarterly, and (d) monthly.

40. *Effective Rate*   Find the effective rate of interest corresponding to a nominal rate of 7.5% per year compounded (a) annually, (b) semiannually, (c) quarterly, and (d) monthly.

41. *Present Value*   How much should be deposited in an account paying 7.2% interest compounded monthly in order to have a balance of $15,503.77 three years from now?

42. *Present Value*   How much should be deposited in an account paying 7.8% interest compounded monthly in order to have a balance of $21,154.03 four years from now?

43. *Future Value*   Find the future value of an $8000 investment if the interest rate is 4.5% compounded monthly for 2 years.

44. *Future Value*   Find the future value of a $6000 investment if the interest rate is 6.25% compounded monthly for 3 years.

45. *Demand*   The demand function for a product is modeled by

$$p = 5000\left(1 - \frac{4}{4 + e^{-0.002x}}\right).$$

Find the price of the product if the quantity demanded is (a) $x = 100$ units and (b) $x = 500$ units. What is the limit of the price as x increases without bound?

46. *Demand*   The demand function for a product is modeled by

$$p = 10,000\left(1 - \frac{3}{3 + te^{-0.001t}}\right).$$

Find the price of the product if the quantity demanded is (a) $x = 1000$ units and (b) $x = 1500$ units. What is the limit of the price as x increases without bound?

47. *Probability* The average time between incoming calls at a switchboard is 3 minutes. If a call has just come in, the probability that the next call will come within the next t minutes is

$$P(t) = 1 - e^{-t/3}.$$

Find the probability of each situation.

(a) A call comes in within $\frac{1}{2}$ minute.

(b) A call comes in within 2 minutes.

(c) A call comes in within 5 minutes.

48. *Consumer Awareness* An automobile gets 28 miles per gallon at speeds of up to and including 50 miles per hour. At speeds greater than 50 miles per hour, the number of miles per gallon drops at the rate of 12% for each 10 miles per hour. If s is the speed (in miles per hour) and y is the number of miles per gallon, then

$$y = 28e^{0.6 - 0.012s}, \quad s > 50.$$

Use this information to complete the table. What can you conclude?

Speed (s)	50	55	60	65	70
Miles per gallon (y)					

49. *Sales* The sales S (in millions of dollars) for Avon Products from 1994 through 2003 can be modeled by

$$S = 3557.12e^{0.0475t}$$

where t is time in years, with $t = 4$ corresponding to 1994. *(Source: Avon Products Inc.)*

(a) Find the sales in 1995, 2000, and 2003.

(b) Using the data points from part (a), would a linear model fit the data? Explain your reasoning.

(c) Use the exponential growth model to estimate when the sales will exceed 10 billion dollars.

50. *Population* The population P (in thousands) of Las Vegas, Nevada from 1970 to 2000 can be modeled by

$$P = 115.49e^{0.0445t}$$

where t is the time in years, with $t = 0$ corresponding to 1970. *(Source: U.S. Census Bureau)*

(a) Find the populations in 1970, 1980, 1990, and 2000.

(b) Explain why the data do not fit a linear model.

(c) Use the model to estimate when the population will exceed 750,000.

 51. *Biology* The population y of a bacterial culture is modeled by the logistic growth function

$$y = \frac{925}{1 + e^{-0.3t}}$$

where t is the time in days.

 (a) Use a graphing utility to graph the model.

(b) Does the population have a limit as t increases without bound? Explain your answer.

(c) How would the limit change if the model were

$$y = \frac{1000}{1 + e^{-0.3t}}?$$

Explain your answer. Draw some conclusions about this type of model.

 52. *Biology: Cell Division* Suppose that you have a single imaginary bacterium able to divide to form two new cells every 30 seconds. Make a table of values for the number of individuals in the population over 30-second intervals up to 5 minutes. Graph the points and use a graphing utility to fit an exponential model to the data. *(Source: Adapted from Levine/Miller, Biology: Discovering Life, Second Edition)*

 53. *Learning Theory* In a learning theory project, the proportion P of correct responses after n trials can be modeled by

$$P = \frac{0.83}{1 + e^{-0.2n}}.$$

(a) Use a graphing utility to estimate the proportion of correct responses after 10 trials. Verify your result analytically.

(b) Use a graphing utility to estimate the number of trials required to have a proportion of correct responses of 0.75.

(c) Does the proportion of correct responses have a limit as n increases without bound? Explain your answer.

54. *Learning Theory* In a typing class, the average number N of words per minute typed after t weeks of lessons can be modeled by

$$N = \frac{95}{1 + 8.5e^{-0.12t}}.$$

(a) Use a graphing utility to estimate the average number of words per minute typed after 10 weeks. Verify your result analytically.

(b) Use a graphing utility to estimate the number of weeks required to achieve an average of 70 words per minute.

(c) Does the number of words per minute have a limit as t increases without bound? Explain your answer.

4.3 DERIVATIVES OF EXPONENTIAL FUNCTIONS

- Find the derivatives of natural exponential functions.
- Use calculus to analyze the graphs of functions that involve the natural exponential function.
- Explore the normal probability density function.

Derivatives of Exponential Functions

In Section 4.2, it was stated that the most convenient base for exponential functions is the irrational number e. The convenience of this base stems primarily from the fact that the function $f(x) = e^x$ *is its own derivative*. You will see that this is not true of other exponential functions of the form $y = a^x$ where $a \neq e$. To verify that $f(x) = e^x$ is its own derivative, notice that the limit

$$\lim_{\Delta x \to 0} (1 + \Delta x)^{1/\Delta x} = e$$

implies that for small values of Δx, $e \approx (1 + \Delta x)^{1/\Delta x}$, or $e^{\Delta x} \approx 1 + \Delta x$. This approximation is used in the following derivation.

$$f'(x) = \lim_{\Delta x \to 0} \frac{f(x + \Delta x) - f(x)}{\Delta x} \qquad \text{Definition of derivative}$$

$$= \lim_{\Delta x \to 0} \frac{e^{x + \Delta x} - e^x}{\Delta x} \qquad \text{Use } f(x) = e^x.$$

$$= \lim_{\Delta x \to 0} \frac{e^x(e^{\Delta x} - 1)}{\Delta x} \qquad \text{Factor numerator.}$$

$$= \lim_{\Delta x \to 0} \frac{e^x[(1 + \Delta x) - 1]}{\Delta x} \qquad \text{Substitute } 1 + \Delta x \text{ for } e^{\Delta x}.$$

$$= \lim_{\Delta x \to 0} \frac{e^x(\Delta x)}{\Delta x} \qquad \text{Divide out like factor.}$$

$$= \lim_{\Delta x \to 0} e^x \qquad \text{Simplify.}$$

$$= e^x \qquad \text{Evaluate limit.}$$

If u is a function of x, you can apply the Chain Rule to obtain the derivative of e^u with respect to x. Both formulas are summarized below.

Derivative of the Natural Exponential Function

Let u be a differentiable function of x.

1. $\dfrac{d}{dx}[e^x] = e^x$ 2. $\dfrac{d}{dx}[e^u] = e^u \dfrac{du}{dx}$

TECHNOLOGY

Let $f(x) = e^x$. Use a graphing utility to evaluate $f(x)$ and the numerical derivative of $f(x)$ at each x-value. Explain the results.

(a) $x = -2$ (b) $x = 0$ (c) $x = 2$

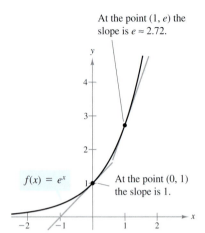

At the point $(1, e)$ the slope is $e \approx 2.72$.

$f(x) = e^x$

At the point $(0, 1)$ the slope is 1.

FIGURE 4.9

EXAMPLE 1 **Interpreting a Derivative Graphically**

Find the slopes of the tangent lines to

$$f(x) = e^x \qquad \text{Original function}$$

at the points $(0, 1)$ and $(1, e)$. What conclusion can you make?

SOLUTION Because the derivative of f is

$$f'(x) = e^x \qquad \text{Derivative}$$

it follows that the slope of the tangent line to the graph of f is

$$f'(0) = e^0 = 1 \qquad \text{Slope at point } (0, 1)$$

at the point $(0, 1)$ and

$$f'(1) = e^1 = e \qquad \text{Slope at point } (1, e)$$

at the point $(1, e)$, as shown in Figure 4.9. From this pattern, you can see that the slope of the tangent line to the graph of $f(x) = e^x$ at any point (x, e^x) is equal to the y-coordinate of the point.

TRY IT 1

Find the equations of the tangent lines to $f(x) = e^x$ at the points $(0, 1)$ and $(1, e)$.

STUDY TIP

In Example 2, notice that when you differentiate an exponential function, the exponent does not change. For instance, the derivative of $y = e^{3x}$ is $y' = 3e^{3x}$. In both the function and its derivative, the exponent is $3x$.

EXAMPLE 2 **Differentiating Exponential Functions**

Differentiate each function.

(a) $f(x) = e^{2x}$ (b) $f(x) = e^{-3x^2}$

(c) $f(x) = 6e^{x^3}$ (d) $f(x) = e^{-x}$

SOLUTION

(a) Let $u = 2x$. Then $du/dx = 2$, and you can apply the Chain Rule.

$$f'(x) = e^u \frac{du}{dx} = e^{2x}(2) = 2e^{2x}$$

(b) Let $u = -3x^2$. Then $du/dx = -6x$, and you can apply the Chain Rule.

$$f'(x) = e^u \frac{du}{dx} = e^{-3x^2}(-6x) = -6xe^{-3x^2}$$

(c) Let $u = x^3$. Then $du/dx = 3x^2$, and you can apply the Chain Rule.

$$f'(x) = 6e^u \frac{du}{dx} = 6e^{x^3}(3x^2) = 18x^2 e^{x^3}$$

(d) Let $u = -x$. Then $du/dx = -1$, and you can apply the Chain Rule.

$$f'(x) = e^u \frac{du}{dx} = e^{-x}(-1) = -e^{-x}$$

TRY IT 2

Differentiate each function.

(a) $f(x) = e^{3x}$

(b) $f(x) = e^{-2x^3}$

(c) $f(x) = 4e^{x^2}$

(d) $f(x) = e^{-2x}$

The differentiation rules that you studied in Chapter 2 can be used with exponential functions, as shown in Example 3.

EXAMPLE 3   **Differentiating Exponential Functions**

Differentiate each function.

(a) $f(x) = xe^x$ (b) $f(x) = \dfrac{e^x - e^{-x}}{2}$ (c) $f(x) = \dfrac{e^x}{x}$ (d) $f(x) = xe^x - e^x$

SOLUTION

(a) $f(x) = xe^x$ Write original function.

 $f'(x) = xe^x + e^x(1)$ Product Rule

   $= xe^x + e^x$ Simplify.

(b) $f(x) = \dfrac{e^x - e^{-x}}{2}$ Write original function.

   $= \tfrac{1}{2}(e^x - e^{-x})$ Rewrite.

 $f'(x) = \tfrac{1}{2}(e^x + e^{-x})$ Constant Multiple Rule

(c) $f(x) = \dfrac{e^x}{x}$ Write original function.

 $f'(x) = \dfrac{xe^x - e^x(1)}{x^2}$ Quotient Rule

   $= \dfrac{e^x(x - 1)}{x^2}$ Simplify.

(d) $f(x) = xe^x - e^x$ Write original function.

 $f'(x) = [xe^x + e^x(1)] - e^x$ Product and Difference Rules

   $= xe^x + e^x - e^x$

   $= xe^x$ Simplify.

TRY IT 3

Differentiate each function.

(a) $f(x) = x^2 e^x$ (b) $f(x) = \dfrac{e^x + e^{-x}}{2}$

(c) $f(x) = \dfrac{e^x}{x^2}$ (d) $f(x) = x^2 e^x - e^x$

TECHNOLOGY

 If you have access to a symbolic differentiation utility, try using it to find the derivatives of the functions in Example 3.

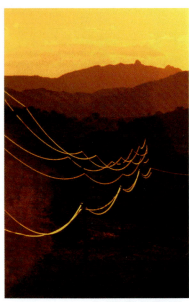

David Buffington/PhotoDisc

Utility wires strung between poles have the shape of a catenary.

Applications

In Chapter 3, you learned how to use derivatives to analyze the graphs of functions. The next example applies those techniques to a function composed of exponential functions. In the example, notice that $e^a = e^b$ implies that $a = b$.

EXAMPLE 4   **Analyzing a Catenary**

When a telephone wire is hung between two poles, the wire forms a U-shaped curve called a **catenary.** For instance, the function

$$y = 30(e^{x/60} + e^{-x/60}), \qquad -30 \le x \le 30$$

models the shape of a telephone wire strung between two poles that are 60 feet apart (x and y are measured in feet). Show that the lowest point on the wire is midway between the two poles. How much does the wire sag between the two poles?

SOLUTION   The derivative of the function is

$$y' = 30\left[e^{x/60}\left(\tfrac{1}{60}\right) + e^{-x/60}\left(-\tfrac{1}{60}\right)\right]$$
$$= \tfrac{1}{2}(e^{x/60} - e^{-x/60}).$$

To find the critical numbers, set the derivative equal to zero.

$\tfrac{1}{2}(e^{x/60} - e^{-x/60}) = 0$	Set derivative equal to 0.
$e^{x/60} - e^{-x/60} = 0$	Multiply each side by 2.
$e^{x/60} = e^{-x/60}$	Add $e^{-x/60}$ to each side.
$\dfrac{x}{60} = -\dfrac{x}{60}$	If $e^a = e^b$, then $a = b$.
$x = -x$	Multiply each side by 60.
$2x = 0$	Add x to each side.
$x = 0$	Divide each side by 2.

Using the First-Derivative Test, you can determine that the critical number $x = 0$ yields a relative minimum of the function. From the graph in Figure 4.10, you can see that this relative minimum is actually a minimum on the interval $[-30, 30]$. To find how much the wire sags between the two poles, you can compare its height at each pole with its height at the midpoint.

$y = 30(e^{-30/60} + e^{-(-30)/60}) \approx 67.7$ feet	Height at left pole
$y = 30(e^{0/60} + e^{-(0)/60}) = 60$ feet	Height at midpoint
$y = 30(e^{30/60} + e^{-(30)/60}) \approx 67.7$ feet	Height at right pole

From this, you can see that the wire sags about 7.7 feet.

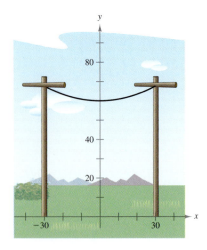

FIGURE 4.10

TRY IT 4

Use a graphing utility to graph the function in Example 4. Verify the minimum value. Use the information in the example to choose an appropriate viewing window.

| ◼ **EXAMPLE 5** | **Finding a Maximum Revenue** |

The demand function for a product is modeled by

$$p = 56e^{-0.000012x} \qquad \text{Demand function}$$

where p is the price per unit (in dollars) and x is the number of units. What price will yield a maximum revenue?

SOLUTION The revenue function is

$$R = xp = 56xe^{-0.000012x}. \qquad \text{Revenue function}$$

To find the maximum revenue *analytically*, you would set the marginal revenue, dR/dx, equal to zero and solve for x. In this problem, it is easier to use a *graphical* approach. After experimenting to find a reasonable viewing window, you can obtain a graph of R that is similar to that shown in Figure 4.11. Using the *zoom* and *trace* features, you can conclude that the maximum revenue occurs when x is about 83,300 units. To find the price that corresponds to this production level, substitute $x \approx 83,300$ into the demand function.

$$p \approx 56e^{-0.000012(83,300)} \approx \$20.61.$$

So, a price of about \$20.61 will yield a maximum revenue.

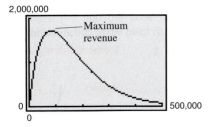

FIGURE 4.11   Use the *zoom* and *trace* features to approximate the x-value that corresponds to the maximum revenue.

───────

TRY IT 5

The demand function for a product is modeled by

$$p = 50e^{-0.0000125x}$$

where p is the price per unit in dollars and x is the number of units. What price will yield a maximum revenue?

❚ **STUDY TIP**

Try solving the problem in Example 5 analytically. When you do this, you obtain

$$\frac{dR}{dx} = 56xe^{-0.000012x}(-0.000012) + e^{-0.000012x}(56) = 0.$$

Explain how you would solve this equation. What is the solution?

PREREQUISITE REVIEW 4.5

The following warm-up exercises involve skills that were covered in earlier sections. You will use these skills in the exercise set for this section.

In Exercises 1–6, expand the logarithmic expression.

1. $\ln(x + 1)^2$

2. $\ln x(x + 1)$

3. $\ln \dfrac{x}{x + 1}$

4. $\ln\left(\dfrac{x}{x - 3}\right)^3$

5. $\ln \dfrac{4x(x - 7)}{x^2}$

6. $\ln x^3(x + 1)$

In Exercises 7 and 8, find dy/dx implicitly.

7. $y^2 + xy = 7$

8. $x^2 y - xy^2 = 3x$

In Exercises 9 and 10, find the second derivative of f.

9. $f(x) = x^2(x + 1) - 3x^3$

10. $f(x) = -\dfrac{1}{x^2}$

EXERCISES 4.5

In Exercises 1–4, find the slope of the tangent line to the graph of the function at the point $(1, 0)$.

1. $y = \ln x^3$

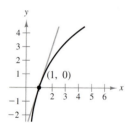

2. $y = \ln x^{5/2}$

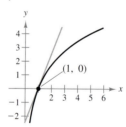

3. $y = \ln x^2$

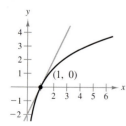

4. $y = \ln x^{1/2}$

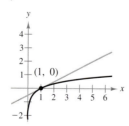

In Exercises 5–26, find the derivative of the function.

5. $y = \ln x^2$

6. $f(x) = \ln 2x$

7. $y = \ln(x^2 + 3)$

8. $f(x) = \ln(1 - x^2)$

9. $y = \ln \sqrt{x^4 - 4x}$

10. $y = \ln(1 - x)^{3/2}$

11. $y = \frac{1}{2}(\ln x)^6$

12. $y = (\ln x^2)^2$

13. $f(x) = x^2 \ln x$

14. $y = \dfrac{\ln x}{x^2}$

15. $y = \ln\left(x\sqrt{x^2 - 1}\right)$

16. $y = \ln \dfrac{x}{x^2 + 1}$

17. $y = \ln \dfrac{x}{x + 1}$

18. $y = \ln \dfrac{x^2}{x^2 + 1}$

19. $y = \ln \sqrt[3]{\dfrac{x - 1}{x + 1}}$

20. $y = \ln \sqrt{\dfrac{x + 1}{x - 1}}$

21. $y = \ln \dfrac{\sqrt{4 + x^2}}{x}$

22. $y = \ln\left(x\sqrt{4 + x^2}\right)$

23. $g(x) = e^{-x} \ln x$

24. $f(x) = x \ln e^{x^2}$

25. $g(x) = \ln \dfrac{e^x + e^{-x}}{2}$

26. $f(x) = \ln \dfrac{1 + e^x}{1 - e^x}$

In Exercises 27–30, write the expression with base e.

27. 2^x

28. 3^x

29. $\log_4 x$

30. $\log_3 x$

In Exercises 31–36, use a calculator to evaluate the logarithm. Round to three decimal places.

31. $\log_2 48$

32. $\log_5 12$

33. $\log_3 \frac{1}{2}$

34. $\log_7 \frac{2}{9}$

35. $\log_{1/5} 31$

36. $\log_{2/3} 32$

In Exercises 37–46, find the derivative of the function.

37. $y = 3^x$

38. $y = \left(\frac{1}{4}\right)^x$

39. $f(x) = \log_2 x$

40. $g(x) = \log_5 x$

41. $h(x) = 4^{2x-3}$

42. $y = 6^{5x}$

43. $y = \log_{10}(x^2 + 6x)$

44. $f(x) = 10^{x^2}$

45. $y = x2^x$

46. $y = x3^{x+1}$

In Exercises 47–50, determine an equation of the tangent line to the function at the given point.

Function	Point
47. $y = x \ln x$	$(1, 0)$
48. $y = \dfrac{\ln x}{x}$	$\left(e, \dfrac{1}{e}\right)$
49. $y = \log_3(3x + 7)$	$\left(\dfrac{2}{3}, 2\right)$
50. $g(x) = \log_2(3x - 1)$	$(11, 5)$

In Exercises 51–54, find dy/dx implicitly.

51. $x^2 - 3 \ln y + y^2 = 10$

52. $\ln xy + 5x = 30$

53. $4x^3 + \ln y^2 + 2y = 2x$

54. $4xy + \ln(x^2 y) = 7$

In Exercises 55–58, find the second derivative of the function.

55. $f(x) = x \ln \sqrt{x} + 2x$

56. $f(x) = 3 + 2 \ln x$

57. $f(x) = 5^x$

58. $f(x) = \log_{10} x$

59. *Sound Intensity* The relationship between the number of decibels β and the intensity of a sound I in watts per square centimeter is given by

$$\beta = 10 \log_{10}\left(\frac{I}{10^{-16}}\right).$$

Find the rate of change in the number of decibels when the intensity is 10^{-4} watts per square centimeter.

60. *Chemistry* The temperatures T (°F) at which water boils at selected pressures p (pounds per square inch) can be modeled by

$$T = 87.97 + 34.96 \ln p + 7.91 \sqrt{p}.$$

Find the rate of change of the temperature when the pressure is 60 pounds per square inch.

In Exercises 61–66, find the slope of the graph at the indicated point. Then write an equation of the tangent line at the point.

61. $f(x) = 1 + 2x \ln x$

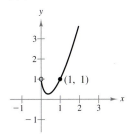

62. $f(x) = 2 \ln x^3$

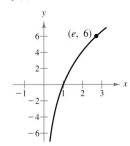

63. $f(x) = \ln \dfrac{5(x + 2)}{x}$

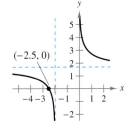

64. $f(x) = \ln\left(x\sqrt{x + 3}\right)$

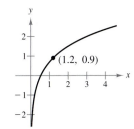

65. $f(x) = x \log_2 x$

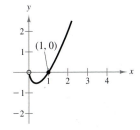

66. $f(x) = x^2 \log_3 x$

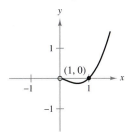

In Exercises 67–72, graph and analyze the function. Include any relative extrema and points of inflection in your analysis. Use a graphing utility to verify your results.

67. $y = x - \ln x$

68. $y = \frac{1}{2}x^2 - \ln x$

69. $y = \dfrac{\ln x}{x}$

70. $y = x \ln x$

71. $y = x^2 \ln x$

72. $y = (\ln x)^2$

Radioactive Decay   In Exercises 11–16, complete the table for each radioactive isotope.

Isotope	Half-life (in years)	Initial quantity	Amount after 1000 years	Amount after 10,000 years
11. ^{226}Ra	1599	10 grams		
12. ^{226}Ra	1599		1.5 grams	
13. ^{14}C	5715			2 grams
14. ^{14}C	5715	3 grams		
15. ^{239}Pu	24,100		2.1 grams	
16. ^{239}Pu	24,100			0.4 gram

17. *Radioactive Decay*   What percent of a present amount of radioactive radium (^{226}Ra) will remain after 900 years?

18. *Radioactive Decay*   Find the half-life of a radioactive material if after 1 year 99.57% of the initial amount remains.

19. *Carbon Dating* ^{14}C dating assumes that the carbon dioxide on the Earth today has the same radioactive content as it did centuries ago. If this is true, then the amount of ^{14}C absorbed by a tree that grew several centuries ago should be the same as the amount of ^{14}C absorbed by a similar tree today. A piece of ancient charcoal contains only 15% as much of the radioactive carbon as a piece of modern charcoal. How long ago was the tree burned to make the ancient charcoal? (The half-life of ^{14}C is 5715 years.)

20. *Carbon Dating*   Repeat Exercise 19 for a piece of charcoal that contains 30% as much radioactive carbon as a modern piece.

In Exercises 21 and 22, find exponential models

$$y_1 = Ce^{k_1 t} \quad \text{and} \quad y_2 = C(2)^{k_2 t}$$

that pass through the points. Compare the values of k_1 and k_2. Briefly explain your results.

21. $(0, 5)$, $(12, 20)$

22. $(0, 8)$, $\left(20, \frac{1}{2}\right)$

23. *Population Growth*   The number of a certain type of bacteria increases continuously at a rate proportional to the number present. There are 150 present at a given time and 450 present 5 hours later.

    (a) How many will there be 10 hours after the initial time?

    (b) How long will it take for the population to double?

    (c) Does the answer to part (b) depend on the starting time? Explain your reasoning.

24. *School Enrollment*   In 1960, the total enrollment in public universities and colleges in the United States was 2.3 million students. By 2000, enrollment had risen to 12.0 million students. Assume enrollment can be modeled by exponential growth. *(Source: U.S. Census Bureau)*

    (a) Estimate the total enrollments in 1970, 1980, and 1990.

    (b) How many years until the enrollment doubles from the 2000 figure?

    (c) By what percent is the enrollment increasing each year?

Compound Interest   In Exercises 25–30, complete the table for an account in which interest is compounded continuously.

	Initial investment	Annual rate	Time to double	Amount after 10 years	Amount after 25 years
25.	$1,000	12%			
26.	$20,000	$10\frac{1}{2}$%			
27.	$750		$7\frac{3}{4}$ years		
28.	$10,000		5 years		
29.	$500			$1292.85	
30.	$2,000				$6008.33

31. *Effective Yield*   The effective yield is the annual rate i that will produce the same interest per year as the nominal rate r compounded n times per year.

    (a) For a rate r that is compounded n times per year, show that the effective yield is

    $$i = \left(1 + \frac{r}{n}\right)^n - 1.$$

    (b) Find the effective yield for a nominal rate of 6%, compounded monthly.

32. *Effective Yield*   The effective yield is the annual rate i that will produce the same interest per year as the nominal rate r.

    (a) For a rate r that is compounded continuously, show that the effective yield is $i = e^r - 1$.

    (b) Find the effective yield for a nominal rate of 6%, compounded continuously.

Effective Yield   In Exercises 33 and 34, use the results of Exercises 31 and 32 to complete the table showing the effective yield for a nominal rate of r.

Number of compoundings per year	4	12	365	Continuous
Effective yield				

33. $r = 5\%$

34. $r = 7\frac{1}{2}\%$

35. *Investment: Rule of 70* Verify that the time necessary for an investment to double its value is approximately $70/r$, where r is the annual interest rate entered as a percent.

36. *Investment: Rule of 70* Use the Rule of 70 from Exercise 35 to approximate the times necessary for an investment to double in value if (a) $r = 10\%$ and (b) $r = 7\%$.

37. *Revenue* The revenues for Sonic Corporation were $83.3 million in 1993 and $446.6 million in 2003. *(Source: Sonic Corporation)*

(a) Use an exponential growth model to estimate the revenue in 2008.

(b) Use a linear model to estimate the 2008 revenue.

 (c) Use a graphing utility to graph the models from parts (a) and (b). Which model is more accurate?

 38. *Sales* The sales (in millions of dollars) for in-line skating and wheel sports in the United States were $150 million in 1990 and $1074 million in 2000. *(Source: National Sporting Goods Association)*

(a) Use the *regression* feature of a graphing utility to find an exponential growth model and a linear model for the data.

(b) Use the exponential growth model to estimate the sales in 2006.

(c) Use the linear model to estimate the sales in 2006.

(d) Use a graphing utility to graph the models from part (a). Which model is more accurate?

39. *Sales* The cumulative sales S (in thousands of units) of a new product after it has been on the market for t years are modeled by

$$S = Ce^{k/t}.$$

During the first year, 5000 units were sold. The saturation point for the market is 30,000 units. That is, the limit of S as $t \to \infty$ is 30,000.

(a) Solve for C and k in the model.

(b) How many units will be sold after 5 years?

 (c) Use a graphing utility to graph the sales function.

40. *Sales* The cumulative sales S (in thousands of units) of a new product after it has been on the market for t years are modeled by

$$S = 30(1 - 3^{kt}).$$

During the first year, 5000 units were sold.

(a) Solve for k in the model.

(b) What is the saturation point for this product?

(c) How many units will be sold after 5 years?

 (d) Use a graphing utility to graph the sales function.

41. *Learning Curve* The management of a factory finds that the maximum number of units a worker can produce in a day is 30. The learning curve for the number of units N produced per day after a new employee has worked t days is modeled by $N = 30(1 - e^{kt})$. After 20 days on the job, a worker is producing 19 units in a day. How many days should pass before this worker is producing 25 units per day?

42. *Learning Curve* The management in Exercise 41 requires that a new employee be producing at least 20 units per day after 30 days on the job.

(a) Find a learning curve model that describes this minimum requirement.

(b) Find the number of days before a minimal achiever is producing 25 units per day.

43. *Profit* Because of a slump in the economy, a company finds that its annual profits have dropped from $742,000 in 1998 to $632,000 in 2000. If the profit follows an exponential pattern of decline, what is the expected profit for 2003? (Let $t = 0$ correspond to 1998.)

44. *Revenue* A small business assumes that the demand function for one of its new products can be modeled by $p = Ce^{kx}$. When $p = \$45$, $x = 1000$ units, and when $p = \$40$, $x = 1200$ units.

(a) Solve for C and k.

(b) Find the values of x and p that will maximize the revenue for this product.

45. *Revenue* Repeat Exercise 44 given that when $p = \$5$, $x = 300$ units, and when $p = \$4$, $x = 400$ units.

46. *Forestry* The value V (in dollars) of a tract of timber can be modeled by $V = 100,000e^{0.75\sqrt{t}}$, where $t = 0$ corresponds to 1990. If money earns interest at a rate of 4%, compounded continuously, then the present value A of the timber at any time t is $A = Ve^{-0.04t}$. Find the year in which the timber should be harvested to maximize the present value.

47. *Forestry* Repeat Exercise 46 using the model

$$V = 100,000e^{0.6\sqrt{t}}.$$

48. *Earthquake Intensity* On the Richter scale, the magnitude R of an earthquake of intensity I is given by

$$R = \frac{\ln I - \ln I_0}{\ln 10}$$

where I_0 is the minimum intensity used for comparison. Assume $I_0 = 1$.

(a) Find the intensity of the 1906 San Francisco earthquake in which $R = 8.3$.

(b) Find the factor by which the intensity is increased when the value of R is doubled.

(c) Find dR/dI.

ALGEBRA REVIEW

Solving Exponential and Logarithmic Equations

To find the extrema or points of inflection of an exponential or logarithmic function, you must know how to solve exponential and logarithmic equations. A few examples are given on page 285. Some additional examples are presented in this Algebra Review.

As with all equations, remember that your basic goal is to isolate the variable on one side of the equation. To do this, you use inverse operations. For instance, to get rid of an exponential expression such as e^{2x}, take the natural log of each side and use the property $\ln e^{2x} = 2x$. Similarly, to get rid of a logarithmic expression such as $\log_2 3x$, exponentiate each side and use the property $2^{\log_2 3x} = 3x$.

EXAMPLE 1   Solving Exponential Equations

Solve each exponential equation.

(a) $25 = 5e^{7t}$ (b) $80{,}000 = 100{,}000e^{k(4)}$ (c) $300 = \left(\dfrac{100}{e^{2k}}\right)e^{4k}$

SOLUTION

(a)
$$25 = 5e^{7t} \qquad \text{Write original equation.}$$
$$5 = e^{7t} \qquad \text{Divide each side by 5.}$$
$$\ln 5 = \ln e^{7t} \qquad \text{Take natural log of each side.}$$
$$\ln 5 = 7t \qquad \text{Apply the property } \ln e^a = a.$$
$$\tfrac{1}{7}\ln 5 = t \qquad \text{Divide each side by 7.}$$

(b)
$$80{,}000 = 100{,}000e^{k(4)} \qquad \text{Example 4, page 304}$$
$$0.8 = e^{4k} \qquad \text{Divide each side by 100,000.}$$
$$\ln 0.8 = \ln e^{4k} \qquad \text{Take natural log of each side.}$$
$$\ln 0.8 = 4k \qquad \text{Apply the property } \ln e^a = a.$$
$$\tfrac{1}{4}\ln 0.8 = k \qquad \text{Divide each side by 4.}$$

(c)
$$300 = \left(\dfrac{100}{e^{2k}}\right)e^{4k} \qquad \text{Example 2, page 301}$$
$$300 = (100)\dfrac{e^{4k}}{e^{2k}} \qquad \text{Rewrite product.}$$
$$300 = 100e^{4k-2k} \qquad \text{To divide powers, subtract exponents.}$$
$$300 = 100e^{2k} \qquad \text{Simplify.}$$
$$3 = e^{2k} \qquad \text{Divide each side by 100.}$$
$$\ln 3 = \ln e^{2k} \qquad \text{Take natural log of each side.}$$
$$\ln 3 = 2k \qquad \text{Apply the property } \ln e^a = a.$$
$$\tfrac{1}{2}\ln 3 = k \qquad \text{Divide each side by 2.}$$

| EXAMPLE 2 | **Solving Logarithmic Equations** |

Solve each logarithmic equation.

(a) $\ln x = 2$ (b) $5 + 2 \ln x = 4$

(c) $2 \ln 3x = 4$ (d) $\ln x - \ln(x - 1) = 1$

SOLUTION

(a) $\ln x = 2$ Write original equation.

$e^{\ln x} = e^2$ Exponentiate each side.

$x = e^2$ Apply the property $e^{\ln a} = a$.

(b) $5 + 2 \ln x = 4$ Write original equation.

$2 \ln x = -1$ Subtract 5 from each side.

$\ln x = -\dfrac{1}{2}$ Divide each side by 2.

$e^{\ln x} = e^{-1/2}$ Exponentiate each side.

$x = e^{-1/2}$ Apply the property $e^{\ln a} = a$.

(c) $2 \ln 3x = 4$ Write original equation.

$\ln 3x = 2$ Divide each side by 2.

$e^{\ln 3x} = e^2$ Exponentiate each side.

$3x = e^2$ Apply the property $e^{\ln a} = a$.

$x = \frac{1}{3} e^2$ Divide each side by 3.

(d) $\ln x - \ln(x - 1) = 1$ Write original equation.

$\ln \dfrac{x}{x - 1} = 1$ $\ln m - \ln n = \ln(m/n)$

$e^{\ln(x/x-1)} = e^1$ Exponentiate each side.

$\dfrac{x}{x - 1} = e^1$ Apply the property $e^{\ln a} = a$.

$x = ex - e$ Multiply each side by $x - 1$.

$x - ex = -e$ Subtract ex from each side.

$x(1 - e) = -e$ Factor.

$x = \dfrac{-e}{1 - e}$ Divide each side by $1 - e$.

$x = \dfrac{e}{e - 1}$ Simplify.

STUDY TIP

Because the domain of a logarithmic function generally does not include all real numbers, be sure to check for extraneous solutions.

4 CHAPTER SUMMARY AND STUDY STRATEGIES

After studying this chapter, you should have acquired the following skills. The exercise numbers are keyed to the Review Exercises that begin on page 312. Answers to odd-numbered Review Exercises are given in the back of the text.*

■ Use the properties of exponents to evaluate and simplify exponential expressions. *(Section 4.1 and Section 4.2)*

Review Exercises 1–16

$$a^0 = 1, \qquad a^x a^y = a^{x+y}, \qquad \frac{a^x}{a^y} = a^{x-y}, \qquad (a^x)^y = a^{xy}$$

$$(ab)^x = a^x b^x, \qquad \left(\frac{a}{b}\right)^x = \frac{a^x}{b^x}, \qquad a^{-x} = \frac{1}{a^x}$$

■ Use properties of exponents to answer questions about real life. *(Section 4.1)*
Review Exercises 17, 18

■ Sketch the graphs of exponential functions. *(Section 4.1 and Section 4.2)*
Review Exercises 19–28

■ Evaluate limits of exponential functions in real life. *(Section 4.2)*
Review Exercises 29, 30

■ Evaluate and graph functions involving the natural exponential function. *(Section 4.2)*
Review Exercises 31–34

■ Graph logistic growth functions. *(Section 4.2)*
Review Exercises 35, 36

■ Solve compound interest problems. *(Section 4.2)*
Review Exercises 37–40

$$A = P(1 + r/n)^{nt}, \quad A = Pe^{rt}$$

■ Solve effective rate of interest problems. *(Section 4.2)*
Review Exercises 41, 42

$$r_{eff} = (1 + r/n)^n - 1$$

■ Solve present value problems. *(Section 4.2)*
Review Exercises 43, 44

$$P = \frac{A}{(1 + r/n)^{nt}}$$

■ Answer questions involving the natural exponential function as a real-life model. *(Section 4.2)*
Review Exercises 45, 46

■ Find the derivatives of natural exponential functions. *(Section 4.3)*
Review Exercises 47–54

$$\frac{d}{dx}[e^x] = e^x, \quad \frac{d}{dx}[e^u] = e^u \frac{du}{dx}$$

■ Use calculus to analyze the graphs of functions that involve the natural exponential function. *(Section 4.3)*
Review Exercises 55–62

■ Use the definition of the natural logarithmic function to write exponential equations in logarithmic form, and vice versa. *(Section 4.4)*
Review Exercises 63–66

$$\ln x = b \quad \text{if and only if} \quad e^b = x.$$

■ Sketch the graphs of natural logarithmic functions. *(Section 4.4)*
Review Exercises 67–70

■ Use properties of logarithms to expand and condense logarithmic expressions. *(Section 4.4)*
Review Exercises 71–76

$$\ln xy = \ln x + \ln y, \quad \ln \frac{x}{y} = \ln x - \ln y, \quad \ln x^n = n \ln x$$

■ Use inverse properties of exponential and logarithmic functions to solve exponential and logarithmic equations. *(Section 4.4)*
Review Exercises 77–92

$$\ln e^x = x, \quad e^{\ln x} = x$$

* Use a wide range of valuable study aids to help you master the material in this chapter. The *Student Solutions Guide* includes step-by-step solutions to all odd-numbered exercises to help you review and prepare. The *HM mathSpace® Student CD-ROM* helps you brush up on your algebra skills. The *Graphing Technology Guide*, available on the Web at *math.college.hmco.com/students*, offers step-by-step commands and instructions for a wide variety of graphing calculators, including the most recent models.

■ Use properties of natural logarithms to answer questions about real life.   *(Section 4.4)*   *Review Exercises 93, 94*

■ Find the derivatives of natural logarithmic functions.   *(Section 4.5)*   *Review Exercises 95–108*

$$\frac{d}{dx}[\ln x] = \frac{1}{x}, \quad \frac{d}{dx}[\ln u] = \frac{1}{u}\frac{du}{dx}$$

■ Use calculus to analyze the graphs of functions that involve the natural logarithmic   *Review Exercises 109–112*
function.   *(Section 4.5)*

■ Use the definition of logarithms to evaluate logarithmic expressions involving other   *Review Exercises 113–116*
bases.   *(Section 4.5)*

$$\log_a x = b \quad \text{if and only if} \quad a^b = x$$

■ Use the change-of-base formula to evaluate logarithmic expressions involving other   *Review Exercises 117–120*
bases.   *(Section 4.5)*

$$\log_a x = \frac{\ln x}{\ln a}$$

■ Find the derivatives of exponential and logarithmic functions involving other bases.   *Review Exercises 121–124*
(Section 4.5)

$$\frac{d}{dx}[a^x] = (\ln a)a^x, \quad \frac{d}{dx}[a^u] = (\ln a)a^u\frac{du}{dx}$$

$$\frac{d}{dx}[\log_a x] = \left(\frac{1}{\ln a}\right)\frac{1}{x}, \quad \frac{d}{dx}[\log_a u] = \left(\frac{1}{\ln a}\right)\left(\frac{1}{u}\right)\frac{du}{dx}$$

■ Use calculus to answer questions about real-life rates of change.   *(Section 4.5)*   *Review Exercises 125, 126*

■ Use exponential growth and decay to model real-life situations.   *(Section 4.6)*   *Review Exercises 127–132*

■ ***Classifying Differentiation Rules***   Differentiation rules fall into two basic classes:
(1) general rules that apply to all differentiable functions; and (2) specific rules that apply
to special types of functions. At this point in the course, you have studied six general
rules: the Constant Rule, the Constant Multiple Rule, the Sum Rule, the Difference Rule,
the Product Rule, and the Quotient Rule. Although these rules were introduced in the
context of algebraic functions, remember that they can also be used with exponential
and logarithmic functions. You have also studied three specific rules: the Power Rule, the
derivative of the natural exponential function, and the derivative of the natural logarithmic
function. Each of these rules comes in two forms: the "simple" version, such as
$D_x[e^x] = e^x$, and the Chain Rule version, such as $D_x[e^u] = e^u(du/dx)$.

■ ***To Memorize or Not to Memorize?***   When studying mathematics, you need to memorize
some formulas and rules. Much of this will come from practice—the formulas that you
use most often will be committed to memory. Some formulas, however, are used only
infrequently. With these, it is helpful to be able to *derive* the formula from a *known*
formula. For instance, knowing the Log Rule for differentiation and the change-of-base
formula, $\log_a x = (\ln x)/(\ln a)$, allows you to derive the formula for the derivative of a
logarithmic function to base a.

Study Tools   *Additional resources that accompany this chapter*

■ **Algebra Review**   (pages 308 and 309)

■ **Chapter Summary and Study Strategies**   (pages 310 and 311)

■ **Review Exercises**   (pages 312–315)

■ **Sample Post-Graduation Exam Questions**   (page 316)

■ **Web Exercises**   (page 289, Exercise 80; page 298, Exercise 83)

■ **Student Solutions Guide**

■ **HM mathSpace® Student CD-ROM**

■ **Graphing Technology Guide**   (*math.college.hmco.com/students*)

4 SAMPLE POST-GRADUATION EXAM QUESTIONS

CPA
GMAT
GRE
Actuarial
CLAST

The following questions represent the types of questions that appear on certified public accountant (CPA) exams, Graduate Management Admission Tests (GMAT), Graduate Records Exams (GRE), actuarial exams, and College-Level Academic Skills Tests (CLAST). The answers to the questions are given in the back of the book.

1. 10^x means that 10 is to be used as a factor x times, and 10^{-x} is equal to

$$\frac{1}{10^x}.$$

   A very large or very small number, therefore, is frequently written as a decimal multiplied by 10^x, where x is an integer. Which, if any, are false?

   (a) $470{,}000 = 4.7 \times 10^5$

   (b) 450 billion $= 4.5 \times 10^{11}$

   (c) $0.0000000075 = 7.5 \times 10^{-10}$

   (d) 86 hundred-thousandths $= 8.6 \times 10^2$

2. The rate of decay of a radioactive substance is proportional to the amount of the substance present. Three years ago there was 6 grams of substance. Now there is 5 grams. How many grams will there be 3 years from now?

   (a) 4   (b) $\frac{25}{6}$ (c) $\frac{125}{36}$ (d) $\frac{75}{36}$

3. In a certain town, 45% of the people have brown hair, 30% have brown eyes, and 15% have both brown hair and brown eyes. What percent of the people in the town have neither brown hair nor brown eyes?

   (a) 25%   (b) 35%   (c) 40%   (d) 50%

4. You deposit $900 in a savings account that is compounded continuously at 4.76%. After 16 years, the amount in the account will be

   (a) $1927.53   (b) $1077.81   (c) $943.88   (d) $2827.53

5. A bookstore orders 75 books. Each book costs the bookstore $29 and is sold for $42. The bookstore must pay a $4 service charge for each unsold book returned. If the bookstore returns seven books, how much profit will the bookstore make?

   (a) $975   (b) $947   (c) $856   (d) $681

Figure for 6–9

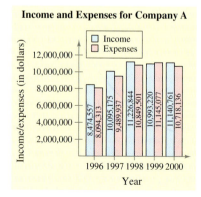

Income and Expenses for Company A

For Questions 6–9, use the data given in the graph.

6. In how many of the years were expenses greater than in the preceding year?

   (a) 2   (b) 4   (c) 1   (d) 3

7. In which year was the profit the greatest?

   (a) 1997   (b) 2000   (c) 1996   (d) 1998

8. In 1999, profits decreased by x percent from 1998 with x equal to

   (a) 60%   (b) 140%   (c) 340%   (d) 40%

9. In 2000, profits increased by y percent from 1999 with y equal to

   (a) 64%   (b) 136%   (c) 178%   (d) 378%

5

Integration and Its Applications

Jerry Alexander/Alamy

Integration can be used to solve physics problems, such as finding the time it takes for a sandbag to fall to the ground when dropped from a hot air balloon.

STRATEGIES FOR SUCCESS

WHAT YOU SHOULD LEARN:

- How to find the antiderivative F of a function—that is, $F'(x) = f(x)$

- How to use the General Power Rule, Exponential Rule, and Log Rule to calculate antiderivatives

- How to evaluate definite integrals and apply the Fundamental Theorem of Calculus to find the area bounded by two graphs

- How to use the Midpoint Rule to approximate definite integrals

- How to use integration to find the volume of a solid of revolution

WHY YOU SHOULD LEARN IT:

Integration has many applications in real life, as demonstrated by the examples below, which represent a small sample of the applications in this chapter.

- Demand Function, Exercises 67–70 on page 327

- Vertical Motion, Exercises 75–78 on page 328

- Marginal Propensity to Consume, Exercises 57 and 58 on page 336

- Annuity, Example 9 on page 352, Exercises 81–84 on page 355

- Capital Accumulation, Exercises 85–88 on page 355

- Consumer and Producer Surpluses, Exercises 41–46 on page 363

- Lorenz Curve, Exercises 58 and 59 on page 364

5.1   ANTIDERIVATIVES AND INDEFINITE INTEGRALS

- Understand the definition of antiderivative.
- Use indefinite integral notation for antiderivatives.
- Use basic integration rules to find antiderivatives.
- Use initial conditions to find particular solutions of indefinite integrals.
- Use antiderivatives to solve real-life problems.

Antiderivatives

Up to this point in the text, you have been concerned primarily with this problem: given a function, find its derivative. Many important applications of calculus involve the inverse problem: given the derivative of a function, find the function. For example, suppose you are given

$$f'(x) = 2, \quad g'(x) = 3x^2, \quad \text{and} \quad s'(t) = 4t.$$

Your goal is to determine the functions f, g, and s. By making educated guesses, you might come up with the following functions.

$$f(x) = 2x \quad \text{because} \quad \frac{d}{dx}[2x] = 2.$$

$$g(x) = x^3 \quad \text{because} \quad \frac{d}{dx}[x^3] = 3x^2.$$

$$s(t) = 2t^2 \quad \text{because} \quad \frac{d}{dt}[2t^2] = 4t.$$

This operation of determining the original function from its derivative is the inverse operation of differentiation. It is called **antidifferentiation.**

Definition of Antiderivative

A function F is an **antiderivative** of a function f if for every x in the domain of f, it follows that $F'(x) = f(x)$.

If $F(x)$ is an antiderivative of $f(x)$, then $F(x) + C$, where C is any constant, is also an antiderivative of $f(x)$. For example,

$$F(x) = x^3, \quad G(x) = x^3 - 5, \quad \text{and} \quad H(x) = x^3 + 0.3$$

are all antiderivatives of $3x^2$ because the derivative of each is $3x^2$. As it turns out, *all* antiderivatives of $3x^2$ are of the form $x^3 + C$. So, the process of antidifferentiation does not determine a single function, but rather a *family* of functions, each differing from the others by a constant.

STUDY TIP

In this text, the phrase "$F(x)$ is an antiderivative of $f(x)$" is used synonymously with "F is an antiderivative of f."

Notation for Antiderivatives and Indefinite Integrals

The antidifferentiation process is also called **integration** and is denoted by the symbol

$$\int$$ Integral sign

which is called an **integral sign.** The symbol

$$\int f(x)\,dx$$ Indefinite integral

is the **indefinite integral** of $f(x)$, and it denotes the family of antiderivatives of $f(x)$. That is, if $F'(x) = f(x)$ for all x, then you can write

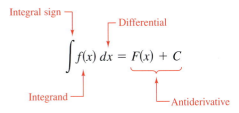

where $f(x)$ is the **integrand** and C is the **constant of integration.** The differential dx in the indefinite integral identifies the variable of integration. That is, the symbol $\int f(x)\,dx$ denotes the "antiderivative of f with respect to x" just as the symbol dy/dx denotes the "derivative of y with respect to x."

Integral Notation of Antiderivatives

The notation

$$\int f(x)\,dx = F(x) + C$$

where C is an arbitrary constant, means that F is an antiderivative of f. That is, $F'(x) = f(x)$ for all x in the domain of f.

EXAMPLE 1   **Notation for Antiderivatives**

Using integral notation, you can write the three antiderivatives from the beginning of this section as shown.

(a) $\displaystyle\int 2\,dx = 2x + C$ (b) $\displaystyle\int 3x^2\,dx = x^3 + C$ (c) $\displaystyle\int 4t\,dt = 2t^2 + C$

TRY IT 1

Rewrite each antiderivative using integral notation.

(a) $\dfrac{d}{dx}[3x] = 3$ (b) $\dfrac{d}{dx}[x^2] = 2x$ (c) $\dfrac{d}{dt}[3t^3] = 9t^2$

DISCOVERY

Verify that $F_1(x) = x^2 - 2x$, $F_2(x) = x^2 - 2x - 1$, and $F_3(x) = (x - 1)^2$ are all anti-derivatives of $f(x) = 2x - 2$. Use a graphing utility to graph F_1, F_2, and F_3 in the same coordinate plane. How are their graphs related? What can you say about the graph of any other antiderivative of f?

Finding Antiderivatives

The inverse relationship between the operations of integration and differentiation can be shown symbolically, as shown.

$$\frac{d}{dx}\left[\int f(x)\,dx\right] = f(x)$$

Differentiation is the inverse of integration.

$$\int f'(x)\,dx = f(x) + C$$

Integration is the inverse of differentiation.

This inverse relationship between integration and differentiation allows you to obtain integration formulas directly from differentiation formulas. The following summary lists the integration formulas that correspond to some of the differentiation formulas you have studied.

Basic Integration Rules

1. $\displaystyle\int k\,dx = kx + C, \quad k$ is a constant.   Constant Rule

2. $\displaystyle\int kf(x)\,dx = k\int f(x)\,dx$ Constant Multiple Rule

3. $\displaystyle\int [f(x) + g(x)]\,dx = \int f(x)\,dx + \int g(x)\,dx$ Sum Rule

4. $\displaystyle\int [f(x) - g(x)]\,dx = \int f(x)\,dx - \int g(x)\,dx$ Difference Rule

5. $\displaystyle\int x^n\,dx = \frac{x^{n+1}}{n+1} + C, \quad n \neq -1$ Simple Power Rule

STUDY TIP

You will study the General Power Rule for integration in Section 5.2 and the Exponential and Log Rules in Section 5.3.

STUDY TIP

In Example 2(b), the integral $\int 1\,dx$ is usually shortened to the form $\int dx$.

Be sure you see that the Simple Power Rule has the restriction that n cannot be -1. So, you *cannot* use the Simple Power Rule to evaluate the integral

$$\int \frac{1}{x}\,dx.$$

To evaluate this integral, you need the Log Rule, which is described in Section 5.3.

TRY IT 2

Find each indefinite integral.

(a) $\displaystyle\int 5\,dx$

(b) $\displaystyle\int -1\,dr$

(c) $\displaystyle\int 2\,dt$

EXAMPLE 2   Finding Indefinite Integrals

Find each indefinite integral.

(a) $\displaystyle\int \frac{1}{2}\,dx$ (b) $\displaystyle\int 1\,dx$ (c) $\displaystyle\int -5\,dt$

SOLUTION

(a) $\displaystyle\int \frac{1}{2}\,dx = \frac{1}{2}x + C$ (b) $\displaystyle\int 1\,dx = x + C$ (c) $\displaystyle\int -5\,dt = -5t + C$

EXAMPLE 3   **Finding an Indefinite Integral**

Find $\int 3x\,dx$.

SOLUTION

$$\int 3x\,dx = 3\int x\,dx \qquad \text{Constant Multiple Rule}$$

$$= 3\int x^1\,dx \qquad \text{Rewrite } x \text{ as } x^1.$$

$$= 3\left(\frac{x^2}{2}\right) + C \qquad \text{Simple Power Rule with } n = 1$$

$$= \frac{3}{2}x^2 + C \qquad \text{Simplify.}$$

TRY IT 3

Find $\int 5x\,dx$.

In finding indefinite integrals, a strict application of the basic integration rules tends to produce cumbersome constants of integration. For instance, in Example 3, you could have written

$$\int 3x\,dx = 3\int x\,dx = 3\left(\frac{x^2}{2} + C\right) = \frac{3}{2}x^2 + 3C.$$

However, because C represents *any* constant, it is unnecessary to write $3C$ as the constant of integration. You can simply write $\frac{3}{2}x^2 + C$.

In Example 3, note that the general pattern of integration is similar to that of differentiation.

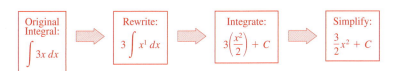

| Original Integral: $\int 3x\,dx$ | Rewrite: $3\int x^1\,dx$ | Integrate: $3\left(\frac{x^2}{2}\right) + C$ | Simplify: $\frac{3}{2}x^2 + C$ |

STUDY TIP

Remember that you can check your answer to an antidifferentiation problem by differentiating. For instance, in Example 4(b), you can check that $\frac{2}{3}x^{3/2}$ is the correct antiderivative by differentiating to obtain

$$\frac{d}{dx}\left[\frac{2}{3}x^{3/2}\right] = \left(\frac{2}{3}\right)\left(\frac{3}{2}\right)x^{1/2}$$

$$= \sqrt{x}.$$

EXAMPLE 4   **Rewriting Before Integrating**

Find each indefinite integral.

(a) $\int \dfrac{1}{x^3}\,dx$

(b) $\int \sqrt{x}\,dx$

SOLUTION

	Original Integral	*Rewrite*	*Integrate*	*Simplify*
(a)	$\int \dfrac{1}{x^3}\,dx$	$\int x^{-3}\,dx$	$\dfrac{x^{-2}}{-2} + C$	$-\dfrac{1}{2x^2} + C$
(b)	$\int \sqrt{x}\,dx$	$\int x^{1/2}\,dx$	$\dfrac{x^{3/2}}{3/2} + C$	$\dfrac{2}{3}x^{3/2} + C$

TRY IT 4

Find each indefinite integral.

(a) $\int \dfrac{1}{x^2}\,dx$ (b) $\int \sqrt[3]{x}\,dx$

With the five basic integration rules, you can integrate *any* polynomial function, as demonstrated in the next example.

TRY IT 5

Find each indefinite integral.

(a) $\displaystyle\int (x + 4)\, dx$

(b) $\displaystyle\int (4x^3 - 5x + 2)\, dx$

EXAMPLE 5   **Integrating Polynomial Functions**

Find each indefinite integral.

(a) $\displaystyle\int (x + 2)\, dx$

(b) $\displaystyle\int (3x^4 - 5x^2 + x)\, dx$

SOLUTION

(a) Use the Sum Rule to integrate each part separately.

$$\int (x + 2)\, dx = \int x\, dx + \int 2\, dx$$

$$= \frac{x^2}{2} + 2x + C$$

(b) Try to identify each basic integration rule used to evaluate this integral.

$$\int (3x^4 - 5x^2 + x)\, dx = 3\left(\frac{x^5}{5}\right) - 5\left(\frac{x^3}{3}\right) + \frac{x^2}{2} + C$$

$$= \frac{3}{5}x^5 - \frac{5}{3}x^3 + \frac{1}{2}x^2 + C$$

STUDY TIP

When integrating quotients, remember *not* to integrate the numerator and denominator separately. For instance, in Example 6, be sure you see that

$$\int \frac{x + 1}{\sqrt{x}}\, dx \neq \frac{\int (x + 1)\, dx}{\int \sqrt{x}\, dx}.$$

EXAMPLE 6   **Rewriting Before Integrating**

Find $\displaystyle\int \frac{x + 1}{\sqrt{x}}\, dx$.

SOLUTION   Begin by rewriting the quotient in the integrand as a sum. Then rewrite each term using rational exponents.

$$\int \frac{x + 1}{\sqrt{x}}\, dx = \int \left(\frac{x}{\sqrt{x}} + \frac{1}{\sqrt{x}}\right) dx \qquad \text{Rewrite as a sum.}$$

$$= \int (x^{1/2} + x^{-1/2})\, dx \qquad \text{Rewrite using rational exponents.}$$

$$= \frac{x^{3/2}}{3/2} + \frac{x^{1/2}}{1/2} + C \qquad \text{Apply Power Rule.}$$

$$= \frac{2}{3}x^{3/2} + 2x^{1/2} + C \qquad \text{Simplify.}$$

TRY IT 6

Find $\displaystyle\int \frac{x + 2}{\sqrt{x}}\, dx$.

ALGEBRA REVIEW

For help on the algebra in Example 6, see Example 1(a) in the *Chapter 5 Algebra Review*, on page 378.

TECHNOLOGY

If you have access to a symbolic integration utility, try using it to solve Example 6. The utility may list the antiderivative as $2\sqrt{x}(x + 3)/3$, which is equivalent to the result listed above.

Particular Solutions

You have already seen that the **differential equation** (an equation that involves x, y, and derivatives of y) $y = \int f(x)\, dx$ has many solutions, each differing from the others by a constant. This means that the graphs of any two antiderivatives of f are vertical translations of each other. For instance, Figure 5.1 shows the graphs of several antiderivatives of the form

$$y = F(x) = \int (3x^2 - 1)\, dx = x^3 - x + C$$

where $F(x) = x^3 - x + C$ is called the **general solution** of the differential equation. Each of these antiderivatives is a solution of $dy/dx = 3x^2 - 1$.

In many applications of integration, you are given enough information to determine a **particular solution.** To do this, you only need to know the value of $F(x)$ for one value of x. (This information is called an **initial condition.**) For example, in Figure 5.1, there is only one curve that passes through the point $(2, 4)$. To find this curve, use the information below.

$F(x) = x^3 - x + C$ General solution

$F(2) = 4$ Initial condition

By using the initial condition in the general solution, you can determine that $F(2) = 2^3 - 2 + C = 4$, which implies that $C = -2$. So, the particular solution is

$F(x) = x^3 - x - 2.$ Particular solution

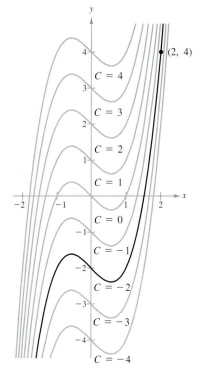

$F(x) = x^3 - x + C$

FIGURE 5.1

EXAMPLE 7   **Finding a Particular Solution**

Find the general solution of

$$F'(x) = 2x - 2$$

and find the particular solution that satisfies the initial condition $F(1) = 2$.

SOLUTION   Begin by integrating to find the general solution.

$F(x) = \displaystyle\int (2x - 2)\, dx$ Integrate $F'(x)$ to obtain $F(x)$.

$\quad\;\; = x^2 - 2x + C$ General solution

Using the initial condition $F(1) = 2$, you can write

$$F(1) = 1^2 - 2(1) + C = 2$$

which implies that $C = 3$. So, the particular solution is

$F(x) = x^2 - 2x + 3.$ Particular solution

This solution is shown graphically in Figure 5.2. Note that each of the gray curves represents a solution of the equation $F'(x) = 2x - 2$. The black curve, however, is the only solution that passes through the point $(1, 2)$, which means that $F(x) = x^2 - 2x + 3$ is the only solution that satisfies the initial condition.

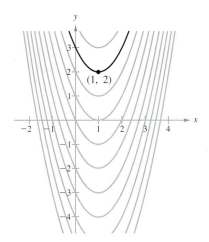

FIGURE 5.2

TRY IT 7

Find the general solution of $F'(x) = 4x + 2$, and find the particular solution that satisfies the initial condition $F(1) = 8$.

Applications

In Chapter 2, you used the general position function (neglecting air resistance) for a falling object

$$s(t) = -16t^2 + v_0t + s_0$$

where $s(t)$ is the height (in feet) and t is the time (in seconds). In the next example, integration is used to *derive* this function.

EXAMPLE 8   **Deriving a Position Function**

A ball is thrown upward with an initial velocity of 64 feet per second from an initial height of 80 feet, as shown in Figure 5.3. Derive the position function giving the height s (in feet) as a function of the time t (in seconds). When does the ball hit the ground?

SOLUTION   Let $t = 0$ represent the initial time. Then the two given conditions can be written as

$$s(0) = 80 \qquad \text{Initial height is 80 feet.}$$
$$s'(0) = 64. \qquad \text{Initial velocity is 64 feet per second.}$$

Because the acceleration due to gravity is -32 feet per second per second, you can integrate the acceleration function to find the velocity function as shown.

$$s''(t) = -32 \qquad \text{Acceleration due to gravity}$$
$$s'(t) = \int -32 \, dt \qquad \text{Integrate } s''(t) \text{ to obtain } s'(t).$$
$$= -32t + C_1 \qquad \text{Velocity function}$$

Using the initial velocity, you can conclude that $C_1 = 64$.

$$s'(t) = -32t + 64 \qquad \text{Velocity function}$$
$$s(t) = \int (-32t + 64) \, dt \qquad \text{Integrate } s'(t) \text{ to obtain } s(t).$$
$$= -16t^2 + 64t + C_2 \qquad \text{Position function}$$

Using the initial height, it follows that $C_2 = 80$. So, the position function is given by

$$s(t) = -16t^2 + 64t + 80. \qquad \text{Position function}$$

To find the time when the ball hits the ground, set the position function equal to 0 and solve for t.

$$-16t^2 + 64t + 80 = 0 \qquad \text{Set } s(t) \text{ equal to zero.}$$
$$-16(t + 1)(t - 5) = 0 \qquad \text{Factor.}$$
$$t = -1, \quad t = 5 \qquad \text{Solve for } t.$$

Because the time must be positive, you can conclude that the ball hits the ground 5 seconds after it is thrown.

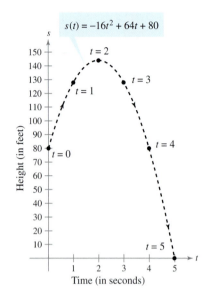

$s(t) = -16t^2 + 64t + 80$

FIGURE 5.3

TRY IT 8

Derive the position function if a ball is thrown upward with an initial velocity of 32 feet per second from an initial height of 48 feet. When does the ball hit the ground? With what velocity does the ball hit the ground?

EXAMPLE 9 **Finding a Cost Function**

The marginal cost for producing x units of a product is modeled by

$$\frac{dC}{dx} = 32 - 0.04x. \qquad \text{Marginal cost}$$

It costs $50 to produce one unit. Find the total cost of producing 200 units.

SOLUTION To find the cost function, integrate the marginal cost function.

$$C = \int (32 - 0.04x)\, dx \qquad \text{Integrate } \frac{dC}{dx} \text{ to obtain } C.$$

$$= 32x - 0.04\left(\frac{x^2}{2}\right) + K$$

$$= 32x - 0.02x^2 + K \qquad \text{Cost function}$$

To solve for K, use the initial condition that $C = 50$ when $x = 1$.

$$50 = 32(1) - 0.02(1)^2 + K \qquad \text{Substitute 50 for } C \text{ and 1 for } x.$$

$$18.02 = K \qquad \text{Solve for } K.$$

So, the total cost function is given by

$$C = 32x - 0.02x^2 + 18.02 \qquad \text{Cost function}$$

which implies that the cost of producing 200 units is

$$C = 32(200) - 0.02(200)^2 + 18.02$$

$$= \$5618.02.$$

TRY IT 9

The marginal cost function for producing x units of a product is modeled by

$$\frac{dC}{dx} = 28 - 0.02x.$$

It costs $40 to produce one unit. Find the cost of producing 200 units.

TAKE ANOTHER LOOK

Investigating Marginal Cost

In Example 9, you were given a marginal cost function of

$$\frac{dC}{dx} = 32 - 0.04x.$$

This means that the cost of making each additional unit decreases by about $0.04. You can confirm this by finding the costs of making different amounts of the product.

x	1	2	3	4	5	6
C	$50.00	$81.94	$113.84	$145.70	$177.52	$209.30

From this, you can see that the first unit cost $50, the second cost $31.94, the third cost $31.90, the fourth cost $31.86, the fifth cost $31.82, and so on. At what production level would this costing scheme cease to make sense? Explain your reasoning.

PREREQUISITE REVIEW 5.1

The following warm-up exercises involve skills that were covered in earlier sections. You will use these skills in the exercise set for this section.

In Exercises 1–6, rewrite the expression using rational exponents.

1. $\dfrac{\sqrt{x}}{x}$

2. $\sqrt[3]{2x}\,(2x)$

3. $\sqrt{5x^3} + \sqrt{x^5}$

4. $\dfrac{1}{\sqrt{x}} + \dfrac{1}{\sqrt[3]{x^2}}$

5. $\dfrac{(x+1)^3}{\sqrt{x+1}}$

6. $\dfrac{\sqrt{x}}{\sqrt[3]{x}}$

In Exercises 7–10, let $(x, y) = (2, 2)$, and solve the equation for C.

7. $y = x^2 + 5x + C$

8. $y = 3x^3 - 6x + C$

9. $y = -16x^2 + 26x + C$

10. $y = -\frac{1}{4}x^4 - 2x^2 + C$

EXERCISES 5.1

In Exercises 1–8, verify the statement by showing that the derivative of the right side is equal to the integrand of the left side.

1. $\displaystyle\int \left(-\frac{9}{x^4}\right) dx = \frac{3}{x^3} + C$

2. $\displaystyle\int \frac{4}{\sqrt{x}}\, dx = 8\sqrt{x} + C$

3. $\displaystyle\int \left(4x^3 - \frac{1}{x^2}\right) dx = x^4 + \frac{1}{x} + C$

4. $\displaystyle\int \left(1 - \frac{1}{\sqrt[3]{x^2}}\right) dx = x - 3\sqrt[3]{x} + C$

5. $\displaystyle\int 2\sqrt{x}\,(x - 3)\, dx = \frac{4x^{3/2}(x-5)}{5} + C$

6. $\displaystyle\int 4\sqrt{x}\,(x^2 - 2)\, dx = \frac{8x^{3/2}(3x^2 - 14)}{21} + C$

7. $\displaystyle\int \frac{x^2 - 1}{x^{3/2}}\, dx = \frac{2(x^2 + 3)}{3\sqrt{x}} + C$

8. $\displaystyle\int \frac{2x - 1}{x^{4/3}}\, dx = \frac{3(x + 1)}{\sqrt[3]{x}} + C$

In Exercises 9–20, find the indefinite integral and check your result by differentiation.

9. $\displaystyle\int 6\, dx$

10. $\displaystyle\int -4\, dx$

11. $\displaystyle\int 5t^2\, dt$

12. $\displaystyle\int 3t^4\, dt$

13. $\displaystyle\int 5x^{-3}\, dx$

14. $\displaystyle\int 4y^{-3}\, dy$

15. $\displaystyle\int du$

16. $\displaystyle\int dr$

17. $\displaystyle\int e\, dt$

18. $\displaystyle\int e^3\, dy$

19. $\displaystyle\int y^{3/2}\, dy$

20. $\displaystyle\int v^{-1/2}\, dv$

In Exercises 21–28, find the indefinite integral using the columns in Example 4 as a model. Use a symbolic integration utility to verify your results.

21. $\displaystyle\int \sqrt[3]{x}\, dx$

22. $\displaystyle\int \frac{1}{x^2}\, dx$

23. $\displaystyle\int \frac{1}{x\sqrt{x}}\, dx$

24. $\displaystyle\int \frac{1}{x^2\sqrt{x}}\, dx$

25. $\displaystyle\int x(x^2 + 3)\, dx$

26. $\displaystyle\int t(t^2 + 2)\, dt$

27. $\displaystyle\int \frac{1}{2x^3}\, dx$

28. $\displaystyle\int \frac{1}{8x^3}\, dx$

In Exercises 29–32, find two functions that have the given derivative, and sketch the graph of each. (There is more than one correct answer.)

29.
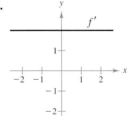

30.
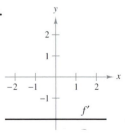

31.

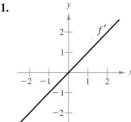

32.

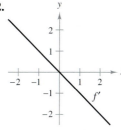

Figure for 55

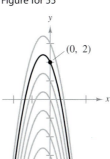

Figure for 56

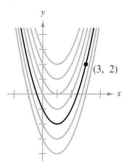

In Exercises 33–42, find the indefinite integral and check your result by differentiation.

33. $\displaystyle\int (x^3 + 2)\,dx$

34. $\displaystyle\int (x^2 - 2x + 3)\,dx$

35. $\displaystyle\int \left(\sqrt[3]{x} - \frac{1}{2\sqrt[3]{x}}\right)dx$

36. $\displaystyle\int \left(\sqrt{x} + \frac{1}{2\sqrt{x}}\right)dx$

37. $\displaystyle\int \left(\sqrt[3]{x^2} + 1\right)dx$

38. $\displaystyle\int \left(\sqrt[4]{x^3} + 1\right)dx$

39. $\displaystyle\int \frac{1}{3x^4}\,dx$

40. $\displaystyle\int \frac{1}{4x^2}\,dx$

41. $\displaystyle\int \frac{2x^3 + 1}{x^3}\,dx$

42. $\displaystyle\int \frac{t^2 + 2}{t^2}\,dt$

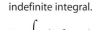 In Exercises 43–48, use a symbolic integration utility to find the indefinite integral.

43. $\displaystyle\int u(3u^2 + 1)\,du$

44. $\displaystyle\int \sqrt{x}(x + 1)\,dx$

45. $\displaystyle\int (x - 1)(6x - 5)\,dx$

46. $\displaystyle\int (2t^2 - 1)^2\,dt$

47. $\displaystyle\int y^2\sqrt{y}\,dy$

48. $\displaystyle\int (1 + 3t)t^2\,dt$

In Exercises 49–54, find the particular solution $y = f(x)$ that satisfies the differential equation and initial condition.

49. $f'(x) = 3\sqrt{x} + 3; \quad f(1) = 4$

50. $f'(x) = \frac{1}{5}x - 2; \quad f(10) = -10$

51. $f'(x) = 6x(x - 1); \quad f(1) = -1$

52. $f'(x) = (2x - 3)(2x + 3); \quad f(3) = 0$

53. $f'(x) = \dfrac{2 - x}{x^3}, \quad x > 0; \quad f(2) = \dfrac{3}{4}$

54. $f'(x) = \dfrac{x^2 - 5}{x^2}, \quad x > 0; \quad f(1) = 2$

In Exercises 55 and 56, you are shown a family of graphs, each of which is a general solution of the given differential equation. Find the equation of the particular solution that passes through the indicated point.

55. $\dfrac{dy}{dx} = -5x - 2$

56. $\dfrac{dy}{dx} = 2(x - 1)$

In Exercises 57 and 58, find the equation of the function f whose graph passes through the point.

Derivative	Point
57. $f'(x) = 6\sqrt{x} - 10$	$(4, 2)$
58. $f'(x) = \dfrac{6}{x^2}$	$(2, 5)$

In Exercises 59–62, find a function f that satisfies the conditions.

59. $f''(x) = 2, \quad f'(2) = 5, \quad f(2) = 10$

60. $f''(x) = x^2, \quad f'(0) = 6, \quad f(0) = 3$

61. $f''(x) = x^{-2/3}, \quad f'(8) = 6, \quad f(0) = 0$

62. $f''(x) = x^{-3/2}, \quad f'(1) = 2, \quad f(9) = -4$

Cost   In Exercises 63–66, find the cost function for the marginal cost and fixed cost.

Marginal Cost	Fixed Cost $(x = 0)$
63. $\dfrac{dC}{dx} = 85$	\$5500
64. $\dfrac{dC}{dx} = \dfrac{1}{50}x + 10$	\$1000
65. $\dfrac{dC}{dx} = \dfrac{1}{20\sqrt{x}} + 4$	\$750
66. $\dfrac{dC}{dx} = \dfrac{\sqrt[4]{x}}{10} + 10$	\$2300

Demand Function   In Exercises 67–70, find the revenue and demand functions for the given marginal revenue. (Use the fact that $R = 0$ when $x = 0$.)

67. $\dfrac{dR}{dx} = 225 - 3x$

68. $\dfrac{dR}{dx} = 310 - 4x$

69. $\dfrac{dR}{dx} = 225 + 2x - x^2$

70. $\dfrac{dR}{dx} = 100 - 6x - 2x^2$

Profit   In Exercises 71–74, find the profit function for the given marginal profit and initial condition.

Marginal Profit	Initial Condition
71. $\dfrac{dP}{dx} = -18x + 1650$	$P(15) = \$22{,}725$
72. $\dfrac{dP}{dx} = -40x + 250$	$P(5) = \$650$
73. $\dfrac{dP}{dx} = -24x + 805$	$P(12) = \$8000$
74. $\dfrac{dP}{dx} = -30x + 920$	$P(8) = \$6500$

Vertical Motion   In Exercises 75–78, use $a(t) = -32$ feet per second per second as the acceleration due to gravity.

75. A ball is thrown vertically upward with an initial velocity of 60 feet per second. How high will the ball go?

76. The Grand Canyon is 6000 feet deep at the deepest part. A rock is dropped from this height. Express the height of the rock as a function of the time t (in seconds). How long will it take the rock to hit the canyon floor?

77. With what initial velocity must an object be thrown upward from the ground to reach the height of the Washington Monument (550 feet)?

78. A balloon, rising vertically with a velocity of 16 feet per second, releases a sandbag at an instant when the balloon is 64 feet above the ground.

(a) How many seconds after its release will the bag strike the ground?

(b) With what velocity will the bag strike the ground?

79. *Cost*   A company produces a product for which the marginal cost of producing x units is modeled by

$$\frac{dC}{dx} = 2x - 12$$

and the fixed costs are $125.

(a) Find the total cost function and the average cost function.

(b) Find the total cost of producing 50 units.

(c) In part (b), how much of the total cost is fixed? How much is variable? Give examples of fixed costs associated with the manufacturing of a product. Give examples of variable costs.

80. *Population Growth*   The growth rate of Horry County in South Carolina can be modeled by

$$\frac{dP}{dt} = 105.74t + 2639.3$$

where t is the time in years, with $t = 0$ corresponding to 1970. The county's population was 196,629 in 2000. *(Source: U.S. Census Bureau)*

(a) Find the model for Horry County's population.

(b) Use the model to predict the population in 2010. Does your answer seem reasonable? Explain your reasoning.

81. *Vital Statistics*   The rate of increase of the number of married couples M (in thousands) in the United States from 1970 to 2000 can be modeled by

$$\frac{dM}{dt} = 0.636t^2 - 28.48t + 632.7$$

where t is the time in years, with $t = 0$ corresponding to 1970. The number of married couples in 2000 was 56,497 thousand. *(Source: U.S. Census Bureau)*

(a) Find the model for the number of married couples in the United States.

(b) Use the model to predict the number of married couples in the United States in 2010. Does your answer seem reasonable? Explain your reasoning.

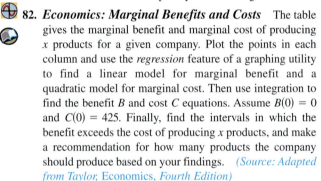

82. *Economics: Marginal Benefits and Costs*   The table gives the marginal benefit and marginal cost of producing x products for a given company. Plot the points in each column and use the *regression* feature of a graphing utility to find a linear model for marginal benefit and a quadratic model for marginal cost. Then use integration to find the benefit B and cost C equations. Assume $B(0) = 0$ and $C(0) = 425$. Finally, find the intervals in which the benefit exceeds the cost of producing x products, and make a recommendation for how many products the company should produce based on your findings. *(Source: Adapted from Taylor, Economics, Fourth Edition)*

Number of products	1	2	3	4	5
Marginal benefit	330	320	290	270	250
Marginal cost	150	120	100	110	120

Number of products	6	7	8	9	10
Marginal benefit	230	210	190	170	160
Marginal cost	140	160	190	250	320

83. *Research Project*   Use your school's library, the Internet, or some other reference source to research a company that markets a natural resource. Find data on the revenue of the company and on the consumption of the resource. Then find a model for each. Is the company's revenue related to the consumption of the resource? Explain your reasoning.

5.2 THE GENERAL POWER RULE

- Use the General Power Rule to find indefinite integrals.
- Use substitution to find indefinite integrals.
- Use the General Power Rule to solve real-life problems.

The General Power Rule

In Section 5.1, you used the Simple Power Rule

$$\int x^n \, dx = \frac{x^{n+1}}{n+1} + C, \qquad n \neq -1$$

to find antiderivatives of functions expressed as powers of x alone. In this section, you will study a technique for finding antiderivatives of more complicated functions.

To begin, consider how you might find the antiderivative of $2x(x^2 + 1)^3$. Because you are hunting for a function whose derivative is $2x(x^2 + 1)^3$, you might discover the antiderivative as shown.

$$\frac{d}{dx}[(x^2 + 1)^4] = 4(x^2 + 1)^3(2x) \qquad \text{Use Chain Rule.}$$

$$\frac{d}{dx}\left[\frac{(x^2 + 1)^4}{4}\right] = (x^2 + 1)^3(2x) \qquad \text{Divide both sides by 4.}$$

$$\frac{(x^2 + 1)^4}{4} + C = \int 2x(x^2 + 1)^3 \, dx \qquad \text{Write in integral form.}$$

The key to this solution is the presence of the factor $2x$ in the integrand. In other words, this solution works because $2x$ is precisely the derivative of $(x^2 + 1)$. Letting $u = x^2 + 1$, you can write

$$\int \overbrace{(x^2 + 1)^3}^{u^3} \underbrace{2x \, dx}_{du} = \int u^3 \, du$$

$$= \frac{u^4}{4} + C.$$

This is an example of the **General Power Rule** for integration.

General Power Rule for Integration

If u is a differentiable function of x, then

$$\int u^n \frac{du}{dx} \, dx = \int u^n \, du = \frac{u^{n+1}}{n+1} + C, \qquad n \neq -1.$$

When using the General Power Rule, you must first identify a factor u of the integrand that is raised to a power. Then, you must show that its derivative du/dx is also a factor of the integrand. This is demonstrated in Example 1.

EXAMPLE 1 **Applying the General Power Rule**

Find each indefinite integral.

(a) $\displaystyle\int 3(3x - 1)^4 \, dx$　　(b) $\displaystyle\int (2x + 1)(x^2 + x) \, dx$

(c) $\displaystyle\int 3x^2 \sqrt{x^3 - 2} \, dx$　　(d) $\displaystyle\int \frac{-4x}{(1 - 2x^2)^2} \, dx$

SOLUTION

STUDY TIP

Example 1(b) illustrates a case of the General Power Rule that is sometimes overlooked—when the power is $n = 1$. In this case, the rule takes the form

$$\int u \frac{du}{dx} \, dx = \frac{u^2}{2} + C.$$

(a) $\displaystyle\int 3(3x - 1)^4 \, dx = \int \overbrace{(3x - 1)^4}^{u^n} \overbrace{(3)}^{\frac{du}{dx}} \, dx$　　Let $u = 3x - 1$.

$$= \frac{(3x - 1)^5}{5} + C$$　　General Power Rule

(b) $\displaystyle\int (2x + 1)(x^2 + x) \, dx = \int \overbrace{(x^2 + x)}^{u^n} \overbrace{(2x + 1)}^{\frac{du}{dx}} \, dx$　　Let $u = x^2 + x$.

$$= \frac{(x^2 + x)^2}{2} + C$$　　General Power Rule

(c) $\displaystyle\int 3x^2 \sqrt{x^3 - 2} \, dx = \int \overbrace{(x^3 - 2)^{1/2}}^{u^n} \overbrace{(3x^2)}^{\frac{du}{dx}} \, dx$　　Let $u = x^3 - 2$.

$$= \frac{(x^3 - 2)^{3/2}}{3/2} + C$$　　General Power Rule

$$= \frac{2}{3}(x^3 - 2)^{3/2} + C$$　　Simplify.

STUDY TIP

Remember that you can verify the result of an indefinite integral by differentiating the function. Check the answer to Example 1(d) by differentiating the function

$$F(x) = -\frac{1}{1 - 2x^2} + C.$$

$$\frac{d}{dx}\left(-\frac{1}{1 - 2x^2} + C \right)$$

$$= \frac{-4x}{(1 - 2x^2)^2}$$

(d) $\displaystyle\int \frac{-4x}{(1 - 2x^2)^2} \, dx = \int \overbrace{(1 - 2x^2)^{-2}}^{u^n} \overbrace{(-4x)}^{\frac{du}{dx}} \, dx$　　Let $u = 1 - 2x^2$.

$$= \frac{(1 - 2x^2)^{-1}}{-1} + C$$　　General Power Rule

$$= -\frac{1}{1 - 2x^2} + C$$　　Simplify.

TRY IT 1

Find each indefinite integral.

(a) $\displaystyle\int (3x^2 + 6)(x^3 + 6x)^2 \, dx$　　(b) $\displaystyle\int 2x \sqrt{x^2 - 2} \, dx$

Many times, part of the derivative du/dx is missing from the integrand, and in *some* cases you can make the necessary adjustments to apply the General Power Rule.

EXAMPLE 2   **Multiplying and Dividing by a Constant**

Find $\displaystyle\int x(3 - 4x^2)^2 \, dx$.

SOLUTION   Let $u = 3 - 4x^2$. To apply the General Power Rule, you need to create $du/dx = -8x$ as a factor of the integrand. You can accomplish this by multiplying and dividing by the constant -8.

$$\int x(3 - 4x^2)^2 \, dx = \int \left(-\frac{1}{8}\right) \overbrace{(3 - 4x^2)^2}^{u^n} \overbrace{(-8x)}^{\frac{du}{dx}} \, dx \qquad \text{Multiply and divide by } -8.$$

$$= -\frac{1}{8}\int (3 - 4x^2)^2 (-8x) \, dx \qquad \text{Factor } -\frac{1}{8} \text{ out of integrand.}$$

$$= \left(-\frac{1}{8}\right)\frac{(3 - 4x^2)^3}{3} + C \qquad \text{General Power Rule}$$

$$= -\frac{(3 - 4x^2)^3}{24} + C \qquad \text{Simplify.}$$

ALGEBRA REVIEW

For help on the algebra in Example 2, see Example 1(b) in the *Chapter 5 Algebra Review*, on page 378.

STUDY TIP

Try using the Chain Rule to check the result of Example 2. After differentiating $-\frac{1}{24}(3 - 4x^2)^3$ and simplifying, you should obtain the original integrand.

TRY IT 2

Find $\displaystyle\int x^3(3x^4 + 1)^2 \, dx$.

EXAMPLE 3   **A Failure of the General Power Rule**

Find $\displaystyle\int -8(3 - 4x^2)^2 \, dx$.

SOLUTION   Let $u = 3 - 4x^2$. As in Example 2, to apply the General Power Rule you must create $du/dx = -8x$ as a factor of the integrand. In Example 2, you could do this by multiplying and dividing by a constant, and then factoring that constant out of the integrand. This strategy doesn't work with variables. That is,

$$\int -8(3 - 4x^2)^2 \, dx \neq \frac{1}{x}\int (3 - 4x^2)^2 (-8x) \, dx.$$

To find this indefinite integral, you can expand the integrand and use the Simple Power Rule.

$$\int -8(3 - 4x^2)^2 \, dx = \int (-72 + 192x^2 - 128x^4) \, dx$$

$$= -72x + 64x^3 - \frac{128}{5}x^5 + C$$

STUDY TIP

In Example 3, be sure you see that you cannot factor variable quantities outside the integral sign. After all, if this were permissible, then you could move the entire integrand outside the integral sign and eliminate the need for all integration rules except the rule $\int dx = x + C$.

TRY IT 3

Find $\displaystyle\int 2(3x^4 + 1)^2 \, dx$.

When an integrand contains an extra constant factor that is not needed as part of du/dx, you can simply move the factor outside the integral sign, as shown in the next example.

EXAMPLE 4 **Applying the General Power Rule**

Find $\int 7x^2\sqrt{x^3 + 1}\, dx$.

SOLUTION Let $u = x^3 + 1$. Then you need to create $du/dx = 3x^2$ by multiplying and dividing by 3. The constant factor $\frac{7}{3}$ is not needed as part of du/dx, and can be moved outside the integral sign.

$$\int 7x^2\sqrt{x^3 + 1}\, dx = \int 7x^2(x^3 + 1)^{1/2}\, dx \qquad \text{Rewrite with rational exponent.}$$

$$= \int \frac{7}{3}(x^3 + 1)^{1/2}(3x^2)\, dx \qquad \text{Multiply and divide by 3.}$$

$$= \frac{7}{3}\int (x^3 + 1)^{1/2}(3x^2)\, dx \qquad \text{Factor } \tfrac{7}{3} \text{ outside integral.}$$

$$= \frac{7}{3}\frac{(x^3 + 1)^{3/2}}{3/2} + C \qquad \text{General Power Rule}$$

$$= \frac{14}{9}(x^3 + 1)^{3/2} + C \qquad \text{Simplify.}$$

TRY IT 4

Find $\int 5x\sqrt{x^2 + 1}\, dx$.

ALGEBRA REVIEW

For help on the algebra in Example 4, see Example 1(c) in the *Chapter 5 Algebra Review*, on page 378.

TECHNOLOGY

If you use a symbolic integration utility to find indefinite integrals, you should be in for some surprises. This is true because integration is not nearly as straightforward as differentiation. By trying different integrands, you should be able to find several that the program cannot solve: in such situations, it may list a new indefinite integral. You should also be able to find several that have horrendous antiderivatives, some with functions that you may not recognize.

Substitution

The integration technique used in Examples 1, 2, and 4 depends on your ability to recognize or create an integrand of the form $u^n \, du/dx$. With more complicated integrands, it is difficult to recognize the steps needed to fit the integrand to a basic integration formula. When this occurs, an alternative procedure called **substitution** or **change of variables** can be helpful. With this procedure, you completely rewrite the integral in terms of u and du. That is, if $u = f(x)$, then $du = f'(x) \, dx$, and the General Power Rule takes the form

$$\int u^n \frac{du}{dx} \, dx = \int u^n \, du. \qquad \text{General Power Rule}$$

EXAMPLE 5   **Integrating by Substitution**

Find $\displaystyle\int \sqrt{1 - 3x} \, dx$.

SOLUTION   Begin by letting $u = 1 - 3x$. Then, $du/dx = -3$ and $du = -3 \, dx$. This implies that $dx = -\tfrac{1}{3} \, du$, and you can find the indefinite integral as shown.

$$\int \sqrt{1 - 3x} \, dx = \int (1 - 3x)^{1/2} \, dx \qquad \text{Rewrite with rational exponent.}$$

$$= \int u^{1/2} \left(-\frac{1}{3} \, du \right) \qquad \text{Substitute } u \text{ and } du.$$

$$= -\frac{1}{3} \int u^{1/2} \, du \qquad \text{Factor } -\tfrac{1}{3} \text{ out of integrand.}$$

$$= -\frac{1}{3} \frac{u^{3/2}}{3/2} + C \qquad \text{Apply Power Rule.}$$

$$= -\frac{2}{9} u^{3/2} + C \qquad \text{Simplify.}$$

$$= -\frac{2}{9}(1 - 3x)^{3/2} + C \qquad \text{Substitute } 1 - 3x \text{ for } u.$$

TRY IT 5

Find $\displaystyle\int \sqrt{1 - 2x} \, dx$ by the method of substitution.

To become efficient at integration, you should learn to use *both* techniques discussed in this section. For simpler integrals, you should use pattern recognition and create du/dx by multiplying and dividing by an appropriate constant. For more complicated integrals, you should use a formal change of variables, as shown in Example 5. (You will learn more about this technique in Chapter 6.) For the integrals in this section's exercise set, try working several of the problems twice—once with pattern recognition and once using formal substitution.

Extended Application: Propensity to Consume

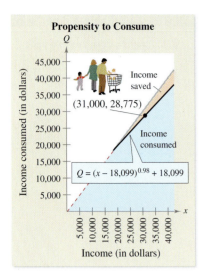

Propensity to Consume

(31,000, 28,775)

Income saved

Income consumed

$Q = (x - 18,099)^{0.98} + 18,099$

Income (in dollars)

FIGURE 5.4

In 2001, the U.S. poverty level for a family of four was about $18,100. Families at or below the poverty level tend to consume 100% of their income—that is, they use all their income to purchase necessities such as food, clothing, and shelter. As income level increases, the average consumption tends to drop below 100%. For instance, a family earning $20,000 may be able to save $400 and so consume only $19,600 (98%) of their income. As the income increases, the ratio of consumption to savings tends to decrease. The rate of change of consumption with respect to income is called the **marginal propensity to consume.**   *(Source: U.S. Census Bureau)*

EXAMPLE 6   **Analyzing Consumption**

For a family of four in 2001, the marginal propensity to consume income x can be modeled by

$$\frac{dQ}{dx} = \frac{0.98}{(x - 18,099)^{0.02}}, \qquad x \geq 18,100$$

where Q represents the income consumed. Use the model to estimate the amount consumed by a family of four whose 2001 income was $31,000.

SOLUTION   Begin by integrating dQ/dx to find a model for the consumption Q. Use the initial condition that $Q = 18,100$ when $x = 18,100$.

$$\frac{dQ}{dx} = \frac{0.98}{(x - 18,099)^{0.02}} \qquad \text{\color{red}Marginal propensity to consume}$$

$$Q = \int \frac{0.98}{(x - 18,099)^{0.02}}\, dx \qquad \text{\color{red}Integrate to obtain } Q.$$

$$= \int 0.98(x - 18,099)^{-0.02}\, dx \qquad \text{\color{red}Rewrite.}$$

$$= (x - 18,099)^{0.98} + C \qquad \text{\color{red}General Power Rule}$$

$$= (x - 18,099)^{0.98} + 18,099 \qquad \text{\color{red}Use initial condition to find } C.$$

Using this model, you can estimate that a family of four with an income of $x = 31,000$ consumed about $28,775. The graph of Q is shown in Figure 5.4.

When you use the initial condition to find the value of C in Example 6, you substitute 18,100 for Q and 18,100 for x.

$$Q = (x - 18,099)^{0.98} + C$$

$$18,100 = (18,100 - 18,099)^{0.98} + C$$

$$18,100 = 1 + C$$

$$18,099 = C$$

TRY IT 6

According to the model in Example 6, at what income level would a family of four consume $28,000?

TAKE ANOTHER LOOK

Propensity to Consume

According to the model in Example 6, at what income level would a family of four consume $36,000? Use the graph in Figure 5.4 to verify your results.

The following warm-up exercises involve skills that were covered in earlier sections. You will use these skills in the exercise set for this section.

In Exercises 1–10, find the indefinite integral.

1. $\int (2x^3 + 1)\, dx$

2. $\int (x^{1/2} + 3x - 4)\, dx$

3. $\int \frac{1}{x^2}\, dx$

4. $\int \frac{1}{3t^3}\, dt$

5. $\int (1 + 2t)t^{3/2}\, dt$

6. $\int \sqrt{x}(2x - 1)\, dx$

7. $\int \frac{5x^3 + 2}{x^2}\, dx$

8. $\int \frac{2x^2 - 5}{x^4}\, dx$

9. $\int (x^2 + 1)^2\, dx$

10. $\int (x^3 - 2x + 1)^2\, dx$

In Exercises 11–14, simplify the expression.

11. $\left(-\frac{5}{4}\right)\frac{(x - 2)^4}{4}$

12. $\left(\frac{1}{6}\right)\frac{(x - 1)^{-2}}{-2}$

13. $(6)\frac{(x^2 + 3)^{2/3}}{2/3}$

14. $\left(\frac{5}{2}\right)\frac{(1 - x^3)^{-1/2}}{-1/2}$

EXERCISES 5.2

In Exercises 1–8, identify u and du/dx for the integral $\int u^n (du/dx)\, dx$.

1. $\int (5x^2 + 1)^2(10x)\, dx$

2. $\int (3 - 4x^2)^3(-8x)\, dx$

3. $\int \sqrt{1 - x^2}(-2x)\, dx$

4. $\int 3x^2 \sqrt{x^3 + 1}\, dx$

5. $\int \left(4 + \frac{1}{x^2}\right)^5\left(\frac{-2}{x^3}\right)\, dx$

6. $\int \frac{1}{(1 + 2x)^2}(2)\, dx$

7. $\int (1 + \sqrt{x})^3\left(\frac{1}{2\sqrt{x}}\right)\, dx$

8. $\int (4 - \sqrt{x})^2\left(\frac{-1}{2\sqrt{x}}\right)\, dx$

In Exercises 9–28, find the indefinite integral and check the result by differentiation.

9. $\int (1 + 2x)^4(2)\, dx$

10. $\int (x^2 - 1)^3(2x)\, dx$

11. $\int \sqrt{5x^2 - 4}\,(10x)\, dx$

12. $\int \sqrt{3 - x^3}\,(3x^2)\, dx$

13. $\int (x - 1)^4\, dx$

14. $\int (x - 3)^{5/2}\, dx$

15. $\int x(x^2 - 1)^7\, dx$

16. $\int x(1 - 2x^2)^3\, dx$

17. $\int \frac{x^2}{(1 + x^3)^2}\, dx$

18. $\int \frac{x^2}{(x^3 - 1)^2}\, dx$

19. $\int \frac{x + 1}{(x^2 + 2x - 3)^2}\, dx$

20. $\int \frac{6x}{(1 + x^2)^3}\, dx$

21. $\int \frac{x - 2}{\sqrt{x^2 - 4x + 3}}\, dx$

22. $\int \frac{4x + 6}{(x^2 + 3x + 7)^3}\, dx$

23. $\int 5u \sqrt[3]{1 - u^2}\, du$

24. $\int u^3 \sqrt{u^4 + 2}\, du$

25. $\int \frac{4y}{\sqrt{1 + y^2}}\, dy$

26. $\int \frac{x^2}{\sqrt{1 - x^3}}\, dx$

27. $\int \frac{-3}{\sqrt{2t + 3}}\, dt$

28. $\int \frac{t + 2t^2}{\sqrt{t}}\, dt$

In Exercises 29–34, use a symbolic integration utility to find the indefinite integral.

29. $\int \frac{x^3}{\sqrt{1 - x^4}}\, dx$

30. $\int \frac{3x}{\sqrt{1 - 4x^2}}\, dx$

31. $\int \left(1 + \frac{4}{t^2}\right)^2\left(\frac{1}{t^3}\right)\, dt$

32. $\int \left(1 + \frac{1}{t}\right)^3\left(\frac{1}{t^2}\right)\, dt$

33. $\int (x^3 + 3x)(x^2 + 1)\, dx$

34. $\int (3 - 2x - 4x^2)(1 + 4x)\, dx$

In Exercises 35–42, use formal substitution (as illustrated in Example 5) to find the indefinite integral.

35. $\displaystyle\int x(6x^2 - 1)^3\, dx$

36. $\displaystyle\int x^2(1 - x^3)^2\, dx$

37. $\displaystyle\int x^2(2 - 3x^3)^{3/2}\, dx$

38. $\displaystyle\int t\sqrt{t^2 + 1}\, dt$

39. $\displaystyle\int \frac{x}{\sqrt{x^2 + 25}}\, dx$

40. $\displaystyle\int \frac{3}{\sqrt{2x + 1}}\, dx$

41. $\displaystyle\int \frac{x^2 + 1}{\sqrt{x^3 + 3x + 4}}\, dx$

42. $\displaystyle\int \sqrt{x}\,(4 - x^{3/2})^2\, dx$

In Exercises 43–46, (a) perform the integration in two ways: once using the Simple Power Rule and once using the General Power Rule. (b) Explain the difference in the results. (c) Which method do you prefer? Explain your reasoning.

43. $\displaystyle\int (2x - 1)^2\, dx$

44. $\displaystyle\int (3 - 2x)^2\, dx$

45. $\displaystyle\int x(x^2 - 1)^2\, dx$

46. $\displaystyle\int x(2x^2 + 1)^2\, dx$

47. Find the equation of the function f whose graph passes through the point $\left(0, \frac{4}{3}\right)$ and whose derivative is
$$f'(x) = x\sqrt{1 - x^2}.$$

48. Find the equation of the function f whose graph passes through the point $\left(0, \frac{7}{3}\right)$ and whose derivative is
$$f'(x) = x\sqrt{1 - x^2}.$$

49. *Cost* The marginal cost of a product is modeled by
$$\frac{dC}{dx} = \frac{4}{\sqrt{x + 1}}.$$
When $x = 15$, $C = 50$.

(a) Find the cost function.

 (b) Use a graphing utility to graph dC/dx and C in the same viewing window.

50. *Cost* The marginal cost of a product is modeled by
$$\frac{dC}{dx} = \frac{12}{\sqrt[3]{12x + 1}}.$$
When $x = 13$, $C = 100$.

(a) Find the cost function.

(b) Use a graphing utility to graph dC/dx and C in the same viewing window.

Supply Function In Exercises 51 and 52, find the supply function $x = f(p)$ that satisfies the initial conditions.

51. $\dfrac{dx}{dp} = p\sqrt{p^2 - 25}$, $\quad x = 600$ when $p = \$13$

52. $\dfrac{dx}{dp} = \dfrac{10}{\sqrt{p - 3}}$, $\quad x = 100$ when $p = \$3$

Demand Function In Exercises 53 and 54, find the demand function $x = f(p)$ that satisfies the initial conditions.

53. $\dfrac{dx}{dp} = -\dfrac{6000p}{(p^2 - 16)^{3/2}}$, $\quad x = 5000$ when $p = \$5$

54. $\dfrac{dx}{dp} = -\dfrac{400}{(0.02p - 1)^3}$, $\quad x = 10{,}000$ when $p = \$100$

55. *Gardening* An evergreen nursery usually sells a type of shrub after 5 years of growth and shaping. The growth rate during those 5 years is approximated by
$$\frac{dh}{dt} = \frac{17.6t}{\sqrt{17.6t^2 + 1}}$$
where t is time in years and h is height in inches. The seedlings are 6 inches tall when planted ($t = 0$).

(a) Find the height function.

(b) How tall are the shrubs when they are sold?

56. *Cash Flow* The rate of disbursement dQ/dt of a $2 million federal grant is proportional to the square of $100 - t$, where t is the time (in days, $0 \le t \le 100$) and Q is the amount that remains to be disbursed. Find the amount that remains to be disbursed after 50 days. Assume that the entire grant will be disbursed after 100 days.

 Marginal Propensity to Consume In Exercises 57 and 58, (a) use the marginal propensity to consume, dQ/dx, to write Q as a function of x, where x is the income (in dollars) and Q is the income consumed (in dollars). Assume that 100% of the income is consumed for families that have annual incomes of \$20,000 or less. (b) Use the result of part (a) to complete the table showing the income consumed and the income saved, $x - Q$, for various incomes. (c) Use a graphing utility to represent graphically the income consumed and saved.

x	20,000	50,000	100,000	150,000
Q				
$x - Q$				

57. $\dfrac{dQ}{dx} = \dfrac{0.95}{(x - 19{,}999)^{0.05}}$, $\quad x \ge 20{,}000$

58. $\dfrac{dQ}{dx} = \dfrac{0.93}{(x - 19{,}999)^{0.07}}$, $\quad x \ge 20{,}000$

In Exercises 59 and 60, use a symbolic integration utility to find the indefinite integral. Verify the result by differentiating.

59. $\displaystyle\int \frac{1}{\sqrt{x} + \sqrt{x + 1}}\, dx$

60. $\displaystyle\int \frac{x}{\sqrt{3x + 2}}\, dx$

5.3   EXPONENTIAL AND LOGARITHMIC INTEGRALS

- Use the Exponential Rule to find indefinite integrals.
- Use the Log Rule to find indefinite integrals.

Using the Exponential Rule

Each of the differentiation rules for exponential functions has its corresponding integration rule.

Integrals of Exponential Functions

Let u be a differentiable function of x.

$$\int e^x \, dx = e^x + C \qquad \text{Simple Exponential Rule}$$

$$\int e^u \frac{du}{dx} \, dx = \int e^u \, du = e^u + C \qquad \text{General Exponential Rule}$$

EXAMPLE 1   **Integrating Exponential Functions**

Find each indefinite integral.

(a) $\displaystyle \int 2e^x \, dx$

(b) $\displaystyle \int 2e^{2x} \, dx$

(c) $\displaystyle \int (e^x + x) \, dx$

SOLUTION

(a) $\displaystyle \int 2e^x \, dx = 2 \int e^x \, dx$ Constant Multiple Rule

$\displaystyle \qquad\qquad = 2e^x + C$ Simple Exponential Rule

(b) $\displaystyle \int 2e^{2x} \, dx = \int e^{2x}(2) \, dx$ Let $u = 2x$, then $\dfrac{du}{dx} = 2$.

$\displaystyle \qquad\qquad = \int e^u \frac{du}{dx} \, dx$

$\displaystyle \qquad\qquad = e^{2x} + C$ General Exponential Rule

(c) $\displaystyle \int (e^x + x) \, dx = \int e^x \, dx + \int x \, dx$ Sum Rule

$\displaystyle \qquad\qquad = e^x + \frac{x^2}{2} + C$ Simple Exponential and Power Rules

You can check each of these results by differentiating.

TRY IT 1

Find each indefinite integral.

(a) $\displaystyle \int 3e^x \, dx$

(b) $\displaystyle \int 5e^{5x} \, dx$

(c) $\displaystyle \int (e^x - x) \, dx$

If you use a symbolic integration utility to find antiderivatives of exponential or logarithmic functions, you can easily obtain results that are beyond the scope of this course. For instance, the antiderivative of e^{x^2} involves the imaginary unit i and the probability function called "ERF." In this course, you are not expected to interpret or use such results. You can simply state that the function cannot be integrated using elementary functions.

EXAMPLE 2 **Integrating an Exponential Function**

Find $\displaystyle\int e^{3x+1}\,dx$.

SOLUTION Let $u = 3x + 1$, then $du/dx = 3$. You can introduce the missing factor of 3 in the integrand by multiplying and dividing by 3.

$$\int e^{3x+1}\,dx = \frac{1}{3}\int e^{3x+1}(3)\,dx \qquad \text{Multiply and divide by 3.}$$

$$= \frac{1}{3}\int e^{u}\frac{du}{dx}\,dx \qquad \text{Substitute } u \text{ and } du/dx.$$

$$= \frac{1}{3}e^{u} + C \qquad \text{General Exponential Rule}$$

$$= \frac{1}{3}e^{3x+1} + C \qquad \text{Substitute for } u.$$

TRY IT 2

Find $\displaystyle\int e^{2x+3}\,dx$.

For help on the algebra in Example 3, see Example 1(d) in the *Chapter 5 Algebra Review*, on page 378.

EXAMPLE 3 **Integrating an Exponential Function**

Find $\displaystyle\int 5xe^{-x^2}\,dx$.

SOLUTION Let $u = -x^2$, then $du/dx = -2x$. You can create the factor $-2x$ in the integrand by multiplying and dividing by -2.

$$\int 5xe^{-x^2}\,dx = \int\left(-\frac{5}{2}\right)e^{-x^2}(-2x)\,dx \qquad \text{Multiply and divide by } -2.$$

$$= -\frac{5}{2}\int e^{-x^2}(-2x)\,dx \qquad \text{Factor } -\tfrac{5}{2} \text{ out of the integrand.}$$

$$= -\frac{5}{2}\int e^{u}\frac{du}{dx}\,dx \qquad \text{Substitute } u \text{ and } \tfrac{du}{dx}.$$

$$= -\frac{5}{2}e^{u} + C \qquad \text{General Exponential Rule}$$

$$= -\frac{5}{2}e^{-x^2} + C \qquad \text{Substitute for } u.$$

STUDY TIP

Remember that you cannot introduce a missing *variable* in the integrand. For instance, you cannot find $\int e^{x^2}\,dx$ by multiplying and dividing by $2x$ and then factoring $1/(2x)$ out of the integral. That is,

$$\int e^{x^2}\,dx \neq \frac{1}{2x}\int e^{x^2}(2x)\,dx.$$

TRY IT 3

Find $\displaystyle\int 4xe^{x^2}\,dx$.

Using the Log Rule

When the Power Rules for integration were introduced in Sections 5.1 and 5.2, you saw that they work for powers other than $n = -1$.

$$\int x^n \, dx = \frac{x^{n+1}}{n+1} + C, \qquad n \neq -1 \qquad \text{Simple Power Rule}$$

$$\int u^n \frac{du}{dx} \, dx = \int u^n \, du = \frac{u^{n+1}}{n+1} + C, \qquad n \neq -1 \qquad \text{General Power Rule}$$

The Log Rules for integration allow you to integrate functions of the form $\int x^{-1} \, dx$ and $\int u^{-1} \, du$.

> ### Integrals of Logarithmic Functions
>
> Let u be a differentiable function of x.
>
> $$\int \frac{1}{x} \, dx = \ln|x| + C \qquad \text{Simple Logarithmic Rule}$$
>
> $$\int \frac{du/dx}{u} \, dx = \int \frac{1}{u} \, du = \ln|u| + C \qquad \text{General Logarithmic Rule}$$

You can verify each of these rules by differentiating. For instance, to verify that $d/dx[\ln|x|] = 1/x$, notice that

$$\frac{d}{dx}[\ln x] = \frac{1}{x} \qquad \text{and} \qquad \frac{d}{dx}[\ln(-x)] = \frac{-1}{-x} = \frac{1}{x}.$$

EXAMPLE 4 Integrating Logarithmic Functions

Find each indefinite integral.

(a) $\displaystyle\int \frac{4}{x} \, dx$ (b) $\displaystyle\int \frac{2x}{x^2} \, dx$ (c) $\displaystyle\int \frac{3}{3x+1} \, dx$

SOLUTION

(a) $\displaystyle\int \frac{4}{x} \, dx = 4 \int \frac{1}{x} \, dx$ Constant Multiple Rule

$\qquad\qquad = 4 \ln|x| + C$ Simple Logarithmic Rule

(b) $\displaystyle\int \frac{2x}{x^2} \, dx = \int \frac{du/dx}{u} \, dx$ Let $u = x^2$, then $\dfrac{du}{dx} = 2x$.

$\qquad\qquad = \ln|u| + C$ General Logarithmic Rule

$\qquad\qquad = \ln x^2 + C$ Substitute for u.

(c) $\displaystyle\int \frac{3}{3x+1} \, dx = \int \frac{du/dx}{u} \, dx$ Let $u = 3x + 1$, then $\dfrac{du}{dx} = 3$.

$\qquad\qquad = \ln|u| + C$ General Logarithmic Rule

$\qquad\qquad = \ln|3x + 1| + C$ Substitute for u.

DISCOVERY

The General Power Rule is not valid for $n = -1$. Can you find an antiderivative for u^{-1}?

STUDY TIP

Notice the absolute values in the Log Rules. For those special cases in which u or x cannot be negative, you can omit the absolute value. For instance, in Example 4(b), it is not necessary to write the antiderivative as $\ln|x^2| + C$ because x^2 cannot be negative.

TRY IT 4

Find each indefinite integral.

(a) $\displaystyle\int \frac{2}{x} \, dx$

(b) $\displaystyle\int \frac{3x^2}{x^3} \, dx$

(c) $\displaystyle\int \frac{2}{2x+1} \, dx$

EXAMPLE 5   **Using the Log Rule**

Find $\displaystyle\int \frac{1}{2x-1}\, dx$.

SOLUTION   Let $u = 2x - 1$, then $du/dx = 2$. You can create the necessary factor of 2 in the integrand by multiplying and dividing by 2.

$$\int \frac{1}{2x-1}\, dx = \frac{1}{2}\int \frac{2}{2x-1}\, dx \qquad \text{Multiply and divide by 2.}$$

$$= \frac{1}{2}\int \frac{du/dx}{u}\, dx \qquad \text{Substitute } u \text{ and } \frac{du}{dx}.$$

$$= \frac{1}{2}\ln|u| + C \qquad \text{General Log Rule}$$

$$= \frac{1}{2}\ln|2x-1| + C \qquad \text{Substitute for } u.$$

TRY IT 5

Find $\displaystyle\int \frac{1}{4x+1}\, dx$.

EXAMPLE 6   **Using the Log Rule**

Find $\displaystyle\int \frac{6x}{x^2+1}\, dx$.

SOLUTION   Let $u = x^2 + 1$, then $du/dx = 2x$. You can create the necessary factor of $2x$ in the integrand by factoring a 3 out of the integrand.

$$\int \frac{6x}{x^2+1}\, dx = 3\int \frac{2x}{x^2+1}\, dx \qquad \text{Factor 3 out of integrand.}$$

$$= 3\int \frac{du/dx}{u}\, dx \qquad \text{Substitute } u \text{ and } \frac{du}{dx}.$$

$$= 3\ln|u| + C \qquad \text{General Log Rule}$$

$$= 3\ln(x^2+1) + C \qquad \text{Substitute for } u.$$

TRY IT 6

Find $\displaystyle\int \frac{3x}{x^2+4}\, dx$.

ALGEBRA REVIEW

For help on the algebra in the integral at the right, see Example 2(d) in the *Chapter 5 Algebra Review*, on page 379.

Integrals to which the Log Rule can be applied are often given in disguised form. For instance, if a rational function has a numerator of degree greater than or equal to that of the denominator, you should use long division to rewrite the integrand. Here is an example.

$$\int \frac{x^2+6x+1}{x^2+1}\, dx = \int \left(1 + \frac{6x}{x^2+1}\right) dx$$

$$= x + 3\ln(x^2+1) + C$$

The next example summarizes some additional situations in which it is helpful to rewrite the integrand in order to recognize the antiderivative.

EXAMPLE 7 **Rewriting Before Integrating**

Find each indefinite integral.

(a) $\displaystyle\int \frac{3x^2 + 2x - 1}{x^2}\, dx$ (b) $\displaystyle\int \frac{1}{1 + e^{-x}}\, dx$ (c) $\displaystyle\int \frac{x^2 + x + 1}{x - 1}\, dx$

SOLUTION

(a) Begin by rewriting the integrand as the sum of three fractions.

$$\int \frac{3x^2 + 2x - 1}{x^2}\, dx = \int \left(\frac{3x^2}{x^2} + \frac{2x}{x^2} - \frac{1}{x^2} \right) dx$$

$$= \int \left(3 + \frac{2}{x} - \frac{1}{x^2} \right) dx$$

$$= 3x + 2\ln|x| + \frac{1}{x} + C$$

(b) Begin by rewriting the integrand by multiplying and dividing by e^x.

$$\int \frac{1}{1 + e^{-x}}\, dx = \int \left(\frac{e^x}{e^x} \right) \frac{1}{1 + e^{-x}}\, dx$$

$$= \int \frac{e^x}{e^x + 1}\, dx$$

$$= \ln(e^x + 1) + C$$

(c) Begin by dividing the numerator by the denominator.

$$\int \frac{x^2 + x + 1}{x - 1}\, dx = \int \left(x + 2 + \frac{3}{x - 1} \right) dx$$

$$= \frac{x^2}{2} + 2x + 3\ln|x - 1| + C$$

ALGEBRA REVIEW

For help on the algebra in Example 7, see Example 2(a)–(c) in the *Chapter 5 Algebra Review*, on page 379.

TRY IT 7

Find each indefinite integral.

(a) $\displaystyle\int \frac{4x^2 - 3x + 2}{x^2}\, dx$

(b) $\displaystyle\int \frac{2}{e^{-x} + 1}\, dx$

(c) $\displaystyle\int \frac{x^2 + 2x + 4}{x + 1}\, dx$

STUDY TIP

The Exponential and Log Rules are necessary to solve certain real-life problems, such as population growth. You will see such problems in the exercise set for this section.

TAKE ANOTHER LOOK

Using the General Log Rule

One of the most common applications of the Log Rule is to find indefinite integrals of the form

$$\int \frac{a}{bx + c}\, dx.$$

Describe a quick way to find this indefinite integral. Then apply your technique to each integral below.

a. $\displaystyle\int \frac{1}{2x - 5}\, dx$ b. $\displaystyle\int \frac{4}{3x + 2}\, dx$ c. $\displaystyle\int \frac{7}{8x - 3}\, dx$

Use a symbolic integration utility to verify your results.

The following warm-up exercises involve skills that were covered in earlier sections. You will use these skills in the exercise set for this section.

In Exercises 1 and 2, find the domain of the function.

1. $y = \ln(2x - 5)$

2. $y = \ln(x^2 - 5x + 6)$

In Exercises 3–6, use long division to rewrite the quotient.

3. $\dfrac{x^2 + 4x + 2}{x + 2}$

4. $\dfrac{x^2 - 6x + 9}{x - 4}$

5. $\dfrac{x^3 + 4x^2 - 30x - 4}{x^2 - 4x}$

6. $\dfrac{x^4 - x^3 + x^2 + 15x + 2}{x^2 + 5}$

In Exercises 7–10, evaluate the integral.

7. $\displaystyle \int \left(x^3 + \frac{1}{x^2} \right) dx$

8. $\displaystyle \int \frac{x^2 + 2x}{x} \, dx$

9. $\displaystyle \int \frac{x^3 + 4}{x^2} \, dx$

10. $\displaystyle \int \frac{x + 3}{x^3} \, dx$

EXERCISES 5.3

In Exercises 1–12, use the Exponential Rule to find the indefinite integral.

1. $\displaystyle \int 2e^{2x} \, dx$

2. $\displaystyle \int -3e^{-3x} \, dx$

3. $\displaystyle \int e^{4x} \, dx$

4. $\displaystyle \int e^{-0.25x} \, dx$

5. $\displaystyle \int 9xe^{-x^2} \, dx$

6. $\displaystyle \int 3xe^{0.5x^2} \, dx$

7. $\displaystyle \int 5x^2 \, e^{x^3} \, dx$

8. $\displaystyle \int (2x + 1)e^{x^2 + x} \, dx$

9. $\displaystyle \int (x^2 + 2x)e^{x^3 + 3x^2 - 1} \, dx$

10. $\displaystyle \int 3(x - 4)e^{x^2 - 8x} \, dx$

11. $\displaystyle \int 5e^{2 - x} \, dx$

12. $\displaystyle \int 3e^{-(x + 1)} \, dx$

In Exercises 13–28, use the Log Rule to find the indefinite integral.

13. $\displaystyle \int \frac{1}{x + 1} \, dx$

14. $\displaystyle \int \frac{1}{x - 5} \, dx$

15. $\displaystyle \int \frac{1}{3 - 2x} \, dx$

16. $\displaystyle \int \frac{1}{6x - 5} \, dx$

17. $\displaystyle \int \frac{2}{3x + 5} \, dx$

18. $\displaystyle \int \frac{5}{2x - 1} \, dx$

19. $\displaystyle \int \frac{x}{x^2 + 1} \, dx$

20. $\displaystyle \int \frac{x^2}{3 - x^3} \, dx$

21. $\displaystyle \int \frac{x^2}{x^3 + 1} \, dx$

22. $\displaystyle \int \frac{x}{x^2 + 4} \, dx$

23. $\displaystyle \int \frac{x + 3}{x^2 + 6x + 7} \, dx$

24. $\displaystyle \int \frac{x^2 + 2x + 3}{x^3 + 3x^2 + 9x + 1} \, dx$

25. $\displaystyle \int \frac{1}{x \ln x} \, dx$

26. $\displaystyle \int \frac{1}{x(\ln x)^2} \, dx$

27. $\displaystyle \int \frac{e^{-x}}{1 - e^{-x}} \, dx$

28. $\displaystyle \int \frac{e^x}{1 + e^x} \, dx$

In Exercises 29–38, use a symbolic integration utility to find the indefinite integral.

29. $\displaystyle \int \frac{1}{x^2} e^{2/x} \, dx$

30. $\displaystyle \int \frac{1}{x^3} e^{1/4x^2} \, dx$

31. $\displaystyle \int \frac{1}{\sqrt{x}} e^{\sqrt{x}} \, dx$

32. $\displaystyle \int \frac{e^{1/\sqrt{x}}}{x^{3/2}} \, dx$

33. $\displaystyle \int (e^x - 2)^2 \, dx$

34. $\displaystyle \int (e^x - e^{-x})^2 \, dx$

35. $\displaystyle \int \frac{e^{-x}}{1 + e^{-x}} \, dx$

36. $\displaystyle \int \frac{3e^x}{2 + e^x} \, dx$

37. $\displaystyle \int \frac{4e^{2x}}{5 - e^{2x}} \, dx$

38. $\displaystyle \int \frac{-e^{3x}}{2 - e^{3x}} \, dx$

In Exercises 39–54, use any basic integration formula or formulas to find the indefinite integral.

39. $\int \dfrac{e^{2x} + 2e^x + 1}{e^x}\, dx$

40. $\int (6x + e^x)\sqrt{3x^2 + e^x}\, dx$

41. $\int e^x \sqrt{1 - e^x}\, dx$

42. $\int \dfrac{2(e^x - e^{-x})}{(e^x + e^{-x})^2}\, dx$

43. $\int \dfrac{1}{(x - 1)^2}\, dx$

44. $\int \dfrac{1}{\sqrt{x} + 1}\, dx$

45. $\int 4e^{2x-1}\, dx$

46. $\int (5e^{-2x} + 1)\, dx$

47. $\int \dfrac{x^3 - 8x}{2x^2}\, dx$

48. $\int \dfrac{x - 1}{4x}\, dx$

49. $\int \dfrac{2}{1 + e^{-x}}\, dx$

50. $\int \dfrac{3}{1 + e^{-3x}}\, dx$

51. $\int \dfrac{x^2 + 2x + 5}{x - 1}\, dx$

52. $\int \dfrac{x - 3}{x + 3}\, dx$

53. $\int \dfrac{1 + e^{-x}}{1 + xe^{-x}}\, dx$

54. $\int \dfrac{5}{e^{-5x} + 7}\, dx$

In Exercises 55 and 56, find the equation of the function f whose graph passes through the point.

55. $f'(x) = \dfrac{x^2 + 4x + 3}{x - 1}$; $\quad (2, 4)$

56. $f'(x) = \dfrac{x^3 - 4x^2 + 3}{x - 3}$; $\quad (4, -1)$

57. *Biology* A population of bacteria is growing at the rate of

$$\frac{dP}{dt} = \frac{3000}{1 + 0.25t}$$

where t is the time in days. When $t = 0$, the population is 1000.

(a) Write an equation that models the population P in terms of the time t.

(b) What is the population after 3 days?

(c) After how many days will the population be 12,000?

58. *Biology* Because of an insufficient oxygen supply, the trout population in a lake is dying. The population's rate of change can be modeled by

$$\frac{dP}{dt} = -125e^{-t/20}$$

where t is the time in days. When $t = 0$, the population is 2500.

(a) Write an equation that models the population P in terms of the time t.

(b) What is the population after 15 days?

(c) According to this model, how long will it take for the entire trout population to die?

 59. *Demand* The marginal price for the demand of a product can be modeled by $dp/dx = 0.1e^{-x/500}$, where x is the quantity demanded. When the demand is 600 units, the price is $30.

(a) Find the demand function, $p = f(x)$.

(b) Use a graphing utility to graph the demand function. Does price increase or decrease as demand increases?

(c) Use the *zoom* and *trace* features of the graphing utility to find the quantity demanded when the price is $22.

 60. *Revenue* The marginal revenue for the sale of a product can be modeled by

$$\frac{dR}{dx} = 50 - 0.02x + \frac{100}{x + 1}$$

where x is the quantity demanded.

(a) Find the revenue function.

(b) Use a graphing utility to graph the revenue function.

(c) Find the revenue when 1500 units are sold.

(d) Use the *zoom* and *trace* features of the graphing utility to find the number of units sold when the revenue is $60,230.

61. *Average Salary* From 1995 through 2002, the average salary for superintendents S (in dollars) in the United States changed at the rate of

$$\frac{dS}{dt} = 2621.7e^{0.07t}$$

where $t = 5$ corresponds to 1995. In 2001, the average salary for superintendents was $118,496. *(Source: Educational Research Service)*

(a) Write a model that gives the average salary for superintendents per year.

(b) Use the model to find the average salary for superintendents in 1999.

62. *Sales* The rate of change in sales for The Yankee Candle Company from 1998 through 2003 can be modeled by

$$\frac{dS}{dt} = 1.04t + \frac{544.694}{t}$$

where S is the sales (in millions) and $t = 8$ corresponds to 1998. In 1999, the sales for The Yankee Candle Company were $256.6 million. *(Source: The Yankee Candle Company)*

(a) Find a model for sales from 1998 through 2003.

(b) Find The Yankee Candle Company's sales in 2002.

True or False? In Exercises 63 and 64, determine whether the statement is true or false. If it is false, explain why or give an example that shows it is false.

63. $(\ln x)^{1/2} = \frac{1}{2}(\ln x)$

64. $\int \ln x = \left(\dfrac{1}{x}\right) + C$

Guidelines for Using the Fundamental Theorem of Calculus

1. The Fundamental Theorem of Calculus describes a way of *evaluating* a definite integral, not a procedure for finding antiderivatives.

2. In applying the Fundamental Theorem, it is helpful to use the notation

$$\int_a^b f(x)\, dx = F(x)\Big]_a^b = F(b) - F(a).$$

3. The constant of integration C can be dropped because

$$\int_a^b f(x)\, dx = \left[F(x) + C\right]_a^b$$
$$= [F(b) + C] - [F(a) + C]$$
$$= F(b) - F(a) + C - C$$
$$= F(b) - F(a).$$

In the development of the Fundamental Theorem of Calculus, f was assumed to be nonnegative on the closed interval $[a, b]$. As such, the definite integral was defined as an area. Now, with the Fundamental Theorem, the definition can be extended to include functions that are negative on all or part of the closed interval $[a, b]$. Specifically, if f is *any* function that is continuous on a closed interval $[a, b]$, then the **definite integral** of $f(x)$ from a to b is defined to be

$$\int_a^b f(x)\, dx = F(b) - F(a)$$

where F is an antiderivative of f. Remember that definite integrals do not necessarily represent areas and can be negative, zero, or positive.

STUDY TIP

Be sure you see the distinction between indefinite and definite integrals. The *indefinite integral*

$$\int f(x)\, dx$$

denotes a family of *functions*, each of which is an antiderivative of f, whereas the *definite integral*

$$\int_a^b f(x)\, dx$$

is a *number*.

Properties of Definite Integrals

Let f and g be continuous on the closed interval $[a, b]$.

1. $\displaystyle\int_a^b kf(x)\, dx = k\int_a^b f(x)\, dx,\quad k$ is a constant.

2. $\displaystyle\int_a^b [f(x) \pm g(x)]\, dx = \int_a^b f(x)\, dx \pm \int_a^b g(x)\, dx$

3. $\displaystyle\int_a^b f(x)\, dx = \int_a^c f(x)\, dx + \int_c^b f(x)\, dx,\qquad a < c < b$

4. $\displaystyle\int_a^a f(x)\, dx = 0$

5. $\displaystyle\int_a^b f(x)\, dx = -\int_b^a f(x)\, dx$

■ **EXAMPLE 2** **Finding Area by the Fundamental Theorem**

Find the area of the region bounded by the x-axis and the graph of

$$f(x) = x^2 - 1, \quad 1 \le x \le 2.$$

SOLUTION   Note that $f(x) \ge 0$ on the interval $1 \le x \le 2$, as shown in Figure 5.9. So, you can represent the area of the region by a definite integral. To find the area, use the Fundamental Theorem of Calculus.

$$\begin{aligned}
\text{Area} &= \int_1^2 (x^2 - 1)\, dx && \text{Definition of definite integral} \\[2mm]
&= \left[\frac{x^3}{3} - x \right]_1^2 && \text{Find antiderivative.} \\[2mm]
&= \left[\frac{(2)^3}{3} - 2 \right] - \left[\frac{(1)^3}{3} - 1 \right] && \text{Apply Fundamental Theorem.} \\[2mm]
&= \frac{4}{3} && \text{Simplify.}
\end{aligned}$$

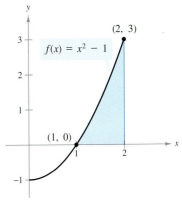

FIGURE 5.9   Area $= \displaystyle\int_1^2 (x^2 - 1)\, dx$

So, the area of the region is $\frac{4}{3}$ square units.

❚ **STUDY TIP**

It is easy to make errors in signs when evaluating definite integrals. To avoid such errors, enclose the values of the antiderivative at the upper and lower limits of integration in separate sets of parentheses, as shown above.

TRY IT 2

Find the area of the region bounded by the x-axis and the graph of $f(x) = x^2 + 1$, $2 \le x \le 3$.

■ **EXAMPLE 3** **Evaluating a Definite Integral**

Evaluate the definite integral

$$\int_0^1 (4t + 1)^2\, dt$$

and sketch the region whose area is represented by the integral.

SOLUTION

$$\begin{aligned}
\int_0^1 (4t + 1)^2\, dt &= \frac{1}{4} \int_0^1 (4t + 1)^2 (4)\, dt && \text{Multiply and divide by 4.} \\[2mm]
&= \frac{1}{4}\left[\frac{(4t + 1)^3}{3} \right]_0^1 && \text{Find antiderivative.} \\[2mm]
&= \frac{1}{4}\left[\left(\frac{5^3}{3} \right) - \left(\frac{1}{3} \right) \right] && \text{Apply Fundamental Theorem.} \\[2mm]
&= \frac{31}{3} && \text{Simplify.}
\end{aligned}$$

The region is shown in Figure 5.10.

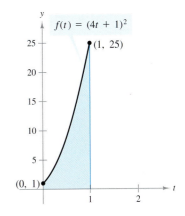

FIGURE 5.10

TRY IT 3

Evaluate $\displaystyle\int_0^1 (2t + 3)^3\, dt.$

EXAMPLE 4   **Evaluating Definite Integrals**

Evaluate each definite integral.

(a) $\displaystyle\int_0^3 e^{2x}\,dx$ (b) $\displaystyle\int_1^2 \frac{1}{x}\,dx$ (c) $\displaystyle\int_1^4 -3\sqrt{x}\,dx$

SOLUTION

(a) $\displaystyle\int_0^3 e^{2x}\,dx = \frac{1}{2}e^{2x}\Big]_0^3 = \frac{1}{2}(e^6 - e^0) \approx 201.21$

(b) $\displaystyle\int_1^2 \frac{1}{x}\,dx = \ln x\Big]_1^2 = \ln 2 - \ln 1 = \ln 2 \approx 0.69$

(c) $\displaystyle\int_1^4 -3\sqrt{x}\,dx = -3\int_1^4 x^{1/2}\,dx$ Rewrite with rational exponent.

$\qquad = -3\left[\frac{x^{3/2}}{3/2}\right]_1^4$ Find antiderivative.

$\qquad = -2x^{3/2}\Big]_1^4$

$\qquad = -2(4^{3/2} - 1^{3/2})$ Apply Fundamental Theorem.

$\qquad = -2(8 - 1)$

$\qquad = -14$ Simplify.

TRY IT 4

Evaluate each definite integral.

(a) $\displaystyle\int_0^1 e^{4x}\,dx$

(b) $\displaystyle\int_2^5 -\frac{1}{x}\,dx$

STUDY TIP

In Example 4(c), note that the value of a definite integral can be negative.

EXAMPLE 5   **Interpreting Absolute Value**

Evaluate $\displaystyle\int_0^2 |2x - 1|\,dx$.

SOLUTION   The region represented by the definite integral is shown in Figure 5.11. From the definition of absolute value, you can write

$$|2x - 1| = \begin{cases} -(2x - 1), & x < \frac{1}{2} \\ 2x - 1, & x \ge \frac{1}{2} \end{cases}.$$

Using Property 3 of definite integrals, you can rewrite the integral as two definite integrals.

$$\int_0^2 |2x - 1|\,dx = \int_0^{1/2} -(2x - 1)\,dx + \int_{1/2}^2 (2x - 1)\,dx$$

$$= \left[-x^2 + x\right]_0^{1/2} + \left[x^2 - x\right]_{1/2}^2$$

$$= \left(-\frac{1}{4} + \frac{1}{2}\right) - (0 + 0) + (4 - 2) - \left(\frac{1}{4} - \frac{1}{2}\right)$$

$$= \frac{5}{2}$$

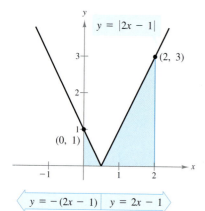

FIGURE 5.11

$y = |2x - 1|$
$(2, 3)$
$(0, 1)$
$y = -(2x - 1)$ $y = 2x - 1$

TRY IT 5

Evaluate $\displaystyle\int_0^5 |x - 2|\,dx$.

Marginal Analysis

You have already studied *marginal analysis* in the context of derivatives and differentials (Sections 2.3 and 3.8). There, you were given a cost, revenue, or profit function, and you used the derivative to approximate the additional cost, revenue, or profit obtained by selling one additional unit. In this section, you will examine the reverse process. That is, you will be given the marginal cost, marginal revenue, or marginal profit and will be asked to use a definite integral to find the exact increase or decrease in cost, revenue, or profit obtained by selling one or several additional units.

For instance, suppose you wanted to find the additional revenue obtained by increasing sales from x_1 to x_2 units. If you knew the revenue function R you could simply subtract $R(x_1)$ from $R(x_2)$. If you didn't know the revenue function, but did know the marginal revenue function, you could still find the additional revenue by using a definite integral, as shown.

$$\int_{x_1}^{x_2} \frac{dR}{dx} \, dx = R(x_2) - R(x_1)$$

EXAMPLE 6   Analyzing a Profit Function

The marginal profit for a product is modeled by $\dfrac{dP}{dx} = -0.0005x + 12.2$.

(a) Find the change in profit when sales increase from 100 to 101 units.

(b) Find the change in profit when sales increase from 100 to 110 units.

SOLUTION

(a) The change in profit obtained by increasing sales from 100 to 101 units is

$$\int_{100}^{101} \frac{dP}{dx} \, dx = \int_{100}^{101} (-0.0005x + 12.2) \, dx$$
$$= \Big[-0.00025x^2 + 12.2x \Big]_{100}^{101}$$
$$\approx \$12.15.$$

(b) The change in profit obtained by increasing sales from 100 to 110 units is

$$\int_{100}^{110} \frac{dP}{dx} \, dx = \int_{100}^{110} (-0.0005x + 12.2) \, dx$$
$$= \Big[-0.00025x^2 + 12.2x \Big]_{100}^{110}$$
$$\approx \$121.48$$

TRY IT 6

The marginal profit for a product is modeled by

$$\frac{dP}{dx} = -0.0002x + 14.2.$$

(a) Find the change in profit when sales increase from 100 to 101 units.

(b) Find the change in profit when sales increase from 100 to 110 units.

TECHNOLOGY

Symbolic integration utilities can be used to evaluate definite integrals as well as indefinite integrals. If you have access to such a program, try using it to evaluate several of the definite integrals in this section.

Average Value

The *average value* of a function on a closed interval is defined below.

> ### Definition of the Average Value of a Function
>
> If f is continuous on $[a, b]$, then the **average value** of f on $[a, b]$ is
>
> $$\text{Average value of } f \text{ on } [a, b] = \frac{1}{b-a} \int_a^b f(x) \, dx.$$

In Section 3.5, you studied the effects of production levels on cost using an average cost function. In the next example, you will study the effects of time on cost by using integration to find the average cost.

EXAMPLE 7   Finding the Average Cost

The cost per unit c of producing CD players over a two-year period is modeled by

$$c = 0.005t^2 + 0.01t + 13.15, \qquad 0 \le t \le 24$$

where t is the time in months. Approximate the average cost per unit over the two-year period.

SOLUTION   The average cost can be found by integrating c over the interval $[0, 24]$.

$$
\begin{aligned}
\text{Average cost per unit} &= \frac{1}{24} \int_0^{24} (0.005t^2 + 0.01t + 13.15) \, dt \\
&= \frac{1}{24} \left[\frac{0.005t^3}{3} + \frac{0.01t^2}{2} + 13.15t \right]_0^{24} \\
&= \frac{1}{24}(341.52) \\
&= \$14.23 \qquad \text{(See Figure 5.12.)}
\end{aligned}
$$

To check the reasonableness of the average value found in Example 7, assume that one unit is produced each month, beginning with $t = 0$ and ending with $t = 24$. When $t = 0$, the cost is

$$
\begin{aligned}
c &= 0.005(0)^2 + 0.01(0) + 13.15 \\
&= \$13.15.
\end{aligned}
$$

Similarly, when $t = 1$, the cost is

$$
\begin{aligned}
c &= 0.005(1)^2 + 0.01(1) + 13.15 \\
&\approx \$13.17.
\end{aligned}
$$

Each month, the cost increases, and the average of the 25 costs is

$$\frac{13.15 + 13.17 + 13.19 + 13.23 + \cdots + 16.27}{25} \approx \$14.25.$$

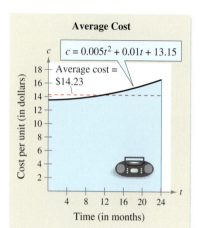

Average Cost

$c = 0.005t^2 + 0.01t + 13.15$

Average cost = $14.23

Cost per unit (in dollars)

Time (in months)

FIGURE 5.12

TRY IT 7

Find the average cost per unit over a two-year period if the cost per unit c of roller blades is given by $c = 0.005t^2 + 0.02t + 12.5$, for $0 \le t \le 24$, where t is the time in months.

Even and Odd Functions

Several common functions have graphs that are symmetric with respect to the y-axis or the origin, as shown in Figure 5.13. If the graph of f is symmetric with respect to the y-axis, as in Figure 5.13(a), then

$$f(-x) = f(x) \qquad \text{Even function}$$

and f is called an **even** function. If the graph of f is symmetric with respect to the origin, as in Figure 5.13(b), then

$$f(-x) = -f(x) \qquad \text{Odd function}$$

and f is called an **odd** function.

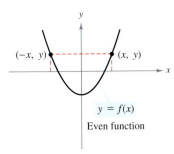

(a) y-axis symmetry

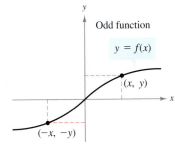

(b) Origin symmetry

FIGURE 5.13

Integration of Even and Odd Functions

1. If f is an *even* function, then $\displaystyle \int_{-a}^{a} f(x)\,dx = 2\int_{0}^{a} f(x)\,dx.$

2. If f is an *odd* function, then $\displaystyle \int_{-a}^{a} f(x)\,dx = 0.$

EXAMPLE 8   **Integrating Even and Odd Functions**

Evaluate each definite integral.

(a) $\displaystyle \int_{-2}^{2} x^2\,dx$ (b) $\displaystyle \int_{-2}^{2} x^3\,dx$

SOLUTION

(a) Because $f(x) = x^2$ is even,

$$\int_{-2}^{2} x^2\,dx = 2\int_{0}^{2} x^2\,dx = 2\left[\frac{x^3}{3}\right]_{0}^{2} = 2\left(\frac{8}{3} - 0\right) = \frac{16}{3}.$$

(b) Because $f(x) = x^3$ is odd,

$$\int_{-2}^{2} x^3\,dx = 0.$$

TRY IT 8

Evaluate each definite integral.

(a) $\displaystyle \int_{-1}^{1} x^4\,dx$

(b) $\displaystyle \int_{-1}^{1} x^5\,dx$

Annuity

A sequence of equal payments made at regular time intervals over a period of time is called an **annuity.** Some examples of annuities are payroll savings plans, monthly home mortgage payments, and individual retirement accounts. The **amount of an annuity** is the sum of the payments plus the interest earned and can be found as shown below.

Amount of an Annuity

If c represents a continuous income function in dollars per year (where t is the time in years), r represents the interest rate compounded continuously, and T represents the term of the annuity in years, then the **amount of an annuity** is

$$\text{Amount of an annuity} = e^{rT} \int_0^T c(t)e^{-rt}\, dt.$$

EXAMPLE 9   **Finding the Amount of an Annuity**

You deposit $2000 each year for 15 years in an individual retirement account (IRA) paying 10% interest. How much will you have in your IRA after 15 years?

SOLUTION   The income function for your deposit is $c(t) = 2000$. So, the amount of the annuity after 15 years will be

$$
\begin{aligned}
\text{Amount of an annuity} &= e^{rT} \int_0^T c(t)e^{-rt}\, dt \\
&= e^{(0.10)(15)} \int_0^{15} 2000e^{-0.10t}\, dt \\
&= 2000e^{1.5}\left[-\frac{e^{-0.10t}}{0.10}\right]_0^{15} \\
&\approx \$69{,}633.78.
\end{aligned}
$$

TRY IT 9

If you deposit $1000 in a savings account every year, paying 8% interest, how much will be in the account after 10 years?

TAKE ANOTHER LOOK

Using Geometry to Evaluate Definite Integrals

When using the Fundamental Theorem of Calculus to evaluate $\int_a^b f(x)\, dx$, remember that you must first be able to find an antiderivative of $f(x)$. If you are unable to find an antiderivative, you cannot use the Fundamental Theorem. In some cases, you can still evaluate the definite integral. For instance, explain how you can use geometry to evaluate

$$\int_{-2}^2 \sqrt{4 - x^2}\, dx.$$

Use a symbolic integration utility to verify your answer.

PREREQUISITE REVIEW 5.4

The following warm-up exercises involve skills that were covered in earlier sections. You will use these skills in the exercise set for this section.

In Exercises 1–4, find the indefinite integral.

1. $\int (3x + 7)\, dx$

2. $\int \left(x^{3/2} + 2\sqrt{x}\right) dx$

3. $\int \dfrac{1}{5x}\, dx$

4. $\int e^{-6x}\, dx$

In Exercises 5 and 6, evaluate the expression when $a = 5$ and $b = 3$.

5. $\left(\dfrac{a}{5} - a\right) - \left(\dfrac{b}{5} - b\right)$

6. $\left(6a - \dfrac{a^3}{3}\right) - \left(6b - \dfrac{b^3}{3}\right)$

In Exercises 7–10, integrate the marginal function.

7. $\dfrac{dC}{dx} = 0.02x^{3/2} + 29{,}500$

8. $\dfrac{dR}{dx} = 9000 + 2x$

9. $\dfrac{dP}{dx} = 25{,}000 - 0.01x$

10. $\dfrac{dC}{dx} = 0.03x^2 + 4600$

EXERCISES 5.4

In Exercises 1–8, sketch the region whose area is represented by the definite integral. Then use a geometric formula to evaluate the integral.

1. $\displaystyle\int_0^2 3\, dx$

2. $\displaystyle\int_0^4 2\, dx$

3. $\displaystyle\int_0^5 (x + 1)\, dx$

4. $\displaystyle\int_0^3 (2x + 1)\, dx$

5. $\displaystyle\int_{-2}^3 |x - 1|\, dx$

6. $\displaystyle\int_{-1}^4 |x - 2|\, dx$

7. $\displaystyle\int_{-3}^3 \sqrt{9 - x^2}\, dx$

8. $\displaystyle\int_0^2 \sqrt{4 - x^2}\, dx$

In Exercises 9 and 10, use the values $\int_0^5 f(x)\, dx = 8$ and $\int_0^5 g(x)\, dx = 3$ to evaluate the definite integral.

9. (a) $\displaystyle\int_0^5 [f(x) + g(x)]\, dx$

   (b) $\displaystyle\int_0^5 [f(x) - g(x)]\, dx$

   (c) $\displaystyle\int_0^5 -4f(x)\, dx$

   (d) $\displaystyle\int_0^5 [f(x) - 3g(x)]\, dx$

10. (a) $\displaystyle\int_0^5 2g(x)\, dx$

    (b) $\displaystyle\int_5^0 f(x)\, dx$

    (c) $\displaystyle\int_5^5 f(x)\, dx$

    (d) $\displaystyle\int_0^5 [f(x) - f(x)]\, dx$

In Exercises 11–18, find the area of the region.

11. $y = x - x^2$

12. $y = 1 - x^4$

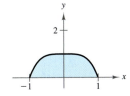

13. $y = \dfrac{1}{x^2}$

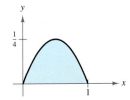

14. $y = \dfrac{2}{\sqrt{x}}$

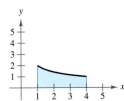

15. $y = 3e^{-x/2}$

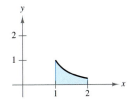
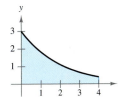

16. $y = 2e^{x/2}$

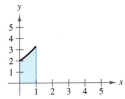

17. $y = \dfrac{x^2 + 4}{x}$

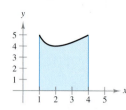

18. $y = \dfrac{x - 2}{x}$

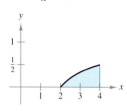

In Exercises 19–42, evaluate the definite integral.

19. $\displaystyle\int_0^1 2x \, dx$

20. $\displaystyle\int_2^7 3v \, dv$

21. $\displaystyle\int_{-1}^0 (2x + 1) \, dx$

22. $\displaystyle\int_2^5 (-3x + 4) \, dx$

23. $\displaystyle\int_{-1}^1 (2t - 1)^2 \, dt$

24. $\displaystyle\int_0^1 (1 - 2x)^2 \, dx$

25. $\displaystyle\int_0^3 (x - 2)^3 \, dx$

26. $\displaystyle\int_2^2 (x - 3)^4 \, dx$

27. $\displaystyle\int_{-1}^1 \left(\sqrt[3]{t} - 2\right) dt$

28. $\displaystyle\int_1^4 \sqrt{\dfrac{2}{x}} \, dx$

29. $\displaystyle\int_1^4 \dfrac{2u - 1}{\sqrt{u}} \, du$

30. $\displaystyle\int_0^1 \dfrac{x - \sqrt{x}}{3} \, dx$

31. $\displaystyle\int_{-1}^0 (t^{1/3} - t^{2/3}) \, dt$

32. $\displaystyle\int_0^4 (x^{1/2} + x^{1/4}) \, dx$

33. $\displaystyle\int_0^4 \dfrac{1}{\sqrt{2x + 1}} \, dx$

34. $\displaystyle\int_0^2 \dfrac{x}{\sqrt{1 + 2x^2}} \, dx$

35. $\displaystyle\int_0^1 e^{-2x} \, dx$

36. $\displaystyle\int_1^2 e^{1 - x} \, dx$

37. $\displaystyle\int_1^3 \dfrac{e^{3/x}}{x^2} \, dx$

38. $\displaystyle\int_{-1}^1 (e^x - e^{-x}) \, dx$

39. $\displaystyle\int_0^1 e^{2x}\sqrt{e^{2x} + 1} \, dx$

40. $\displaystyle\int_0^1 \dfrac{e^{-x}}{\sqrt{e^{-x} + 1}} \, dx$

41. $\displaystyle\int_0^2 \dfrac{x}{1 + 4x^2} \, dx$

42. $\displaystyle\int_0^1 \dfrac{e^{2x}}{e^{2x} + 1} \, dx$

In Exercises 43–46, evaluate the definite integral by the most convenient method. Explain your approach.

43. $\displaystyle\int_{-1}^1 |4x| \, dx$

44. $\displaystyle\int_0^3 |2x - 3| \, dx$

45. $\displaystyle\int_0^4 (2 - |x - 2|) \, dx$

46. $\displaystyle\int_{-4}^4 (4 - |x|) \, dx$

 In Exercises 47–50, evaluate the definite integral by hand. Then use a symbolic integration utility to evaluate the definite integral. Briefly explain any differences in your results.

47. $\displaystyle\int_{-1}^2 \dfrac{x}{x^2 - 9} \, dx$

48. $\displaystyle\int_2^3 \dfrac{x + 1}{x^2 + 2x - 3} \, dx$

49. $\displaystyle\int_0^3 \dfrac{2e^x}{2 + e^x} \, dx$

50. $\displaystyle\int_1^2 \dfrac{(2 + \ln x)^3}{x} \, dx$

 In Exercises 51–56, evaluate the definite integral by hand. Then use a graphing utility to graph the region whose area is represented by the integral.

51. $\displaystyle\int_1^3 (4x - 3) \, dx$

52. $\displaystyle\int_0^2 (x + 4) \, dx$

53. $\displaystyle\int_0^1 (x - x^3) \, dx$

54. $\displaystyle\int_0^1 \sqrt{x}(1 - x) \, dx$

55. $\displaystyle\int_2^4 \dfrac{3x^2}{x^3 - 1} \, dx$

56. $\displaystyle\int_0^{\ln 6} \dfrac{e^x}{2} \, dx$

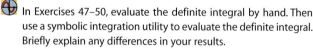 In Exercises 57–60, find the area of the region bounded by the graphs. Use a graphing utility to graph the region and verify your results.

57. $y = 3x^2 + 1$, $y = 0$, $x = 0$, and $x = 2$

58. $y = 1 + \sqrt{x}$, $y = 0$, $x = 0$, and $x = 4$

59. $y = (x + 5)/x$, $y = 0$, $x = 1$, and $x = 5$

60. $y = 3e^x$, $y = 0$, $x = -2$, and $x = 1$

In Exercises 61–68, use a graphing utility to graph the function over the interval. Find the average value of the function over the interval. Then find all x-values in the interval for which the function is equal to its average value.

	Function	*Interval*
61.	$f(x) = 6 - x^2$	$[-2, 2]$
62.	$f(x) = x - 2\sqrt{x}$	$[0, 4]$
63.	$f(x) = 5e^{0.2(x - 10)}$	$[0, 10]$
64.	$f(x) = 2e^{x/4}$	$[0, 4]$
65.	$f(x) = x\sqrt{4 - x^2}$	$[0, 2]$
66.	$f(x) = \dfrac{1}{(x - 3)^2}$	$[0, 2]$
67.	$f(x) = \dfrac{5x}{x^2 + 1}$	$[0, 7]$
68.	$f(x) = \dfrac{4x}{x^2 + 1}$	$[0, 1]$

In Exercises 69–72, state whether the function is even, odd, or neither.

69. $f(x) = 3x^4$

70. $g(x) = x^3 - 2x$

71. $g(t) = 2t^5 - 3t^2$

72. $f(t) = 5t^4 + 1$

73. Use the value $\int_0^2 x^2 \, dx = \frac{8}{3}$ to evaluate each definite integral. Explain your reasoning.

(a) $\displaystyle\int_{-2}^0 x^2 \, dx$ (b) $\displaystyle\int_{-2}^2 x^2 \, dx$ (c) $\displaystyle\int_0^2 -x^2 \, dx$

74. Use the value $\int_0^2 x^3 \, dx = 4$ to evaluate each definite integral. Explain your reasoning.

(a) $\displaystyle\int_{-2}^0 x^3 \, dx$ (b) $\displaystyle\int_{-2}^2 x^3 \, dx$ (c) $\displaystyle\int_0^2 3x^3 \, dx$

Marginal Analysis   In Exercises 75–80, find the change in cost C, revenue R, or profit P, for the given marginal. In each case, assume that the number of units x increases by 3 from the specified value of x.

Marginal	*Number of Units, x*
75. $\dfrac{dC}{dx} = 2.25$	$x = 100$
76. $\dfrac{dC}{dx} = \dfrac{20{,}000}{x^2}$	$x = 10$
77. $\dfrac{dR}{dx} = 48 - 3x$	$x = 12$
78. $\dfrac{dR}{dx} = 75\left(20 + \dfrac{900}{x}\right)$	$x = 500$
79. $\dfrac{dP}{dx} = \dfrac{400 - x}{150}$	$x = 200$
80. $\dfrac{dP}{dx} = 12.5\left(40 - 3\sqrt{x}\right)$	$x = 125$

Annuity   In Exercises 81–84, find the amount of an annuity with income function $c(t)$, interest rate r, and term T.

81. $c(t) = \$250$, $r = 8\%$, $T = 6$ years

82. $c(t) = \$500$, $r = 9\%$, $T = 4$ years

83. $c(t) = \$1500$, $r = 2\%$, $T = 10$ years

84. $c(t) = \$2000$, $r = 3\%$, $T = 15$ years

Capital Accumulation   In Exercises 85–88, you are given the rate of investment dI/dt. Find the capital accumulation over a five-year period by evaluating the definite integral

$$\text{Capital accumulation} = \int_0^5 \frac{dI}{dt}\, dt$$

where t is the time in years.

85. $\dfrac{dI}{dt} = 500$

86. $\dfrac{dI}{dt} = 100t$

87. $\dfrac{dI}{dt} = 500\sqrt{t + 1}$

88. $\dfrac{dI}{dt} = \dfrac{12{,}000t}{(t^2 + 2)^2}$

89. *Cost*   The total cost of purchasing and maintaining a piece of equipment for x years can be modeled by

$$C = 5000\left(25 + 3\int_0^x t^{1/4}\, dt\right).$$

Find the total cost after (a) 1 year, (b) 5 years, and (c) 10 years.

90. *Depreciation*   A company purchases a new machine for which the rate of depreciation can be modeled by

$$\frac{dV}{dt} = 10{,}000(t - 6),\quad 0 \le t \le 5$$

where V is the value of the machine after t years. Set up and evaluate the definite integral that yields the total loss of value of the machine over the first 3 years.

91. *Compound Interest*   A deposit of $2250 is made in a savings account at an annual interest rate of 12%, compounded continuously. Find the average balance in the account during the first 5 years.

92. *Mortgage Debt*   The rate of change of mortgage debt outstanding for one- to four-family homes in the United States from 1993 through 2002 can be modeled by

$$\frac{dM}{dt} = 5.4399t^2 + 6603.7e^{-t}$$

where M is the mortgage debt outstanding (in billions of dollars) and $t = 3$ corresponds to 1993. In 1993, the mortgage debt outstanding in the United States was $3119 billion. *(Source: Board of Governors of the Federal Reserve System)*

(a) Write a model for the debt as a function of t.

(b) What was the average mortgage debt outstanding for 1993 through 2002?

93. *Medicine*   The velocity v of blood at a distance r from the center of an artery of radius R can be modeled by

$$v = k(R^2 - r^2)$$

where k is a constant. Find the average velocity along a radius of the artery. (Use 0 and R as the limits of integration.)

94. *Biology*   The rate of change in the number of coyotes $N(t)$ in a population is directly proportional to $650 - N(t)$, where t is time in years.

$$\frac{dN}{dt} = k[650 - N(t)]$$

When $t = 0$, the population is 300, and when $t = 2$, the population has increased to 500.

(a) Find the population function.

(b) Find the average number of coyotes over the first 5 years.

 In Exercises 95–98, use a symbolic integration utility to evaluate the definite integral.

95. $\displaystyle\int_3^6 \frac{x}{3\sqrt{x^2 - 8}}\, dx$

96. $\displaystyle\int_{1/2}^1 (x + 1)\sqrt{1 - x}\, dx$

97. $\displaystyle\int_2^5 \left(\frac{1}{x^2} - \frac{1}{x^3}\right) dx$

98. $\displaystyle\int_0^1 x^3(x^3 + 1)^3\, dx$

In Exercises 1–4, simplify the expression.

1. $(-x^2 + 4x + 3) - (x + 1)$

2. $(-2x^2 + 3x + 9) - (-x + 5)$

3. $(-x^3 + 3x^2 - 1) - (x^2 - 4x + 4)$

4. $(3x + 1) - (-x^3 + 9x + 2)$

In Exercises 5–10, find the points of intersection of the graphs.

5. $f(x) = x^2 - 4x + 4$, $g(x) = 4$

6. $f(x) = -3x^2$, $g(x) = 6 - 9x$

7. $f(x) = x^2$, $g(x) = -x + 6$

8. $f(x) = \frac{1}{2}x^3$, $g(x) = 2x$

9. $f(x) = x^2 - 3x$, $g(x) = 3x - 5$

10. $f(x) = e^x$, $g(x) = e$

EXERCISES 5.5

In Exercises 1–8, find the area of the region.

1. $f(x) = x^2 - 6x$
$g(x) = 0$

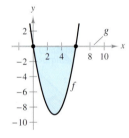

2. $f(x) = x^2 + 2x + 1$
$g(x) = 2x + 5$

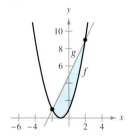

3. $f(x) = x^2 - 4x + 3$
$g(x) = -x^2 + 2x + 3$

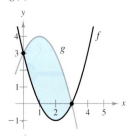

4. $f(x) = x^2$
$g(x) = x^3$

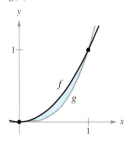

5. $f(x) = 3(x^3 - x)$
$g(x) = 0$

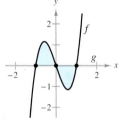

6. $f(x) = (x - 1)^3$
$g(x) = x - 1$

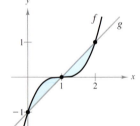

7. $f(x) = e^x - 1$
$g(x) = 0$

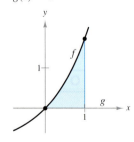

8. $f(x) = -x + 3$
$g(x) = 2x^{-1}$

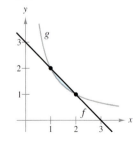

In Exercises 9–14, sketch the region whose area is represented by the definite integral.

9. $\int_0^4 \left[(x+1) - \frac{1}{2}x\right] dx$

10. $\int_{-1}^1 \left[(1-x^2) - (x^2-1)\right] dx$

11. $\int_{-2}^2 \left[2x^2 - (x^4 - 2x^2)\right] dx$

12. $\int_{-4}^0 \left[(x-6) - (x^2 + 5x - 6)\right] dx$

13. $\int_{-1}^2 \left[(y^2 + 2) - 1\right] dy$

14. $\int_{-2}^3 \left[(y+6) - y^2\right] dy$

In Exercises 15–26, sketch the region bounded by the graphs of the functions and find the area of the region.

15. $y = \dfrac{1}{x^2}, y = 0, x = 1, x = 5$

16. $y = x^3 - 2x + 1, y = -2x, x = 1$

17. $f(x) = \sqrt[3]{x}, g(x) = x$

18. $f(x) = \sqrt{3x} + 1, g(x) = x + 1$

19. $y = x^2 - 4x + 3, y = 3 + 4x - x^2$

20. $y = 4 - x^2, y = x^2$

21. $y = xe^{-x^2}, y = 0, x = 0, x = 1$

22. $y = \dfrac{e^{1/x}}{x^2}, y = 0, x = 1, x = 3$

23. $y = \dfrac{8}{x}, y = x^2, y = 0, x = 1, x = 4$

24. $y = \dfrac{1}{x}, y = x^3, x = \dfrac{1}{2}, x = 1$

25. $f(x) = e^{0.5x}, g(x) = -\dfrac{1}{x}, x = 1, x = 2$

26. $f(x) = \dfrac{1}{x}, g(x) = -e^x, x = \dfrac{1}{2}, x = 1$

In Exercises 27–30, sketch the region bounded by the graphs of the functions and find the area of the region.

27. $f(y) = y^2, g(y) = y + 2$

28. $f(y) = y(2-y), g(y) = -y$

29. $f(y) = \sqrt{y}, y = 9, x = 0$

30. $f(y) = y^2 + 1, g(y) = 4 - 2y$

 In Exercises 31–34, use a graphing utility to graph the region bounded by the graphs of the functions. Write the definite integrals that represent the area of the region. (*Hint:* Multiple integrals may be necessary.)

31. $f(x) = 2x, g(x) = 4 - 2x, h(x) = 0$

32. $f(x) = x(x^2 - 3x + 3), g(x) = x^2$

33. $y = \dfrac{4}{x}, y = x, x = 1, x = 4$

34. $y = x^3 - 4x^2 + 1, y = x - 3$

 In Exercises 35–38, use a graphing utility to graph the region bounded by the graphs of the functions, and find the area of the region.

35. $f(x) = x^2 - 4x, g(x) = 0$

36. $f(x) = 3 - 2x - x^2, g(x) = 0$

37. $f(x) = x^2 + 2x + 1, g(x) = x + 1$

38. $f(x) = -x^2 + 4x + 2, g(x) = x + 2$

In Exercises 39 and 40, use integration to find the area of the triangular region having the given vertices.

39. $(0, 0), (4, 0), (4, 4)$

40. $(0, 0), (4, 0), (6, 4)$

Consumer and Producer Surpluses   In Exercises 41–46, find the consumer and producer surpluses.

Demand Function	*Supply Function*
41. $p_1(x) = 50 - 0.5x$	$p_2(x) = 0.125x$
42. $p_1(x) = 300 - x$	$p_2(x) = 100 + x$
43. $p_1(x) = 200 - 0.02x^2$	$p_2(x) = 100 + x$
44. $p_1(x) = 1000 - 0.4x^2$	$p_2(x) = 42x$
45. $p_1(x) = \dfrac{10{,}000}{\sqrt{x + 100}}$	$p_2(x) = 100\sqrt{0.05x + 10}$
46. $p_1(x) = \sqrt{25 - 0.1x}$	$p_2(x) = \sqrt{9 + 0.1x} - 2$

47. *Writing*   Describe the characteristics of typical demand and supply functions.

48. *Writing*   Suppose that the demand and supply functions for a product do not intersect. What can you conclude?

Revenue   In Exercises 49 and 50, two models, R_1 and R_2, are given for revenue (in billions of dollars per year) for a large corporation. Both models are estimates of revenues for 2004–2008, with $t = 4$ corresponding to 2004. Which model is projecting the greater revenue? How much more total revenue does that model project over the four-year period?

49. $R_1 = 7.21 + 0.58t, R_2 = 7.21 + 0.45t$

50. $R_1 = 7.21 + 0.26t + 0.02t^2, R_2 = 7.21 + 0.1t + 0.01t^2$

51. *Fuel Cost*   The projected fuel cost C (in millions of dollars per year) for an airline company from 2004 through 2010 is $C_1 = 568.5 + 7.15t$, where $t = 4$ corresponds to 2004. If the company purchases more efficient airplane engines, fuel cost is expected to decrease and to follow the model $C_2 = 525.6 + 6.43t$. How much can the company save with the more efficient engines? Explain your reasoning.

52. *Health*   An epidemic was spreading such that t weeks after its outbreak it had infected

$$N_1(t) = 0.1t^2 + 0.5t + 150, \quad 0 \le t \le 50$$

people. Twenty-five weeks after the outbreak, a vaccine was developed and administered to the public. At that point, the number of people infected was governed by the model

$$N_2(t) = -0.2t^2 + 6t + 200.$$

Approximate the number of people that the vaccine prevented from becoming ill during the epidemic.

53. *Consumer Trends*   For the years 1990 through 2001, the per capita consumption of tomatoes (in pounds per year) in the United States can be modeled by

$$C(t) = \begin{cases} 0.085t^3 - 0.309t^2 + 0.13t + 15.5, \\ 0 \le t \le 4 \\ 0.01515t^4 - 0.5348t^3 + 6.864t^2 - 37.68t + 91.4 \\ 4 < t \le 11 \end{cases}$$

where $t = 0$ corresponds to 1990. *(Source: U.S. Department of Agriculture)*

(a) Use a graphing utility to graph this model.

(b) Suppose the tomato consumption from 1995 through 2001 had continued to follow the model for 1990 through 1994. How many more or fewer pounds of tomatoes would have been consumed from 1995 through 2001?

54. *Consumer and Producer Surpluses*   Factory orders for an air conditioner are about 6000 units per week when the price is $331 and about 8000 units per week when the price is $303. The supply function is given by $p = 0.0275x$. Find the consumer and producer surpluses. (Assume the demand function is linear.)

55. *Consumer and Producer Surpluses*   Repeat Exercise 54 with a demand of about 6000 units per week when the price is $325 and about 8000 units per week when the price is $300. Find the consumer and producer surpluses. (Assume the demand function is linear.)

56. *Cost, Revenue, and Profit*   The revenue from a manufacturing process (in millions of dollars per year) is projected to follow the model $R = 100$ for 10 years. Over the same period of time, the cost (in millions of dollars per year) is projected to follow the model $C = 60 + 0.2t^2$, where t is the time (in years). Approximate the profit over the 10-year period.

57. *Cost, Revenue, and Profit*   Repeat Exercise 56 for revenue and cost models given by $R = 100 + 0.08t$ and $C = 60 + 0.2t^2$.

58. *Lorenz Curve*   Economists use *Lorenz curves* to illustrate the distribution of income in a country. Letting x represent the percent of families in a country and y the percent of total income, the model $y = x$ would represent a country in which each family had the same income. The Lorenz curve, $y = f(x)$, represents the actual income distribution. The area between these two models, for $0 \le x \le 100$, indicates the "income inequality" of a country. In 2001, the Lorenz curve for the United States could be modeled by

$$y = (0.00059x^2 + 0.0233x + 1.731)^2, \quad 0 \le x \le 100$$

where x is measured from the poorest to the wealthiest families. Find the income inequality for the United States in 2001. *(Source: U.S. Census Bureau)*

59. *Income Distribution*   Using the Lorenz curve in Exercise 58, complete the table, which lists the percent of total income earned by each quintile in the United States in 2001.

Quintile	Lowest	2nd	3rd	4th	Highest
Percent					

60. *Research Project*   Use your school's library, the Internet, or some other reference source to research a small company similar to that described above. Describe the impact of different factors, such as start-up capital and market conditions, on a company's revenue.

5.6   THE DEFINITE INTEGRAL AS THE LIMIT OF A SUM

■ Use the Midpoint Rule to approximate definite integrals.
■ Use a symbolic integration utility to approximate definite integrals.

The Midpoint Rule

In Section 5.4, you learned that you cannot use the Fundamental Theorem of Calculus to evaluate a definite integral unless you can find an antiderivative of the integrand. In cases where this cannot be done, you can approximate the value of the integral using an approximation technique. One such technique is called the **Midpoint Rule.** (Two other techniques are discussed in Section 6.5.)

EXAMPLE 1   **Approximating the Area of a Plane Region**

Use the five rectangles in Figure 5.22 to approximate the area of the region bounded by the graph of $f(x) = -x^2 + 5$, the x-axis, and the lines $x = 0$ and $x = 2$.

SOLUTION   You can find the heights of the five rectangles by evaluating f at the midpoint of each of the following intervals.

$$\left[0, \frac{2}{5}\right], \quad \left[\frac{2}{5}, \frac{4}{5}\right], \quad \left[\frac{4}{5}, \frac{6}{5}\right], \quad \left[\frac{6}{5}, \frac{8}{5}\right], \quad \left[\frac{8}{5}, \frac{10}{5}\right]$$

Evaluate f at the midpoints of these intervals.

The width of each rectangle is $\frac{2}{5}$. So, the sum of the five areas is

$$\text{Area} \approx \frac{2}{5}f\left(\frac{1}{5}\right) + \frac{2}{5}f\left(\frac{3}{5}\right) + \frac{2}{5}f\left(\frac{5}{5}\right) + \frac{2}{5}f\left(\frac{7}{5}\right) + \frac{2}{5}f\left(\frac{9}{5}\right)$$

$$= \frac{2}{5}\left[f\left(\frac{1}{5}\right) + f\left(\frac{3}{5}\right) + f\left(\frac{5}{5}\right) + f\left(\frac{7}{5}\right) + f\left(\frac{9}{5}\right)\right]$$

$$= \frac{2}{5}\left(\frac{124}{25} + \frac{116}{25} + \frac{100}{25} + \frac{76}{25} + \frac{44}{25}\right)$$

$$= \frac{920}{125}$$

$$= 7.36.$$

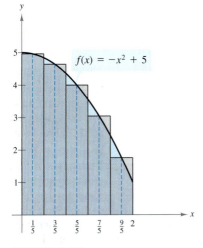

FIGURE 5.22

TRY IT 1

Use four rectangles to approximate the area of the region bounded by the graph of $f(x) = x^2 + 1$, the x-axis, $x = 0$ and $x = 2$.

For the region in Example 1, you can find the exact area with a definite integral. That is,

$$\text{Area} = \int_0^2 (-x^2 + 5)\, dx = \frac{22}{3} \approx 7.33.$$

The easiest way to use the Midpoint Rule to approximate the definite integral $\int_a^b f(x)\,dx$ is to program it into a computer or programmable calculator. For instance, the pseudocode below will help you write a program to evaluate the Midpoint Rule. (Appendix E lists this program for several models of graphing utilities.)

Program

- *Prompt for value of a.*
- *Input value of a.*
- *Prompt for value of b.*
- *Input value of b.*
- *Prompt for value of n.*
- *Input value of n.*
- *Initialize sum of areas.*
- *Calculate width of subinterval.*
- *Initialize counter.*
- *Begin loop.*
- *Calculate left endpoint.*
- *Calculate right endpoint.*
- *Calculate midpoint of subinterval.*
- *Add area to sum.*
- *Test counter.*
- *End loop.*
- *Display approximation.*

Before executing the program, enter the function. When the program is executed, you will be prompted to enter the lower and upper limits of integration and the number of subintervals you want to use.

With most integrals, you can determine the accuracy of the approximation by using increasingly larger values of *n*. For example, if you use the program to approximate the value of $\int_0^2 \sqrt{x^3 + 1}\,dx$, you will obtain an approximation of 3.241, which is accurate to three decimal places.

The approximation procedure used in Example 1 is the **Midpoint Rule.** You can use the Midpoint Rule to approximate *any* definite integral—not just those representing area. The basic steps are summarized below.

Guidelines for Using the Midpoint Rule

To approximate the definite integral $\int_a^b f(x)\,dx$ with the Midpoint Rule, use the steps below.

1. Divide the interval $[a, b]$ into *n* subintervals, each of width

$$\Delta x = \frac{b - a}{n}.$$

2. Find the midpoint of each subinterval.

$$\text{Midpoints} = \{x_1, x_2, x_3, \dots, x_n\}$$

3. Evaluate *f* at each midpoint and form the sum as shown.

$$\int_a^b f(x)\,dx \approx \frac{b - a}{n}[f(x_1) + f(x_2) + f(x_3) + \cdots + f(x_n)]$$

An important characteristic of the Midpoint Rule is that the approximation tends to improve as *n* increases. The table below shows the approximations for the area of the region described in Example 1 for various values of *n*. For example, for $n = 10$, the Midpoint Rule yields

$$\int_0^2 (-x^2 + 5)\,dx \approx \frac{2}{10}\left[f\left(\frac{1}{10}\right) + f\left(\frac{3}{10}\right) + \cdots + f\left(\frac{19}{10}\right)\right]$$
$$= 7.34.$$

n	5	10	15	20	25	30
Approximation	7.3600	7.3400	7.3363	7.3350	7.3344	7.3341

Note that as *n* increases, the approximation gets closer and closer to the exact value of the integral, which was found to be

$$\frac{22}{3} \approx 7.3333.$$

STUDY TIP

In Example 1, the Midpoint Rule is used to approximate an integral whose exact value can be found with the Fundamental Theorem of Calculus. This was done to illustrate the accuracy of the rule. In practice, of course, you would use the Midpoint Rule to approximate the values of definite integrals for which you cannot find an antiderivative. Examples 2 and 3 illustrate such integrals.

EXAMPLE 2 **Using the Midpoint Rule**

Use the Midpoint Rule with $n = 5$ to approximate $\int_0^1 \frac{1}{x^2 + 1}\, dx$.

SOLUTION With $n = 5$, the interval $[0, 1]$ is divided into five subintervals.

$$\left[0, \frac{1}{5}\right], \quad \left[\frac{1}{5}, \frac{2}{5}\right], \quad \left[\frac{2}{5}, \frac{3}{5}\right], \quad \left[\frac{3}{5}, \frac{4}{5}\right], \quad \left[\frac{4}{5}, 1\right]$$

The midpoints of these intervals are $\frac{1}{10}, \frac{3}{10}, \frac{5}{10}, \frac{7}{10}$, and $\frac{9}{10}$. Because each subinterval has a width of $\Delta x = (1 - 0)/5 = \frac{1}{5}$, you can approximate the value of the definite integral as shown.

$$\int_0^1 \frac{1}{x^2 + 1}\, dx \approx \frac{1}{5}\left(\frac{1}{1.01} + \frac{1}{1.09} + \frac{1}{1.25} + \frac{1}{1.49} + \frac{1}{1.81}\right)$$
$$\approx 0.786$$

The region whose area is represented by the definite integral is shown in Figure 5.23. The actual area of this region is $\pi/4 \approx 0.785$. So, the approximation is off by only 0.001.

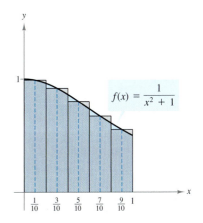

FIGURE 5.23

EXAMPLE 3 **Using the Midpoint Rule**

Use the Midpoint Rule with $n = 10$ to approximate $\int_1^3 \sqrt{x^2 + 1}\, dx$.

SOLUTION Begin by dividing the interval $[1, 3]$ into 10 subintervals. The midpoints of these intervals are

$$\frac{11}{10}, \quad \frac{13}{10}, \quad \frac{3}{2}, \quad \frac{17}{10}, \quad \frac{19}{10}, \quad \frac{21}{10}, \quad \frac{23}{10}, \quad \frac{5}{2}, \quad \frac{27}{10}, \quad \text{and} \quad \frac{29}{10}.$$

Because each subinterval has a width of $\Delta x = (3 - 1)/10 = \frac{1}{5}$, you can approximate the value of the definite integral as shown.

$$\int_1^3 \sqrt{x^2 + 1}\, dx \approx \frac{1}{5}\left[\sqrt{(1.1)^2 + 1} + \sqrt{(1.3)^2 + 1} + \cdots + \sqrt{(2.9)^2 + 1}\right]$$
$$\approx 4.504$$

The region whose area is represented by the definite integral is shown in Figure 5.24. Using techniques that are not within the scope of this course, it can be shown that the actual area is

$$\tfrac{1}{2}\left[3\sqrt{10} + \ln\left(3 + \sqrt{10}\right) - \sqrt{2} - \ln\left(1 + \sqrt{2}\right)\right] \approx 4.505.$$

So, the approximation is off by only 0.001.

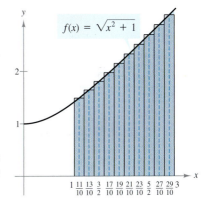

FIGURE 5.24

The Definite Integral as the Limit of a Sum

Consider the closed interval $[a, b]$, divided into n subintervals whose midpoints are x_i and whose widths are $\Delta x = (b - a)/n$. In this section, you have seen that the midpoint approximation

$$\int_a^b f(x)\, dx \approx f(x_1)\, \Delta x + f(x_2)\, \Delta x + f(x_3)\, \Delta x + \cdots + f(x_n)\, \Delta x$$

$$= [f(x_1) + f(x_2) + f(x_3) + \cdots + f(x_n)]\, \Delta x$$

becomes better and better as n increases. In fact, the limit of this sum as n approaches infinity is exactly equal to the definite integral. That is,

$$\int_a^b f(x)\, dx = \lim_{n \to \infty} [f(x_1) + f(x_2) + f(x_3) + \cdots + f(x_n)]\, \Delta x.$$

It can be shown that this limit is valid as long as x_i is *any* point in the ith interval.

EXAMPLE 4   Approximating a Definite Integral

Use a computer, programmable calculator, or symbolic integration utility to approximate the definite integral

$$\int_0^1 e^{-x^2}\, dx.$$

SOLUTION   Using the program on page 366, with $n = 10, 20, 30, 40,$ and 50, it appears that the value of the integral is approximately 0.7468. If you have access to a computer or calculator with a built-in program for approximating definite integrals, try using it to approximate this integral. When a computer with such a built-in program approximated the integral, it returned a value of 0.746824.

TRY IT 4

Use a computer, programmable calculator, or symbolic integration utility to approximate the definite integral

$$\int_0^1 e^{x^2}\, dx.$$

TAKE ANOTHER LOOK

A Failure of the Midpoint Rule

Suppose you use the Midpoint Rule to approximate the definite integral

$$\int_0^1 \frac{1}{x^2}\, dx.$$

Use a program similar to the one on page 366 to complete the table for $n = 10, 20, 30, 40, 50,$ and 60.

n	10	20	30	40	50	60
Approximation						

Why are the approximations getting larger? Why isn't the Midpoint Rule working?

PREREQUISITE REVIEW 5.6

The following warm-up exercises involve skills that were covered in earlier sections. You will use these skills in the exercise set for this section.

In Exercises 1–6, find the midpoint of the interval.

1. $\left[0, \frac{1}{3}\right]$
2. $\left[\frac{1}{10}, \frac{2}{10}\right]$
3. $\left[\frac{3}{20}, \frac{4}{20}\right]$
4. $\left[1, \frac{7}{6}\right]$
5. $\left[2, \frac{31}{15}\right]$
6. $\left[\frac{26}{9}, 3\right]$

In Exercises 7–10, find the limit.

7. $\lim\limits_{x \to \infty} \dfrac{2x^2 + 4x - 1}{3x^2 - 2x}$

8. $\lim\limits_{x \to \infty} \dfrac{4x + 5}{7x - 5}$

9. $\lim\limits_{x \to \infty} \dfrac{x - 7}{x^2 + 1}$

10. $\lim\limits_{x \to \infty} \dfrac{5x^3 + 1}{x^3 + x^2 + 4}$

EXERCISES 5.6

In Exercises 1–4, use the Midpoint Rule with $n = 4$ to approximate the area of the region. Compare your result with the exact area obtained with a definite integral.

1. $f(x) = -2x + 3, \quad [0, 1]$
2. $f(x) = \dfrac{1}{x}, \quad [1, 5]$

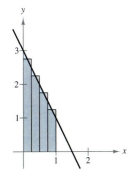

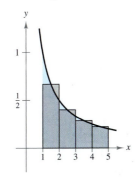

3. $f(x) = \sqrt{x}, \quad [0, 1]$
4. $f(x) = 1 - x^2, \quad [-1, 1]$

In Exercises 5–12, use the Midpoint Rule with $n = 4$ to approximate the area of the region bounded by the graph of f and the x-axis over the interval. Compare your result with the exact area. Sketch the region.

	Function	*Interval*
5.	$f(x) = x^2 + 2$	$[-1, 1]$
6.	$f(x) = 4 - x^2$	$[0, 2]$
7.	$f(x) = 2x^2$	$[1, 3]$
8.	$f(x) = 2x - x^3$	$[0, 1]$
9.	$f(x) = x^2 - x^3$	$[0, 1]$
10.	$f(x) = x^2 - x^3$	$[-1, 0]$
11.	$f(x) = x(1 - x)^2$	$[0, 1]$
12.	$f(x) = x^2(3 - x)$	$[0, 3]$

 In Exercises 13–16, use a program similar to that on page 366 to approximate the area of the region. How large must n be to obtain an approximation that is correct to within 0.01?

13. $\displaystyle\int_0^4 (2x^2 + 3)\, dx$

14. $\displaystyle\int_0^4 (2x^3 + 3)\, dx$

15. $\displaystyle\int_1^2 (2x^2 - x + 1)\, dx$

16. $\displaystyle\int_1^2 (x^3 - 1)\, dx$

In Exercises 17–20, use the Midpoint Rule with $n = 4$ to approximate the area of the region. Compare your result with the exact area obtained with a definite integral.

17. $f(y) = \frac{1}{4}y$, $[2, 4]$

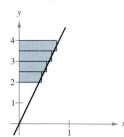

18. $f(y) = 2y$, $[0, 2]$

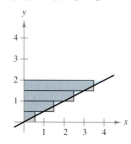

19. $f(y) = y^2 + 1$, $[0, 4]$

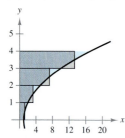

20. $f(y) = 4y - y^2$, $[0, 4]$

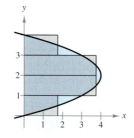

Trapezoidal Rule In Exercises 21 and 22, use the Trapezoidal Rule with $n = 8$ to approximate the definite integral. Compare the result with the exact value and the approximation obtained with $n = 8$ and the Midpoint Rule. Which approximation technique appears to be better? Let f be continuous on $[a, b]$ and let n be the number of equal subintervals (see figure). Then the Trapezoidal Rule for approximating $\int_a^b f(x)\, dx$ is

$$\frac{b - a}{2n}[f(x_0) + 2f(x_1) + \cdots + 2f(x_{n-1}) + f(x_n)].$$

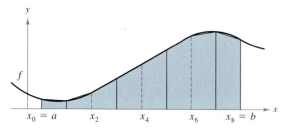

21. $\displaystyle\int_0^2 x^3\, dx$

22. $\displaystyle\int_1^3 \frac{1}{x^2}\, dx$

In Exercises 23–26, use the Trapezoidal Rule with $n = 4$ to approximate the definite integral.

23. $\displaystyle\int_0^2 \frac{1}{x + 1}\, dx$

24. $\displaystyle\int_0^4 \sqrt{1 + x^2}\, dx$

25. $\displaystyle\int_{-1}^1 \frac{1}{x^2 + 1}\, dx$

26. $\displaystyle\int_1^5 \frac{\sqrt{x - 1}}{x}\, dx$

 In Exercises 27 and 28, use a computer or programmable calculator to approximate the definite integral using the Midpoint Rule and the Trapezoidal Rule for $n = 4, 8, 12, 16,$ and 20.

27. $\displaystyle\int_0^4 \sqrt{2 + 3x^2}\, dx$

28. $\displaystyle\int_0^2 \frac{5}{x^3 + 1}\, dx$

In Exercises 29 and 30, use the Trapezoidal Rule with $n = 10$ to approximate the area of the region bounded by the graphs of the equations.

29. $y = \sqrt{\dfrac{x^3}{4 - x}}$, $y = 0$, $x = 3$

30. $y = x\sqrt{\dfrac{4 - x}{4 + x}}$, $y = 0$, $x = 4$

31. *Velocity and Acceleration* The table lists the velocity v (in feet per second) of an accelerating car over a 20-second interval. Use the Trapezoidal Rule to approximate the distance in feet that the car travels during the 20 seconds. (The distance is given by $s = \int_0^{20} v\, dt$.)

Time, t	0	5	10	15	20
Velocity, v	0.0	29.3	51.3	66.0	73.3

32. *Surface Area* To estimate the surface area of a pond, a surveyor takes several measurements, as shown in the figure. Estimate the surface area of the pond using (a) the Midpoint Rule and (b) the Trapezoidal Rule.

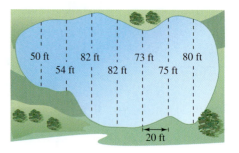
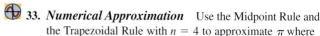

 33. *Numerical Approximation* Use the Midpoint Rule and the Trapezoidal Rule with $n = 4$ to approximate π where

$$\pi = \int_0^1 \frac{4}{1 + x^2}\, dx.$$

Then use a graphing utility to evaluate the definite integral. Compare all of your results.

5.7   VOLUMES OF SOLIDS OF REVOLUTION

- Use the Disk Method to find volumes of solids of revolution.
- Use the Washer Method to find volumes of solids of revolution with holes.
- Use solids of revolution to solve real-life problems.

The Disk Method

Another important application of the definite integral is its use in finding the volume of a three-dimensional solid. In this section, you will study a particular type of three-dimensional solid—one whose cross sections are similar. You will begin with solids of revolution. These solids, such as axles, funnels, pills, bottles, and pistons, are used commonly in engineering and manufacturing.

As shown in Figure 5.25, a **solid of revolution** is formed by revolving a plane region about a line. The line is called the **axis of revolution.**

To develop a formula for finding the volume of a solid of revolution, consider a continuous function f that is nonnegative on the interval $[a, b]$. Suppose that the area of the region is approximated by n rectangles, each of width Δx, as shown in Figure 5.26. By revolving the rectangles about the x-axis, you obtain n circular disks, each with a volume of $\pi[f(x_i)]^2\Delta x$. The volume of the solid formed by revolving the region about the x-axis is approximately equal to the sum of the volumes of the n disks. Moreover, by taking the limit as n approaches infinity, you can see that the exact volume is given by a definite integral. This result is called the **Disk Method.**

Plane region

Axis of revolution

FIGURE 5.25

The Disk Method

The volume of the solid formed by revolving the region bounded by the graph of f and the x-axis ($a \le x \le b$) about the x-axis is

$$\text{Volume} = \pi \int_a^b [f(x)]^2 \, dx.$$

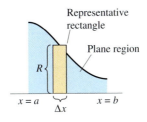

Representative rectangle

Plane region

R

$x = a$ Δx $x = b$

Approximation by n rectangles

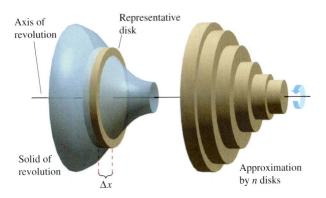

Axis of revolution

Representative disk

Solid of revolution

Δx

Approximation by n disks

FIGURE 5.26

EXAMPLE 1   **Finding the Volume of a Solid of Revolution**

Find the volume of the solid formed by revolving the region bounded by the graph of $f(x) = -x^2 + x$ and the x-axis about the x-axis.

SOLUTION   Begin by sketching the region bounded by the graph of f and the x-axis. As shown in Figure 5.27(a), sketch a representative rectangle whose height is $f(x)$ and whose width is Δx. From this rectangle, you can see that the radius of the solid is

$$\text{Radius} = f(x) = -x^2 + x.$$

Using the Disk Method, you can find the volume of the solid of revolution.

$$\text{Volume} = \pi \int_0^1 [f(x)]^2 \, dx \qquad \text{Disk Method}$$

$$= \pi \int_0^1 (-x^2 + x)^2 \, dx \qquad \text{Substitute for } f(x).$$

$$= \pi \int_0^1 (x^4 - 2x^3 + x^2) \, dx \qquad \text{Expand integrand.}$$

$$= \pi \left[\frac{x^5}{5} - \frac{x^4}{2} + \frac{x^3}{3} \right]_0^1 \qquad \text{Find antiderivative.}$$

$$= \frac{\pi}{30} \qquad \text{Apply Fundamental Theorem.}$$

$$\approx 0.105 \qquad \text{Round to three decimal places.}$$

So, the volume of the solid is about 0.105 cubic unit.

TECHNOLOGY

Try using the integration capabilities of a graphing utility to verify the solution in Example 1. Consult your user's manual for specific keystrokes.

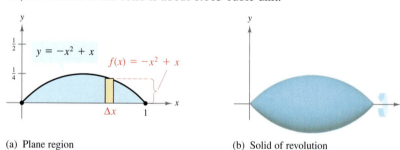

(a) Plane region   (b) Solid of revolution

FIGURE 5.27

TRY IT 1

Find the volume of the solid formed by revolving the region bounded by the graph of $f(x) = -x^2 + 4$ and the x-axis about the x-axis.

STUDY TIP

In Example 1, the entire problem was solved *without* referring to the three-dimensional sketch given in Figure 5.27(b). In general, to set up the integral for calculating the volume of a solid of revolution, a sketch of the plane region is more useful than a sketch of the solid, because the radius is more readily visualized in the plane region.

The Washer Method

You can extend the Disk Method to find the volume of a solid of revolution with a *hole*. Consider a region that is bounded by the graphs of f and g, as shown in Figure 5.28(a). If the region is revolved about the x-axis, then the volume of the resulting solid can be found by applying the Disk Method to f and g and subtracting the results.

$$\text{Volume} = \pi \int_a^b [f(x)]^2 \, dx - \pi \int_a^b [g(x)]^2 \, dx$$

Writing this as a single integral produces the **Washer Method.**

The Washer Method

Let f and g be continuous and nonnegative on the closed interval $[a, b]$, as shown in Figure 5.28(a). If $g(x) \le f(x)$ for all x in the interval, then the volume of the solid formed by revolving the region bounded by the graphs of f and g ($a \le x \le b$) about the x-axis is

$$\text{Volume} = \pi \int_a^b \{[f(x)]^2 - [g(x)]^2\} \, dx.$$

$f(x)$ is the **outer radius** and $g(x)$ is the **inner radius.**

In Figure 5.28(b), note that the solid of revolution has a hole. Moreover, the radius of the hole is $g(x)$, the inner radius.

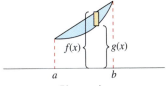

$f(x)$ $g(x)$

a b

Plane region

(a)

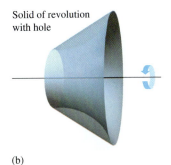

Solid of revolution with hole

(b)

FIGURE 5.28

EXAMPLE 2   **Using the Washer Method**

Find the volume of the solid formed by revolving the region bounded by the graphs of

$$f(x) = \sqrt{25 - x^2} \text{ and } g(x) = 3$$

about the x-axis (see Figure 5.29).

SOLUTION   First find the points of intersection of f and g by setting f(x) equal to g(x) and solving for x.

$$f(x) = g(x)$$ Set $f(x)$ equal to $g(x)$.

$$\sqrt{25 - x^2} = 3$$ Substitute for $f(x)$ and $g(x)$.

$$25 - x^2 = 9$$ Square each side.

$$16 = x^2$$

$$\pm 4 = x$$ Solve for x.

Using f(x) as the outer radius and g(x) as the inner radius, you can find the volume of the solid as shown.

$$\text{Volume} = \pi \int_{-4}^{4} \{[f(x)]^2 - [g(x)]^2\} \, dx$$ Washer Method

$$= \pi \int_{-4}^{4} \left[\left(\sqrt{25 - x^2} \right)^2 - (3)^2 \right] dx$$ Substitute for $f(x)$ and $g(x)$.

$$= \pi \int_{-4}^{4} (16 - x^2) \, dx$$ Simplify.

$$= \pi \left[16x - \frac{x^3}{3} \right]_{-4}^{4}$$ Find antiderivative.

$$= \frac{256\pi}{3}$$ Apply Fundamental Theorem.

$$\approx 268.08$$ Round to two decimal places.

So, the volume of the solid is about 268.08 cubic inches.

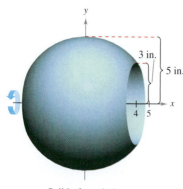

Solid of revolution

(b)

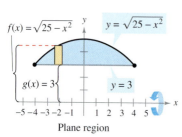

Plane region

(a)

TRY IT 2

Find the volume of the solid formed by revolving the region bounded by the graphs of $f(x) = 5 - x^2$ and $g(x) = 1$ about the x-axis.

FIGURE 5.29

Application

© Jessica Rinaldi/Stringer/Reuters/CORBIS

| EXAMPLE 3 | **Finding a Football's Volume**

A regulation-size football can be modeled as a solid of revolution formed by revolving the graph of

$$f(x) = -0.0944x^2 + 3.4, \quad -5.5 \le x \le 5.5$$

about the x-axis, as shown in Figure 5.30. Use this model to find the volume of a football. (In the model, x and y are measured in inches.)

SOLUTION   To find the volume of the solid of revolution, use the Disk Method.

$$\text{Volume} = \pi \int_{-5.5}^{5.5} [f(x)]^2 \, dx \qquad \text{Disk Method}$$

$$= \pi \int_{-5.5}^{5.5} (-0.0944x^2 + 3.4)^2 \, dx \qquad \text{Substitute for } f(x).$$

$$\approx 232 \text{ cubic inches} \qquad \text{Volume}$$

So, the volume of the football is about 232 cubic inches.

American football, in its modern form, is a twentieth-century invention. In the 1800s a rough, soccer-like game was played with a "round football." In 1905, at the request of President Theodore Roosevelt, the Intercollegiate Athletic Association (which became the NCAA in 1910) was formed. With the introduction of the forward pass in 1906, the shape of the ball was altered to make it easier to grip.

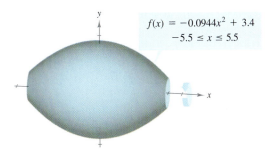

$$f(x) = -0.0944x^2 + 3.4$$
$$-5.5 \le x \le 5.5$$

FIGURE 5.30   A football-shaped solid is formed by revolving a parabolic segment about the x-axis.

TRY IT 3

A soup bowl can be modeled as a solid of revolution formed by revolving the graph of

$$f(x) = \sqrt{x} + 1, \quad 0 \le x \le 3$$

about the x-axis. Use this model, where x and y are measured in inches, to find the volume of the soup bowl.

TAKE ANOTHER LOOK

Testing the Reasonableness of an Answer

A football is about 11 inches long and has a diameter of about 7 inches. In Example 3, the volume of a football was approximated to be 232 cubic inches. Explain how you can determine whether this answer is reasonable.

The following warm-up exercises involve skills that were covered in earlier sections. You will use these skills in the exercise set for this section.

In Exercises 1–6, solve for x.

1. $x^2 = 2x$

2. $-x^2 + 4x = x^2$

3. $x = -x^3 + 5x$

4. $x^2 + 1 = x + 3$

5. $-x + 4 = \sqrt{4x - x^2}$

6. $\sqrt{x - 1} = \frac{1}{2}(x - 1)$

In Exercises 7–10, evaluate the integral.

7. $\displaystyle\int_0^2 2e^{2x}\, dx$

8. $\displaystyle\int_{-1}^3 \frac{2x + 1}{x^2 + x + 2}\, dx$

9. $\displaystyle\int_0^2 x\sqrt{x^2 + 1}\, dx$

10. $\displaystyle\int_1^5 \frac{(\ln x)^2}{x}\, dx$

EXERCISES 5.7

In Exercises 1–16, find the volume of the solid formed by revolving the region bounded by the graph(s) of the equation(s) about the x-axis.

1. $y = \sqrt{4 - x^2}$

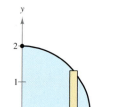

2. $y = x^2$

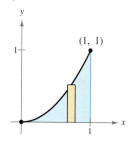

3. $y = \sqrt{x}$

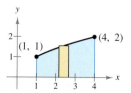

4. $y = \sqrt{4 - x^2}$

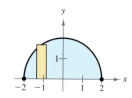

5. $y = 4 - x^2$, $y = 0$

6. $y = x$, $y = 0$, $x = 4$

7. $y = 1 - \frac{1}{4}x^2$, $y = 0$

8. $y = x^2 + 1$, $y = 5$

9. $y = -x + 1$, $y = 0$, $x = 0$

10. $y = x$, $y = e^{x-1}$, $x = 0$

11. $y = \sqrt{x} + 1$, $y = 0$, $x = 0$, $x = 9$

12. $y = \sqrt{x}$, $y = 0$, $x = 4$

13. $y = 2x^2$, $y = 0$, $x = 2$

14. $y = \dfrac{1}{x}$, $y = 0$, $x = 1$, $x = 3$

15. $y = e^x$, $y = 0$, $x = 0$, $x = 1$

16. $y = x^2$, $y = 4x - x^2$

In Exercises 17–24, find the volume of the solid formed by revolving the region bounded by the graph(s) of the equation(s) about the y-axis.

17. $y = x^2$, $y = 4$, $0 \le x \le 2$

18. $y = \sqrt{16 - x^2}$, $y = 0$, $0 \le x \le 4$

19. $x = 1 - \frac{1}{2}y$, $x = 0$, $y = 0$

20. $x = y(y - 1)$, $x = 0$

21. $y = x^{2/3}$

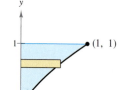

22. $x = -y^2 + 4y$

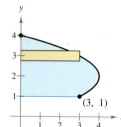

23. $y = \sqrt{4 - x}$, $y = 0$, $x = 0$

24. $y = 4$, $y = 0$, $x = 2$, $x = 0$

25. **Volume** The line segment from $(0, 0)$ to $(6, 3)$ is revolved about the x-axis to form a cone. What is the volume of the cone?

26. **Volume** The line segment from $(0, 0)$ to $(4, 2)$ is revolved about the y-axis to form a cone. What is the volume of the cone?

27. **Volume** Use the Disk Method to verify that the volume of a right circular cone is $\frac{1}{3}\pi r^2 h$, where r is the radius of the base and h is the height.

28. **Volume** Use the Disk Method to verify that the volume of a sphere of radius r is $\frac{4}{3}\pi r^3$.

29. **Volume** The right half of the ellipse

$$9x^2 + 25y^2 = 225$$

is revolved about the y-axis to form an oblate spheroid (shaped like an M&M candy). Find the volume of the spheroid.

30. **Volume** The upper half of the ellipse

$$9x^2 + 16y^2 = 144$$

is revolved about the x-axis to form a prolate spheroid (shaped like a football). Find the volume of the spheroid.

31. **Volume** A tank on the wing of a jet airplane is modeled by revolving the region bounded by the graph of $y = \frac{1}{8}x^2\sqrt{2 - x}$ and the x-axis about the x-axis, where x and y are measured in meters (see figure). Find the volume of the tank.

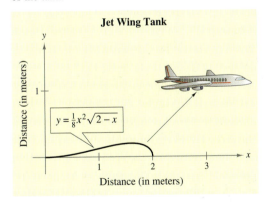

Jet Wing Tank

$y = \frac{1}{8}x^2\sqrt{2 - x}$

Distance (in meters)

Distance (in meters)

32. **Volume** A soup bowl can be modeled as a solid of revolution formed by revolving the graph of

$$y = \sqrt{\frac{x}{2} + 1}, \quad 0 \le x \le 4$$

about the x-axis. Use this model, where x and y are measured in inches, to find the volume of the soup bowl.

33. **Biology** A pond is to be stocked with a species of fish. The food supply in 500 cubic feet of pond water can adequately support one fish. The pond is nearly circular, is 20 feet deep at its center, and has a radius of 200 feet. The bottom of the pond can be modeled by

$$y = 20[(0.005x)^2 - 1].$$

(a) How much water is in the pond?

(b) How many fish can the pond support?

34. **Modeling a Body of Water** A pond is approximately circular, with a diameter of 400 feet (see figure). Starting at the center, the depth of the water is measured every 25 feet and recorded in the table.

x	0	25	50	75	100
Depth	20	19	19	17	15

x	125	150	175	200
Depth	14	10	6	0

(a) Use a graphing utility to plot the depths and graph the model of the pond's depth, $y = 20 - 0.00045x^2$.

(b) Use the model in part (a) to find the pond's volume.

(c) Use the result of part (b) to approximate the number of gallons of water in the pond (1 ft³ ≈ 7.48 gal).

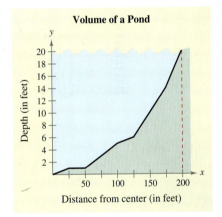

Volume of a Pond

Depth (in feet)

Distance from center (in feet)

In Exercises 35 and 36, use a program similar to the one on page 366 to approximate the volume of a solid generated by revolving the region bounded by the graphs of the equations about the x-axis.

35. $y = \sqrt[3]{x + 1}$, $y = 0$, $x = 0$, $x = 7$

36. $y = \dfrac{10}{x^2 + 1}$, $y = 0$, $x = 0$, $x = 3$

ALGEBRA REVIEW

"Unsimplifying" an Algebraic Expression

In algebra it is often helpful to write an expression in simplest form. In this chapter, you have seen that the reverse is often true in integration. That is, to fit an integrand to an integration formula, it often helps to "unsimplify" the expression. To do this, you use the same algebraic rules, but your goal is different. Here are some examples.

EXAMPLE 1   Rewriting an Algebraic Expression

Rewrite each algebraic expression as indicated in the example.

(a) $\dfrac{x+1}{\sqrt{x}}$ Example 6, page 322   (b) $x(3-4x^2)^2$ Example 2, page 331

(c) $7x^2\sqrt{x^3+1}$ Example 4, page 332   (d) $5xe^{-x^2}$ Example 3, page 338

SOLUTION

(a) $\dfrac{x+1}{\sqrt{x}} = \dfrac{x}{\sqrt{x}} + \dfrac{1}{\sqrt{x}}$ Example 6, page 322
Rewrite as two fractions.

$\qquad = \dfrac{x^1}{x^{1/2}} + \dfrac{1}{x^{1/2}}$ Rewrite with rational exponents.

$\qquad = x^{1-1/2} + x^{-1/2}$ Properties of exponents

$\qquad = x^{1/2} + x^{-1/2}$ Simplify exponent.

(b) $x(3-4x^2)^2 = \dfrac{-8}{-8}x(3-4x^2)^2$ Example 2, page 331
Multiply and divide by -8.

$\qquad = \left(-\dfrac{1}{8}\right)(-8)x(3-4x^2)^2$ Regroup.

$\qquad = \left(-\dfrac{1}{8}\right)(3-4x^2)^2(-8x)$ Regroup.

(c) $7x^2\sqrt{x^3+1} = 7x^2(x^3+1)^{1/2}$ Example 4, page 332
Rewrite with rational exponent.

$\qquad = \dfrac{3}{3}(7x^2)(x^3+1)^{1/2}$ Multiply and divide by 3.

$\qquad = \dfrac{7}{3}(3x^2)(x^3+1)^{1/2}$ Regroup.

$\qquad = \dfrac{7}{3}(x^3+1)^{1/2}(3x^2)$ Regroup.

(d) $5xe^{-x^2} = \dfrac{-2}{-2}(5x)e^{-x^2}$ Example 3, page 338
Multiply and divide by -2.

$\qquad = \left(-\dfrac{5}{2}\right)(-2x)e^{-x^2}$ Regroup.

$\qquad = \left(-\dfrac{5}{2}\right)e^{-x^2}(-2x)$ Regroup.

| EXAMPLE 2 | **Rewriting an Algebraic Expression** |

Rewrite each algebraic expression.

(a) $\dfrac{3x^2 + 2x - 1}{x^2}$ (b) $\dfrac{1}{1 + e^{-x}}$

(c) $\dfrac{x^2 + x + 1}{x - 1}$ (d) $\dfrac{x^2 + 6x + 1}{x^2 + 1}$

SOLUTION

(a) $\dfrac{3x^2 + 2x - 1}{x^2} = \dfrac{3x^2}{x^2} + \dfrac{2x}{x^2} - \dfrac{1}{x^2}$ Example 7(a), page 341
Rewrite as separate fractions.

$= 3 + \dfrac{2}{x} - x^{-2}$ Properties of exponents.

$= 3 + 2\left(\dfrac{1}{x}\right) - x^{-2}$ Regroup.

(b) $\dfrac{1}{1 + e^{-x}} = \left(\dfrac{e^x}{e^x}\right)\dfrac{1}{1 + e^{-x}}$ Example 7(b), page 341
Multiply and divide by e^x.

$= \dfrac{e^x}{e^x + e^x(e^{-x})}$ Multiply.

$= \dfrac{e^x}{e^x + e^{x-x}}$ Property of exponents

$= \dfrac{e^x}{e^x + e^0}$ Simplify exponent.

$= \dfrac{e^x}{e^x + 1}$ $e^0 = 1$

(c) $\dfrac{x^2 + x + 1}{x - 1} = x + 2 + \dfrac{3}{x - 1}$ Example 7(c), page 341
Use long division as shown below.

$$
\begin{array}{r}
x + 2 \\
x - 1 \overline{\smash{)}\, x^2 + x + 1} \\
\underline{x^2 - x} \\
2x + 1 \\
\underline{2x - 2} \\
3
\end{array}
$$

(d) $\dfrac{x^2 + 6x + 1}{x^2 + 1} = 1 + \dfrac{6x}{x^2 + 1}$ Bottom of page 340.
Use long division as shown below.

$$
\begin{array}{r}
1 \\
x^2 + 1 \overline{\smash{)}\, x^2 + 6x + 1} \\
\underline{x^2 + 1} \\
6x
\end{array}
$$

5 CHAPTER SUMMARY AND STUDY STRATEGIES

After studying this chapter, you should have acquired the following skills. The exercise numbers are keyed to the Review Exercises that begin on page 382. Answers to odd-numbered Review Exercises are given in the back of the text.*

■ Use basic integration rules to find indefinite integrals. *(Section 5.1)*   *Review Exercises 1–10*

$$\int k \, dx = kx + C \qquad\qquad \int [f(x) - g(x)] \, dx = \int f(x) \, dx - \int g(x) \, dx$$

$$\int kf(x) \, dx = k \int f(x) \, dx \qquad\qquad \int x^n \, dx = \frac{x^{n+1}}{n+1} + C, \quad n \neq -1$$

$$\int [f(x) + g(x)] \, dx = \int f(x) \, dx + \int g(x) \, dx$$

■ Use initial conditions to find particular solutions of indefinite integrals. *(Section 5.1)*   *Review Exercises 11–14*

■ Use antiderivatives to solve real-life problems. *(Section 5.1)*   *Review Exercises 15, 16*

■ Use the General Power Rule to find indefinite integrals. *(Section 5.2)*   *Review Exercises 17–24*

$$\int u^n \frac{du}{dx} \, dx = \int u^n \, du = \frac{u^{n+1}}{n+1} + C, \quad n \neq -1$$

■ Use the General Power Rule to solve real-life problems. *(Section 5.2)*   *Review Exercises 25, 26*

■ Use the Exponential and Log Rules to find indefinite integrals. *(Section 5.3)*   *Review Exercises 27–32*

$$\int e^x \, dx = e^x + C \qquad\qquad \int \frac{1}{x} \, dx = \ln|x| + C$$

$$\int e^u \frac{du}{dx} \, dx = \int e^u \, du = e^u + C \qquad\qquad \int \frac{du/dx}{u} \, dx = \int \frac{1}{u} \, du = \ln|u| + C$$

■ Use a symbolic integration utility to find indefinite integrals. *(Section 5.3)*   *Review Exercises 33, 34*

■ Find the areas of regions bounded by the graph of a function and the x-axis. *(Section 5.4)*   *Review Exercises 35–40*

■ Use the Fundamental Theorem of Calculus to evaluate definite integrals. *(Section 5.4)*   *Review Exercises 41–52*

$$\int_a^b f(x) \, dx = F(x) \Big]_a^b = F(b) - F(a), \quad \text{where} \quad F'(x) = f(x)$$

■ Use definite integrals to solve marginal analysis problems. *(Section 5.4)*   *Review Exercises 53, 54*

■ Find average values of functions over closed intervals. *(Section 5.4)*   *Review Exercises 55–58*

$$\text{Average value} = \frac{1}{b-a} \int_a^b f(x) \, dx$$

■ Use average values to solve real-life problems. *(Section 5.4)*   *Review Exercises 59–62*

■ Find amounts of annuities. *(Section 5.4)*   *Review Exercises 63, 64*

■ Use properties of even and odd functions to help evaluate definite integrals. *(Section 5.4)*   *Review Exercises 65–68*

    Even function: $f(-x) = f(x)$ *Odd function:* $f(-x) = -f(x)$

■ Find areas of regions bounded by two (or more) graphs. *(Section 5.5)*   *Review Exercises 69–76*

$$A = \int_a^b [f(x) - g(x)] \, dx$$

■ Find consumer and producer surpluses. *(Section 5.5)*   *Review Exercises 77, 78*

■ Use the areas of regions bounded by two graphs to solve real-life problems.   *Review Exercises 79–82*
(Section 5.5)

■ Use the Midpoint Rule to approximate values of definite integrals. *(Section 5.6)*   *Review Exercises 83–86*

$$\int_a^b f(x) \, dx \approx \frac{b-a}{n} [f(x_1) + f(x_2) + f(x_3) + \cdots + f(x_n)]$$

■ Use the Disk Method to find volumes of solids of revolution. *(Section 5.7)*   *Review Exercises 87–90*

$$\text{Volume} = \pi \int_a^b [f(x)]^2 \, dx$$

■ Use the Washer Method to find volumes of solids of revolution with holes.   *Review Exercises 91–94*
(Section 5.7)

$$\text{Volume} = \pi \int_a^b \{[f(x)]^2 - [g(x)]^2\} \, dx$$

■ Use solids of revolution to solve real-life problems. *(Section 5.7)*   *Review Exercises 95, 96*

■ ***Indefinite and Definite Integrals***   When evaluating integrals, remember that an indefinite integral is a *family of antiderivatives*, each differing by a constant C, whereas a definite integral is a number.

■ ***Checking Antiderivatives by Differentiating***   When finding an antiderivative, remember that you can check your result by differentiating. For example, you can check that the antiderivative

$$\int (3x^3 - 4x) \, dx = \frac{3}{4}x^4 - 2x^2 + C$$

is correct by differentiating to obtain

$$\frac{d}{dx}\left[\frac{3}{4}x^4 - 2x^2 + C\right] = 3x^3 - 4x.$$

Because the derivative is equal to the original integrand, you know that the antiderivative is correct.

■ ***Grouping Symbols and the Fundamental Theorem***   When using the Fundamental Theorem of Calculus to evaluate a definite integral, you can avoid sign errors by using grouping symbols. Here is an example.

$$\int_1^3 (x^3 - 9x) \, dx = \left[\frac{x^4}{4} - \frac{9x^2}{2}\right]_1^3 = \left[\frac{3^4}{4} - \frac{9(3^2)}{2}\right] - \left[\frac{1^4}{4} - \frac{9(1^2)}{2}\right] = \frac{81}{4} - \frac{81}{2} - \frac{1}{4} + \frac{9}{2} = -16$$

Study Tools   *Additional resources that accompany this chapter*

■ **Algebra Review**   (pages 378 and 379)
■ **Chapter Summary and Study Strategies** (pages 380 and 381)
■ **Review Exercises**   (pages 382–385)
■ **Sample Post-Graduation Exam Questions**   (page 386)

■ **Web Exercises**   (page 328, Exercise 83; page 364, Exercise 60)
■ **Student Solutions Guide**
■ **HM mathSpace® Student CD-ROM**
■ **Graphing Technology Guide** (*math.college.hmco.com/students*)

5   CHAPTER REVIEW EXERCISES

In Exercises 1–10, find the indefinite integral.

1. $\displaystyle\int 16\,dx$

2. $\displaystyle\int \tfrac{3}{5}x\,dx$

3. $\displaystyle\int (2x^2 + 5x)\,dx$

4. $\displaystyle\int (5 - 6x^2)\,dx$

5. $\displaystyle\int \frac{2}{3\sqrt[3]{x}}\,dx$

6. $\displaystyle\int 6x^2\sqrt{x}\,dx$

7. $\displaystyle\int \left(\sqrt[3]{x^4} + 3x\right)dx$

8. $\displaystyle\int \left(\frac{4}{\sqrt{x}} + \sqrt{x}\right)dx$

9. $\displaystyle\int \frac{2x^4 - 1}{\sqrt{x}}\,dx$

10. $\displaystyle\int \frac{1 - 3x}{x^2}\,dx$

In Exercises 11–14, find the particular solution, $y = f(x)$, that satisfies the conditions.

11. $f'(x) = 3x + 1, \quad f(2) = 6$

12. $f'(x) = x^{-1/3} - 1, \quad f(8) = 4$

13. $f''(x) = 2x^2, \quad f'(3) = 10, \quad f(3) = 6$

14. $f''(x) = \dfrac{6}{\sqrt{x}} + 3, \quad f'(1) = 12, \quad f(4) = 56$

15. Vertical Motion   An object is projected upward from the ground with an initial velocity of 80 feet per second.

(a) How long does it take the object to rise to its maximum height?

(b) What is the maximum height?

(c) When is the velocity of the object half of its initial velocity?

(d) What is the height of the object when its velocity is one-half the initial velocity?

16. Revenue   The weekly revenue for a new product has been increasing. The rate of change of the revenue can be modeled by

$$\frac{dR}{dt} = 0.675t^{3/2}, \quad 0 \le t \le 225$$

where t is the time (in weeks). When $t = 0$, $R = 0$.

(a) Find a model for the revenue function.

(b) When will the weekly revenue be $27,000?

In Exercises 17–24, find the indefinite integral.

17. $\displaystyle\int (1 + 5x)^2\,dx$

18. $\displaystyle\int (x - 6)^{4/3}\,dx$

19. $\displaystyle\int \frac{1}{\sqrt{5x - 1}}\,dx$

20. $\displaystyle\int \frac{4x}{\sqrt{1 - 3x^2}}\,dx$

21. $\displaystyle\int x(1 - 4x^2)\,dx$

22. $\displaystyle\int \frac{x^2}{(x^3 - 4)^2}\,dx$

23. $\displaystyle\int (x^4 - 2x)(2x^3 - 1)\,dx$

24. $\displaystyle\int \frac{\sqrt{x}}{(1 - x^{3/2})^3}\,dx$

25. Production   The output P (in board-feet) of a small sawmill changes according to the model

$$\frac{dP}{dt} = 2t(0.001t^2 + 0.5)^{1/4}, \quad 0 \le t \le 40$$

where t is measured in hours. Find the numbers of board-feet produced in (a) 6 hours and (b) 12 hours.

26. Cost   The marginal cost for a catering service to cater to x people can be modeled by

$$\frac{dC}{dx} = \frac{5x}{\sqrt{x^2 + 1000}}.$$

When $x = 225$, the cost is $1136.06. Find the costs of catering to (a) 500 people and (b) 1000 people.

In Exercises 27–32, find the indefinite integral.

27. $\displaystyle\int 3e^{-3x}\,dx$

28. $\displaystyle\int (2t - 1)e^{t^2 - t}\,dt$

29. $\displaystyle\int (x - 1)e^{x^2 - 2x}\,dx$

30. $\displaystyle\int \frac{4}{6x - 1}\,dx$

31. $\displaystyle\int \frac{x^2}{1 - x^3}\,dx$

32. $\displaystyle\int \frac{x - 4}{x^2 - 8x}\,dx$

In Exercises 33 and 34, use a symbolic integration utility to find the indefinite integral.

33. $\displaystyle\int \frac{(\sqrt{x} + 1)^2}{\sqrt{x}}\,dx$

34. $\displaystyle\int \frac{e^{5x}}{5 + e^{5x}}\,dx$

In Exercises 35–40, find the area of the region.

35. $f(x) = 4 - 2x$

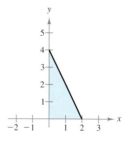

36. $f(x) = 4 - x^2$

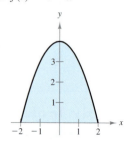

Fi

37. $f(y) = (y - 2)^2$

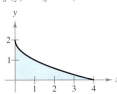

38. $f(x) = \sqrt{9 - x^2}$

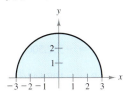

39. $f(x) = \dfrac{2}{x + 1}$

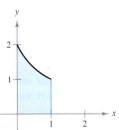

40. $f(x) = 2xe^{x^2 - 4}$

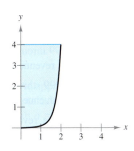

In Exerc
imate t
or com
Compa

83. $\displaystyle\int_0^2$

85. $\displaystyle\int_0^1$

In Exerc
solid of
x-axis.

87. $y =$

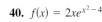

In Exercises 41–52, use the Fundamental Theorem of Calculus to evaluate the definite integral.

41. $\displaystyle\int_0^4 (2 + x)\, dx$

42. $\displaystyle\int_{-1}^1 (t^2 + 2)\, dt$

43. $\displaystyle\int_{-1}^1 (4t^3 - 2t)\, dt$

44. $\displaystyle\int_1^4 2x\sqrt{x}\, dx$

45. $\displaystyle\int_0^3 \frac{1}{\sqrt{1 + x}}\, dx$

46. $\displaystyle\int_3^6 \frac{x}{3\sqrt{x^2 - 8}}\, dx$

47. $\displaystyle\int_1^2 \left(\frac{1}{x^2} - \frac{1}{x^3} \right) dx$

48. $\displaystyle\int_0^1 x^2(x^3 + 1)^3\, dx$

49. $\displaystyle\int_1^3 \frac{(3 + \ln x)}{x}\, dx$

50. $\displaystyle\int_0^{\ln 5} e^{x/5}\, dx$

51. $\displaystyle\int_{-1}^1 3xe^{x^2 - 1}\, dx$

52. $\displaystyle\int_1^3 \frac{1}{x(\ln x + 2)^2}\, dx$

53. Cost The marginal cost of serving a typical additional client at a law firm can be modeled by

$$\frac{dC}{dx} = 675 + 0.5x$$

where x is the number of clients. How does the cost C change when x increases from 50 to 51 clients?

54. Profit The marginal profit obtained by selling x dollars of automobile insurance can be modeled by

$$\frac{dP}{dx} = 0.4\left(1 - \frac{5000}{x} \right), \quad x \geq 5000.$$

Find the change in the profit when x increases from $75,000 to $100,000.

89. $y =$

In Exercises 55–58, find the average value of the function on the closed interval. Then find all x-values in the interval for which the function is equal to its average value.

55. $f(x) = \dfrac{4}{\sqrt{x - 1}}$, $[5, 10]$

56. $f(x) = \dfrac{20 \ln x}{x}$, $[2, 10]$

57. $f(x) = e^{5 - x}$, $[2, 5]$

58. $f(x) = x^3$, $[0, 2]$

59. Compound Interest An interest-bearing checking account yields 4% interest compounded continuously. If you deposit $500 in such an account, and never write checks, what will the average value of the account be over a period of 2 years? Explain your reasoning.

60. Consumer Awareness Suppose that the price p of gasoline can be modeled by

$$p = 0.12 + 0.203t - 0.0057t^2$$

where $t = 5$ corresponds to January 1, 1995. Find the cost of gasoline for an automobile that is driven 15,000 miles per year and gets 33 miles per gallon from 1995 through 2000.

61. Consumer Trends The rates of change of beef prices (in dollars per pound) in the United States from 1998 through 2002 can be modeled by

$$\frac{dB}{dt} = -0.0486t + 0.564$$

where t is the year, with $t = 8$ corresponding to 1998. The price of 1 pound of beef in 2000 was $1.63. *(Source: U.S. Bureau of Labor Statistics)*

(a) Find the price function in terms of the year.

(b) If the price of beef per pound continues to change at this rate, in what year will the price of beef per pound surpass $2.00? Explain your reasoning.

62. Medical Science The volume V (in liters) of air in the lungs during a five-second respiratory cycle is approximated by the model

$$V = 0.1729t + 0.1522t^2 - 0.0374t^3$$

where t is time in seconds.

(a) Use a graphing utility to graph the equation on the interval $[0, 5]$.

(b) Determine the intervals on which the function is increasing and decreasing.

(c) Determine the maximum volume during the respiratory cycle.

(d) Determine the average volume of air in the lungs during one cycle.

(e) Briefly explain your results for parts (a) through (d).

Annuity In Exercises 63 and 64, find the amount of an annuity with income function $c(t)$, interest rate r, and term T.

63. $c(t) = \$3000$, $r = 6\%$, $T = 5$ years

64. $c(t) = \$1200$, $r = 7\%$, $T = 8$ years

5 SAMPLE POST-GRADUATION EXAM QUESTIONS

| CPA |
| GMAT |
| GRE |
| Actuarial |
| CLAST |

The following questions represent the types of questions that appear on certified public accountant (CPA) exams, Graduate Management Admission Tests (GMAT), Graduate Records Exams (GRE), actuarial exams, and College-Level Academic Skills Tests (CLAST). The answers to the questions are given in the back of the book.

For Questions 1–3, use the data shown in the table.

Table for 1–3

Number of students	Number of correct answers
12	45 to 50
11	40 to 44
14	35 to 39
10	30 to 34
7	0 to 29

1. To pass a 50-question exam, a student must correctly answer 75% of the questions. What is the maximum number of students that could have passed the exam?

(a) 23　(b) 47　(c) 30　(d) 37

2. What percent of the class answered 40 or more questions correctly?

(a) 69%　(b) 43%　(c) 22%　(d) 31%

3. The number of students who answered 35 to 39 questions correctly is y times the number who answered 29 or fewer correctly, where y is

(a) $\frac{1}{2}$　(b) 1　(c) 2　(d) $\frac{7}{5}$

4. For which of the following statements is $x = -2$ a solution?

I. $x^2 + 4x + 4 \leq 0$

II. $|3x + 6| = 0$

III. $x^2 + 7x + 10 > 0$

(a) I, II, and III　(b) I and II　(c) II and III　(d) I and III

5. The area of the shaded region in the graph is A square units, with A equal to

(a) $\frac{50}{3}$　(b) $\frac{25}{3}$　(c) 50　(d) $\frac{4}{3}$

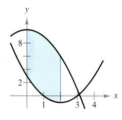

For Questions 6–8, use the data shown below.

56, 58, 54, 54, 59, 56, 55, 57, 56, 62

6. The mean of the data is

(a) 55.5　(b) 56.7　(c) 56.5　(d) 56

7. The median of the data is

(a) 55.5　(b) 56.7　(c) 56.5　(d) 56

8. The mode of the data is

(a) 55.5　(b) 56.7　(c) 56.5　(d) 56

9. If the number 52 were added to these data, which of these would change?

(a) Mean　(b) Median　(c) Mode　(d) All of these

Techniques of Integration

Photodisc/Getty Images

Integration is used to find the present value of an income over several years.

WHAT YOU SHOULD LEARN:

- How to find indefinite and definite integrals using integration by substitution

- How to evaluate integrals by parts, by using partial fractions, and by using a table of integrals

- How to use the Trapezoidal Rule and Simpson's Rule to approximate definite integrals

- How to evaluate improper integrals with infinite limits of integration and with infinite integrands

WHY YOU SHOULD LEARN IT:

The various integration techniques can be used in many applications in real life, as demonstrated by the examples below, which represent a small sample of the applications in this chapter.

- Present and Future Value, Exercises 63–74 on page 405

- Health: Epidemic, Exercise 54 on page 414

- Consumer and Producer Surpluses, Exercise 64 on page 425

- Surveying, Exercises 29 and 30 on page 434

- Capitalized Cost, Exercises 41 and 42 on page 445

PREREQUISITE REVIEW 6.1

The following warm-up exercises involve skills that were covered in earlier sections. You will use these skills in the exercise set for this section.

In Exercises 1–8, evaluate the indefinite integral.

1. $\int 5\,dx$

2. $\int \frac{1}{3}\,dx$

3. $\int x^{3/2}\,dx$

4. $\int x^{2/3}\,dx$

5. $\int 2x(x^2 + 1)^3\,dx$

6. $\int 3x^2(x^3 - 1)^2\,dx$

7. $\int 6e^{6x}\,dx$

8. $\int \frac{2}{2x + 1}\,dx$

In Exercises 9–12, simplify the expression.

9. $2x(x - 1)^2 + x(x - 1)$

10. $6x(x + 4)^3 - 3x^2(x + 4)^2$

11. $3(x + 7)^{1/2} - 2x(x + 7)^{-1/2}$

12. $(x + 5)^{1/3} - 5(x + 5)^{-2/3}$

EXERCISES 6.1

In Exercises 1–38, find the indefinite integral.

1. $\int (x - 2)^4\,dx$

2. $\int (x + 5)^{3/2}\,dx$

3. $\int \frac{2}{(t - 9)^2}\,dt$

4. $\int \frac{4}{(1 - t)^3}\,dt$

5. $\int \frac{2t - 1}{t^2 - t + 2}\,dt$

6. $\int \frac{2y^3}{y^4 + 1}\,dy$

7. $\int \sqrt{1 + x}\,dx$

8. $\int (3 + x)^{5/2}\,dx$

9. $\int \frac{12x + 2}{3x^2 + x}\,dx$

10. $\int \frac{6x^2 + 2}{x^3 + x}\,dx$

11. $\int \frac{1}{(5x + 1)^3}\,dx$

12. $\int \frac{1}{(3x + 1)^2}\,dx$

13. $\int \frac{1}{\sqrt{x + 1}}\,dx$

14. $\int \frac{1}{\sqrt{5x + 1}}\,dx$

15. $\int \frac{e^{3x}}{1 - e^{3x}}\,dx$

16. $\int \frac{4e^{2x}}{1 + e^{2x}}\,dx$

17. $\int \frac{2x}{e^{3x^2}}\,dx$

18. $\int \frac{e^{\sqrt{x+1}}}{\sqrt{x + 1}}\,dx$

19. $\int \frac{x^2}{x - 1}\,dx$

20. $\int \frac{2x}{x - 4}\,dx$

21. $\int x\sqrt{x^2 + 4}\,dx$

22. $\int \frac{t}{\sqrt{1 - t^2}}\,dt$

23. $\int e^{5x}\,dx$

24. $\int te^{t^2 + 1}\,dt$

25. $\int \frac{e^{-x}}{e^{-x} + 2}\,dx$

26. $\int \frac{e^x}{1 + e^x}\,dx$

27. $\int \frac{x}{(x + 1)^4}\,dx$

28. $\int \frac{x^2}{(x + 1)^3}\,dx$

29. $\int \frac{x}{(3x - 1)^2}\,dx$

30. $\int \frac{5x}{(x - 4)^3}\,dx$

31. $\int \frac{1}{\sqrt{t - 1}}\,dt$

32. $\int \frac{1}{\sqrt{x + 1}}\,dx$

33. $\int \frac{2\sqrt{t} + 1}{t}\,dt$

34. $\int \frac{6x + \sqrt{x}}{x}\,dx$

35. $\int \frac{x}{\sqrt{2x + 1}}\,dx$

36. $\int \frac{x^2}{\sqrt{x - 1}}\,dx$

37. $\int t^2\sqrt{1 - t}\,dt$

38. $\int y^2\sqrt[3]{y + 1}\,dy$

In Exercises 39–46, evaluate the definite integral.

39. $\int_0^4 \sqrt{2x + 1}\,dx$

40. $\int_2^4 \sqrt{4x + 1}\,dx$

41. $\int_0^1 3xe^{x^2}\,dx$

42. $\int_0^2 e^{-2x}\,dx$

43. $\int_0^4 \frac{x}{(x + 4)^2}\,dx$

44. $\int_0^1 x(x + 5)^4\,dx$

45. $\int_0^{0.5} x(1 - x)^3\,dx$

46. $\int_0^{0.5} x^2(1 - x)^3\,dx$

In Exercises 47–54, find the area of the region bounded by the graphs of the equations. Then use a graphing utility to graph the region and verify your answer.

47. $y = x\sqrt{x - 3}$, $y = 0$, $x = 7$

48. $y = x\sqrt{2x + 1}$, $y = 0$, $x = 4$

49. $y = x^2\sqrt{1 - x}$, $y = 0$, $x = -3$

50. $y = x^2\sqrt{x + 2}$, $y = 0$, $x = 7$

51. $y = \dfrac{x^2 - 1}{\sqrt{2x - 1}}$, $y = 0$, $x = 1$, $x = 5$

52. $y = \dfrac{2x - 1}{\sqrt{x + 3}}$, $y = 0$, $x = \dfrac{1}{2}$, $x = 6$

53. $y = x\sqrt[3]{x + 1}$, $y = 0$, $x = 0$, $x = 7$

54. $y = x\sqrt[3]{x - 2}$, $y = 0$, $x = 2$, $x = 10$

In Exercises 55–58, find the area of the region bounded by the graphs of the equations.

55. $y = -x\sqrt{x + 2}$, $y = 0$ **56.** $y = x\sqrt[3]{1 - x}$, $y = 0$

 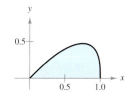

57. $y^2 = x^2(1 - x^2)$

(*Hint:* Find the area of the region bounded by $y = x\sqrt{1 - x^2}$ and $y = 0$. Then multiply by 4.)

58. $y = 1/(1 + \sqrt{x})$, $y = 0$, $x = 0$, $x = 4$

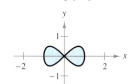 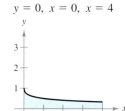

In Exercises 59 and 60, find the volume of the solid generated by revolving the region bounded by the graph(s) of the equation(s) about the *x*-axis.

59. $y = x\sqrt{1 - x^2}$

60. $y = \sqrt{x}(1 - x)^2$, $y = 0$

In Exercises 61 and 62, find the average amount by which the function *f* exceeds the function *g* on the interval.

61. $f(x) = \dfrac{1}{x + 1}$, $g(x) = \dfrac{x}{(x + 1)^2}$, $[0, 1]$

62. $f(x) = x\sqrt{4x + 1}$, $g(x) = 2\sqrt{x^3}$, $[0, 2]$

63. *Probability* The probability of recall in an experiment is modeled by

$$P(a \le x \le b) = \int_a^b \frac{15}{4}x\sqrt{1 - x}\,dx$$

where *x* is the percent of recall (see figure).

(a) What is the probability of recalling between 40% and 80%?

(b) What is the median percent recall? That is, for what value of *b* is $P(0 \le x \le b) = 0.5$?

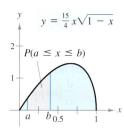   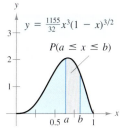

Figure for 63   Figure for 64

64. *Probability* The probability of finding between *a* and *b* percent iron in ore samples is modeled by

$$P(a \le x \le b) = \int_a^b \frac{1155}{32}x^3(1 - x)^{3/2}\,dx$$

(see figure). Find the probabilities that a sample will contain between (a) 0% and 25% and (b) 50% and 100% iron.

65. *Meteorology* During a two-week period in March in a small town near Lake Erie, the measurable snowfall *S* (in inches) on the ground can be modeled by

$$S(t) = t\sqrt{14 - t}, \quad 0 \le t \le 14$$

where *t* represents the day.

 (a) Use a graphing utility to graph the function.

(b) Find the average amount of snow on the ground during the two-week period.

(c) Find the total snowfall over the two-week period.

66. *Revenue* A company sells a seasonal product that generates a daily revenue *R* (in dollars per day) modeled by

$$R = 0.06t^2(365 - t)^{1/2} + 1250, \quad 0 \le t \le 365$$

where *t* represents the day.

(a) Find the average daily revenue over a period of 1 year.

(b) Describe a product whose seasonal sales pattern resembles the model. Explain your reasoning.

 In Exercises 67 and 68, use a program similar to the Midpoint Rule program on page 366 with $n = 10$ to approximate the area of the region bounded by the graph(s) of the equation(s).

67. $y = \sqrt[3]{x}\sqrt{4 - x}$, $y = 0$

68. $y^2 = x^2(1 - x^2)$

In Exercises 39–42, find the indefinite integral using each speci-fied method. Then write a brief statement explaining which method you prefer.

39. $\int 2x\sqrt{2x-3}\,dx$

    (a) By parts, letting $dv = \sqrt{2x-3}\,dx$

    (b) By substitution, letting $u = \sqrt{2x-3}$

40. $\int x\sqrt{4+x}\,dx$

    (a) By parts, letting $dv = \sqrt{4+x}\,dx$

    (b) By substitution, letting $u = \sqrt{4+x}$

41. $\int \dfrac{x}{\sqrt{4+5x}}\,dx$

    (a) By parts, letting $dv = \dfrac{1}{\sqrt{4+5x}}\,dx$

    (b) By substitution, letting $u = \sqrt{4+5x}$

42. $\int x\sqrt{4-x}\,dx$

    (a) By parts, letting $dv = \sqrt{4-x}\,dx$

    (b) By substitution, letting $u = \sqrt{4-x}$

In Exercises 43 and 44, use integration by parts to verify the formula.

43. $\int x^n \ln x\,dx = \dfrac{x^{n+1}}{(n+1)^2}\left[-1 + (n+1)\ln x\right] + C,$

    $n \neq -1$

44. $\int x^n e^{ax}\,dx = \dfrac{x^n e^{ax}}{a} - \dfrac{n}{a}\int x^{n-1} e^{ax}\,dx$

In Exercises 45–48, use the results of Exercises 43 and 44 to find the indefinite integral.

45. $\int x^2 e^{5x}\,dx$

46. $\int xe^{-3x}\,dx$

47. $\int x^{-2}\ln x\,dx$

48. $\int x^{1/2}\ln x\,dx$

In Exercises 49–52, find the area of the region bounded by the graphs of the given equations.

49. $y = xe^{-x},\ y = 0,\ x = 4$

50. $y = \frac{1}{9}xe^{-x/3},\ y = 0,\ x = 0,\ x = 3$

51. $y = x\ln x,\ y = 0,\ x = e$

52. $y = x^{-3}\ln x,\ y = 0,\ x = e$

53. Given the region bounded by the graphs of $y = 2\ln x$, $y = 0$, and $x = e$ (see figure), find

    (a) the area of the region.

    (b) the volume of the solid generated by revolving the region about the x-axis.

Figure for 53

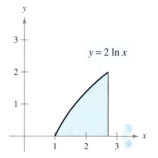

54. Given the region bounded by the graphs of $y = xe^x$, $y = 0$, $x = 0$, and $x = 1$ (see figure), find

    (a) the area of the region.

    (b) the volume of the solid generated by revolving the region about the x-axis.

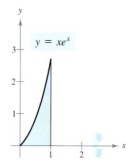

In Exercises 55–58, use a symbolic integration utility to evaluate the integral.

55. $\displaystyle\int_0^2 t^3 e^{-4t}\,dt$

56. $\displaystyle\int_1^4 \ln x(x^2 + 4)\,dx$

57. $\displaystyle\int_0^5 x^4(25 - x^2)^{3/2}\,dx$

58. $\displaystyle\int_1^e x^9 \ln x\,dx$

59. *Demand* A manufacturing company forecasts that the demand x (in units per year) for its product over the next 10 years can be modeled by $x = 500(20 + te^{-0.1t})$ for $0 \le t \le 10$, where t is the time in years.

    (a) Use a graphing utility to decide whether the company is forecasting an increase or a decrease in demand over the decade.

    (b) According to the model, what is the total demand over the next 10 years?

    (c) Find the average annual demand during the 10-year period.

60. *Capital Campaign* The board of trustees of a college is planning a five-year capital gifts campaign to raise money for the college. The goal is to have an annual gift income I that is modeled by $I = 2000(375 + 68te^{-0.2t})$ for $0 \le t \le 5$, where t is the time in years.

(a) Use a graphing utility to decide whether the board of trustees expects the gift income to increase or decrease over the five-year period.

(b) Find the expected total gift income over the five-year period.

(c) Determine the average annual gift income over the five-year period. Compare the result with the income given when $t = 3$.

61. *Learning Theory* A model for the ability M of a child to memorize, measured on a scale from 0 to 10, is

$$M = 1 + 1.6t \ln t, \qquad 0 < t \le 4$$

where t is the child's age in years. Find the average value of this model between

(a) the child's first and second birthdays.

(b) the child's third and fourth birthdays.

62. *Revenue* A company sells a seasonal product. The revenue R (in dollars per year) generated by sales of the product can be modeled by

$$R = 410.5t^2 e^{-t/30} + 25,000, \qquad 0 \le t \le 365$$

where t is the time in days.

(a) Find the average daily receipts during the first quarter, which is given by $0 \le t \le 90$.

(b) Find the average daily receipts during the fourth quarter, which is given by $274 \le t \le 365$.

(c) Find the total daily receipts during the year.

Present Value In Exercises 63–68, find the present value of the income c (measured in dollars) over t_1 years at the given annual inflation rate r.

63. $c = 5000, \ r = 5\%, \ t_1 = 4$ years

64. $c = 450, \ r = 4\%, \ t_1 = 10$ years

65. $c = 150,000 + 2500t, \ r = 4\%, \ t_1 = 10$ years

66. $c = 30,000 + 500t, \ r = 7\%, \ t_1 = 6$ years

67. $c = 1000 + 50e^{t/2}, \ r = 6\%, \ t_1 = 4$ years

68. $c = 5000 + 25te^{t/10}, \ r = 6\%, \ t_1 = 10$ years

69. *Present Value* A company expects its income c during the next 4 years to be modeled by

$$c = 150,000 + 75,000t.$$

(a) Find the actual income for the business over the 4 years.

(b) Assuming an annual inflation rate of 4%, what is the present value of this income?

70. *Present Value* A professional athlete signs a three-year contract in which the earnings can be modeled by

$$c = 300,000 + 125,000t.$$

(a) Find the actual value of the athlete's contract.

(b) Assuming an annual inflation rate of 5%, what is the present value of the contract?

Future Value In Exercises 71 and 72, find the future value of the income (in dollars) given by $f(t)$ over t_1 years at the annual interest rate of r. If the function f represents a continuous investment over a period of t_1 years at an annual interest rate of r (compounded continuously), then the future value of the investment is given by

$$\text{Future value} = e^{rt_1} \int_0^{t_1} f(t)e^{-rt}\, dt.$$

71. $f(t) = 3000, \ r = 8\%, \ t_1 = 10$ years

72. $f(t) = 3000e^{0.05t}, \ r = 10\%, \ t_1 = 5$ years

 73. *Finance: Future Value* Use the equation from Exercises 71 and 72 to calculate the following. *(Source: Adapted from Garman/Forgue, Personal Finance, Fifth Edition)*

(a) The future value of $1200 saved each year for 10 years earning 7% interest.

(b) A person who wishes to invest $1200 each year finds one investment choice that is expected to pay 9% interest per year and another, riskier choice that may pay 10% interest per year. What is the difference in return (future value) if the investment is made for 15 years?

74. *Consumer Awareness* In 2004, the total cost to attend Pennsylvania State University for 1 year was estimated to be $19,843. If your grandparents had continuously invested in a college fund according to the model

$$f(t) = 400t$$

for 18 years, at an annual interest rate of 10%, would the fund have grown enough to allow you to cover 4 years of expenses at Pennsylvania State University? *(Source: Pennsylvania State University)*

 75. Use a program similar to the Midpoint Rule program on page 366 with $n = 10$ to approximate

$$\int_1^4 \frac{4}{\sqrt{x} + \sqrt[3]{x}}\, dx.$$

 76. Use a program similar to the Midpoint Rule program on page 366 with $n = 12$ to approximate the volume of the solid generated by revolving the region bounded by the graphs of

$$y = \frac{10}{\sqrt{xe^x}}, \ y = 0, \ x = 1, \ \text{and} \ x = 4$$

about the x-axis.

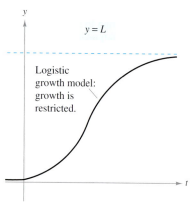

FIGURE 6.6

Logistic Growth Function

In Section 4.6, you saw that exponential growth occurs in situations for which the rate of growth is proportional to the quantity present at any given time. That is, if y is the quantity at time t, then

$$\frac{dy}{dt} = ky \qquad \frac{dy}{dt} \text{ is proportional to } y.$$

$$y = Ce^{kt}. \qquad \text{Exponential growth function}$$

Exponential growth is unlimited. As long as C and k are positive, the value of Ce^{kt} can be made arbitrarily large by choosing sufficiently large values of t.

In many real-life situations, however, the growth of a quantity is limited and cannot increase beyond a certain size L, as shown in Figure 6.6. The **logistic growth** model assumes that the rate of growth is proportional to both the quantity y and the difference between the quantity and the limit L. That is

$$\frac{dy}{dt} = ky(L - y). \qquad \frac{dy}{dt} \text{ is proportional to } y \text{ and } (L - y).$$

The solution of this *differential equation* is given in Example 4.

STUDY TIP

The logistic growth model in Example 4 is simplified by assuming that the limit of the quantity y is 1. If the limit were L, then the solution would be

$$y = \frac{L}{1 + be^{-kt}}.$$

In the fourth step of the solution, notice that partial fractions are used to integrate the left side of the equation.

EXAMPLE 4   **Deriving the Logistic Growth Function**

Solve the equation

$$\frac{dy}{dt} = ky(1 - y).$$

Assume $y > 0$ and $1 - y > 0$.

SOLUTION

$$\frac{dy}{dt} = ky(1 - y) \qquad \text{Write differential equation.}$$

$$\frac{1}{y(1 - y)} dy = k\, dt \qquad \text{Write in differential form.}$$

$$\int \frac{1}{y(1 - y)} dy = \int k\, dt \qquad \text{Integrate each side.}$$

$$\int \left(\frac{1}{y} + \frac{1}{1 - y} \right) dy = \int k\, dt \qquad \text{Rewrite using partial fractions.}$$

$$\ln y - \ln(1 - y) = kt + C_1 \qquad \text{Find antiderivative.}$$

$$\ln \frac{y}{1 - y} = kt + C_1 \qquad \text{Simplify.}$$

$$\frac{y}{1 - y} = Ce^{kt} \qquad \text{Exponentiate and let } e^{C_1} = C.$$

Solving this equation for y produces

$$y = \frac{1}{1 + be^{-kt}} \qquad \text{Logistic growth function}$$

where $b = 1/C$.

TRY IT 4

Show that if

$$y = \frac{1}{1 + be^{-kt}}, \text{ then}$$

$$\frac{dy}{dt} = ky(1 - y).$$

[*Hint:* First find $ky(1 - y)$ in terms of t, then find dy/dt and show that they are equivalent.]

| **EXAMPLE 5** | **Comparing Logistic Growth Functions** | |

Use a graphing utility to investigate the effects of the values of L, b, and k on the graph of

$$y = \frac{L}{1 + be^{-kt}}.$$ Logistic growth function $(L > 0, b > 0, k > 0)$

SOLUTION   The value of L determines the horizontal asymptote of the graph to the right. In other words, as t increases without bound, the graph approaches a limit of L (see Figure 6.7).

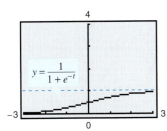 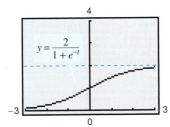 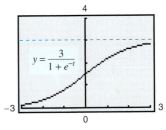

FIGURE 6.7

The value of b determines the point of inflection of the graph. When $b = 1$, the point of inflection occurs when $t = 0$. If $b > 1$, the point of inflection is to the right of the y-axis. If $0 < b < 1$, the point of inflection is to the left of the y-axis (see Figure 6.8).

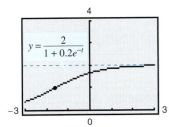 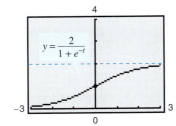 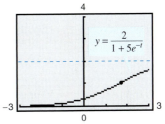

FIGURE 6.8

The value of k determines the rate of growth of the graph. For fixed values of b and L, larger values of k correspond to higher rates of growth (see Figure 6.9).

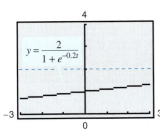 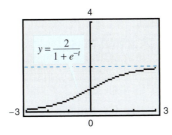 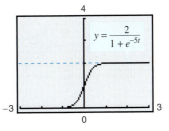

FIGURE 6.9

TRY IT 5

Find the horizontal asymptote of the graph of $y = \dfrac{4}{1 + 5e^{-6t}}$.

© Galen Rowell/CORBIS

The American peregrine falcon was removed from the endangered species list in 1999 due to its recovery from 324 nesting pairs in North America in 1975 to 1650 pairs in the United States and Canada. The peregrine was put on the endangered species list in 1970 because of the use of the chemical pesticide DDT. The Fish and Wildlife Service, state wildlife agencies, and many other organizations contributed to the recovery by setting up protective breeding programs among other efforts.

EXAMPLE 6   **Modeling a Population**

The state game commission releases 100 deer into a game preserve. During the first 5 years, the population increases to 432 deer. The commission believes that the population can be modeled by logistic growth with a limit of 2000 deer. Write the logistic growth model for this population. Then use the model to create a table showing the size of the deer population over the next 30 years.

SOLUTION   Let y represent the number of deer in year t. Assuming a logistic growth model means that the rate of change in the population is proportional to both y and $(2000 - y)$. That is

$$\frac{dy}{dt} = ky(2000 - y), \qquad 100 \le y \le 2000.$$

The solution of this equation is

$$y = \frac{2000}{1 + be^{-kt}}.$$

Using the fact that $y = 100$ when $t = 0$, you can solve for b.

$$100 = \frac{2000}{1 + be^{-k(0)}} \qquad \Longrightarrow \qquad b = 19$$

Then, using the fact that $y = 432$ when $t = 5$, you can solve for k.

$$432 = \frac{2000}{1 + 19e^{-k(5)}} \qquad \Longrightarrow \qquad k \approx 0.33106$$

So, the logistic growth model for the population is

$$y = \frac{2000}{1 + 19e^{-0.33106t}}. \qquad \text{Logistic growth model}$$

The population, in five-year intervals, is shown in the table.

Time, t	0	5	10	15	20	25	30
Population, y	100	432	1181	1766	1951	1990	1998

TRY IT 6

Write the logistic growth model for the population of deer in Example 6 if the game preserve could contain a limit of 4000 deer.

TAKE ANOTHER LOOK

Logistic Growth

Analyze the graph of the logistic growth function in Example 6. During which years is the *rate of growth* of the herd increasing? During which years is the *rate of growth* of the herd decreasing? How would these answers change if, instead of a limit of 2000 deer, the game preserve could contain a limit of 3000 deer?

PREREQUISITE REVIEW 6.3

The following warm-up exercises involve skills that were covered in earlier sections. You will use these skills in the exercise set for this section.

In Exercises 1–8, factor the expression.

1. $x^2 - 16$

2. $x^2 - 25$

3. $x^2 - x - 12$

4. $x^2 + x - 6$

5. $x^3 - x^2 - 2x$

6. $x^3 - 4x^2 + 4x$

7. $x^3 - 4x^2 + 5x - 2$

8. $x^3 - 5x^2 + 7x - 3$

In Exercises 9–14, rewrite the improper rational expression as the sum of a proper rational expression and a polynomial.

9. $\dfrac{x^2 - 2x + 1}{x - 2}$

10. $\dfrac{2x^2 - 4x + 1}{x - 1}$

11. $\dfrac{x^3 - 3x^2 + 2}{x - 2}$

12. $\dfrac{x^3 + 2x - 1}{x + 1}$

13. $\dfrac{x^3 + 4x^2 + 5x + 2}{x^2 - 1}$

14. $\dfrac{x^3 + 3x^2 - 4}{x^2 - 1}$

EXERCISES 6.3

In Exercises 1–12, write the partial fraction decomposition for the expression.

1. $\dfrac{2(x + 20)}{x^2 - 25}$

2. $\dfrac{3x + 11}{x^2 - 2x - 3}$

3. $\dfrac{8x + 3}{x^2 - 3x}$

4. $\dfrac{10x + 3}{x^2 + x}$

5. $\dfrac{4x - 13}{x^2 - 3x - 10}$

6. $\dfrac{7x + 5}{6(2x^2 + 3x + 1)}$

7. $\dfrac{3x^2 - 2x - 5}{x^3 + x^2}$

8. $\dfrac{3x^2 - x + 1}{x(x + 1)^2}$

9. $\dfrac{x + 1}{3(x - 2)^2}$

10. $\dfrac{3x - 4}{(x - 5)^2}$

11. $\dfrac{8x^2 + 15x + 9}{(x + 1)^3}$

12. $\dfrac{6x^2 - 5x}{(x + 2)^3}$

In Exercises 13–32, find the indefinite integral.

13. $\displaystyle\int \dfrac{1}{x^2 - 1}\,dx$

14. $\displaystyle\int \dfrac{9}{x^2 - 9}\,dx$

15. $\displaystyle\int \dfrac{-2}{x^2 - 16}\,dx$

16. $\displaystyle\int \dfrac{-4}{x^2 - 4}\,dx$

17. $\displaystyle\int \dfrac{1}{3x^2 - x}\,dx$

18. $\displaystyle\int \dfrac{3}{x^2 - 3x}\,dx$

19. $\displaystyle\int \dfrac{1}{2x^2 + x}\,dx$

20. $\displaystyle\int \dfrac{5}{x^2 + x - 6}\,dx$

21. $\displaystyle\int \dfrac{3}{x^2 + x - 2}\,dx$

22. $\displaystyle\int \dfrac{1}{4x^2 - 9}\,dx$

23. $\displaystyle\int \dfrac{5 - x}{2x^2 + x - 1}\,dx$

24. $\displaystyle\int \dfrac{x + 1}{x^2 + 4x + 3}\,dx$

25. $\displaystyle\int \dfrac{x^2 + 12x + 12}{x^3 - 4x}\,dx$

26. $\displaystyle\int \dfrac{3x^2 - 7x - 2}{x^3 - x}\,dx$

27. $\displaystyle\int \dfrac{x + 2}{x^2 - 4x}\,dx$

28. $\displaystyle\int \dfrac{4x^2 + 2x - 1}{x^3 + x^2}\,dx$

29. $\displaystyle\int \dfrac{4 - 3x}{(x - 1)^2}\,dx$

30. $\displaystyle\int \dfrac{x^4}{(x - 1)^3}\,dx$

31. $\displaystyle\int \dfrac{3x^2 + 3x + 1}{x(x^2 + 2x + 1)}\,dx$

32. $\displaystyle\int \dfrac{3x}{x^2 - 6x + 9}\,dx$

In Exercises 33–40, evaluate the definite integral.

33. $\displaystyle\int_4^5 \dfrac{1}{9 - x^2}\,dx$

34. $\displaystyle\int_0^1 \dfrac{3}{2x^2 + 5x + 2}\,dx$

35. $\displaystyle\int_1^5 \dfrac{x - 1}{x^2(x + 1)}\,dx$

36. $\displaystyle\int_0^1 \dfrac{x^2 - x}{x^2 + x + 1}\,dx$

37. $\int_0^1 \dfrac{x^3}{x^2 - 2}\, dx$

38. $\int_0^1 \dfrac{x^3 - 1}{x^2 - 4}\, dx$

39. $\int_1^2 \dfrac{x^3 - 4x^2 - 3x + 3}{x^2 - 3x}\, dx$

40. $\int_2^4 \dfrac{x^4 - 4}{x^2 - 1}\, dx$

In Exercises 41–44, find the area of the shaded region.

41. $y = \dfrac{14}{16 - x^2}$

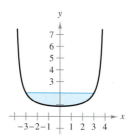

42. $y = \dfrac{-4}{x^2 - x - 6}$

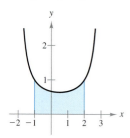

43. $y = \dfrac{x + 1}{x^2 - x}$

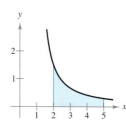

44. $y = \dfrac{x^2 + 2x - 1}{x^2 - 4}$

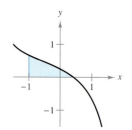

In Exercises 45–48, write the partial fraction decomposition for the rational expression. Check your result algebraically. Then assign a value to the constant a and use a graphing utility to check the result graphically.

45. $\dfrac{1}{a^2 - x^2}$

46. $\dfrac{1}{x(x + a)}$

47. $\dfrac{1}{x(a - x)}$

48. $\dfrac{1}{(x + 1)(a - x)}$

In Exercises 49–52, use a graphing utility to graph the function. Then find the volume of the solid generated by revolving the region bounded by the graphs of the given equations about the x-axis by using the integration capabilities of a graphing utility and by integrating by hand using partial fraction decomposition.

49. $y = \dfrac{10}{x(x + 10)}$, $y = 0$, $x = 1$, $x = 5$

50. $y = \dfrac{-4}{(x + 1)(x - 4)}$, $y = 0$, $x = 0$, $x = 3$

51. $y = \dfrac{2}{x^2 - 4}$, $x = 1$, $x = -1$, $y = 0$

52. $y = \dfrac{25x}{x^2 + x - 6}$, $x = -2$, $x = 0$, $y = 0$

53. *Biology*   A conservation organization releases 100 animals of an endangered species into a game preserve. The organization believes that the preserve has a capacity of 1000 animals and that the herd will grow according to a logistic growth model. That is, the size y of the herd will follow the equation

$$\int \frac{1}{y(1000 - y)}\, dy = \int k\, dt$$

where t is measured in years. Find this logistic curve. (To solve for the constant of integration C and the proportionality constant k, assume $y = 100$ when $t = 0$ and $y = 134$ when $t = 2$.) Use a graphing utility to graph your solution.

54. *Health: Epidemic*   A single infected individual enters a community of 500 individuals susceptible to the disease. The disease spreads at a rate proportional to the product of the total number infected and the number of susceptible individuals not yet infected. A model for the time it takes for the disease to spread to x individuals is

$$t = 5010 \int \frac{1}{(x + 1)(500 - x)}\, dx$$

where t is the time in hours.

(a) Find the time it takes for 75% of the population to become infected (when $t = 0$, $x = 1$).

(b) Find the number of people infected after 100 hours.

55. *Marketing*   After test-marketing a new menu item, a fast-food restaurant predicts that sales of the new item will grow according to the model

$$\frac{dS}{dt} = \frac{2t}{(t + 4)^2}$$

where t is the time in weeks and S is the sales (in thousands of dollars). Find the sales of the menu item at 10 weeks.

56. *Biology*   One gram of a bacterial culture is present at time $t = 0$, and 10 grams is the upper limit of the culture's weight. The time required for the culture to grow to y grams is modeled by

$$kt = \int \frac{1}{y(10 - y)}\, dy$$

where y is the weight of the culture (in grams) and t is the time in hours.

(a) Verify that the weight of the culture at time t is modeled by

$$y = \frac{10}{1 + 9e^{-10kt}}.$$

Use the fact that $y = 1$ when $t = 0$.

(b) Use the graph to determine the constant k.

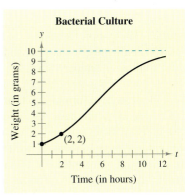

Bacterial Culture

$(2, 2)$

Time (in hours)

Weight (in grams)

57. Revenue The revenue R (in millions of dollars per year) for Symantec Corporation from 1995 through 2003 can be modeled by

$$R = \frac{410t^2 + 28{,}490t + 28{,}080}{-6t^2 + 94t + 100}$$

where $t = 5$ corrresponds to 1995. Find the total revenue from 1995 through 2003. Then find the average revenue during this time period. *(Source: Symantec Corporation)*

58. Medicine On a college campus, 50 students return from semester break with a contagious flu virus. The virus has a history of spreading at a rate of

$$\frac{dN}{dt} = \frac{100e^{-0.1t}}{(1 + 4e^{-0.1t})^2}$$

where N is the number of students infected after t days.

(a) Find the model giving the number of students infected with the virus in terms of the number of days since returning from semester break.

(b) If nothing is done to stop the virus from spreading, will the virus spread to infect half the student population of 1000 students? Explain your answer.

59. Biology A conservation organization releases 100 animals of an endangered species into a game preserve. The organization believes the population of the species will increase at a rate of

$$\frac{dN}{dt} = \frac{125e^{-0.125t}}{(1 + 9e^{-0.125t})^2}$$

where N is the population and t is the time in months.

(a) Use the fact that $N = 100$ when $t = 0$ to find the population after 2 years.

(b) Find the limiting size of the population as time increases without bound.

 60. Biology: Population Growth The graph shows the logistic growth curves for two species of the single-celled *Paramecium* in a laboratory culture. During which time intervals is the rate of growth of each species increasing? During which time intervals is the rate of growth of each species decreasing? Which species has a higher limiting population under these conditions? *(Source: Adapted from Levine/Miller,* Biology: Discovering Life, *Second Edition)*

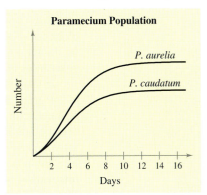

Paramecium Population

P. aurelia

P. caudatum

Number

Days

BUSINESS CAPSULE

Courtesy of Susie Wang/Aqua Dessa

While a math communications major at the University of California at Berkeley, Susie Wang began researching the idea of selling natural skin-care products. She used $10,000 to start her company, Aqua Dessa, and uses word-of-mouth as an advertising tactic. Aqua Dessa products are used and sold at spas and exclusive cosmetics counters throughout the United States.

61. Research Project Use your school's library, the Internet, or some other reference source to research the opportunity cost of attending graduate school for 2 years to receive a Masters of Business Administration (MBA) degree rather than working for 2 years with a bachelor's degree. Write a short paper describing these costs.

6.4 INTEGRATION TABLES AND COMPLETING THE SQUARE

- Use integration tables to find indefinite integrals.
- Use reduction formulas to find indefinite integrals.
- Use completing the square to find indefinite integrals.

Integration Tables

So far in this chapter, you have studied three integration techniques to be used along with the basic integration formulas. Certainly these techniques and formulas do not cover every possible method for finding an antiderivative, but they do cover most of the important ones.

In this section, you will expand the list of integration formulas to form a table of integrals. As you add new integration formulas to the basic list, two effects occur. On one hand, it becomes increasingly difficult to memorize, or even become familiar with, the entire list of formulas. On the other hand, with a longer list you need fewer techniques for fitting an integral to one of the formulas on the list. The procedure of integrating by means of a long list of formulas is called **integration by tables.** (The table in this section constitutes only a partial listing of integration formulas. Much longer lists exist, some of which contain several hundred formulas.)

Integration by tables should not be considered a trivial task. It requires considerable thought and insight, and it often requires substitution. Many people find a table of integrals to be a valuable supplement to the integration techniques discussed in the first three sections of this chapter. We encourage you to gain competence in the use of integration tables, as well as to continue to improve in the use of the various integration techniques. In doing so, you should find that a combination of techniques and tables is the most versatile approach to integration.

Each integration formula in the table on the next three pages can be developed using one or more of the techniques you have studied. You should try to verify several of the formulas. For instance, Formula 4

$$\int \frac{u}{(a + bu)^2}\, du = \frac{1}{b^2}\left(\frac{a}{a + bu} + \ln|a + bu|\right) + C \qquad \text{Formula 4}$$

can be verified using partial fractions, Formula 17

$$\int \frac{\sqrt{a + bu}}{u}\, du = 2\sqrt{a + bu} + a\int \frac{1}{u\sqrt{a + bu}}\, du \qquad \text{Formula 17}$$

can be verified using integration by parts, and Formula 37

$$\int \frac{1}{1 + e^u}\, du = u - \ln(1 + e^u) + C \qquad \text{Formula 37}$$

can be verified using substitution.

STUDY TIP

A symbolic integration utility consists, in part, of a database of integration tables. The primary difference between using a symbolic integration utility and using a table of integrals is that with a symbolic integration utility the computer searches through the database to find a fit. With a table of integrals, *you* must do the searching.

In the table of integrals below and on the next two pages, the formulas have been grouped into eight different types according to the form of the integrand.

Forms involving u^n

Forms involving $a + bu$

Forms involving $\sqrt{a + bu}$

Forms involving $\sqrt{u^2 \pm a^2}$

Forms involving $u^2 - a^2$

Forms involving $\sqrt{a^2 - u^2}$

Forms involving e^u

Forms involving $\ln u$

Table of Integrals

Forms involving u^n

1. $\displaystyle \int u^n \, du = \frac{u^{n+1}}{n+1} + C, \quad n \neq -1$

2. $\displaystyle \int \frac{1}{u} \, du = \ln|u| + C$

Forms involving $a + bu$

3. $\displaystyle \int \frac{u}{a + bu} \, du = \frac{1}{b^2}(bu - a \ln|a + bu|) + C$

4. $\displaystyle \int \frac{u}{(a + bu)^2} \, du = \frac{1}{b^2}\left(\frac{a}{a + bu} + \ln|a + bu|\right) + C$

5. $\displaystyle \int \frac{u}{(a + bu)^n} \, du = \frac{1}{b^2}\left[\frac{-1}{(n - 2)(a + bu)^{n-2}} + \frac{a}{(n - 1)(a + bu)^{n-1}}\right] + C, \quad n \neq 1, 2$

6. $\displaystyle \int \frac{u^2}{a + bu} \, du = \frac{1}{b^3}\left[-\frac{bu}{2}(2a - bu) + a^2 \ln|a + bu|\right] + C$

7. $\displaystyle \int \frac{u^2}{(a + bu)^2} \, du = \frac{1}{b^3}\left(bu - \frac{a^2}{a + bu} - 2a \ln|a + bu|\right) + C$

8. $\displaystyle \int \frac{u^2}{(a + bu)^3} \, du = \frac{1}{b^3}\left[\frac{2a}{a + bu} - \frac{a^2}{2(a + bu)^2} + \ln|a + bu|\right] + C$

9. $\displaystyle \int \frac{u^2}{(a + bu)^n} \, du = \frac{1}{b^3}\left[\frac{-1}{(n - 3)(a + bu)^{n-3}} + \frac{2a}{(n - 2)(a + bu)^{n-2}} - \frac{a^2}{(n - 1)(a + bu)^{n-1}}\right] + C, \quad n \neq 1, 2, 3$

10. $\displaystyle \int \frac{1}{u(a + bu)} \, du = \frac{1}{a} \ln\left|\frac{u}{a + bu}\right| + C$

11. $\displaystyle \int \frac{1}{u(a + bu)^2} \, du = \frac{1}{a}\left(\frac{1}{a + bu} + \frac{1}{a} \ln\left|\frac{u}{a + bu}\right|\right) + C$

12. $\displaystyle \int \frac{1}{u^2(a + bu)} \, du = -\frac{1}{a}\left(\frac{1}{u} + \frac{b}{a} \ln\left|\frac{u}{a + bu}\right|\right) + C$

13. $\displaystyle \int \frac{1}{u^2(a + bu)^2} \, du = -\frac{1}{a^2}\left[\frac{a + 2bu}{u(a + bu)} + \frac{2b}{a} \ln\left|\frac{u}{a + bu}\right|\right] + C$

Table of Integrals (*continued*)

Forms involving $\sqrt{a + bu}$

14. $\displaystyle\int u^n \sqrt{a + bu}\, du = \frac{2}{b(2n + 3)}\left[u^n(a + bu)^{3/2} - na \int u^{n-1}\sqrt{a + bu}\, du \right]$

15. $\displaystyle\int \frac{1}{u\sqrt{a + bu}}\, du = \frac{1}{\sqrt{a}} \ln\left| \frac{\sqrt{a + bu} - \sqrt{a}}{\sqrt{a + bu} + \sqrt{a}} \right| + C, \quad a > 0$

16. $\displaystyle\int \frac{1}{u^n \sqrt{a + bu}}\, du = \frac{-1}{a(n - 1)}\left[\frac{\sqrt{a + bu}}{u^{n-1}} + \frac{(2n - 3)b}{2} \int \frac{1}{u^{n-1}\sqrt{a + bu}}\, du \right], \quad n \neq 1$

17. $\displaystyle\int \frac{\sqrt{a + bu}}{u}\, du = 2\sqrt{a + bu} + a \int \frac{1}{u\sqrt{a + bu}}\, du$

18. $\displaystyle\int \frac{\sqrt{a + bu}}{u^n}\, du = \frac{-1}{a(n - 1)}\left[\frac{(a + bu)^{3/2}}{u^{n-1}} + \frac{(2n - 5)b}{2} \int \frac{\sqrt{a + bu}}{u^{n-1}}\, du \right], \quad n \neq 1$

19. $\displaystyle\int \frac{u}{\sqrt{a + bu}}\, du = -\frac{2(2a - bu)}{3b^2}\sqrt{a + bu} + C$

20. $\displaystyle\int \frac{u^n}{\sqrt{a + bu}}\, du = \frac{2}{(2n + 1)b}\left(u^n\sqrt{a + bu} - na \int \frac{u^{n-1}}{\sqrt{a + bu}}\, du \right)$

Forms involving $\sqrt{u^2 \pm a^2}, \quad a > 0$

21. $\displaystyle\int \sqrt{u^2 \pm a^2}\, du = \frac{1}{2}\left(u\sqrt{u^2 \pm a^2} \pm a^2 \ln\left| u + \sqrt{u^2 \pm a^2} \right| \right) + C$

22. $\displaystyle\int u^2 \sqrt{u^2 \pm a^2}\, du = \frac{1}{8}\left[u(2u^2 \pm a^2)\sqrt{u^2 \pm a^2} - a^4 \ln\left| u + \sqrt{u^2 \pm a^2} \right| \right] + C$

23. $\displaystyle\int \frac{\sqrt{u^2 + a^2}}{u}\, du = \sqrt{u^2 + a^2} - a \ln\left| \frac{a + \sqrt{u^2 + a^2}}{u} \right| + C$

24. $\displaystyle\int \frac{\sqrt{u^2 \pm a^2}}{u^2}\, du = \frac{-\sqrt{u^2 \pm a^2}}{u} + \ln\left| u + \sqrt{u^2 \pm a^2} \right| + C$

25. $\displaystyle\int \frac{1}{\sqrt{u^2 \pm a^2}}\, du = \ln\left| u + \sqrt{u^2 \pm a^2} \right| + C$

26. $\displaystyle\int \frac{1}{u\sqrt{u^2 + a^2}}\, du = -\frac{1}{a} \ln\left| \frac{a + \sqrt{u^2 + a^2}}{u} \right| + C$

27. $\displaystyle\int \frac{u^2}{\sqrt{u^2 \pm a^2}}\, du = \frac{1}{2}\left(u\sqrt{u^2 \pm a^2} \mp a^2 \ln\left| u + \sqrt{u^2 \pm a^2} \right| \right) + C$

28. $\displaystyle\int \frac{1}{u^2\sqrt{u^2 \pm a^2}}\, du = \mp\frac{\sqrt{u^2 \pm a^2}}{a^2 u} + C$

Table of Integrals (*continued*)

Forms involving $u^2 - a^2$, $a > 0$

29. $\displaystyle\int \frac{1}{u^2 - a^2}\, du = -\int \frac{1}{a^2 - u^2}\, du = \frac{1}{2a} \ln\left|\frac{u - a}{u + a}\right| + C$

30. $\displaystyle\int \frac{1}{(u^2 - a^2)^n}\, du = \frac{-1}{2a^2(n - 1)}\left[\frac{u}{(u^2 - a^2)^{n-1}} + (2n - 3)\int \frac{1}{(u^2 - a^2)^{n-1}}\, du\right], \quad n \neq 1$

Forms involving $\sqrt{a^2 - u^2}$, $a > 0$

31. $\displaystyle\int \frac{\sqrt{a^2 - u^2}}{u}\, du = \sqrt{a^2 - u^2} - a \ln\left|\frac{a + \sqrt{a^2 - u^2}}{u}\right| + C$

32. $\displaystyle\int \frac{1}{u\sqrt{a^2 - u^2}}\, du = -\frac{1}{a} \ln\left|\frac{a + \sqrt{a^2 - u^2}}{u}\right| + C$

33. $\displaystyle\int \frac{1}{u^2\sqrt{a^2 - u^2}}\, du = \frac{-\sqrt{a^2 - u^2}}{a^2 u} + C$

Forms involving e^u

34. $\displaystyle\int e^u\, du = e^u + C$

35. $\displaystyle\int u e^u\, du = (u - 1)e^u + C$

36. $\displaystyle\int u^n e^u\, du = u^n e^u - n\int u^{n-1} e^u\, du$

37. $\displaystyle\int \frac{1}{1 + e^u}\, du = u - \ln(1 + e^u) + C$

38. $\displaystyle\int \frac{1}{1 + e^{nu}}\, du = u - \frac{1}{n}\ln(1 + e^{nu}) + C$

Forms involving $\ln u$

39. $\displaystyle\int \ln u\, du = u(-1 + \ln u) + C$

40. $\displaystyle\int u \ln u\, du = \frac{u^2}{4}(-1 + 2 \ln u) + C$

41. $\displaystyle\int u^n \ln u\, du = \frac{u^{n+1}}{(n + 1)^2}[-1 + (n + 1)\ln u] + C, \quad n \neq -1$

42. $\displaystyle\int (\ln u)^2\, du = u[2 - 2 \ln u + (\ln u)^2] + C$

43. $\displaystyle\int (\ln u)^n\, du = u(\ln u)^n - n\int (\ln u)^{n-1}\, du$

PREREQUISITE REVIEW 6.4

The following warm-up exercises involve skills that were covered in earlier sections. You will use these skills in the exercise set for this section.

In Exercises 1–4, expand the expression.

1. $(x + 4)^2$

2. $(x - 1)^2$

3. $\left(x + \frac{1}{2}\right)^2$

4. $\left(x - \frac{1}{3}\right)^2$

In Exercises 5–8, write the partial fraction decomposition for the expression.

5. $\dfrac{4}{x(x + 2)}$

6. $\dfrac{3}{x(x - 4)}$

7. $\dfrac{x + 4}{x^2(x - 2)}$

8. $\dfrac{3x^2 + 4x - 8}{x(x - 2)(x + 1)}$

In Exercises 9 and 10, use integration by parts to find the indefinite integral.

9. $\displaystyle\int 2xe^x\, dx$

10. $\displaystyle\int 3x^2 \ln x\, dx$

EXERCISES 6.4

In Exercises 1–8, use the indicated formula from the table of integrals in this section to find the indefinite integral.

1. $\displaystyle\int \frac{x}{(2 + 3x)^2}\, dx$, Formula 4

2. $\displaystyle\int \frac{1}{x(2 + 3x)^2}\, dx$, Formula 11

3. $\displaystyle\int \frac{x}{\sqrt{2 + 3x}}\, dx$, Formula 19

4. $\displaystyle\int \frac{4}{x^2 - 9}\, dx$, Formula 29

5. $\displaystyle\int \frac{2x}{\sqrt{x^4 - 9}}\, dx$, Formula 25

6. $\displaystyle\int x^2\sqrt{x^2 + 9}\, dx$, Formula 22

7. $\displaystyle\int x^3 e^{x^2}\, dx$, Formula 35

8. $\displaystyle\int \frac{x}{1 + e^{x^2}}\, dx$, Formula 37

In Exercises 9–34, use the table of integrals in this section to find the indefinite integral.

9. $\displaystyle\int \frac{1}{x(1 + x)}\, dx$

10. $\displaystyle\int \frac{1}{x(1 + x)^2}\, dx$

11. $\displaystyle\int \frac{1}{x\sqrt{x^2 + 9}}\, dx$

12. $\displaystyle\int \frac{1}{\sqrt{x^2 - 1}}\, dx$

13. $\displaystyle\int \frac{1}{x\sqrt{4 - x^2}}\, dx$

14. $\displaystyle\int \frac{\sqrt{x^2 - 9}}{x^2}\, dx$

15. $\displaystyle\int x \ln x\, dx$

16. $\displaystyle\int x^2(\ln x^3)^2\, dx$

17. $\displaystyle\int \frac{6x}{1 + e^{3x^2}}\, dx$

18. $\displaystyle\int \frac{1}{1 + e^x}\, dx$

19. $\displaystyle\int x\sqrt{x^4 - 4}\, dx$

20. $\displaystyle\int \frac{x}{x^4 - 9}\, dx$

21. $\displaystyle\int \frac{t^2}{(2 + 3t)^3}\, dt$

22. $\displaystyle\int \frac{\sqrt{3 + 4t}}{t}\, dt$

23. $\displaystyle\int \frac{s}{s^2\sqrt{3 + s}}\, ds$

24. $\displaystyle\int \sqrt{3 + x^2}\, dx$

25. $\displaystyle\int \frac{x^2}{(3 + 2x)^5}\,dx$

26. $\displaystyle\int \frac{1}{x^2\sqrt{x^2 - 4}}\,dx$

27. $\displaystyle\int \frac{1}{x^2\sqrt{1 - x^2}}\,dx$

28. $\displaystyle\int \frac{2x}{(1 - 3x)^2}\,dx$

29. $\displaystyle\int x^2 \ln x\,dx$

30. $\displaystyle\int xe^{x^2}\,dx$

31. $\displaystyle\int \frac{x^2}{(3x - 5)^2}\,dx$

32. $\displaystyle\int \frac{1}{2x^2(2x - 1)^2}\,dx$

33. $\displaystyle\int \frac{\ln x}{x(4 + 3\ln x)}\,dx$

34. $\displaystyle\int (\ln x)^3\,dx$

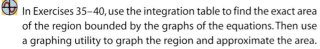 In Exercises 35–40, use the integration table to find the exact area of the region bounded by the graphs of the equations. Then use a graphing utility to graph the region and approximate the area.

35. $y = \dfrac{x}{\sqrt{x + 1}}$, $y = 0$, $x = 8$

36. $y = \dfrac{2}{1 + e^{4x}}$, $y = 0$, $x = 0$, $x = 1$

37. $y = \dfrac{x}{1 + e^{x^2}}$, $y = 0$, $x = 2$

38. $y = \dfrac{-e^x}{1 - e^{2x}}$, $y = 0$, $x = 1$, $x = 2$

39. $y = x^2\sqrt{x^2 + 4}$, $y = 0$, $x = \sqrt{5}$

40. $y = \dfrac{1}{\sqrt{x}(1 + 2\sqrt{x})}$, $y = 0$, $x = 1$, $x = 4$

In Exercises 41–44, evaluate the definite integral.

41. $\displaystyle\int_0^5 \frac{x}{\sqrt{5 + 2x}}\,dx$

42. $\displaystyle\int_0^5 \frac{x}{(4 + x)^2}\,dx$

43. $\displaystyle\int_0^4 \frac{6}{1 + e^{0.5x}}\,dx$

44. $\displaystyle\int_1^4 x \ln x\,dx$

In Exercises 45–48, find the indefinite integral (a) using the integration table and (b) using the specified method.

Integral	Method
45. $\displaystyle\int x^2 e^x\,dx$	Integration by parts
46. $\displaystyle\int x^4 \ln x\,dx$	Integration by parts
47. $\displaystyle\int \dfrac{1}{x^2(x + 1)}\,dx$	Partial fractions
48. $\displaystyle\int \dfrac{1}{x^2 - 75}\,dx$	Partial fractions

In Exercises 49–52, complete the square to express each polynomial as the sum or difference of squares.

49. (a) $x^2 + 6x$
(b) $x^2 - 8x + 9$
(c) $x^4 + 2x^2 - 5$
(d) $3 - 2x - x^2$

50. (a) $x^2 + 4x$
(b) $x^2 + 16x - 1$
(c) $x^4 + 8x^2 + 1$
(d) $9x^2 + 36x - 1$

51. (a) $4x^2 + 12x + 15$
(b) $3x^2 - 12x - 9$
(c) $x^2 - 2x$
(d) $9 + 8x - x^2$

52. (a) $16x^2 - 96x + 3$
(b) $x^2 + 4x - 1$
(c) $1 - 8x - x^2$
(d) $6x - x^2$

In Exercises 53–60, complete the square and then use the integration table to find the indefinite integral.

53. $\displaystyle\int \frac{1}{x^2 + 6x - 8}\,dx$

54. $\displaystyle\int \frac{1}{x^2 + 4x - 5}\,dx$

55. $\displaystyle\int \frac{1}{(x - 1)\sqrt{x^2 - 2x + 2}}\,dx$

56. $\displaystyle\int \sqrt{x^2 - 6x}\,dx$

57. $\displaystyle\int \frac{1}{2x^2 - 4x - 6}\,dx$

58. $\displaystyle\int \frac{\sqrt{7 - 6x - x^2}}{x + 3}\,dx$

59. $\displaystyle\int \frac{x}{\sqrt{x^4 + 2x^2 + 2}}\,dx$

60. $\displaystyle\int \frac{x\sqrt{x^4 + 4x^2 + 5}}{x^2 + 2}\,dx$

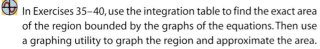 ***Population Growth***   In Exercises 61 and 62, use a graphing utility to graph the growth function. Use the table of integrals to find the average value of the growth function over the interval, where N is the size of a population and t is the time in days.

61. $N = \dfrac{50}{1 + e^{4.8 - 1.9t}}$, $[3, 4]$

62. $N = \dfrac{375}{1 + e^{4.20 - 0.25t}}$, $[21, 28]$

63. ***Revenue***   The revenue (in dollars per year) for a new product is modeled by

$$R = 10{,}000\left[1 - \frac{1}{(1 + 0.1t^2)^{1/2}}\right]$$

where t is the time in years. Estimate the total revenue from sales of the product over its first 2 years on the market.

64. ***Consumer and Producer Surpluses***   Find the consumer surplus and the producer surplus for a product with the given demand and supply functions.

$$\text{Demand: } p = \frac{60}{\sqrt{x^2 + 81}}, \quad \text{Supply: } p = \frac{x}{3}$$

65. ***Profit***   The net profits P (in billions of dollars per year) for Hershey Foods from 2000 through 2003 can be modeled by

$$P = \sqrt{0.04t - 0.3}, \quad 10 \le t \le 13$$

where t is the time in years, with $t = 10$ corresponding to 2000. Find the average net profit over that time period. *(Source: Hershey Foods Corp.)*

In Exercises 13–20, approximate the integral using (a) the Trapezoidal Rule and (b) Simpson's Rule. (Round your answers to three significant digits.)

Definite Integral	Subdivisions
13. $\int_0^1 \dfrac{1}{1 + x^2}\,dx$	$n = 4$
14. $\int_0^2 \dfrac{1}{\sqrt{1 + x^3}}\,dx$	$n = 4$
15. $\int_0^1 \sqrt{1 - x^2}\,dx$	$n = 4$
16. $\int_0^1 \sqrt{1 - x^2}\,dx$	$n = 8$
17. $\int_0^2 e^{-x^2}\,dx$	$n = 2$
18. $\int_0^2 e^{-x^2}\,dx$	$n = 4$
19. $\int_0^3 \dfrac{1}{2 - 2x + x^2}\,dx$	$n = 6$
20. $\int_0^3 \dfrac{x}{2 + x + x^2}\,dx$	$n = 6$

 Present Value   In Exercises 21 and 22, use a program similar to the Simpson's Rule program on page 430 with $n = 8$ to approximate the present value of the income $c(t)$ over t_1 years at the given annual interest rate r. Then use the integration capabilities of a graphing utility to approximate the present value. Compare the results. (Present value is defined in Section 6.2.)

21. $c(t) = 6000 + 200\sqrt{t}$, $r = 7\%$, $t_1 = 4$

22. $c(t) = 200{,}000 + 15{,}000\sqrt[3]{t}$, $r = 10\%$, $t_1 = 8$

 Marginal Analysis   In Exercises 23 and 24, use a program similar to the Simpson's Rule program on page 430 with $n = 4$ to approximate the change in revenue from the marginal revenue function dR/dx. In each case, assume that the number of units sold x increases from 14 to 16.

23. $\dfrac{dR}{dx} = 5\sqrt{8000 - x^3}$

24. $\dfrac{dR}{dx} = 50\sqrt{x}\sqrt{20 - x}$

 Probability   In Exercises 25–28, use a program similar to the Simpson's Rule program on page 430 with $n = 6$ to approximate the indicated normal probability. The standard normal probability density function is

$$f(x) = \dfrac{1}{\sqrt{2\pi}}e^{-x^2/2}.$$

If x is chosen at random from a population with this density, then the probability that x lies in the interval $[a, b]$ is

$$P(a \le x \le b) = \int_a^b f(x)\,dx.$$

25. $P(0 \le x \le 1)$

26. $P(0 \le x \le 2)$

27. $P(0 \le x \le 4)$

28. $P(0 \le x \le 1.5)$

 Surveying   In Exercises 29 and 30, use a program similar to the Simpson's Rule program on page 430 to estimate the number of square feet of land in the lot, where x and y are measured in feet, as shown in the figures. In each case, the land is bounded by a stream and two straight roads.

29.

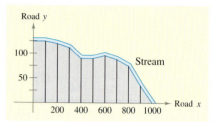

x	0	100	200	300	400	500
y	125	125	120	112	90	90

x	600	700	800	900	1000
y	95	88	75	35	0

30.

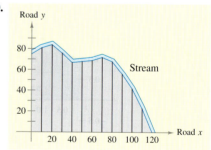

x	0	10	20	30	40	50	60
y	75	81	84	76	67	68	69

x	70	80	90	100	110	120
y	72	68	56	42	23	0

In Exercises 31–34, use the error formulas to find bounds for the error in approximating the integral using (a) the Trapezoidal Rule and (b) Simpson's Rule. (Let $n = 4$.)

31. $\displaystyle\int_0^2 x^4 \, dx$

32. $\displaystyle\int_0^1 \frac{1}{x+1} \, dx$

33. $\displaystyle\int_0^1 e^{x^3} \, dx$

34. $\displaystyle\int_0^1 e^{-x^2} \, dx$

In Exercises 35–38, use the error formulas to find n such that the error in the approximation of the definite integral is less than 0.0001 using (a) the Trapezoidal Rule and (b) Simpson's Rule.

35. $\displaystyle\int_0^1 x^4 \, dx$

36. $\displaystyle\int_1^3 \frac{1}{x} \, dx$

37. $\displaystyle\int_1^3 e^{2x} \, dx$

38. $\displaystyle\int_3^5 \ln x \, dx$

 In Exercises 39–42, use the program for Simpson's Rule given on page 430 to approximate the integral. Use $n = 100$.

39. $\displaystyle\int_1^4 x\sqrt{x+4} \, dx$

40. $\displaystyle\int_1^4 x^2\sqrt{x+4} \, dx$

41. $\displaystyle\int_2^5 10xe^{-x} \, dx$

42. $\displaystyle\int_2^5 10x^2e^{-x} \, dx$

43. Prove that Simpson's Rule is exact when used to approximate the integral of a cubic polynomial function, and demonstrate the result for

$$\int_0^1 x^3 \, dx, \qquad n = 2.$$

 44. Use a program similar to the Simpson's Rule program on page 430 with $n = 4$ to find the volume of the solid generated by revolving the region bounded by the graphs of

$$y = x\sqrt[3]{x+4}, \quad y = 0, \quad \text{and} \quad x = 4$$

about the x-axis.

In Exercises 45 and 46, use the definite integral below to find the required arc length. If f has a continuous derivative, then the arc length of f between the points $(a, f(a))$ and $(b, f(b))$ is

$$\int_b^a \sqrt{1 + [f'(x)]^2} \, dx.$$

 45. Arc Length The suspension cable on a bridge that is 400 feet long is in the shape of a parabola whose equation is

$$y = \frac{x^2}{800} \text{ (see figure).}$$

Use a program similar to the Simpson's Rule program on page 430 with $n = 12$ to approximate the length of the cable. Compare this result with the length obtained by using the table of integrals in Section 6.4 to perform the integration.

Figure for 45

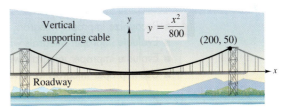

46. Arc Length A fleeing hare leaves its burrow $(0, 0)$ and moves due north (up the y-axis). At the same time, a pursuing lynx leaves from 1 yard east of the burrow $(1, 0)$ and always moves toward the fleeing hare (see figure). If the lynx's speed is twice that of the hare's, the equation of the lynx's path is

$$y = \frac{1}{3}(x^{3/2} - 3x^{1/2} + 2).$$

Find the distance traveled by the lynx by integrating over the interval $[0, 1]$.

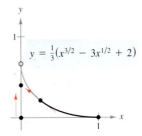

47. Medicine A body assimilates a 12-hour cold tablet at a rate modeled by

$$\frac{dC}{dt} = 8 - \ln(t^2 - 2t + 4), \qquad 0 \le t \le 12$$

where dC/dt is measured in milligrams per hour and t is the time in hours. Find the total amount of the drug absorbed into the body during the 12 hours.

48. Medicine The concentration M (in grams per liter) of a 6-hour allergy medicine in a body is modeled by

$$M = 12 - 4\ln(t^2 - 4t + 6), \qquad 0 \le t \le 6$$

where t is the time in hours since the allergy medication was taken. Find the average level of concentration in the body over the six-hour period.

49. Consumer Trends The rate of change S in the number of subscribers to a newly introduced magazine is modeled by

$$\frac{dS}{dt} = 1000t^2e^{-t}, \qquad 0 \le t \le 6$$

where t is the time in years. Find the total increase in the number of subscribers during the first 6 years.

PREREQUISITE REVIEW 6.6

The following warm-up exercises involve skills that were covered in earlier sections. You will use these skills in the exercise set for this section.

In Exercises 1–6, find the limit.

1. $\lim\limits_{x \to 2} (2x + 5)$

2. $\lim\limits_{x \to 1} \left(\dfrac{1}{x} + 2x^2 \right)$

3. $\lim\limits_{x \to -4} \dfrac{x + 4}{x^2 - 16}$

4. $\lim\limits_{x \to 0} \dfrac{x^2 - 2x}{x^3 + 3x^2}$

5. $\lim\limits_{x \to 1} \dfrac{1}{\sqrt{x - 1}}$

6. $\lim\limits_{x \to -3} \dfrac{x^2 + 2x - 3}{x + 3}$

In Exercises 7–10, evaluate the expression (a) when $x = b$ and (b) when $x = 0$.

7. $\dfrac{4}{3}(2x - 1)^3$

8. $\dfrac{1}{x - 5} + \dfrac{3}{(x - 2)^2}$

9. $\ln(5 - 3x^2) - \ln(x + 1)$

10. $e^{3x^2} + e^{-3x^2}$

EXERCISES 6.6

In Exercises 1–14, determine whether or not the improper integral converges. If it does, evaluate the integral.

1. $\displaystyle\int_0^\infty e^{-x}\,dx$

2. $\displaystyle\int_{-\infty}^0 e^{2x}\,dx$

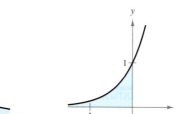

3. $\displaystyle\int_1^\infty \dfrac{1}{x^2}\,dx$

4. $\displaystyle\int_1^\infty \dfrac{1}{\sqrt{x}}\,dx$

5. $\displaystyle\int_0^\infty e^{x/3}\,dx$

6. $\displaystyle\int_0^\infty \dfrac{5}{e^{2x}}\,dx$

7. $\displaystyle\int_5^\infty \dfrac{x}{\sqrt{x^2 - 16}}\,dx$

8. $\displaystyle\int_{1/2}^\infty \dfrac{1}{\sqrt{2x - 1}}\,dx$

9. $\displaystyle\int_{-\infty}^0 e^{-x}\,dx$

10. $\displaystyle\int_{-\infty}^{-1} \dfrac{1}{x^2}\,dx$

11. $\displaystyle\int_1^\infty \dfrac{e^{\sqrt{x}}}{\sqrt{x}}\,dx$

12. $\displaystyle\int_{-\infty}^0 \dfrac{x}{x^2 + 1}\,dx$

13. $\displaystyle\int_{-\infty}^\infty 2xe^{-3x^2}\,dx$

14. $\displaystyle\int_{-\infty}^\infty x^2 e^{-x^3}\,dx$

In Exercises 15–18, determine the divergence or convergence of the improper integral. Evaluate the integral if it converges.

15. $\displaystyle\int_0^4 \dfrac{1}{\sqrt{x}}\,dx$

16. $\displaystyle\int_3^4 \dfrac{1}{\sqrt{x - 3}}\,dx$

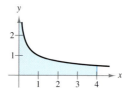

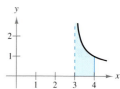

17. $\int_0^2 \dfrac{1}{(x-1)^{2/3}}\, dx$

18. $\int_0^2 \dfrac{1}{(x-1)^2}\, dx$

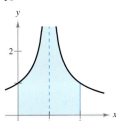
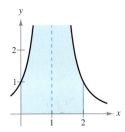

In Exercises 19–28, evaluate the improper integral.

19. $\int_0^1 \dfrac{1}{1-x}\, dx$

20. $\int_0^{27} \dfrac{5}{\sqrt[3]{x}}\, dx$

21. $\int_0^9 \dfrac{1}{\sqrt{9-x}}\, dx$

22. $\int_0^2 \dfrac{x}{\sqrt{4-x^2}}\, dx$

23. $\int_0^1 \dfrac{1}{x^2}\, dx$

24. $\int_0^1 \dfrac{1}{x}\, dx$

25. $\int_0^2 \dfrac{1}{\sqrt[3]{x-1}}\, dx$

26. $\int_0^2 \dfrac{1}{(x-1)^{4/3}}\, dx$

27. $\int_3^4 \dfrac{1}{\sqrt{x^2-9}}\, dx$

28. $\int_3^5 \dfrac{1}{x^2\sqrt{x^2-9}}\, dx$

In Exercises 29 and 30, (a) find the area of the region bounded by the graphs of the given equations and (b) find the volume of the solid generated by revolving the region about the x-axis.

29. $y = \dfrac{1}{x^2}, y = 0, x \geq 1$

30. $y = e^{-x}, y = 0, x \geq 0$

In Exercises 31–34, complete the table for the specified values of a and n to demonstrate that

$$\lim_{x \to \infty} x^n e^{-ax} = 0, \quad a > 0, n > 0.$$

x	1	10	25	50
$x^n e^{-ax}$				

31. $a = 1, n = 1$

32. $a = 2, n = 4$

33. $a = \tfrac{1}{2}, n = 2$

34. $a = \tfrac{1}{2}, n = 5$

In Exercises 35–38, use the results of Exercises 31–34 to evaluate the improper integral.

35. $\int_0^\infty x^2 e^{-x}\, dx$

36. $\int_0^\infty (x-1)e^{-x}\, dx$

37. $\int_0^\infty xe^{-2x}\, dx$

38. $\int_0^\infty xe^{-x}\, dx$

39. *Present Value* A business is expected to yield a continuous flow of profit at the rate of $500,000 per year. If money will earn interest at the nominal rate of 9% per year compounded continuously, what is the present value of the business (a) for 20 years and (b) forever? (Present value is defined in Section 6.2.)

40. *Present Value* Repeat Exercise 39 for a farm that is expected to produce a profit of $75,000 per year. Assume that money will earn interest at the nominal rate of 8% compounded continuously. (Present value is defined in Section 6.2.)

Capitalized Cost In Exercises 41 and 42, find the capitalized cost C of an asset (a) for $n = 5$ years, (b) for $n = 10$ years, and (c) forever. The capitalized cost is given by

$$C = C_0 + \int_0^n c(t)e^{-rt}\,dt$$

where C_0 is the original investment, t is the time in years, r is the annual interest rate compounded continuously, and $c(t)$ is the annual cost of maintenance. [*Hint:* For part (c), see Exercises 31–34.]

41. $C_0 = \$650{,}000, c(t) = 25{,}000, r = 10\%$

42. $C_0 = \$650{,}000, c(t) = \$25{,}000(1 + 0.08t), r = 12\%$

43. *Women's Height* The mean height of American women between the ages of 25 and 34 is 64.5 inches, and the standard deviation is 2.4 inches. Find the probability that a 25- to 34-year-old woman chosen at random is

(a) between 5 and 6 feet tall.

(b) 5 feet 8 inches or taller.

(c) 6 feet or taller.

(Source: U.S. National Center for Health Statistics)

44. *Quality Control* A company manufactures wooden yardsticks. The lengths of the yardsticks are normally distributed with a mean of 36 inches and a standard deviation of 0.2 inch. Find the probability that a yardstick is

(a) longer than 35.5 inches.

(b) longer than 35.9 inches.

ALGEBRA REVIEW

Algebra and Integration Techniques

Integration techniques involve many different algebraic skills. Study the examples in this Algebra Review. Be sure that you understand the algebra used in each step.

EXAMPLE 1   **Algebra and Integration Techniques**

Perform each operation and simplify.

(a) $\dfrac{28}{9}(1 - u^3)(u)(-3u^2)$ (b) $\dfrac{2}{x - 3} - \dfrac{1}{x + 2}$ (c) $\dfrac{6}{x} - \dfrac{1}{x + 1} + \dfrac{9}{(x + 1)^2}$

SOLUTION

(a) $\dfrac{28}{9}(1 - u^3)(u)(-3u^2)$ Example 6, page 393

$= \dfrac{(3)(28)}{9}(1 - u^3)(-u^3)$ Regroup factors.

$= \dfrac{28}{3}(1 - u^3)(-u^3)$ Simplify fraction.

$= \dfrac{28}{3}(u^6 - u^3)$ Multiply.

(b) $\dfrac{2}{x - 3} - \dfrac{1}{x + 2}$ Example 1, page 407

$= \dfrac{2(x + 2)}{(x - 3)(x + 2)} - \dfrac{(x - 3)}{(x - 3)(x + 2)}$ Rewrite with common denominator.

$= \dfrac{2(x + 2) - (x - 3)}{(x - 3)(x + 2)}$ Rewrite as single fraction.

$= \dfrac{2x + 4 - x + 3}{x^2 - x - 6}$ Multiply factors.

$= \dfrac{x + 7}{x^2 - x - 6}$ Combine like terms.

(c) $\dfrac{6}{x} - \dfrac{1}{x + 1} + \dfrac{9}{(x + 1)^2}$ Example 2, page 408

$= \dfrac{6(x + 1)^2}{x(x + 1)^2} - \dfrac{x(x + 1)}{x(x + 1)^2} + \dfrac{9x}{x(x + 1)^2}$ Rewrite with common denominator.

$= \dfrac{6(x + 1)^2 - x(x + 1) + 9x}{x(x + 1)^2}$ Rewrite as single fraction.

$= \dfrac{6x^2 + 12x + 6 - x^2 - x + 9x}{x^3 + 2x^2 + x}$ Multiply factors.

$= \dfrac{5x^2 + 20x + 6}{x^3 + 2x^2 + x}$ Combine like terms.

<table>
<tr><td>

EXAMPLE 2

</td><td>

Algebra and Integration Techniques

</td></tr>
</table>

Perform each operation and simplify.

(a) $x + 1 + \dfrac{1}{x^3} + \dfrac{1}{x-1}$

(b) $x^2 e^x - 2(x-1)e^x$

(c) $6\ln|x| - \ln|x+1| + 9\dfrac{(x+1)^{-1}}{-1}$

SOLUTION

(a) $x + 1 + \dfrac{1}{x^3} + \dfrac{1}{x-1}$ Example 3, page 409

$= \dfrac{(x+1)(x^3)(x-1)}{x^3(x-1)} + \dfrac{x-1}{x^3(x-1)} + \dfrac{x^3}{x^3(x-1)}$ Rewrite with common denominator.

$= \dfrac{(x+1)(x^3)(x-1) + (x-1) + x^3}{x^3(x-1)}$ Rewrite as single fraction.

$= \dfrac{(x^2-1)(x^3) + x - 1 + x^3}{x^3(x-1)}$ $(x+1)(x-1) = x^2 - 1$

$= \dfrac{x^5 - x^3 + x - 1 + x^3}{x^4 - x^3}$ Multiply factors.

$= \dfrac{x^5 + x - 1}{x^4 - x^3}$ Combine like terms.

(b) $x^2 e^x - 2(x-1)e^x$ Example 5, page 422

$= x^2 e^x - 2(xe^x - e^x)$ Multiply factors.

$= x^2 e^x - 2xe^x + 2e^x$ Multiply factors.

$= e^x(x^2 - 2x + 2)$ Factor.

(c) $6\ln|x| - \ln|x+1| + 9\dfrac{(x+1)^{-1}}{-1}$ Example 2, page 408

$= \ln|x|^6 - \ln|x+1| + 9\dfrac{(x+1)^{-1}}{-1}$ $m\ln n = \ln n^m$

$= \ln|x^6| - \ln|x+1| + 9\dfrac{(x+1)^{-1}}{-1}$ Property of absolute value

$= \ln\dfrac{|x^6|}{|x+1|} + 9\dfrac{(x+1)^{-1}}{-1}$ $\ln m - \ln n = \ln\dfrac{m}{n}$

$= \ln\left|\dfrac{x^6}{x+1}\right| + 9\dfrac{(x+1)^{-1}}{-1}$ $\dfrac{|a|}{|b|} = \left|\dfrac{a}{b}\right|$

$= \ln\left|\dfrac{x^6}{x+1}\right| - 9(x+1)^{-1}$ Rewrite sum as difference.

$= \ln\left|\dfrac{x^6}{x+1}\right| - \dfrac{9}{x+1}$ Rewrite with positive exponent.

6   CHAPTER SUMMARY AND STUDY STRATEGIES

After studying this chapter, you should have acquired the following skills. The exercise numbers are keyed to the Review Exercises that begin on page 450. Answers to odd-numbered Review Exercises are given in the back of the text.*

■ Use the basic integration formulas to find indefinite integrals.   *(Section 6.1)*   *Review Exercises 1–12*

Constant Rule: $\int k\,dx = kx + C$

Power Rules: $\int x^n\,dx = \dfrac{x^{n+1}}{n+1} + C, \quad \int u^n\dfrac{du}{dx}\,dx = \dfrac{u^{n+1}}{n+1} + C, \quad n \neq -1$

Exponential Rules: $\int e^x\,dx = e^x + C, \quad \int e^u\dfrac{du}{dx}\,dx = \int e^u\,du = e^u + C$

Log Rules: $\int \dfrac{1}{x}\,dx = \ln|x| + C, \quad \int \dfrac{du/dx}{u}\,dx = \int \dfrac{1}{u}\,du = \ln|u| + C$

■ Use substitution to find indefinite integrals.   *(Section 6.1)*   *Review Exercises 13–20*

■ Use substitution to evaluate definite integrals.   *(Section 6.1)*   *Review Exercises 21–24*

■ Use integration to solve real-life problems.   *(Section 6.1)*   *Review Exercises 25–28*

■ Use integration by parts to find indefinite integrals.   *(Section 6.2)*   *Review Exercises 29–32*

$\int u\,dv = uv - \int v\,du$

■ Use integration by parts repeatedly to find indefinite integrals.   *(Section 6.2)*   *Review Exercises 33, 34*

■ Find the present value of future income.   *(Section 6.2)*   *Review Exercises 35–42*

■ Use partial fractions to find indefinite integrals.   *(Section 6.3)*   *Review Exercises 43–48*

■ Use logistic growth functions to model real-life situations.   *(Section 6.3)*   *Review Exercises 49, 50*

■ Use integration tables to find indefinite and definite integrals.   *(Section 6.4)*   *Review Exercises 51–56*

■ Use reduction formulas to find indefinite integrals.   *(Section 6.4)*   *Review Exercises 57–60*

■ Use completing the square to find indefinite integrals.   *(Section 6.4)*   *Review Exercises 61–64*

■ Use the Trapezoidal Rule to approximate definite integrals.   *(Section 6.5)*   *Review Exercises 65–68*

$\int_a^b f(x)\,dx \approx \left(\dfrac{b-a}{2n}\right)[f(x_0) + 2f(x_1) + \cdots + 2f(x_{n-1}) + f(x_n)]$

■ Use Simpson's Rule to approximate definite integrals.   *(Section 6.5)*   *Review Exercises 69–72*

$\int_a^b f(x)\,dx \approx \left(\dfrac{b-a}{3n}\right)[f(x_0) + 4f(x_1) + 2f(x_2) + 4f(x_3) + \cdots + 4f(x_{n-1}) + f(x_n)]$

■ Analyze the sizes of the errors when approximating definite integrals with the Trapezoidal Rule.   *(Section 6.5)*   *Review Exercises 73, 74*

$|E| \leq \dfrac{(b-a)^3}{12n^2}[\max|f''(x)|], \quad a \leq x \leq b$

■ Analyze the sizes of the errors when approximating definite integrals with Simpson's Rule.   *(Section 6.5)*   *Review Exercises 75, 76*

$|E| \leq \dfrac{(b-a)^5}{180n^4}[\max|f^{(4)}(x)|], \quad a \leq x \leq b$

* Use a wide range of valuable study aids to help you master the material in this chapter. The *Student Solutions Guide* includes step-by-step solutions to all odd-numbered exercises to help you review and prepare. The *HM mathSpace® Student CD-ROM* helps you brush up on your algebra skills. The *Graphing Technology Guide*, available on the Web at *math.college.hmco.com/students*, offers step-by-step commands and instructions for a wide variety of graphing calculators, including the most recent models.

■ Evaluate improper integrals with infinite limits of integration. *(Section 6.6)*   *Review Exercises 77–80*

$$\int_a^\infty f(x)\,dx = \lim_{b\to\infty} \int_a^b f(x)\,dx, \qquad \int_{-\infty}^b f(x)\,dx = \lim_{a\to-\infty} \int_a^b f(x)\,dx,$$

$$\int_{-\infty}^\infty f(x)\,dx = \int_{-\infty}^c f(x)\,dx + \int_c^\infty f(x)\,dx$$

■ Evaluate improper integrals with infinite integrands. *(Section 6.6)*   *Review Exercises 81–84*

$$\int_a^b f(x)\,dx = \lim_{c\to b^-} \int_a^c f(x)\,dx, \qquad \int_a^b f(x)\,dx = \lim_{c\to a^+} \int_c^b f(x)\,dx,$$

$$\int_a^b f(x)\,dx = \int_a^c f(x)\,dx + \int_c^b f(x)\,dx$$

■ Use improper integrals to solve real-life problems. *(Section 6.6)*   *Review Exercises 85–87*

■ ***Use a Variety of Approaches***   To be efficient at finding antiderivatives, you need to use a variety of approaches.

1. Check to see whether the integral fits one of the basic integration formulas—you should have these formulas memorized.

2. Try an integration technique such as substitution, integration by parts, partial fractions, or completing the square to rewrite the integral in a form that fits one of the basic integration formulas.

3. Use a table of integrals.

4. Use a symbolic integration utility.

■ ***Use Numerical Integration***   When solving a definite integral, remember that you cannot apply the Fundamental Theorem of Calculus unless you can find an antiderivative of the integrand. This is not always possible—even with a symbolic integration utility. In such cases, you can use a numerical technique such as the Midpoint Rule, the Trapezoidal Rule, or Simpson's Rule to approximate the value of the integral.

■ ***Improper Integrals***   When solving integration problems, remember that the symbols used to denote definite integrals are the same as those used to denote improper integrals. Evaluating an improper integral as a definite integral can lead to an incorrect value. For instance, if you evaluated the integral

$$\int_{-2}^1 \frac{1}{x^2}\,dx$$

as though it were a definite integral, you would obtain a value of $-\frac{3}{2}$. This is not, however, correct. This integral is actually a divergent improper integral. If you have access to a symbolic integration utility, try using it to evaluate this integral—it will probably make the same mistake.

Study Tools   *Additional resources that accompany this chapter*

■ **Algebra Review** (pages 446 and 447)

■ **Chapter Summary and Study Strategies** (pages 448 and 449)

■ **Review Exercises** (pages 450–453)

■ **Sample Post-Graduation Exam Questions** (page 454)

■ **Web Exercise** (page 415, Exercise 61)

■ **Student Solutions Guide**

■ **HM mathSpace® Student CD-ROM**

■ **Graphing Technology Guide** (*math.college.hmco.com/students*)

6 SAMPLE POST-GRADUATION EXAM QUESTIONS

CPA
GMAT
GRE
Actuarial
CLAST

The following questions represent the types of questions that appear on certified public accountant (CPA) exams, Graduate Management Admission Tests (GMAT), Graduate Records Exams (GRE), actuarial exams, and College-Level Academic Skills Tests (CLAST). The answers to the questions are given in the back of the book.

For Questions 1–4, the total 2000 population of Florida was 15,982,000 and the total 2002 population was 16,713,000. Also use the data shown in the graphs.   *(Source: U.S. Census Bureau)*

Figure for 1–4

Resident Population by Age
State of Florida, 2000

75 and over

9%   0 – 17   23%
55 – 74   19%
18 – 34   21%
35 – 54   28%

Resident Population by Age
State of Florida, 2002

75 and over

8%   0 – 17   23%
55 – 74   19%
18 – 34   21%
35 – 54   29%

1. Find the number of people aged 75 and over for the year 2000.

(a) 1,438,380   (b) 1,278,560   (c) 1,504,170   (d) 1,337,040

2. Find the increase in population of 35- to 54-year-olds from 2000 to 2002.

(a) 204,680   (b) 448,600   (c) 211,900   (d) 371,810

3. In 2002, how many people were 54 years old or younger?

(a) 12,200,490   (b) 12,033,360   (c) 11,666,860   (d) 11,507,040

4. In what age group did the population and the population percent decrease between 2000 and 2002?

(a) 0–17   (b) 75 and over   (c) 18–34   (d) 55–74

5. $\displaystyle\int_0^3 \sqrt{x+1}\,dx =$

(a) $-\frac{14}{3}$ (b) $2\sqrt{3}$ (c) $\frac{14}{3}$ (d) $\frac{15}{2}$

6. $\displaystyle\int_1^6 \frac{x}{\sqrt{x+3}}\,dx =$

(a) $\frac{10}{3}$ (b) $-\frac{20}{3}$ (c) -4 (d) $\frac{20}{3}$

7. $\displaystyle\lim_{x\to\infty} \frac{2x^2 + 5x - 16}{5x^2 - 15x + 48} =$

(a) 0 (b) $\frac{2}{5}$ (c) ∞ (d) $\frac{1}{3}$

8. The city council is planning to construct a new street, as shown in the figure. Construction costs for the new street are $110 per linear foot. What is the projected cost for constructing the new street?

New street
Maple Drive
3 miles
4 miles
East Avenue

(a) $2,904,000   (b) $4,065,600   (c) $1,904,000   (d) $3,864,000

9. The following information pertains to Varn Co.:

Sales $1,000,000,   Variable costs $200,000,   Fixed costs $50,000

What is Varn's break-even point in sales dollars?

(a) $40,000   (b) $250,000   (c) $62,500   (d) $200,000

Functions of Several Variables

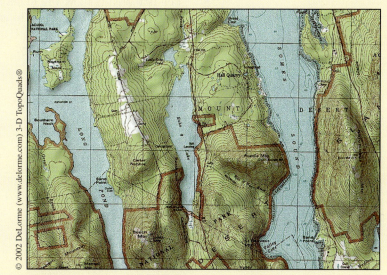

© 2002 DeLorme (www.delorme.com) 3-D TopoQuads®

In addition to distance and location, contour maps indicate other variables such as elevation, population, and climate.

STRATEGIES FOR SUCCESS

WHAT YOU SHOULD LEARN:

- How to analyze surfaces and graph functions of two variables on the three-dimensional coordinate system
- How to calculate partial derivatives and find extrema of functions of several variables
- How to use Lagrange multipliers to solve constrained optimization problems
- How to use least squares regression for mathematical modeling
- How to evaluate double integrals and use them to find area and volume

WHY YOU SHOULD LEARN IT:

Functions of several variables have many applications in real life, such as the examples below, which represent a small sample of the applications in this chapter.

- Cobb-Douglas Production Function, Exercises 41 and 42 on page 481
- Geology: A Contour Map, Exercise 49 on page 482
- Complementary and Substitute Products, Exercise 71 on page 493
- Hardy-Weinberg Law, Exercise 45 on page 503
- Investment Strategy, Exercises 49 and 50 on page 513

7.1   THE THREE-DIMENSIONAL COORDINATE SYSTEM

■ Plot points in space.
■ Find distances between points in space and find midpoints of line segments in space.
■ Write the standard forms of the equations of spheres and find the centers and radii of spheres.
■ Sketch the coordinate plane traces of surfaces.

The Three-Dimensional Coordinate System

Recall that the Cartesian plane is determined by two perpendicular number lines called the x-axis and the y-axis. These axes together with their point of intersection (the origin) allow you to develop a two-dimensional coordinate system for identifying points in a plane. To identify a point in space, you must introduce a third dimension to the model. The geometry of this three-dimensional model is called **solid analytic geometry.**

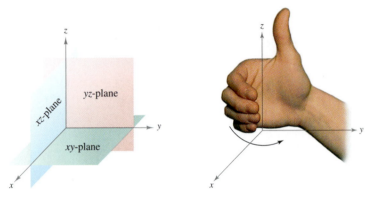

FIGURE 7.1   **FIGURE 7.2**

You can construct a **three-dimensional coordinate system** by passing a z-axis perpendicular to both the x- and y-axes at the origin. Figure 7.1 shows the positive portion of each coordinate axis. Taken as pairs, the axes determine three **coordinate planes:** the **xy-plane,** the **xz-plane,** and the **yz-plane.** These three coordinate planes separate the three-dimensional coordinate system into eight **octants.** The first octant is the one for which all three coordinates are positive. In this three-dimensional system, a point P in space is determined by an ordered triple (x, y, z), where x, y, and z are as shown.

x = directed distance from yz-plane to P

y = directed distance from xz-plane to P

z = directed distance from xy-plane to P

A three-dimensional coordinate system can have either a **left-handed** or a **right-handed** orientation. In this text, you will work exclusively with right-handed systems, as shown in Figure 7.2.

■ **EXAMPLE 1**   **Plotting Points in Space**

Plot each point in space.

(a) $(2, -3, 3)$

(b) $(-2, 6, 2)$

(c) $(1, 4, 0)$

(d) $(2, 2, -3)$

SOLUTION   To plot the point $(2, -3, 3)$, notice that $x = 2$, $y = -3$, and $z = 3$. To help visualize the point (see Figure 7.3), locate the point $(2, -3)$ in the xy-plane (denoted by a cross). The point $(2, -3, 3)$ lies three units above the cross. The other three points are also shown in the figure.

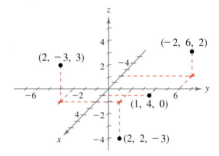

FIGURE 7.3

TRY IT 1

Plot each point on the three-dimensional coordinate system.

(a) $(2, 5, 1)$

(b) $(-2, -4, 3)$

(c) $(4, 0, -5)$

The Distance and Midpoint Formulas

Many of the formulas established for the two-dimensional coordinate system can be extended to three dimensions. For example, to find the distance between two points in space, you can use the Pythagorean Theorem twice, as shown in Figure 7.4. By doing this, you will obtain the formula for the distance between two points in space.

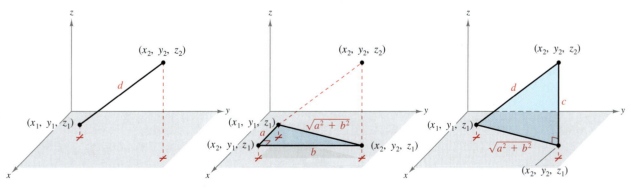

d = distance between two points $a = |x_2 - x_1|$, $b = |y_2 - y_1|$

$c = |z_2 - z_1|$
$d = \sqrt{a^2 + b^2 + c^2}$
$\quad = \sqrt{(x_2 - x_1)^2 + (y_2 - y_1)^2 + (z_2 - z_1)^2}$

FIGURE 7.4

Distance Formula in Space

The distance between the points (x_1, y_1, z_1) and (x_2, y_2, z_2) is

$$d = \sqrt{(x_2 - x_1)^2 + (y_2 - y_1)^2 + (z_2 - z_1)^2}.$$

EXAMPLE 2   **Finding the Distance Between Two Points**

Find the distance between $(1, 0, 2)$ and $(2, 4, -3)$.

SOLUTION

$$\begin{aligned}
d &= \sqrt{(x_2 - x_1)^2 + (y_2 - y_1)^2 + (z_2 - z_1)^2} &&\text{Write Distance Formula.}\\
&= \sqrt{(2 - 1)^2 + (4 - 0)^2 + (-3 - 2)^2} &&\text{Substitute.}\\
&= \sqrt{1 + 16 + 25} &&\text{Simplify.}\\
&= \sqrt{42}. &&\text{Simplify.}
\end{aligned}$$

TRY IT 2

Find the distance between $(2, 3, -1)$ and $(0, 5, 3)$.

Notice the similarity between the Distance Formula in the plane and the Distance Formula in space. The Midpoint Formulas in the plane and in space are also similar.

Midpoint Formula in Space

The midpoint of the line segment joining the points (x_1, y_1, z_1) and (x_2, y_2, z_2) is

$$\text{Midpoint} = \left(\frac{x_1 + x_2}{2}, \frac{y_1 + y_2}{2}, \frac{z_1 + z_2}{2}\right).$$

EXAMPLE 3   **Using the Midpoint Formula**

Find the midpoint of the line segment joining $(5, -2, 3)$ and $(0, 4, 4)$.

SOLUTION   Using the Midpoint Formula, the midpoint is

$$\left(\frac{5 + 0}{2}, \frac{-2 + 4}{2}, \frac{3 + 4}{2}\right) = \left(\frac{5}{2}, 1, \frac{7}{2}\right)$$

as shown in Figure 7.5.

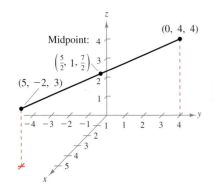

FIGURE 7.5

TRY IT 3

Find the midpoint of the line segment joining $(3, -2, 0)$ and $(-8, 6, -4)$.

The Equation of a Sphere

A **sphere** with center at (h, k, l) and radius r is defined to be the set of all points (x, y, z) such that the distance between (x, y, z) and (h, k, l) is r, as shown in Figure 7.6. Using the Distance Formula, this condition can be written as

$$\sqrt{(x - h)^2 + (y - k)^2 + (z - l)^2} = r.$$

By squaring both sides of this equation, you obtain the standard equation of a sphere.

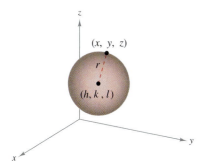

FIGURE 7.6 Sphere: Radius r, Center (h, k, l)

Standard Equation of a Sphere

The **standard equation of a sphere** whose center is (h, k, l) and whose radius is r is

$$(x - h)^2 + (y - k)^2 + (z - l)^2 = r^2.$$

EXAMPLE 4 **Finding the Equation of a Sphere**

Find the standard equation for the sphere whose center is $(2, 4, 3)$ and whose radius is 3. Does this sphere intersect the xy-plane?

SOLUTION

$$(x - h)^2 + (y - k)^2 + (z - l)^2 = r^2 \qquad \text{Write standard equation.}$$
$$(x - 2)^2 + (y - 4)^2 + (z - 3)^2 = 3^2 \qquad \text{Substitute.}$$
$$(x - 2)^2 + (y - 4)^2 + (z - 3)^2 = 9 \qquad \text{Simplify.}$$

From the graph shown in Figure 7.7, you can see that the center of the sphere lies three units above the xy-plane. Because the sphere has a radius of 3, you can conclude that it does intersect the xy-plane—at the point $(2, 4, 0)$. ——

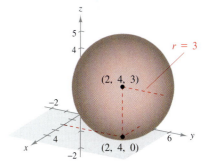

FIGURE 7.7

TRY IT 4

Find the standard equation of the sphere whose center is $(4, 3, 2)$ and whose radius is 5.

EXAMPLE 5 **Finding the Equation of a Sphere**

Find the equation of the sphere that has the points $(3, -2, 6)$ and $(-1, 4, 2)$ as endpoints of a diameter.

SOLUTION By the Midpoint Formula, the center of the sphere is

$$(h, k, l) = \left(\frac{3 + (-1)}{2}, \frac{-2 + 4}{2}, \frac{6 + 2}{2} \right) \qquad \text{Apply Midpoint Formula.}$$

$$= (1, 1, 4). \qquad \text{Simplify.}$$

By the Distance Formula, the radius is

$$r = \sqrt{(3 - 1)^2 + (-2 - 1)^2 + (6 - 4)^2} \qquad \text{Distance Formula using } (3, -2, 6) \text{ and } (1, 1, 4)$$

$$= \sqrt{17}. \qquad \text{Simplify.}$$

So, the standard equation of the sphere is

$$(x - h)^2 + (y - k)^2 + (z - l)^2 = r^2 \qquad \text{Write formula for a sphere.}$$

$$(x - 1)^2 + (y - 1)^2 + (z - 4)^2 = 17. \qquad \text{Substitute.}$$

> **TRY IT 5**
>
> Find the equation of the sphere that has the points $(-2, 5, 7)$ and $(4, 1, -3)$ as endpoints of a diameter.

EXAMPLE 6 **Finding the Center and Radius of a Sphere**

Find the center and radius of the sphere whose equation is

$$x^2 + y^2 + z^2 - 2x + 4y - 6z + 8 = 0.$$

SOLUTION You can obtain the standard equation of the sphere by completing the square. To do this, begin by grouping terms with the same variable. Then add "the square of half the coefficient of each linear term" to each side of the equation. For instance, to complete the square of $(x^2 - 2x)$, add $\left[\frac{1}{2}(-2)\right]^2 = 1$ to each side.

$$x^2 + y^2 + z^2 - 2x + 4y - 6z + 8 = 0$$

$$(x^2 - 2x + \quad) + (y^2 + 4y + \quad) + (z^2 - 6z + \quad) = -8$$

$$(x^2 - 2x + 1) + (y^2 + 4y + 4) + (z^2 - 6z + 9) = -8 + 1 + 4 + 9$$

$$(x - 1)^2 + (y + 2)^2 + (z - 3)^2 = 6$$

So, the center of the sphere is $(1, -2, 3)$, and its radius is $\sqrt{6}$, as shown in Figure 7.8.

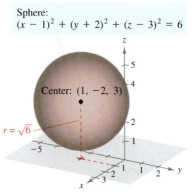

Sphere:
$(x - 1)^2 + (y + 2)^2 + (z - 3)^2 = 6$

Center: $(1, -2, 3)$

$r = \sqrt{6}$

FIGURE 7.8

> **TRY IT 6**
>
> Find the center and radius of the sphere whose equation is
>
> $$x^2 + y^2 + z^2 + 6x - 8y + 2z - 10 = 0.$$

Note in Example 6 that the points satisfying the equation of the sphere are "surface points," not "interior points." In general, the collection of points satisfying an equation involving x, y, and z is called a **surface in space.**

Traces of Surfaces

Finding the intersection of a surface with one of the three coordinate planes (or with a plane parallel to one of the three coordinate planes) helps visualize the surface. Such an intersection is called a **trace** of the surface. For example, the *xy*-trace of a surface consists of all points that are common to both the surface *and* the *xy*-plane. Similarly, the *xz*-trace of a surface consists of all points that are common to both the surface and the *xz*-plane.

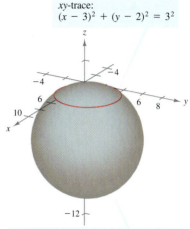

xy-trace:
$(x - 3)^2 + (y - 2)^2 = 3^2$

| **EXAMPLE 7** | **Finding a Trace of a Surface** |

Sketch the *xy*-trace of the sphere whose equation is

$$(x - 3)^2 + (y - 2)^2 + (z + 4)^2 = 5^2.$$

SOLUTION   To find the *xy*-trace of this surface, use the fact that every point in the *xy*-plane has a *z*-coordinate of zero. This means that if you substitute $z = 0$ into the original equation, the resulting equation will represent the intersection of the surface with the *xy*-plane.

$(x - 3)^2 + (y - 2)^2 + (z + 4)^2 = 5^2$ Write original equation.

$(x - 3)^2 + (y - 2)^2 + (0 + 4)^2 = 25$ Let $z = 0$ to find *xy*-trace.

$(x - 3)^2 + (y - 2)^2 + 16 = 25$

$(x - 3)^2 + (y - 2)^2 = 9$

$(x - 3)^2 + (y - 2)^2 = 3^2$ Equation of circle

From this equation, you can see that the *xy*-trace is a circle of radius 3, as shown in Figure 7.9.

Sphere:
$(x - 3)^2 + (y - 2)^2 + (z + 4)^2 = 5^2$

FIGURE 7.9

TRY IT 7

Find the equation of the *xy*-trace of the sphere whose equation is

$$(x + 1)^2 + (y - 2)^2 + (z + 3)^2 = 5^2.$$

TAKE ANOTHER LOOK

Comparing Two and Three Dimensions

In this section, you saw similarities between formulas in two-dimensional coordinate geometry and formulas in three-dimensional coordinate geometry. In two-dimensional coordinate geometry, the graph of the equation

$ax + by + c = 0$

is a line. In three-dimensional coordinate geometry, what is the graph of the equation

$ax + by + c = 0?$

Is it a line? Explain your reasoning. Use a three-dimensional graphing utility to verify your result.

In Exercises 1–4, find the distance between the points.

1. $(5, 1), (3, 5)$

2. $(2, 3), (-1, -1)$

3. $(-5, 4), (-5, -4)$

4. $(-3, 6), (-3, -2)$

In Exercises 5–8, find the midpoint of the line segment connecting the points.

5. $(2, 5), (6, 9)$

6. $(-1, -2), (3, 2)$

7. $(-6, 0), (6, 6)$

8. $(-4, 3), (2, -1)$

In Exercises 9 and 10, write the standard equation of the circle.

9. Center: $(2, 3)$; radius: 2

10. Endpoints of a diameter: $(4, 0), (-2, 8)$

EXERCISES 7.1

In Exercises 1–4, plot the points on the same three-dimensional coordinate system.

1. (a) $(2, 1, 3)$

  (b) $(-1, 2, 1)$

2. (a) $(3, -2, 5)$

  (b) $\left(\frac{3}{2}, 4, -2\right)$

3. (a) $(5, -2, 2)$

  (b) $(5, -2, -2)$

4. (a) $(0, 4, -5)$

  (b) $(4, 0, 5)$

In Exercises 5–8, find the distance between the two points.

5. $(4, 1, 5), (8, 2, 6)$

6. $(-4, -1, 1), (2, -1, 5)$

7. $(-1, -5, 7), (-3, 4, -4)$

8. $(8, -2, 2), (8, -2, 4)$

In Exercises 9–12, find the coordinates of the midpoint of the line segment joining the two points.

9. $(6, -9, 1), (-2, -1, 5)$

10. $(4, 0, -6), (8, 8, 20)$

11. $(-5, -2, 5), (6, 3, -7)$

12. $(0, -2, 5), (4, 2, 7)$

In Exercises 13–16, find (x, y, z).

13.

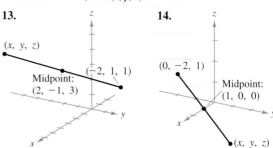

(x, y, z)

Midpoint: $(2, -1, 3)$

$(-2, 1, 1)$

14.

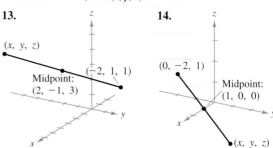

$(0, -2, 1)$

Midpoint: $(1, 0, 0)$

(x, y, z)

15.

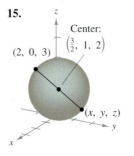

Center: $\left(\frac{3}{2}, 1, 2\right)$

$(2, 0, 3)$

(x, y, z)

16.

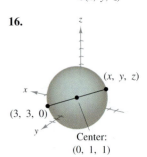

(x, y, z)

$(3, 3, 0)$

Center: $(0, 1, 1)$

In Exercises 17–20, find the lengths of the sides of the triangle with the given vertices, and determine whether the triangle is a right triangle, an isosceles triangle, or neither of these.

17. $(0, 0, 0), (2, 2, 1), (2, -4, 4)$

18. $(5, 3, 4), (7, 1, 3), (3, 5, 3)$

19. $(-2, 2, 4), (-2, 2, 6), (-2, 4, 8)$

20. $(5, 0, 0), (0, 2, 0), (0, 0, -3)$

In Exercises 21–30, find the standard form of the equation of the sphere.

21.
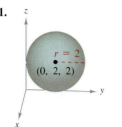

22.
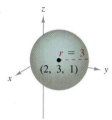

23.
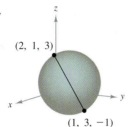

24.
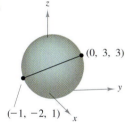

25. Center: $(1, 1, 5)$; radius: 3

26. Center: $(4, -1, 1)$; radius: 5

27. Endpoints of diameter: $(2, 0, 0), (0, 6, 0)$

28. Endpoints of diameter: $(1, 0, 0), (0, 5, 0)$

29. Center: $(-2, 1, 1)$; tangent to the xy-coordinate plane

30. Center: $(1, 2, 0)$; tangent to the yz-coordinate plane

In Exercises 31–36, find the sphere's center and radius.

31. $x^2 + y^2 + z^2 - 5x = 0$

32. $x^2 + y^2 + z^2 - 8y = 0$

33. $x^2 + y^2 + z^2 - 2x + 6y + 8z + 1 = 0$

34. $x^2 + y^2 + z^2 - 4y + 6z + 4 = 0$

35. $2x^2 + 2y^2 + 2z^2 - 4x - 12y - 8z + 3 = 0$

36. $4x^2 + 4y^2 + 4z^2 - 8x + 16y + 11 = 0$

In Exercises 37–40, sketch the xy-trace of the sphere.

37. $(x - 1)^2 + (y - 3)^2 + (z - 2)^2 = 25$

38. $(x + 1)^2 + (y + 2)^2 + (z - 2)^2 = 16$

39. $x^2 + y^2 + z^2 - 6x - 10y + 6z + 30 = 0$

40. $x^2 + y^2 + z^2 - 4y + 2z - 60 = 0$

In Exercises 41 and 42, sketch the yz-trace of the sphere.

41. $x^2 + y^2 + z^2 - 4x - 4y - 6z - 12 = 0$

42. $x^2 + y^2 + z^2 - 6x - 10y + 6z + 30 = 0$

In Exercises 43–46, sketch the trace of the intersection of each plane with the given sphere.

43. $x^2 + y^2 + z^2 = 25$

(a) $z = 3$ (b) $x = 4$

44. $x^2 + y^2 + z^2 = 169$

(a) $x = 5$ (b) $y = 12$

45. $x^2 + y^2 + z^2 - 4x - 6y + 9 = 0$

(a) $x = 2$ (b) $y = 3$

46. $x^2 + y^2 + z^2 - 8x - 6z + 16 = 0$

(a) $x = 4$ (b) $z = 3$

47. *Geology* Crystals are classified according to their symmetry. Crystals shaped like cubes are classified as isometric. Suppose you have mapped the vertices of a crystal onto a three-dimensional coordinate system. Determine (x, y, z) if the crystal is isometric.

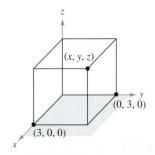

Halite crystals (rock salt) are classified as isometric.

48. *Physical Science* Assume that Earth is a sphere with a radius of 3963 miles. If the center of Earth is placed at the origin of a three-dimensional coordinate system, what is the equation of the sphere? Lines of longitude that run north-south could be represented by what trace(s)? What shape would each of these traces form? Why? Lines of latitude that run east-west could be represented by what trace(s)? Why? What shape would each of these traces form? Why?

7.2 SURFACES IN SPACE

- Sketch planes in space.
- Draw planes in space with different numbers of intercepts.
- Classify quadric surfaces in space.

Equations of Planes in Space

In Section 7.1, you studied one type of surface in space—a sphere. In this section, you will study a second type—a plane in space. The **general equation of a plane** in space is

$$ax + by + cz = d. \qquad \text{General equation of a plane}$$

Note the similarity of this equation to the general equation of a line in the plane. In fact, if you intersect the plane represented by this equation with each of the three coordinate planes, you will obtain traces that are lines, as shown in Figure 7.10.

In Figure 7.10, the points where the plane intersects the three coordinate axes are the x-, y-, and z-intercepts of the plane. By connecting these three points, you can form a triangular region, which helps you visualize the plane in space.

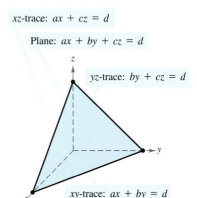

xz-trace: $ax + cz = d$

Plane: $ax + by + cz = d$

yz-trace: $by + cz = d$

xy-trace: $ax + by = d$

FIGURE 7.10

EXAMPLE 1 Sketching a Plane in Space

Find the x-, y-, and z-intercepts of the plane given by

$$3x + 2y + 4z = 12.$$

Then sketch the plane.

SOLUTION To find the x-intercept, let both y and z be zero.

$$3x + 2(0) + 4(0) = 12 \qquad \text{Substitute 0 for } y \text{ and } z.$$
$$3x = 12 \qquad \text{Simplify.}$$
$$x = 4 \qquad \text{Solve for } x.$$

So, the x-intercept is $(4, 0, 0)$. To find the y-intercept, let x and z be zero and conclude that $y = 6$. So, the y-intercept is $(0, 6, 0)$. Similarly, by letting x and y be zero, you can determine that $z = 3$ and that the z-intercept is $(0, 0, 3)$. Figure 7.11 shows the triangular portion of the plane formed by connecting the three intercepts.

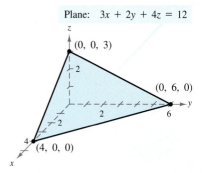

Plane: $3x + 2y + 4z = 12$

$(0, 0, 3)$

$(0, 6, 0)$

$(4, 0, 0)$

FIGURE 7.11 Sketch Made by Connecting Intercepts: $(4, 0, 0), (0, 6, 0), (0, 0, 3)$

> **TRY IT 1**
>
> Find the x-, y-, and z-intercepts of the plane given by
>
> $$2x + 4y + z = 8.$$
>
> Then sketch the plane.

Drawing Planes in Space

The planes shown in Figures 7.10 and 7.11 have three intercepts. When this occurs, we suggest that you draw the plane by sketching the triangular region formed by connecting the three intercepts.

It is possible for a plane in space to have fewer than three intercepts. This occurs when one or more of the coefficients in the equation $ax + by + cz = d$ is zero. Figure 7.12 shows some planes in space that have only one intercept, and Figure 7.13 shows some that have only two intercepts. In each figure, note the use of dashed lines and shading to give the illusion of three dimensions.

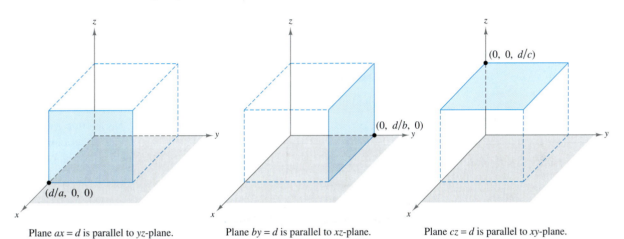

Plane $ax = d$ is parallel to yz-plane. Plane $by = d$ is parallel to xz-plane. Plane $cz = d$ is parallel to xy-plane.

FIGURE 7.12 Planes Parallel to Coordinate Planes

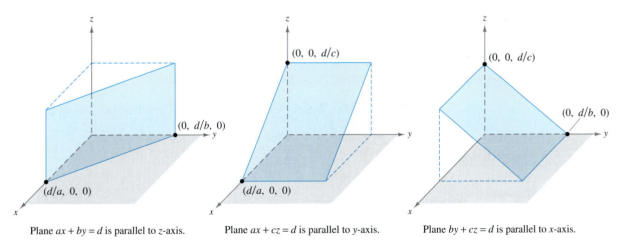

Plane $ax + by = d$ is parallel to z-axis. Plane $ax + cz = d$ is parallel to y-axis. Plane $by + cz = d$ is parallel to x-axis.

FIGURE 7.13 Planes Parallel to Coordinate Axes

DISCOVERY

What is the equation of each plane?

(a) xy-plane   (b) xz-plane   (c) yz-plane

Quadric Surfaces

A third common type of surface in space is a **quadric surface.** Every quadric surface has an equation of the form

$$Ax^2 + By^2 + Cz^2 + Dx + Ey + Fz + G = 0.$$ Second-degree equation

There are six basic types of quadric surfaces.

1. Elliptic cone

2. Elliptic paraboloid

3. Hyperbolic paraboloid

4. Ellipsoid

5. Hyperboloid of one sheet

6. Hyperboloid of two sheets

The six types are summarized on pages 468 and 469. Notice that each surface is pictured with two types of three-dimensional sketches. The computer-generated sketches use traces with hidden lines to give the illusion of three dimensions. The artist-rendered sketches use shading to create the same illusion.

All of the quadric surfaces on pages 468 and 469 are centered at the origin and have axes along the coordinate axes. Moreover, only one of several possible orientations of each surface is shown. If the surface has a different center or is oriented along a different axis, then its standard equation will change accordingly. For instance, the ellipsoid

$$\frac{x^2}{1^2} + \frac{y^2}{3^2} + \frac{z^2}{2^2} = 1$$

has $(0, 0, 0)$ as its center, but the ellipsoid

$$\frac{(x - 2)^2}{1^2} + \frac{(y + 1)^2}{3^2} + \frac{(z - 4)^2}{2^2} = 1$$

has $(2, -1, 4)$ as its center. A computer-generated graph of the first ellipsoid is shown in Figure 7.14.

DISCOVERY

One way to help visualize a quadric surface is to determine the intercepts of the surface with the coordinate axes. What are the intercepts of the ellipsoid in Figure 7.14?

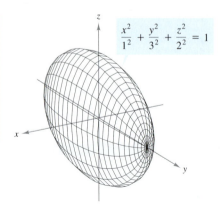

$$\frac{x^2}{1^2} + \frac{y^2}{3^2} + \frac{z^2}{2^2} = 1$$

FIGURE 7.14

Using a Three-Dimensional Graphing Utility

Most three-dimensional graphing utilities represent surfaces by sketching several traces of the surface. The traces are usually taken in equally spaced parallel planes. Depending on the graphing utility, the sketch can be made with one set, two sets, or three sets of traces. For instance, the two sketches shown below use two sets of traces: in one set the traces are parallel to the xz-plane, and in the other set the traces are parallel to the yz-plane.

To sketch the graph of an equation involving x, y, and z with a three-dimensional "function grapher," you must first solve the equation for z. After entering the equation, you need to specify a rectangular viewing box (the three-dimensional analog of a viewing window).

The two sketches shown below were generated by *Derive* for Windows. If you have access to a three-dimensional graphing utility, try using it to duplicate these graphs.

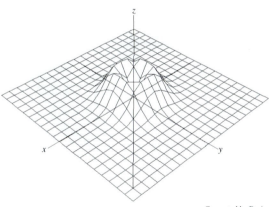

Equation: $z = (x^2 + y^2)e^{1-x^2-y^2}$
Grid: 20 traces by 20 traces
Viewing Box: $-5 \le x \le 5$
$-5 \le y \le 5$
$-5 \le z \le 5$

Generated by Derive

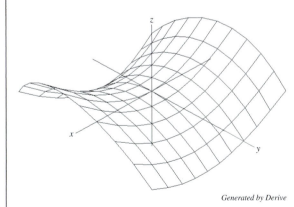

Equation: $z = \frac{1}{4}x^2 - \frac{1}{4}y^2$
Grid: 10 traces by 10 traces
Viewing Box: $-4 \le x \le 4$
$-4 \le y \le 4$
$-4 \le z \le 4$

Generated by Derive

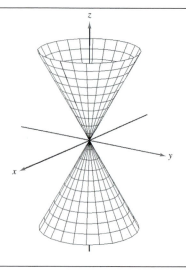

Elliptic Cone

$$\frac{x^2}{a^2} + \frac{y^2}{b^2} - \frac{z^2}{c^2} = 0$$

Trace	Plane
Ellipse	Parallel to xy-plane
Hyperbola	Parallel to xz-plane
Hyperbola	Parallel to yz-plane

The axis of the cone corresponds to the variable whose coefficient is negative. The traces in the coordinate planes parallel to this axis are intersecting lines.

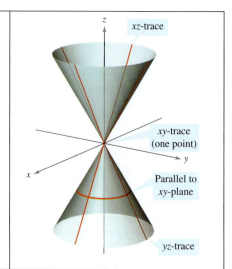

Elliptic Paraboloid

$$z = \frac{x^2}{a^2} + \frac{y^2}{b^2}$$

Trace	Plane
Ellipse	Parallel to xy-plane
Parabola	Parallel to xz-plane
Parabola	Parallel to yz-plane

The axis of the paraboloid corresponds to the variable raised to the first power.

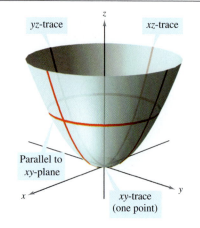

Hyperbolic Paraboloid

$$z = \frac{y^2}{b^2} - \frac{x^2}{a^2}$$

Trace	Plane
Hyperbola	Parallel to xy-plane
Parabola	Parallel to xz-plane
Parabola	Parallel to yz-plane

The axis of the paraboloid corresponds to the variable raised to the first power.

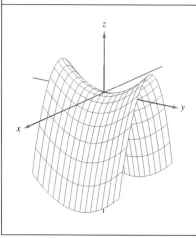

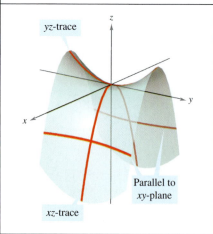

Ellipsoid

$$\frac{x^2}{a^2} + \frac{y^2}{b^2} + \frac{z^2}{c^2} = 1$$

Trace	Plane
Ellipse	Parallel to xy-plane
Ellipse	Parallel to xz-plane
Ellipse	Parallel to yz-plane

The surface is a sphere if the coefficients a, b, and c are equal and nonzero.

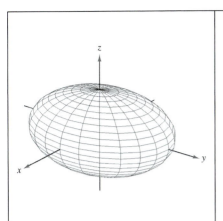

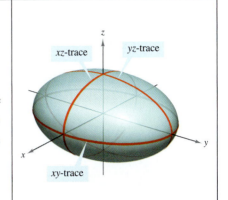

Hyperboloid of One Sheet

$$\frac{x^2}{a^2} + \frac{y^2}{b^2} - \frac{z^2}{c^2} = 1$$

Trace	Plane
Ellipse	Parallel to xy-plane
Hyperbola	Parallel to xz-plane
Hyperbola	Parallel to yz-plane

The axis of the hyperboloid corresponds to the variable whose coefficient is negative.

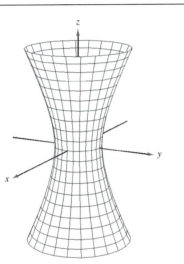

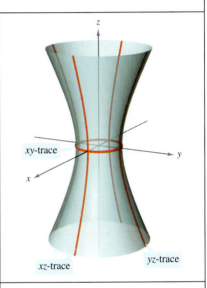

Hyperboloid of Two Sheets

$$\frac{z^2}{c^2} - \frac{x^2}{a^2} - \frac{y^2}{b^2} = 1$$

Trace	Plane
Ellipse	Parallel to xy-plane
Hyperbola	Parallel to xz-plane
Hyperbola	Parallel to yz-plane

The axis of the hyperboloid corresponds to the variable whose coefficient is positive. There is no trace in the coordinate plane perpendicular to this axis.

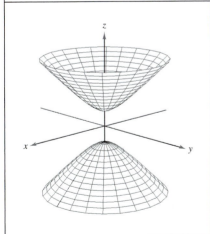

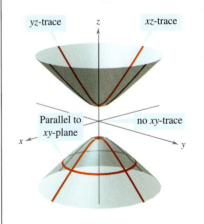

When classifying quadric surfaces, note that the two types of paraboloids have one variable raised to the first power. The other four types of quadric surfaces have equations that are of second degree in *all* three variables.

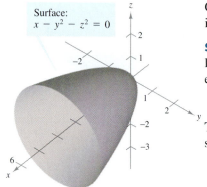

Surface:
$x - y^2 - z^2 = 0$

FIGURE 7.15   Elliptic Paraboloid

| **EXAMPLE 2** | **Classifying a Quadric Surface** |

Classify the surface given by $x - y^2 - z^2 = 0$. Describe the traces of the surface in the xy-plane, the xz-plane, and the plane given by $x = 1$.

SOLUTION   Because x is raised only to the first power, the surface is a paraboloid whose axis is the x-axis, as shown in Figure 7.15. In standard form, the equation is

$$x = y^2 + z^2.$$

The traces in the xy-plane, the xz-plane, and the plane given by $x = 1$ are as shown.

Trace in xy-plane ($z = 0$):	$x = y^2$	Parabola
Trace in xz-plane ($y = 0$):	$x = z^2$	Parabola
Trace in plane $x = 1$:	$y^2 + z^2 = 1$	Circle

These three traces are shown in Figure 7.16. From the traces, you can see that the surface is an elliptic (or circular) paraboloid. If you have access to a three-dimensional graphing utility, try using it to graph this surface. If you do this, you will discover that sketching surfaces in space is not a simple task—even with a graphing utility.

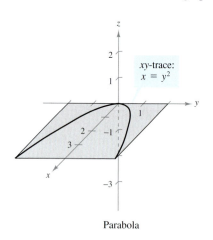

xy-trace:
$x = y^2$

Parabola

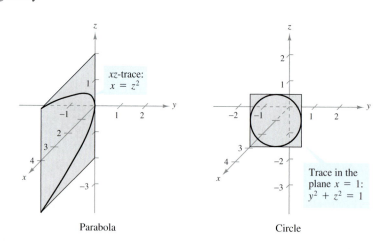

xz-trace:
$x = z^2$

Parabola

Trace in the plane $x = 1$:
$y^2 + z^2 = 1$

Circle

FIGURE 7.16

TRY IT 2

Classify the surface given by $x^2 + y^2 - z^2 = 1$. Describe the traces of the surface in the xy-plane, the yz-plane, the xz-plane, and the plane given by $z = 3$.

EXAMPLE 3 **Classifying Quadric Surfaces**

Write each quadric surface in standard form and classify each equation.

(a) $x^2 - 4y^2 - 4z^2 - 4 = 0$

(b) $x^2 + 4y^2 + z^2 - 4 = 0$

SOLUTION

(a) The equation $x^2 - 4y^2 - 4z^2 - 4 = 0$ can be written in standard form as

$$\frac{x^2}{4} - y^2 - z^2 = 1. \qquad \text{Standard form}$$

From the standard form, you can see that the graph is a hyperboloid of two sheets, with the x-axis as its axis, as shown in Figure 7.17(a).

(b) The equation $x^2 + 4y^2 + z^2 - 4 = 0$ can be written in standard form as

$$\frac{x^2}{4} + y^2 + \frac{z^2}{4} = 1. \qquad \text{Standard form}$$

From the standard form, you can see that the graph is an ellipsoid, as shown in Figure 7.17(b).

Surface:
$x^2 - 4y^2 - 4z^2 - 4 = 0$

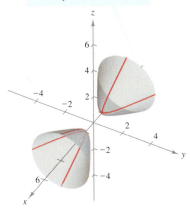

(a)

Surface:
$x^2 + 4y^2 + z^2 - 4 = 0$

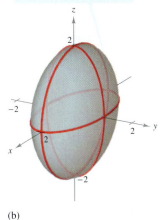

(b)

FIGURE 7.17

TRY IT 3

Write each quadric surface in standard form and classify each equation.

(a) $4x^2 + 9y^2 - 36z = 0$

(b) $36x^2 + 16y^2 - 144z^2 = 0$

TAKE ANOTHER LOOK

Classifying Quadric Surfaces

Classify each quadric surface. Use a three-dimensional graphing utility to verify your results.

a. $\dfrac{x^2}{2^2} - \dfrac{y^2}{4^2} + \dfrac{z^2}{3^2} = 0$ b. $\dfrac{x^2}{2^2} + \dfrac{y^2}{4^2} + \dfrac{z^2}{3^2} = 1$ c. $\dfrac{x^2}{2^2} - \dfrac{y^2}{4^2} + \dfrac{z^2}{3^2} = 1$

The following warm-up exercises involve skills that were covered in earlier sections. You will use these skills in the exercise set for this section.

In Exercises 1–4, find the x- and y-intercepts of the function.

1. $3x + 4y = 12$

2. $6x + y = -8$

3. $-2x + y = -2$

4. $-x - y = 5$

In Exercises 5–8, rewrite the expression by completing the square.

5. $x^2 + y^2 + z^2 - 2x - 4y - 6z + 15 = 0$

6. $x^2 + y^2 - z^2 - 8x + 4y - 6z + 11 = 0$

7. $z - 2 = x^2 + y^2 + 2x - 2y$

8. $x^2 + y^2 + z^2 - 6x + 10y + 26z = -202$

In Exercises 9 and 10, write the expression in standard form.

9. $16x^2 - 16y^2 + 16z^2 = 4$

10. $9x^2 - 9y^2 + 9z^2 = 36$

EXERCISES 7.2

In Exercises 1–12, find the intercepts and sketch the graph of the plane.

1. $4x + 2y + 6z = 12$

2. $3x + 6y + 2z = 6$

3. $3x + 3y + 5z = 15$

4. $x + y + z = 3$

5. $2x - y + 3z = 8$

6. $2x - y + z = 4$

7. $z = 3$

8. $y = -4$

9. $y + z = 5$

10. $x + 2y = 8$

11. $x + y - z = 0$

12. $x - 3z = 3$

In Exercises 13–22, determine whether the planes $a_1x + b_1y + c_1z = d_1$ and $a_2x + b_2y + c_2z = d_2$ are parallel, perpendicular, or neither. The planes are parallel if there exists a nonzero constant k such that $a_1 = ka_2, b_1 = kb_2, c_1 = kc_2$, and perpendicular if $a_1a_2 + b_1b_2 + c_1c_2 = 0$.

13. $5x - 3y + z = 4,\ x + 4y + 7z = 1$

14. $3x + y - 4z = 3,\ -9x - 3y + 12z = 4$

15. $x - 5y - z = 1,\ 5x - 25y - 5z = -3$

16. $x + 3y + 2z = 6,\ 4x - 12y + 8z = 24$

17. $x + 2y = 3,\ 4x + 8y = 5$

18. $x + 3y + z = 7,\ x - 5z = 0$

19. $2x + y = 3,\ 3x - 5z = 0$

20. $2x - z = 1,\ 4x + y + 8z = 10$

21. $x = 6,\ y = -1$

22. $x = -2,\ y = 4$

In Exercises 23–30, find the distance between the point and the plane (see figure). The distance D between a point (x_0, y_0, z_0) and the plane $ax + by + cz + d = 0$ is

$$D = \frac{|ax_0 + by_0 + cz_0 + d|}{\sqrt{a^2 + b^2 + c^2}}.$$

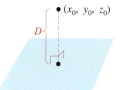

Plane:
$ax + by + cz + d = 0$

23. $(0, 0, 0),\ 2x + 3y + z = 12$

24. $(0, 0, 0),\ x - 3y + 4z = 6$

25. $(1, 5, -4),\ 3x - y + 2z = 6$

26. $(1, 2, 3),\ 2x - y + z = 4$

27. $(1, 0, -1),\ 2x - 4y + 3z = 12$

28. $(2, -1, 0),\ 3x + 3y + 2z = 6$

29. $(3, 2, -1),\ 2x - 3y + 4z = 24$

30. $(-2, 1, 0),\ 2x + 5y - z = 20$

In Exercises 31–38, match the given equation with the correct graph. [The graphs are labeled (a)–(h).]

31. $\dfrac{x^2}{9} + \dfrac{y^2}{16} + \dfrac{z^2}{9} = 1$

32. $x^2 - \dfrac{4y^2}{15} + z^2 = -\dfrac{4}{15}$

33. $4x^2 + 4y^2 - z^2 = 4$

34. $y^2 = 4x^2 + 9z^2$

35. $4x^2 - 4y + z^2 = 0$

36. $12z = -3y^2 + 4x^2$

37. $4x^2 - y^2 + 4z = 0$

38. $x^2 + y^2 + z^2 = 9$

(a)

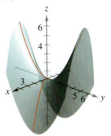

(b)

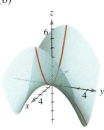

(c)

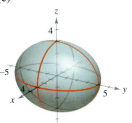

(d)

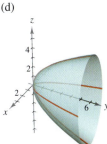

(e)

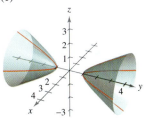

(f)

(g)

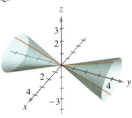

(h)

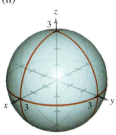

In Exercises 39–42, describe the traces of the surface in the given planes.

Surface	Planes
39. $x^2 - y - z^2 = 0$	xy-plane, $y = 1$, yz-plane
40. $y = x^2 + z^2$	xy-plane, $y = 1$, yz-plane
41. $\dfrac{x^2}{4} + y^2 + z^2 = 1$	xy-plane, xz-plane, yz-plane
42. $y^2 + z^2 - x^2 = 1$	xy-plane, xz-plane, yz-plane

In Exercises 43–56, identify the quadric surface.

43. $x^2 + \dfrac{y^2}{4} + z^2 = 1$

44. $\dfrac{x^2}{9} + \dfrac{y^2}{16} + \dfrac{z^2}{16} = 1$

45. $25x^2 + 25y^2 - z^2 = 5$

46. $9x^2 + 4y^2 - 8z^2 = 72$

47. $x^2 - y + z^2 = 0$

48. $z = 4x^2 + y^2$

49. $x^2 - y^2 + z = 0$

50. $z^2 - x^2 - \dfrac{y^2}{4} = 1$

51. $4x^2 - y^2 + 4z^2 = -16$

52. $z^2 = x^2 + \dfrac{y^2}{4}$

53. $z^2 = 9x^2 + y^2$

54. $4y = x^2 + z^2$

55. $3z = -y^2 + x^2$

56. $z^2 = 2x^2 + 2y^2$

 In Exercises 57–60, use a three-dimensional graphing utility to graph the function.

57. $z = y^2 - x^2 + 1$

58. $z = x^2 + y^2 + 1$

59. $z = \dfrac{x^2}{2} + \dfrac{y^2}{4}$

60. $z = \dfrac{1}{12}\sqrt{144 - 16x^2 - 9y^2}$

61. *Physical Science* Because of the forces caused by its rotation, Earth is actually an oblate ellipsoid rather than a sphere. The equatorial radius is 3963 miles and the polar radius is 3950 miles. Find an equation of the ellipsoid. Assume that the center of Earth is at the origin and the xy-trace ($z = 0$) corresponds to the equator.

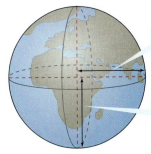

Equatorial radius = 3963 mi

Polar radius = 3950 mi

7.3   FUNCTIONS OF SEVERAL VARIABLES

- Evaluate functions of several variables.
- Find the domains and ranges of functions of several variables.
- Read contour maps and sketch level curves of functions of two variables.
- Use functions of several variables to answer questions about real-life situations.

Functions of Several Variables

In the first six chapters of this text, you studied functions of a single independent variable. Many quantities in science, business, and technology, however, are functions not of one, but of two or more variables. For instance, the demand function for a product is often dependent on the price *and* the advertising, rather than on the price alone.

The notation for functions of two or more variables is similar to that used for functions of a single variable. For example,

$$f(x, y) = x^2 + xy \quad \text{and} \quad g(x, y) = e^{x+y}$$

2 variables   2 variables

are functions of two variables, and

$$f(x, y, z) = x + 2y - 3z$$

3 variables

is a function of three variables.

Definition of a Function of Two Variables

Let D be a set of ordered pairs of real numbers. If to each ordered pair (x, y) in D there corresponds a unique real number $z = f(x, y)$, then f is called a **function of x and y.** The set D is the **domain** of f, and the corresponding set of z-values is the **range** of f. Functions of three, four, or more variables are defined similarly.

EXAMPLE 1   **Evaluating Functions of Several Variables**

(a) For $f(x, y) = 2x^2 - y^2$, you can evaluate $f(2, 3)$ as shown.

$$f(2, 3) = 2(2)^2 - (3)^2$$
$$= 8 - 9$$
$$= -1$$

(b) For $f(x, y, z) = e^x(y + z)$, you can evaluate $f(0, -1, 4)$ as shown.

$$f(0, -1, 4) = e^0(-1 + 4)$$
$$= (1)(3)$$
$$= 3$$

TRY IT 1

Find the function values of $f(x, y)$.

(a) For $f(x, y) = x^2 + 2xy$, find $f(2, -1)$.

(b) For $f(x, y, z) = \dfrac{2x^2z}{y^3}$, find $f(-3, 2, 1)$.

The Graph of a Function of Two Variables

A function of two variables can be represented graphically as a surface in space by letting $z = f(x, y)$. When sketching the graph of a function of x and y, remember that even though the graph is three-dimensional, the domain of the function is two-dimensional—it consists of the points in the xy-plane for which the function is defined. As with functions of a single variable, unless specifically restricted, the domain of a function of two variables is assumed to be the set of all points (x, y) for which the defining equation has meaning. In other words, to each point (x, y) in the domain of f there corresponds a point (x, y, z) on the surface, and conversely, to each point (x, y, z) on the surface there corresponds a point (x, y) in the domain of f.

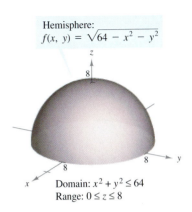

Hemisphere:
$f(x, y) = \sqrt{64 - x^2 - y^2}$

Domain: $x^2 + y^2 \leq 64$
Range: $0 \leq z \leq 8$

FIGURE 7.18

EXAMPLE 2   **Finding the Domain and Range of a Function**

Find the domain and range of the function

$$f(x, y) = \sqrt{64 - x^2 - y^2}.$$

SOLUTION   Because no restrictions are given, the domain is assumed to be the set of all points for which the defining equation makes sense.

$64 - x^2 - y^2 \geq 0$ Quantity inside radical must be nonnegative.

$\quad\quad x^2 + y^2 \leq 64$ Domain of the function

So, the domain is the set of all points that lie on or inside the circle given by $x^2 + y^2 = 8^2$. The range of f is the set

$0 \leq z \leq 8.$ Range of the function

As shown in Figure 7.18, the graph of the function is a hemisphere. ───

TRY IT 2

Find the domain and range of the function

$$f(x, y) = \sqrt{9 - x^2 - y^2}.$$

TECHNOLOGY

Some three-dimensional graphing utilities can graph equations in x, y, and z. Others are programmed to graph only functions of x and y. A surface in space represents the graph of a function of x and y only if each vertical line intersects the surface at most once. For instance, the surface shown in Figure 7.18 passes this vertical line test, but the surface at the right (drawn by *Mathematica*) does not represent the graph of a function of x and y.

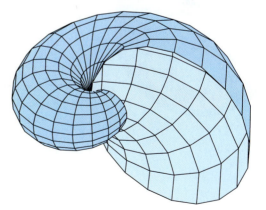

Some vertical lines intersect this surface more than once. So, the surface does not pass the vertical line test and is not a function of x and y.

Contour Maps and Level Curves

A **contour map** of a surface is created by *projecting* traces, taken in evenly spaced planes that are parallel to the xy-plane, onto the xy-plane. Each projection is a **level curve** of the surface.

Contour maps are used to create weather maps, topographical maps, and population density maps. For instance, Figure 7.19(a) shows a graph of a "mountain and valley" surface given by $z = f(x, y)$. Each of the level curves in Figure 7.19(b) represents the intersection of the surface $z = f(x, y)$ with a plane $z = c$, where $c = 828, 830, \ldots, 854$.

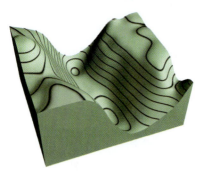

(a) Surface

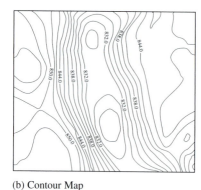

(b) Contour Map

FIGURE 7.19

EXAMPLE 3   **Reading a Contour Map**

The "contour map" in Figure 7.20 was computer generated using data collected by satellite instrumentation. The map uses color to represent levels of chlorine nitrate in the atmosphere. Chlorine nitrate contributes to the ozone depletion in the Earth's atmosphere. The red areas represent the highest level of chlorine nitrate and the dark blue areas represent the lowest level. Describe the areas that have the highest levels of chlorine nitrate.   *(Source: Lockheed Missiles and Space Company)*

SOLUTION   The highest levels of chlorine nitrate are in the Antarctic Ocean, surrounding Antarctica. Although chlorine nitrate is not itself harmful to ozone, it has a tendency to convert to chlorine monoxide, which *is* harmful to ozone. Once the chlorine nitrate is converted to chlorine monoxide, it no longer shows on the contour map. So, Antarctica itself shows little chlorine nitrate—the nitrate has been converted to monoxide. If you have seen maps showing the "ozone hole" in Earth's atmosphere, you know that the hole occurs over Antarctica.

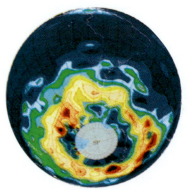

Lockheed Missiles and Space Company

FIGURE 7.20

TRY IT 3

When the level curves of a contour map are close together, is the surface of the contour map steep or nearly level? When the level curves of a contour map are far apart, is the surface of the contour map steep or nearly level?

| **EXAMPLE 4** | **Reading a Contour Map** | |

The contour map shown in Figure 7.21 represents the economy in the United States. Discuss the use of color to represent the level curves. *(Source: U.S. Census Bureau)*

SOLUTION You can see from the key that the light yellow regions are mainly used in crop production. The gray areas represent regions that are unproductive. Manufacturing centers are denoted by large red dots and mineral deposits are denoted by small black dots.

One advantage of such a map is that it allows you to "see" the components of the country's economy at a glance. From the map it is clear that the Midwest is responsible for most of the crop production in the United States.

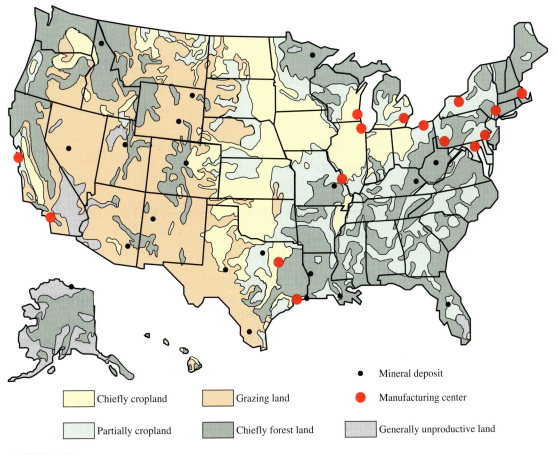

Chiefly cropland	
Partially cropland	
Grazing land	
Chiefly forest land	
• Mineral deposit	
🔴 Manufacturing center	
Generally unproductive land	

FIGURE 7.21

TRY IT 4

Use Figure 7.21 to describe how Alaska contributes to the U.S. economy. Does Alaska contain any manufacturing centers? Does Alaska contain any mineral deposits?

In Exercises 1–4, evaluate the function when $x = -3$.

1. $f(x) = 5 - 2x$

2. $f(x) = -x^2 + 4x + 5$

3. $y = \sqrt{4x^2 - 3x + 4}$

4. $y = \sqrt[3]{34 - 4x + 2x^2}$

In Exercises 5–8, find the domain of the function.

5. $f(x) = 5x^2 + 3x - 2$

6. $g(x) = \dfrac{1}{2x} - \dfrac{2}{x + 3}$

7. $h(y) = \sqrt{y - 5}$

8. $f(y) = \sqrt{y^2 - 5}$

In Exercises 9 and 10, evaluate the expression.

9. $(476)^{0.65}$

10. $(251)^{0.35}$

EXERCISES 7.3

In Exercises 1–14, find the function values.

1. $f(x, y) = \dfrac{x}{y}$

(a) $f(3, 2)$ (b) $f(-1, 4)$

(c) $f(30, 5)$ (d) $f(5, y)$

(e) $f(x, 2)$ (f) $f(5, t)$

2. $f(x, y) = 4 - x^2 - 4y^2$

(a) $f(0, 0)$ (b) $f(0, 1)$

(c) $f(2, 3)$ (d) $f(1, y)$

(e) $f(x, 0)$ (f) $f(t, 1)$

3. $f(x, y) = xe^y$

(a) $f(5, 0)$ (b) $f(3, 2)$

(c) $f(2, -1)$ (d) $f(5, y)$

(e) $f(x, 2)$ (f) $f(t, t)$

4. $g(x, y) = \ln|x + y|$

(a) $g(2, 3)$ (b) $g(5, 6)$

(c) $g(e, 0)$ (d) $g(0, 1)$

(e) $g(2, -3)$ (f) $g(e, e)$

5. $h(x, y, z) = \dfrac{xy}{z}$

(a) $h(2, 3, 9)$ (b) $h(1, 0, 1)$

6. $f(x, y, z) = \sqrt{x + y + z}$

(a) $f(0, 5, 4)$ (b) $f(6, 8, -3)$

7. $V(r, h) = \pi r^2 h$

(a) $V(3, 10)$ (b) $V(5, 2)$

8. $F(r, N) = 500\left(1 + \dfrac{r}{12}\right)^N$

(a) $F(0.09, 60)$ (b) $F(0.14, 240)$

9. $A(P, r, t) = P\left[\left(1 + \dfrac{r}{12}\right)^{12t} - 1\right]\left(1 + \dfrac{12}{r}\right)$

(a) $A(100, 0.10, 10)$ (b) $A(275, 0.0925, 40)$

10. $A(P, r, t) = Pe^{rt}$

(a) $A(500, 0.10, 5)$ (b) $A(1500, 0.12, 20)$

11. $f(x, y) = \displaystyle\int_x^y (2t - 3)\, dt$

(a) $f(1, 2)$ (b) $f(1, 4)$

12. $g(x, y) = \displaystyle\int_x^y \dfrac{1}{t}\, dt$

(a) $g(4, 1)$ (b) $g(6, 3)$

13. $f(x, y) = x^2 - 2y$

(a) $f(x + \Delta x, y)$

(b) $\dfrac{f(x, y + \Delta y) - f(x, y)}{\Delta y}$

14. $f(x, y) = 3xy + y^2$

(a) $f(x + \Delta x, y)$

(b) $\dfrac{f(x, y + \Delta y) - f(x, y)}{\Delta y}$

In Exercises 15–18, describe the region R in the xy-coordinate plane that corresponds to the domain of the function, and find the range of the function.

15. $f(x, y) = \sqrt{16 - x^2 - y^2}$

16. $f(x, y) = x^2 + y^2 - 1$

17. $f(x, y) = e^{x/y}$

18. $f(x, y) = \ln(x + y)$

In Exercises 19–28, describe the region R in the xy-coordinate plane that corresponds to the domain of the function.

19. $f(x, y) = \sqrt{9 - 9x^2 - y^2}$

20. $f(x, y) = \sqrt{x^2 + y^2 - 1}$

21. $f(x, y) = \dfrac{x}{y}$

22. $f(x, y) = \dfrac{4y}{x - 1}$

23. $f(x, y) = \dfrac{1}{xy}$

24. $g(x, y) = \dfrac{1}{x - y}$

25. $h(x, y) = x\sqrt{y}$

26. $f(x, y) = \sqrt{xy}$

27. $g(x, y) = \ln(4 - x - y)$

28. $f(x, y) = ye^{1/x}$

In Exercises 29–32, match the graph of the surface with one of the contour maps. [The contour maps are labeled (a)–(d).]

29. $f(x, y) = x^2 + \dfrac{y^2}{4}$

30. $f(x, y) = e^{1 - x^2 + y^2}$

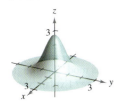

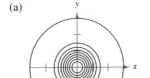

31. $f(x, y) = e^{1 - x^2 - y^2}$

32. $f(x, y) = \ln|y - x^2|$

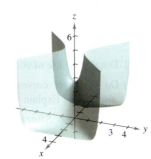

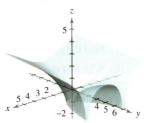

(a)

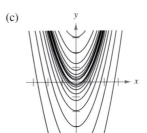

(b)

(c)

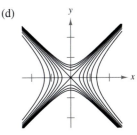

(d)

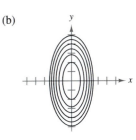

In Exercises 33–40, describe the level curves of the function. Sketch the level curves for the given c-values.

Function	*c-Values*
33. $z = x + y$	$c = -1, 0, 2, 4$
34. $z = 6 - 2x - 3y$	$c = 0, 2, 4, 6, 8, 10$
35. $z = \sqrt{16 - x^2 - y^2}$	$c = 0, 1, 2, 3, 4$
36. $f(x, y) = x^2 + y^2$	$c = 0, 2, 4, 6, 8$
37. $f(x, y) = xy$	$c = \pm 1, \pm 2, \dots, \pm 6$
38. $z = e^{xy}$	$c = 1, 2, 3, 4, \frac{1}{2}, \frac{1}{3}, \frac{1}{4}$
39. $f(x, y) = \dfrac{x}{x^2 + y^2}$	$c = \pm\frac{1}{2}, \pm 1, \pm\frac{3}{2}, \pm 2$
40. $f(x, y) = \ln(x - y)$	$c = 0, \pm\frac{1}{2}, \pm 1, \pm\frac{3}{2}, \pm 2$

41. *Cobb-Douglas Production Function* A manufacturer estimates the Cobb-Douglas production function to be given by

$$f(x, y) = 100x^{0.75}y^{0.25}.$$

Estimate the production levels when $x = 1500$ and $y = 1000$.

42. *Cobb-Douglas Production Function* Use the Cobb-Douglas production function (Example 5) to show that if both the number of units of labor and the number of units of capital are doubled, the production level is also doubled.

43. *Cost* A company manufactures two types of woodburning stoves: a freestanding model and a fireplace-insert model. The cost function for producing x freestanding stoves and y fireplace-insert stoves is given by

$$C(x, y) = 27\sqrt{xy} + 195x + 215y + 980.$$

Find the cost when $x = 80$ and $y = 20$.

A function of two variables has two first partial derivatives and four second partial derivatives. For a function of three variables, there are three first partials

$$f_x, f_y, \quad \text{and} \quad f_z$$

and nine second partials

$$f_{xx}, f_{xy}, f_{xz}, f_{yx}, f_{yy}, f_{yz}, f_{zx}, f_{zy}, \quad \text{and} \quad f_{zz}$$

of which six are mixed partials. To find partial derivatives of order three and higher, follow the same pattern used to find second partial derivatives. For instance, if $z = f(x, y)$, then

$$z_{xxx} = \frac{\partial}{\partial x}\left(\frac{\partial^2 f}{\partial x^2}\right) = \frac{\partial^3 f}{\partial x^3} \quad \text{and} \quad z_{xxy} = \frac{\partial}{\partial y}\left(\frac{\partial^2 f}{\partial x^2}\right) = \frac{\partial^3 f}{\partial y \partial x^2}.$$

EXAMPLE 7   **Finding Second Partial Derivatives**

Find the second partial derivatives of

$$f(x, y, z) = ye^x + x \ln z.$$

SOLUTION   Begin by finding the first partial derivatives.

$$f_x(x, y, z) = ye^x + \ln z, \qquad f_y(x, y, z) = e^x, \qquad f_z(x, y, z) = \frac{x}{z}$$

Then, differentiate with respect to x, y, and z to find the nine second partial derivatives.

$$f_{xx}(x, y, z) = ye^x, \qquad f_{xy}(x, y, z) = e^x, \qquad f_{xz}(x, y, z) = \frac{1}{z}$$

$$f_{yx}(x, y, z) = e^x, \qquad f_{yy}(x, y, z) = 0, \qquad f_{yz}(x, y, z) = 0$$

$$f_{zx}(x, y, z) = \frac{1}{z}, \qquad f_{zy}(x, y, z) = 0, \qquad f_{zz}(x, y, z) = -\frac{x}{z^2}$$

TRY IT 7

Find the second partial derivatives of

$$f(x, y, z) = xe^y + 2xz + y^2.$$

TAKE ANOTHER LOOK

Interpreting a Partial Derivative

The upper half of the sphere shown at the right is given by

$$f(x, y) = \sqrt{1 - x^2 - y^2}.$$

a. Describe the values of $f_x(x, 0)$. Can you find a value of x for which this partial derivative is 1? Use a three-dimensional graphing utility to illustrate your answer graphically.

b. Describe the values of $f_y(0, y)$. Can you find a value of y for which this partial derivative is 1? Use a three-dimensional graphing utility to illustrate your answer graphically.

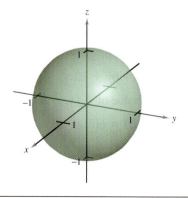

PREREQUISITE
REVIEW 7.4

The following warm-up exercises involve skills that were covered in earlier sections. You will use these skills in the exercise set for this section.

In Exercises 1–8, find the derivative of the function.

1. $f(x) = \sqrt{x^2 + 3}$

2. $g(x) = (3 - x^2)^3$

3. $g(t) = te^{2t+1}$

4. $f(x) = e^{2x}\sqrt{1 - e^{2x}}$

5. $f(x) = \ln(3 - 2x)$

6. $u(t) = \ln\sqrt{t^3 - 6t}$

7. $g(x) = \dfrac{5x^2}{(4x - 1)^2}$

8. $f(x) = \dfrac{(x + 2)^3}{(x^2 - 9)^2}$

In Exercises 9 and 10, evaluate the derivative at the point $(2, 4)$.

9. $f(x) = x^2 e^{x-2}$

10. $g(x) = x\sqrt{x^2 - x + 2}$

EXERCISES 7.4

In Exercises 1–14, find the first partial derivatives with respect to x and with respect to y.

1. $f(x, y) = 2x - 3y + 5$

2. $f(x, y) = x^2 - 3y^2 + 7$

3. $f(x, y) = 5\sqrt{x} - 6y^2$

4. $f(x, y) = x^{-1/2} + 4y^{3/2}$

5. $f(x, y) = \dfrac{x}{y}$

6. $z = x\sqrt{y}$

7. $f(x, y) = \sqrt{x^2 + y^2}$

8. $f(x, y) = \dfrac{xy}{x^2 + y^2}$

9. $z = x^2 e^{2y}$

10. $z = xe^{x+y}$

11. $h(x, y) = e^{-(x^2+y^2)}$

12. $g(x, y) = e^{x/y}$

13. $z = \ln\dfrac{x - y}{(x + y)^2}$

14. $g(x, y) = \ln\sqrt{x^2 + y^2}$

In Exercises 15–20, let $f(x, y) = 3x^2 y e^{x-y}$ and $g(x, y) = 3xy^2 e^{y-x}$. Find each of the following.

15. $f_x(x, y)$

16. $f_y(x, y)$

17. $g_x(x, y)$

18. $g_y(x, y)$

19. $f_x(1, 1)$

20. $g_x(-2, -2)$

In Exercises 21–28, evaluate f_x and f_y at the point.

Function	Point
21. $f(x, y) = 3x^2 + xy - y^2$	$(2, 1)$
22. $f(x, y) = x^2 - 3xy + y^2$	$(1, -1)$
23. $f(x, y) = e^{3xy}$	$(0, 4)$
24. $f(x, y) = e^x y^2$	$(0, 2)$
25. $f(x, y) = \dfrac{xy}{x - y}$	$(2, -2)$
26. $f(x, y) = \dfrac{4xy}{\sqrt{x^2 + y^2}}$	$(1, 0)$
27. $f(x, y) = \ln(x^2 + y^2)$	$(1, 0)$
28. $f(x, y) = \ln\sqrt{xy}$	$(-1, -1)$

In Exercises 29–32, find the first partial derivatives with respect to $x, y,$ and z.

29. $w = 3x^2 y - 5xyz + 10yz^2$

30. $w = \sqrt{x^2 + y^2 + z^2}$

31. $w = \dfrac{xy}{x + y + z}$

32. $w = \dfrac{1}{\sqrt{1 - x^2 - y^2 - z^2}}$

In Exercises 33–38, evaluate w_x, w_y, and w_z at the point.

Function	Point
33. $w = \sqrt{x^2 + y^2 + z^2}$	$(2, -1, 2)$
34. $w = \dfrac{xy}{x + y + z}$	$(1, 2, 0)$
35. $w = \ln\sqrt{x^2 + y^2 + z^2}$	$(3, 0, 4)$
36. $w = \dfrac{1}{\sqrt{1 - x^2 - y^2 - z^2}}$	$(0, 0, 0)$
37. $w = 2xz^2 + 3xyz - 6y^2 z$	$(1, -1, 2)$
38. $w = xye^{z^2}$	$(2, 1, 0)$

In Exercises 39–42, find values of x and y such that $f_x(x, y) = 0$ and $f_y(x, y) = 0$ simultaneously.

39. $f(x, y) = x^2 + 4xy + y^2 - 4x + 16y + 3$

40. $f(x, y) = 3x^3 - 12xy + y^3$

41. $f(x, y) = \dfrac{1}{x} + \dfrac{1}{y} + xy$

42. $f(x, y) = \ln(x^2 + y^2 + 1)$

In Exercises 43–50, find the slope of the surface at the given point in (a) the x-direction and (b) the y-direction.

Function	Point
43. $z = 2x - 3y + 5$	$(2, 1, 6)$
44. $z = xy$	$(1, 2, 2)$
45. $z = x^2 - 9y^2$	$(3, 1, 0)$
46. $z = x^2 + 4y^2$	$(2, 1, 8)$
47. $z = \sqrt{25 - x^2 - y^2}$	$(3, 0, 4)$
48. $z = \dfrac{x}{y}$	$(3, 1, 3)$
49. $z = 4 - x^2 - y^2$	$(1, 1, 2)$

50. $z = x^2 - y^2$ $(-2, 1, 3)$

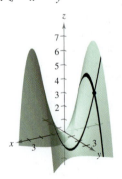

In Exercises 51–54, show that $\partial^2 z/(\partial x \partial y) = \partial^2 z/(\partial y \partial x)$.

51. $z = x^2 - 2xy + 3y^2$

52. $z = x^4 - 3x^2y^2 + y^4$

53. $z = \dfrac{e^{2xy}}{4x}$

54. $z = \dfrac{x^2 - y^2}{2xy}$

In Exercises 55–62, find the second partial derivatives

$$\frac{\partial^2 z}{\partial x^2}, \quad \frac{\partial^2 z}{\partial y^2}, \quad \frac{\partial^2 z}{\partial y \partial x}, \quad \text{and} \quad \frac{\partial^2 z}{\partial x \partial y}.$$

55. $z = x^3 - 4y^2$

56. $z = 3x^2 - xy + 2y^3$

57. $z = 4x^3 + 3xy^2 - 4y^3$

58. $z = \sqrt{9 - x^2 - y^2}$

59. $z = \dfrac{xy}{x - y}$

60. $z = \dfrac{x}{x + y}$

61. $z = xe^{-y^2}$

62. $z = xe^y + ye^x$

In Exercises 63–66, evaluate the second partial derivatives f_{xx}, f_{xy}, f_{yy}, and f_{yx} at the point.

Function	Point
63. $f(x, y) = x^4 - 3x^2y^2 + y^2$	$(1, 0)$
64. $f(x, y) = \sqrt{x^2 + y^2}$	$(0, 2)$
65. $f(x, y) = \ln(x - y)$	$(2, 1)$
66. $f(x, y) = x^2e^y$	$(-1, 0)$

67. *Marginal Cost*   A company manufactures two models of bicycles: a mountain bike and a racing bike. The cost function for producing x mountain bikes and y racing bikes is given by

$$C = 10\sqrt{xy} + 149x + 189y + 675.$$

Find the marginal costs ($\partial C/\partial x$ and $\partial C/\partial y$) when $x = 120$ and $y = 160$.

68. *Marginal Revenue*   A pharmaceutical corporation has two plants that produce the same over-the-counter medicine. If x_1 and x_2 are the numbers of units produced at plant 1 and plant 2, respectively, then the total revenue for the product is given by

$$R = 200x_1 + 200x_2 - 4x_1^2 - 8x_1x_2 - 4x_2^2.$$

If $x_1 = 4$ and $x_2 = 12$, find the following.

(a) The marginal revenue for plant 1, $\partial R/\partial x_1$

(b) The marginal revenue for plant 2, $\partial R/\partial x_2$

69. *Marginal Productivity* Let $x = 1000$ and $y = 500$ in the Cobb-Douglas production function given by

$$f(x, y) = 100x^{0.6}y^{0.4}.$$

(a) Find the marginal productivity of labor, $\partial f/\partial x$.

(b) Find the marginal productivity of capital, $\partial f/\partial y$.

70. *Marginal Productivity* Repeat Exercise 69 for the production function given by $f(x, y) = 100x^{0.75}y^{0.25}$.

71. *Complementary and Substitute Products* Using the notation of Example 4 in this section, let x_1 and x_2 be the demands for products 1 and 2, respectively, and p_1 and p_2 the prices of products 1 and 2, respectively. Determine whether the demand functions below describe complementary or substitute product relationships.

(a) $x_1 = 150 - 2p_1 - \frac{5}{2}p_2$, $x_2 = 350 - \frac{3}{2}p_1 - 3p_2$

(b) $x_1 = 150 - 2p_1 + 1.8p_2$, $x_2 = 350 + \frac{3}{4}p_1 - 1.9p_2$

(c) $x_1 = \dfrac{1000}{\sqrt{p_1 p_2}}$, $x_2 = \dfrac{750}{p_2\sqrt{p_1}}$

72. *Psychology* Early in the twentieth century, an intelligence test called the *Stanford-Binet Test* (more commonly known as the *IQ test*) was developed. In this test, an individual's mental age M is divided by the individual's chronological age C and the quotient is multiplied by 100. The result is the individual's *IQ*.

$$IQ(M, C) = \frac{M}{C} \times 100$$

Find the partial derivatives of *IQ* with respect to M and with respect to C. Evaluate the partial derivatives at the point $(12, 10)$ and interpret the result. *(Source: Adapted from Bernstein/Clark-Stewart/Roy/Wickens, Psychology, Fourth Edition)*

73. *Education* Let N be the number of applicants to a university, p the charge for food and housing at the university, and t the tuition. Suppose that N is a function of p and t such that $\partial N/\partial p < 0$ and $\partial N/\partial t < 0$. How would you interpret the fact that both partials are negative?

74. *Chemistry* The temperature at any point (x, y) in a steel plate is given by

$$T = 500 - 0.6x^2 - 1.5y^2$$

where x and y are measured in meters. At the point $(2, 3)$, find the rate of change of the temperature with respect to the distance moved along the plate in the directions of the x- and y-axes.

75. *Chemistry* A measure of what hot weather feels like to two average persons is the Apparent Temperature Index. A model for this index is

$$A = 0.885t - 78.7h + 1.20th + 2.70$$

where A is the apparent temperature, t is the air temperature, and h is the relative humidity in decimal form. *(Source: The UMAP Journal)*

(a) Find $\partial A/\partial t$ and $\partial A/\partial h$ when $t = 90°\text{F}$ and $h = 0.80$.

(b) Which has a greater effect on A, air temperature or humidity? Explain your reasoning.

76. *Marginal Utility* The utility function $U = f(x, y)$ is a measure of the utility (or satisfaction) derived by a person from the consumption of two goods x and y. Suppose the utility function is given by

$$U = -5x^2 + xy - 3y^2.$$

(a) Determine the marginal utility of good x.

(b) Determine the marginal utility of good y.

(c) When $x = 2$ and $y = 3$, should a person consume one more unit of good x or one more unit of good y? Explain your reasoning.

(d) Use a three-dimensional graphing utility to graph the function. Interpret the marginal utilities of goods x and y graphically.

BUSINESS CAPSULE

© Charles Nix/www.nixphoto.com

Fred and Richard Calloway of Augusta, Georgia, cofounded Male Care, which provides barber, dry cleaning, and car wash services. Among the many advertising techniques used by the Calloways to attract new clients are coupons, customer referrals, and radio advertising. They also feel that their prime location provides them with a strong customer base. Eighty percent of their 1800 monthly clients are repeat customers.

77. *Research Project* Use your school's library, the Internet, or some other reference source to research a company that increased the demand for its product by creative advertising. Write a paper about the company. Use graphs to show how a change in demand is related to a change in the marginal utility of a product or service.

Lagrange Multipliers with Two Constraints

In Examples 1 through 5, each of the optimization problems contained only one constraint. When an optimization problem has two constraints, you need to introduce a second Lagrange multiplier. The customary symbol for this second multiplier is μ, the Greek letter mu.

EXAMPLE 6 Using Lagrange Multipliers: Two Constraints

Find the minimum value of

$$f(x, y, z) = x^2 + y^2 + z^2 \qquad \text{Objective function}$$

subject to the constraints

$$x + y - 3 = 0 \qquad \text{Constraint 1}$$
$$x + z - 5 = 0. \qquad \text{Constraint 2}$$

SOLUTION Begin by forming the function

$$F(x, y, z, \lambda, \mu) = x^2 + y^2 + z^2 - \lambda(x + y - 3) - \mu(x + z - 5).$$

Next, set the five partial derivatives equal to zero, and solve the resulting system of equations for x, y, and z.

$$F_x(x, y, z, \lambda, \mu) = 2x - \lambda - \mu = 0 \qquad \text{Equation 1}$$
$$F_y(x, y, z, \lambda, \mu) = 2y - \lambda = 0 \qquad \text{Equation 2}$$
$$F_z(x, y, z, \lambda, \mu) = 2z - \mu = 0 \qquad \text{Equation 3}$$
$$F_\lambda(x, y, z, \lambda, \mu) = -x - y + 3 = 0 \qquad \text{Equation 4}$$
$$F_\mu(x, y, z, \lambda, \mu) = -x - z + 5 = 0 \qquad \text{Equation 5}$$

Solving this system of equations produces $x = \frac{8}{3}$, $y = \frac{1}{3}$, and $z = \frac{7}{3}$. So, the minimum value of $f(x, y, z)$ is

$$f\left(\frac{8}{3}, \frac{1}{3}, \frac{7}{3}\right) = \left(\frac{8}{3}\right)^2 + \left(\frac{1}{3}\right)^2 + \left(\frac{7}{3}\right)^2$$
$$= \frac{38}{3}.$$

TRY IT 6

Find the minimum value of
$f(x, y, z) = x^2 + y^2 + z^2$
subject to the constraints

$$x + y - 2 = 0$$
$$x + z - 4 = 0.$$

TAKE ANOTHER LOOK

Solving a System of Equations

Explain how you would solve the system of equations shown in Example 6. Illustrate your description by actually solving the system.

Suppose that in Example 6 you had been asked to find the maximum value of $f(x, y, z)$ subject to the two constraints. How would your answer have been different? Use a three-dimensional graphing utility to verify your conclusion.

The following warm-up exercises involve skills that were covered in earlier sections. You will use these skills in the exercise set for this section.

In Exercises 1–6, solve the system of linear equations.

1. $\begin{cases} 4x - 6y = 3 \\ 2x + 3y = 2 \end{cases}$

2. $\begin{cases} 6x - 6y = 5 \\ -3x - y = 1 \end{cases}$

3. $\begin{cases} 5x - y = 25 \\ x - 5y = 15 \end{cases}$

4. $\begin{cases} 4x - 9y = 5 \\ -x + 8y = -2 \end{cases}$

5. $\begin{cases} 2x - y + z = 3 \\ 2x + 2y + z = 4 \\ -x + 2y + 3z = -1 \end{cases}$

6. $\begin{cases} -x - 4y + 6z = -2 \\ x - 3y - 3z = 4 \\ 3x + y + 3z = 0 \end{cases}$

In Exercises 7–10, find all first partial derivatives.

7. $f(x, y) = x^2y + xy^2$

8. $f(x, y) = 25(xy + y^2)^2$

9. $f(x, y, z) = x(x^2 - 2xy + yz)$

10. $f(x, y, z) = z(xy + xz + yz)$

EXERCISES 7.6

In Exercises 1–12, use Lagrange multipliers to find the given extremum. In each case, assume that x and y are positive.

Objective Function	Constraint
1. Maximize $f(x, y) = xy$	$x + y = 10$
2. Maximize $f(x, y) = xy$	$2x + y = 4$
3. Minimize $f(x, y) = x^2 + y^2$	$x + y - 4 = 0$
4. Minimize $f(x, y) = x^2 + y^2$	$-2x - 4y + 5 = 0$
5. Maximize $f(x, y) = x^2 - y^2$	$y - x^2 = 0$
6. Minimize $f(x, y) = x^2 - y^2$	$x - 2y + 6 = 0$
7. Maximize $f(x, y) = 3x + xy + 3y$	$x + y = 25$
8. Maximize $f(x, y) = 3x + y + 10$	$x^2y = 6$
9. Maximize $f(x, y) = \sqrt{6 - x^2 - y^2}$	$x + y - 2 = 0$
10. Minimize $f(x, y) = \sqrt{x^2 + y^2}$	$2x + 4y - 15 = 0$
11. Maximize $f(x, y) = e^{xy}$	$x^2 + y^2 - 8 = 0$
12. Minimize $f(x, y) = 2x + y$	$xy = 32$

In Exercises 13–18, use Lagrange multipliers to find the given extremum. In each case, assume that x, y, and z are positive.

13. Minimize $f(x, y, z) = 2x^2 + 3y^2 + 2z^2$

Constraint: $x + y + z - 24 = 0$

14. Maximize $f(x, y, z) = xyz$

Constraint: $x + y + z - 6 = 0$

15. Minimize $f(x, y, z) = x^2 + y^2 + z^2$

Constraint: $x + y + z = 1$

16. Minimize $f(x, y) = x^2 - 8x + y^2 - 12y + 48$

Constraint: $x + y = 8$

17. Maximize $f(x, y, z) = x + y + z$

Constraint: $x^2 + y^2 + z^2 = 1$

18. Maximize $f(x, y, z) = x^2y^2z^2$

Constraint: $x^2 + y^2 + z^2 = 1$

In Exercises 19 and 20, use Lagrange multipliers with the objective function

$$f(x, y, z, w) = 2x^2 + y^2 + z^2 + 2w^2$$

and with the given constraints to find the given extremum. In each case, assume that $x, y, z,$ and w are nonnegative.

19. Maximize $f(x, y, z, w)$

Constraint: $2x + 2y + z + w = 2$

20. Maximize $f(x, y, z, w)$

Constraint: $x + y + 2z + 2w = 4$

In Exercises 21–24, use Lagrange multipliers to find the given extremum of f subject to two constraints. In each case, assume that $x, y,$ and z are nonnegative.

21. Maximize $f(x, y, z) = xyz$

Constraints: $x + y + z = 24,\ x - y + z = 12$

22. Minimize $f(x, y, z) = x^2 + y^2 + z^2$

Constraints: $x + 2z = 4,\ x + y = 8$

23. Maximize $f(x, y, z) = xyz$

Constraints: $x^2 + z^2 = 5,\ x - 2y = 0$

24. Maximize $f(x, y, z) = xy + yz$

Constraints: $x + 2y = 6,\ x - 3z = 0$

In Exercises 25 and 26, use a spreadsheet to find the given extremum. In each case, assume that $x, y,$ and z are nonnegative.

25. Maximize $f(x, y, z) = xyz$

Constraints: $x + 3y = 6,\ x - 2z = 0$

26. Minimize $f(x, y, z) = x^2 + y^2 + z^2$

Constraints: $x + 2y = 8,\ x + z = 4$

In Exercises 27–30, find three positive numbers $x, y,$ and z that satisfy the given conditions.

27. The sum is 120 and the product is maximum.

28. The sum is 120 and the sum of the squares is minimum.

29. The sum is S and the product is maximum.

30. The sum is S and the sum of the squares is minimum.

In Exercises 31–34, find the minimum distance from the curve or surface to the given point. (*Hint:* Start by minimizing the square of the distance.)

31. Line: $x + 2y = 5,\ (0, 0)$

Minimize $d^2 = x^2 + y^2$

32. Circle: $(x - 4)^2 + y^2 = 4,\ (0, 10)$

Minimize $d^2 = x^2 + (y - 10)^2$

33. Plane: $x + y + z = 1,\ (2, 1, 1)$

Minimize $d^2 = (x - 2)^2 + (y - 1)^2 + (z - 1)^2$

34. Cone: $z = \sqrt{x^2 + y^2},\ (4, 0, 0)$

Minimize $d^2 = (x - 4)^2 + y^2 + z^2$

35. *Volume*   Find the dimensions of the rectangular package of largest volume subject to the constraint that the sum of the length and the girth cannot exceed 108 inches (see figure).   (*Hint:* Maximize $V = xyz$ subject to the constraint $x + 2y + 2z = 108$.)

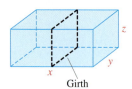

Girth

36. *Cost*   In redecorating an office, the cost for new carpeting is five times the cost of wallpapering a wall. Find the dimensions of the largest office that can be redecorated for a fixed cost C (see figure).   (*Hint:* Maximize $V = xyz$ subject to $5xy + 2xz + 2yz = C$.)

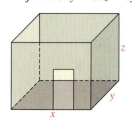

37. *Cost*   A cargo container (in the shape of a rectangular solid) must have a volume of 480 cubic feet. Use Lagrange multipliers to find the dimensions of the container of this size that has a minimum cost, if the bottom will cost $5 per square foot to construct and the sides and top will cost $3 per square foot to construct.

38. *Cost*   A manufacturer has an order for 1000 units of fine paper that can be produced at two locations. Let x_1 and x_2 be the numbers of units produced at the two plants. Find the number of units that should be produced at each plant to minimize the cost if the cost function is given by

$$C = 0.25x_1^2 + 25x_1 + 0.05x_2^2 + 12x_2.$$

39. *Cost*   A manufacturer has an order for 2000 units of all-terrain vehicle tires that can be produced at two locations. Let x_1 and x_2 be the numbers of units produced at the two plants. The cost function is modeled by

$$C = 0.25x_1^2 + 10x_1 + 0.15x_2^2 + 12x_2.$$

Find the number of units that should be produced at each plant to minimize the cost.

40. *Hardy-Weinberg Law*   Repeat Exercise 45 in Section 7.5 using Lagrange multipliers—that is, maximize

$$P(p, q, r) = 2pq + 2pr + 2qr$$

subject to the constraint

$$p + q + r = 1.$$

41. Least-Cost Rule The production function for a company is given by

$$f(x, y) = 100x^{0.25}y^{0.75}$$

where x is the number of units of labor and y is the number of units of capital. Suppose that labor costs $48 per unit, capital costs $36 per unit, and management sets a production goal of 20,000 units.

(a) Find the numbers of units of labor and capital needed to meet the production goal while minimizing the cost.

(b) Show that the conditions of part (a) are met when

$$\frac{\text{Marginal productivity of labor}}{\text{Marginal productivity of capital}} = \frac{\text{unit price of labor}}{\text{unit price of capital}}.$$

This proportion is called the *Least-Cost Rule* (or *Equimarginal Rule*).

42. Least-Cost Rule Repeat Exercise 41 for the production function given by

$$f(x, y) = 100x^{0.6}y^{0.4}.$$

43. Production The production function for a company is given by

$$f(x, y) = 100x^{0.25}y^{0.75}$$

where x is the number of units of labor and y is the number of units of capital. Suppose that labor costs $48 per unit and capital costs $36 per unit. The total cost of labor and capital is limited to $100,000.

(a) Find the maximum production level for this manufacturer.

(b) Find the marginal productivity of money.

(c) Use the marginal productivity of money to find the maximum number of units that can be produced if $125,000 is available for labor and capital.

44. Production Repeat Exercise 43 for the production function given by

$$f(x, y) = 100x^{0.06}y^{0.04}.$$

45. Biology A microbiologist must prepare a culture medium in which to grow a certain type of bacteria. The percent of salt contained in this medium is given by

$$S = 12xyz$$

where x, y, and z are the nutrient solutions to be mixed in the medium. For the bacteria to grow, the medium must be 13% salt. Nutrient solutions x, y, and z cost $1, $2, and $3 per liter, respectively. How much of each nutrient solution should be used to minimize the cost of the culture medium?

46. Biology Repeat Exercise 45 for a salt-content model of

$$S = 0.01x^2y^2z^2.$$

47. Construction A rancher plans to use an existing stone wall and the side of a barn as a boundary for two adjacent rectangular corrals. Fencing for the perimeter costs $10 per foot. To separate the corrals, a fence that costs $4 per foot will divide the region. The total area of the two corrals is to be 6000 square feet.

(a) Use Lagrange multipliers to find the dimensions that will minimize the cost of the fencing.

(b) What is the minimum cost?

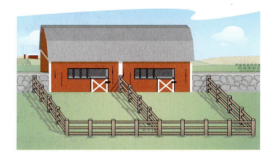

48. Area Use Lagrange multipliers to show that the maximum area of a rectangle with dimensions x and y and a given perimeter P is $\frac{1}{16}P^2$.

49. Investment Strategy An investor is considering three different stocks in which to invest $300,000. The average annual dividends for the stocks are

General Mills (G)	2.5%
Pepsico, Inc. (P)	1.4%
Sara Lee (S)	3.1%.

The amount invested in Pepsico, Inc. must follow the equation

$$3000(G) - 3000(S) + P^2 = 0.$$

How much should be invested in each stock to yield a maximum of dividends?

50. Investment Strategy An investor is considering three different stocks in which to invest $20,000. The average annual dividends for the stocks are

General Motors (G)	5.2%
Campbell Soup (C)	2.7%
Kellogg Co. (K)	3.2%.

The amount invested in Campbell Soup must follow the equation

$$1000(K) - 1000(G) + C^2 = 0.$$

How much should be invested in each stock to yield a maximum of dividends?

51. Research Project Use your school's library, the Internet, or some other reference source to write a paper about two different types of available investment options. Find examples of each type and find the data about their dividends for the past 10 years. What are the similarities and differences between the two types?

The following warm-up exercises involve skills that were covered in earlier sections. You will use these skills in the exercise set for this section.

In Exercises 1–12, evaluate the definite integral.

1. $\displaystyle\int_0^1 dx$

2. $\displaystyle\int_0^2 3\,dy$

3. $\displaystyle\int_1^4 2x^2\,dx$

4. $\displaystyle\int_0^1 2x^3\,dx$

5. $\displaystyle\int_1^2 (x^3 - 2x + 4)\,dx$

6. $\displaystyle\int_0^2 (4 - y^2)\,dy$

7. $\displaystyle\int_1^2 \frac{2}{7x^2}\,dx$

8. $\displaystyle\int_1^4 \frac{2}{\sqrt{x}}\,dx$

9. $\displaystyle\int_0^2 \frac{2x}{x^2 + 1}\,dx$

10. $\displaystyle\int_2^e \frac{1}{y - 1}\,dy$

11. $\displaystyle\int_0^2 xe^{x^2 + 1}\,dx$

12. $\displaystyle\int_0^1 e^{-2y}\,dy$

In Exercises 13–16, sketch the region bounded by the graphs of the equations.

13. $y = x,\ y = 0,\ x = 3$

14. $y = x,\ y = 3,\ x = 0$

15. $y = 4 - x^2,\ y = 0,\ x = 0$

16. $y = x^2,\ y = 4x$

EXERCISES 7.8

In Exercises 1–10, evaluate the partial integral.

1. $\displaystyle\int_0^x (2x - y)\,dy$

2. $\displaystyle\int_x^{x^2} \frac{y}{x}\,dy$

3. $\displaystyle\int_1^{2y} \frac{y}{x}\,dx$

4. $\displaystyle\int_0^{e^y} y\,dx$

5. $\displaystyle\int_0^{\sqrt{9 - x^2}} x^2 y\,dy$

6. $\displaystyle\int_{x^2}^{\sqrt{x}} (x^2 + y^2)\,dy$

7. $\displaystyle\int_{e^y}^{y} \frac{y \ln x}{x}\,dx$

8. $\displaystyle\int_{-\sqrt{1 - y^2}}^{\sqrt{1 - y^2}} (x^2 + y^2)\,dx$

9. $\displaystyle\int_0^{x^3} ye^{-y/x}\,dy$

10. $\displaystyle\int_y^3 \frac{xy}{\sqrt{x^2 + 1}}\,dx$

In Exercises 11–24, evaluate the double integral.

11. $\displaystyle\int_0^2\int_0^1 (x - y)\,dy\,dx$

12. $\displaystyle\int_0^2\int_0^2 (6 - x^2)\,dy\,dx$

13. $\displaystyle\int_0^4\int_0^3 xy\,dy\,dx$

14. $\displaystyle\int_0^1\int_0^x \sqrt{1 - x^2}\,dy\,dx$

15. $\displaystyle\int_0^1\int_0^{\sqrt{1 - y^2}} (x + y)\,dx\,dy$

16. $\displaystyle\int_0^2\int_{3y^2 - 6y}^{2y - y^2} 3y\,dx\,dy$

17. $\displaystyle\int_1^2\int_0^4 (x^2 - 2y^2 + 1)\,dx\,dy$

18. $\displaystyle\int_0^1\int_y^{2y} (1 + 2x^2 + 2y^2)\,dx\,dy$

19. $\displaystyle\int_0^2\int_0^{\sqrt{1 - y^2}} -5xy\,dx\,dy$

20. $\displaystyle\int_0^4\int_0^x \frac{2}{(x + 1)(y + 1)}\,dy\,dx$

21. $\displaystyle\int_0^2\int_0^{4 - x^2} x^3\,dy\,dx$

22. $\displaystyle\int_0^a\int_0^{a - x} (x^2 + y^2)\,dy\,dx$

23. $\displaystyle\int_0^\infty\int_0^\infty e^{-(x + y)/2}\,dy\,dx$

24. $\displaystyle\int_0^\infty\int_0^\infty xye^{-(x^2 + y^2)}\,dx\,dy$

In Exercises 25–32, sketch the region R whose area is given by the double integral. Then change the order of integration and show that both orders yield the same area.

25. $\int_0^1 \int_0^2 dy \, dx$

26. $\int_1^2 \int_2^4 dx \, dy$

27. $\int_0^1 \int_{2y}^2 dx \, dy$

28. $\int_0^4 \int_0^{\sqrt{x}} dy \, dx$

29. $\int_0^2 \int_{x/2}^1 dy \, dx$

30. $\int_0^4 \int_{\sqrt{x}}^2 dx \, dy$

31. $\int_0^1 \int_{y^2}^{\sqrt[3]{y}} dx \, dy$

32. $\int_{-2}^2 \int_0^{4-y^2} dx \, dy$

In Exercises 33 and 34, evaluate the double integral. Note that it is necessary to change the order of integration.

33. $\int_0^3 \int_y^3 e^{x^2} dx \, dy$

34. $\int_0^2 \int_x^2 e^{-y^2} dy \, dx$

In Exercises 35–40, use a double integral to find the area of the specified region.

35.

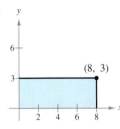

36.

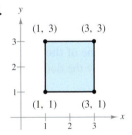

37.

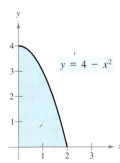

38.

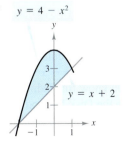

39.

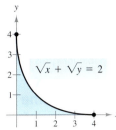

40.

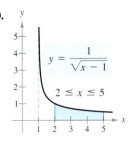

In Exercises 41–46, use a double integral to find the area of the region bounded by the graphs of the equations.

41. $y = 25 - x^2, \ y = 0$

42. $y = x^{3/2}, \ y = x$

43. $5x - 2y = 0, \ x + y = 3, \ y = 0$

44. $xy = 9, \ y = x, \ y = 0, \ x = 9$

45. $y = x, \ y = 2x, \ x = 2$

46. $y = x^2 + 2x + 1, \ y = 3(x + 1)$

In Exercises 47–54, use a symbolic integration utility to evaluate the double integral.

47. $\int_0^1 \int_0^2 e^{-x^2-y^2} dx \, dy$

48. $\int_0^2 \int_{x^2}^{2x} (x^3 + 3y^2) \, dy \, dx$

49. $\int_1^2 \int_0^x e^{xy} dy \, dx$

50. $\int_1^2 \int_y^{2y} \ln(x + y) \, dx \, dy$

51. $\int_0^1 \int_x^1 \sqrt{1 - x^2} \, dy \, dx$

52. $\int_0^3 \int_0^{x^2} \sqrt{x}\sqrt{1 + x} \, dy \, dx$

53. $\int_0^2 \int_{\sqrt{4-x^2}}^{4-x^2/4} \frac{xy}{x^2 + y^2 + 1} \, dy \, dx$

54. $\int_0^4 \int_0^y \frac{2}{(x + 1)(y + 1)} \, dx \, dy$

True or False? In Exercises 55 and 56, determine whether the statement is true or false. If it is false, explain why or give an example that shows it is false.

55. Changing the order of integration will sometimes change the value of a double integral.

56. $\int_2^5 \int_1^6 x \, dy \, dx = \int_1^6 \int_2^5 x \, dx \, dy$

In Exercises 1–4, sketch the region that is described.

1. $0 \le x \le 2$, $0 \le y \le 1$

2. $1 \le x \le 3$, $2 \le y \le 3$

3. $0 \le x \le 4$, $0 \le y \le 2x - 1$

4. $0 \le x \le 2$, $0 \le y \le x^2$

In Exercises 5–10, evaluate the double integral.

5. $\int_0^1 \int_1^2 dy\, dx$

6. $\int_0^3 \int_1^3 dx\, dy$

7. $\int_0^1 \int_0^x x\, dy\, dx$

8. $\int_0^4 \int_1^y y\, dx\, dy$

9. $\int_1^3 \int_x^{x^2} 2\, dy\, dx$

10. $\int_0^1 \int_x^{-x^2+2} dy\, dx$

EXERCISES 7.9

In Exercises 1–8, sketch the region of integration and evaluate the double integral.

1. $\int_0^2 \int_0^1 (3x + 4y)\, dy\, dx$

2. $\int_0^3 \int_0^1 (2x + 6y)\, dy\, dx$

3. $\int_0^1 \int_y^{\sqrt{y}} x^2 y^2\, dx\, dy$

4. $\int_0^6 \int_{y/2}^3 (x + y)\, dx\, dy$

5. $\int_0^1 \int_0^{\sqrt{1-x^2}} y\, dy\, dx$

6. $\int_0^2 \int_0^{4-x^2} xy^2\, dy\, dx$

7. $\int_{-a}^a \int_{-\sqrt{a^2-x^2}}^{\sqrt{a^2-x^2}} dy\, dx$

8. $\int_0^a \int_0^{\sqrt{a^2-x^2}} dy\, dx$

In Exercises 9–12, set up the integral for both orders of integration and use the more convenient order to evaluate the integral over the region R.

9. $\int_R \int xy\, dA$

    R: rectangle with vertices at $(0, 0)$, $(0, 5)$, $(3, 5)$, $(3, 0)$

10. $\int_R \int x\, dA$

    R: semicircle bounded by $y = \sqrt{25 - x^2}$ and $y = 0$

11. $\int_R \int \dfrac{y}{x^2 + y^2}\, dA$

    R: triangle bounded by $y = x$, $y = 2x$, $x = 2$

12. $\int_R \int \dfrac{y}{1 + x^2}\, dA$

    R: region bounded by $y = 0$, $y = \sqrt{x}$, $x = 4$

In Exercises 13 and 14, evaluate the double integral. Note that it is necessary to change the order of integration.

13. $\int_0^1 \int_{y/2}^{1/2} e^{-x^2}\, dx\, dy$

14. $\int_0^{\ln 10} \int_{e^x}^{10} \dfrac{1}{\ln y}\, dy\, dx$

In Exercises 15–26, use a double integral to find the volume of the specified solid.

15.

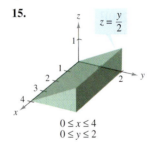

$0 \le x \le 4$
$0 \le y \le 2$

16.

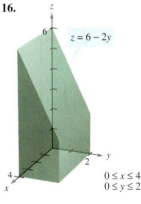

$0 \le x \le 4$
$0 \le y \le 2$

17.

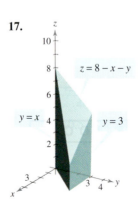

18.

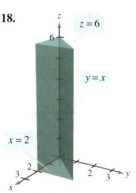

19.

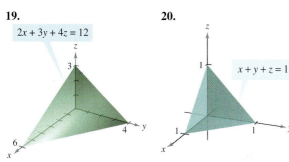

$2x + 3y + 4z = 12$

20.

$x + y + z = 1$

21.

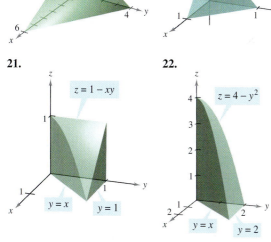

$z = 1 - xy$

$y = x$ $y = 1$

22.

$z = 4 - y^2$

$y = x$ $y = 2$

23.

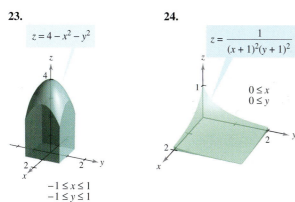

$z = 4 - x^2 - y^2$

$-1 \le x \le 1$
$-1 \le y \le 1$

24.

$z = \dfrac{1}{(x + 1)^2 (y + 1)^2}$

$0 \le x$
$0 \le y$

25.

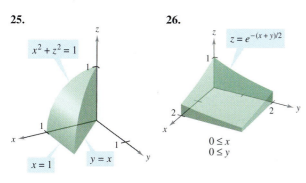

$x^2 + z^2 = 1$

$x = 1$ $y = x$

26.

$z = e^{-(x + y)/2}$

$0 \le x$
$0 \le y$

In Exercises 27–30, use a double integral to find the volume of the solid bounded by the graphs of the equations.

27. $z = xy$, $z = 0$, $y = 0$, $y = 4$, $x = 0$, $x = 1$

28. $z = x$, $z = 0$, $y = x$, $y = 0$, $x = 0$, $x = 4$

29. $z = x^2$, $z = 0$, $x = 0$, $x = 2$, $y = 0$, $y = 4$

30. $z = x + y$, $x^2 + y^2 = 4$ (first octant)

31. *Population Density* The population density (in people per square mile) for a coastal town can be modeled by

$$f(x, y) = \frac{120,000}{(2 + x + y)^3}$$

where x and y are measured in miles. What is the population inside the rectangular area defined by the vertices $(0, 0)$, $(2, 0)$, $(0, 2)$, and $(2, 2)$?

32. *Population Density* The population density (in people per square mile) for a coastal town on an island can be modeled by

$$f(x, y) = \frac{5000xe^y}{1 + 2x^2}$$

where x and y are measured in miles. What is the population inside the rectangular area defined by the vertices $(0, 0)$, $(4, 0)$, $(0, -2)$, and $(4, -2)$?

In Exercises 33–36, find the average value of $f(x, y)$ over the region R.

Integral	*Region R*
33. $f(x, y) = x$	Rectangle with vertices $(0, 0)$, $(4, 0)$, $(4, 2)$, $(0, 2)$
34. $f(x, y) = xy$	Rectangle with vertices $(0, 0)$, $(4, 0)$, $(4, 2)$, $(0, 2)$
35. $f(x, y) = x^2 + y^2$	Square with vertices $(0, 0)$, $(2, 0)$, $(2, 2)$, $(0, 2)$
36. $f(x, y) = e^{x+y}$	Triangle with vertices $(0, 0)$, $(0, 1)$, $(1, 1)$

37. *Average Revenue* A company sells two products whose demand functions are given by

$$x_1 = 500 - 3p_1 \quad \text{and} \quad x_2 = 750 - 2.4p_2.$$

So, the total revenue is given by

$$R = x_1 p_1 + x_2 p_2.$$

Estimate the average revenue if the price p_1 varies between $50 and $75 and the price p_2 varies between $100 and $150.

38. *Average Weekly Profit* A firm's weekly profit in marketing two products is given by

$$P = 192x_1 + 576x_2 - x_1^2 - 5x_2^2 - 2x_1 x_2 - 5000$$

where x_1 and x_2 represent the numbers of units of each product sold weekly. Estimate the average weekly profit if x_1 varies between 40 and 50 units and x_2 varies between 45 and 50 units.

ALGEBRA REVIEW

Solving Systems of Equations

Nonlinear System in Two Variables

$$\begin{cases} 4x + 3y = 6 \\ x^2 - y = 4 \end{cases}$$

Linear System in Three Variables

$$\begin{cases} -x + 2y + 4z = 2 \\ 2x - y + z = 0 \\ 6x + 2z = 3 \end{cases}$$

Three of the sections in this chapter (7.5, 7.6, and 7.7) involve solutions of systems of equations. These systems can be linear or nonlinear, as shown at the left.

There are many techniques for solving a system of linear equations. Two of the more common ones are listed here.

1. *Substitution*: Solve for one of the variables in one of the equations and substitute the value into another equation.

2. *Elimination*: Add multiples of one equation to a second equation to eliminate a variable in the second equation.

EXAMPLE 1 Solving a System of Equations

Solve each system of equations.

(a) $\begin{cases} y - x^3 = 0 \\ x - y^3 = 0 \end{cases}$

(b) $\begin{cases} -400p_1 + 300p_2 = -25 \\ 300p_1 - 360p_2 = -535 \end{cases}$

SOLUTION

(a) Example 3, page 498

$$\begin{cases} y - x^3 = 0 \\ x - y^3 = 0 \end{cases}$$
Equation 1
Equation 2

$$y = x^3$$ Solve for y in Equation 1.

$$x - (x^3)^3 = 0$$ Substitute x^3 for y in Equation 2.

$$x - x^9 = 0$$ $(x^m)^n = x^{mn}$

$$x(x - 1)(x + 1)(x^2 + 1)(x^4 + 1) = 0$$ Factor.

$$x = 0$$ Set factors equal to zero.

$$x = 1$$ Set factors equal to zero.

$$x = -1$$ Set factors equal to zero.

(b) Example 4, page 499

$$\begin{cases} -400p_1 + 300p_2 = -25 \\ 300p_1 - 360p_2 = -535 \end{cases}$$
Equation 1
Equation 2

$$p_2 = \tfrac{1}{12}(16p_1 - 1)$$ Solve for p_2 in Equation 1.

$$300p_1 - 360\left(\tfrac{1}{12}\right)(16p_1 - 1) = -535$$ Substitute for p_2 in Equation 2.

$$300p_1 - 30(16p_1 - 1) = -535$$ Multiply factors.

$$-180p_1 = -565$$ Combine like terms.

$$p_1 = \tfrac{113}{36} \approx 3.14$$ Divide each side by -180.

$$p_2 = \tfrac{1}{12}\left[16\left(\tfrac{113}{36}\right) - 1\right]$$ Find p_2 by substituting p_1.

$$p_2 \approx 4.10$$ Solve for p_2.

EXAMPLE 2   Solving a System of Equations

Solve each system of equations.

(a) $\begin{cases} y(24 - 12x - 4y) = 0 \\ x(24 - 6x - 8y) = 0 \end{cases}$ (b) $\begin{cases} 28a - 4b = 10 \\ -4a + 8b = 12 \end{cases}$

SOLUTION

(a) Example 5, page 500

Before solving this system of equations, factor 4 out of the first equation and factor 2 out of the second equation.

$\begin{cases} y(24 - 12x - 4y) = 0 \\ x(24 - 6x - 8y) = 0 \end{cases}$ Original Equation 1

                                         Original Equation 2

$\begin{cases} y(4)(6 - 3x - y) = 0 \\ x(2)(12 - 3x - 4y) = 0 \end{cases}$ Factor 4 out of Equation 1.

                                         Factor 2 out of Equation 2.

$\begin{cases} y(6 - 3x - y) = 0 \\ x(12 - 3x - 4y) = 0 \end{cases}$ Equation 1

                                         Equation 2

In each equation, either factor can be 0, so you obtain four different linear systems. For the first system, substitute $y = 0$ into the second equation to obtain $x = 4$.

$\begin{cases} y = 0 \\ 12 - 3x - 4y = 0 \end{cases}$ $(4, 0)$ is a solution.

You can solve the second system by the method of elimination.

$\begin{cases} 6 - 3x - y = 0 \\ 12 - 3x - 4y = 0 \end{cases}$ $\left(\frac{4}{3}, 2\right)$ is a solution.

The third system is already solved.

$\begin{cases} y = 0 \\ x = 0 \end{cases}$ $(0, 0)$ is a solution.

You can solve the last system by substituting $x = 0$ into the first equation to obtain $y = 6$.

$\begin{cases} 6 - 3x - y = 0 \\ x = 0 \end{cases}$ $(0, 6)$ is a solution.

(b) Example 2, page 516

$\begin{cases} 28a - 4b = 10 \\ -4a + 8b = 12 \end{cases}$ Equation 1

                                         Equation 2

$-2a + 4b = 6$ Divide Equation 2 by 2.

$26a = 16$ Add new equation to Equation 1.

$a = \frac{8}{13}$ Divide each side by 26.

$28\left(\frac{8}{13}\right) - 4b = 10$ Substitute for a in Equation 1.

$b = \frac{47}{26}$ Solve for b.

7 CHAPTER SUMMARY AND STUDY STRATEGIES

After studying this chapter, you should have acquired the following skills. The exercise numbers are keyed to the Review Exercises that begin on page 544. Answers to odd-numbered Review Exercises are given in the back of the text. *

■ Plot points in space. *(Section 7.1)*   *Review Exercises 1, 2*

■ Find the distance between two points in space. *(Section 7.1)*   *Review Exercises 3, 4*

$$d = \sqrt{(x_2 - x_1)^2 + (y_2 - y_1)^2 + (z_2 - z_1)^2}$$

■ Find the midpoints of line segments in space. *(Section 7.1)*   *Review Exercises 5, 6*

$$\text{Midpoint} = \left(\frac{x_1 + x_2}{2}, \frac{y_1 + y_2}{2}, \frac{z_1 + z_2}{2}\right)$$

■ Write the standard forms of the equations of spheres. *(Section 7.1)*   *Review Exercises 7–10*

$$(x - h)^2 + (y - k)^2 + (z - l)^2 = r^2$$

■ Find the centers and radii of spheres. *(Section 7.1)*   *Review Exercises 11, 12*

■ Sketch the coordinate plane traces of spheres. *(Section 7.1)*   *Review Exercises 13, 14*

■ Sketch planes in space. *(Section 7.2)*   *Review Exercises 15–18*

■ Classify quadric surfaces in space. *(Section 7.2)*   *Review Exercises 19–26*

■ Evaluate functions of several variables. *(Section 7.3)*   *Review Exercises 27, 28*

■ Find the domains and ranges of functions of several variables. *(Section 7.3)*   *Review Exercises 29, 30*

■ Sketch the level curves of functions of two variables. *(Section 7.3)*   *Review Exercises 31–34*

■ Use functions of several variables to answer questions about real-life situations.   *Review Exercises 35–40*
(Section 7.3)

■ Find the first partial derivatives of functions of several variables. *(Section 7.4)*   *Review Exercises 41–50*

$$\frac{\partial z}{\partial x} = \lim_{\Delta x \to 0} \frac{f(x + \Delta x, y) - f(x, y)}{\Delta x}$$

$$\frac{\partial z}{\partial y} = \lim_{\Delta y \to 0} \frac{f(x, y + \Delta y) - f(x, y)}{\Delta y}$$

■ Find the slopes of surfaces in the *x*- and *y*-directions. *(Section 7.4)*   *Review Exercises 51–54*

■ Find the second partial derivatives of functions of several variables. *(Section 7.4)*   *Review Exercises 55–58*

■ Use partial derivatives to answer questions about real-life situations. *(Section 7.4)* *Review Exercises 59–62*

■ Find the relative extrema of functions of two variables. *(Section 7.5)*   *Review Exercises 63–70*

■ Use relative extrema to answer questions about real-life situations. *(Section 7.5)*   *Review Exercises 71, 72*

■ Use Lagrange multipliers to find extrema of functions of several variables.   *Review Exercises 73–78*
(Section 7.6)

■ Use a spreadsheet to find the indicated extremum. *(Section 7.6)*   *Review Exercises 79, 80*

■ Use Lagrange multipliers to answer questions about real-life situations. *(Section 7.6)* *Review Exercises 81, 82*

---

* Use a wide range of valuable study aids to help you master the material in this chapter. The *Student Solutions Guide* includes step-by-step solutions to all odd-numbered exercises to help you review and prepare. The *HM mathSpace®* Student CD-ROM helps you brush up on your algebra skills. The *Graphing Technology Guide*, available on the Web at *math.college.hmco.com/students*, offers step-by-step commands and instructions for a wide variety of graphing calculators, including the most recent models.

■ Find the least squares regression lines, $y = ax + b$, for data and calculate the sum of the squared errors for data. *(Section 7.7)*   *Review Exercises 83, 84*

$$a = \left[n\sum_{i=1}^{n} x_i y_i - \sum_{i=1}^{n} x_i \sum_{i=1}^{n} y_i \right] \bigg/ \left[n\sum_{i=1}^{n} x_i^2 - \left(\sum_{i=1}^{n} x_i\right)^2 \right], \qquad b = \frac{1}{n}\left(\sum_{i=1}^{n} y_i - a\sum_{i=1}^{n} x_i \right)$$

■ Use least squares regression lines to model real-life data. *(Section 7.7)*   *Review Exercises 85, 86*

■ Find the least squares regression quadratics for data. *(Section 7.7)*   *Review Exercises 87, 88*

■ Evaluate double integrals. *(Section 7.8)*   *Review Exercises 89–92*

■ Use double integrals to find the areas of regions. *(Section 7.8)*   *Review Exercises 93–96*

■ Use double integrals to find the volumes of solids. *(Section 7.9)*   *Review Exercises 97, 98*

$$\text{Volume} = \int_R \int f(x, y)\, dA$$

■ Use double integrals to find the average values of real-life models. *(Section 7.9)*   *Review Exercises 99, 100*

$$\text{Average value} = \frac{1}{A}\int_R \int f(x, y)\, dA$$

■ ***Comparing Two Dimensions with Three Dimensions***   Many of the formulas and techniques in this chapter are generalizations of formulas and techniques used in earlier chapters in the text. Here are several examples.

Two-Dimensional Coordinate System	Three-Dimensional Coordinate System
Distance Formula $d = \sqrt{(x_2 - x_1)^2 + (y_2 - y_1)^2}$	*Distance Formula* $d = \sqrt{(x_2 - x_1)^2 + (y_2 - y_1)^2 + (z_2 - z_1)^2}$
Midpoint Formula $\text{Midpoint} = \left(\dfrac{x_1 + x_2}{2}, \dfrac{y_1 + y_2}{2} \right)$	*Midpoint Formula* $\text{Midpoint} = \left(\dfrac{x_1 + x_2}{2}, \dfrac{y_1 + y_2}{2}, \dfrac{z_1 + z_2}{2} \right)$
Equation of Circle $(x - h)^2 + (y - k)^2 = r^2$	*Equation of Sphere* $(x - h)^2 + (y - k)^2 + (z - l)^2 = r^2$
Equation of Line $ax + by = c$	*Equation of Plane* $ax + by + cz = d$
Derivative of $y = f(x)$ $\dfrac{dy}{dx} = \lim_{\Delta x \to 0} \dfrac{f(x + \Delta x) - f(x)}{\Delta x}$	*Partial Derivative of $z = f(x, y)$* $\dfrac{\partial z}{\partial x} = \lim_{\Delta x \to 0} \dfrac{f(x + \Delta x, y) - f(x, y)}{\Delta x}$
Area of Region $A = \displaystyle\int_a^b f(x)\, dx$	*Volume of Region* $V = \displaystyle\int_R \int f(x, y)\, dA$

Study Tools   *Additional resources that accompany this chapter*

■ **Algebra Review** (pages 540 and 541)

■ **Chapter Summary and Study Strategies** (pages 542 and 543)

■ **Review Exercises** (pages 544–547)

■ **Sample Post-Graduation Exam Questions**   (page 548)

■ **Web Exercises**   (page 493, Exercise 77; page 513, Exercise 51)

■ **Student Solutions Guide**

■ **HM mathSpace® Student CD-ROM**

■ **Graphing Technology Guide** (*math.college.hmco.com/students*)

60. *Marginal Revenue* At a baseball stadium, souvenir caps are sold at two locations. If x_1 and x_2 are the numbers of baseball caps sold at location 1 and location 2, respectively, then the total revenue for the caps is modeled by

$$R = 15x_1 + 16x_2 - \frac{1}{10}x_1^2 - \frac{1}{10}x_2^2 - \frac{1}{100}x_1x_2.$$

Given that $x_1 = 50$ and $x_2 = 40$, find the marginal revenues at location 1 and at location 2.

61. *Medical Science* The surface area A of an average human body in square centimeters can be approximated by the model

$$A(w, h) = 101.4w^{0.425}h^{0.725}$$

where w is the weight in pounds and h is the height in inches.

(a) Determine the partial derivatives of A with respect to w and with respect to h.

(b) Evaluate dA/dw at $(180, 70)$. Explain your result.

62. *Medicine* In order to treat a certain bacterial infection, a combination of two drugs is being tested. Studies have shown that the duration of the infection in laboratory tests can be modeled by

$$D(x, y) = x^2 + 2y^2 - 18x - 24y + 2xy + 120$$

where x is the dose in hundreds of milligrams of the first drug and y is the dose in hundreds of milligrams of the second drug. Evaluate $D(5, 2.5)$ and $D(7.5, 8)$ and interpret your results.

In Exercises 63–70, find any critical points and relative extrema of the function.

63. $f(x, y) = x^2 + 2xy + y^2$

64. $f(x, y) = x^3 - 3xy + y^2$

65. $f(x, y) = x^2 + 6xy + 3y^2 + 6x + 8$

66. $f(x, y) = x + y^2 - e^x$

67. $f(x, y) = x^3 + y^2 - xy$

68. $f(x, y) = y^2 + xy + 3y - 2x + 5$

69. $f(x, y) = x^3 + y^3 - 3x - 3y + 2$

70. $f(x, y) = y^2 - x^2$

71. *Revenue* A company manufactures and sells two products. The demand functions for the products are given by

$$p_1 = 100 - x_1$$

and

$$p_2 = 200 - 0.5x_2.$$

(a) Find the total revenue functions for x_1 and x_2.

(b) Find x_1 and x_2 such that the revenue is maximized.

(c) What is the maximum revenue?

72. *Profit* A company manufactures a product at two locations. The costs of manufacturing x_1 units at plant 1 and x_2 units at plant 2 are modeled by

$$C_1 = 0.03x_1^2 + 4x_1 + 300 \quad \text{and}$$
$$C_2 = 0.05x_2^2 + 7x_2 + 175$$

respectively. If the product sells for $10 per unit, find x_1 and x_2 such that the profit, $P = 10(x_1 + x_2) - C_1 - C_2$, is maximized.

In Exercises 73–78, locate any extrema of the function by using Lagrange multipliers.

73. $f(x, y) = x^2y$

Constraint: $x + 2y = 2$

74. $f(x, y) = x^2 + y^2$

Constraint: $x + y = 4$

75. $f(x, y, z) = xyz$

Constraint: $x + 2y + z - 4 = 0$

76. $f(x, y, z) = xz + yz$

Constraint: $x + y + z = 6$

77. $f(x, y, z) = x^2 + y^2 + z^2$

Constraints: $x + z = 6, y + z = 8$

78. $f(x, y, z) = xyz$

Constraints: $x + y + z = 32, x - y + z = 0$

In Exercises 79 and 80, use a spreadsheet to find the indicated extremum. In each case, assume that $x, y,$ and z are nonnegative.

79. Maximize $f(x, y, z) = xy$

Constraints: $x^2 + y^2 = 16, x - 2z = 0$

80. Minimize $f(x, y, z) = x^2 + y^2 + z^2$

Constraints: $x - 2z = 4, x + y = 8$

81. *Maximum Production Level* The production function for a manufacturer is given by $f(x, y) = 4x + xy + 2y$. Assume that the total amount available for labor x and capital y is $2000 and that units of labor and capital cost $20 and $4, respectively. Find the maximum production level for this manufacturer.

82. *Minimum Cost* A manufacturer has an order for 1500 units that can be produced at two locations. Let x_1 and x_2 be the numbers of units produced at the two locations. Find the number that should be produced at each location to meet the order and minimize cost if the cost function is given by

$$C = 0.20x_1^2 + 10x_1 + 0.15x_2^2 + 12x_2.$$

In Exercises 83 and 84, (a) use the method of least squares to find the least squares regression line and (b) calculate the sum of the squared errors.

83. $(-2, -3), (-1, -1), (1, 2), (3, 2)$

84. $(-3, -1), (-2, -1), (0, 0), (1, 1), (2, 1)$

85. *Agriculture* An agronomist used four test plots to determine the relationship between the wheat yield (in bushels per acre) and the amount of fertilizer (in hundreds of pounds per acre). The results are listed in the table.

Fertilizer, x	1.0	1.5	2.0	2.5
Yield, y	32	41	48	53

(a) Use a graphing utility with a least squares regression program to find the least squares regression line for the data.

(b) Estimate the yield for a fertilizer application of 20 pounds per acre.

86. *Work Force* The table gives the percents and numbers (in millions) of women in the work force for selected years. *(Source: U.S. Bureau of Labor Statistics)*

Year	1970	1975	1980	1985
Percent, x	43.3	46.3	51.5	54.5
Number, y	31.5	37.5	45.5	51.1

Year	1990	1995	2000
Percent, x	57.5	58.9	59.9
Number, y	56.8	60.9	66.3

(a) Use a spreadsheet software program with a least squares regression program to find the least squares regression line for the data.

(b) According to this model, approximately how many women enter the labor force for each one-point increase in the percent of women in the labor force?

In Exercises 87 and 88, find the least squares regression quadratic for the given points. Plot the points and sketch the least squares regression quadratic.

87. $(-1, 9), (0, 7), (1, 5), (2, 6), (4, 23)$

88. $(0, 10), (2, 9), (3, 7), (4, 4), (5, 0)$

In Exercises 89–92, evaluate the double integral.

89. $\displaystyle\int_0^1 \int_0^{1+x} (3x + 2y)\, dy\, dx$

90. $\displaystyle\int_{-2}^2 \int_0^4 (x - y^2)\, dx\, dy$

91. $\displaystyle\int_1^2 \int_1^{2y} \frac{x}{y^2}\, dx\, dy$

92. $\displaystyle\int_0^4 \int_0^{\sqrt{16-x^2}} 2x\, dy\, dx$

In Exercises 93–96, use a double integral to find the area of the region.

93.

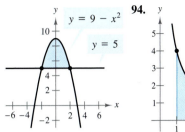

94.

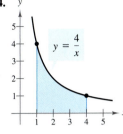

95.

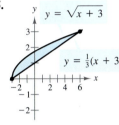

96.

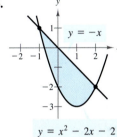

97. Find the volume of the solid bounded by the graphs of $z = (xy)^2$, $z = 0$, $y = 0$, $y = 4$, $x = 0$, and $x = 4$.

98. Find the volume of the solid bounded by the graphs of $z = x + y$, $z = 0$, $x = 0$, $x = 3$, $y = x$, and $y = 0$.

99. *Average Elevation* In a triangular coastal area, the elevation in miles above sea level at the point (x, y) is modeled by

$$f(x, y) = 0.25 - 0.025x - 0.01y$$

where x and y are measured in miles (see figure). Find the average elevation of the triangular area.

Figure for 99   Figure for 100

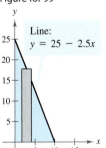   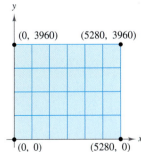

100. *Real Estate* The average value of real estate (in dollars per square foot) for a rectangular section of a city is given by

$$f(x, y) = 2.5x^{3/2}y^{3/4}$$

(see figure). Find the average value of real estate for this section.

7 SAMPLE POST-GRADUATION EXAM QUESTIONS

CPA
GMAT
GRE
Actuarial
CLAST

The following questions represent the types of questions that appear on certified public accountant (CPA) exams, Graduate Management Admission Tests (GMAT), Graduate Records Exams (GRE), actuarial exams, and College-Level Academic Skills Tests (CLAST). The answers to the questions are given in the back of the book.

1. What is the derivative of $f(x, y) = y^2(x + y)^3$ with respect to y?

   (a) $6y(x + y)^2$ (b) $y(x + y)^2(2x + 5y)$

   (c) $3y^2(x + y)^2$ (d) $y(x + y)^2(5x + 2y)$

2. Let $f(x, y) = x^2 + y^2 + 6x - 2y + 4$. At which point does f have a relative minimum?

   (a) $(-3, 1, -13)$ (b) $(-3, 1, -6)$ (c) $(3, 1, -2)$ (d) $(-3, -1, -2)$

For Questions 3 and 4, use the following excerpts from the 2003 Tax Rate Schedules.

SCHEDULE X—Use if your filing status is **Single**

If the amount on Form 1040, line 40, is over—	But not over—	Enter on Form 1040, line 41	of the amount over—
$0	$7000	.............10%	$0
7000	28,400	**$700.00 + 15%**	7000
28,400	68,800	3910.00 + 25%	28,400
68,800	143,500	14,010.00 + 28%	68,800
143,500	311,950	34,926.00 + 33%	143,500
311,950	....	90,514.50 + 35%	311,950

SCHEDULE Y-1—Use if your filing status is **Married filing jointly** or **Qualifying widow(er)**

If the amount on Form 1040, line 40, is over—	But not over—	Enter on Form 1040, line 41	of the amount over—
$0	$14,000	.............10%	$0
14,000	56,800	**$1400.00 + 15%**	14,000
56,800	114,650	7820.00 + 25%	56,800
114,650	174,700	22,282.50 + 28%	114,650
174,700	311,950	39,096.50 + 33%	174,700
311,950	....	84,389.00 + 35%	311,950

3. The tax for a married couple filing jointly whose amount on Form 1040, line 40, is $125,480 is

   (a) $29,880.40   (b) $55,339.10   (c) $25,314.90   (d) $54,384.50

4. The tax for a single person whose amount on Form 1040, line 40, is $1,000,000 is

   (a) $199,697.00   (b) $325,206.50   (c) $343,630.08   (d) $331,332.00

5. If x, y, and z are chosen from the three numbers $\frac{1}{3}$, -2, and 4, what is the largest possible value of the expression $x^2 + z/y^2$?

   (a) 126   (b) 40   (c) 24   (d) 72

6. If $xz = 4y$, then

   $$\frac{x^3}{2}$$

   equals

   (a) $\dfrac{2y^3}{z^3}$ (b) $\dfrac{16y^2}{z^3}$ (c) $\dfrac{32y^3}{z^3}$ (d) $\dfrac{64y^3}{z^3}$

7. Mave Co. calculated the following ratios for one of its profit centers: gross margin 31%, return on sales 26%, capital turnover 0.5 time.

   What is Mave's return on investment for this profit center?

   (a) 8.5%   (b) 13%   (c) 15.5%   (d) 26%

8

Trigonometric Functions

© Jose Luis Pelaez, Inc./CORBIS

The amounts of precipitation during certain times of the year can be modeled by trigonometric functions.

STRATEGIES FOR SUCCESS

WHAT YOU SHOULD LEARN:

- How to measure angles using both degrees and radians
- How to evaluate trigonometric functions and solve trigonometric equations
- How to graph trigonometric functions
- How to find the derivative of a trigonometric function and use trigonometric derivatives in applications
- How to find the integral of a trigonometric function and use trigonometric integrals in applications
- How to use L'Hôpital's Rule to find limits of indeterminate forms

WHY YOU SHOULD LEARN IT:

Trigonometric functions have many applications in real life, such as the examples below, which represent a small sample of the applications in this chapter.

- Predator-Prey Cycles, Exercises 63 and 64 on page 577
- Biorhythms, Exercises 67 and 68 on page 577
- Finance: Cyclical Stocks, Exercise 77 on page 578
- Meteorology, Exercise 59 on page 595, Exercise 60 on page 596
- Consumer Trends, Exercise 117 on page 613

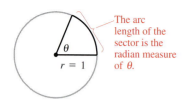

The arc length of the sector is the radian measure of θ.

FIGURE 8.6

Radian Measure

A second way to measure angles is in terms of radians. To assign a radian measure to an angle θ, consider θ to be the central angle of a circular sector of radius 1, as shown in Figure 8.6. The radian measure of θ is then defined to be the length of the arc of the sector. Recall that the circumference of a circle is given by

$$\text{Circumference} = (2\pi)(\text{radius}).$$

So, the circumference of a circle of radius 1 is simply 2π, and you can conclude that the radian measure of an angle measuring $360°$ is 2π. In other words

$$360° = 2\pi \text{ radians}$$

or

$$180° = \pi \text{ radians}.$$

Figure 8.7 gives the radian measures of several common angles.

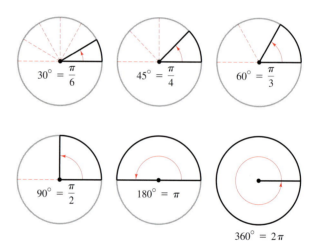

FIGURE 8.7   Radian Measures of Several Common Angles

It is important for you to be able to convert back and forth between the degree and radian measures of an angle. You should remember the conversions for the common angles shown in Figure 8.7. For other conversions, you can use the conversion rule below.

Angle Measure Conversion Rule

The degree measure and radian measure of an angle are related by the equation

$$180° = \pi \text{ radians}.$$

EXAMPLE 2 **Converting from Degrees to Radians**

Convert each degree measure to radian measure.

(a) 135°   (b) 40°   (c) 540°   (d) −270°

SOLUTION To convert from degree measure to radian measure, multiply the degree measure by $(\pi \text{ radians})/180°$.

(a) $135° = (135 \text{ degrees})\left(\dfrac{\pi \text{ radians}}{180 \text{ degrees}}\right) = \dfrac{3\pi}{4} \text{ radians}$

(b) $40° = (40 \text{ degrees})\left(\dfrac{\pi \text{ radians}}{180 \text{ degrees}}\right) = \dfrac{2\pi}{9} \text{ radian}$

(c) $540° = (540 \text{ degrees})\left(\dfrac{\pi \text{ radians}}{180 \text{ degrees}}\right) = 3\pi \text{ radians}$

(d) $-270° = (-270 \text{ degrees})\left(\dfrac{\pi \text{ radians}}{180 \text{ degrees}}\right) = -\dfrac{3\pi}{2} \text{ radians}$

TRY IT 2

Convert each degree measure to radian measure.

(a) 225°   (b) −45°   (c) 240°   (d) 150°

Although it is common to list radian measure in multiples of π, this is not necessary. For instance, if the degree measure of an angle is 79.3°, the radian measure is

$$79.3° = (79.3 \text{ degrees})\left(\dfrac{\pi \text{ radians}}{180 \text{ degrees}}\right) \approx 1.384 \text{ radians.}$$

EXAMPLE 3 **Converting from Radians to Degrees**

Convert each radian measure to degree measure.

(a) $-\dfrac{\pi}{2}$ (b) $\dfrac{7\pi}{4}$ (c) $\dfrac{11\pi}{6}$ (d) $\dfrac{9\pi}{2}$

SOLUTION To convert from radian measure to degree measure, multiply the radian measure by $180°/(\pi \text{ radians})$.

(a) $-\dfrac{\pi}{2} \text{ radians} = \left(-\dfrac{\pi}{2} \text{ radians}\right)\left(\dfrac{180 \text{ degrees}}{\pi \text{ radians}}\right) = -90°$

(b) $\dfrac{7\pi}{4} \text{ radians} = \left(\dfrac{7\pi}{4} \text{ radians}\right)\left(\dfrac{180 \text{ degrees}}{\pi \text{ radians}}\right) = 315°$

(c) $\dfrac{11\pi}{6} \text{ radians} = \left(\dfrac{11\pi}{6} \text{ radians}\right)\left(\dfrac{180 \text{ degrees}}{\pi \text{ radians}}\right) = 330°$

(d) $\dfrac{9\pi}{2} \text{ radians} = \left(\dfrac{9\pi}{2} \text{ radians}\right)\left(\dfrac{180 \text{ degrees}}{\pi \text{ radians}}\right) = 810°$

TRY IT 3

Convert each radian measure to degree measure.

(a) $\dfrac{5\pi}{3}$ (b) $\dfrac{7\pi}{6}$

(c) $\dfrac{3\pi}{2}$ (d) $-\dfrac{3\pi}{4}$

Triangles

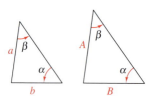

FIGURE 8.8 $a^2 + b^2 = c^2$

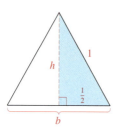

FIGURE 8.9 $\dfrac{a}{b} = \dfrac{A}{B}$

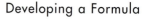

FIGURE 8.10

A Summary of Rules About Triangles

1. The sum of the angles of a triangle is 180°.
2. The sum of the two acute angles of a right triangle is 90°.
3. **Pythagorean Theorem** The sum of the squares of the legs of a right triangle is equal to the square of the hypotenuse, as shown in Figure 8.8.
4. **Similar Triangles** If two triangles are similar (have the same angle measures), then the ratios of the corresponding sides are equal, as shown in Figure 8.9.
5. The area of a triangle is equal to one-half the base times the height. That is, $A = \frac{1}{2}bh$.
6. Each angle of an equilateral triangle measures 60°.
7. Each acute angle of an isosceles right triangle measures 45°.
8. The altitude of a triangle bisects its base.

EXAMPLE 4   **Finding the Area of a Triangle**

Find the area of an equilateral triangle with one-foot sides.

SOLUTION To use the formula $A = \frac{1}{2}bh$, you must first find the height of the triangle, as shown in Figure 8.10. To do this, apply the Pythagorean Theorem to the shaded portion of the triangle.

$$h^2 + \left(\frac{1}{2}\right)^2 = 1^2 \qquad \textcolor{red}{\text{Pythagorean Theorem}}$$

$$h^2 = \frac{3}{4} \qquad \textcolor{red}{\text{Simplify.}}$$

$$h = \frac{\sqrt{3}}{2} \qquad \textcolor{red}{\text{Solve for } h.}$$

So, the area of the triangle is

$$A = \frac{1}{2}bh = \frac{1}{2}(1)\left(\frac{\sqrt{3}}{2}\right)$$

$$= \frac{\sqrt{3}}{4} \text{ square foot.}$$

TRY IT 4

Find the area of an isosceles right triangle with a hypotenuse of $\sqrt{2}$ feet.

TAKE ANOTHER LOOK

Developing a Formula

Develop a formula for the area of an equilateral triangle whose sides each have a length of s.

PREREQUISITE REVIEW 8.1

The following warm-up exercises involve skills that were covered in earlier sections. You will use these skills in the exercise set for this section.

In Exercises 1 and 2, find the area of the triangle.

1. Base: 10 cm; height: 7 cm

2. Base: 4 in.; height: 6 in.

In Exercises 3–6, let a and b represent the lengths of the legs, and let c represent the length of the hypotenuse, of a right triangle. Solve for the missing side length.

3. $a = 5, b = 12$

4. $a = 3, c = 5$

5. $a = 8, c = 17$

6. $b = 8, c = 10$

In Exercises 7–10, let a, b, and c represent the side lengths of a triangle. Use the information below to determine whether the figure is a right triangle, an isosceles triangle, or an equilateral triangle.

7. $a = 4, b = 4, c = 4$

8. $a = 1, b = 1, c = 2$

9. $a = 12, b = 16, c = 20$

10. $a = 1, b = 1, c = \sqrt{2}$

EXERCISES 8.1

In Exercises 1–4, determine two coterminal angles (one positive and one negative) for the given angle. Give the answers in degrees.

1. (a)

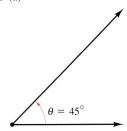

$\theta = 45°$

(b)

$\theta = -41°$

2. (a)

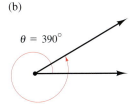

$\theta = -120°$

(b)

$\theta = 390°$

3. (a)

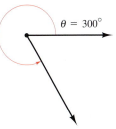

$\theta = 300°$

(b)

$\theta = 740°$

4. (a)

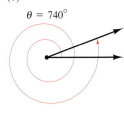

$\theta = -420°$

(b)

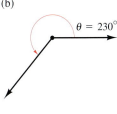

$\theta = 230°$

In Exercises 5–8, determine two coterminal angles (one positive and one negative) for each angle. Give the answers in radians.

5. (a)   (b)

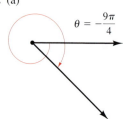

$\theta = \dfrac{\pi}{9}$ $\theta = \dfrac{2\pi}{3}$

6. (a)   (b)

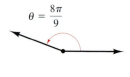

$\theta = \dfrac{11\pi}{6}$ $\theta = -\dfrac{7\pi}{6}$

7. (a)   (b)

$\theta = -\dfrac{9\pi}{4}$ $\theta = -\dfrac{2\pi}{15}$

8. (a)   (b)

$\theta = \dfrac{8\pi}{9}$ $\theta = \dfrac{8\pi}{45}$

In Exercises 9–20, express the angle in radian measure as a multiple of π. Use a calculator to verify your result.

9. 30°   **10.** 150°
11. 225°   **12.** 210°
13. 315°   **14.** 120°
15. −30°   **16.** −240°
17. −270°   **18.** −330°
19. 390°   **20.** 405°

In Exercises 21–30, express the angle in degree measure. Use a calculator to verify your result.

21. $\dfrac{3\pi}{2}$ **22.** $\dfrac{7\pi}{6}$

23. $\dfrac{11\pi}{6}$ **24.** $\dfrac{7\pi}{4}$

25. $-\dfrac{5\pi}{3}$ **26.** $-\dfrac{3\pi}{4}$

27. $\dfrac{9\pi}{4}$ **28.** $\dfrac{5\pi}{2}$

29. $\dfrac{19\pi}{6}$ **30.** $\dfrac{8\pi}{3}$

In Exercises 31–34, find the indicated measure of the angle. Express radian measure as a multiple of π.

Degree Measure	Radian Measure
31. −270°	
32.	$\dfrac{\pi}{9}$
33. 144°	
34.	$-\dfrac{7\pi}{12}$

In Exercises 35–42, solve the triangle for the indicated side and/or angle.

35.   **36.**

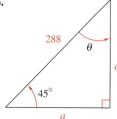

37.   **38.**

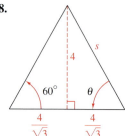

39.   **40.**

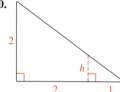

41.   **42.**

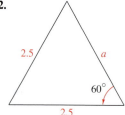

In Exercises 43–46, find the area of the equilateral triangle with sides of length s.

43. $s = 6$ in.   **44.** $s = 8$ m

45. $s = 5$ ft   **46.** $s = 12$ cm

47. Height   A person 6 feet tall standing 16 feet from a streetlight casts a shadow 8 feet long (see figure). What is the height of the streetlight?

Figure for 47   Figure for 48

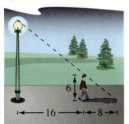
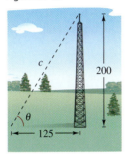

48. Length   A guy wire is stretched from a broadcasting tower at a point 200 feet above the ground to an anchor 125 feet from the base (see figure). How long is the wire?

49. Let r represent the radius of a circle, θ the central angle (measured in radians), and s the length of the arc interceped by the angle (see figure). Use the relationship $\theta = s/r$ to complete the table.

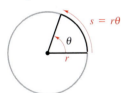

$$s = r\theta$$

r	8 ft	15 in.	85 cm		
s	12 ft			96 in.	8642 mi
θ		1.6	$\dfrac{3\pi}{4}$	4	$\dfrac{2\pi}{3}$

50. Arc Length   The minute hand on a clock is $3\frac{1}{2}$ inches long (see figure). Through what distance does the tip of the minute hand move in 25 minutes?

51. Distance   A man bends his elbow through $75°$. The distance from his elbow to the tip of his index finger is $18\frac{3}{4}$ inches (see figure).

(a) Find the radian measure of this angle.

(b) Find the distance the tip of the index finger moves.

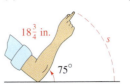

52. Distance   A tractor tire that is 5 feet in diameter d is partially filled with a liquid ballast for additional traction. To check the air pressure, the tractor operator rotates the tire until the valve stem is at the top so that the liquid will not enter the gauge. On a given occasion, the operator notes that the tire must be rotated $80°$ to have the stem in the proper position (see figure).

(a) Find the radian measure of this rotation.

(b) How far must the tractor be moved to get the valve stem in the proper position?

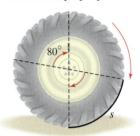

53. Speed of Revolution   A compact disc can have an angular speed up to 3142 radians per minute.

(a) At this angular speed, how many revolutions per minute would the CD make?

(b) How long would it take the CD to make 10,000 revolutions?

54. Speed of Revolution   The radius of the magnetic disk in a 3.5-inch diskette is 1.68 inches. Find the linear speed of a point on the circumference of the disk if it rotates at a speed of 360 revolutions per minute.

True or False?   In Exercises 55–58, determine whether the statement is true or false. If it is false, explain why or give an example that shows it is false.

55. An angle whose measure is $75°$ is obtuse.

56. $\theta = -35°$ is coterminal to $325°$.

57. A right triangle can have one angle whose measure is $89°$.

58. An angle whose measure is π radians is a straight angle.

8.2   THE TRIGONOMETRIC FUNCTIONS

- Recognize trigonometric functions.
- Use trigonometric identities.
- Evaluate trigonometric functions and solve right triangles.
- Solve trigonometric equations.

The Trigonometric Functions

There are two common approaches to the study of trigonometry. In one case the trigonometric functions are defined as ratios of two sides of a right triangle. In the other case these functions are defined in terms of a point on the terminal side of an arbitrary angle. The first approach is the one generally used in surveying, navigation, and astronomy, where a typical problem involves a triangle, three of whose six parts (sides and angles) are known and three of which are to be determined. The second approach is the one normally used in science and economics, where the periodic nature of the trigonometric functions is emphasized. In the definitions below, the six trigonometric functions are defined from both viewpoints.

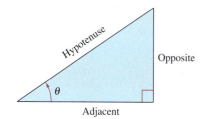

FIGURE 8.11

Definitions of the Trigonometric Functions

Right Triangle Definition: $0 < \theta < \dfrac{\pi}{2}$. (See Figure 8.11.)

$$\sin \theta = \frac{\text{opp.}}{\text{hyp.}} \qquad \csc \theta = \frac{\text{hyp.}}{\text{opp.}}$$

$$\cos \theta = \frac{\text{adj.}}{\text{hyp.}} \qquad \sec \theta = \frac{\text{hyp.}}{\text{adj.}}$$

$$\tan \theta = \frac{\text{opp.}}{\text{adj.}} \qquad \cot \theta = \frac{\text{adj.}}{\text{opp.}}$$

Circular Function Definition: θ is any angle in standard position and (x, y) is a point on the terminal ray of the angle. (See Figure 8.12.)

$$\sin \theta = \frac{y}{r} \qquad \csc \theta = \frac{r}{y}$$

$$\cos \theta = \frac{x}{r} \qquad \sec \theta = \frac{r}{x}$$

$$\tan \theta = \frac{y}{x} \qquad \cot \theta = \frac{x}{y}$$

The full names of the trigonometric functions are **sine, cosecant, cosine, secant, tangent,** and **cotangent.**

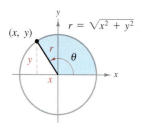

FIGURE 8.12

Trigonometric Identities

In the circular function definition of the six trigonometric functions, the value of r is always positive. From this, it follows that the signs of the trigonometric functions are determined from the signs of x and y, as shown in Figure 8.13.

The trigonometric reciprocal identities below are also direct consequences of the definitions.

$$\sin\theta = \frac{1}{\csc\theta} \qquad \cos\theta = \frac{1}{\sec\theta} \qquad \tan\theta = \frac{\sin\theta}{\cos\theta} = \frac{1}{\cot\theta}$$

$$\csc\theta = \frac{1}{\sin\theta} \qquad \sec\theta = \frac{1}{\cos\theta} \qquad \cot\theta = \frac{\cos\theta}{\sin\theta} = \frac{1}{\tan\theta}$$

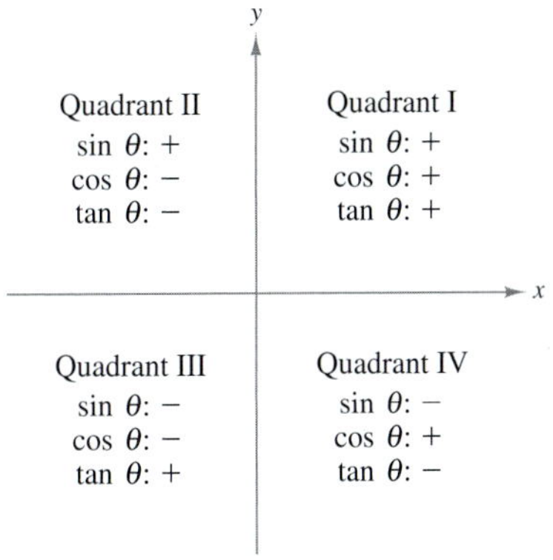
y
x
Quadrant II
sin θ: +
cos θ: −
tan θ: −
Quadrant I
sin θ: +
cos θ: +
tan θ: +
Quadrant III
sin θ: −
cos θ: −
tan θ: +
Quadrant IV
sin θ: −
cos θ: +
tan θ: −

FIGURE 8.13

Furthermore, because

$$\begin{aligned}\sin^2\theta + \cos^2\theta &= \left(\frac{y}{r}\right)^2 + \left(\frac{x}{r}\right)^2 \\ &= \frac{x^2 + y^2}{r^2} \\ &= \frac{r^2}{r^2} \\ &= 1\end{aligned}$$

you can obtain the Pythagorean Identity $\sin^2\theta + \cos^2\theta = 1$. Other trigonometric identities are listed below. In the list, ϕ is the lowercase Greek letter phi.

Trigonometric Identities

Pythagorean Identities

$\sin^2\theta + \cos^2\theta = 1$

$\tan^2\theta + 1 = \sec^2\theta$

$\cot^2\theta + 1 = \csc^2\theta$

Sum or Difference of Two Angles

$\sin(\theta \pm \phi) = \sin\theta\cos\phi \pm \cos\theta\sin\phi$

$\cos(\theta \pm \phi) = \cos\theta\cos\phi \mp \sin\theta\sin\phi$

$\tan(\theta \pm \phi) = \dfrac{\tan\theta \pm \tan\phi}{1 \mp \tan\theta\tan\phi}$

Reduction Formulas

$\sin(-\theta) = -\sin\theta$

$\cos(-\theta) = \cos\theta$

$\tan(-\theta) = -\tan\theta$

$\sin\theta = -\sin(\theta - \pi)$

$\cos\theta = -\cos(\theta - \pi)$

$\tan\theta = \tan(\theta - \pi)$

Double Angle

$\sin 2\theta = 2\sin\theta\cos\theta$

$\cos 2\theta = 2\cos^2\theta - 1 = 1 - 2\sin^2\theta$

Half Angle

$\sin^2\theta = \frac{1}{2}(1 - \cos 2\theta)$

$\cos^2\theta = \frac{1}{2}(1 + \cos 2\theta)$

STUDY TIP

The symbol $\sin^2\theta$ is used to represent $(\sin\theta)^2$.

Although an angle can be measured in either degrees or radians, radian measure is preferred in calculus. So, all angles in the remainder of this chapter are assumed to be measured in radians unless otherwise indicated. In other words, sin 3 means the sine of 3 radians, and sin 3° means the sine of 3 degrees.

Evaluating Trigonometric Functions

There are two common methods of evaluating trigonometric functions: decimal approximations using a calculator and exact evaluations using trigonometric identities and formulas from geometry. The next three examples illustrate the second method.

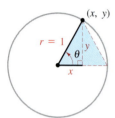

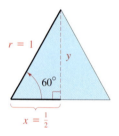

FIGURE 8.14

> **EXAMPLE 1**   **Evaluating Trigonometric Functions**

Evaluate the sine, cosine, and tangent of $\pi/3$.

SOLUTION   Begin by drawing the angle $\theta = \pi/3$ in standard position, as shown in Figure 8.14. Because $\pi/3$ radians is 60°, you can imagine an equilateral triangle with sides of length 1 and with θ as one of its angles. Because the altitude of the triangle bisects its base, you know that $x = \frac{1}{2}$. So, using the Pythagorean Theorem, you have

$$y = \sqrt{r^2 - x^2} = \sqrt{1^2 - \left(\frac{1}{2}\right)^2} = \sqrt{\frac{3}{4}} = \frac{\sqrt{3}}{2}.$$

Now, using $x = \frac{1}{2}$, $y = \frac{1}{2}\sqrt{3}$, and $r = 1$, you can find the values of the sine, cosine, and tangent as shown.

$$\sin \frac{\pi}{3} = \frac{y}{r} = \frac{\frac{1}{2}\sqrt{3}}{1} = \frac{\sqrt{3}}{2}$$

$$\cos \frac{\pi}{3} = \frac{x}{r} = \frac{\frac{1}{2}}{1} = \frac{1}{2}$$

$$\tan \frac{\pi}{3} = \frac{y}{x} = \frac{\frac{1}{2}\sqrt{3}}{\frac{1}{2}} = \sqrt{3}$$

TRY IT 1

Evaluate the sine, cosine, and tangent of $\dfrac{\pi}{6}$.

STUDY TIP

Learning the table of values at the right is worth the effort. Doing so will increase both your efficiency and your confidence. Here is a pattern for the sine function that may help you remember the values.

θ	0°	30°	45°	60°	90°
$\sin \theta$	$\frac{\sqrt{0}}{2}$	$\frac{\sqrt{1}}{2}$	$\frac{\sqrt{2}}{2}$	$\frac{\sqrt{3}}{2}$	$\frac{\sqrt{4}}{2}$

Reverse the order to get cosine values of the same angles.

The sines, cosines, and tangents of several common angles are listed in the table below. You should remember, or be able to derive, these values.

Table of Common Trigonometric Function Values

Degree measure of θ	0	30°	45°	60°	90°
Radian measure of θ	0	$\frac{\pi}{6}$	$\frac{\pi}{4}$	$\frac{\pi}{3}$	$\frac{\pi}{2}$
$\sin \theta$	0	$\frac{1}{2}$	$\frac{\sqrt{2}}{2}$	$\frac{\sqrt{3}}{2}$	1
$\cos \theta$	1	$\frac{\sqrt{3}}{2}$	$\frac{\sqrt{2}}{2}$	$\frac{1}{2}$	0
$\tan \theta$	0	$\frac{1}{\sqrt{3}}$	1	$\sqrt{3}$	Undefined

To extend the use of the values in the table on page 560 to angles in quadrants other than the first quadrant, you can use the concept of a reference angle, as shown in Figure 8.15, together with the appropriate quadrant sign. The **reference angle θ'** for an angle θ is the smallest positive angle between the terminal side of θ and the x-axis. For instance, the reference angle for $135°$ is $45°$ and the reference angle for $210°$ is $30°$.

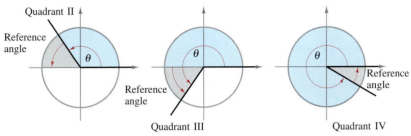

Reference angle: $\pi - \theta$ Reference angle: $\theta - \pi$ Reference angle: $2\pi - \theta$

FIGURE 8.15

To find the value of a trigonometric function of any angle θ, first determine the function value for the associated reference angle θ'. Then, depending on the quadrant in which θ lies, prefix the appropriate sign to the function value.

EXAMPLE 2 **Evaluating Trigonometric Functions**

Evaluate each trigonometric function.

(a) $\sin \dfrac{3\pi}{4}$ (b) $\tan 330°$ (c) $\cos \dfrac{7\pi}{6}$

SOLUTION

(a) Because the reference angle for $3\pi/4$ is $\pi/4$ and the sine is positive in the second quadrant, you can write

$$\sin \frac{3\pi}{4} = \sin \frac{\pi}{4} \qquad \text{Reference angle}$$

$$= \frac{\sqrt{2}}{2}. \qquad \text{See Figure 8.16(a).}$$

(b) Because the reference angle for $330°$ is $30°$ and the tangent is negative in the fourth quadrant, you can write

$$\tan 330° = -\tan 30° \qquad \text{Reference angle}$$

$$= -\frac{1}{\sqrt{3}}. \qquad \text{See Figure 8.16(b).}$$

(c) Because the reference angle for $7\pi/6$ is $\pi/6$ and the cosine is negative in the third quadrant, you can write

$$\cos \frac{7\pi}{6} = -\cos \frac{\pi}{6} \qquad \text{Reference angle}$$

$$= -\frac{\sqrt{3}}{2}. \qquad \text{See Figure 8.16(c).}$$

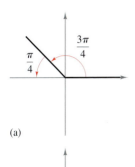

(a)

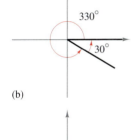

(b)

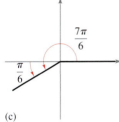

(c)

FIGURE 8.16

TRY IT 2

Evaluate each trigonometric function.

(a) $\sin \dfrac{5\pi}{6}$

(b) $\cos 135°$

(c) $\tan \dfrac{5\pi}{3}$

EXAMPLE 3 **Evaluating Trigonometric Functions**

Evaluate each trigonometric function.

(a) $\sin\left(-\dfrac{\pi}{3}\right)$ (b) $\sec 60°$ (c) $\cos 15°$ (d) $\sin 2\pi$ (e) $\cot 0°$ (f) $\tan \dfrac{9\pi}{4}$

SOLUTION

(a) By the reduction formula $\sin(-\theta) = -\sin\theta$,

$$\sin\left(-\frac{\pi}{3}\right) = -\sin\frac{\pi}{3} = -\frac{\sqrt{3}}{2}.$$

(b) By the reciprocal formula $\sec\theta = 1/\cos\theta$,

$$\sec 60° = \frac{1}{\cos 60°} = \frac{1}{1/2} = 2.$$

(c) By the difference formula $\cos(\theta - \phi) = \cos\theta\cos\phi + \sin\theta\sin\phi$,

$$\cos 15° = \cos(45° - 30°)$$
$$= (\cos 45°)(\cos 30°) + (\sin 45°)(\sin 30°)$$
$$= \left(\frac{\sqrt{2}}{2}\right)\left(\frac{\sqrt{3}}{2}\right) + \left(\frac{\sqrt{2}}{2}\right)\left(\frac{1}{2}\right)$$
$$= \frac{\sqrt{6} + \sqrt{2}}{4}.$$

(d) Because the reference angle for 2π is 0,

$$\sin 2\pi = \sin 0 = 0.$$

(e) Using the reciprocal formula $\cot\theta = 1/\tan\theta$ and the fact that $\tan 0° = 0$, you can conclude that $\cot 0°$ is undefined.

(f) Because the reference angle for $9\pi/4$ is $\pi/4$ and the tangent is positive in the first quadrant,

$$\tan\frac{9\pi}{4} = \tan\frac{\pi}{4} = 1.$$

TRY IT 3

Evaluate each trigonometric function.

(a) $\sin\left(-\dfrac{\pi}{6}\right)$ (b) $\csc 45°$

(c) $\cos 75°$ (d) $\cos 2\pi$

(e) $\sec 0°$ (f) $\cot \dfrac{13\pi}{4}$

TECHNOLOGY

Examples 1, 2, and 3 all involve standard angles such as $\pi/6$ and $\pi/3$. To evaluate trigonometric functions involving nonstandard angles, you should use a calculator. When doing this, remember to set the calculator to the proper mode—either *degree* mode or *radian* mode. Furthermore, most calculators have only three trigonometric functions: sine, cosine, and tangent. To evaluate the other three functions, you should combine these keys with the reciprocal key. For instance, you can evaluate the secant of $\pi/7$ as shown:

Function	*Calculator Steps*	*Display*
$\sec\dfrac{\pi}{7}$	(COS π ÷ 7)) 1/x	1.109916264.

EXAMPLE 4 **Solving a Right Triangle**

A surveyor is standing 50 feet from the base of a large tree. The surveyor measures the angle of elevation to the top of the tree as 71.5°. How tall is the tree?

SOLUTION Referring to Figure 8.17, you can see that

$$\tan 71.5° = \frac{y}{x}$$

where $x = 50$ and y is the height of the tree. So, you can determine the height of the tree to be

$$y = (x)(\tan 71.5°) \approx (50)(2.98868) \approx 149.4 \text{ feet.}$$

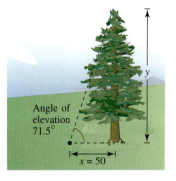

FIGURE 8.17

TRY IT 4

Find the height of a building that casts a 75-foot shadow when the angle of elevation of the sun is 35°.

EXAMPLE 5 **Calculating Peripheral Vision**

To measure the extent of your peripheral vision, stand 1 foot from the corner of a room, facing the corner. Have a friend move an object along the wall until you can just barely see it. If the object is 2 feet from the corner, as shown in Figure 8.18, what is the total angle of your peripheral vision?

SOLUTION Let α represent the total angle of your peripheral vision. As shown in Figure 8.19, you can model the physical situation with an isosceles right triangle whose legs are $\sqrt{2}$ feet and whose hypotenuse is 2 feet. In the triangle, the angle θ is given by

$$\tan \theta = \frac{y}{x} = \frac{\sqrt{2}}{\sqrt{2} - 1} \approx 3.414.$$

Using the inverse tangent function of a calculator, you can determine that $\theta \approx 73.7°$. So, $\alpha/2 \approx 180° - 73.7° = 106.3°$, which implies that $\alpha \approx 212.6°$. In other words, the total angle of your peripheral vision is about 212.6°.

© George Hall/CORBIS

Some occupations, such as that of a fighter pilot, require excellent vision, including good depth perception and good peripheral vision.

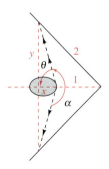

FIGURE 8.18

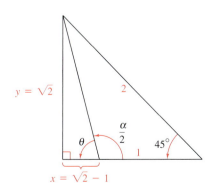

FIGURE 8.19

TRY IT 5

If the object in Example 5 is 4 feet from the corner, find the total angle of your peripheral vision.

Solving Trigonometric Equations

ALGEBRA REVIEW

For more examples of the algebra involved in solving trigonometric equations, see the *Chapter 8 Algebra Review* on pages 606 and 607.

An important part of the study of trigonometry is learning how to solve trigonometric equations. For example, consider the equation

$$\sin\theta = 0.$$

You know that $\theta = 0$ is one solution. Also, in Example 3(d), you saw that $\theta = 2\pi$ is another solution. But these are not the only solutions. In fact, this equation has infinitely many solutions. Any one of the values of θ shown below will work.

$$\ldots, -3\pi, -2\pi, -\pi, 0, \pi, 2\pi, 3\pi, \ldots$$

To simplify the situation, the search for solutions is usually restricted to the interval $0 \le \theta \le 2\pi$, as shown in Example 6.

EXAMPLE 6 Solving Trigonometric Equations

Solve for θ in each equation. Assume $0 \le \theta \le 2\pi$.

(a) $\sin\theta = -\dfrac{\sqrt{3}}{2}$ (b) $\cos\theta = 1$ (c) $\tan\theta = 1$

SOLUTION

(a) To solve the equation $\sin\theta = -\sqrt{3}/2$, first remember that

$$\sin\frac{\pi}{3} = \frac{\sqrt{3}}{2}.$$

Because the sine is negative in the third and fourth quadrants, it follows that you are seeking values of θ in these quadrants that have a reference angle of $\pi/3$. The two angles fitting these criteria are

$$\theta = \pi + \frac{\pi}{3} = \frac{4\pi}{3} \quad \text{and} \quad \theta = 2\pi - \frac{\pi}{3} = \frac{5\pi}{3}$$

as indicated in Figure 8.20.

(b) To solve $\cos\theta = 1$, remember that $\cos 0 = 1$ and note that in the interval $[0, 2\pi]$, the only angles whose reference angles are zero are zero, π, and 2π. Of these, zero and 2π have cosines of 1. (The cosine of π is -1.) So, the equation has two solutions:

$$\theta = 0 \quad \text{and} \quad \theta = 2\pi.$$

(c) Because $\tan \pi/4 = 1$ and the tangent is positive in the first and third quadrants, it follows that the two solutions are

$$\theta = \frac{\pi}{4} \quad \text{and} \quad \theta = \pi + \frac{\pi}{4} = \frac{5\pi}{4}.$$

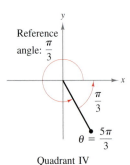

Reference angle: $\dfrac{\pi}{3}$

$\dfrac{\pi}{3}$

$\theta = \dfrac{4\pi}{3}$

Quadrant III

Reference angle: $\dfrac{\pi}{3}$

$\dfrac{\pi}{3}$

$\theta = \dfrac{5\pi}{3}$

Quadrant IV

FIGURE 8.20

TRY IT 6

Solve for θ in each equation. Assume $0 \le \theta \le 2\pi$.

(a) $\cos\theta = \dfrac{\sqrt{2}}{2}$ (b) $\tan\theta = -\sqrt{3}$ (c) $\sin\theta = -\dfrac{1}{2}$

EXAMPLE 7 **Solving a Trigonometric Equation**

Solve the equation for θ.

$$\cos 2\theta = 2 - 3 \sin \theta, \quad 0 \leq \theta \leq 2\pi$$

SOLUTION You can use the double-angle identity $\cos 2\theta = 1 - 2 \sin^2\theta$ to rewrite the original equation, as shown.

$$\cos 2\theta = 2 - 3 \sin \theta$$
$$1 - 2 \sin^2\theta = 2 - 3 \sin \theta$$
$$0 = 2 \sin^2\theta - 3 \sin \theta + 1$$
$$0 = (2 \sin \theta - 1)(\sin \theta - 1)$$

For $2 \sin \theta - 1 = 0$, you have $\sin \theta = \frac{1}{2}$, which has solutions of

$$\theta = \frac{\pi}{6} \quad \text{and} \quad \theta = \frac{5\pi}{6}.$$

For $\sin \theta - 1 = 0$, you have $\sin \theta = 1$, which has a solution of

$$\theta = \frac{\pi}{2}.$$

So, for $0 \leq \theta \leq 2\pi$, the three solutions are

$$\theta = \frac{\pi}{6}, \quad \frac{\pi}{2}, \quad \text{and} \quad \frac{5\pi}{6}.$$

ALGEBRA REVIEW

In Example 7, note that the expression $2 \sin^2\theta - 3 \sin \theta + 1$ is a quadratic in $\sin \theta$, and as such can be factored. For instance, if you let $x = \sin \theta$, then the quadratic factors as

$$2x^2 - 3x + 1 = (2x - 1)(x - 1).$$

TRY IT 7

Solve the equation for θ.

$$\sin 2\theta + \sin \theta = 0, \quad 0 \leq \theta \leq 2\pi$$

TAKE ANOTHER LOOK

Solving Trigonometric Equations

A calculator can be used to solve equations of the form

$$\sin \theta = a, \quad -\frac{\pi}{2} \leq \theta \leq \frac{\pi}{2}$$

$$\cos \theta = a, \quad 0 \leq \theta \leq \pi$$

$$\tan \theta = a, \quad -\frac{\pi}{2} < \theta < \frac{\pi}{2}.$$

For instance, you can solve the equation $\cos \theta = \frac{1}{2}$ by first setting the calculator to *radian* mode. When you use the inverse trigonometric function of your calculator, you obtain 1.047197551 (which is a decimal approximation of $\pi/3$). Note, however, that the calculator displays only one solution—the solution that occurs in the interval $0 \leq \theta \leq \pi$. Use your knowledge of trigonometry to obtain the other solutions of $\cos \theta = \frac{1}{2}$ in the interval $0 \leq \theta \leq 2\pi$.

The following warm-up exercises involve skills that were covered in earlier sections. You will use these skills in the exercise set for this section.

In Exercises 1–8, convert the angle to radian measure.

1. $135°$

2. $315°$

3. $-210°$

4. $-300°$

5. $-120°$

6. $-225°$

7. $540°$

8. $390°$

In Exercises 9–16, solve for x.

9. $x^2 - x = 0$

10. $2x^2 + x = 0$

11. $2x^2 - x = 1$

12. $x^2 - 2x = 3$

13. $x^2 - 2x = -1$

14. $2x^2 + x = 1$

15. $x^2 - 5x = -6$

16. $x^2 + x = 2$

In Exercises 17–20, solve for t.

17. $\dfrac{2\pi}{24}(t - 4) = \dfrac{\pi}{2}$

18. $\dfrac{2\pi}{12}(t - 2) = \dfrac{\pi}{4}$

19. $\dfrac{2\pi}{365}(t - 10) = \dfrac{\pi}{4}$

20. $\dfrac{2\pi}{12}(t - 4) = \dfrac{\pi}{2}$

EXERCISES 8.2

In Exercises 1–6, determine all six trigonometric functions for the angle θ.

1.

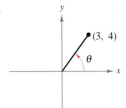

2.

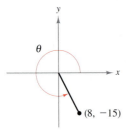

3.

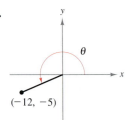

4.

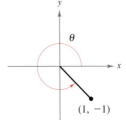

5.

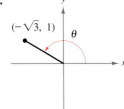

6.

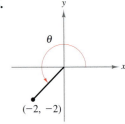

In Exercises 7–12, find the indicated trigonometric function from the given function.

7. Given $\sin \theta = \frac{1}{2}$, find $\csc \theta$.

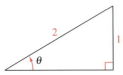

8. Given $\sin \theta = \frac{1}{3}$, find $\tan \theta$.

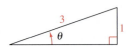

9. Given $\cos \theta = \frac{4}{5}$, find $\cot \theta$.

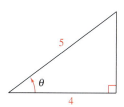

10. Given $\sec \theta = \frac{13}{5}$, find $\cot \theta$.

11. Given $\cot \theta = \frac{15}{8}$, find $\sec \theta$.

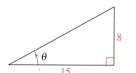

12. Given $\tan \theta = \frac{1}{2}$, find $\sin \theta$.

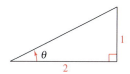

In Exercises 13–18, sketch a right triangle corresponding to the trigonometric function of the angle θ and find the other five trigonometric functions of θ.

13. $\sin \theta = \frac{1}{3}$ **14.** $\cot \theta = 5$

15. $\sec \theta = \frac{3}{2}$ **16.** $\cos \theta = \frac{5}{7}$

17. $\tan \theta = 3.5$ **18.** $\csc \theta = 4.25$

In Exercises 19–24, determine the quadrant in which θ lies.

19. $\sin \theta < 0$, $\cos \theta > 0$

20. $\sin \theta > 0$, $\cos \theta < 0$

21. $\sin \theta > 0$, $\sec \theta > 0$

22. $\cot \theta < 0$, $\cos \theta > 0$

23. $\csc \theta > 0$, $\tan \theta < 0$

24. $\cos \theta > 0$, $\tan \theta < 0$

In Exercises 25–32, evaluate the sines, cosines, and tangents of the angles *without* using a calculator.

25. (a) $60°$ (b) $-\dfrac{2\pi}{3}$

26. (a) $\dfrac{\pi}{4}$ (b) $\dfrac{5\pi}{4}$

27. (a) $-\dfrac{\pi}{6}$ (b) $150°$

28. (a) $-\dfrac{\pi}{2}$ (b) $\dfrac{\pi}{2}$

29. (a) $225°$ (b) $-225°$

30. (a) $300°$ (b) $330°$

31. (a) $750°$ (b) $510°$

32. (a) $\dfrac{10\pi}{3}$ (b) $\dfrac{17\pi}{3}$

In Exercises 33–40, use a calculator to evaluate the trigonometric functions to four decimal places.

33. (a) $\sin 12°$ (b) $\csc 12°$

34. (a) $\sec 225°$ (b) $\sec 135°$

35. (a) $\tan \dfrac{\pi}{9}$ (b) $\tan \dfrac{10\pi}{9}$

36. (a) $\cot 1.35$ (b) $\tan 1.35$

37. (a) $\cos(-110°)$ (b) $\cos 250°$

38. (a) $\tan 240°$ (b) $\cot 210°$

39. (a) $\csc 2.62$ (b) $\csc 150°$

40. (a) $\sin(-0.65)$ (b) $\sin 5.63$

In Exercises 41–46, find two values of θ corresponding to each function. List the measure of θ in radians ($0 \le \theta \le 2\pi$). Do not use a calculator.

41. (a) $\sin \theta = \frac{1}{2}$ (b) $\sin \theta = -\frac{1}{2}$

42. (a) $\cos \theta = \dfrac{\sqrt{2}}{2}$ (b) $\cos \theta = -\dfrac{\sqrt{2}}{2}$

43. (a) $\csc \theta = \dfrac{2\sqrt{3}}{3}$ (b) $\cot \theta = -1$

44. (a) $\sec \theta = 2$ (b) $\sec \theta = -2$

45. (a) $\tan \theta = -1$ (b) $\cot \theta = -\sqrt{3}$

46. (a) $\sin \theta = \dfrac{\sqrt{3}}{2}$ (b) $\sin \theta = -\dfrac{\sqrt{3}}{2}$

In Exercises 47–56, solve the equation for θ ($0 \le \theta \le 2\pi$). For some of the equations you should use the trigonometric identities listed in this section. Use the *trace* feature of a graphing utility to verify your results.

47. $2 \sin^2 \theta = 1$ **48.** $\tan^2 \theta = 3$

49. $\tan^2 \theta - \tan \theta = 0$ **50.** $2 \cos^2 \theta - \cos \theta = 1$

51. $\sin 2\theta - \cos \theta = 0$ **52.** $\cos 2\theta + 3 \cos \theta + 2 = 0$

53. $\sin \theta = \cos \theta$ **54.** $\sec \theta \csc \theta = 2 \csc \theta$

55. $\cos^2 \theta + \sin \theta = 1$ **56.** $\cos \dfrac{\theta}{2} - \cos \theta = 1$

In Exercises 57–62, solve for x, y, or r as indicated.

57. Solve for y.

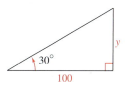

58. Solve for x.

59. Solve for *x*.

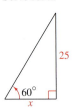

60. Solve for *r*.

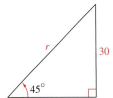

61. Solve for *r*.

62. Solve for *x*.

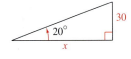

63. *Length*   A 20-foot ladder leaning against the side of a house makes a 75° angle with the ground (see figure). How far up the side of the house does the ladder reach?

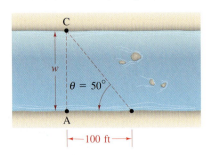

64. *Width*   A biologist wants to know the width *w* of a river in order to set instruments to study the pollutants in the water. From point A the biologist walks downstream 100 feet and sights to point C. From this sighting it is determined that $\theta = 50°$ (see figure). How wide is the river?

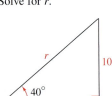

65. *Distance*   From a 150-foot observation tower on the coast, a Coast Guard officer sights a boat in difficulty. The angle of depression of the boat is 3° (see figure). How far is the boat from the shoreline?

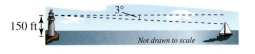

66. *Angle Measure*   A ramp $17\frac{1}{2}$ feet in length rises to a loading platform that is $3\frac{1}{2}$ feet off the ground (see figure). Find the angle that the ramp makes with the ground.

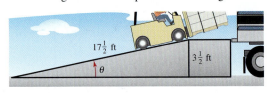

67. *Medicine*   The temperature *T* in degrees Fahrenheit of a patient *t* hours after arriving at the emergency room of a hospital at 10:00 P.M. is given by

$$T(t) = 98.6 + 4\cos\frac{\pi t}{36}, \quad 0 \le t \le 18.$$

Find the patient's temperature at each time.

(a) 10:00 P.M.

(b) 4:00 A.M.

(c) 10:00 A.M.

At what time do you expect the patient's temperature to return to normal? Explain your reasoning.

68. *Sales*   A company that produces a window and door insulating kit forecasts monthly sales over the next 2 years to be

$$S = 23.1 + 0.442t + 4.3\sin\frac{\pi t}{6}$$

where *S* is measured in thousands of units and *t* is the time in months, with $t = 1$ corresponding to January 2006. Use a graphing utility to estimate sales for each month.

(a) February 2006   (b) February 2007

(c) September 2006   (d) September 2007

In Exercises 69 and 70, use a graphing utility to complete the table. Then graph the function.

x	0	2	4	6	8	10
f(x)						

69. $f(x) = \frac{2}{5}x + 2\sin\frac{\pi x}{5}$

70. $f(x) = \frac{1}{2}(5 - x) + 3\cos\frac{\pi x}{5}$

8.3 GRAPHS OF TRIGONOMETRIC FUNCTIONS

- Sketch graphs of trigonometric functions.
- Evaluate limits of trigonometric functions.
- Use trigonometric functions to model real-life situations.

Graphs of Trigonometric Functions

When you are sketching the graph of a trigonometric function, it is common to use x (rather than θ) as the independent variable. On the simplest level, you can sketch the graph of a function such as

$$f(x) = \sin x$$

by constructing a table of values, plotting the resulting points, and connecting them with a smooth curve, as shown in Figure 8.21. Some examples of values are shown in the table below.

x	0	$\dfrac{\pi}{6}$	$\dfrac{\pi}{4}$	$\dfrac{\pi}{3}$	$\dfrac{\pi}{2}$	$\dfrac{2\pi}{3}$	$\dfrac{3\pi}{4}$	$\dfrac{5\pi}{6}$	π
$\sin x$	0.00	0.50	0.71	0.87	1.00	0.87	0.71	0.50	0.00

In Figure 8.21, note that the maximum value of $\sin x$ is 1 and the minimum value is -1. The **amplitude** of the sine function (or the cosine function) is defined to be half of the difference between its maximum and minimum values. So, the amplitude of $f(x) = \sin x$ is 1.

The periodic nature of the sine function becomes evident when you observe that as x increases beyond 2π, the graph repeats itself over and over, continuously oscillating about the x-axis. The **period** of the function is the distance (on the x-axis) between successive cycles. So, the period of $f(x) = \sin x$ is 2π.

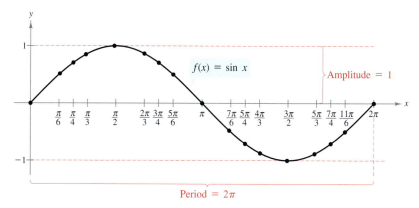

FIGURE 8.21

Figure 8.22 shows the graphs of at least one cycle of all six trigonometric functions.

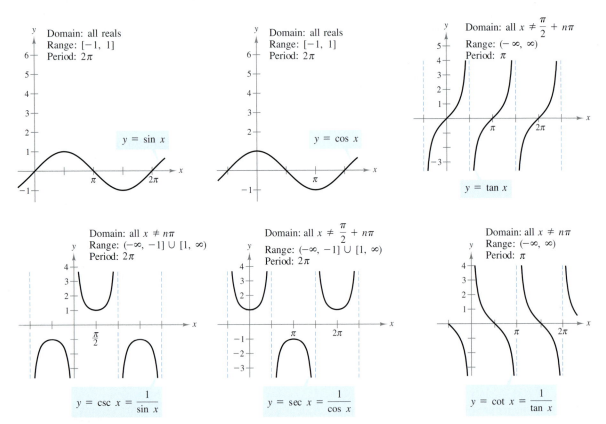

FIGURE 8.22   Graphs of the Six Trigonometric Functions

Familiarity with the graphs of the six basic trigonometric functions allows you to sketch graphs of more general functions such as

$$y = a \sin bx$$

and

$$y = a \cos bx.$$

Note that the function $y = a \sin bx$ oscillates between $-a$ and a and so has an amplitude of

$$|a|. \qquad \textcolor{red}{\text{Amplitude of } y = a \sin bx}$$

Furthermore, because $bx = 0$ when $x = 0$ and $bx = 2\pi$ when $x = 2\pi/b$, it follows that the function $y = a \sin bx$ has a period of

$$\frac{2\pi}{|b|}. \qquad \textcolor{red}{\text{Period of } y = a \sin bx}$$

EXAMPLE 1 **Graphing a Trigonometric Function**

Sketch the graph of $f(x) = 4 \sin x$.

SOLUTION The graph of $f(x) = 4 \sin x$ has the characteristics below.

    Amplitude: 4

       Period: 2π

Three cycles of the graph are shown in Figure 8.23, starting with the point $(0, 0)$.

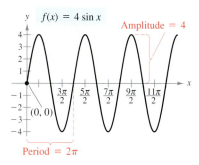

FIGURE 8.23

TRY IT 1

Sketch the graph of $g(x) = 2 \cos x$.

EXAMPLE 2 **Graphing a Trigonometric Function**

Sketch the graph of $f(x) = 3 \cos 2x$.

SOLUTION The graph of $f(x) = 3 \cos 2x$ has the characteristics below.

    Amplitude: 3

       Period: $\dfrac{2\pi}{2} = \pi$

Almost three cycles of the graph are shown in Figure 8.24, starting with the maximum point $(0, 3)$.

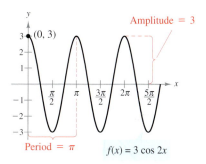

FIGURE 8.24

TRY IT 2

Sketch the graph of $g(x) = 2 \sin 4x$.

EXAMPLE 3 **Graphing a Trigonometric Function**

Sketch the graph of $f(x) = -2 \tan 3x$.

SOLUTION The graph of this function has a period of $\pi/3$. The vertical asymptotes of this tangent function occur at

$$x = \ldots, -\frac{\pi}{6}, \underbrace{\frac{\pi}{6}, \frac{\pi}{2}, \frac{5\pi}{6}}_{\text{Period} = \frac{\pi}{3}}, \ldots.$$

Several cycles of the graph are shown in Figure 8.25, starting with the vertical asymptote $x = -\pi/6$.

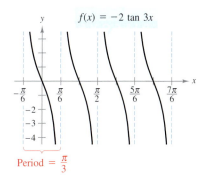

FIGURE 8.25

TRY IT 3

Sketch the graph of $g(x) = \tan 4x$.

EXAMPLE 6   Modeling Biorhythms

A popular theory that attempts to explain the ups and downs of everyday life states that each of us has three cycles, which begin at birth. These three cycles can be modeled by sine waves

$$\text{Physical (23 days):}\qquad P = \sin\frac{2\pi t}{23}, \quad t \ge 0$$

$$\text{Emotional (28 days):}\qquad E = \sin\frac{2\pi t}{28}, \quad t \ge 0$$

$$\text{Intellectual (33 days):}\qquad I = \sin\frac{2\pi t}{33}, \quad t \ge 0$$

where t is the number of days since birth. Describe the biorhythms during the month of September 2004, for a person who was born on July 20, 1984.

SOLUTION   Figure 8.28 shows the person's biorhythms during the month of September 2004. Note that September 1, 2004 was the 7348th day of the person's life.

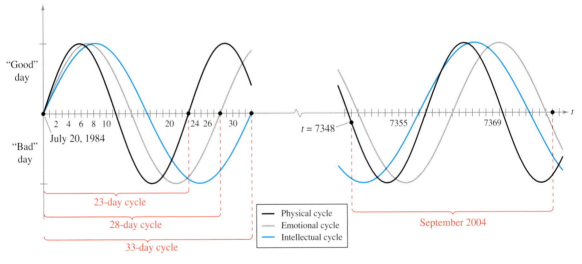

FIGURE 8.28

TRY IT 6

Use a graphing utility to describe the biorhythms of the person in Example 6 during the month of January 2004. Assume that January 1, 2004, is the 7104th day of the person's life.

TAKE ANOTHER LOOK

Biorhythms

Use a graphing utility to describe the August 2004 biorhythms for the person in Example 6. Assume that August 1, 2004, is the 7317th day of the person's life.

In Exercises 1 and 2, find the limit.

1. $\lim_{x\to2} (x^2 + 4x + 2)$

2. $\lim_{x\to3} (x^3 - 2x^2 + 1)$

In Exercises 3–10, evaluate the trigonometric function without using a calculator.

3. $\cos \dfrac{\pi}{2}$ **4.** $\sin \pi$

5. $\tan \dfrac{5\pi}{4}$ **6.** $\cot \dfrac{2\pi}{3}$

7. $\sin \dfrac{11\pi}{6}$ **8.** $\cos \dfrac{5\pi}{6}$

9. $\cos \dfrac{5\pi}{3}$ **10.** $\sin \dfrac{4\pi}{3}$

In Exercises 11–18, use a calculator to evaluate the trigonometric function to four decimal places.

11. $\cos 15°$ **12.** $\sin 220°$

13. $\sin 275°$ **14.** $\cos 310°$

15. $\sin 103°$ **16.** $\cos 72°$

17. $\tan 327°$ **18.** $\tan 140°$

EXERCISES 8.3

In Exercises 1–14, determine the period and amplitude of the function.

1. $y = 2 \sin 2x$

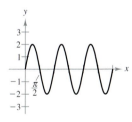

2. $y = 3 \cos 3x$

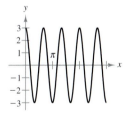

5. $y = \dfrac{1}{2} \cos \pi x$

6. $y = \dfrac{5}{2} \cos \dfrac{\pi x}{2}$

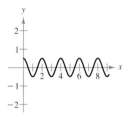

3. $y = \dfrac{3}{2} \cos \dfrac{x}{2}$

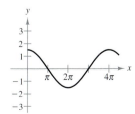

4. $y = -2 \sin \dfrac{x}{3}$

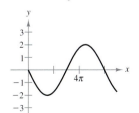

7. $y = -2 \sin x$

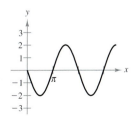

8. $y = -\cos \dfrac{2x}{3}$

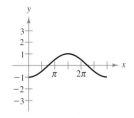

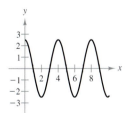

9. $y = -2 \sin 10x$

10. $y = \frac{1}{3} \sin 8x$

11. $y = \frac{1}{2} \sin \frac{2x}{3}$

12. $y = 5 \cos \frac{x}{4}$

13. $y = 3 \sin 4\pi x$

14. $y = \frac{2}{3} \cos \frac{\pi x}{10}$

In Exercises 15–20, find the period of the function.

15. $y = 5 \tan 2x$

16. $y = 7 \tan 2\pi x$

17. $y = 3 \sec 5x$

18. $y = \csc 4x$

19. $y = \cot \frac{\pi x}{6}$

20. $y = 5 \tan \frac{2\pi x}{3}$

In Exercises 21–26, match the trigonometric function with the correct graph and give the period of the function. [The graphs are labeled (a)–(f).]

(a)
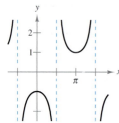

(b)
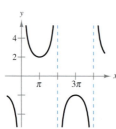

(c)
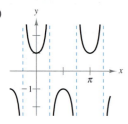

(d)
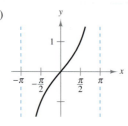

(e)
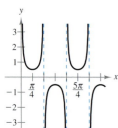

(f)
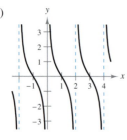

21. $y = \sec 2x$

22. $y = \frac{1}{2} \csc 2x$

23. $y = \cot \frac{\pi x}{2}$

24. $y = -\sec x$

25. $y = 2 \csc \frac{x}{2}$

26. $y = \tan \frac{x}{2}$

In Exercises 27–36, sketch the graph of the function by hand. Use a graphing utility to verify your sketch.

27. $y = \sin \frac{x}{2}$

28. $y = 4 \sin \frac{x}{3}$

29. $y = 2 \cos \frac{2x}{3}$

30. $y = \frac{3}{2} \cos \frac{2x}{3}$

31. $y = -2 \sin 6x$

32. $y = -3 \cos 4x$

33. $y = \cos 2\pi x$

34. $y = \frac{3}{2} \sin \frac{\pi x}{4}$

35. $y = 2 \tan x$

36. $y = 2 \cot x$

In Exercises 37–46, sketch the graph of the function.

37. $y = -\sin \frac{2\pi x}{3}$

38. $y = 10 \cos \frac{\pi x}{6}$

39. $y = \cot 2x$

40. $y = 3 \tan \pi x$

41. $y = \csc \frac{2x}{3}$

42. $y = \csc \frac{x}{3}$

43. $y = 2 \sec 2x$

44. $y = \sec \pi x$

45. $y = \csc 2\pi x$

46. $y = -\tan x$

 In Exercises 47–56, complete the table (using a graphing utility set in *radian* mode) to estimate $\lim\limits_{x \to 0} f(x)$.

x	-0.1	-0.01	-0.001	0.001	0.01	0.1
$f(x)$						

47. $f(x) = \dfrac{1 - \cos 2x}{2x}$

48. $f(x) = \dfrac{\sin 2x}{\sin 3x}$

49. $f(x) = \dfrac{\sin x}{5x}$

50. $f(x) = \dfrac{1 - \cos 2x}{x}$

51. $f(x) = \dfrac{3(1 - \cos x)}{x}$

52. $f(x) = \dfrac{2 \sin(x/4)}{x}$

53. $f(x) = \dfrac{\tan 2x}{x}$

54. $f(x) = \dfrac{\tan 4x}{3x}$

55. $f(x) = \dfrac{1 - \cos^2 4x}{x}$

56. $f(x) = \dfrac{1 - \cos^2 x}{2x}$

In Exercises 57 and 58, determine the values of $a, b, c,$ and d such that the graph of $y = a \sin(bx + c) + d$ is shown.

57.

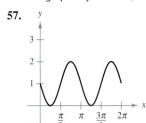

58.

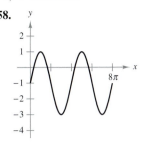

59. Health For a person at rest, the velocity v (in liters per second) of air flow into and out of the lungs during a respiratory cycle is given by

$$v = 0.9 \sin \frac{\pi t}{3}$$

where t is the time in seconds. Inhalation occurs when $v > 0$, and exhalation occurs when $v < 0$.

(a) Find the time for one full respiratory cycle.

(b) Find the number of cycles per minute.

 (c) Use a graphing utility to graph the velocity function.

 60. Health After exercising for a few minutes, a person has a respiratory cycle for which the velocity of air flow is approximated by

$$y = 1.75 \sin \frac{\pi t}{2}.$$

Use this model to repeat Exercise 59.

61. Music When tuning a piano, a technician strikes a tuning fork for the A above middle C and sets up wave motion that can be approximated by

$$y = 0.001 \sin 880 \pi t$$

where t is the time in seconds.

(a) What is the period p of this function?

(b) What is the frequency f of this note $(f = 1/p)$?

 (c) Use a graphing utility to graph this function.

62. Health The function

$$P = 100 - 20 \cos \frac{5\pi t}{3}$$

approximates the blood pressure P (in millimeters of mercury) at time t in seconds for a person at rest.

(a) Find the period of the function.

(b) Find the number of heartbeats per minute.

(c) Use a graphing utility to graph the pressure function.

63. Biology: Predator-Prey Cycle The population P of a predator at time t (in months) is modeled by

$$P = 8000 + 2500 \sin \frac{2\pi t}{24}$$

and the population p of its prey is modeled by

$$p = 12{,}000 + 4000 \cos \frac{2\pi t}{24}.$$

 (a) Use a graphing utility to graph both models in the same viewing window.

(b) Explain the oscillations in the size of each population.

64. Biology: Predator-Prey Cycle The population P of a predator at time t (in months) is modeled by

$$P = 5700 + 1200 \sin \frac{2\pi t}{24}$$

and the population p of its prey is modeled by

$$p = 9800 + 2750 \cos \frac{2\pi t}{24}.$$

 (a) Use a graphing utility to graph both models in the same viewing window.

(b) Explain the oscillations in the size of each population.

Sales In Exercises 65 and 66, sketch the graph of the sales function over 1 year where S is sales in thousands of units and t is the time in months, with $t = 1$ corresponding to January.

65. $S = 22.3 - 3.4 \cos \dfrac{\pi t}{6}$ **66.** $S = 74.50 + 43.75 \sin \dfrac{\pi t}{6}$

67. Biorhythms For the person born on July 20, 1984, use the biorhythm cycles given in Example 6 to calculate this person's three energy levels on December 31, 2008. Assume this is the 8930th day of the person's life.

68. Biorhythms Use your birthday and the concept of biorhythms as given in Example 6 to calculate your three energy levels on December 31, 2008. Use the internet or some other reference source to calculate the number of days between your birthday and December 31, 2008.

 In Exercises 69 and 70, use a graphing utility to graph the functions in the same viewing window where $0 \le x \le 2$.

69. (a) $y = \dfrac{4}{\pi} \sin \pi x$

(b) $y = \dfrac{4}{\pi}\left(\sin \pi x + \dfrac{1}{3} \sin 3\pi x\right)$

70. (a) $y = \dfrac{1}{2} - \dfrac{4}{\pi^2} \cos \pi x$

(b) $y = \dfrac{1}{2} - \dfrac{4}{\pi^2}\left(\cos \pi x + \dfrac{1}{9} \cos 3\pi x\right)$

 In Exercises 71–74, use a graphing utility to graph the function f and find $\lim\limits_{x \to 0} f(x)$.

71. $f(x) = \dfrac{\sin x}{x}$ **72.** $f(x) = \dfrac{\sin 5x}{2x}$

73. $f(x) = \dfrac{\sin 5x}{\sin 2x}$ **74.** $f(x) = \dfrac{\tan 2x}{3x}$

75. Sales　The sales S (in thousands of units) of snowmobiles are modeled by

$$S = 58.3 + 32.5 \cos \frac{\pi t}{6}$$

where t is the time in months, with $t = 1$ corresponding to January and $t = 12$ corresponding to December.

 (a) Use a graphing utility to graph S.

(b) Determine the months when sales exceed 75,000 units.

76. Meteorology　The normal monthly high temperatures for Erie, Pennsylvania are approximated by

$$H(t) = 54.33 - 20.38 \cos \frac{\pi t}{6} - 15.69 \sin \frac{\pi t}{6}$$

and the normal monthly low temperatures are approximated by

$$L(t) = 39.36 - 15.70 \cos \frac{\pi t}{6} - 14.16 \sin \frac{\pi t}{6}$$

where t is the time in months, with $t = 1$ corresponding to January. *(Source: NOAA)* Use the figure to answer the questions below.

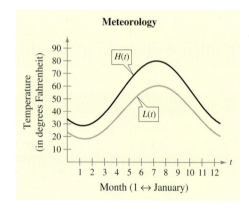

Meteorology

(a) During what part of the year is the difference between the normal high and low temperatures greatest? When is it smallest?

(b) The sun is the farthest north in the sky around June 21, but the graph shows the highest temperatures at a later date. Approximate the lag time of the temperatures relative to the position of the sun.

 77. Finance: Cyclical Stocks　The term "cyclical stock" describes the stock of a company whose profits are greatly influenced by changes in the economic business cycle. The market prices of cyclical stocks mirror the general state of the economy and reflect the various phases of the business cycle. Give a description and sketch of the graph of a given corporation's stock prices during recurrent periods of prosperity and recession. *(Source: Adapted from Garman/ Forgue, Personal Finance, Fifth Edition)*

 78. Physics　Use the graphs below to answer each question.

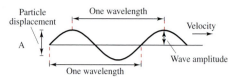

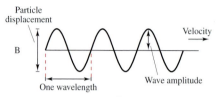

(a) Which graph (A or B) has a longer wavelength, or period?

(b) Which graph (A or B) has a greater amplitude?

(c) The frequency of a graph is the number of oscillations or cycles that occur during a given period of time. Which graph (A or B) has a greater frequency?

(d) Based on the definition of frequency, how are frequency and period related?

(Source: Adapted from Shipman/Wilson/Todd, An Introduction to Physical Science, Tenth Edition)

 79. Biology: Predator-Prey Cycles　The graph below demonstrates snowshoe hare and lynx population fluctuations. The cycles of each population follow a periodic pattern. Describe several factors that could be contributing to the cyclical patterns. *(Source: Adapted from Levine/ Miller, Biology: Discovering Life, Second Edition)*

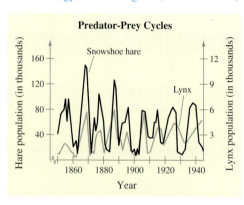

True or False?　In Exercises 80–83, determine whether the statement is true or false. If it is false, explain why or give an example that shows it is false.

80. The amplitude of $f(x) = -3 \cos 2x$ is -3.

81. The period of $f(x) = 5 \cot\left(-\frac{4x}{3}\right)$ is $\frac{3\pi}{2}$.

82. $\lim\limits_{x \to 0} \dfrac{\sin 5x}{3x} = \dfrac{5}{3}$

83. One solution of $\tan \dfrac{x}{2} = 1$ is $\dfrac{5\pi}{4}$.

8.4 DERIVATIVES OF TRIGONOMETRIC FUNCTIONS

- Find derivatives of trigonometric functions.
- Find the relative extrema of trigonometric functions.
- Use derivatives of trigonometric functions to answer questions about real-life situations.

Derivatives of Trigonometric Functions

In Example 4 and Try It 4 in the preceding section, you looked at two important trigonometric limits:

$$\lim_{\Delta x \to 0} \frac{\sin \Delta x}{\Delta x} = 1 \quad \text{and} \quad \lim_{\Delta x \to 0} \frac{1 - \cos \Delta x}{\Delta x} = 0.$$

These two limits are used in the development of the derivative of the sine function.

$$\frac{d}{dx}[\sin x] = \lim_{\Delta x \to 0} \frac{\sin(x + \Delta x) - \sin x}{\Delta x}$$

$$= \lim_{\Delta x \to 0} \frac{\sin x \cos \Delta x + \cos x \sin \Delta x - \sin x}{\Delta x}$$

$$= \lim_{\Delta x \to 0} \left(\cos x \frac{\sin \Delta x}{\Delta x} - \sin x \frac{1 - \cos \Delta x}{\Delta x} \right)$$

$$= \cos x \left(\lim_{\Delta x \to 0} \frac{\sin \Delta x}{\Delta x} \right) - \sin x \left(\lim_{\Delta x \to 0} \frac{1 - \cos \Delta x}{\Delta x} \right)$$

$$= (\cos x)(1) - (\sin x)(0)$$

$$= \cos x.$$

This differentiation rule is illustrated graphically in Figure 8.29. Note that the *slope* of the sine curve determines the *value* of the cosine curve. If u is a function of x, the Chain Rule version of this differentiation rule is

$$\frac{d}{dx}[\sin u] = \cos u \frac{du}{dx}.$$

The Chain Rule versions of the differentiation rules for all six trigonometric functions are listed below.

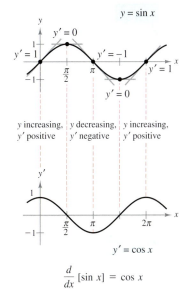

$$\frac{d}{dx}[\sin x] = \cos x$$

FIGURE 8.29

Derivatives of Trigonometric Functions

$$\frac{d}{dx}[\sin u] = \cos u \frac{du}{dx} \qquad \frac{d}{dx}[\cos u] = -\sin u \frac{du}{dx}$$

$$\frac{d}{dx}[\tan u] = \sec^2 u \frac{du}{dx} \qquad \frac{d}{dx}[\cot u] = -\csc^2 u \frac{du}{dx}$$

$$\frac{d}{dx}[\sec u] = \sec u \tan u \frac{du}{dx} \qquad \frac{d}{dx}[\csc u] = -\csc u \cot u \frac{du}{dx}$$

STUDY TIP

To help you remember these differentiation rules, note that each trigonometric function that begins with a "c" has a negative sign in its derivative.

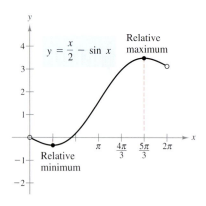

FIGURE 8.30

TRY IT 7

Find the relative extrema of

$$y = \frac{x}{2} - \cos x$$

on the interval $(0, 2\pi)$.

Relative Extrema of Trigonometric Functions

| **EXAMPLE 7** | **Finding Relative Extrema** |

Find the relative extrema of

$$y = \frac{x}{2} - \sin x$$

on the interval $(0, 2\pi)$.

SOLUTION   To find the relative extrema of the function, begin by finding its critical numbers. The derivative of y is

$$\frac{dy}{dx} = \frac{1}{2} - \cos x.$$

By setting the derivative equal to zero, you obtain $\cos x = \frac{1}{2}$. So, in the interval $(0, 2\pi)$, the critical numbers are $x = \pi/3$ and $x = 5\pi/3$. Using the First-Derivative Test, you can conclude that $\pi/3$ yields a relative minimum and $5\pi/3$ yields a relative maximum, as shown in Figure 8.30.

STUDY TIP

Recall that the critical numbers of a function $y = f(x)$ are the x-values for which $f'(x) = 0$ or $f'(x)$ is undefined.

| **EXAMPLE 8** | **Finding Relative Extrema** |

Find the relative extrema of $f(x) = 2 \sin x - \cos 2x$ on the interval $(0, 2\pi)$.

SOLUTION

$$f(x) = 2 \sin x - \cos 2x \qquad \text{Write original function.}$$
$$f'(x) = 2 \cos x + 2 \sin 2x \qquad \text{Differentiate.}$$
$$0 = 2 \cos x + 2 \sin 2x \qquad \text{Set derivative equal to 0.}$$
$$0 = 2 \cos x + 4 \cos x \sin x \qquad \text{Identity: } \sin 2x = 2 \cos x \sin x$$
$$0 = 2(\cos x)(1 + 2 \sin x) \qquad \text{Factor.}$$

From this, you can see that the critical numbers occur when $\cos x = 0$ and when $\sin x = -\frac{1}{2}$. So, in the interval $(0, 2\pi)$, the critical numbers are

$$x = \frac{\pi}{2}, \frac{3\pi}{2}, \frac{7\pi}{6}, \frac{11\pi}{6}.$$

Using the First-Derivative Test, you can determine that $(\pi/2, 3)$ and $(3\pi/2, -1)$ are relative maxima, and $\left(7\pi/6, -\frac{3}{2}\right)$ and $\left(11\pi/6, -\frac{3}{2}\right)$ are relative minima, as shown in Figure 8.31.

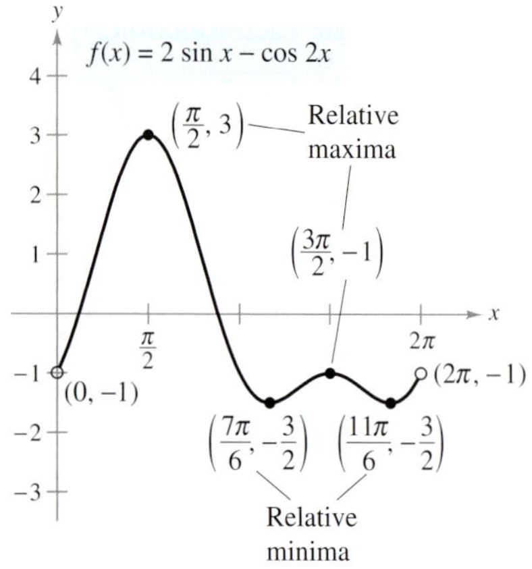

FIGURE 8.31

TRY IT 8

Find the relative extrema of $y = \frac{1}{2} \sin 2x + \cos x$ on the interval $(0, 2\pi)$.

Applications

EXAMPLE 9   **Modeling Seasonal Sales**

A fertilizer manufacturer finds that the sales of one of its fertilizer brands follows a seasonal pattern that can be modeled by

$$F = 100{,}000\left[1 + \sin\frac{2\pi(t - 60)}{365}\right], \quad t \geq 0$$

where F is the amount sold (in pounds) and t is the time (in days), with $t = 1$ corresponds to January 1. On which day of the year is the maximum amount of fertilizer sold?

SOLUTION   The derivative of the model is

$$\frac{dF}{dt} = 100{,}000\left(\frac{2\pi}{365}\right)\cos\frac{2\pi(t - 60)}{365}.$$

Setting this derivative equal to zero produces

$$\cos\frac{2\pi(t - 60)}{365} = 0.$$

Because cosine is zero at $\pi/2$ and $3\pi/2$, you can find the critical numbers as shown.

$$\frac{2\pi(t - 60)}{365} = \frac{\pi}{2} \qquad\qquad \frac{2\pi(t - 60)}{365} = \frac{3\pi}{2}$$

$$t - 60 = \frac{365}{4} \qquad\qquad t - 60 = \frac{3(365)}{4}$$

$$t = \frac{365}{4} + 60 \approx 151 \qquad\qquad t = \frac{3(365)}{4} + 60 \approx 334$$

The 151st day of the year is May 31 and the 334th day of the year is November 30. From the graph in Figure 8.32, you can see that, according to the model, the maximum sales occur on May 31.

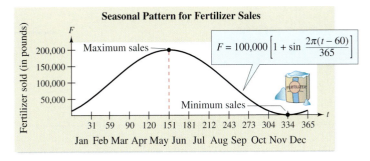

Seasonal Pattern for Fertilizer Sales

$$F = 100{,}000\left[1 + \sin\frac{2\pi(t - 60)}{365}\right]$$

FIGURE 8.32

Graphing Trigonometric Functions

 Because of the difficulty of solving some trigonometric equations, it can be difficult to find the critical numbers of a trigonometric function. For example, consider the function $f(x) = 2\sin x - \cos 3x$. Setting the derivative of this function equal to zero produces

$$f'(x) = 2\cos x + 3\sin 3x = 0.$$

This equation is difficult to solve analytically. So, it is difficult to find the relative extrema of f analytically. With a graphing utility, however, you can easily graph the function and use the *zoom* feature to estimate the relative extrema. In the graph shown below, notice that the function has three relative minima and three relative maxima in the interval $(0, 2\pi)$.

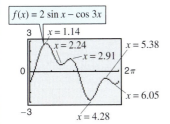

Once you have obtained rough approximations for the relative extrema, you can further refine the approximations by applying other approximation techniques, such as Newton's Method, which is discussed in Section 10.6, to the equation

$$f'(x) = 2\cos x + 3\sin 3x = 0.$$

Try using technology to locate the relative extrema of the function

$$f(x) = 2\sin x - \cos 4x.$$

How many relative extrema does this function have in the interval $(0, 2\pi)$?

TRY IT 9

Using the model from Example 9, find the rate at which sales are changing when $t = 59$.

EXAMPLE 10   **Modeling Temperature Change**

The temperature T (in degrees Fahrenheit) during a given 24-hour period can be modeled by

$$T = 70 + 15 \sin \frac{\pi(t - 8)}{12}, \quad t \geq 0$$

where t is the time (in hours), with $t = 0$ corresponding to midnight, as shown in Figure 8.33. Find the rate at which the temperature is changing at 6 A.M.

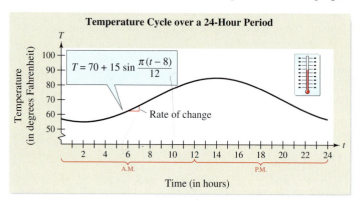

FIGURE 8.33

SOLUTION

The rate of change of the temperature is given by the derivative

$$\frac{dT}{dt} = \frac{15\pi}{12} \cos \frac{\pi(t - 8)}{12}.$$

Because 6 A.M. corresponds to $t = 6$, the rate of change at 6 A.M. is

$$\frac{15\pi}{12} \cos \left(\frac{-2\pi}{12} \right) = \frac{5\pi}{4} \cos \left(-\frac{\pi}{6} \right)$$

$$= \frac{5\pi}{4} \left(\frac{\sqrt{3}}{2} \right)$$

$$\approx 3.4° \text{ per hour.}$$

TRY IT 10

In Example 10, find the rate at which the temperature is changing at 8 P.M.

TAKE ANOTHER LOOK

Modeling Temperature Changes

In Example 10, when does the minimum temperature occur? When does the maximum temperature occur? Suppose that each occurred 1 hour later. How would this change the model?

PREREQUISITE REVIEW 8.4

The following warm-up exercises involve skills that were covered in earlier sections. You will use these skills in the exercise set for this section.

In Exercises 1–4, find the derivative of the function.

1. $f(x) = 3x^3 - 2x^2 + 4x - 7$

2. $g(x) = (x^3 + 4)^4$

3. $f(x) = (x - 1)(x^2 + 2x + 3)$

4. $g(x) = \dfrac{2x}{x^2 + 5}$

In Exercises 5 and 6, find the relative extrema of the function.

5. $f(x) = x^2 + 4x + 1$

6. $f(x) = \frac{1}{3}x^3 - 4x + 2$

In Exercises 7–10, solve the trigonometric equation for x where $0 \le x \le 2\pi$.

7. $\sin x = \dfrac{\sqrt{3}}{2}$

8. $\cos x = -\dfrac{1}{2}$

9. $\cos \dfrac{x}{2} = 0$

10. $\sin \dfrac{x}{2} = -\dfrac{\sqrt{2}}{2}$

EXERCISES 8.4

In Exercises 1–26, find the derivative of the function.

1. $y = \frac{1}{2} - 3\sin x$

2. $y = 5 + \sin x$

3. $y = x^2 - \cos x$

4. $g(t) = \pi \cos t - \dfrac{1}{t^2}$

5. $f(x) = 4\sqrt{x} + 3\cos x$

6. $f(x) = \sin x + \cos x$

7. $f(t) = t^2 \cos t$

8. $f(x) = (x + 1)\cos x$

9. $g(t) = \dfrac{\cos t}{t}$

10. $f(x) = \dfrac{\sin x}{x}$

11. $y = \tan x + x^2$

12. $y = x + \cot x$

13. $y = e^{x^2} \sec x$

14. $y = e^{-x} \sin x$

15. $y = \cos 3x + \sin^2 x$

16. $y = \csc^2 x - \cos 2x$

17. $y = \sin \pi x$

18. $y = \frac{1}{2} \csc 2x$

19. $y = x \sin \dfrac{1}{x}$

20. $y = x^2 \sin \dfrac{1}{x}$

21. $y = 3 \tan 4x$

22. $y = \tan e^x$

23. $y = 2 \tan^2 4x$

24. $y = -\sin^4 2x$

25. $y = e^{2x} \sin 2x$

26. $y = e^{-x} \cos \dfrac{x}{2}$

In Exercises 27–38, find the derivative of the function and simplify your answer by using the trigonometric identities listed in Section 8.2.

27. $y = \cos^2 x$

28. $y = \frac{1}{4} \sin^2 2x$

29. $y = \cos^2 x - \sin^2 x$

30. $y = \dfrac{x}{2} + \dfrac{\sin 2x}{4}$

31. $y = \ln|\sin x|$

32. $y = -\ln|\cos x|$

33. $y = \ln|\csc x^2 - \cot x^2|$

34. $y = \ln|\sec x + \tan x|$

35. $y = \tan x - x$

36. $y = \dfrac{\sec^7 x}{7} - \dfrac{\sec^5 x}{5}$

37. $y = \ln(\sin^2 x)$

38. $y = \frac{1}{2}(x \tan x - \sec x)$

In Exercises 39–46, find an equation of the tangent line to the graph of the function at the given point.

Function	Point
39. $y = \tan x$	$\left(-\dfrac{\pi}{4}, -1\right)$
40. $y = \sec x$	$\left(\dfrac{\pi}{3}, 2\right)$
41. $y = \sin 4x$	$(\pi, 0)$
42. $y = \csc^2 x$	$\left(\dfrac{\pi}{2}, 1\right)$
43. $y = \dfrac{\cos x}{\sin x}$	$\left(\dfrac{3\pi}{4}, -1\right)$
44. $y = \sin x \cos x$	$\left(\dfrac{3\pi}{2}, 0\right)$
45. $y = \ln|\cot x|$	$\left(\dfrac{\pi}{4}, 0\right)$
46. $y = \sqrt{\sin x}$	$\left(\dfrac{\pi}{6}, \dfrac{\sqrt{2}}{2}\right)$

In Exercises 47 and 48, use implicit differentiation to find dy/dx and evaluate the derivative at the given point.

Function	Point
47. $\sin x + \cos 2y = 1$	$\left(\dfrac{\pi}{2}, \dfrac{\pi}{4}\right)$
48. $\tan(x + y) = x$	$(0, 0)$

In Exercises 49–52, show that the function satisfies the differential equation.

49. $y = 2 \sin x + 3 \cos x$

$y'' + y = 0$

50. $y = \dfrac{10 - \cos x}{x}$

$xy' + y = \sin x$

51. $y = \cos 2x + \sin 2x$

$y'' + 4y = 0$

52. $y = e^x \left(\cos \sqrt{2}x + \sin \sqrt{2}x\right)$

$y'' - 2y' + 3y = 0$

In Exercises 53–58, find the slope of the tangent line to the given sine function at the origin. Compare this value with the number of complete cycles in the interval $[0, 2\pi]$.

53. $y = \sin \dfrac{5x}{4}$ **54.** $y = \sin \dfrac{5x}{2}$

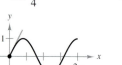

55. $y = \sin 2x$ **56.** $y = \sin \dfrac{3x}{2}$

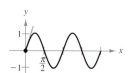 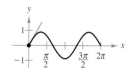

57. $y = \sin x$ **58.** $y = \sin \dfrac{x}{2}$

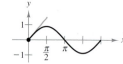 

In Exercises 59–64, determine the relative extrema of the function on the interval $(0, 2\pi)$. Use a graphing utility to confirm your result.

59. $y = 2 \sin x + \sin 2x$ **60.** $y = 2 \sin x + \cos 2x$

61. $y = x - 2 \sin x$ **62.** $y = e^{-x} \sin x$

63. $y = e^{-x} \cos x$ **64.** $y = \sec \dfrac{x}{2}$

65. *Biology*   Plants do not grow at constant rates during a normal 24-hour period because their growth is affected by sunlight. Suppose that the growth of a certain plant species in a controlled environment is given by the model

$$h = 0.2t + 0.03 \sin 2\pi t$$

where h is the height of the plant in inches and t is the time in days, with $t = 0$ corresponding to midnight of day 1 (see figure). During what time of day is the rate of growth of this plant

(a) a maximum?   (b) a minimum?

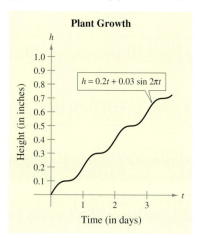

66. *Meteorology*   The normal average daily temperature in degrees Fahrenheit for a city is given by

$$T = 55 - 21 \cos \dfrac{2\pi(t - 32)}{365}$$

where t is the time in days, with $t = 1$ corresponding to January 1. Find the expected date of

(a) the warmest day.   (b) the coldest day.

67. *Physics*   An amusement park ride is constructed such that its height h in feet above ground in terms of the horizontal distance x in feet from the starting point can be modeled by

$$h = 50 + 45 \sin \dfrac{\pi x}{150}, \quad 0 \le x \le 300.$$

(a) Use a graphing utility to graph the model. Be sure to choose an appropriate viewing window.

(b) Determine dh/dx and evaluate for $x = 50, 150, 200,$ and 250. Interpret these values of dh/dx.

(c) Find the maximum height and the minimum height of the ride.

(d) Find the distance from the starting point at which the ride's rate of change is the greatest.

68. *Meteorology* The number of hours of daylight D in New Orleans can be modeled by

$$D = 12.2 - 1.9 \cos \frac{\pi(t + 0.2)}{6}, \quad 0 \le t \le 12$$

where t represents the month, with $t = 0$ corresponding to January. Find the month t when New Orleans has the maximum number of daylight hours. What is this maximum number of daylight hours?

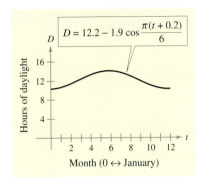

69. For $f(x) = \sec^2 x$ and $g(x) = \tan^2 x$, show that $f'(x) = g'(x)$.

70. For $f(x) = \sin^2 x$ and $g(x) = \cos^2 x$, show that $f'(x) = -g'(x)$.

71. *Physics* A 15-centimeter pendulum moves according to the equation

$$\theta = 0.2 \cos 8t$$

where θ is the angular displacement from the vertical in radians and t is the time in seconds.

(a) Determine the maximum angular displacement.

(b) Find the rate of change of θ when $t = 3$ seconds.

72. *Tides* Throughout the day, the depth of water D in meters at the end of a dock varies with the tides. The depth for one particular day can be modeled by

$$D = 3.5 + 1.5 \cos \frac{\pi t}{6}, \quad 0 \le t \le 24$$

where $t = 0$ represents midnight.

(a) Determine dD/dt.

(b) Evaluate dD/dt for $t = 4$ and $t = 20$ and interpret your results.

(c) Find the time(s) when the water depth is the greatest and the time(s) when the water depth is the least.

(d) What is the greatest depth? What is the least depth? Did you have to use calculus to determine these depths? Explain your reasoning.

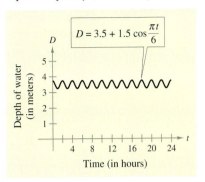

 In Exercises 73–78, use a graphing utility (a) to graph f and f' on the same coordinate axes over the specified interval, (b) to find the critical numbers of f, and (c) to find the interval(s) on which f' is positive and the interval(s) on which it is negative. Note the behavior of f in relation to the sign of f'.

Function	*Interval*
73. $f(t) = t^2 \sin t$	$(0, 2\pi)$
74. $f(x) = \dfrac{x}{2} + \cos \dfrac{x}{2}$	$(0, 4\pi)$
75. $f(x) = \sin x - \frac{1}{3} \sin 3x + \frac{1}{5} \sin 5x$	$(0, \pi)$
76. $f(x) = x \sin x$	$(0, \pi)$
77. $f(x) = \sqrt{2x} \sin x$	$(0, 2\pi)$
78. $f(x) = 4e^{-0.5x} \sin \pi x$	$(0, 4)$

 In Exercises 79–84, use a graphing utility to find the relative extrema of the trigonometric function. Let $0 < x < 2\pi$.

79. $f(x) = \dfrac{x}{\sin x}$

80. $f(x) = \dfrac{x^2 - 2}{\sin x} - 5x$

81. $f(x) = \ln x \cos x$

82. $f(x) = \ln x \sin x$

83. $f(x) = \sin(0.1x^2)$

84. $f(x) = \sin \sqrt{x}$

True or False? In Exercises 85–88, determine whether the statement is true or false. If it is false, explain why or give an example that shows it is false.

85. If $y = (1 - x)^{1/2}$, then $y' = \frac{1}{2}(1 - x)^{-1/2}$.

86. If $f(x) = \sin^2(2x)$, then $f'(x) = 2(\sin 2x)(\cos 2x)$.

87. If $y = x \sin^3 x$, then $y' = 3x \sin^2 x$.

88. The maximum value of $y = 3 \sin x + 2 \cos x$ is 5.

8.5   INTEGRALS OF TRIGONOMETRIC FUNCTIONS

- Find the six basic trigonometric integrals.
- Solve trigonometric integrals.
- Use trigonometric integrals to solve real-life problems.

The Six Basic Trigonometric Integrals

For each trigonometric differentiation rule, there is a corresponding integration rule. For instance, corresponding to the differentiation rule

$$\frac{d}{dx}[\cos u] = -\sin u \frac{du}{dx}$$

is the integration rule

$$\int \sin u \, du = -\cos u + C.$$

The list below contains the integration formulas that correspond to the six basic trigonometric differentiation rules.

Integrals Involving Trigonometric Functions

Differentiation Rule	*Integration Rule*
$\dfrac{d}{dx}[\sin u] = \cos u \dfrac{du}{dx}$	$\displaystyle\int \cos u \, du = \sin u + C$
$\dfrac{d}{dx}[\cos u] = -\sin u \dfrac{du}{dx}$	$\displaystyle\int \sin u \, du = -\cos u + C$
$\dfrac{d}{dx}[\tan u] = \sec^2 u \dfrac{du}{dx}$	$\displaystyle\int \sec^2 u \, du = \tan u + C$
$\dfrac{d}{dx}[\sec u] = \sec u \tan u \dfrac{du}{dx}$	$\displaystyle\int \sec u \tan u \, du = \sec u + C$
$\dfrac{d}{dx}[\cot u] = -\csc^2 u \dfrac{du}{dx}$	$\displaystyle\int \csc^2 u \, du = -\cot u + C$
$\dfrac{d}{dx}[\csc u] = -\csc u \cot u \dfrac{du}{dx}$	$\displaystyle\int \csc u \cot u \, du = -\csc u + C$

STUDY TIP

Note that this list gives you formulas for integrating only two of the six trigonometric functions: the sine function and the cosine function. The list does not show you how to integrate the other four trigonometric functions. Rules for integrating those functions are discussed later in this section.

EXAMPLE 1 **Integrating a Trigonometric Function**

Find $\displaystyle\int 2\cos x\,dx$.

SOLUTION Let $u = x$. Then $du = dx$.

$$\int 2\cos x\,dx = 2\int \cos x\,dx \qquad \text{Apply Constant Multiple Rule.}$$

$$= 2\int \cos u\,du \qquad \text{Substitute for } x \text{ and } dx.$$

$$= 2\sin u + C \qquad \text{Integrate.}$$

$$= 2\sin x + C \qquad \text{Substitute for } u.$$

> **TRY IT 1**
>
> Find $\displaystyle\int 5\sin x\,dx$.

EXAMPLE 2 **Integrating a Trigonometric Function**

Find $\displaystyle\int 3x^2 \sin x^3\,dx$.

SOLUTION Let $u = x^3$. Then $du = 3x^2\,dx$.

$$\int 3x^2 \sin x^3\,dx = \int (\sin x^3)3x^2\,dx \qquad \text{Rewrite integrand.}$$

$$= \int \sin u\,du \qquad \text{Substitute for } x^3 \text{ and } 3x^2\,dx.$$

$$= -\cos u + C \qquad \text{Integrate.}$$

$$= -\cos x^3 + C \qquad \text{Substitute for } u.$$

> **TRY IT 2**
>
> Find $\displaystyle\int 4x^3 \cos x^4\,dx$.

EXAMPLE 3 **Integrating a Trigonometric Function**

Find $\displaystyle\int \sec 3x \tan 3x\,dx$.

SOLUTION Let $u = 3x$. Then $du = 3\,dx$.

$$\int \sec 3x \tan 3x\,dx = \frac{1}{3}\int (\sec 3x \tan 3x)3\,dx \qquad \text{Multiply and divide by 3.}$$

$$= \frac{1}{3}\int \sec u \tan u\,du \qquad \text{Substitute for } 3x \text{ and } 3\,dx.$$

$$= \frac{1}{3}\sec u + C \qquad \text{Integrate.}$$

$$= \frac{1}{3}\sec 3x + C \qquad \text{Substitute for } u.$$

> **TECHNOLOGY**
>
> If you have access to a symbolic integration utility, try using it to integrate the functions in Examples 1, 2, and 3. Does your utility give the same results that are given in the examples?

> **TRY IT 3**
>
> Find $\displaystyle\int \sec^2 5x\,dx$.

| EXAMPLE 4 | **Integrating a Trigonometric Function** |

Find $\displaystyle\int e^x \sec^2 e^x\, dx$.

SOLUTION   Let $u = e^x$. Then $du = e^x\, dx$.

$$\int e^x \sec^2 e^x\, dx = \int (\sec^2 e^x)e^x\, dx \qquad \text{Rewrite integrand.}$$

$$= \int \sec^2 u\, du \qquad \text{Substitute for } e^x \text{ and } e^x\, dx.$$

$$= \tan u + C \qquad \text{Integrate.}$$

$$= \tan e^x + C \qquad \text{Substitute for } u. \quad \rule{2cm}{0.5mm}$$

The next two examples use the General Power Rule for integration and the Log Rule for integration. Recall from Chapter 5 that these rules are

$$\int u^n \frac{du}{dx}\, dx = \frac{u^{n+1}}{n+1} + C, \quad n \neq -1 \qquad \text{General Power Rule}$$

and

$$\int \frac{du/dx}{u}\, dx = \ln|u| + C. \qquad \text{Log Rule}$$

The key to using these two rules is identifying the proper substitution for u. For instance, in the next example, the proper choice for u is $\sin 4x$.

TRY IT 4

Find $\displaystyle\int 2\csc 2x \cot 2x\, dx$.

STUDY TIP

It is a good idea to check your answers to integration problems by differentiating. In Example 5, for instance, try differentiating the answer

$$y = \frac{1}{12}\sin^3 4x + C.$$

You should obtain the original integrand, as shown.

$$y' = \frac{1}{12}\, 3(\sin 4x)^2(\cos 4x)4$$

$$= \sin^2 4x \cos 4x$$

| EXAMPLE 5 | **Using the General Power Rule** |

Find $\displaystyle\int \sin^2 4x \cos 4x\, dx$.

SOLUTION   Let $u = \sin 4x$. Then $du/dx = 4\cos 4x$.

$$\int \sin^2 4x \cos 4x\, dx = \frac{1}{4}\int \overbrace{(\sin 4x)^2}^{u^2} \overbrace{(4\cos 4x)}^{du/dx}\, dx \qquad \text{Rewrite integrand.}$$

$$= \frac{1}{4}\int u^2\, du \qquad \text{Substitute for } \sin 4x \text{ and } 4\cos 4x\, dx.$$

$$= \frac{1}{4}\frac{u^3}{3} + C \qquad \text{Integrate.}$$

$$= \frac{1}{4}\frac{(\sin 4x)^3}{3} + C \qquad \text{Substitute for } u.$$

$$= \frac{1}{12}\sin^3 4x + C \qquad \text{Simplify.}$$

TRY IT 5

Find $\displaystyle\int \cos^3 2x \sin 2x\, dx$.

EXAMPLE 6 Using the Log Rule

Find $\displaystyle\int \frac{\sin x}{\cos x}\, dx$.

SOLUTION Let $u = \cos x$. Then $du/dx = -\sin x$.

$$\int \frac{\sin x}{\cos x}\, dx = -\int \frac{-\sin x}{\cos x}\, dx \qquad \text{Rewrite integrand.}$$

$$= -\int \frac{du/dx}{u}\, dx \qquad \text{Substitute for } \cos x \text{ and } -\sin x.$$

$$= -\ln|u| + C \qquad \text{Apply Log Rule.}$$

$$= -\ln|\cos x| + C \qquad \text{Substitute for } u.$$

TRY IT 6

Find $\displaystyle\int \frac{\cos x}{\sin x}\, dx$.

EXAMPLE 7 Evaluating a Definite Integral

Evaluate $\displaystyle\int_0^{\pi/4} \cos 2x\, dx$.

SOLUTION

$$\int_0^{\pi/4} \cos 2x\, dx = \left[\frac{1}{2}\sin 2x\right]_0^{\pi/4}$$

$$= \frac{1}{2} - 0 = \frac{1}{2}$$

TRY IT 7

Find $\displaystyle\int_0^{\pi/2} \sin 2x\, dx$.

EXAMPLE 8 Finding Area by Integration

Find the area of the region bounded by the x-axis and one arc of the graph of $y = \sin x$.

SOLUTION As indicated in Figure 8.34, this area is given by

$$\text{Area} = \int_0^{\pi} \sin x\, dx$$

$$= \left[-\cos x\right]_0^{\pi}$$

$$= -(-1) - (-1)$$

$$= 2.$$

So, the region has an area of 2 square units.

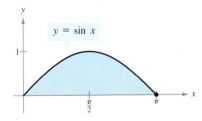

$y = \sin x$

FIGURE 8.34

TRY IT 8

Find the area of the region bounded by the graphs of $y = \cos x$ and $y = 0$ for

$$0 \le x \le \frac{\pi}{2}.$$

Other Trigonometric Integrals

At the beginning of this section, the integration rules for the sine and cosine functions were listed. Now, using the result of Example 6, you have an integration rule for the tangent function. That rule is

$$\int \tan x \, dx = \int \frac{\sin x}{\cos x} \, dx = -\ln|\cos x| + C.$$

Integration formulas for the other three trigonometric functions can be developed in a similar way. For instance, to obtain the integration formula for the secant function, you can integrate as shown.

$$\int \sec x \, dx = \int \frac{\sec x(\sec x + \tan x)}{\sec x + \tan x} \, dx$$

$$= \int \frac{\sec^2 x + \sec x \tan x}{\sec x + \tan x} \, dx \qquad \text{Use substitution with } u = \sec x + \tan x$$

$$= \ln|\sec x + \tan x| + C$$

These formulas, and integration formulas for the other two trigonometric functions, are summarized below.

Integrals of Trigonometric Functions

$$\int \tan u \, du = -\ln|\cos u| + C \qquad \int \sec u \, du = \ln|\sec u + \tan u| + C$$

$$\int \cot u \, du = \ln|\sin u| + C \qquad \int \csc u \, du = \ln|\csc u - \cot u| + C$$

EXAMPLE 9   Integrating a Trigonometric Function

Find $\int \tan 4x \, dx$.

SOLUTION   Let $u = 4x$. Then $du = 4 \, dx$.

$$\int \tan 4x \, dx = \frac{1}{4} \int (\tan 4x) 4 \, dx \qquad \text{Rewrite integrand.}$$

$$= \frac{1}{4} \int \tan u \, du \qquad \text{Substitute for } 4x \text{ and } 4 \, dx.$$

$$= -\frac{1}{4} \ln|\cos u| + C \qquad \text{Apply Tangent Differentiation Rule.}$$

$$= -\frac{1}{4} \ln|\cos 4x| + C \qquad \text{Substitute for } u.$$

TRY IT 9

Find $\int \sec 2x \, dx$.

Application

In the next example, recall that the average value of a function f over an interval $[a, b]$ is given by

$$\frac{1}{b - a} \int_a^b f(x) \, dx.$$

EXAMPLE 10 **Finding an Average Temperature**

The temperature T (in degrees Fahrenheit) during a 24-hour period can be modeled by

$$T = 72 + 18 \sin \frac{\pi(t - 8)}{12}$$

where t is the time (in hours), with $t = 0$ corresponding to midnight. Find the average temperature during the four-hour period from noon to 4 P.M.

SOLUTION   To find the average temperature A, use the formula for the average value of a function over an interval.

$$A = \frac{1}{4} \int_{12}^{16} \left[72 + 18 \sin \frac{\pi(t - 8)}{12} \right] dt$$

$$= \frac{1}{4} \left[72t + 18 \left(\frac{12}{\pi} \right) \left(-\cos \frac{\pi(t - 8)}{12} \right) \right]_{12}^{16}$$

$$= \frac{1}{4} \left[72(16) + 18 \left(\frac{12}{\pi} \right) \left(\frac{1}{2} \right) - 72(12) + 18 \left(\frac{12}{\pi} \right) \left(\frac{1}{2} \right) \right]$$

$$= \frac{1}{4} \left(288 + \frac{216}{\pi} \right)$$

$$= 72 + \frac{54}{\pi} \approx 89.2°$$

So, the average temperature is 89.2°, as indicated in Figure 8.35. ——— **FIGURE 8.35**

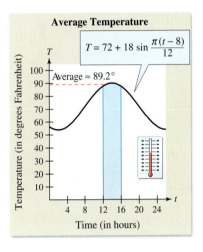

Average Temperature

TRY IT 10

Use the function in Example 10 to find the average temperature from 9 A.M. to noon.

TAKE ANOTHER LOOK

Finding Average Temperature

Use the temperature model in Example 10 to find the average temperature during each time interval below. What can you conclude?

a. From midnight to the following midnight

b. From midnight to noon

c. From noon to midnight

The following warm-up exercises involve skills that were covered in earlier sections. You will use these skills in the exercise set for this section.

In Exercises 1–8, evaluate the trigonometric function.

1. $\cos \dfrac{5\pi}{4}$

2. $\sin \dfrac{7\pi}{6}$

3. $\sin\left(-\dfrac{\pi}{3}\right)$

4. $\cos\left(-\dfrac{\pi}{6}\right)$

5. $\tan \dfrac{5\pi}{6}$

6. $\cot \dfrac{5\pi}{3}$

7. $\sec \pi$

8. $\cos \dfrac{\pi}{2}$

In Exercises 9–16, simplify the expression using the trigonometric identities.

9. $\sin x \sec x$

10. $\csc x \cos x$

11. $\cos^2 x(\sec^2 x - 1)$

12. $\sin^2 x(\csc^2 x - 1)$

13. $\sec x \sin\left(\dfrac{\pi}{2} - x\right)$

14. $\cot x \cos\left(\dfrac{\pi}{2} - x\right)$

15. $\cot x \sec x$

16. $\cot x(\sin^2 x)$

In Exercises 17–20, evaluate the definite integral.

17. $\displaystyle\int_0^4 (x^2 + 3x - 4)\, dx$

18. $\displaystyle\int_{-1}^1 (1 - x^2)\, dx$

19. $\displaystyle\int_0^2 x(4 - x^2)\, dx$

20. $\displaystyle\int_0^1 x(9 - x^2)\, dx$

EXERCISES 8.5

In Exercises 1–34, evaluate the integral.

1. $\displaystyle\int (2 \sin x + 3 \cos x)\, dx$

2. $\displaystyle\int (t^2 - \sin t)\, dt$

3. $\displaystyle\int (1 - \csc t \cot t)\, dt$

4. $\displaystyle\int (\theta^2 + \sec^2 \theta)\, d\theta$

5. $\displaystyle\int (\csc^2 \theta - \cos \theta)\, d\theta$

6. $\displaystyle\int (\sec y \tan y - \sec^2 y)\, dy$

7. $\displaystyle\int \sin 2x\, dx$

8. $\displaystyle\int \cos 6x\, dx$

9. $\displaystyle\int x \cos x^2\, dx$

10. $\displaystyle\int x \sin x^2\, dx$

11. $\displaystyle\int \sec^2 \dfrac{x}{2}\, dx$

12. $\displaystyle\int \csc^2 \dfrac{x}{2}\, dx$

13. $\displaystyle\int \tan 3x\, dx$

14. $\displaystyle\int \csc 2x \cot 2x\, dx$

15. $\displaystyle\int \tan^3 x \sec^2 x\, dx$

16. $\displaystyle\int \sqrt{\cot x}\, \csc^2 x\, dx$

17. $\displaystyle\int \cot \pi x\, dx$

18. $\displaystyle\int \tan 5x\, dx$

19. $\displaystyle\int \csc 2x\, dx$

20. $\displaystyle\int \sec \dfrac{x}{2}\, dx$

21. $\displaystyle\int \dfrac{\sec^2 2x}{\tan 2x}\, dx$

22. $\displaystyle\int \dfrac{\sin x}{\cos^2 x}\, dx$

23. $\displaystyle\int \dfrac{\sec x \tan x}{\sec x - 1}\, dx$

24. $\displaystyle\int \dfrac{\cos t}{1 + \sin t}\, dt$

25. $\displaystyle\int \dfrac{\sin x}{1 + \cos x}\, dx$

26. $\displaystyle\int \dfrac{\sin \sqrt{x}}{\sqrt{x}}\, dx$

27. $\displaystyle\int \dfrac{\csc^2 x}{\cot^3 x}\, dx$

28. $\displaystyle\int \dfrac{1 - \cos \theta}{\theta - \sin \theta}\, d\theta$

29. $\displaystyle\int e^x \sin e^x\, dx$

30. $\displaystyle\int e^{-x} \tan e^{-x}\, dx$

31. $\displaystyle\int e^{\sin x} \cos x\, dx$

32. $\displaystyle\int e^{\sec x} \sec x \tan x\, dx$

33. $\displaystyle\int (\sin 2x + \cos 2x)^2 \, dx$ **34.** $\displaystyle\int (\csc 2\theta - \cot 2\theta)^2 \, d\theta$

In Exercises 35–38, evaluate the integral.

35. $\displaystyle\int x \cos x \, dx$ **36.** $\displaystyle\int x \sin x \, dx$

37. $\displaystyle\int x \sec^2 x \, dx$ **38.** $\displaystyle\int \theta \sec \theta \tan \theta \, d\theta$

In Exercises 39–46, evaluate the definite integral. Use a symbolic integration utility to verify your results.

39. $\displaystyle\int_0^{\pi/4} \cos \frac{4x}{3} \, dx$ **40.** $\displaystyle\int_0^{\pi/2} \sin 2x \, dx$

41. $\displaystyle\int_{\pi/2}^{2\pi/3} \sec^2 \frac{x}{2} \, dx$ **42.** $\displaystyle\int_0^{\pi/2} (x + \cos x) \, dx$

43. $\displaystyle\int_{\pi/12}^{\pi/4} \csc 2x \cot 2x \, dx$ **44.** $\displaystyle\int_0^{\pi/8} \sin 2x \cos 2x \, dx$

45. $\displaystyle\int_0^1 \tan(1 - x) \, dx$ **46.** $\displaystyle\int_0^{\pi/4} \sec x \tan x \, dx$

In Exercises 47–52, determine the area of the region.

47. $y = \cos \dfrac{x}{4}$

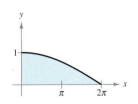

48. $y = \tan x$

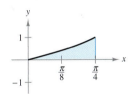

49. $y = x + \sin x$

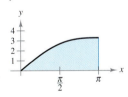

50. $y = 2 \sin x + \sin 2x$

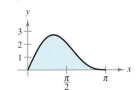

51. $y = \sin x + \cos 2x$

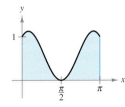

52. $y = x \sin x$

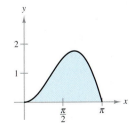

In Exercises 53 and 54, find the volume of the solid generated by revolving the region bounded by the graphs of the given equations about the x-axis.

53. $y = \sec x, \ y = 0, \ x = 0, \ x = \dfrac{\pi}{4}$

54. $y = \csc x, \ y = 0, \ x = \dfrac{\pi}{6}, \ x = \dfrac{5\pi}{6}$

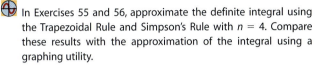 In Exercises 55 and 56, approximate the definite integral using the Trapezoidal Rule and Simpson's Rule with $n = 4$. Compare these results with the approximation of the integral using a graphing utility.

55. $\displaystyle\int_0^{\pi/2} f(x) \, dx, \qquad f(x) = \begin{cases} \dfrac{\sin x}{x}, & x > 0 \\ 1, & x = 0 \end{cases}$

56. $\displaystyle\int_0^1 \cos x^2 \, dx$

57. *Inventory* The minimum stockpile level of gasoline in the United States can be approximated by the model

$$Q = 217 + 13 \cos \frac{\pi(t - 3)}{6}$$

where Q is measured in millions of barrels of gasoline and t is the time in months, with $t = 1$ corresponding to January. Find the average minimum level given by this model during

(a) the first quarter $(0 \le t \le 3)$

(b) the second quarter $(3 \le t \le 6)$

(c) the entire year $(0 \le t \le 12)$.

58. *Seasonal Sales* The sales of a software product are given by the model

$$S = 74.50 + 43.75 \sin \frac{\pi t}{6}$$

where S is measured in thousands of units per month and t is the time in months, with $t = 1$ corresponding to January. Use a graphing utility to estimate average sales during

(a) the first quarter $(0 \le t \le 3)$

(b) the second quarter $(3 \le t \le 6)$

(c) the entire year $(0 \le t \le 12)$.

59. *Meteorology* The average monthly precipitation P in inches, including rain, snow, and ice, for Sacramento, California can be modeled by

$$P = 2.34 \sin(0.38t + 1.91) + 2.22$$

where t is the time in months, with $t = 1$ corresponding to January. Find the total annual precipitation for Sacramento. *(Source: U.S. National Oceanic and Atmospheric Administration)*

60. *Meteorology* The average monthly precipitation P in inches, including rain, snow, and ice, for Bismarck, North Dakota can be modeled by

$$P = 1.07 \sin(0.59t + 3.94) + 1.52$$

where t is the time in months, with $t = 1$ corresponding to January. (*Source: National Oceanic and Atmospheric Administration*)

(a) Find the maximum and minimum precipitation and the month in which each occurs.

(b) Determine the average monthly precipitation for the year.

(c) Find the total annual precipitation for Bismarck.

 61. *Cost* Suppose that the temperature in degrees Fahrenheit is given by

$$T = 72 + 12 \sin \frac{\pi(t - 8)}{12}$$

where t is the time in hours, with $t = 0$ corresponding to midnight. Furthermore, suppose that it costs $0.10 to cool a particular house 1° for 1 hour.

(a) Use the integration capabilities of a graphing utility to find the cost C of cooling this house between 8 A.M. and 8 P.M., if the thermostat is set at 72° (see figure) and the cost is given by

$$C = 0.1 \int_{8}^{20} \left[72 + 12 \sin \frac{\pi(t - 8)}{12} - 72 \right] dt.$$

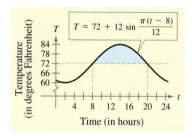

(b) Use the integration capabilities of a graphing utility to find the savings realized by resetting the thermostat to 78° (see figure) by evaluating the integral

$$C = 0.1 \int_{10}^{18} \left[72 + 12 \sin \frac{\pi(t - 8)}{12} - 78 \right] dt.$$

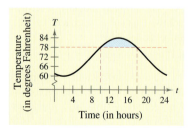

62. *Health* For a person at rest, the velocity v (in liters per second) of air flow into and out of the lungs during a respiratory cycle is approximated by

$$v = 0.9 \sin \frac{\pi t}{3}$$

where t is the time in seconds. Find the volume in liters of air inhaled during one cycle by integrating this function over the interval $[0, 3]$.

63. *Health* After exercising for a few minutes, a person has a respiratory cycle for which the velocity of air flow is approximated by

$$v = 1.75 \sin \frac{\pi t}{2}.$$

How much does the lung capacity of a person increase as a result of exercising? Use the results of Exercise 62 to determine how much more air is inhaled during a cycle after exercising than is inhaled during a cycle at rest. (Note that the cycle is shorter and you must integrate over the interval $[0, 2]$.)

64. *Sales* In Example 9 in Section 8.4, the sales of a seasonal product were approximated by the model

$$F = 100,000 \left[1 + \sin \frac{2\pi(t - 60)}{365} \right], \quad t \geq 0$$

where F was measured in pounds and t was the time in days, with $t = 1$ corresponding to January 1. The manufacturer of this product wants to set up a manufacturing schedule to produce a uniform amount each day. What should this amount be? (Assume that there are 200 production days during the year.)

 In Exercises 65–68, use a graphing utility and Simpson's Rule to approximate the integral.

Integral		n
65. $\int_{0}^{\pi/2} \sqrt{x} \sin x \, dx$		8
66. $\int_{0}^{\pi/2} \cos \sqrt{x} \, dx$		8
67. $\int_{0}^{\pi} \sqrt{1 + \cos^2 x} \, dx$		20
68. $\int_{0}^{2} (4 + x + \sin \pi x) \, dx$		20

True or False? In Exercises 69–71, determine whether the statement is true or false. If it is false, explain why or give an example that shows it is false.

69. $\displaystyle\int_{a}^{b} \sin x \, dx = \int_{a}^{b + 2\pi} \sin x \, dx$

70. $\displaystyle 4 \int \sin x \cos x \, dx = -\cos 2x + C$

71. $\displaystyle \int \sin^2 2x \cos 2x \, dx = \frac{1}{3} \sin^3 2x + C$

8.6   L'HÔPITAL'S RULE

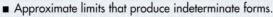

■ Approximate limits that produce indeterminate forms.
■ Use L'Hôpital's Rule to evaluate limits.

Indeterminate Forms and Limits

In Sections 1.5 and 3.6, you studied limits such as

$$\lim_{x \to 1} \frac{x^2 - 1}{x - 1}$$

and

$$\lim_{x \to \infty} \frac{2x + 1}{x + 1}.$$

In those sections, you discovered that direct substitution can produce an **indeterminate form** such as $0/0$ or ∞/∞. For instance, if you substitute $x = 1$ into the first limit, you obtain

$$\lim_{x \to 1} \frac{x^2 - 1}{x - 1} = \frac{0}{0} \qquad \text{Indeterminate form}$$

which tells you nothing about the limit. To find the limit, you can factor and divide out like factors, as shown.

$$\lim_{x \to 1} \frac{x^2 - 1}{x - 1} = \lim_{x \to 1} \frac{(x - 1)(x + 1)}{x - 1} \qquad \text{Factor.}$$

$$= \lim_{x \to 1} \frac{(x - 1)(x + 1)}{x - 1} \qquad \text{Divide out like factors.}$$

$$= \lim_{x \to 1} (x + 1) \qquad \text{Simplify.}$$

$$= 1 + 1 \qquad \text{Direct substitution}$$

$$= 2 \qquad \text{Simplify.}$$

For the second limit, direct substitution produces the indeterminate form ∞/∞, which again tells you nothing about the limit. To evaluate this limit, you can divide the numerator and denominator by x. Then you can use the fact that the limit of $1/x$, as $x \to \infty$, is 0.

$$\lim_{x \to \infty} \frac{2x + 1}{x + 1} = \lim_{x \to \infty} \frac{2 + (1/x)}{1 + (1/x)} \qquad \text{Divide numerator and denominator by } x.$$

$$= \frac{2 + 0}{1 + 0} \qquad \text{Evaluate limits.}$$

$$= 2 \qquad \text{Simplify.}$$

Algebraic techniques such as these tend to work well as long as the function itself is algebraic. To find the limits of other types of functions, such as exponential functions or trigonometric functions, you generally need to use a different approach.

L'Hôpital's Rule can be used to compare the rates of growth of two functions. For instance, consider the limit in Example 5

$$\lim_{x \to \infty} \frac{e^x}{e^{2x} + 1} = 0.$$

Both of the functions $f(x) = e^x$ and $g(x) = e^{2x} + 1$ approach infinity as $x \to \infty$. However, because the quotient $f(x)/g(x)$ approaches 0 as $x \to \infty$, it follows that the denominator is growing much more rapidly than the numerator.

STUDY TIP

L'Hôpital's Rule is necessary to solve certain real-life problems such as compound interest problems and other business applications.

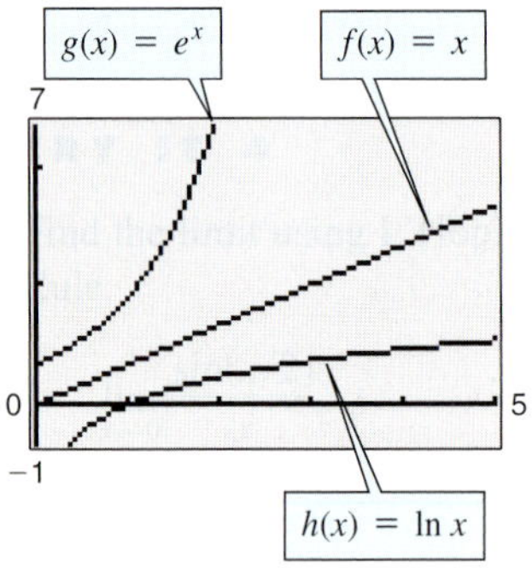

EXAMPLE 7   **Comparing Rates of Growth**

Each of the functions below approaches infinity as x approaches infinity. Which function has the highest rate of growth?

(a) $f(x) = x$ (b) $g(x) = e^x$ (c) $h(x) = \ln x$

SOLUTION   Using L'Hôpital's Rule, you can show that each of the limits is zero.

$$\lim_{x \to \infty} \frac{x}{e^x} = \lim_{x \to \infty} \frac{\frac{d}{dx}[x]}{\frac{d}{dx}[e^x]} = \lim_{x \to \infty} \frac{1}{e^x} = 0$$

$$\lim_{x \to \infty} \frac{\ln x}{x} = \lim_{x \to \infty} \frac{\frac{d}{dx}[\ln x]}{\frac{d}{dx}[x]} = \lim_{x \to \infty} \frac{1}{x} = 0$$

$$\lim_{x \to \infty} \frac{\ln x}{e^x} = \lim_{x \to \infty} \frac{\frac{d}{dx}[\ln x]}{\frac{d}{dx}[e^x]} = \lim_{x \to \infty} \frac{1}{xe^x} = 0$$

From this, you can conclude that $h(x) = \ln x$ has the lowest rate of growth, $f(x) = x$ has a higher rate of growth, and $g(x) = e^x$ has the highest rate of growth. This conclusion is confirmed graphically in Figure 8.38.

FIGURE 8.38

$g(x) = e^x$ $f(x) = x$
$h(x) = \ln x$

TRY IT 7

Use a graphing utility to order the functions according to the rate of growth of each function as x approaches infinity.

(a) $f(x) = e^{2x}$

(b) $g(x) = x^2$

(c) $h(x) = \ln x^2$

TAKE ANOTHER LOOK

Comparing Rates of Growth

Each of the functions below approaches infinity as x approaches infinity. In each pair, which function has the higher rate of growth? Use a graphing utility to verify your choice.

a. $f(x) = e^x$, $g(x) = e^{\sqrt{x}}$ b. $f(x) = \sqrt{x^3 + 1}$, $g(x) = x$ c. $f(x) = \tan \frac{\pi x}{2x + 1}$, $g(x) = x$

PREREQUISITE REVIEW 8.6

The following warm-up exercises involve skills that were covered in earlier sections. You will use these skills in the exercise set for this section.

In Exercises 1–8, find the limit, if it exists.

1. $\lim\limits_{x\to\infty} (x^2 - 10x + 1)$

2. $\lim\limits_{x\to\infty} \dfrac{2x + 5}{3}$

3. $\lim\limits_{x\to\infty} \dfrac{x + 1}{x^2}$

4. $\lim\limits_{x\to\infty} \dfrac{5x - 2}{x^2 + 1}$

5. $\lim\limits_{x\to\infty} \dfrac{2x^2 - 5x + 7}{3x^2 + 12x + 4}$

6. $\lim\limits_{x\to\infty} \dfrac{x^2 - 1}{x^2 + 4}$

7. $\lim\limits_{x\to\infty} \dfrac{x^2 - 5}{2x + 7}$

8. $\lim\limits_{x\to\infty} \dfrac{x^2 - 2x + 1}{x - 4}$

In Exercises 9–12, find the first derivative of the function.

9. $f(x) = \cos x^2$

10. $f(x) = \sin(5x - 1)$

11. $f(x) = \sec 4x$

12. $f(x) = \tan(x^2 - 2)$

In Exercises 13–16, find the second derivative of the function.

13. $f(x) = \sin(2x + 3)$

14. $f(x) = \cos \dfrac{x}{2}$

15. $f(x) = \tan x$

16. $f(x) = \cot x$

EXERCISES 8.6

In Exercises 1–6, decide whether the limit produces an indeterminate form.

1. $\lim\limits_{x\to 0} \dfrac{2x + \sqrt{x}}{x}$

2. $\lim\limits_{x\to\infty} \dfrac{x^2 + 4x - 3}{7x^2 + 2}$

3. $\lim\limits_{x\to -\infty} \dfrac{4}{x^2 + e^x}$

4. $\lim\limits_{x\to 0} \dfrac{\sin x}{e^x}$

5. $\lim\limits_{x\to\infty} \dfrac{2xe^{2x}}{3e^x}$

6. $\lim\limits_{x\to\infty} \dfrac{\ln x}{x}$

In Exercises 7–10, complete the table to estimate the limit numerically.

7. $\lim\limits_{x\to 0} \dfrac{e^{-x} - 1}{3x}$

x	-0.1	-0.01	-0.001	0	0.001	0.01	0.1
$f(x)$				?			

8. $\lim\limits_{x\to 3} \dfrac{x^2 + x - 12}{x^2 - 9}$

x	2.9	2.99	2.999	3	3.001	3.01	3.1
$f(x)$				?			

9. $\lim\limits_{x\to 0} \dfrac{\sin x}{5x}$

x	-0.1	-0.01	-0.001	0	0.001	0.01	0.1
$f(x)$				?			

10. $\lim\limits_{x\to 0} \dfrac{\cos 2x - 1}{6x}$

x	-0.1	-0.01	-0.001	0	0.001	0.01	0.1
$f(x)$				?			

 In Exercises 11–16, use a graphing utility to find the indicated limit graphically.

11. $\lim\limits_{x\to\infty} \dfrac{\ln x}{x^2}$

12. $\lim\limits_{x\to\infty} \dfrac{e^x}{x^2}$

13. $\lim\limits_{x\to 1} \dfrac{\ln(2 - x)}{x - 1}$

14. $\lim\limits_{x\to 1} \dfrac{e^{x-1}}{x^2 - 1}$

15. $\lim\limits_{x\to 2} \dfrac{x^2 - x - 2}{x^2 - 5x + 6}$

16. $\lim\limits_{x\to 1} \dfrac{x^2 + 3x - 4}{x^2 + 2x - 3}$

In Exercises 17–48, use L'Hôpital's Rule to find the limit. You may need to use L'Hôpital's Rule repeatedly.

17. $\lim\limits_{x\to 0} \dfrac{e^{-x} - 1}{x}$

18. $\lim\limits_{x\to 3} \dfrac{x^2 + x - 12}{x^2 - 9}$

19. $\lim\limits_{x\to 0} \dfrac{\sin x}{5x}$

20. $\lim\limits_{x\to 0} \dfrac{\cos 2x - 1}{6x}$

21. $\lim\limits_{x\to \infty} \dfrac{\ln x}{x^2}$

22. $\lim\limits_{x\to \infty} \dfrac{e^x}{x^2}$

23. $\lim\limits_{x\to 2} \dfrac{x^2 - x - 2}{x^2 - 5x + 6}$

24. $\lim\limits_{x\to 1} \dfrac{x^2 + 3x - 4}{x^2 + 2x - 3}$

25. $\lim\limits_{x\to 0} \dfrac{2x + 1 - e^x}{x}$

26. $\lim\limits_{x\to 0} \dfrac{2x - 1 + e^{-x}}{3x}$

27. $\lim\limits_{x\to \infty} \dfrac{\ln x}{e^x}$

28. $\lim\limits_{x\to \infty} \dfrac{3x}{e^x}$

29. $\lim\limits_{x\to \infty} \dfrac{4x^2 + 2x - 1}{3x^2 - 7}$

30. $\lim\limits_{x\to \infty} \dfrac{3x^2 + 5}{2x^2 - 11}$

31. $\lim\limits_{x\to \infty} \dfrac{1 - x}{e^x}$

32. $\lim\limits_{x\to 0} \dfrac{e^{3x} - 1}{2x}$

33. $\lim\limits_{x\to 1} \dfrac{\ln x}{x^2 - 1}$

34. $\lim\limits_{x\to 2} \dfrac{\ln(x - 1)}{x - 2}$

35. $\lim\limits_{x\to 0} \dfrac{\sin 2x}{\sin 5x}$

36. $\lim\limits_{x\to 0} \dfrac{\tan 2x}{x}$

37. $\lim\limits_{x\to \pi/2} \dfrac{2 \cos x}{3 \cos x}$

38. $\lim\limits_{x\to 0} \dfrac{\cos 2x - 1}{x}$

39. $\lim\limits_{x\to 0} \dfrac{\sin x}{e^x - 1}$

40. $\lim\limits_{x\to 1} \dfrac{\ln x}{1 - x^2}$

41. $\lim\limits_{x\to \pi} \dfrac{\cos(x/2)}{x - \pi}$

42. $\lim\limits_{x\to 1} \dfrac{x - 1}{\sqrt{x} - 1}$

43. $\lim\limits_{x\to \infty} \dfrac{x}{\sqrt{x + 1}}$

44. $\lim\limits_{x\to 0} \dfrac{\sqrt{x + 4} - 2}{x}$

45. $\lim\limits_{x\to 0} \dfrac{\sqrt{4 - x^2} - 2}{x}$

46. $\lim\limits_{x\to \infty} \dfrac{x^2 + 2}{e^{x^2}}$

47. $\lim\limits_{x\to \infty} \dfrac{e^{3x}}{x^3}$

48. $\lim\limits_{x\to \infty} \dfrac{x^2}{e^x \ln x}$

In Exercises 49–56, find the limit. (*Hint:* L'Hôpital's Rule does not apply in every case.)

49. $\lim\limits_{x\to \infty} \dfrac{x^2 + 2x + 1}{x^2 + 3}$

50. $\lim\limits_{x\to \infty} \dfrac{3x^3 - 7x + 5}{4x^3 + 2x - 11}$

51. $\lim\limits_{x\to -1} \dfrac{x^3 + 3x^2 - 6x - 8}{2x^3 - 3x^2 - 5x + 6}$

52. $\lim\limits_{x\to 0} \dfrac{2x^3 - x^2 - 3x}{x^4 + 2x^3 - 9x^2 - 18x}$

53. $\lim\limits_{x\to 3} \dfrac{\ln(x - 2)}{x - 2}$

54. $\lim\limits_{x\to -1} \dfrac{\ln(x + 2)}{x + 2}$

55. $\lim\limits_{x\to 1} \dfrac{2 \ln x}{e^x}$

56. $\lim\limits_{x\to 0} \dfrac{\sin \pi x}{x}$

In Exercises 57–62, use L'Hôpital's Rule to compare the rates of growth of the numerator and the denominator.

57. $\lim\limits_{x\to \infty} \dfrac{x^2}{e^{4x}}$

58. $\lim\limits_{x\to \infty} \dfrac{x^3}{e^{2x}}$

59. $\lim\limits_{x\to \infty} \dfrac{(\ln x)^4}{x}$

60. $\lim\limits_{x\to \infty} \dfrac{(\ln x)^2}{x^4}$

61. $\lim\limits_{x\to \infty} \dfrac{(\ln x)^n}{x^m}$

62. $\lim\limits_{x\to \infty} \dfrac{x^m}{e^{nx}}$

63. Complete the table to show that x eventually "overpowers" $(\ln x)^5$.

x	10	10^2	10^3	10^4	10^5	10^6
$\dfrac{(\ln x)^5}{x}$						

64. Complete the table to show that e^x eventually "overpowers" x^6.

x	1	2	4	8	12	20	30
$\dfrac{e^x}{x^6}$							

In Exercises 65–68, L'Hôpital's Rule is used incorrectly. Describe the error.

65. $\lim\limits_{x\to 0} \dfrac{e^{3x} - 1}{e^x} = \lim\limits_{x\to 0} \dfrac{3e^{3x}}{e^x} = \lim\limits_{x\to 0} 3e^{2x} = 3$ ✕

66. $\lim\limits_{x\to 0} \dfrac{\sin \pi x + 1}{x} = \lim\limits_{x\to 0} \dfrac{\pi \cos \pi x}{1} = \pi$ ✕

67. $\lim\limits_{x\to 1} \dfrac{e^x - 1}{\ln x} = \lim\limits_{x\to 1} \dfrac{e^x}{(1/x)} = \lim\limits_{x\to 1} xe^x = e$ ✕

68. $\lim\limits_{x\to \infty} \dfrac{e^{-x}}{1 - e^{-x}} = \lim\limits_{x\to \infty} \dfrac{-e^{-x}}{e^{-x}} = \lim\limits_{x\to \infty} -1 = -1$ ✕

In Exercises 69–72, use a graphing utility to (a) graph the function and (b) find the limit (if it exists).

69. $\lim\limits_{x \to 2} \dfrac{x - 2}{\ln(3x - 5)}$

70. $\lim\limits_{x \to \infty} \dfrac{x^4}{e^x}$

71. $\lim\limits_{x \to -2} \dfrac{\sqrt{x^2 - 4} - 5}{x + 2}$

72. $\lim\limits_{x \to 0} \dfrac{3 \tan x}{5x}$

73. Show that L'Hôpital's Rule fails for the limit

$$\lim_{x \to \infty} \frac{x}{\sqrt{x^2 + 1}}.$$

What is the limit?

  74. Use a graphing utility to graph

$$f(x) = \frac{x^k - 1}{k}$$

for $k = 1, 0.1,$ and 0.01. Then find

$$\lim_{k \to 0^+} \frac{x^k - 1}{k}.$$

75. *Sales* The sales (in millions of dollars per year) for the years 1994 through 2003 for two major drugstore chains are modeled by

$$f(t) = -1673.4 + 449.78t^2 - 28.548t^3 + 0.0113e^t$$

Rite Aid

$$g(t) = 5947.9 - 3426.94t + 841.872t^2 - 35.4110t^3$$

CVS

where t is the year, with $t = 4$ corresponding to 1994. *(Source: Rite Aid Corporation; CVS Corporation)*

  (a) Use a graphing utility to graph both models for $4 \le t \le 13$.

(b) Use your graph to determine which company had the larger rate of growth of sales for $4 \le t \le 13$.

(c) Use your knowledge of functions to predict which company ultimately will have the larger rate of growth. Explain your reasoning.

(d) If the models continue to be valid after 2004, when will one company's sales surpass the other company's sales and always be greater?

76. *Compound Interest* The formula for the amount A in a savings account compounded n times per year for t years at an interest rate r and an initial deposit of P is given by

$$A = P\left(1 + \frac{r}{n}\right)^{nt}.$$

Use L'Hôpital's Rule to show that the limiting formula as the number of compoundings per year becomes infinite is $A = Pe^{rt}$.

77. *Research Project* Use your school's library, the Internet, or some other reference source to gather data on the sales growth of two competing companies. Model the data for both companies, and determine which company is growing at a higher rate. Write a short paper that describes your findings, including the factors that account for the growth of the faster-growing company.

True or False? In Exercises 78–81, determine whether the statement is true or false. If it is false, explain why or give an example that shows it is false.

78. $\lim\limits_{x \to 0} \dfrac{x^2 + 3x - 1}{x + 1} = \lim\limits_{x \to 0} \dfrac{2x + 3}{1} = 3$

79. $\lim\limits_{x \to \infty} \dfrac{x}{1 - x} = \lim\limits_{x \to \infty} \dfrac{1}{-1} = 1$

80. If

$$\lim_{x \to \infty} \frac{f(x)}{g(x)} = 0,$$

then $g(x)$ has a higher growth rate than $f(x)$.

81. If $\lim\limits_{x \to \infty} \dfrac{f(x)}{g(x)} = 1$, then $g(x) = f(x)$.

ALGEBRA REVIEW

Solving Trigonometric Equations

Solving a trigonometric equation requires the use of trigonometry, but it also requires the use of algebra. Some examples of solving trigonometric equations were presented on pages 564 and 565. Here are several others.

EXAMPLE 1   **Solving a Trigonometric Equation**

Solve each trigonometric equation.

(a) $\sin x + \sqrt{2} = -\sin x$

(b) $3 \tan^2 x = 1$

(c) $\cot x \cos^2 x = 2 \cot x$

SOLUTION

(a) $\sin x + \sqrt{2} = -\sin x$ Write original equation.

$\sin x + \sin x = -\sqrt{2}$ Add $\sin x$ to, and subtract $\sqrt{2}$ from, each side.

$2 \sin x = -\sqrt{2}$ Combine like terms.

$\sin x = -\dfrac{\sqrt{2}}{2}$ Divide each side by 2.

$x = \dfrac{5\pi}{4}, \dfrac{7\pi}{4}, \quad 0 \le x \le 2\pi$

(b) $3 \tan^2 x = 1$ Write original equation.

$\tan^2 x = \dfrac{1}{3}$ Divide each side by 3.

$\tan x = \pm \dfrac{1}{\sqrt{3}}$ Extract square roots.

$x = \dfrac{\pi}{6}, \dfrac{5\pi}{6}, \dfrac{7\pi}{6}, \dfrac{11\pi}{6}, \quad 0 \le x \le 2\pi$

(c) $\cot x \cos^2 x = 2 \cot x$ Write original equation.

$\cot x \cos^2 x - 2 \cot x = 0$ Subtract $2 \cot x$ from each side.

$\cot x (\cos^2 x - 2) = 0$ Factor.

Setting each factor equal to zero, you obtain the solutions in the interval $0 \le x \le 2\pi$ as shown.

$\cot x = 0 \qquad$ and $\quad \cos^2 x - 2 = 0$

$x = \dfrac{\pi}{2}, \dfrac{3\pi}{2} \qquad\qquad \cos^2 x = 2$

$\cos x = \pm \sqrt{2}$

No solution is obtained from $\cos x = \pm \sqrt{2}$ because $\pm \sqrt{2}$ are outside the range of the cosine function.

| **EXAMPLE 2** | **Solving a Trigonometric Equation** |

Solve each trigonometric equation.

(a) $2 \sin^2 x - \sin x - 1 = 0$

(b) $2 \sin^2 x + 3 \cos x - 3 = 0$

(c) $2 \cos 3t - 1 = 0$

SOLUTION

(a)　　$2 \sin^2 x - \sin x - 1 = 0$　　　　　Write original equation.

　　　$(2 \sin x + 1)(\sin x - 1) = 0$　　　　　Factor.

Setting each factor equal to zero, you obtain the solutions in the interval $[0, 2\pi]$ as shown.

$2 \sin x + 1 = 0$　　　and　$\sin x - 1 = 0$

$\sin x = -\dfrac{1}{2}$　　　　　$\sin x = 1$

$x = \dfrac{7\pi}{6}, \dfrac{11\pi}{6}$　　　　$x = \dfrac{\pi}{2}$

(b)　　　　$2 \sin^2 x + 3 \cos x - 3 = 0$　　　Write original equation.

　　$2(1 - \cos^2 x) + 3 \cos x - 3 = 0$　　　Pythagorean Identity

　　　　$-2 \cos^2 x + 3 \cos x - 1 = 0$　　　Combine like terms.

　　　　$2 \cos^2 x - 3 \cos x + 1 = 0$　　　Multiply each side by -1.

　　　$(2 \cos x - 1)(\cos x - 1) = 0$　　　Factor.

Setting each factor equal to zero, you obtain the solutions in the interval $[0, 2\pi]$ as shown.

$2 \cos x - 1 = 0$　　　and　$\cos x - 1 = 0$

$2 \cos x = 1$　　　　　　$\cos x = 1$

$\cos x = \dfrac{1}{2}$　　　　　$x = 0, 2\pi$

$x = \dfrac{\pi}{3}, \dfrac{5\pi}{3}$

(c)　$2 \cos 3t - 1 = 0$　　　　　　　　Write original equation.

　　　$2 \cos 3t = 1$　　　　　　　　Add 1 to each side.

　　　$\cos 3t = \dfrac{1}{2}$　　　　　　　　Divide each side by 2.

　　　$3t = \dfrac{\pi}{3}, \dfrac{5\pi}{3}, \quad 0 \le 3t \le 2\pi$

　　　$t = \dfrac{\pi}{9}, \dfrac{5\pi}{9}, \quad 0 \le t \le \dfrac{2}{3}\pi$

In the interval $0 \le t \le 2\pi$, there are four other solutions.

8   CHAPTER SUMMARY AND STUDY STRATEGIES

After studying this chapter, you should have acquired the following skills. The exercise numbers are keyed to the Review Exercises that begin on page 610. Answers to odd-numbered Review Exercises are given in the back of the text.*

■ Find coterminal angles.   *(Section 8.1)*   *Review Exercises 1–8*

■ Convert from degree to radian measure and from radian to degree measure.   *Review Exercises 9–20*
(Section 8.1)

π radians $= 180°$

■ Use formulas relating to triangles.   *(Section 8.1)*   *Review Exercises 21–24*

■ Use formulas relating to triangles to solve real-life problems.   *(Section 8.1)*   *Review Exercises 25, 26*

■ Find the reference angles for given angles.   *(Section 8.2)*   *Review Exercises 27–34*

■ Evaluate trigonometric functions exactly.   *(Section 8.2)*   *Review Exercises 35–46*

Right Triangle Definition: $0 < \theta < \dfrac{\pi}{2}$.

$$\sin \theta = \frac{\text{opp.}}{\text{hyp.}} \qquad \cos \theta = \frac{\text{adj.}}{\text{hyp.}} \qquad \tan \theta = \frac{\text{opp.}}{\text{adj.}}$$

$$\csc \theta = \frac{\text{hyp.}}{\text{opp.}} \qquad \sec \theta = \frac{\text{hyp.}}{\text{adj.}} \qquad \cot \theta = \frac{\text{adj.}}{\text{opp.}}$$

Circular Function Definition: θ is any angle in standard position and (x, y) is a point on the terminal ray of the angle.

$$\sin \theta = \frac{y}{r} \qquad \cos \theta = \frac{x}{r} \qquad \tan \theta = \frac{y}{x}$$

$$\csc \theta = \frac{r}{y} \qquad \sec \theta = \frac{r}{x} \qquad \cot \theta = \frac{x}{y}$$

■ Use a calculator to approximate values of trigonometric functions.   *(Section 8.2)*   *Review Exercises 47–54*

■ Solve right triangles.   *(Section 8.2)*   *Review Exercises 55–58*

■ Solve trigonometric equations.   *(Section 8.2)*   *Review Exercises 59–64*

■ Use right triangles to solve real-life problems.   *(Section 8.2)*   *Review Exercises 65, 66*

■ Sketch graphs of trigonometric functions.   *(Section 8.3)*   *Review Exercises 67–74*

■ Use trigonometric functions to model real-life situations.   *(Section 8.3)*   *Review Exercises 75, 76*

■ Find derivatives of trigonometric functions.   *(Section 8.4)*   *Review Exercises 77–88*

$$\frac{d}{dx}[\sin u] = \cos u \frac{du}{dx} \qquad\qquad \frac{d}{dx}[\cos u] = -\sin u \frac{du}{dx}$$

$$\frac{d}{dx}[\tan u] = \sec^2 u \frac{du}{dx} \qquad\qquad \frac{d}{dx}[\cot u] = -\csc^2 u \frac{du}{dx}$$

$$\frac{d}{dx}[\sec u] = \sec u \tan u \frac{du}{dx} \qquad\qquad \frac{d}{dx}[\csc u] = -\csc u \cot u \frac{du}{dx}$$

■ Find the equations of tangent lines to graphs of trigonometric functions.   *(Section 8.4)*   *Review Exercises 89–94*

■ Analyze the graphs of trigonometric functions.   *(Section 8.4)*   *Review Exercises 95–98*

---

* Use a wide range of valuable study aids to help you master the material in this chapter. The *Student Solutions Guide* includes step-by-step solutions to all odd-numbered exercises to help you review and prepare. The *HM mathSpace® Student CD-ROM* helps you brush up on your algebra skills. The *Graphing Technology Guide*, available on the Web at *math.college.hmco.com/students*, offers step-by-step commands and instructions for a wide variety of graphing calculators, including the most recent models.

- Use relative extrema to solve real-life problems. *(Section 8.4)*   *Review Exercises 99, 100*
- Solve trigonometric integrals. *(Section 8.5)*   *Review Exercises 101–112*

$$\int \cos u \, du = \sin u + C \qquad\qquad \int \sin u \, du = -\cos u + C$$

$$\int \sec^2 u \, du = \tan u + C \qquad\qquad \int \sec u \tan u \, du = \sec u + C$$

$$\int \csc^2 u \, du = -\cot u + C \qquad\qquad \int \csc u \cot u \, du = -\csc u + C$$

$$\int \tan u \, du = -\ln|\cos u| + C \qquad\qquad \int \sec u \, du = \ln|\sec u + \tan u| + C$$

$$\int \cot u \, du = \ln|\sin u| + C \qquad\qquad \int \csc u \, du = \ln|\csc u - \cot u| + C$$

- Find the areas of regions in the plane. *(Section 8.5)*   *Review Exercises 113–116*
- Use trigonometric integrals to solve real-life problems. *(Section 8.5)*   *Review Exercises 117–120*
- Use L'Hôpital's Rule to evaluate limits. *(Section 8.6)*   *Review Exercises 121–134*
- Compare the rates of growth of two functions. *(Section 8.6)*   *Review Exercises 135, 136*

- ***Degree and Radian Modes***   When using a computer or calculator to evaluate or graph a trigonometric function, be sure that you use the proper mode—*radian* mode or *degree* mode.

- ***Checking the Form of an Answer***   Because of the abundance of trigonometric identities, solutions of problems in this chapter can take a variety of forms. For instance, the expressions $-\ln|\cot x| + C$ and $\ln|\tan x| + C$ are equivalent. So, when you are checking your solutions with those given in the back of the text, remember that your solution might be correct, even if its form doesn't agree precisely with that given in the text.

- ***Using Technology***   Throughout this chapter, remember that technology can help you *graph* trigonometric functions, *evaluate* trigonometric functions, *differentiate* trigonometric functions, and *integrate* trigonometric functions. Consider, for instance, the difficulty of sketching the graph of the function below *without* using a graphing utility.

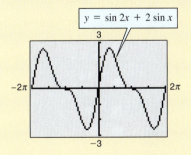

$y = \sin 2x + 2 \sin x$

Study Tools   *Additional resources that accompany this chapter*

- **Algebra Review**   (pages 606 and 607)
- **Chapter Summary and Study Strategies**   (pages 608 and 609)
- **Review Exercises**   (pages 610–613)
- **Sample Post-Graduation Exam Questions**   (page 614)

- **Web Exercises**   (page 605, Exercise 77)
- **Student Solutions Guide**
- **HM mathSpace® Student CD-ROM**
- **Graphing Technology Guide**   (*math.college.hmco.com/students*)

8   CHAPTER REVIEW EXERCISES

In Exercises 1–8, sketch the angle in standard position and list one positive and one negative coterminal angle.

1. $\dfrac{7\pi}{4}$

2. $\dfrac{11\pi}{6}$

3. $\dfrac{3\pi}{2}$

4. $\dfrac{\pi}{2}$

5. $135°$

6. $210°$

7. $-405°$

8. $-330°$

In Exercises 9–16, convert the degree measure to radian measure. Use a calculator to verify your results.

9. $210°$

10. $300°$

11. $-60°$

12. $-30°$

13. $-480°$

14. $-540°$

15. $135°$

16. $315°$

In Exercises 17–20, convert the radian measure to degree measure. Use a calculator to verify your results.

17. $\dfrac{4\pi}{3}$

18. $\dfrac{5\pi}{6}$

19. $-\dfrac{2\pi}{3}$

20. $-\dfrac{11\pi}{6}$

In Exercises 21–24, solve the triangle for the indicated side and/or angle.

21.
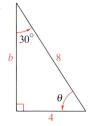

22.
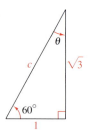

23.
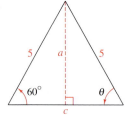

24.
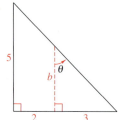

25. *Height*   A ladder of length 16 feet leans against the side of a house. The bottom of the ladder is 4.4 feet from the house (see figure). Find the height h of the top of the ladder.

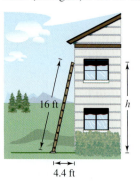

26. *Length*   To stabilize a 75-foot tower for a radio antenna, a guy wire must be attached from the top of the tower to an anchor 50 feet from the base. How long is the wire?

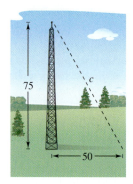

In Exercises 27–34, find the reference angle for the given angle.

27. $\dfrac{2\pi}{3}$

28. $\dfrac{5\pi}{6}$

29. $-\dfrac{5\pi}{6}$

30. $-\dfrac{5\pi}{3}$

31. $240°$

32. $300°$

33. $420°$

34. $480°$

In Exercises 35–46, evaluate the trigonometric function without using a calculator.

35. $\cos 45°$

36. $\sin 240°$

37. $\tan \dfrac{\pi}{3}$

38. $\csc \dfrac{\pi}{4}$

39. $\sin \dfrac{5\pi}{3}$

40. $\cos \dfrac{7\pi}{6}$

41. $\cot\left(-\dfrac{5\pi}{6}\right)$

42. $\tan\left(-\dfrac{5\pi}{3}\right)$

43. $\sec(-180°)$

44. $\csc(-270°)$

45. $\cos\left(-\dfrac{4\pi}{3}\right)$

46. $\sin\left(-\dfrac{11\pi}{6}\right)$

 In Exercises 47–54, use a calculator to evaluate the trigonometric function. Round to four decimal places.

47. $\tan 33°$

48. $\cot 216°$

49. $\sec \dfrac{12\pi}{5}$

50. $\csc \dfrac{2\pi}{9}$

51. $\sin\left(-\dfrac{\pi}{9}\right)$

52. $\cos\left(-\dfrac{3\pi}{7}\right)$

53. $\cos 105°$

54. $\sin 224°$

In Exercises 55–58, solve for x, y, or r as indicated.

55.

56.

57.

58.

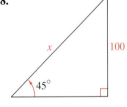

In Exercises 59–64, solve the trigonometric equation.

59. $2 \cos x + 1 = 0$

60. $2 \sin^2 x = 1$

61. $2 \sin^2 x + 3 \sin x + 1 = 0$

62. $\cos^3 x = \cos x$

63. $\sec^2 x - \sec x - 2 = 0$

64. $2 \sec^2 x + \tan^2 x - 3 = 0$

65. *Height*   The length of a shadow of a tree is 125 feet when the angle of elevation of the sun is 33° (see figure). Approximate the height h of the tree.

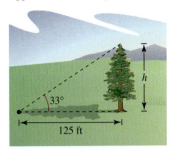

66. *Distance*   A passenger in an airplane flying at 35,000 feet sees two towns directly to the left of the airplane. The angles of depression to the towns are 32° and 76° (see figure). How far apart are the towns?

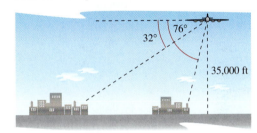

In Exercises 67–74, sketch a graph of the trigonometric function.

67. $y = 2 \cos 6x$

68. $y = \sin 2\pi x$

69. $y = \dfrac{1}{3} \tan x$

70. $y = \cot \dfrac{x}{2}$

71. $y = 3 \sin \dfrac{2x}{5}$

72. $y = 8 \cos\left(-\dfrac{x}{4}\right)$

73. $y = \sec 2\pi x$

74. $y = 3 \csc 2x$

 75. *Seasonal Sales*   The daily revenue from sales S (in thousands of units per day) of jet skis is modeled by

$$S = 74 + \dfrac{3}{365}t - 40 \sin \dfrac{2\pi t}{365}$$

where t is the time in days, with $t = 1$ corresponding to January 1. Use a graphing utility to graph this model over a one-year period.

 76. *Seasonal Sales*   The daily revenue from sales S (in hundreds of units per day) of gas stoves is modeled by

$$S = 25 + \dfrac{2}{365}t + 20 \sin \dfrac{2\pi t}{365}$$

where t is the time in days, with $t = 1$ corresponding to January 1. Use a graphing utility to graph this model over a one-year period.

In Exercises 77–88, find the derivative of the function.

77. $y = \sin 5\pi x$

78. $y = \tan(4x - \pi)$

79. $y = -x \tan x$

80. $y = \csc 3x + \cot 3x$

81. $y = \dfrac{\cos x}{x^2}$

82. $y = \dfrac{\cos(x - 1)}{x - 1}$

83. $y = 3 \sin^2 4x + x$

84. $y = x \cos x - \sin x$

85. $y = 2 \csc^3 x$

86. $y = \sec^2 2x$

87. $y = e^x \cot x$

88. $y = \frac{1}{2} e^{\sin 2x}$

In Exercises 89–94, find an equation of the tangent line to the graph of the function at the given point.

Function	*Point*
89. $y = 4 \sin 2x$	$(\pi, 0)$
90. $y = -x \cos x$	$(0, 0)$
91. $y = \dfrac{1}{4} \sin^2 2x$	$\left(\dfrac{\pi}{4}, \dfrac{1}{4}\right)$
92. $y = \dfrac{2x}{\cos x}$	$(0, 0)$
93. $y = e^x \tan 2x$	$(0, 0)$
94. $y = x^2 \cot x$	$\left(\dfrac{\pi}{4}, \dfrac{\pi^2}{16}\right)$

In Exercises 95–98, find the relative extrema of the function on the interval $(0, 2\pi)$.

95. $f(x) = \dfrac{x}{2} + \cos x$

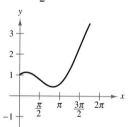

96. $f(x) = \sin x \cos x$

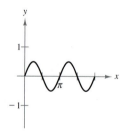

97. $f(x) = \sin^2 x + \sin x$

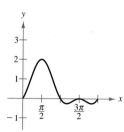

98. $f(x) = \dfrac{\cos x}{1 + \sin^2 x}$

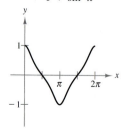

 99. *Seasonal Sales* Refer to the model given in Exercise 75.

(a) Use a graphing utility to find the maximum daily sales of jet skis. On what day of the year does the maximum daily revenue occur?

(b) Use a graphing utility to find the minimum daily sales of jet skis. On what day of the year does the minimum daily revenue occur?

100. *Seasonal Sales* Refer to the model given in Exercise 76.

(a) Use a graphing utility to find the maximum daily sales of gas stoves. On what day of the year do the maximum daily sales occur?

(b) Use a graphing utility to find the minimum daily sales of gas stoves. On what day of the year do the minimum daily sales occur?

In Exercises 101–112, evaluate the integral.

101. $\displaystyle\int (3 \sin x - 2 \cos x)\, dx$

102. $\displaystyle\int \csc 2x \cot 2x\, dx$

103. $\displaystyle\int \sin^3 x \cos x\, dx$

104. $\displaystyle\int x \cos x^2\, dx$

105. $\displaystyle\int_0^{\pi} (1 + \sin x)\, dx$

106. $\displaystyle\int_{-\pi}^{\pi} (x - \cos^2 x)\, dx$

107. $\displaystyle\int_{-\pi/6}^{\pi/6} \sec^2 x\, dx$

108. $\displaystyle\int_{\pi/6}^{\pi/2} \csc^2 x\, dx$

109. $\displaystyle\int_{-\pi/3}^{\pi/3} 4 \sec x \tan x\, dx$

110. $\displaystyle\int -2 \csc \dfrac{x}{2} \cot \dfrac{x}{2}\, dx$

111. $\displaystyle\int_{-\pi/2}^{\pi/2} (2x + \cos x)\, dx$

112. $\displaystyle\int_0^{\pi} 2x \sin x^2\, dx$

In Exercises 113–116, find the area of the region.

113. $y = \sin 2x$

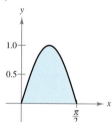

114. $y = \cot x$

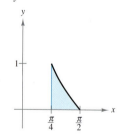

115. $y = 2 \sin x + \cos 3x$

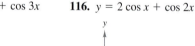

116. $y = 2 \cos x + \cos 2x$

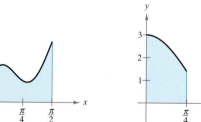

117. *Consumer Trends* Domestic energy consumption in the United States is seasonal. Assume that consumption is approximated by the model

$$Q = 6.9 + \cos \frac{\pi(2t - 1)}{12}$$

where Q is the total consumption (in quadrillion Btu) and t is the time in months, with $t = 1$ corresponding to January (see figure). Find the average consumption rate of domestic energy during a year.

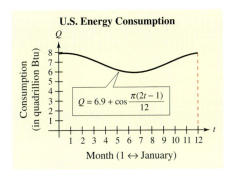

U.S. Energy Consumption

$Q = 6.9 + \cos \dfrac{\pi(2t - 1)}{12}$

Consumption (in quadrillion Btu)

Month (1 ↔ January)

118. *Seasonal Sales* The monthly sales (in millions of units) of snow blowers can be modeled by

$$S = 15 + 6 \sin \frac{\pi(t - 8)}{6}$$

where t is the time in months, with $t = 1$ corresponding to January (see figure). Find the average monthly sales

(a) during a year.

(b) from July through December.

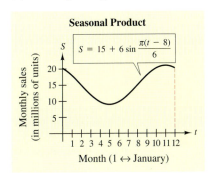

Seasonal Product

$S = 15 + 6 \sin \dfrac{\pi(t - 8)}{6}$

Monthly sales (in millions of units)

Month (1 ↔ January)

119. *Meteorology* The average monthly precipitation P in inches, including rain, snow, and ice, for San Francisco, California can be modeled by

$$P = 2.77 \sin(0.39t + 1.91) + 2.52$$

where t is the time in months, with $t = 1$ corresponding to January. Find the total annual precipitation for San Francisco. *(Source: U.S. National Oceanic and Atmospheric Administration)*

 120. *Sales* The sales S (in millions of dollars per year) for 7-Eleven for the years 1994 through 2003 can be modeled by

$$S = 2180.74 \sin(0.35t + 2.90) + 8794.58,$$

$$4 \le t \le 13$$

where t is the year, with $t = 4$ corresponding to 1994. *(Source: 7-Eleven, Inc.)*

(a) Use a graphing utility to graph the model.

(b) Find the rates at which the sales were changing in 1997, 2000, and 2003. Explain your results.

(c) Determine 7-Eleven's total sales from 1994 through 2003.

In Exercises 121–134, find the limit, if it exists. Use L'Hôpital's Rule when necessary.

121. $\lim\limits_{x \to 1} \dfrac{3x - 1}{5x + 5}$

122. $\lim\limits_{x \to 0} \dfrac{2x - 1}{x^2 + 1}$

123. $\lim\limits_{x \to 2} \dfrac{x^2 + 2x + 11}{x^3 + 4}$

124. $\lim\limits_{x \to -1} \dfrac{x^2 + 2x + 1}{x^2 - 1}$

125. $\lim\limits_{x \to 1} \dfrac{x^3 - x^2 + 4x - 4}{x^3 - 6x^2 + 5x}$

126. $\lim\limits_{x \to 2} \dfrac{4x^2 + 13x - 42}{3x^2 - 23x + 34}$

127. $\lim\limits_{x \to 0} \dfrac{\sin \pi x}{\sin 2\pi x}$

128. $\lim\limits_{x \to 0} \dfrac{\tan 2x}{5x}$

129. $\lim\limits_{x \to 0} \dfrac{\cos^2 x - 1}{\sin x}$

130. $\lim\limits_{x \to 0} \dfrac{1 - \cos 2x}{x}$

131. $\lim\limits_{x \to 0} \dfrac{\sin^2 x}{e^x}$

132. $\lim\limits_{x \to 0} \dfrac{\sin 2x}{e^x - 1}$

133. $\lim\limits_{x \to \infty} \dfrac{e^{5x}}{x^2}$

134. $\lim\limits_{x \to 1} \dfrac{\ln(x^2)}{x - 1}$

135. Consider

$$\lim_{x \to \infty} \frac{f(x)}{g(x)} = 0.$$

What can you say about the relative rates of growth of f and g?

136. Consider

$$\lim_{x \to \infty} \frac{f(x)}{g(x)} = \infty.$$

What can you say about the relative rates of growth of f and g?

8 SAMPLE POST-GRADUATION EXAM QUESTIONS

CPA
GMAT
GRE
Actuarial
CLAST

The following questions represent the types of questions that appear on certified public accountant (CPA) exams, Graduate Management Admission Tests (GMAT), Graduate Records Exams (GRE), actuarial exams, and College-Level Academic Skills Tests (CLAST). The answers to the questions are given in the back of the book.

For Question 1, use the bar graph at the left. *(Source: U.S. Energy Information Administration)*

Figure for 1

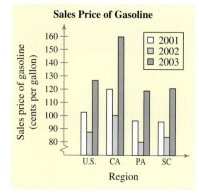

Sales Price of Gasoline

1. Which of the following can be concluded from the data presented in the bar graph?

   I. For the years 2001 through 2003, the sales price of gasoline in California was more than the national average, whereas in Pennsylvania and South Carolina the sales price of gasoline was less than the national average.

   II. The sales price of gasoline in California in 2003 was approximately 125% of the national average in 2003.

   III. The sales price of gasoline in South Carolina in 2002 was approximately $\frac{4}{5}$ of the price in California in 2002.

   (a) I only   (b) II and III only   (c) III only   (d) I, II, and III

For Question 2, use the figure below.

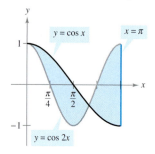

2. The area of the closed region in the xy-plane bounded by the graphs of $y = \cos x$, $y = \cos 2x$, and $x = \pi$, as shown in the shaded portion of the diagram, is equal to

   (a) $1 - \dfrac{\sqrt{3}}{2}$ (b) $\sqrt{3}$ (c) $\sqrt{3} + \dfrac{1}{2}$ (d) $\dfrac{3\sqrt{3}}{2}$ (e) $3\sqrt{3}$

3. The real value of x for which $e^x - 3e^{-x} = 2$ is

   (a) -1 (b) 0   (c) $\frac{1}{4} \ln 2$ (d) $\ln 3$ (e) 3

4. The line $y = x + 1$ and the curve $y = x^2$ intersect at two points P and Q. The y-coordinate of the midpoint of the line segment PQ is

   (a) $\dfrac{1}{2}$ (b) $\dfrac{\sqrt{5}}{2}$ (c) $\dfrac{3}{2}$ (d) $\sqrt{5}$ (e) 3

5. An official baseball diamond is a square, 90 feet on each side. The shortest distance (to the nearest foot) from third base to first base is

   (a) 90 feet

   (b) 127 feet

   (c) 135 feet

   (d) 180 feet

   (e) less than 90 feet

chapter

9

Probability and Calculus

© John Gillmoure/CORBIS

Calculus can be used to find the probability that doubles are thrown on any given roll of a pair of dice.

STRATEGIES FOR SUCCESS

WHAT YOU SHOULD LEARN:

- How to calculate the probability of a discrete random variable
- How to find the expected values, variances, and standard deviations of both discrete and continuous random variables
- How to use integration to calculate probabilities for a probability density function of a continuous random variable
- How to use special probability density functions in applications

WHY YOU SHOULD LEARN IT:

Probability theory has many applications in real life, such as the examples below, which represent a small sample of the applications in this chapter.

- Biology: Genetics, Exercise 19 on page 623
- Personal Income, Exercise 28 on page 624
- Demand, Exercises 33 and 34 on page 632
- Learning Theory, Exercise 37 on page 642
- Consumer Trends, Exercise 48 on page 643
- Dice Toss, Review Exercise 11 on page 647
- Chemistry: Hydrogen Orbitals, Review Exercise 51 on page 649

9.1   DISCRETE PROBABILITY

- Describe sample spaces for experiments.
- Assign values to, and form frequency distributions for, discrete random variables.
- Find the probabilities of events for discrete random variables.
- Find the expected values or means of discrete random variables.
- Find the variances and standard deviations of discrete random variables.

Sample Spaces

PhotoDisc

When a weather forecaster states that there is a 50% chance of thunderstorms, it means that thunderstorms have occurred on half of all days that have had similar weather conditions.

When assigning measurements to the uncertainties of everyday life, people often use ambiguous terminology such as "fairly certain," "probable," and "highly unlikely." Probability theory allows you to remove this ambiguity by assigning a number to the likelihood of the occurrence of an event. This number is called the **probability** that the event will occur. For example, if you toss a fair coin, the probability that it will land heads up is one-half or 0.5.

In probability theory, any happening whose result is uncertain is called an **experiment.** The possible results of the experiment are the **outcomes,** a collection of outcomes is an **event,** and the collection of all possible outcomes is the **sample space** of the experiment. For instance, consider an experiment in which a coin is tossed. The sample space of this experiment consists of two outcomes: either the coin will land heads up (denoted by H) or it will land tails up (denoted by T). So, the sample space S is

$$S = \{H, T\}. \qquad \text{Sample space}$$

In this text, all outcomes of a sample space are assumed to be equally likely. For instance, when a coin is tossed, H and T are assumed to be equally likely.

EXAMPLE 1   **Finding a Sample Space**

An experiment consists of tossing a six-sided die.

(a) What is the sample space?

(b) Describe the event corresponding to a number greater than 2 turning up.

SOLUTION

(a) The sample space S consists of six outcomes, which can be represented by the numbers 1 through 6. That is

$$S = \{1, 2, 3, 4, 5, 6\}. \qquad \text{Sample space}$$

Note that each of the outcomes in the sample space is equally likely.

(b) The event E corresponding to a number greater than 2 turning up is a subset of S. That is

$$E = \{3, 4, 5, 6\}. \qquad \text{Event}$$

TRY IT 1

An experiment consists of tossing two six-sided dice.

(a) What is the sample space?

(b) Describe the event corresponding to a sum greater than or equal to seven points when the dice are tossed.

Discrete Random Variables

A function that assigns a numerical value to each of the outcomes in a sample space is called a **random variable.** For instance, in the sample space $S = \{HH, HT, TH, TT\}$, the outcomes could be assigned the numbers 2, 1, and 0, depending on the number of heads in the outcome.

Definition of Discrete Random Variable

Let S be a sample space. A **random variable** is a function x that assigns a numerical value to each outcome in S. If the set of values taken on by the random variable is finite, then the random variable is **discrete.** The number of times a specific value of x occurs is the **frequency** of x and is denoted by $n(x)$.

EXAMPLE 2 | **Finding Frequencies**

Three coins are tossed. A random variable assigns the number 0, 1, 2, or 3 to each possible outcome, depending on the number of heads that turn up.

$$S = \{HHH, HHT, HTH, HTT, THH, THT, TTH, TTT\}$$

$$\quad\;\; 3 \qquad 2 \qquad 2 \qquad 1 \qquad 2 \qquad 1 \qquad 1 \qquad 0$$

Find the frequencies of 0, 1, 2, and 3. Then use a bar graph to represent the result.

SOLUTION   To find the frequencies, simply count the number of occurrences of each value of the random variable, as shown in the table.

Random variable, x	0	1	2	3
Frequency of x, $n(x)$	1	3	3	1

This table is called a **frequency distribution** of the random variable. The result is shown graphically by the bar graph in Figure 9.1.

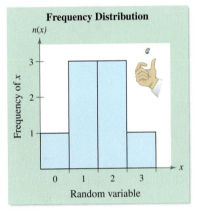

FIGURE 9.1

TRY IT 2

Use a graphing utility to create a bar graph similar to the one shown in Figure 9.1, representing the frequency for tossing two six-sided dice. Let the random variable be the sum of the points when the dice are tossed.

STUDY TIP

In Example 2, note that the sample space consists of *eight* outcomes, each of which is *equally likely.* The sample space does *not* consist of the outcomes "zero heads," "one head," "two heads," and "three heads." You cannot consider these events to be outcomes because they are not equally likely.

Discrete Probability

The probability of a random variable x is

$$P(x) = \frac{\text{Frequency of } x}{\text{Number of outcomes in } S}$$

$$= \frac{n(x)}{n(S)}$$ Probability

where $n(S)$ is the number of equally likely outcomes in the sample space. By this definition, it follows that the probability of an event must be a number between 0 and 1. That is, $0 \le P(x) \le 1$.

The collection of probabilities corresponding to the values of the random variable is called the **probability distribution** of the random variable. If the range of a discrete random variable consists of m different values $\{x_1, x_2, x_3, \ldots, x_m\}$, then the sum of the probabilities of x_i is 1. This can be written as

$$P(x_1) + P(x_2) + P(x_3) + \cdots + P(x_m) = 1.$$

EXAMPLE 3   **Finding a Probability Distribution**

Five coins are tossed. Graph the probability distribution for the random variable giving the number of heads that turn up.

SOLUTION

x	Event	$n(x)$
0	TTTTT	1
1	HTTTT, THTTT, TTHTT, TTTHT, TTTTH	5
2	HHTTT, HTHTT, HTTHT, HTTTH, THHTT	10
	THTHT, THTTH, TTHHT, TTHTH, TTTHH	
3	HHHTT, HHTHT, HHTTH, HTHHT, HTHTH	10
	HTTHH, THHHT, THHTH, THTHH, TTHHH	
4	HHHHT, HHHTH, HHTHH, HTHHH, THHHH	5
5	HHHHH	1

The number of outcomes in the sample space is $n(S) = 32$. The probability of each value of the random variable is shown in the table.

Random variable, x	0	1	2	3	4	5
Probability, $P(x)$	$\frac{1}{32}$	$\frac{5}{32}$	$\frac{10}{32}$	$\frac{10}{32}$	$\frac{5}{32}$	$\frac{1}{32}$

A graph of this probability distribution is shown in Figure 9.2. Note that values of the random variable are represented by intervals on the x-axis. Observe that the sum of the probabilities is 1.

TRY IT 3

Two six-sided dice are tossed. Graph the probability distribution for the random variable giving the sum of the points when the dice are tossed.

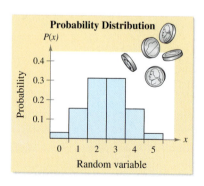

FIGURE 9.2

Expected Value

Suppose you repeated the coin-tossing experiment in Example 3 several times. On the average, how many heads would you expect to turn up? From Figure 9.2, it seems reasonable that the average number of heads would be $2\frac{1}{2}$. This "average" is the **expected value** of the random variable.

Definition of Expected Value

If the range of a discrete random variable consists of m different values $\{x_1, x_2, x_3, \ldots, x_m\}$, then the **expected value** of the random variable is

$$E(x) = x_1 P(x_1) + x_2 P(x_2) + x_3 P(x_3) + \cdots + x_m P(x_m).$$

The expected value is also called the **mean** of the random variable.

STUDY TIP

Although the expected value of x is denoted by $E(x)$, the mean of x is usually denoted by μ. Because the mean often occurs near the center of the values in the range of the random variable, it is called a **measure of central tendency**.

EXAMPLE 4   **Finding an Expected Value**

Five coins are tossed. Find the expected value for the number of heads that will turn up.

SOLUTION   Using the results of Example 3, you obtain the expected value as shown.

$\quad$ 0 Heads $\quad$ 1 Head $\quad$ 2 Heads $\quad$ 3 Heads $\quad$ 4 Heads $\quad$ 5 Heads

$$E(x) = (0)\left(\tfrac{1}{32}\right) + (1)\left(\tfrac{5}{32}\right) + (2)\left(\tfrac{10}{32}\right) + (3)\left(\tfrac{10}{32}\right) + (4)\left(\tfrac{5}{32}\right) + (5)\left(\tfrac{1}{32}\right)$$

$$= \tfrac{80}{32} = 2.5$$

EXAMPLE 5   **Finding an Expected Value**

Over a period of 1 year (225 selling days), a sales representative sold from zero to eight units per day, as shown in Figure 9.3. From these data, what is the average number of units per day the sales representative should expect to sell?

SOLUTION   One way to answer this question is to calculate the expected value of the number of units.

$$E(x) = (0)\left(\tfrac{33}{225}\right) + (1)\left(\tfrac{45}{225}\right) + (2)\left(\tfrac{52}{225}\right) + (3)\left(\tfrac{46}{225}\right) + (4)\left(\tfrac{24}{225}\right) +$$

$$(5)\left(\tfrac{11}{225}\right) + (6)\left(\tfrac{8}{225}\right) + (7)\left(\tfrac{5}{225}\right) + (8)\left(\tfrac{1}{225}\right)$$

$$= \tfrac{529}{225} \approx 2.35 \text{ units per day}$$

TRY IT 4

Two six-sided dice are tossed. Find the expected value for a sum of seven points.

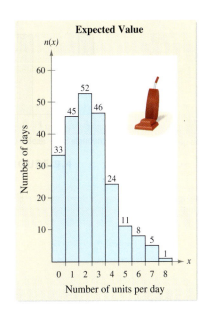

FIGURE 9.3

TRY IT 5

Over a period of 1 year, a salesperson worked 6 days a week (312 selling days) and sold from zero to six units per day. Using the data from the table shown below, what is the average number of units per day the sales representative should expect to sell?

Number of units per day	0	1	2	3	4	5	6
Number of days		39	60	75	62	48	18	10

20. *Experiment* Use a computer to conduct the following experiment. The four letters A, R, S, and T are placed in a box. The letters are then chosen one at a time and placed in the order in which they were chosen. What is the probability that the letters spell an English word in the order they are chosen? Conduct this procedure 100 times. How many times did an English word occur?

In Exercises 21–24, find $E(x)$, $V(x)$, and σ for the given probability distribution.

21.

x	1	2	3	4	5
$P(x)$	$\frac{1}{16}$	$\frac{3}{16}$	$\frac{8}{16}$	$\frac{3}{16}$	$\frac{1}{16}$

22.

x	1	2	3	4	5
$P(x)$	$\frac{4}{10}$	$\frac{2}{10}$	$\frac{2}{10}$	$\frac{1}{10}$	$\frac{1}{10}$

23.

x	-3	-1	0	3	5
$P(x)$	$\frac{1}{5}$	$\frac{1}{5}$	$\frac{1}{5}$	$\frac{1}{5}$	$\frac{1}{5}$

24.

x	$-\$5000$	$-\$2500$	$\$300$
$P(x)$	0.008	0.052	0.940

In Exercises 25 and 26, find the mean and variance of the discrete random variable x.

25. *Die Toss* x is (a) the number of points when a four-sided die is tossed once and (b) the sum of the points when the four-sided die is tossed twice.

26. *Coin Toss* x is the number of heads when a coin is tossed four times.

27. *Revenue* A publishing company introduces a new weekly magazine that sells for $2.95 on the newsstand. The marketing group of the company estimates that sales x in thousands will be approximated by the following probability function.

x	10	15	20	30	40
$P(x)$	0.25	0.30	0.25	0.15	0.05

(a) Find $E(x)$ and σ.

(b) Find the expected revenue.

28. *Personal Income* The probability distribution of the random variable x, the annual income of a family, in thousands of dollars, in a certain section of a large city, is shown in the table.

x	30	40	50	60	80
$P(x)$	0.10	0.20	0.50	0.15	0.05

Find $E(x)$ and σ.

29. *Insurance* An insurance company needs to determine the annual premium required to break even on fire protection policies with a face value of $90,000. If x is the claim size on these policies and the analysis is restricted to the losses $30,000, $60,000, and $90,000, then the probability distribution of x is shown in the table.

x	0	30,000	60,000	90,000
$P(x)$	0.995	0.0036	0.0011	0.0003

What premium should customers be charged for the company to break even?

30. *Insurance* An insurance company needs to determine the annual premium required to break even for collision protection for cars with a value of $10,000. If x is the claim size on these policies and the analysis is restricted to the losses $1000, $5000, and $10,000, then the probability distribution of x is shown in the table.

x	0	1000	5,000	10,000
$P(x)$	0.936	0.040	0.020	0.004

What premium should customers be charged for the company to break even?

Games of Chance If x is the net gain to a player in a game of chance, then $E(x)$ is usually negative. This value gives the average amount per game the player can expect to lose over the long run. In Exercises 31 and 32, find the expected net gain to the player for one play of the specified game.

31. In roulette, the wheel has the 38 numbers 00, 0, 1, 2, . . . , 34, 35, and 36, marked on equally spaced slots. If a player bets $1 on a number and wins, then the player keeps the dollar and receives an additional $35. Otherwise, the dollar is lost.

32. A service organization is selling $2 raffle tickets as part of a fund-raising program. The first prize is a boat valued at $2950, and the second prize is a camping tent valued at $400. In addition to the first and second prizes, there are 25 $20 gift certificates to be awarded. The number of tickets sold is 3000.

33. ***Market Analysis*** After considerable market study, a sporting goods company has decided on two possible cities in which to open a new store. Management estimates that city 1 will yield $20 million in revenues if successful and will lose $4 million if not, whereas city 2 will yield $50 million in revenues if successful and lose $9 million if not. City 1 has a 0.3 probability of being successful and city 2 has a 0.2 probability of being successful. In which city should the sporting goods company open the new store with respect to the expected return from each store?

34. Repeat Exercise 33 if the probabilities of city 1 and city 2 being successful are 0.4 and 0.25, respectively.

35. ***Health*** The table shown below gives the probability distribution of the numbers of deaths due to cancer in the United States in 2001 by age group. *(Source: U.S. National Center for Health Statistics)*

Age, a	14 and under	15–24	25–34	35–44	45–54
$P(a)$	0.0018	0.0036	0.0072	0.0307	0.0903

Age, a	55–64	65–74	75–84	85 and over
$P(a)$	0.1625	0.2653	0.2978	0.1408

(a) Sketch the probability distribution.

(b) Find the probability that an individual who died of cancer was between 35 and 74 years of age.

(c) Find the probability that an individual who died of cancer was over 45 years of age.

(d) Find the probability that an individual who died of cancer was under 24 years of age.

36. ***Education*** The table gives the probability distribution of the educational attainments of people in the United States in 2002, ages 25 to 34, where $x = 0$ represents no high school degree, $x = 1$ represents a high school degree, $x = 2$ represents some college, $x = 3$ represents an associate's degree, $x = 4$ represents a bachelor's degree, and $x = 5$ represents an advanced degree. *(Source: U.S. Census Bureau)*

x	0	1	2	3	4	5
$P(x)$	0.131	0.284	0.191	0.087	0.232	0.074

(a) Sketch the probability distribution.

(b) Determine $E(x)$, $V(x)$, and σ. Explain the meanings of these values.

37. ***Athletics*** A baseball fan examined the record of a favorite baseball player's performance during his last 50 games. The numbers of games in which the player had zero, one, two, three, and four hits are recorded in the table shown below.

Number of hits	0	1	2	3	4
Frequency	14	26	7	2	1

(a) Complete the table below, where x is the number of hits.

x	0	1	2	3	4
$P(x)$					

(b) Use the table in part (a) to sketch the graph of the probability distribution.

(c) Use the table in part (a) to find $P(1 \leq x \leq 3)$.

(d) Determine $E(x)$, $V(x)$, and σ. Explain your results.

38. ***Economics: Investment*** Suppose you are trying to make a decision about how to invest $10,000 over the next year. One option is a low-risk bank deposit paying 5% interest per year. The other is a high-risk corporate stock with a 5% dividend, plus a 50% chance of a 30% price decline and a 50% chance of a 30% price increase. Determine the expected value of each option and choose one of the options. Explain your choice. How would your decision change if the corporate stock offered a 20% dividend instead of a 5% dividend? *(Source: Adapted from Taylor, Economics, Fourth Edition)*

39. ***Employment*** Use a graphing utility to find the mean and standard deviation of the data shown below, which give the numbers of hours of overtime for 60 employees of a particular company during a given week.

0.0, 0.0, 0.0, 0.0, 0.0, 0.5, 0.5, 1.0, 1.0, 1.0, 2.0, 2.0, 2.0, 2.0, 2.0, 2.0, 2.0, 2.5, 2.5, 2.5, 3.0, 3.0, 3.0, 3.0, 3.0, 3.0, 3.0, 4.0, 4.0, 4.0, 4.0, 4.5, 4.5, 5.0, 5.0, 5.0, 5.0, 5.0, 6.0, 6.0, 6.5, 6.5, 6.5, 7.0, 7.0, 7.0, 7.0, 8.0, 8.0, 8.0, 8.0, 8.0, 8.0, 8.0, 8.0, 8.5, 8.5, 8.5, 9.0, 9.5

9.2   CONTINUOUS RANDOM VARIABLES

- Verify continuous probability density functions and use continuous probability density functions to find probabilities.
- Use continuous probability density functions to answer questions about real-life situations.

Continuous Random Variables

In many applications of probability, it is useful to consider a random variable whose range is an interval on the real line. Such a random variable is called **continuous.** For instance, the random variable that measures the height of a person in a population is continuous.

To define the probability of an event involving a continuous random variable, you cannot simply count the number of ways the event can occur (as you can with a discrete random variable). Rather, you need to define a function f called a **probability density function.**

Definition of Probability Density Function

Consider a function f of a continuous random variable x whose range is the interval $[a, b]$. The function is a **probability density function** if it is nonnegative and continuous on the interval $[a, b]$ and if

$$\int_a^b f(x)\, dx = 1.$$

The probability that x lies in the interval $[c, d]$ is

$$P(c \le x \le d) = \int_c^d f(x)\, dx$$

as shown in Figure 9.5. If the range of the continuous random variable is an infinite interval, then the integrals are improper integrals.

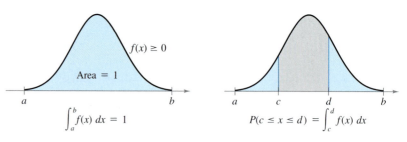

$$\int_a^b f(x)\, dx = 1 \qquad\qquad P(c \le x \le d) = \int_c^d f(x)\, dx$$

FIGURE 9.5   Probability Density Function

EXAMPLE 1   **Verifying a Probability Density Function**

Show that

$$f(x) = 12x(1 - x)^2$$

is a probability density function over the interval $[0, 1]$.

SOLUTION   Begin by observing that f is continuous and nonnegative on the interval $[0, 1]$.

$$f(x) = 12x(1 - x)^2 \geq 0, \quad 0 \leq x \leq 1$$ *$f(x)$ is nonnegative on $[0, 1]$.*

Next, evaluate the integral below.

$$\int_0^1 12x(1 - x)^2 \, dx = 12 \int_0^1 (x^3 - 2x^2 + x) \, dx$$ *Expand polynomial.*

$$= 12\left[\frac{x^4}{4} - \frac{2x^3}{3} + \frac{x^2}{2}\right]_0^1$$ *Integrate.*

$$= 12\left(\frac{1}{4} - \frac{2}{3} + \frac{1}{2}\right)$$ *Apply Fundamental Theorem of Calculus.*

$$= 1$$ *Simplify.*

Because this value is 1, you can conclude that f is a probability density function over the interval $[0, 1]$. The graph of f is shown in Figure 9.6.

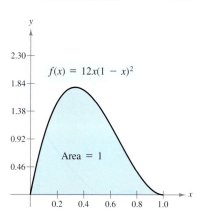

FIGURE 9.6

TRY IT 1

Show that $f(x) = \frac{1}{2}x$ is a probability density function over the interval $[0, 2]$.

The next example deals with an infinite interval and its corresponding improper integral.

EXAMPLE 2   **Verifying a Probability Density Function**

Show that

$$f(t) = 0.1e^{-0.1t}$$

is a probability density function over the infinite interval $[0, \infty)$.

SOLUTION   Begin by observing that f is continuous and nonnegative on the interval $[0, \infty)$.

$$f(t) = 0.1e^{-0.1t} \geq 0, \quad 0 \leq t$$ *$f(t)$ is nonnegative on $[0, 1]$.*

Next, evaluate the integral below.

$$\int_0^\infty 0.1e^{-0.1t} \, dt = \lim_{b \to \infty} \left[-e^{-0.1t}\right]_0^b$$ *Improper integral*

$$= \lim_{b \to \infty} (-e^{-0.1b} + 1)$$ *Evaluate limit.*

$$= 1$$

Because this value is 1, you can conclude that f is a probability density function over the interval $[0, \infty)$. The graph of f is shown in Figure 9.7.

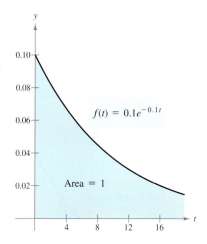

FIGURE 9.7

TRY IT 2

Show that $f(x) = 2e^{-2x}$ is a probability density function over the interval $[0, \infty)$.

EXAMPLE 6   **Modeling Weekly Demand**

The weekly demand for a product is modeled by the probability density function

$$f(x) = \frac{1}{36}(-x^2 + 6x), \quad 0 \le x \le 6$$

where x is the number of units sold (in thousands). What are the minimum and maximum weekly sales? Find the probability that the sales for a randomly chosen week will be between 2000 and 4000 units.

SOLUTION   Because $0 \le x \le 6$, the weekly sales vary from a minimum of 0 to a maximum of 6000 units. The probability is given by the integral

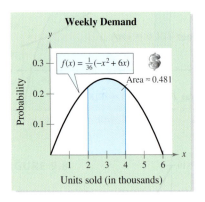

Weekly Demand

$f(x) = \frac{1}{36}(-x^2 + 6x)$

Area ≈ 0.481

Units sold (in thousands)

FIGURE 9.10

$$P(2 \le x \le 4) = \frac{1}{36}\int_2^4 (-x^2 + 6x)\, dx \qquad \text{Integrate } f(x) \text{ over } [2, 4].$$

$$= \frac{1}{36}\left[-\frac{x^3}{3} + 3x^2 \right]_2^4 \qquad \text{Find antiderivative.}$$

$$= \frac{1}{36}\left(-\frac{64}{3} + 48 + \frac{8}{3} - 12 \right) \qquad \begin{array}{l}\text{Apply Fundamental Theorem} \\ \text{of Calculus.}\end{array}$$

$$= \frac{13}{27} \qquad \text{Simplify.}$$

$$\approx 0.481. \qquad \text{Approximate.}$$

So, the probability that the weekly sales will be between 2000 and 4000 units is about 0.481 or 48.1%, as indicated in Figure 9.10. ▬▬▬▬

TRY IT 6

Find the probabilities that the sales for a randomly chosen week for the product in Example 6 will be (a) less than 2000 units and (b) more than 4000 units. Explain how you can find these probabilities without integration.

TAKE ANOTHER LOOK

Modeling Population Heights

Which of the probability density functions below do you think is the best model of the heights of American women (ages 18–24)? Explain your reasoning.

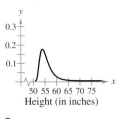

Height (in inches)

a.

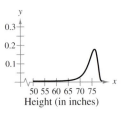

Height (in inches)

b.

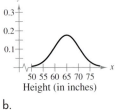

Height (in inches)

c.

Use the model to estimate the average height of American women.

PREREQUISITE REVIEW 9.2   The following warm-up exercises involve skills that were covered in earlier sections. You will use these skills in the exercise set for this section.

In Exercises 1–4, determine whether f is continuous and nonnegative on the given interval.

1. $f(x) = \dfrac{1}{x}$, $[1, 4]$

2. $f(x) = x^2 - 1$, $[0, 1]$

3. $f(x) = 3 - x$, $[1, 5]$

4. $f(x) = e^{-x}$, $[0, 1]$

In Exercises 5–10, evaluate the definite integral.

5. $\displaystyle\int_0^4 \frac{1}{4}\, dx$

6. $\displaystyle\int_1^3 \frac{1}{4}\, dx$

7. $\displaystyle\int_0^2 \frac{2 - x}{2}\, dx$

8. $\displaystyle\int_1^2 \frac{2 - x}{2}\, dx$

9. $\displaystyle\int_0^\infty 0.4e^{-0.4t}\, dt$

10. $\displaystyle\int_0^\infty 3e^{-3t}\, dt$

EXERCISES 9.2

 In Exercises 1–14, use a graphing utility to graph the function. Then verify that f is a probability density function over the given interval.

1. $f(x) = \frac{1}{8}$, $[0, 8]$

2. $f(x) = \frac{1}{5}$, $[0, 5]$

3. $f(x) = \dfrac{4 - x}{8}$, $[0, 4]$

4. $f(x) = \dfrac{x}{18}$, $[0, 6]$

5. $f(x) = 6x(1 - x)$, $[0, 1]$

6. $f(x) = \dfrac{x(6 - x)}{36}$, $[0, 6]$

7. $f(x) = \frac{1}{5}e^{-x/5}$, $[0, \infty)$

8. $f(x) = \frac{1}{6}e^{-x/6}$, $[0, \infty)$

9. $f(x) = \dfrac{3x}{8}\sqrt{4 - x^2}$, $[0, 2]$

10. $f(x) = 12x^2(1 - x)$, $[0, 1]$

11. $f(x) = \frac{4}{27}x^2(3 - x)$, $[0, 3]$

12. $f(x) = \frac{2}{9}x(3 - x)$, $[0, 3]$

13. $f(x) = \frac{1}{3}e^{-x/3}$, $[0, \infty)$

14. $f(x) = \frac{1}{4}$, $[8, 12]$

In Exercises 15–20, find the constant k such that the function f is a probability density function over the given interval.

15. $f(x) = kx$, $[1, 4]$

16. $f(x) = kx^3$, $[0, 4]$

17. $f(x) = k(4 - x^2)$, $[-2, 2]$

18. $f(x) = k\sqrt{x}(1 - x)$, $[0, 1]$

19. $f(x) = ke^{-x/2}$, $[0, \infty)$

20. $f(x) = \dfrac{k}{b - a}$, $[a, b]$

In Exercises 21–26, sketch the graph of the probability density function over the indicated interval and find the indicated probabilities.

21. $f(x) = \frac{1}{10}$, $[0, 10]$

    (a) $P(0 < x < 6)$ (b) $P(4 < x < 6)$

    (c) $P(8 < x < 10)$ (d) $P(x \geq 2)$

22. $f(x) = \dfrac{x}{50}$, $[0, 10]$

    (a) $P(0 < x < 6)$ (b) $P(4 < x < 6)$

    (c) $P(8 < x < 10)$ (d) $P(x \geq 2)$

23. $f(x) = \frac{3}{16}\sqrt{x}, \quad [0, 4]$

  (a) $P(0 < x < 2)$ (b) $P(2 < x < 4)$

  (c) $P(1 < x < 3)$ (d) $P(x \le 3)$

24. $f(x) = \dfrac{5}{4(x + 1)^2}, \quad [0, 4]$

  (a) $P(0 < x < 2)$ (b) $P(2 < x < 4)$

  (c) $P(1 < x < 3)$ (d) $P(x \le 3)$

25. $f(t) = \frac{1}{3}e^{-t/3}, \quad [0, \infty)$

  (a) $P(t < 2)$ (b) $P(t \ge 2)$

  (c) $P(1 < t < 4)$ (d) $P(t = 3)$

26. $f(t) = \frac{3}{256}(16 - t^2), \quad [-4, 4]$

  (a) $P(t < -2)$ (b) $P(t > 2)$

  (c) $P(-1 < t < 1)$ (d) $P(t > -2)$

27. **Waiting Time**   Buses arrive and depart from a college every 30 minutes. The probability density function for the waiting time t (in minutes) for a person arriving at the bus stop is

$$f(t) = \frac{1}{30}, \quad [0, 30].$$

Find the probabilities that the person will wait (a) no more than 5 minutes and (b) at least 18 minutes.

28. **Learning Theory**   The time t (in hours) required for a new employee successfully to learn to operate a machine in a manufacturing process is described by the probability density function

$$f(t) = \frac{5}{324}t\sqrt{9 - t}, \quad [0, 9].$$

Find the probabilities that a new employee will learn to operate the machine (a) in less than 3 hours and (b) in more than 4 hours but less than 8 hours.

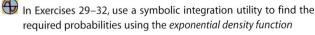 In Exercises 29–32, use a symbolic integration utility to find the required probabilities using the *exponential density function*

$$f(t) = \frac{1}{\lambda}e^{-t/\lambda}, \quad [0, \infty).$$

29. **Waiting Time**   The waiting time (in minutes) for service at the checkout at a grocery store is exponentially distributed with $\lambda = 3$. Find the probabilities of waiting (a) less than 2 minutes, (b) more than 2 minutes but less than 4 minutes, and (c) at least 2 minutes.

30. **Useful Life**   The lifetime (in years) of a battery is exponentially distributed with $\lambda = 5$. Find the probabilities that the lifetime of a given battery will be (a) less than 6 years, (b) more than 2 years but less than 6 years, and (c) more than 8 years.

31. **Waiting Time**   The length of time (in hours) required to unload trucks at a depot is exponentially distributed with $\lambda = \frac{3}{4}$. What proportion of the trucks can be unloaded in less than 1 hour?

32. **Useful Life**   The time (in years) until failure of a component in a machine is exponentially distributed with $\lambda = 3.5$. A manufacturer has a large number of these machines and plans to replace the components in all the machines during regularly scheduled maintenance periods. How much time should elapse between maintenance periods if at least 90% of the components are to remain working throughout the period?

33. **Demand**   The weekly demand x (in tons) for a certain product is a continuous random variable with the density function

$$f(x) = \frac{1}{36}xe^{-x/6}, \quad [0, \infty).$$

Find the probabilities.

  (a) $P(x < 6)$

  (b) $P(6 < x < 12)$

  (c) $P(x > 12) = 1 - P(x \le 12)$

34. **Demand**   Given the conditions of Exercise 33, determine the number of tons that should be ordered each week so that the demand can be met for 90% of the weeks.

35. **Meteorology**   A meteorologist predicts that the amount of rainfall (in inches) expected for a certain coastal community during a hurricane has the probability density function

$$f(x) = \frac{\pi}{30}\sin\frac{\pi x}{15}, \quad 0 \le x \le 15.$$

Find and interpret the probabilities.

  (a) $P(0 \le x \le 10)$ (b) $P(10 \le x \le 15)$

  (c) $P(0 \le x < 5)$ (d) $P(12 \le x \le 15)$

36. **Metallurgy**   The probability density function for the percent of iron in ore samples taken from a certain region is

$$f(x) = \frac{1155}{32}x^3(1 - x)^{3/2}, \quad [0, 1].$$

Find the probabilities that a sample will contain (a) from 0% to 25% iron and (b) from 50% to 100% iron.

37. **Labor Force**   The probability density function for women in the U.S. civilian labor force in 2002, ages 16 and over, is

$$f(x) = 4.7 + 0.61x - 4.15 \ln x - 5.14/x,$$
$$1 \le x \le 5$$

where $x = 1$ represents 16 to 24 years, $x = 2$ represents 25 to 44 years, $x = 3$ represents 45 to 54 years, $x = 4$ represents 55 to 64 years, and $x = 5$ represents 65 years and over. Find the probabilities that a woman in the labor force is (a) between the ages of 16 and 64 and (b) between the ages of 25 and 54.   *(Source: U.S. Bureau of Labor Statistics)*

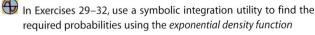 **38.** **Coin Toss**   The probability of obtaining 49, 50, or 51 heads when a fair coin is tossed 100 times is

$$P(49 \le x \le 51) \approx \int_{48.5}^{51.5} \frac{1}{5\sqrt{2\pi}}e^{-(x-50)^2/50} \, dx.$$

Use a computer or graphing utility and Simpson's Rule (with $n = 12$) to approximate this integral.

9.3   EXPECTED VALUE AND VARIANCE

■ Find the expected values or means of continuous probability density functions.
■ Find the variances and standard deviations of continuous probability density functions.
■ Find the medians of continuous probability density functions.
■ Use special probability density functions to answer questions about real-life situations.

Expected Value

In Section 9.1, you studied the concepts of expected value, mean, variance, and standard deviation for *discrete* random variables. In this section, you will extend these concepts to *continuous* random variables.

Definition of Expected Value

If f is a probability density function for a continuous random variable x over the interval $[a, b]$, then the **expected value** or **mean** of x is

$$\mu = E(x) = \int_a^b x f(x)\, dx.$$

EXAMPLE 1   **Finding Average Weekly Demand**

In Example 6 in Section 9.2, the weekly demand for a product was modeled by the probability density function

$$f(x) = \frac{1}{36}(-x^2 + 6x), \quad 0 \le x \le 6.$$

Find the expected weekly demand for this product.

SOLUTION

$$\mu = E(x) = \frac{1}{36}\int_0^6 x(-x^2 + 6x)\, dx$$

$$= \frac{1}{36}\int_0^6 (-x^3 + 6x^2)\, dx$$

$$= \frac{1}{36}\left[\frac{-x^4}{4} + 2x^3\right]_0^6$$

$$= 3$$

In Figure 9.11, you can see that an expected value of 3 seems reasonable because the region is symmetric about the line $x = 3$.

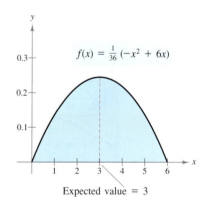

Expected value = 3

FIGURE 9.11

TRY IT 1

Find the expected value of the probability density function $f(x) = \frac{1}{32}3x(4 - x)$ on the interval $[0, 4]$.

9 SAMPLE POST-GRADUATION EXAM QUESTIONS

CPA
GMAT
GRE
Actuarial
CLAST

The following questions represent the types of questions that appear on certified public accountant (CPA) exams, Graduate Management Admission Tests (GMAT), Graduate Records Exams (GRE), actuarial exams, and College-Level Academic Skills Tests (CLAST). The answers to the questions are given in the back of the book.

In Questions 1 and 2, use the circle graph below.

Living Vietnam Veterans (in thousands)
by Age as of 2002

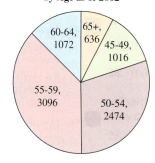

(Source: U.S. Dept. of Veterans Affairs)

1. What is the measure of the central angle of the sector for living Vietnam veterans ages 50–54?

   (a) 120°   (b) 130°   (c) 107°   (d) 128°   (e) 124°

2. The number of living Vietnam veterans age 60 or older is what percent of the total number of living Vietnam veterans?

   (a) 7.7%   (b) 20.6%   (c) 12.9%   (d) 17.0%   (e) 6.4%

3. If the mean of six numbers is 4.5, what is the sum of the numbers?

   (a) 4.5   (b) 24   (c) 27   (d) 30   (e) Cannot be determined

4. Given that a and b are real numbers, let $f(a, b) = ab$ and let $g(a) = a^2 + 2$. Then $f(3, g(3)) =$

   (a) $3a^2 + 2$ (b) $3a^2 + 6$ (c) 27   (d) 29   (e) 33

5. There are 97 men and 3 women in an organization. A committee of five people is chosen at random, and one of these people is randomly designated as chairperson. The probability that the committee includes all three women and has one of the women as chairperson is

   (a) $\dfrac{3(4!\,97!)}{2(100!)}$ (b) $\dfrac{3(5!\,97!)}{2(100!)}$ (c) $\dfrac{5!\,97!}{2(100!)}$ (d) $\dfrac{3!\,5!\,97!}{100!}$ (e) $\dfrac{3^3\,97^2}{100^5}$

6. Let x be a continuous random variable with the density function

$$f(x) = 6x(1 - x), \quad 0 < x < 1.$$

The value of $P\left(\left|x - \tfrac{1}{2}\right| > \tfrac{1}{4}\right)$ is

   (a) 0.0521   (b) 0.1563   (c) 0.3125   (d) 0.5000   (e) 0.8000

chapter

10

Series and Taylor Polynomials

- **10.1** Sequences
- **10.2** Series and Convergence
- **10.3** *p*-Series and the Ratio Test
- **10.4** Power Series and Taylor's Theorem
- **10.5** Taylor Polynomials
- **10.6** Newton's Method

Newton's Method can be applied to approximate the number of years required to maximize the present value for harvesting a tract of timber.

TOM BEER KATE WATSON/Peter Arnold, Inc.

STRATEGIES FOR SUCCESS

WHAT YOU SHOULD LEARN:

- How to find the limit of a sequence of numbers
- How to determine the convergence or divergence of an infinite series
- How to use the Ratio Test and the Convergence Test for *p*-series
- How to use Taylor's Theorem to determine the Taylor and Maclaurin series of simple functions
- How to use Taylor polynomials for approximation
- How to approximate the zeros of a function using Newton's Method

WHY YOU SHOULD LEARN IT:

Series and Taylor polynomials have many applications in real life, such as the examples below, which represent a small sample of the applications in this chapter.

- Budget Analysis, Exercise 69 on page 660
- Consumer Trends: Multiplier Effect, Exercise 51 on page 672
- Finance: Annuity, Exercise 59 on page 672
- Forestry, Exercises 45 and 46 on page 707
- Depreciation, Review Exercise 44 on page 713

The following warm-up exercises involve skills that were covered in earlier sections. You will use these skills in the exercise set for this section.

In Exercises 1 and 2, add the fractions.

1. $\frac{1}{2} + \frac{1}{3} + \frac{1}{4} + \frac{1}{5}$

2. $1 + \frac{3}{4} + \frac{4}{6} + \frac{5}{8}$

In Exercises 3–6, evaluate the expression.

3. $\dfrac{1 - \left(\frac{1}{2}\right)^5}{1 - \frac{1}{2}}$

4. $\dfrac{3\left[1 - \left(\frac{1}{3}\right)^4\right]}{1 - \frac{1}{3}}$

5. $\dfrac{2\left[1 - \left(\frac{1}{4}\right)^3\right]}{1 - \frac{1}{4}}$

6. $\dfrac{\frac{1}{2}\left[1 - \left(\frac{1}{2}\right)^5\right]}{1 - \frac{1}{2}}$

In Exercises 7–10, find the limit.

7. $\lim\limits_{n\to\infty} \dfrac{3n}{4n + 1}$

8. $\lim\limits_{n\to\infty} \dfrac{3n}{n^2 + 1}$

9. $\lim\limits_{n\to\infty} \dfrac{n!}{n! - 3}$

10. $\lim\limits_{n\to\infty} \dfrac{2n! + 1}{4n! - 1}$

EXERCISES 10.2

In Exercises 1–4, find the first five terms of the sequence of partial sums.

1. $\sum\limits_{n=1}^{\infty} \dfrac{1}{n^2} = 1 + \dfrac{1}{4} + \dfrac{1}{9} + \dfrac{1}{16} + \dfrac{1}{25} + \cdots$

2. $\sum\limits_{n=1}^{\infty} (-1)^{n+1}\dfrac{3^n}{2^{n-1}} = 3 - \dfrac{9}{2} + \dfrac{27}{4} - \dfrac{81}{8} + \dfrac{243}{16} - \cdots$

3. $\sum\limits_{n=1}^{\infty} \dfrac{3}{2^{n-1}} = 3 + \dfrac{3}{2} + \dfrac{3}{4} + \dfrac{3}{8} + \dfrac{3}{16} + \cdots$

4. $\sum\limits_{n=1}^{\infty} \dfrac{(-1)^{n+1}}{n!} = 1 - \dfrac{1}{2} + \dfrac{1}{6} - \dfrac{1}{24} + \dfrac{1}{120} - \cdots$

In Exercises 5–12, verify that the infinite series diverges.

5. $\sum\limits_{n=1}^{\infty} \dfrac{n}{n + 1} = \dfrac{1}{2} + \dfrac{2}{3} + \dfrac{3}{4} + \dfrac{4}{5} + \cdots$

6. $\sum\limits_{n=1}^{\infty} \dfrac{n}{2n + 3} = \dfrac{1}{5} + \dfrac{2}{7} + \dfrac{3}{9} + \dfrac{4}{11} + \cdots$

7. $\sum\limits_{n=1}^{\infty} \dfrac{n^2}{n^2 + 1} = \dfrac{1}{2} + \dfrac{4}{5} + \dfrac{9}{10} + \dfrac{16}{17} + \cdots$

8. $\sum\limits_{n=1}^{\infty} \dfrac{n}{\sqrt{n^2 + 1}} = \dfrac{1}{\sqrt{2}} + \dfrac{2}{\sqrt{5}} + \dfrac{3}{\sqrt{10}} + \dfrac{4}{\sqrt{17}} + \cdots$

9. $\sum\limits_{n=0}^{\infty} 3\left(\dfrac{3}{2}\right)^n = 3 + \dfrac{9}{2} + \dfrac{27}{4} + \dfrac{81}{8} + \cdots$

10. $\sum\limits_{n=0}^{\infty} 5\left(-\dfrac{3}{2}\right)^n = 5 - \dfrac{15}{2} + \dfrac{45}{4} - \dfrac{135}{8} + \cdots$

11. $\sum\limits_{n=0}^{\infty} 1000(1.055)^n = 1000 + 1055 + 1113.025 + \cdots$

12. $\sum\limits_{n=0}^{\infty} 2(-1.03)^n = 2 - 2.06 + 2.1218 - \cdots$

In Exercises 13–16, verify that the geometric series converges.

13. $\sum\limits_{n=0}^{\infty} 2\left(\dfrac{3}{4}\right)^n = 2 + \dfrac{3}{2} + \dfrac{9}{8} + \dfrac{27}{32} + \dfrac{81}{128} + \cdots$

14. $\sum\limits_{n=0}^{\infty} 2\left(-\dfrac{1}{2}\right)^n = 2 - 1 + \dfrac{1}{2} - \dfrac{1}{4} + \dfrac{1}{8} - \cdots$

15. $\sum\limits_{n=0}^{\infty} (0.9)^n = 1 + 0.9 + 0.81 + 0.729 + \cdots$

16. $\sum\limits_{n=0}^{\infty} (-0.6)^n = 1 - 0.6 + 0.36 - 0.216 + \cdots$

In Exercises 17–20, use a symbolic algebra utility to find the sum of the convergent series.

17. $\sum_{n=0}^{\infty} \left(\frac{1}{2}\right)^n = 1 + \frac{1}{2} + \frac{1}{4} + \frac{1}{8} + \cdots$

18. $\sum_{n=0}^{\infty} 2\left(\frac{2}{3}\right)^n = 2 + \frac{4}{3} + \frac{8}{9} + \frac{16}{27} + \cdots$

19. $\sum_{n=0}^{\infty} \left(-\frac{1}{2}\right)^n = 1 - \frac{1}{2} + \frac{1}{4} - \frac{1}{8} + \cdots$

20. $\sum_{n=0}^{\infty} 2\left(-\frac{2}{3}\right)^n = 2 - \frac{4}{3} + \frac{8}{9} - \frac{16}{27} + \cdots$

In Exercises 21–30, find the sum of the convergent series.

21. $\sum_{n=0}^{\infty} 2\left(\frac{1}{\sqrt{2}}\right)^n = 2 + \sqrt{2} + 1 + \frac{1}{\sqrt{2}} + \cdots$

22. $\sum_{n=0}^{\infty} 4\left(\frac{1}{4}\right)^n = 4 + 1 + \frac{1}{4} + \frac{1}{16} + \cdots$

23. $1 + 0.1 + 0.01 + 0.001 + \cdots$

24. $8 + 6 + \frac{9}{2} + \frac{27}{8} + \cdots$

25. $2 - \frac{2}{3} + \frac{2}{9} - \frac{2}{27} + \cdots$

26. $4 - 2 + 1 - \frac{1}{2} + \cdots$

27. $\sum_{n=0}^{\infty} \left(\frac{1}{2^n} - \frac{1}{3^n}\right)$

28. $\sum_{n=0}^{\infty} [(0.7)^n + (0.9)^n]$

29. $\sum_{n=0}^{\infty} \left(\frac{1}{3^n} + \frac{1}{4^n}\right)$

30. $\sum_{n=0}^{\infty} [(0.4)^n - (0.8)^n]$

In Exercises 31–40, determine the convergence or divergence of the series. Use a symbolic algebra utility to verify your result.

31. $\sum_{n=1}^{\infty} \frac{n + 10}{10n + 1}$

32. $\sum_{n=0}^{\infty} \frac{4}{2^n}$

33. $\sum_{n=1}^{\infty} \frac{n! + 1}{n!}$

34. $\sum_{n=1}^{\infty} \frac{n + 1}{2n - 1}$

35. $\sum_{n=1}^{\infty} \frac{3n - 1}{2n + 1}$

36. $\sum_{n=0}^{\infty} \frac{1}{4^n}$

37. $\sum_{n=0}^{\infty} (1.075)^n$

38. $\sum_{n=1}^{\infty} \frac{2^n}{100}$

39. $\sum_{n=0}^{\infty} \frac{3}{4^n}$

40. $\sum_{n=0}^{\infty} n!$

In Exercises 41–44, the repeating decimal is expressed as a geometric series. Find the sum of the geometric series and write the decimal as the ratio of two integers.

41. $0.\overline{6} = 0.6 + 0.06 + 0.006 + 0.0006 + \cdots$

42. $0.\overline{23} = 0.23 + 0.0023 + 0.000023 + \cdots$

43. $0.\overline{81} = 0.81 + 0.0081 + 0.000081 + \cdots$

44. $0.\overline{21} = 0.21 + 0.0021 + 0.000021 + \cdots$

45. *Sales*   A company produces a new product for which it estimates the annual sales to be 8000 units. Suppose that in any given year 10% of the units (regardless of age) will become inoperative.

(a) How many units will be in use after n years?

(b) Find the market stabilization level of the product.

46. *Sales*   Repeat Exercise 45 with the assumption that 25% of the units will become inoperative each year.

47. *Physical Science*   A ball is dropped from a height of 16 feet. Each time it drops h feet, it rebounds $0.64h$ feet. Find the total vertical distance traveled by the ball.

48. *Physical Science*   The ball in Exercise 47 takes the times listed below for each fall. (t is measured in seconds.)

$s_1 = -16t^2 + 16$	$s_1 = 0$ if $t = 1$
$s_2 = -16t^2 + 16(0.64)$	$s_2 = 0$ if $t = 0.8$
$s_3 = -16t^2 + 16(0.64)^2$	$s_3 = 0$ if $t = (0.8)^2$
$s_4 = -16t^2 + 16(0.64)^3$	$s_4 = 0$ if $t = (0.8)^3$
$\vdots$	$\vdots$
$s_n = -16t^2 + 16(0.64)^{n-1}$	$s_n = 0$ if $t = (0.8)^{n-1}$

Beginning with s_2, the ball takes the same amount of time to bounce up as to fall, and so the total elapsed time before the ball comes to rest is

$$t = 1 + 2\sum_{n=1}^{\infty} (0.8)^n.$$

Find the total time it takes for the ball to come to rest.

 49. *Annuity*   A deposit of $100 is made at the beginning of each month for 5 years in an account that pays 10% interest, compounded monthly. Use a symbolic algebra utility to find the balance A in the account at the end of the 5 years.

$$A = 100\left(1 + \frac{0.10}{12}\right) + \cdots + 100\left(1 + \frac{0.10}{12}\right)^{60}$$

50. *Annuity*   A deposit of P dollars is made every month for t years in an account that pays an annual interest rate of $r\%$, compounded monthly. Let $N = 12t$ be the total number of deposits. Show that the balance in the account after t years is

$$A = P\left[\left(1 + \frac{r}{12}\right)^N - 1\right]\left(1 + \frac{12}{r}\right), \quad t > 0.$$

51. *Consumer Trends: Multiplier Effect* The annual spending by tourists in a resort city is 100 million dollars. Approximately 75% of that revenue is again spent in the resort city, and of that amount approximately 75% is again spent in the resort city. If this pattern continues, write the geometric series that gives the total amount of spending generated by the 100 million dollars and find the sum of the series.

52. *Area* Find the fraction of the total area of the square that is eventually shaded blue if the pattern of shading shown in the accompanying figure is continued. (Note that the lengths of the sides of the shaded corner squares are one-fourth the lengths of the sides of the squares in which they are placed.)

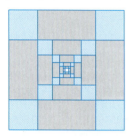

53. *Wages* Suppose that an employer offered to pay you 1¢ the first day, and then double your wages each day thereafter. Find your total wages for working 20 days.

54. *Probability: Coin Toss* A fair coin is tossed until a head appears. The probability that the first head appears on the nth toss is given by $P = \left(\frac{1}{2}\right)^n$, where $n \geq 1$. Show that

$$\sum_{n=1}^{\infty} \left(\frac{1}{2}\right)^n = 1.$$

55. *Probability: Coin Toss* Use a symbolic algebra utility to estimate the expected number of tosses required until the first head occurs in the experiment in Exercise 54.

56. *Profit* The annual profits for eBay from 1998 through 2003 can be approximated by the model

$$a_n = 5.981e^{0.722n}, \qquad n = 1, 2, 3, \ldots, 6$$

where a_n is the annual net profit (in millions of dollars) and n is the year, with $n = 1$ corresponding to 1998. Use the formula for the sum of a geometric sequence to approximate the total net profits earned during this 6-year period. *(Source: eBay, Inc.)*

57. *Environment* A factory is polluting a river such that at every mile down river from the factory an environmental expert finds 15% less pollutant than at the preceding mile. If the pollutant's concentration is 500 ppm at the factory, what is its concentration 12 miles down river?

58. *Physical Science* In a certain brand of CD player, after the STOP function is activated, the disc, during each second after the first second, makes 85% fewer revolutions than it made during the preceding second. In coming to rest, how many revolutions does the disc make if it makes 5.5 revolutions during the first second after the STOP function is activated?

59. *Finance: Annuity* The simplest kind of annuity is a straight-line annuity, which pays a fixed amount per month until the annuitant dies. Suppose that, when he turns 65, Bob wants to purchase a straight-line annuity that has a premium of $100,000 and pays $880 per month. Use sigma notation to represent each scenario below, and give the numerical amount that the summation represents. *(Source: Adapted from Garman/Forgue,* Personal Finance, *Fifth Edition)*

(a) Suppose Bob dies 10 months after he takes out the annuity. How much will he have collected up to that point?

(b) Suppose Bob lives the average number of months beyond age 65 for a man (168 months). How much more or less than the $100,000 will he have collected?

60. *Area* A news reporter is being televised in such a way that an inset of the same camera view is shown in the corner of the screen above the reporter's left shoulder. Another inset view appears in this inset view, and so forth for all subsequent inset views. If in the principal camera view, the reporter occupies approximately 80 square inches of the screen, and if each repeating corner view is $\frac{1}{8}$ of the area of the preceding view, what is the total area of the screen occupied by the news reporter?

In Exercises 61–66, use a symbolic algebra utility to evaluate the summation.

61. $\displaystyle\sum_{n=1}^{\infty} n^2 \left(\frac{1}{2}\right)^n$

62. $\displaystyle\sum_{n=1}^{\infty} 2n^3 \left(\frac{1}{5}\right)^n$

63. $\displaystyle\sum_{n=1}^{\infty} \frac{1}{(2n)!}$

64. $\displaystyle\sum_{n=1}^{\infty} n \left(\frac{4}{11}\right)^n$

65. $\displaystyle\sum_{n=1}^{\infty} e^2 \left(\frac{1}{e}\right)^n$

66. $\displaystyle\sum_{n=1}^{\infty} \ln 2 \left(\frac{1}{8}\right)^{2n}$

True or False? In Exercises 67 and 68, determine whether the statement is true or false. If it is false, explain why or give an example that shows it is false.

67. If $\displaystyle\lim_{n\to\infty} a_n = 0$, then $\displaystyle\sum_{n=1}^{\infty} a_n$ converges.

68. If $|r| < 1$, then $\displaystyle\sum_{n=1}^{\infty} ar^n = \frac{a}{1-r}$.

10.3 *p*-SERIES AND THE RATIO TEST

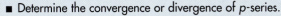

- Determine the convergence or divergence of *p*-series.
- Use the Ratio Test to determine the convergence or divergence of series.

p-Series

In Section 10.2, you studied geometric series. In this section you will study another common type of series called a ***p*-series.**

> **Definition of *p*-Series**
>
> Let *p* be a positive constant. An infinite series of the form
>
> $$\sum_{n=1}^{\infty} \frac{1}{n^p} = \frac{1}{1^p} + \frac{1}{2^p} + \frac{1}{3^p} + \cdots$$
>
> is called a ***p*-series.** If $p = 1$, then the series
>
> $$\sum_{n=1}^{\infty} \frac{1}{n} = 1 + \frac{1}{2} + \frac{1}{3} + \cdots$$
>
> is called the **harmonic series.**

EXAMPLE 1 Classifying *p*-Series

Classify each infinite series.

(a) $\displaystyle\sum_{n=1}^{\infty} \frac{1}{n^3}$

(b) $\displaystyle\sum_{n=1}^{\infty} \frac{1}{\sqrt{n}}$

(c) $\displaystyle\sum_{n=1}^{\infty} \frac{1}{3^n}$

SOLUTION

(a) The infinite series

$$\sum_{n=1}^{\infty} \frac{1}{n^3} = \frac{1}{1^3} + \frac{1}{2^3} + \frac{1}{3^3} + \cdots$$

is a *p*-series with $p = 3$.

(b) The infinite series

$$\sum_{n=1}^{\infty} \frac{1}{\sqrt{n}} = \frac{1}{1^{1/2}} + \frac{1}{2^{1/2}} + \frac{1}{3^{1/2}} + \cdots$$

is a *p*-series with $p = \frac{1}{2}$.

(c) The infinite series

$$\sum_{n=1}^{\infty} \frac{1}{3^n} = \frac{1}{3^1} + \frac{1}{3^2} + \frac{1}{3^3} + \cdots$$

is *not* a *p*-series. It is a geometric series.

> **TRY IT 1**
>
> Classify each infinite series.
>
> (a) $\displaystyle\sum_{n=1}^{\infty} \frac{1}{n^\pi}$
>
> (b) $\displaystyle\sum_{n=1}^{\infty} \frac{1}{n\sqrt{n}}$
>
> (c) $\displaystyle\sum_{n=1}^{\infty} \frac{1}{2^n}$

Some infinite p-series converge and others diverge. With the test below, you can determine the convergence or divergence of a p-series.

Test for Convergence of a p-Series

Consider the p-series

$$\sum_{n=1}^{\infty} \frac{1}{n^p} = \frac{1}{1^p} + \frac{1}{2^p} + \frac{1}{3^p} + \cdots .$$

1. The series diverges if $0 < p \le 1$.

2. The series converges if $p > 1$.

DISCOVERY

It may seem surprising that the harmonic series

$$\sum_{n=1}^{\infty} \frac{1}{n} = 1 + \frac{1}{2} + \frac{1}{3} + \cdots$$

diverges even though its individual terms approach zero. Verify that this series diverges by showing that

$$1 + \frac{1}{2} > \frac{1}{2}$$

$$\frac{1}{3} + \frac{1}{4} > \frac{1}{2}$$

$$\frac{1}{5} + \cdots + \frac{1}{8} > \frac{1}{2}$$

$$\frac{1}{9} + \cdots + \frac{1}{16} > \frac{1}{2}.$$

EXAMPLE 2   **Determining Convergence or Divergence**

Determine whether each p-series converges or diverges.

(a) $\displaystyle\sum_{n=1}^{\infty} \frac{1}{n^{0.9}}$ (b) $\displaystyle\sum_{n=1}^{\infty} \frac{1}{n}$ (c) $\displaystyle\sum_{n=1}^{\infty} \frac{1}{n^{1.1}}$

SOLUTION

(a) For the p-series

$$\sum_{n=1}^{\infty} \frac{1}{n^{0.9}} = \frac{1}{1^{0.9}} + \frac{1}{2^{0.9}} + \frac{1}{3^{0.9}} + \cdots$$

$p = 0.9$. Because $p \le 1$, you can conclude that the series diverges.

(b) For the p-series

$$\sum_{n=1}^{\infty} \frac{1}{n} = \frac{1}{1} + \frac{1}{2} + \frac{1}{3} + \cdots$$

$p = 1$, which means that the series is the *harmonic* series. Because $p \le 1$, you can conclude that the series diverges.

(c) For the p-series

$$\sum_{n=1}^{\infty} \frac{1}{n^{1.1}} = \frac{1}{1^{1.1}} + \frac{1}{2^{1.1}} + \frac{1}{3^{1.1}} + \cdots$$

$p = 1.1$. Because $p > 1$, you can conclude that the series converges.

In Example 2, notice that the p-Series Test tells you only whether the series diverges or converges. It does not give a formula for the sum of a convergent p-series. To approximate such a sum, you can use a computer to evaluate several partial sums. More is said about this on page 675.

TRY IT 2

Determine whether each p-series converges or diverges.

(a) $\displaystyle\sum_{n=1}^{\infty} \frac{1}{n\sqrt{n}}$

(b) $\displaystyle\sum_{n=1}^{\infty} \frac{1}{n^{2.5}}$

(c) $\displaystyle\sum_{n=1}^{\infty} \frac{1}{n^{1/10}}$

TECHNOLOGY

Approximating the Sum of a p-Series

 It can be shown that the sum S of a convergent p-series differs from its Nth partial sum by no more than

$$\frac{1}{(p-1)N^{p-1}}.$$ Maximum error

When p is greater than or equal to 3, this means that you can approximate the sum of the convergent p-series by adding several of the terms. For instance, because

$$\sum_{n=1}^{10} \frac{1}{n^3} \approx 1.19753$$ Tenth partial sum

and

$$\frac{1}{(2)(10^2)} = 0.005$$

you can conclude that the sum of the infinite p-series with $p = 3$ is between 1.19753 and $(1.19753 + 0.005)$ or 1.20253. For convergent p-series in which p is less than 3, you need to add more and more terms of the series to obtain a reasonable approximation of the sum. For instance, when $p = 2$, you can use a computer to find the sums below.

$$\sum_{n=1}^{10} \frac{1}{n^2} \approx 1.54977$$ Tenth partial sum

$$\sum_{n=1}^{100} \frac{1}{n^2} \approx 1.63498$$ 100th partial sum

$$\sum_{n=1}^{1000} \frac{1}{n^2} \approx 1.64393$$ 1000th partial sum

Because

$$\frac{1}{(1)(1000)} = 0.001$$

you can conclude that the partial sum S_{1000} is within 0.001 of the actual sum of the series.

When p is close to 1, approximating the sum of the series becomes difficult. For instance, consider the partial sums below.

$$\sum_{n=1}^{10} \frac{1}{n^{1.1}} \approx 2.680155$$ Tenth partial sum

$$\sum_{n=1}^{100} \frac{1}{n^{1.1}} \approx 4.278024$$ 100th partial sum

$$\sum_{n=1}^{1000} \frac{1}{n^{1.1}} \approx 5.572827$$ 1000th partial sum

Because $1/[(0.1)(1000^{0.1})] \approx 5$, you can see that even the partial sum S_{1000} is not very close to the actual sum of the series.

The Ratio Test

At this point, you have studied two convergence tests: one for a geometric series and one for a *p*-series. The next test is more general: it can be applied to infinite series that do not happen to be geometric series or *p*-series.

The Ratio Test

Let $\displaystyle\sum_{n=1}^{\infty} a_n$ be an infinite series with nonzero terms.

1. The series converges if $\displaystyle\lim_{n\to\infty} \left| \frac{a_{n+1}}{a_n} \right| < 1.$

2. The series diverges if $\displaystyle\lim_{n\to\infty} \left| \frac{a_{n+1}}{a_n} \right| > 1.$

3. The test is inconclusive if $\displaystyle\lim_{n\to\infty} \left| \frac{a_{n+1}}{a_n} \right| = 1.$

The Ratio Test is particularly useful for series that converge rapidly. Series involving factorial or exponential functions are frequently of this type.

EXAMPLE 3   **Using the Ratio Test**

Determine the convergence or divergence of the infinite series

$$\sum_{n=0}^{\infty} \frac{2^n}{n!} = \frac{1}{1} + \frac{2}{1} + \frac{4}{2} + \frac{8}{6} + \frac{16}{24} + \frac{32}{120} + \cdots.$$

SOLUTION   Using the Ratio Test with $a_n = 2^n/n!$, you obtain

$$\lim_{n\to\infty} \left| \frac{a_{n+1}}{a_n} \right| = \lim_{n\to\infty} \left[\frac{2^{n+1}}{(n+1)!} \div \frac{2^n}{n!} \right]$$

$$= \lim_{n\to\infty} \left[\frac{2^{n+1}}{(n+1)!} \cdot \frac{n!}{2^n} \right]$$

$$= \lim_{n\to\infty} \frac{2}{n+1}$$

$$= 0.$$

Because this limit is less than 1, you can apply the Ratio Test to conclude that the series converges. Using a computer, you can approximate the sum of the series to be $S \approx S_{10} \approx 7.39$.

TRY IT 3

Determine the convergence or divergence of the infinite series

$$\sum_{n=0}^{\infty} \frac{3^n}{n!}.$$

Example 3 tells you something about the rates at which the sequences $\{2^n\}$ and $\{n!\}$ increase as n approaches infinity. For example, in the table below, you can see that although the factorial sequence $\{n!\}$ has a slow start, it quickly overpowers the exponential sequence $\{2^n\}$.

n	0	1	2	3	4	5	6	7	8	9
2^n	1	2	4	8	16	32	64	128	256	512
$n!$	1	1	2	6	24	120	720	5040	40,320	362,880

From this table, you can also see that the sequence $\{n\}$ approaches infinity more slowly than the sequence $\{2^n\}$. This is further demonstrated in Example 4.

EXAMPLE 4   Using the Ratio Test

Determine the convergence or divergence of the infinite series

$$\sum_{n=1}^{\infty} \frac{n}{2^n} = \frac{1}{2} + \frac{2}{4} + \frac{3}{8} + \frac{4}{16} + \frac{5}{32} + \frac{6}{64} + \cdots .$$

SOLUTION   Using the Ratio Test with $a_n = n/2^n$, you obtain

$$\lim_{n\to\infty} \left| \frac{a_{n+1}}{a_n} \right| = \lim_{n\to\infty} \left(\frac{n+1}{2^{n+1}} \div \frac{n}{2^n} \right)$$

$$= \lim_{n\to\infty} \left(\frac{n+1}{2^{n+1}} \cdot \frac{2^n}{n} \right)$$

$$= \lim_{n\to\infty} \frac{n+1}{2n} = \frac{1}{2}.$$

Because this limit is less than 1, you can apply the Ratio Test to conclude that the series converges. Using a computer, you can determine that the sum of the series is $S = 2$.

> **TRY IT 4**
>
> Determine the convergence or divergence of the infinite series
>
> $$\sum_{n=1}^{\infty} \frac{5^n}{n^2}.$$

When applying the Ratio Test, remember that if the limit of

$$\left| \frac{a_{n+1}}{a_n} \right|$$

as $n \to \infty$ is 1, then the test does not tell you whether the series converges or diverges. This type of result often occurs with series that converge or diverge slowly. For instance, when you apply the Ratio Test to the harmonic series in which

$$a_n = \frac{1}{n}$$

you obtain

$$\lim_{n\to\infty} \left| \frac{a_{n+1}}{a_n} \right| = \lim_{n\to\infty} \frac{1/(n+1)}{1/n} = \lim_{n\to\infty} \frac{n}{n+1} = 1.$$

So, from the Ratio Test, you cannot conclude that the harmonic series diverges. From the *p*-Series Test, you know that it diverges.

■ **EXAMPLE 5**   **Using the Ratio Test**

Determine the convergence or divergence of the infinite series

$$\sum_{n=1}^{\infty} \frac{2^n}{n^2} = \frac{2}{1} + \frac{4}{4} + \frac{8}{9} + \frac{16}{16} + \frac{32}{25} + \frac{64}{36} + \cdots.$$

SOLUTION   Using the Ratio Test with $a_n = 2^n/n^2$, you obtain

$$\lim_{n\to\infty} \left| \frac{a_{n+1}}{a_n} \right| = \lim_{n\to\infty} \left[\frac{2^{n+1}}{(n+1)^2} \div \frac{2^n}{n^2} \right]$$

$$= \lim_{n\to\infty} \left[\frac{2^{n+1}}{(n+1)^2} \cdot \frac{n^2}{2^n} \right]$$

$$= \lim_{n\to\infty} 2 \left(\frac{n}{n+1} \right)^2$$

$$= 2.$$

Because this limit is greater than 1, you can apply the Ratio Test to conclude that the series diverges.

TRY IT 5

Determine the convergence or divergence of the infinite series

$$\sum_{n=1}^{\infty} \frac{n!}{10^n}.$$

Summary of Tests of Series

Test	Series	Converges	Diverges
nth-Term	$\sum_{n=1}^{\infty} a_n$	No test	$\lim_{n\to\infty} a_n \neq 0$
Geometric	$\sum_{n=0}^{\infty} ar^n$	$\|r\| < 1$	$\|r\| \geq 1$
p-Series	$\sum_{n=1}^{\infty} \frac{1}{n^p}$	$p > 1$	$0 < p \leq 1$
Ratio	$\sum_{n=1}^{\infty} a_n$	$\lim_{n\to\infty} \left\| \frac{a_{n+1}}{a_n} \right\| < 1$	$\lim_{n\to\infty} \left\| \frac{a_{n+1}}{a_n} \right\| > 1$

TAKE ANOTHER LOOK

Comparing Tests

Determine whether each infinite series converges or diverges. Identify the test used and explain your reasoning. If the series converges, use a computer to confirm your reasoning by finding the 1000th partial sum.

a. $\sum_{n=1}^{\infty} \frac{5}{n!}$

b. $\sum_{n=1}^{\infty} \frac{5}{2^n}$

c. $\sum_{n=1}^{\infty} \frac{5}{n^2}$

PREREQUISITE REVIEW 10.3

The following warm-up exercises involve skills that were covered in earlier sections. You will use these skills in the exercise set for this section.

In Exercises 1–4, simplify the expression.

1. $\dfrac{n!}{(n + 1)!}$

2. $\dfrac{(n + 1)!}{n!}$

3. $\dfrac{3^{n+1}}{n + 1} \cdot \dfrac{n}{3^n}$

4. $\dfrac{(n + 1)^2}{(n + 1)!} \cdot \dfrac{n!}{n^2}$

In Exercises 5–8, find the limit.

5. $\lim\limits_{n \to \infty} \dfrac{(n + 1)^2}{n^2}$

6. $\lim\limits_{n \to \infty} \dfrac{5^{n+1}}{5^n}$

7. $\lim\limits_{n \to \infty} \left(\dfrac{5}{n + 1} \div \dfrac{5}{n} \right)$

8. $\lim\limits_{n \to \infty} \left(\dfrac{(n + 1)^3}{3^{n+1}} \div \dfrac{n^3}{3^n} \right)$

In Exercises 9 and 10, decide whether the series is geometric.

9. $\displaystyle\sum_{n=1}^{\infty} \dfrac{1}{4^n}$

10. $\displaystyle\sum_{n=1}^{\infty} \dfrac{1}{n^4}$

EXERCISES 10.3

In Exercises 1–6, determine whether the series is a *p*-series.

1. $\displaystyle\sum_{n=1}^{\infty} \dfrac{1}{n^2}$

2. $\displaystyle\sum_{n=1}^{\infty} \dfrac{1}{\sqrt{n}}$

3. $\displaystyle\sum_{n=1}^{\infty} \dfrac{1}{3^n}$

4. $\displaystyle\sum_{n=1}^{\infty} n^{-3/4}$

5. $\displaystyle\sum_{n=1}^{\infty} \dfrac{1}{n^n}$

6. $\displaystyle\sum_{n=1}^{\infty} \dfrac{1}{n + 1}$

In Exercises 7–16, determine the convergence or divergence of the *p*-series.

7. $\displaystyle\sum_{n=1}^{\infty} \dfrac{1}{n^3}$

8. $\displaystyle\sum_{n=1}^{\infty} \dfrac{1}{n^{1/3}}$

9. $\displaystyle\sum_{n=1}^{\infty} \dfrac{1}{\sqrt[3]{n}}$

10. $\displaystyle\sum_{n=1}^{\infty} \dfrac{1}{n^{4/3}}$

11. $\displaystyle\sum_{n=1}^{\infty} \dfrac{1}{n^{1.03}}$

12. $\displaystyle\sum_{n=1}^{\infty} \dfrac{1}{n^{\pi}}$

13. $1 + \dfrac{1}{\sqrt{2}} + \dfrac{1}{\sqrt{3}} + \dfrac{1}{\sqrt{4}} + \cdots$

14. $1 + \dfrac{1}{4} + \dfrac{1}{9} + \dfrac{1}{16} + \dfrac{1}{25} + \cdots$

15. $1 + \dfrac{1}{2\sqrt{2}} + \dfrac{1}{3\sqrt{3}} + \dfrac{1}{4\sqrt{4}} + \cdots$

16. $1 + \dfrac{1}{2} + \dfrac{1}{3} + \dfrac{1}{4} + \dfrac{1}{5} + \cdots$

In Exercises 17–30, use the Ratio Test to determine the convergence or divergence of the series.

17. $\displaystyle\sum_{n=0}^{\infty} \dfrac{3^n}{n!}$

18. $\displaystyle\sum_{n=1}^{\infty} n \left(\dfrac{2}{3} \right)^n$

19. $\displaystyle\sum_{n=0}^{\infty} \dfrac{n!}{3^n}$

20. $\displaystyle\sum_{n=1}^{\infty} n \left(\dfrac{3}{2} \right)^n$

21. $\displaystyle\sum_{n=1}^{\infty} \dfrac{n}{4^n}$

22. $\displaystyle\sum_{n=1}^{\infty} \dfrac{n^2}{2^n}$

23. $\displaystyle\sum_{n=1}^{\infty} \dfrac{2^n}{n^5}$

24. $\displaystyle\sum_{n=0}^{\infty} (-1)^n e^{-n}$

25. $\displaystyle\sum_{n=0}^{\infty} \frac{(-1)^n 2^n}{n!}$

26. $\displaystyle\sum_{n=0}^{\infty} \frac{4n}{n!}$

27. $\displaystyle\sum_{n=0}^{\infty} \frac{4^n}{3^n + 1}$

28. $\displaystyle\sum_{n=0}^{\infty} \frac{3^n}{n + 1}$

29. $\displaystyle\sum_{n=0}^{\infty} \frac{n5^n}{n!}$

30. $\displaystyle\sum_{n=1}^{\infty} \frac{2n!}{n^5}$

In Exercises 31–34, approximate the sum of the convergent series using the indicated number of terms. Estimate the maximum error of your approximation.

31. $\displaystyle\sum_{n=1}^{\infty} \frac{1}{n^3}$, four terms

32. $\displaystyle\sum_{n=1}^{\infty} \frac{1}{n^4}$, four terms

33. $\displaystyle\sum_{n=1}^{\infty} \frac{1}{n^{3/2}}$, 10 terms

34. $\displaystyle\sum_{n=1}^{\infty} \frac{1}{n^{10}}$, three terms

In Exercises 35 and 36, verify that the Ratio Test is inconclusive for the p-series.

35. $\displaystyle\sum_{n=1}^{\infty} \frac{1}{n^{3/2}}$

36. $\displaystyle\sum_{n=1}^{\infty} \frac{1}{n^3}$

In Exercises 37–40, match the series with the graph of its sequence of partial sums. [The graphs are labeled (a)–(d).] Determine the convergence or divergence of the series.

(a)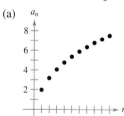

(b)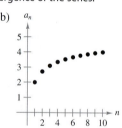

(c)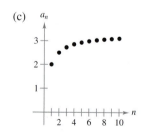

(d)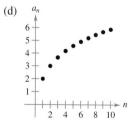

37. $\displaystyle\sum_{n=1}^{\infty} \frac{2}{\sqrt[4]{n^3}}$

38. $\displaystyle\sum_{n=1}^{\infty} \frac{2}{n}$

39. $\displaystyle\sum_{n=1}^{\infty} \frac{2}{n\sqrt{n}}$

40. $\displaystyle\sum_{n=1}^{\infty} \frac{2}{n^2}$

In Exercises 41–56, test the series for convergence or divergence using any appropriate test from this chapter. Identify the test used and explain your reasoning. If the series is convergent, find the sum whenever possible.

41. $\displaystyle\sum_{n=1}^{\infty} \frac{2n}{n + 1}$

42. $\displaystyle\sum_{n=1}^{\infty} \frac{10}{3\sqrt[3]{n^2}}$

43. $\displaystyle\sum_{n=1}^{\infty} \frac{1}{n\sqrt[3]{n}}$

44. $\displaystyle\sum_{n=0}^{\infty} \left(\frac{5}{6}\right)^n$

45. $\displaystyle\sum_{n=0}^{\infty} \frac{(-1)^n 2^n}{3^n}$

46. $\displaystyle\sum_{n=2}^{\infty} \ln n$

47. $\displaystyle\sum_{n=1}^{\infty} \left(\frac{1}{n^2} - \frac{1}{n^3}\right)$

48. $\displaystyle\sum_{n=1}^{\infty} \frac{n3^n}{n!}$

49. $\displaystyle\sum_{n=0}^{\infty} \left(\frac{5}{4}\right)^n$

50. $\displaystyle\sum_{n=1}^{\infty} n(0.4)^n$

51. $\displaystyle\sum_{n=1}^{\infty} \frac{n!}{3^{n-1}}$

52. $\displaystyle\sum_{n=1}^{\infty} \frac{1}{n^{0.95}}$

53. $\displaystyle\sum_{n=1}^{\infty} \frac{2^n}{2^{n+1} + 1}$

54. $\displaystyle\sum_{n=1}^{\infty} 2e^{-n}$

55. $\displaystyle\sum_{n=1}^{\infty} \frac{2^n}{5^{n-1}}$

56. $\displaystyle\sum_{n=1}^{\infty} \frac{n^2}{n^2 + 1}$

In Exercises 57 and 58, use a computer to confirm the sum of the convergent series.

57. $\displaystyle\sum_{n=1}^{\infty} \frac{1}{n^2} = \frac{\pi^2}{6}$

58. $\displaystyle\sum_{n=1}^{\infty} \frac{1}{(2n)^2} = \frac{\pi^2}{24}$

59. Expenses The table lists the research and development expenses at universities for life sciences (in billions of dollars) for the years 1996 through 2001. *(Source: U.S. National Science Foundation)*

Year	1996	1997	1998	1999	2000	2001
Amount	12.70	13.61	14.55	15.59	17.46	19.19

(a) Find a quadratic model for the data with $n = 1$ corresponding to 1996. Use this quadratic model to find an infinite series to model the data.

(b) Can you determine whether the series is converging or diverging by using the Ratio Test? Explain.

60. Research Project Use your school's library, the Internet, or some other reference source to find data about a college or university that can be modeled by a series. Write a summary of your findings, including a description of the convergence or divergence of the series and how it was determined.

10.4   POWER SERIES AND TAYLOR'S THEOREM

- Recognize power series.
- Find the radii of convergence of power series.
- Use Taylor's Theorem to find power series for functions.
- Use the basic list of power series to find power series for functions.

Power Series

In the preceding two sections, you studied infinite series whose terms are constants. In this section, you will study infinite series that have variable terms. Specifically, you will study a type of infinite series that is called a **power series.** Informally, you can think of a power series as a "very long" polynomial.

Definition of Power Series

An infinite series of the form

$$\sum_{n=0}^{\infty} a_n x^n = a_0 + a_1 x + a_2 x^2 + a_3 x^3 + \cdots + a_n x^n + \cdots$$

is called a **power series** in x. An infinite series of the form

$$\sum_{n=0}^{\infty} a_n (x - c)^n = a_0 + a_1 (x - c) + a_2 (x - c)^2 + \cdots +$$

$$a_n (x - c)^n + \cdots$$

is called a **power series centered at c.**

> **STUDY TIP**
>
> The index of a power series usually begins with $n = 0$. In such cases, it is agreed that $(x - c)^0 = 1$ even if $x = c$.

EXAMPLE 1   **Writing Power Series**

(a) The power series

$$\sum_{n=0}^{\infty} \frac{x^n}{n!} = 1 + x + \frac{x^2}{2!} + \frac{x^3}{3!} + \cdots$$

is centered at zero.

(b) The power series

$$\sum_{n=1}^{\infty} \frac{(x - 1)^n}{n} = (x - 1) + \frac{(x - 1)^2}{2} + \frac{(x - 1)^3}{3} + \cdots$$

is centered at 1.

(c) The power series

$$\sum_{n=1}^{\infty} \frac{(x + 1)^n}{n} = (x + 1) + \frac{(x + 1)^2}{2} + \frac{(x + 1)^3}{3} + \cdots$$

is centered at -1.

> **TRY IT 1**
>
> Identify the center of each power series.
>
> (a) $\displaystyle\sum_{n=1}^{\infty} \frac{(x - 2)^n}{n^2}$
>
> (b) $\displaystyle\sum_{n=1}^{\infty} \frac{(x + 3)^n}{3^n}$
>
> (c) $\displaystyle\sum_{n=1}^{\infty} \frac{(-1)^n x^{2n}}{(2n)!}$

Radius of Convergence of a Power Series

A power series centered at c can be viewed as a function of x

$$f(x) = \sum_{n=0}^{\infty} a_n(x - c)^n$$

where the **domain** of f is the set of all x for which the power series converges. Determining this domain is one of the primary problems associated with power series. Every power series converges at its center c, because

$$f(c) = \sum_{n=0}^{\infty} a_n(c - c)^n$$

$$= a_0(1) + 0 + 0 + \cdots$$

$$= a_0.$$

So, c is always in the domain of f. It can happen that the domain of f consists only of the single number c. In fact, this is one of the three basic types of domains of power series, as indicated in Figure 10.5.

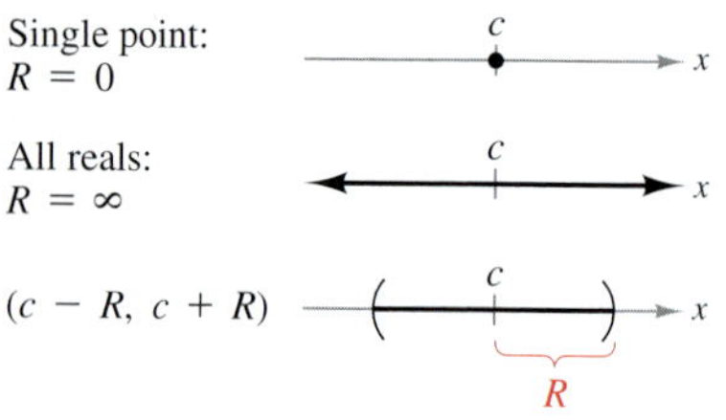

Single point:
$R = 0$

All reals:
$R = \infty$

$(c - R, c + R)$

FIGURE 10.5   Three Types of Domains

Convergence of a Power Series

For a power series centered at c, precisely one of the statements below is true.

1. The series converges only at $x = c$.

2. The series converges for all x.

3. There exists a positive real number R such that the series converges for $|x - c| < R$ and diverges for $|x - c| > R$.

In the third type of domain, the number R is called the **radius of convergence** of the power series. For the first and second types of domains, the radii of convergence are considered to be 0 and ∞, respectively.

In the third case, the series converges in the interval

$$(c - R, c + R)$$

and diverges in the intervals $(-\infty, c - R)$ and $(c + R, \infty)$. Determining the convergence or divergence at the endpoints $c - R$ and $c + R$ can be difficult, and, except for simple cases, the endpoint question is left open. To find the radius of convergence of a power series, use the Ratio Test, as illustrated in Examples 2 and 3.

EXAMPLE 2   **Finding the Radius of Convergence**

Find the radius of convergence of the power series

$$\sum_{n=0}^{\infty} \frac{x^n}{n!}.$$

SOLUTION   For this power series, $a_n = 1/n!$. So, you have

$$\lim_{n \to \infty} \left| \frac{a_{n+1} x^{n+1}}{a_n x^n} \right| = \lim_{n \to \infty} \left| \frac{x^{n+1}/(n+1)!}{x^n/n!} \right|$$

$$= \lim_{n \to \infty} \left| \frac{x}{n+1} \right|$$

$$= 0.$$

So, by the Ratio Test, this series converges for all x, and the radius of convergence is infinite.

TRY IT 2

Find the radius of convergence of the power series $\sum_{n=0}^{\infty} \frac{x^{2n}}{(2n)!}.$

EXAMPLE 3   **Finding the Radius of Convergence**

Find the radius of convergence of the power series

$$\sum_{n=0}^{\infty} \frac{(-1)^n (x+1)^n}{2^n}.$$

SOLUTION   For this power series, $a_n = (-1)^n/2^n$. So, you have

$$\lim_{n \to \infty} \left| \frac{a_{n+1}(x+1)^{n+1}}{a_n(x+1)^n} \right| = \lim_{n \to \infty} \left| \frac{(-1)^{n+1}(x+1)^{n+1}/2^{n+1}}{(-1)^n(x+1)^n/2^n} \right|$$

$$= \lim_{n \to \infty} \left| \frac{(-1)(x+1)}{2} \right|$$

$$= \lim_{n \to \infty} \left| \frac{x+1}{2} \right|$$

$$= \left| \frac{x+1}{2} \right|.$$

By the Ratio Test, this series will converge as long as $|(x+1)/2| < 1$. Because the center of convergence is $x = -1$, the series converges in the interval $(-3, 1)$, and the radius of convergence is $R = 2$.

DISCOVERY

It is possible to show that the power series in Example 3 diverges at both of the endpoints of its interval of convergence. Explain.

TRY IT 3

Find the radius of convergence of the power series $\sum_{n=0}^{\infty} \frac{(-1)^n (x-2)^n}{3^n}.$

EXAMPLE 8 **Using the Basic List of Power Series**

Find the power series for each function.

(a) $f(x) = 1 + 2e^x$, centered at 0

(b) $f(x) = e^{2x+1}$, centered at 0

(c) $f(x) = \ln 2x$, centered at 1

SOLUTION

(a) To find the power series for this function, you can multiply the series for e^x by 2 and add 1.

$$1 + 2e^x = 1 + 2\left(1 + x + \frac{x^2}{2!} + \frac{x^3}{3!} + \frac{x^4}{4!} + \cdots\right)$$

$$= 1 + 2 + 2x + \frac{2x^2}{2!} + \frac{2x^3}{3!} + \frac{2x^4}{4!} + \cdots$$

$$= 1 + \sum_{n=0}^{\infty} \frac{2x^n}{n!}$$

(b) To find the power series for $e^{2x+1} = e^{2x}e$, you can substitute $2x$ for x in the series for e^x and multiply the result by e.

$$e^{2x+1} = e^{2x}e$$

$$= e\left[1 + (2x) + \frac{(2x)^2}{2!} + \frac{(2x)^3}{3!} + \frac{(2x)^4}{4!} + \cdots\right]$$

$$= e\sum_{n=0}^{\infty} \frac{2^n x^n}{n!}$$

(c) To find the power series for $\ln 2x$ centered at 1, use the properties of logarithms.

$$\ln 2x = \ln 2 + \ln x$$

$$= \ln 2 + (x - 1) - \frac{(x-1)^2}{2} + \frac{(x-1)^3}{3} - \frac{(x-1)^4}{4} + \cdots$$

$$= \ln 2 + \sum_{n=1}^{\infty} \frac{(-1)^{n-1}(x-1)^n}{n}$$

TAKE ANOTHER LOOK

Using the Basic List of Power Series

Explain how to use a power series from the basic list to find the power series for each function.

a. $f(x) = 3 - 5e^{-x}$, centered at 0

b. $f(x) = \dfrac{2}{x-2}$, centered at 3

c. $f(x) = \ln \dfrac{4}{x}$, centered at 1

PREREQUISITE REVIEW 10.4 The following warm-up exercises involve skills that were covered in earlier sections. You will use these skills in the exercise set for this section.

In Exercises 1–4, find $f(g(x))$ and $g(f(x))$.

1. $f(x) = x^2$, $g(x) = x - 1$

2. $f(x) = 3x$, $g(x) = 2x + 1$

3. $f(x) = \sqrt{x + 4}$, $g(x) = x^2$

4. $f(x) = e^x$, $g(x) = x^2$

In Exercises 5–8, find $f'(x)$, $f''(x)$, $f'''(x)$, and $f^{(4)}(x)$.

5. $f(x) = 5e^x$

6. $f(x) = \ln x$

7. $f(x) = 3e^{2x}$

8. $f(x) = \ln 2x$

In Exercises 9 and 10, simplify the expression.

9. $\dfrac{3^n}{n!} \div \dfrac{3^{n+1}}{(n+1)!}$

10. $\dfrac{n!}{(n+2)!} \div \dfrac{(n+1)!}{(n+3)!}$

EXERCISES 10.4

In Exercises 1–4, write the first five terms of the power series.

1. $\displaystyle\sum_{n=0}^{\infty} \left(\frac{x}{4}\right)^n$

2. $\displaystyle\sum_{n=1}^{\infty} \frac{(-1)^n (x-2)^n}{3^n}$

3. $\displaystyle\sum_{n=0}^{\infty} \frac{(-1)^{n+1}(x+1)^n}{n!}$

4. $\displaystyle\sum_{n=1}^{\infty} \frac{(-1)^n x^n}{(n-1)!}$

In Exercises 5–24, find the radius of convergence for the series.

5. $\displaystyle\sum_{n=0}^{\infty} \left(\frac{x}{2}\right)^n$

6. $\displaystyle\sum_{n=0}^{\infty} \left(\frac{x}{k}\right)^n$

7. $\displaystyle\sum_{n=1}^{\infty} \frac{(-1)^n x^n}{3n}$

8. $\displaystyle\sum_{n=1}^{\infty} (-1)^{n+1} n x^n$

9. $\displaystyle\sum_{n=0}^{\infty} \frac{(-1)^n x^n}{n!}$

10. $\displaystyle\sum_{n=0}^{\infty} \frac{(3x)^n}{n!}$

11. $\displaystyle\sum_{n=0}^{\infty} n! \left(\frac{x}{2}\right)^n$

12. $\displaystyle\sum_{n=0}^{\infty} \frac{(-1)^n x^n}{(n+1)(n+2)}$

13. $\displaystyle\sum_{n=1}^{\infty} \frac{(-1)^{n+1} x^n}{4^n}$

14. $\displaystyle\sum_{n=0}^{\infty} \frac{(-1)^n n! (x-4)^n}{3^n}$

15. $\displaystyle\sum_{n=1}^{\infty} \frac{(-1)^{n+1}(x-5)^n}{n 5^n}$

16. $\displaystyle\sum_{n=0}^{\infty} \frac{(x-2)^{n+1}}{(n+1)3^{n+1}}$

17. $\displaystyle\sum_{n=0}^{\infty} \frac{(-1)^{n+1}(x-1)^{n+1}}{n+1}$

18. $\displaystyle\sum_{n=1}^{\infty} \frac{(-1)^{n+1}(x-c)^n}{n c^n}$

19. $\displaystyle\sum_{n=1}^{\infty} \frac{(x-c)^{n-1}}{c^{n-1}}$, $0 < c$

20. $\displaystyle\sum_{n=1}^{\infty} \frac{(-1)^{n+1} x^{2n-1}}{2n-1}$

21. $\displaystyle\sum_{n=1}^{\infty} \frac{n}{(n+1)!} (-2x)^{n-1}$

22. $\displaystyle\sum_{n=0}^{\infty} \frac{(-1)^n x^{2n}}{n!}$

23. $\displaystyle\sum_{n=0}^{\infty} \frac{x^{2n+1}}{(2n+1)!}$

24. $\displaystyle\sum_{n=0}^{\infty} \frac{n! x^n}{(n+1)!}$

In Exercises 25–32, apply Taylor's Theorem to find the power series (centered at c) for the function, and find the radius of convergence.

Function	Center
25. $f(x) = e^x$	$c = 0$
26. $f(x) = e^{-x}$	$c = 0$
27. $f(x) = e^{2x}$	$c = 0$
28. $f(x) = e^{-2x}$	$c = 0$
29. $f(x) = \dfrac{1}{x+1}$	$c = 0$
30. $f(x) = \dfrac{1}{2-x}$	$c = 0$
31. $f(x) = \sqrt{x}$	$c = 1$
32. $f(x) = \sqrt{x}$	$c = 4$

In Exercises 33–36, apply Taylor's Theorem to find the binomial series (centered at $c = 0$) for the function, and find the radius of convergence.

33. $f(x) = \dfrac{1}{(1+x)^3}$

34. $f(x) = \sqrt{1+x}$

35. $f(x) = \dfrac{1}{\sqrt{1+x}}$

36. $f(x) = \sqrt[3]{1+x}$

In Exercises 37–40, find the radius of convergence of (a) $f(x)$, (b) $f'(x)$, (c) $f''(x)$, and (d) $\int f(x)\, dx$.

37. $f(x) = \displaystyle\sum_{n=0}^{\infty} \left(\frac{x}{2}\right)^n$

38. $f(x) = \displaystyle\sum_{n=1}^{\infty} \frac{x^n}{n5^n}$

39. $f(x) = \displaystyle\sum_{n=0}^{\infty} \frac{(x+1)^{n+1}}{n+1}$

40. $f(x) = \displaystyle\sum_{n=1}^{\infty} \frac{(-1)^{n+1}(x-1)^n}{n}$

In Exercises 41–52, find the power series for the function using the suggested method. Use the basic list of power series for elementary functions on page 687.

41. Use the power series for e^x.

$$f(x) = e^{x^3}$$

42. Use the series found in Exercise 41.

$$f(x) = e^{-x^3}$$

43. Differentiate the series found in Exercise 41.

$$f(x) = 3x^2 e^{x^3}$$

44. Use the power series for e^x and e^{-x}.

$$f(x) = \frac{e^x + e^{-x}}{2}$$

45. Use the power series for $\dfrac{1}{x+1}$.

$$f(x) = \frac{1}{x^4 + 1}$$

46. Use the power series for $\dfrac{1}{x+1}$.

$$f(x) = \frac{2x}{x+1}$$

47. Use the power series for $\dfrac{1}{x+1}$.

$$f(x) = \ln(x^2 + 1)$$

48. Integrate the series for $\dfrac{1}{x+1}$.

$$f(x) = \ln(x+1)$$

49. Integrate the series for $\dfrac{1}{x}$.

$$f(x) = \ln x$$

50. Differentiate the series found in Exercise 44.

$$f(x) = \frac{e^x - e^{-x}}{2}$$

51. Differentiate the series for $-\dfrac{1}{x}$.

$$f(x) = \frac{1}{x^2}$$

52. Differentiate the power series for e^x term by term and use the resulting series to show that

$$\frac{d}{dx}[e^x] = e^x.$$

In Exercises 53 and 54, use the Taylor series for the exponential function to approximate the expression to four decimal places.

53. $e^{1/2}$

54. e^{-1}

In Exercises 55–58, use a symbolic algebra utility and 50 terms of the series to approximate the function

$$f(x) = \sum_{n=1}^{\infty} \frac{(-1)^{n+1}(x-1)^n}{n}, \quad 0 < x \le 2.$$

55. $f(0.5)$ (Actual sum is $\ln 0.5$.)

56. $f(1.5)$ (Actual sum is $\ln 1.5$.)

57. $f(0.1)$ (Actual sum is $\ln 0.1$.)

58. $f(1.95)$ (Actual sum is $\ln 1.95$.)

10.5 TAYLOR POLYNOMIALS

■ Find Taylor polynomials for functions.
■ Use Taylor polynomials to determine the maximum errors of approximations and to approximate definite integrals.
■ Use Taylor polynomials to model probabilities.

Taylor Polynomials and Approximation

In Section 10.4, you saw that it is sometimes possible to obtain an *exact* power series representation of a function. For example, the function

$$f(x) = e^{-x}$$

can be represented exactly by the power series

$$e^{-x} = \sum_{n=0}^{\infty} \frac{(-1)^n}{n!} x^n$$

$$= 1 - x + \frac{x^2}{2!} - \frac{x^3}{3!} + \frac{x^4}{4!} - \cdots + \frac{(-1)^n x^n}{n!} + \cdots.$$

The problem with using this power series is that the exactness of its representation depends on the summation of an infinite number of terms. In practice, this is not feasible, and you must be content with a finite summation that approximates the function rather than representing it exactly. For instance, consider the sequence of partial sums below.

$$S_0(x) = 1 \qquad\qquad\qquad\qquad\qquad \text{Zeroth-degree Taylor polynomial}$$

$$S_1(x) = 1 - x \qquad\qquad\qquad\qquad \text{First-degree Taylor polynomial}$$

$$S_2(x) = 1 - x + \frac{x^2}{2!} \qquad\qquad\qquad \text{Second-degree Taylor polynomial}$$

$$S_3(x) = 1 - x + \frac{x^2}{2!} - \frac{x^3}{3!} \qquad\qquad \text{Third-degree Taylor polynomial}$$

$$S_4(x) = 1 - x + \frac{x^2}{2!} - \frac{x^3}{3!} + \frac{x^4}{4!} \qquad \text{Fourth-degree Taylor polynomial}$$

$$S_5(x) = 1 - x + \frac{x^2}{2!} - \frac{x^3}{3!} + \frac{x^4}{4!} - \frac{x^5}{5!} \qquad \text{Fifth-degree Taylor polynomial}$$

$$\vdots$$

$$S_n(x) = 1 - x + \frac{x^2}{2!} - \cdots + \frac{(-1)^n x^n}{n!} \qquad n\text{th-degree Taylor polynomial}$$

Each of these polynomial approximations of e^{-x} is a **Taylor polynomial** for e^{-x}. As n approaches infinity, the graphs of these Taylor polynomials become closer and closer approximations of the graph of

$$f(x) = e^{-x}.$$

Finding Taylor Polynomials

Symbolic differentiation utilities are programmed to find Taylor polynomials. Try using a symbolic differentiation utility to compute the fifth-degree Taylor polynomial for $f(x) = e^{-x}$. After finding the Taylor polynomial for $f(x) = e^{-x}$, you can use a graphing utility to compare graphs of this Taylor polynomial with the graph of $f(x) = e^{-x}$, as shown below.

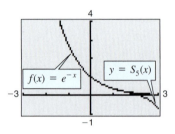

When you use technology to find a Taylor polynomial whose center is not zero, you can obtain polynomials whose forms differ from the standard form. Try using a symbolic differentiation utility to compute the fifth-degree Taylor polynomial for $f(x) = \ln x$, centered at 1. The graph of the fifth-degree Taylor polynomial for $f(x) = \ln x$ is compared with the graph of $f(x) = \ln x$ below.

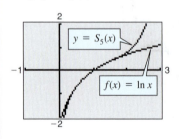

For example, Figure 10.7 shows the graphs of S_1, S_2, S_3, and S_4. Notice in the figure that the closer x is to the center of convergence ($x = 0$ in this case), the better the polynomial approximates e^{-x}. This conclusion is reinforced by the table below. From Section 10.4, you know that the power series for e^{-x} converges for *all* x. From the figure and table, however, you can see that the farther x is from zero, the more terms you need to obtain a good approximation.

x	0	0.5	1.0	1.5	2.0
$S_1 = 1 - x$	1	0.5000	0	-0.5000	-1.0
$S_2 = 1 - x + \dfrac{x^2}{2!}$	1	0.6250	0.5000	0.6250	1.0
$S_3 = 1 - x + \dfrac{x^2}{2!} - \dfrac{x^3}{3!}$	1	0.6042	0.3333	0.0625	-0.3333
$S_4 = 1 - x + \dfrac{x^2}{2!} - \dfrac{x^3}{3!} + \dfrac{x^4}{4!}$	1	0.6068	0.3750	0.2734	0.3333
e^{-x}	1	0.6065	0.3679	0.2231	0.1353

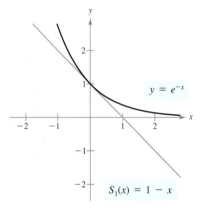

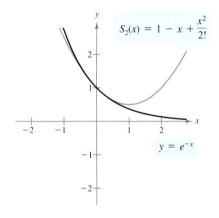

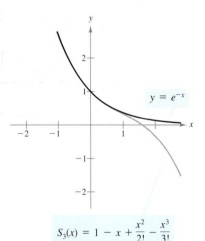

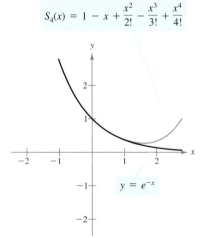

FIGURE 10.7

EXAMPLE 1 **Finding Taylor Polynomials**

Find a Taylor polynomial that is a reasonable approximation of

$$f(x) = e^{-x^2}$$

in the interval $[-1, 1]$. What degree of Taylor polynomial should you use?

SOLUTION To begin, you need to define what constitutes a "reasonable" approximation. Suppose you decide that you want the values of $S_n(x)$ and e^{-x^2} to differ by no more than 0.01 in the interval $[-1, 1]$. To answer the question, you can compute Taylor polynomials of higher and higher degree and then graphically compare them with $f(x) = e^{-x^2}$. For instance, the fourth-, sixth-, and eighth-degree Taylor polynomials for f are as shown.

$$S_4(x) = 1 - x^2 + \frac{x^4}{2}$$ Fourth-degree Taylor polynomial

$$S_6(x) = 1 - x^2 + \frac{x^4}{2} - \frac{x^6}{6}$$ Sixth-degree Taylor polynomial

$$S_8(x) = 1 - x^2 + \frac{x^4}{2} - \frac{x^6}{6} + \frac{x^8}{24}$$ Eighth-degree Taylor polynomial

To compare $S_4(x)$ and e^{-x^2} graphically, use a graphing utility to graph both equations in the same viewing window, as shown in Figure 10.8. Then, use the *trace* feature of the graphing utility to compare the y-values at -1 and 1. When you do this, you will find that the y-values differ by more than 0.01.

Next, perform the same comparison with $S_6(x)$ and e^{-x^2}. With these two graphs, the y-values still differ by more than 0.01 in the interval $[-1, 1]$. Finally, by graphically comparing $S_8(x)$ and e^{-x^2}, you can determine that their graphs differ by less than 0.01 in the interval $[-1, 1]$. To convince yourself of this, try evaluating e^{-x^2} and $S_8(x)$ when $x = \pm 1$. You should obtain the values 0.368 and 0.375, which differ by 0.007.

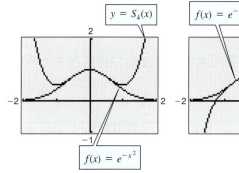
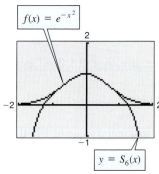
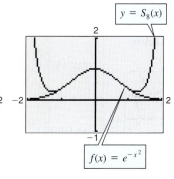

FIGURE 10.8

TRY IT 1

Find the 12th-degree Taylor polynomial for $f(x) = e^{2x^3}$, centered at zero. Then use a graphing utility to graph the function and the polynomial in the same viewing window.

Applications of Probability

Many applications of probability involve experiments in which the sample space is

$$S = \{0, 1, 2, 3, 4, \ldots\}. \qquad \text{\textcolor{red}{Sample space}}$$

As is always true in probability, the sum of the probabilities of the various outcomes is 1. That is, if $P(n)$ is the probability that n will occur, then

$$\sum_{n=0}^{\infty} P(n) = P(0) + P(1) + P(2) + P(3) + \cdots$$

$$= 1.$$

The next example shows how this concept is used in a coin-tossing experiment.

EXAMPLE 4   **Finding Probabilities**

A fair coin is tossed until a head turns up. The possible outcomes are

$$\overset{0}{\frown} \ \overset{1}{\frown} \ \overset{2}{\frown} \ \overset{3}{\frown} \ \overset{4}{\frown}$$

$$\{H, TH, TTH, TTTH, TTTTH, \ldots\}$$

where the random variable assigned to each outcome is the number of tails that have consecutively appeared before a head is tossed (which automatically ends the experiment). Show that the sum of the probabilities in this experiment is 1.

SOLUTION   The probability that no tails will appear is $\frac{1}{2}$. The probability that exactly one tail will appear is $\left(\frac{1}{2}\right)^2$. The probability that exactly two tails will appear is $\left(\frac{1}{2}\right)^3$, and so on. So, the sum of the probabilities is given by a geometric series.

$$\sum_{n=0}^{\infty} P(n) = P(0) + P(1) + P(2) + P(3) + \cdots$$

$$= \sum_{n=0}^{\infty} \left(\frac{1}{2}\right)^{n+1} \qquad \text{\textcolor{red}{Geometric series}}$$

$$= \sum_{n=0}^{\infty} \frac{1}{2}\left(\frac{1}{2}\right)^{n} \qquad \text{\textcolor{red}{$a = \frac{1}{2}$ and $r = \frac{1}{2}$}}$$

$$= \frac{a}{1-r} \qquad \text{\textcolor{red}{Sum of a geometric series}}$$

$$= \frac{1/2}{1 - 1/2}$$

$$= 1$$

So, the sum of the probabilities is 1. This result is shown graphically in Figure 10.10.

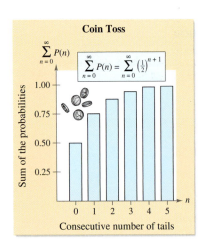

Coin Toss

$$\sum_{n=0}^{\infty} P(n) = \sum_{n=0}^{\infty} \left(\tfrac{1}{2}\right)^{n+1}$$

FIGURE 10.10

TRY IT 4

Use the geometric series in Example 4 to find the probability that exactly 10 tails will appear. How many times must the coin be tossed until the probability that exactly one head appears is less than $\frac{1}{2}\%$?

The **expected value** of a random variable whose range is $\{0, 1, 2, \ldots\}$ is

$$\text{Expected value} = \sum_{n=0}^{\infty} nP(n) = 0P(0) + 1P(1) + 2P(2) + \cdots.$$

This formula is demonstrated in Example 5.

EXAMPLE 5 **Finding an Expected Value**

Find the expected value for the coin-tossing experiment in Example 4.

SOLUTION Using the probabilities from Example 4, you can write

$$\text{Expected value} = \sum_{n=0}^{\infty} nP(n)$$

$$= \sum_{n=0}^{\infty} n\left(\tfrac{1}{2}\right)^{n+1}$$

$$= 0\left(\tfrac{1}{2}\right) + 1\left(\tfrac{1}{2}\right)^2 + 2\left(\tfrac{1}{2}\right)^3 + 3\left(\tfrac{1}{2}\right)^4 + \cdots.$$

This series is not geometric, so to find its sum you need to resort to a different tactic. Using the binomial series from Section 10.4, you can write

$$(1 - x)^{-2} = 1 + 2x + 3x^2 + 4x^3 + \cdots$$

which implies that

$$\left(1 - \tfrac{1}{2}\right)^{-2} = 1 + 2\left(\tfrac{1}{2}\right) + 3\left(\tfrac{1}{2}\right)^2 + 4\left(\tfrac{1}{2}\right)^3 + \cdots$$

$$= \left(\tfrac{1}{2}\right)^{-2} = 4.$$

So, you can conclude that the expected value is

$$\text{Expected value} = 1\left(\tfrac{1}{2}\right)^2 + 2\left(\tfrac{1}{2}\right)^3 + 3\left(\tfrac{1}{2}\right)^4 + \cdots$$

$$= \left(\tfrac{1}{2}\right)^2\left[1 + 2\left(\tfrac{1}{2}\right) + 3\left(\tfrac{1}{2}\right)^2 + 4\left(\tfrac{1}{2}\right)^3 + \cdots\right]$$

$$= \tfrac{1}{4}(4) = 1.$$

This means that if the experiment were conducted many times, the average number of tails that would occur would be 1.

Euan Myles/Stone

An experiment consists of tossing a fair coin repeatedly until the coin finally lands "heads up." Each time the experiment is performed, the number of tosses necessary to obtain "heads" is counted. If this experiment is performed hundreds of times, the average number of tosses per experiment will be 2.

TRY IT 5

Find the expected value for the probability distribution represented by

$$\sum_{n=0}^{\infty} \frac{1}{4}\left(\frac{1}{2}\right)^{n-1}.$$

TAKE ANOTHER LOOK

Expected Value

In Example 5, suppose that the coin is unbalanced so that the probability of landing heads up is 0.6. How does this alter the expected value?

The following warm-up exercises involve skills that were covered in earlier sections. You will use these skills in the exercise set for this section.

In Exercises 1–6, find a power series representation for the function.

1. $f(x) = e^{3x}$

2. $f(x) = e^{-3x}$

3. $f(x) = \dfrac{4}{x}$

4. $f(x) = \ln 5x$

5. $f(x) = (1 + x)^{1/4}$

6. $f(x) = \sqrt{1 + x}$

In Exercises 7–10, evaluate the definite integral.

7. $\displaystyle\int_0^1 (1 - x + x^2 - x^3 + x^4)\,dx$

8. $\displaystyle\int_0^{1/2} \left(1 + \frac{x}{3} - \frac{x^2}{9} + \frac{5x^3}{27}\right) dx$

9. $\displaystyle\int_1^2 \left[(x - 1) - \frac{(x-1)^2}{2} + \frac{(x-1)^3}{3}\right] dx$

10. $\displaystyle\int_1^{3/2} \left[1 - (x - 1) + (x - 1)^2 - (x - 1)^3\right] dx$

EXERCISES 10.5

In Exercises 1–6, find the Taylor polynomials (centered at zero) of degrees (a) 1, (b) 2, (c) 3, and (d) 4.

1. $f(x) = e^x$

2. $f(x) = \ln(x + 1)$

3. $f(x) = \sqrt{x + 1}$

4. $f(x) = \dfrac{1}{(x + 1)^2}$

5. $f(x) = \dfrac{x}{x + 1}$

6. $f(x) = \dfrac{4}{x + 1}$

In Exercises 7 and 8, complete the table using the indicated Taylor polynomials as approximations of the function f.

7.

x	0	$\frac{1}{4}$	$\frac{1}{2}$	$\frac{3}{4}$	1
$f(x) = e^{x/2}$					
$1 + \dfrac{x}{2}$					
$1 + \dfrac{x}{2} + \dfrac{x^2}{8}$					
$1 + \dfrac{x}{2} + \dfrac{x^2}{8} + \dfrac{x^3}{48}$					
$1 + \dfrac{x}{2} + \dfrac{x^2}{8} + \dfrac{x^3}{48} + \dfrac{x^4}{384}$					

8.

x	0	$\frac{1}{4}$	$\frac{1}{2}$	$\frac{3}{4}$
$f(x) = \ln(x^2 + 1)$				
x^2				
$x^2 - \dfrac{x^4}{2}$				
$x^2 - \dfrac{x^4}{2} + \dfrac{x^6}{3}$				
$x^2 - \dfrac{x^4}{2} + \dfrac{x^6}{3} - \dfrac{x^8}{4}$				

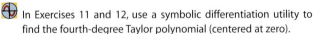 In Exercises 9 and 10, use a symbolic differentiation utility to find the Taylor polynomials (centered at zero) of degrees (a) 2, (b) 4, (c) 6, and (d) 8.

9. $f(x) = \dfrac{1}{1 + x^2}$

10. $f(x) = e^{-x^2}$

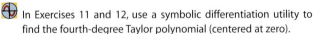 In Exercises 11 and 12, use a symbolic differentiation utility to find the fourth-degree Taylor polynomial (centered at zero).

11. $f(x) = \dfrac{1}{x^2 + 1}$

12. $f(x) = xe^x$

In Exercises 13–16, match the Taylor polynomial approximation of the function $f(x) = e^{-x^2/2}$ with its graph. [The graphs are labeled (a)–(d).] Use a graphing utility to verify your results.

(a)
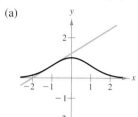

(b)
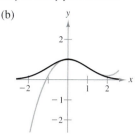

(c)
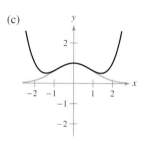

(d)
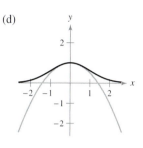

13. $y = -\frac{1}{2}x^2 + 1$

14. $y = \frac{1}{8}x^4 - \frac{1}{2}x^2 + 1$

15. $y = e^{-1/2}[(x + 1) + 1]$

16. $y = e^{-1/2}\left[\frac{1}{3}(x - 1)^3 - (x - 1) + 1\right]$

In Exercises 17–20, use a sixth-degree Taylor polynomial centered at c for the function f to obtain the required approximation.

Function	Approximation
17. $f(x) = e^{-x}$, $c = 0$	$f\left(\frac{1}{2}\right)$
18. $f(x) = x^2 e^{-x}$, $c = 0$	$f\left(\frac{1}{4}\right)$
19. $f(x) = \ln x$, $c = 2$	$f\left(\frac{3}{2}\right)$
20. $f(x) = \sqrt{x}$, $c = 4$	$f(5)$

In Exercises 21–24, use a sixth-degree Taylor polynomial centered at zero to approximate the integral.

Function	Approximation
21. $f(x) = e^{-x^2}$	$\int_0^1 e^{-x^2}\,dx$
22. $f(x) = \ln(x^2 + 1)$	$\int_{-1/4}^{1/4} \ln(x^2 + 1)\,dx$
23. $f(x) = \dfrac{1}{\sqrt{1 + x^2}}$	$\int_0^{1/2} \dfrac{1}{\sqrt{1 + x^2}}\,dx$
24. $f(x) = \dfrac{1}{\sqrt[3]{1 + x^2}}$	$\int_0^{1/2} \dfrac{1}{\sqrt[3]{1 + x^2}}\,dx$

In Exercises 25 and 26, determine the degree of the Taylor polynomial centered at c required to approximate f in the given interval to an accuracy of ± 0.001.

Function	Interval
25. $f(x) = e^x$, $c = 1$	$[0, 2]$
26. $f(x) = \dfrac{1}{x}$, $c = 1$	$\left[1, \dfrac{3}{2}\right]$

In Exercises 27 and 28, determine the maximum error guaranteed by Taylor's Theorem with Remainder when the fifth-degree Taylor polynomial is used to approximate f in the given interval.

27. $f(x) = e^{-x}$, $[0, 1]$,   centered at 0

28. $f(x) = \dfrac{1}{x}$, $\left[1, \dfrac{3}{2}\right]$,   centered at 1

29. *Profit* Let n be a random variable representing the number of units of a certain commodity sold per day in a certain store. The probability distribution of n is shown in the table.

n	0	1	2	3	4, . . .
$P(n)$	$\frac{1}{2}$	$\left(\frac{1}{2}\right)^2$	$\left(\frac{1}{2}\right)^3$	$\left(\frac{1}{2}\right)^4$	$\left(\frac{1}{2}\right)^5$, . . .

(a) Show that

$$\sum_{n=0}^{\infty} P(n) = 1.$$

(b) Find the expected value of the random variable n.

(c) If there is a \$10 profit on each unit sold, what is the expected daily profit on this commodity?

30. *Profit* Repeat Exercise 29 for the probability distribution for n that is shown in the table below.

n	0	1	2	3, . . .
$P(n)$	$\frac{1}{2}\left(\frac{2}{3}\right)$	$\frac{1}{2}\left(\frac{2}{3}\right)^2$	$\frac{1}{2}\left(\frac{2}{3}\right)^3$	$\frac{1}{2}\left(\frac{2}{3}\right)^4$, . . .

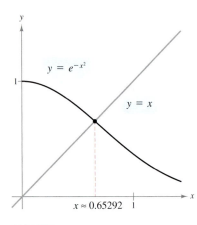

FIGURE 10.17

STUDY TIP

Newton's Method is necessary for working with models of real-life data, such as models of average cost. You will see such models in the exercise set for this section

TRY IT 4

Estimate the point of intersection of the graphs of $y = e^{-x}$ and $y = x$. Use Newton's Method and continue the iterations until two successive approximations differ by less than 0.001.

EXAMPLE 4   **Finding a Point of Intersection**

Estimate the point of intersection of the graphs of

$$y = e^{-x^2} \quad \text{and} \quad y = x$$

as shown in Figure 10.17. Use Newton's Method and continue the iterations until two successive approximations differ by less than 0.0001.

SOLUTION   The point of intersection of the two graphs occurs when $e^{-x^2} = x$, which implies that

$$0 = x - e^{-x^2}.$$

To use Newton's Method, let

$$f(x) = x - e^{-x^2}$$

and employ the iterative formula

$$x_{n+1} = x_n - \frac{f(x_n)}{f'(x_n)}$$

$$= x_n - \frac{x_n - e^{-x_n^2}}{1 + 2x_n e^{-x_n^2}}.$$

The table shows three iterations of Newton's Method beginning with an initial approximation of $x_1 = 0.5$.

n	x_n	$f(x_n)$	$f'(x_n)$	$\dfrac{f(x_n)}{f'(x_n)}$	$x_n - \dfrac{f(x_n)}{f'(x_n)}$
1	0.500000	−0.27880	1.77880	−0.15673	0.65673
2	0.65673	0.00706	1.85331	0.00381	0.65292
3	0.65292	0.00000	1.85261	0.00000	0.65292
4	0.65292				

So, you can estimate that the point of intersection occurs when $x \approx 0.65292$.

TAKE ANOTHER LOOK

Exploring Newton's Method Graphically

Use a graphing utility to graph

$$f(x) = x^{1/3}.$$

In the same viewing window, graph the tangent line to the graph of f at the point $(1, 1)$. If you used $x_1 = 1$ as an initial approximation, what value would Newton's Method yield as the second approximation? Answer the question graphically.

In Exercises 1–4, evaluate f and f' at the given x-value.

1. $f(x) = x^2 - 2x - 1$, $x = 2.4$

2. $f(x) = x^3 - 2x^2 + 1$, $x = -0.6$

3. $f(x) = e^{2x} - 2$, $x = 0.35$

4. $f(x) = e^{x^2} - 7x + 3$, $x = 1.4$

In Exercises 5–8, solve for x.

5. $|x - 5| \leq 0.1$

6. $|4 - 5x| \leq 0.01$

7. $\left|2 - \dfrac{x}{3}\right| \leq 0.01$

8. $|2x + 7| \leq 0.01$

In Exercises 9 and 10, find the point(s) of intersection of the graphs of the two equations.

9. $y = x^2 - x - 2$
   $y = 2x - 1$

10. $y = x^2$
   $y = x + 1$

EXERCISES 10.6

In Exercises 1 and 2, complete one iteration of Newton's Method for the function using the given initial estimate.

Function	Initial Estimate
1. $f(x) = x^2 - 5$	$x_1 = 2.2$
2. $f(x) = 3x^2 - 2$	$x_1 = 1$

In Exercises 3–12, approximate the indicated zero(s) of the function. Use Newton's Method, continuing until two successive approximations differ by less than 0.001.

3. $f(x) = x^3 + x - 1$

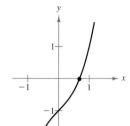

4. $f(x) = x^5 + x - 1$

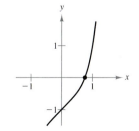

5. $y = 5\sqrt{x - 1} - 2x$

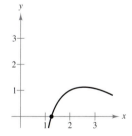

6. $f(x) = \ln x - \dfrac{1}{x}$

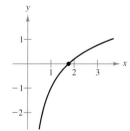

7. $f(x) = \ln x + x$

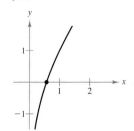

8. $y = e^{3x}(3 - x) - 1$

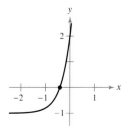

ALGEBRA REVIEW

EXAMPLE 1   **Simplifying Factorial Expressions**

Simplify each expression.

(a) $\dfrac{8!}{2! \cdot 6!}$ (b) $\dfrac{2! \cdot 6!}{3! \cdot 5!}$ (c) $\dfrac{n!}{(n-1)!}$ (d) $\dfrac{(2n+2)!}{(2n+4)!}$

(e) $\dfrac{x^{n+1}}{(n+1)!} \div \dfrac{x^n}{n!}$ (f) $\dfrac{2^{n+1}x^{n+1}}{(n+1)!} \div \dfrac{2^n x^n}{n!}$

SOLUTION

(a) $\dfrac{8!}{2! \cdot 6!} = \dfrac{1 \cdot 2 \cdot 3 \cdot 4 \cdot 5 \cdot 6 \cdot 7 \cdot 8}{1 \cdot 2 \cdot 1 \cdot 2 \cdot 3 \cdot 4 \cdot 5 \cdot 6}$ Factor.

$\qquad = \dfrac{7 \cdot 8}{1 \cdot 2}$ Divide out like factors.

$\qquad = \dfrac{56}{2}$ Multiply.

$\qquad = 28$ Divide.

(b) $\dfrac{2! \cdot 6!}{3! \cdot 5!} = \dfrac{1 \cdot 2 \cdot 1 \cdot 2 \cdot 3 \cdot 4 \cdot 5 \cdot 6}{1 \cdot 2 \cdot 3 \cdot 1 \cdot 2 \cdot 3 \cdot 4 \cdot 5}$ Factor.

$\qquad = \dfrac{6}{3}$ Divide out like factors.

$\qquad = 2$ Divide.

(c) $\dfrac{n!}{(n-1)!} = \dfrac{1 \cdot 2 \cdot 3 \cdots (n-1) \cdot n}{1 \cdot 2 \cdot 3 \cdots (n-1)}$ Factor.

$\qquad = n$ Divide out like factors.

(d) $\dfrac{(2n+2)!}{(2n+4)!} = \dfrac{(2n+2)!}{(2n+2)!(2n+3)(2n+4)}$ Factor.

$\qquad = \dfrac{1}{(2n+3)(2n+4)}$ Divide out like factors.

(e) $\dfrac{x^{n+1}}{(n+1)!} \div \dfrac{x^n}{n!} = \dfrac{x^{n+1}}{(n+1)!} \cdot \dfrac{n!}{x^n}$ Multiply by reciprocal.

$\qquad = \dfrac{x \cdot x^n}{n!(n+1)} \cdot \dfrac{n!}{x^n}$ Factor.

$\qquad = \dfrac{x}{n+1}$ Divide out like factors.

(f) $\dfrac{2^{n+1}x^{n+1}}{(n+1)!} \div \dfrac{2^n x^n}{n!} = \dfrac{2^{n+1}x^{n+1}}{(n+1)!} \cdot \dfrac{n!}{2^n x^n}$ Multiply by reciprocal.

$\qquad = \dfrac{2 \cdot 2^n \cdot x \cdot x^n \cdot n!}{n!(n+1) \cdot 2^n \cdot x^n}$ Factor.

$\qquad = \dfrac{2x}{n+1}$ Divide out like factors.

| EXAMPLE 2 | **Rewriting Expressions with Sigma Notation**

Rewrite each expression.

(a) $\displaystyle\sum_{i=1}^{5} 3i$ (b) $\displaystyle\sum_{k=3}^{6} (1 + k^2)$ (c) $\displaystyle\sum_{i=0}^{8} \frac{1}{i!}$

(d) $\displaystyle\sum_{n=1}^{\infty} 3\left(\frac{1}{2}\right)^n$ (e) $\displaystyle\sum_{n=1}^{20} 6(1.01)^n$ (f) $\displaystyle\sum_{n=1}^{10} \left(n + \frac{1}{3^n}\right)$

SOLUTION

(a) $\displaystyle\sum_{i=1}^{5} 3i = 3(1) + 3(2) + 3(3) + 3(4) + 3(5)$

$= 3(1 + 2 + 3 + 4 + 5)$

$= 3(15) = 45$

(b) $\displaystyle\sum_{k=3}^{6} (1 + k^2) = (1 + 3^2) + (1 + 4^2) + (1 + 5^2) + (1 + 6^2)$

$= 10 + 17 + 26 + 37 = 90$

(c) $\displaystyle\sum_{i=0}^{8} \frac{1}{i!} = \frac{1}{0!} + \frac{1}{1!} + \frac{1}{2!} + \frac{1}{3!} + \frac{1}{4!} + \frac{1}{5!} + \frac{1}{6!} + \frac{1}{7!} + \frac{1}{8!}$

$= 1 + 1 + \frac{1}{2} + \frac{1}{6} + \frac{1}{24} + \frac{1}{120} + \frac{1}{720} + \frac{1}{5040} + \frac{1}{40,320}$

≈ 2.71828

(d) $\displaystyle\sum_{n=1}^{\infty} 3\left(\frac{1}{2}\right)^n = \sum_{n=0}^{\infty} 3\left(\frac{1}{2}\right)^{n+1}$

$= \displaystyle\sum_{n=0}^{\infty} 3\left(\frac{1}{2}\right)\left(\frac{1}{2}\right)^n$

$= \dfrac{3}{2} \displaystyle\sum_{n=0}^{\infty} \left(\frac{1}{2}\right)^n$

(e) $\displaystyle\sum_{n=1}^{20} 6(1.01)^n = 6(1.01)^1 + 6(1.01)^2 + \cdots + 6(1.01)^{20}$

$= -6 + 6 + 6(1.01)^1 + 6(1.01)^2 + \cdots + 6(1.01)^{20}$

$= -6 + 6(1.01)^0 + 6(1.01)^1 + 6(1.01)^2 + \cdots + 6(1.01)^{20}$

$= -6 + \displaystyle\sum_{n=0}^{20} 6(1.01)^n$

(f) $\displaystyle\sum_{n=1}^{10} \left(n + \frac{1}{3^n}\right) = \sum_{n=1}^{10} n + \sum_{n=1}^{10} \frac{1}{3^n}$

$= \displaystyle\sum_{n=1}^{10} n + \sum_{n=1}^{10} \left(\frac{1}{3}\right)^n$

$= \displaystyle\sum_{n=1}^{10} n - 1 + \sum_{n=0}^{10} \left(\frac{1}{3}\right)^n$

10   CHAPTER SUMMARY AND STUDY STRATEGIES

After studying this chapter, you should have acquired the following skills. The exercise numbers are keyed to the Review Exercises that begin on page 712. Answers to odd-numbered Review Exercises are given in the back of the text.

■ Find the terms of sequences.   *(Section 10.1)*   *Review Exercises 1–4*

■ Determine the convergence or divergence of sequences and find the limits of   *Review Exercises 5–12*
convergent sequences.   *(Section 10.1)*

■ Find the *n*th terms of sequences.   *(Section 10.1)*   *Review Exercises 13–16*

■ Use sequences to answer questions about real-life situations.   *(Section 10.1)*   *Review Exercises 17–20*

■ Find the terms of sequences.   *(Section 10.2)*   *Review Exercises 21–24*

$$S_1 = a_1, \quad S_2 = a_1 + a_2, \quad S_3 = a_1 + a_2 + a_3, \quad \ldots$$

■ Determine the convergence or divergence of infinite series.   *(Section 10.2)*   *Review Exercises 25–28*

■ Use the *n*th-Term Test to show that series diverge.   *(Section 10.2)*   *Review Exercises 29–32*

The series $\displaystyle\sum_{n=1}^{\infty} a_n$ diverges if $\displaystyle\lim_{n\to\infty} a_n \neq 0$.

■ Find the *N*th partial sums of geometric series.   *(Section 10.2)*   *Review Exercises 33–36*

$$S_N = \frac{a(1 - r^{N+1})}{1 - r}, \quad r \neq 1.$$

■ Determine the convergence or divergence of geometric series.   *(Section 10.2)*   *Review Exercises 37–40*

If $|r| \geq 1$, the series *diverges*.

If $|r| < 1$, the series *converges* and its sum is $\dfrac{a}{1 - r}$.

■ Use geometric series to model real-life situations.   *(Section 10.2)*   *Review Exercises 41, 42*

■ Use sequences to solve real-life problems.   *(Section 10.2)*   *Review Exercises 43, 44*

■ Determine the convergence or divergence of *p*-series.   *(Section 10.3)*   *Review Exercises 45–48*

The series diverges if $0 < p \leq 1$.

The series converges if $p > 1$.

■ Match sequences with their graphs.   *(Section 10.3)*   *Review Exercises 49–52*

■ Approximate the sums of convergent *p*-series.   *(Section 10.3)*   *Review Exercises 53–56*

■ Use the Ratio Test to determine the convergence or divergence of series.   *(Section 10.3)*   *Review Exercises 57–62*

The series converges if $\displaystyle\lim_{n\to\infty} \left| \frac{a_{n+1}}{a_n} \right| < 1$.

The series diverges if $\displaystyle\lim_{n\to\infty} \left| \frac{a_{n+1}}{a_n} \right| > 1$.

■ Find the radii of convergence of power series.   *(Section 10.4)*   *Review Exercises 63–66*

■ Use Taylor's Theorem with Remainder to find power series for functions.   *(Section 10.4)*   *Review Exercises 67–70*

$$f(x) = \sum_{n=0}^{\infty} \frac{f^{(n)}(c)(x - c)^n}{n!}$$

$$= f(c) + \frac{f'(c)(x - c)}{1!} + \frac{f''(c)(x - c)^2}{2!} + \cdots$$

* Use a wide range of valuable study aids to help you master the material in this chapter. The *Student Solutions Guide* includes step-by-step solutions to all odd-numbered exercises to help you review and prepare. The *HM mathSpace® Student CD-ROM* helps you brush up on your algebra skills. The *Graphing Technology Guide*, available on the Web at *math.college.hmco.com/students*, offers step-by-step commands and instructions for a wide variety of graphing calculators, including the most recent models.

■ Use the basic list of power series to find power series for functions.   *(Section 10.4)*   *Review Exercises 71–78*

■ Find Taylor polynomials for functions.   *(Section 10.5)*   *Review Exercises 79–82*

■ Use Taylor polynomials to approximate the values of functions at points.   *Review Exercises 83–86*
(Section 10.5)

■ Determine the maximum errors of approximations using Taylor polynomials.   *Review Exercises 87, 88*
(Section 10.5)

■ Approximate definite integrals using Taylor polynomials.   *(Section 10.5)*   *Review Exercises 89–92*

■ Use Taylor polynomials to model probabilities.   *(Section 10.5)*   *Review Exercises 93, 94*

■ Use Newton's Method to approximate the zeros of functions.   *(Section 10.6)*   *Review Exercises 95–98*

$$x_{n+1} = x_n - \frac{f(x_n)}{f'(x_n)}$$

■ Use Newton's Method to approximate points of intersection of graphs.   *(Section 10.6)*   *Review Exercises 99–102*

■ ***Using the List of Basic Power Series***   To be efficient at finding Taylor series or
Taylor polynomials, learn how to use the basic list below.

$$\frac{1}{x} = 1 - (x-1) + (x-1)^2 - (x-1)^3 + (x-1)^4 - \cdots + (-1)^n(x-1)^n + \cdots, \quad 0 < x < 2$$

$$\frac{1}{x+1} = 1 - x + x^2 - x^3 + x^4 - x^5 + \cdots + (-1)^n x^n + \cdots, \quad -1 < x < 1$$

$$\ln x = (x-1) - \frac{(x-1)^2}{2} + \frac{(x-1)^3}{3} - \frac{(x-1)^4}{4} + \cdots + \frac{(-1)^{n-1}(x-1)^n}{n} + \cdots, \quad 0 < x \leq 2$$

$$e^x = 1 + x + \frac{x^2}{2!} + \frac{x^3}{3!} + \frac{x^4}{4!} + \cdots + \frac{x^n}{n!} + \cdots, \quad -\infty < x < \infty$$

$$(1+x)^k = 1 + kx + \frac{k(k-1)x^2}{2!} + \frac{k(k-1)(k-2)x^3}{3!} + \frac{k(k-1)(k-2)(k-3)x^4}{4!} + \cdots, \quad -1 < x < 1$$

■ ***Using Technology to Approximate Zeros***   Newton's Method is only one way that
technology can be used to approximate the zeros of a function. Another way is to use the
zoom and *trace* features of a graphing utility, as shown below. (Compare this with the
procedure described on page 703.)

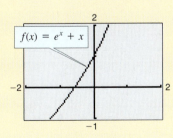

Original screen

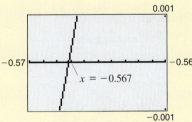

Screen after zooming five times

Study Tools   *Additional resources that accompany this chapter*

■ **Algebra Review**   (pages 708 and 709)

■ **Chapter Summary and Study Strategies** (pages 710 and 711)

■ **Review Exercises**   (pages 712–714)

■ **Sample Post-Graduation Exam Questions**   (page 715)

■ **Web Exercise**   (page 680, Exercise 60)

■ **Student Solutions Guide**

■ **HM mathSpace® Student CD-ROM**

■ **Graphing Technology Guide** (*math.college.hmco.com/students*)

10 CHAPTER REVIEW EXERCISES

In Exercises 1–4, write out the first five terms of the specified sequence. (Begin with $n = 1$.)

1. $a_n = \left(-\frac{1}{3}\right)^n$

2. $a_n = \dfrac{n - 2}{n^2 + 2}$

3. $a_n = \dfrac{4^n}{n!}$

4. $a_n = \dfrac{(-1)^n}{n^3}$

In Exercises 5–8, determine the convergence or divergence of the given sequence. If the sequence converges, find its limit.

5. $a_n = \dfrac{2n + 3}{n^2}$

6. $a_n = \dfrac{1}{\sqrt{n}}$

7. $a_n = \dfrac{n^3}{n^2 + 1}$

8. $a_n = 10e^{-n}$

 In Exercises 9–12, determine the convergence or divergence of the given sequence. If the sequence converges, use a symbolic algebra utility to find its limit.

9. $a_n = 5 + \dfrac{1}{3^n}$

10. $a_n = \dfrac{n}{\sqrt{n^2 + 1}}$

11. $a_n = \dfrac{1}{n^{4/3}}$

12. $a_n = \dfrac{(n - 1)!}{(n + 1)!}$

In Exercises 13–16, find the nth term of the sequence.

13. $\frac{1}{3}, \frac{2}{6}, \frac{3}{9}, \frac{4}{12}, \ldots$

14. $\frac{1}{2}, \frac{2}{5}, \frac{3}{10}, \frac{4}{17}, \ldots$

15. $\frac{1}{3}, -\frac{2}{9}, \frac{4}{27}, -\frac{8}{81}, \ldots$

16. $1, \frac{2}{5}, \frac{4}{25}, \frac{8}{125}, \frac{16}{625}, \ldots$

17. *Sales*   A mail-order company sells $15,000 worth of products during its first year. The company's goal is to increase sales by $10,000 each year for 9 years.

Mail-Order Products

(a) Write an expression for the amount of sales during the nth year.

(b) Compute the total sales for the first 5 years that the mail-order company is in business.

18. *Number of Logs*   Logs are stacked in a pile, as shown in the figure. The top row has 15 logs and the bottom row has 21 logs.

(a) Write an expression for the number of logs in the nth row.

(b) Use the expression in part (a) to verify the number of logs in each row.

 19. *Finance: Compound Interest*   A deposit of $1 is made in an account that earns 7% interest, compounded annually. Find the first 10 terms of the sequence that represents the account balance.   *(Source: Adapted from Garman/Forgue, Personal Finance, Fifth Edition)*

20. *Compound Interest*   A deposit of $5000 is made in an account that earns 5% interest, compounded quarterly. The balance in the account after n quarters is given by

$$A_n = 5000\left(1 + \frac{0.05}{4}\right)^n, \qquad n = 1, 2, 3, \ldots\ .$$

(a) Compute the first eight terms of the sequence.

 (b) Use a symbolic algebra utility to find the balance in the account after 10 years by computing the 40th term of the sequence.

In Exercises 21–24, find the first five terms of the sequence of partial sums for the infinite series.

21. $\displaystyle\sum_{n=0}^{\infty} \left(\frac{3}{2}\right)^n$

22. $\displaystyle\sum_{n=1}^{\infty} \frac{(-1)^{n+1}}{2n}$

23. $\displaystyle\sum_{n=1}^{\infty} \frac{(-1)^{n+1}}{(2n)!}$

24. $\displaystyle\sum_{n=1}^{\infty} \frac{1}{n^2}$

In Exercises 25–28, determine the convergence or divergence of the infinite series.

25. $\displaystyle\sum_{n=1}^{\infty} \frac{n^2 + 1}{n(n + 1)}$

26. $\displaystyle\sum_{n=0}^{\infty} \left(\frac{1}{3}\right)^n$

27. $\displaystyle\sum_{n=0}^{\infty} 2(0.25)^{n+1}$

28. $\displaystyle\sum_{n=1}^{\infty} \frac{\sqrt{n^3}}{n}$

In Exercises 29–32, use the nth-Term Test to verify that the series diverges.

29. $\displaystyle\sum_{n=1}^{\infty} \frac{2n}{n+5}$

30. $\displaystyle\sum_{n=2}^{\infty} \frac{n^3}{1-n^3}$

31. $\displaystyle\sum_{n=0}^{\infty} \left(\frac{5}{4}\right)^n$

32. $\displaystyle\sum_{n=0}^{\infty} 3\left(-\frac{4}{3}\right)^n$

In Exercises 33–36, find the Nth partial sum of the series.

33. $\displaystyle\sum_{n=0}^{\infty} \left(\frac{1}{5}\right)^n$

34. $\displaystyle\sum_{n=0}^{\infty} 3\left(\frac{1}{6}\right)^n$

35. $\displaystyle\sum_{n=0}^{\infty} \left(\frac{1}{2^n} + \frac{1}{4^n}\right)$

36. $\displaystyle\sum_{n=0}^{\infty} 2\left(\frac{1}{\sqrt{3}}\right)^n$

In Exercises 37–40, decide whether the series converges or diverges. If it converges, find its sum.

37. $\displaystyle\sum_{n=0}^{\infty} \frac{1}{4}(4)^n$

38. $\displaystyle\sum_{n=0}^{\infty} 4\left(-\frac{1}{4}\right)^n$

39. $\displaystyle\sum_{n=0}^{\infty} [(0.5)^n + (0.2)^n]$

40. $\displaystyle\sum_{n=0}^{\infty} [(1.5)^n + (0.2)^n]$

41. _Physical Science_ A ball is dropped from a height of 8 feet. Each time it drops h feet, it rebounds $0.7h$ feet.

(a) Write a model for the vertical distance traveled by the ball.

(b) Find the total vertical distance traveled by the ball.

42. _Market Stabilization_ A company estimates the annual sales of a new product to be 8000 units. Each year, 15% of the units that have been sold become inoperative. So, after 1 year 8000 units are in use, after 2 years $[8000 + 0.85(8000)]$ units are in use, and so on. How many units will be in use after n years?

43. _Compound Interest_ The holder of a winning $1,000,000 lottery ticket has the choice of receiving a lump sum payment of $500,000 or receiving an annuity of $40,000 for 25 years. Find the interest rate necessary if the winner wants to deposit the $500,000 in a savings account in order to have $1,000,000 in 25 years. Assume that the interest is compounded quarterly.

44. _Depreciation_ A company buys a machine for $120,000. During the next 5 years it will depreciate at a rate of 30% per year. (That is, at the end of each year the depreciated value will be 70% of what it was at the beginning of the year.)

(a) Find the formula for the nth term of a sequence that gives the value V of the machine t full years after it was purchased.

(b) Find the depreciated value of the machine at the end of 5 full years.

In Exercises 45–48, determine the convergence or divergence of the p-series.

45. $\displaystyle\sum_{n=1}^{\infty} \frac{1}{n^4}$

46. $\displaystyle\sum_{n=1}^{\infty} 2n^{-2/3}$

47. $\displaystyle\sum_{n=1}^{\infty} \frac{1}{n\sqrt[4]{n}}$

48. $\displaystyle\sum_{n=1}^{\infty} \frac{1}{n^e}$

In Exercises 49–52, match the sequence with its graph. [The graphs are labeled (a)–(d).]

(a)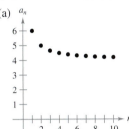

(b)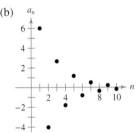

(c)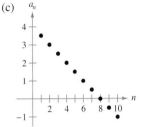

(d)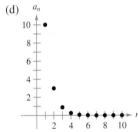

49. $a_n = 4 + \dfrac{2}{n}$

50. $a_n = 4 - \dfrac{1}{2}n$

51. $a_n = 10(0.3)^{n-1}$

52. $a_n = 6\left(-\frac{2}{3}\right)^{n-1}$

In Exercises 53–56, approximate the sum of the given convergent series using the indicated number of terms. Include an estimate of the maximum error for your approximation.

53. $\displaystyle\sum_{n=1}^{\infty} \frac{1}{n^6}$, four terms

54. $\displaystyle\sum_{n=1}^{\infty} \frac{1}{n^3}$, six terms

55. $\displaystyle\sum_{n=1}^{\infty} \frac{1}{n^{5/4}}$, six terms

56. $\displaystyle\sum_{n=1}^{\infty} \frac{1}{n^{12}}$, four terms

In Exercises 57–62, use the Ratio Test to determine the convergence or divergence of the series.

57. $\displaystyle\sum_{n=1}^{\infty} \frac{n4^n}{n!}$

58. $\displaystyle\sum_{n=0}^{\infty} \frac{n!}{4^n}$

59. $\displaystyle\sum_{n=1}^{\infty} \frac{(-1)^n 3^n}{n}$

60. $\displaystyle\sum_{n=1}^{\infty} \frac{2^n}{n}$

61. $\displaystyle\sum_{n=1}^{\infty} \frac{n2^n}{n!}$

62. $\displaystyle\sum_{n=1}^{\infty} \frac{2n}{1-4^n}$

In Exercises 63–66, find the radius of convergence of the power series.

63. $\displaystyle\sum_{n=0}^{\infty} \frac{(-1)^n(x-2)^n}{(n+1)^2}$

64. $\displaystyle\sum_{n=0}^{\infty} (2x)^n$

65. $\displaystyle\sum_{n=0}^{\infty} n!(x-3)^n$

66. $\displaystyle\sum_{n=0}^{\infty} \frac{(x-2)^n}{2^n}$

In Exercises 67–70, use a symbolic differentiation utility to apply Taylor's Theorem with Remainder to find the power series for $f(x)$ centered at c.

67. $f(x) = e^{-0.5x}, \ c = 0$

68. $f(x) = \dfrac{1}{\sqrt{x}}, \ c = 1$

69. $f(x) = \dfrac{1}{x}, \ c = -1$

70. $f(x) = \sqrt[4]{1+x}, \ c = 0$

In Exercises 71–78, use the basic list of power series for elementary functions on page 687 to find the series representation of $f(x)$.

71. $f(x) = \ln(x+2)$

72. $f(x) = e^{2x+1}$

73. $f(x) = (1+x^2)^2$

74. $f(x) = \dfrac{1}{x^3+1}$

75. $f(x) = x^2 e^x$

76. $f(x) = \dfrac{e^x}{x^2}$

77. $f(x) = \dfrac{x^2}{x+1}$

78. $f(x) = \dfrac{\sqrt{x}}{x+1}$

In Exercises 79–82, use a sixth-degree Taylor polynomial to approximate the function in the indicated interval.

79. $f(x) = \dfrac{1}{(x+3)^2}, \ [-1, 1]$

80. $f(x) = e^{x+1}, \ [-2, 2]$

81. $f(x) = \ln(x+2), \ [0, 2]$

82. $f(x) = \sqrt{x+2}, \ [-1, 1]$

In Exercises 83–86, use a sixth-degree Taylor polynomial centered at c for the function f to obtain the desired approximation.

83. $f(x) = e^{x^2}, \ c = 1,$ approximate $f(1.25)$

84. $f(x) = \dfrac{1}{\sqrt{x}}, \ c = 1,$ approximate $f(1.15)$

85. $f(x) = \ln(1+x), \ c = 1,$ approximate $f(1.5)$

86. $f(x) = e^{x-1}, \ c = 0,$ approximate $f(1.75)$

In Exercises 87 and 88, determine the maximum error guaranteed by Taylor's Theorem when the fifth-degree polynomial is used to approximate f in the indicated interval.

87. $f(x) = \dfrac{2}{x}, \ \left[1, \dfrac{3}{2}\right]$

$$P_5(x) = 2[1 - (x-1) + (x-1)^2 - (x-1)^3 + (x-1)^4 - (x-1)^5]$$

88. $f(x) = e^{-2x}, \ [0, 1]$

$$P_5(x) = 1 - 2x + \frac{(2x)^2}{2!} - \frac{(2x)^3}{3!} + \frac{(2x)^4}{4!} - \frac{(2x)^5}{5!}$$

In Exercises 89–92, use a sixth-degree Taylor polynomial centered at zero to approximate the definite integral.

89. $\displaystyle\int_0^{0.3} \sqrt{1+x^3}\,dx$

90. $\displaystyle\int_0^{0.5} e^{-x^2}\,dx$

91. $\displaystyle\int_0^{0.75} \ln(x^2+1)\,dx$

92. $\displaystyle\int_0^{0.5} \sqrt{1+x}\,dx$

Production In Exercises 93 and 94, let n be a random variable representing the number of units produced per hour from one of three machines. Find the expected value of the random variable n, and determine the expected production costs if each unit costs $23.00 to make.

93. $P(n) = 2\left(\frac{1}{3}\right)^{n+1}$

94. $P(n) = \frac{1}{2}\left(\frac{2}{3}\right)^{n+1}$

 In Exercises 95–98, use Newton's Method to approximate to three decimal places the zero(s) of the function.

95. $f(x) = 2x^3 + 3x - 1$

96. $f(x) = x^3 + 2x + 1$

97. $f(x) = \ln 3x + x$

98. $f(x) = e^x - 3$

In Exercises 99–102, use a program similar to the one discussed on page 703 to apply Newton's Method to approximate to three decimal places the x-value of the point of intersection of the graphs of the equations.

99. $f(x) = x^5, \ g(x) = x+3$

100. $f(x) = 2 - x, \ g(x) = x^5 + 2$

101. $f(x) = x^3, \ g(x) = e^{-x}$

102. $f(x) = 2x^2, \ g(x) = 5e^{-x}$

10 SAMPLE POST-GRADUATION EXAM QUESTIONS

CPA
GMAT
GRE
Actuarial
CLAST

The following questions represent the types of questions that appear on certified public accountant (CPA) exams, Graduate Management Admission Tests (GMAT), Graduate Records Exams (GRE), actuarial exams, and College-Level Academic Skills Tests (CLAST). The answers to the questions are given in the back of the book.

For Questions 1 and 2, use the data below.

Territorial Expansion

1 Original 13 States and territories, 1783	888,685
2 Louisiana Purchase, 1803	827,192
3 Florida and Other, 1819	72,003
4 Texas, 1845	390,143
5 Oregon Territory, 1846	285,580
6 Mexican Cession, 1848	529,017
7 Gadsden Purchase, 1853	29,640
8 Alaska, 1867	589,757
9 Hawaii, 1898	6,450
United States Total	3,618,467

GROSS AREA Land and Water, square miles

1. With the Louisiana Purchase, the area of the United States
   (a) roughly tripled   (b) roughly doubled   (c) increased slightly
   (d) stayed the same   (e) decreased slightly

2. Which of the following is closest to the percent of the U.S. total area that is Alaska?
   (a) 10   (b) 15   (c) 20   (d) 25   (e) 30

Figure for Question 3

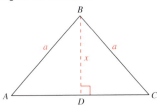

3. In the figure at the left, BD is perpendicular to AC. BA and BC have length a. The area of triangle ABC is
   (a) $2x\sqrt{a^2 - x^2}$ (b) $x\sqrt{a^2 - x^2}$ (c) $a\sqrt{a^2 - x^2}$
   (d) $2a\sqrt{x^2 - a^2}$ (e) $x\sqrt{x^2 - a^2}$

4. Let $f(x) = \dfrac{x^2 + 3x - 4}{x^4 - 1}$. Determine the number of values of k for which $\lim\limits_{x \to k} f(x)$ does not exist.
   (a) 0   (b) 1   (c) 2   (d) 3   (e) 4

5. Let f be a twice-differentiable function such that $f(0) = 4$, $f(3) = 5$, and $f'(3) = 6$. What is $\displaystyle\int_0^3 xf''(x)\, dx$?
   (a) 9   (b) 12   (c) 17   (d) 18   (e) 27

6. The power series
   $$\sum_{n=1}^{\infty} a_n(x - 2)^n \quad \text{and} \quad \sum_{n=1}^{\infty} b_n(x - 1)^n$$
   both converge at $x = 5$. The largest interval over which both series must converge is
   (a) $-5 < x < 5$ (b) $-3 < x < 5$ (c) $-1 \le x \le 5$
   (d) $-1 < x \le 5$ (e) $-3 < x < 6$

PREREQUISITE REVIEW C.1   The following warm-up exercises involve skills that were covered in earlier sections. You will use these skills in the exercise set for this section.

In Exercises 1–4, find the first and second derivatives of the function.

1. $y = 3x^2 + 2x + 1$

2. $y = -2x^3 - 8x + 4$

3. $y = -3e^{2x}$

4. $y = -3e^{x^2}$

In Exercises 5–8, use implicit differentiation to find dy/dx.

5. $x^2 + y^2 = 2x$

6. $2x - y^3 = 4y$

7. $xy^2 = 3$

8. $3xy + x^2y^2 = 10$

In Exercises 9 and 10, solve for k.

9. $0.5 = 9 - 9e^{-k}$

10. $14.75 = 25 - 25e^{-2k}$

EXERCISES C.1

In Exercises 1–10, verify that the function is a solution of the differential equation.

Solution	Differential Equation
1. $y = x^3 + 5$	$y' = 3x^2$
2. $y = 2x^3 - x + 1$	$y' = 6x^2 - 1$
3. $y = e^{-2x}$	$y' + 2y = 0$
4. $y = 3e^{x^2}$	$y' - 2xy = 0$
5. $y = 2x^3$	$y' - \dfrac{3}{x}y = 0$
6. $y = 4x^2$	$y' - \dfrac{2}{x}y = 0$
7. $y = x^2$	$x^2y'' - 2y = 0$
8. $y = \dfrac{1}{x}$	$xy'' + 2y' = 0$
9. $y = 2e^{2x}$	$y'' - y' - 2y = 0$
10. $y = e^{x^3}$	$y'' - 3x^2y' - 6xy = 0$

In Exercises 11–28, verify that the function is a solution of the differential equation for any value of C.

Solution	Differential Equation
11. $y = \dfrac{1}{x} + C$	$\dfrac{dy}{dx} = -\dfrac{1}{x^2}$
12. $y = \sqrt{4 - x^2} + C$	$\dfrac{dy}{dx} = -\dfrac{x}{\sqrt{4 - x^2}}$
13. $y = Ce^{4x}$	$\dfrac{dy}{dx} = 4y$
14. $y = Ce^{-4x}$	$\dfrac{dy}{dx} = -4y$
15. $y = Ce^{-t/3} + 7$	$3\dfrac{dy}{dt} + y - 7 = 0$
16. $y = Ce^{-t} + 10$	$y' + y - 10 = 0$
17. $y = Cx^2 - 3x$	$xy' - 3x - 2y = 0$
18. $y = x \ln x^2 + 2x^{3/2} + Cx$	$y' - \dfrac{y}{x} = 2 + \sqrt{x}$
19. $y = x^2 + 2x + \dfrac{C}{x}$	$xy' + y = x(3x + 4)$
20. $y = C_1 + C_2e^x$	$y'' - y' = 0$

Solution	Differential Equation
21. $y = C_1 e^{x/2} + C_2 e^{-2x}$	$2y'' + 3y' - 2y = 0$
22. $y = C_1 e^{4x} + C_2 e^{-x}$	$y'' - 3y' - 4y = 0$
23. $y = \dfrac{bx^4}{4-a} + Cx^a$	$y' - \dfrac{ay}{x} = bx^3$
24. $y = \dfrac{x^3}{5} - x + C\sqrt{x}$	$2xy' - y = x^3 - x$
25. $y = \dfrac{2}{1 + Ce^{x^2}}$	$y' + 2xy = xy^2$
26. $y = Ce^{x-x^2}$	$y' + (2x - 1)y = 0$
27. $y = x \ln x + Cx + 4$	$x(y' - 1) - (y - 4) = 0$
28. $y = x(\ln x + C)$	$x + y - xy' = 0$

In Exercises 29–32, use implicit differentiation to verify that the equation is a solution of the differential equation for any value of C.

Solution	Differential Equation
29. $x^2 + y^2 = Cy$	$y' = \dfrac{2xy}{x^2 - y^2}$
30. $y^2 + 2xy - x^2 = C$	$(x + y)y' - x + y = 0$
31. $x^2 + xy = C$	$x^2 y'' - 2(x + y) = 0$
32. $x^2 - y^2 = C$	$y^3 y'' + x^2 - y^2 = 0$

In Exercises 33–36, determine whether the function is a solution of the differential equation $y^{(4)} - 16y = 0$.

33. $y = e^{-2x}$

34. $y = 5 \ln x$

35. $y = \dfrac{4}{x}$

36. $y = 4e^{2x}$

In Exercises 37–40, determine whether the function is a solution of the differential equation $y''' - 3y' + 2y = 0$.

37. $y = \frac{2}{9} xe^{-2x}$

38. $y = 4e^x + \frac{2}{9} xe^{-2x}$

39. $y = xe^x$

40. $y = x \ln x$

In Exercises 41–48, verify that the general solution satisfies the differential equation. Then find the particular solution that satisfies the initial condition.

41. General solution: $y = Ce^{-2x}$
Differential equation: $y' + 2y = 0$
Initial condition: $y = 3$ when $x = 0$

42. General solution: $2x^2 + 3y^2 = C$
Differential equation: $2x + 3yy' = 0$
Initial condition: $y = 2$ when $x = 1$

43. General solution: $y = C_1 + C_2 \ln|x|$, $x > 0$
Differential equation: $xy'' + y' = 0$
Initial condition: $y = 5$ and $y' = 0.5$ when $x = 1$

44. General solution: $y = C_1 x + C_2 x^3$
Differential equation: $x^2 y'' - 3xy' + 3y = 0$
Initial condition: $y = 0$ and $y' = 4$ when $x = 2$

45. General solution: $y = C_1 e^{4x} + C_2 e^{-3x}$
Differential equation: $y'' - y' - 12y = 0$
Initial condition: $y = 5$ and $y' = 6$ when $x = 0$

46. General solution: $y = Ce^{x-x^2}$
Differential equation: $y' + (2x - 1)y = 0$
Initial condition: $y = 2$ when $x = 1$

47. General solution: $y = e^{2x/3}(C_1 + C_2 x)$
Differential equation: $9y'' - 12y' + 4y = 0$
Initial condition: $y = 4$ when $x = 0$
$y = 0$ when $x = 3$

48. General solution: $y = \left(C_1 + C_2 x + \frac{1}{12} x^4\right)e^{2x}$
Differential equation: $y'' - 4y' + 4y = x^2 e^{2x}$
Initial condition: $y = 2$ and $y' = 1$ when $x = 0$

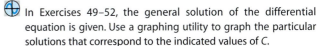 In Exercises 49–52, the general solution of the differential equation is given. Use a graphing utility to graph the particular solutions that correspond to the indicated values of C.

General Solution	Differential Equation	C-Values
49. $y = Cx^2$	$xy' - 2y = 0$	$1, 2, 4$
50. $4y^2 - x^2 = C$	$4yy' - x = 0$	$0, \pm 1, \pm 4$
51. $y = C(x + 2)^2$	$(x + 2)y' - 2y = 0$	$0, \pm 1, \pm 2$
52. $y = Ce^{-x}$	$y' + y = 0$	$0, \pm 1, \pm 2$

In Exercises 53–60, use integration to find the general solution of the differential equation.

53. $\dfrac{dy}{dx} = 3x^2$

54. $\dfrac{dy}{dx} = \dfrac{1}{1 + x}$

55. $\dfrac{dy}{dx} = \dfrac{x + 3}{x}$

56. $\dfrac{dy}{dx} = \dfrac{x - 2}{x}$

57. $\dfrac{dy}{dx} = \dfrac{1}{x^2 - 1}$

58. $\dfrac{dy}{dx} = \dfrac{x}{1 + x^2}$

59. $\dfrac{dy}{dx} = x\sqrt{x - 3}$

60. $\dfrac{dy}{dx} = xe^x$

In Exercises 61–64, you are shown the graphs of some of the solutions of the differential equation. Find the particular solution whose graph passes through the indicated point.

61. $y^2 = Cx^3$

$2xy' - 3y = 0$

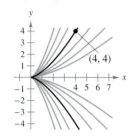

62. $2x^2 - y^2 = C$

$yy' - 2x = 0$

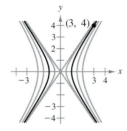

63. $y = Ce^x$

$y' - y = 0$

64. $y^2 = 2Cx$

$2xy' - y = 0$

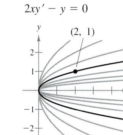

65. Biology　The limiting capacity of the habitat of a wildlife herd is 750. The growth rate dN/dt of the herd is proportional to the unutilized opportunity for growth, as described by the differential equation

$$\frac{dN}{dt} = k(750 - N).$$

The general solution of this differential equation is

$$N = 750 - Ce^{-kt}.$$

When $t = 0$, the population of the herd is 100. After 2 years, the population has grown to 160.

(a) Write the population function N as a function of t.

(b) Use a graphing utility to graph the population function.

(c) What is the population of the herd after 4 years?

66. Investment　The rate of growth of an investment is proportional to the amount in the investment at any time t. That is,

$$\frac{dA}{dt} = kA.$$

The initial investment is $1000, and after 10 years the balance is $3320.12. The general solution is

$$A = Ce^{kt}.$$

What is the particular solution?

67. Marketing　You are working in the marketing department of a computer software company. Your marketing team determines that a maximum of 30,000 units of a new product can be sold in a year. You hypothesize that the rate of growth of the sales x is proportional to the difference between the maximum sales and the current sales. That is,

$$\frac{dx}{dt} = k(30,000 - x).$$

The general solution of this differential equation is

$$x = 30,000 - Ce^{-kt}$$

where t is the time in years. During the first year, 2000 units are sold. Complete the table showing the numbers of units sold in subsequent years.

Year, t	2	4	6	8	10
Units, x					

68. Marketing　In Exercise 67, suppose that the maximum annual sales are 50,000 units. How does this change the sales shown in the table?

69. Safety　Assume that the rate of change in the number of miles s of road cleared per hour by a snowplow is inversely proportional to the depth h of the snow. This rate of change is described by the differential equation

$$\frac{ds}{dh} = \frac{k}{h}.$$

Show that

$$s = 25 - \frac{13}{\ln 3} \ln \frac{h}{2}$$

is a solution of this differential equation.

70. Show that $y = a + Ce^{k(1-b)t}$ is a solution of the differential equation

$$y = a + b(y - a) + \left(\frac{1}{k}\right)\left(\frac{dy}{dt}\right)$$

where k is a constant.

71. The function $y = Ce^{kx}$ is a solution of the differential equation

$$\frac{dy}{dx} = 0.07y.$$

Is it possible to determine C or k from the information given? If so, find its value.

True or False?　In Exercises 72 and 73, determine whether the statement is true or false. If it is false, explain why or give an example that shows it is false.

72. A differential equation can have more than one solution.

73. If $y = f(x)$ is a solution of a differential equation, then $y = f(x) + C$ is also a solution.

C.2   SEPARATION OF VARIABLES

Use separation of variables to solve differential equations. • **Use differential equations to model and solve real-life problems.**

Separation of Variables

The simplest type of differential equation is one of the form $y' = f(x)$. You know that this type of equation can be solved by integration to obtain

$$y = \int f(x)\, dx.$$

In this section, you will learn how to use integration to solve another important family of differential equations—those in which the variables can be separated. This technique is called **separation of variables.**

Separation of Variables
If f and g are continuous functions, then the differential equation $$\frac{dy}{dx} = f(x)g(y)$$ has a general solution of $$\int \frac{1}{g(y)}\, dy = \int f(x)\, dx + C.$$

TECHNOLOGY

You can use a symbolic integration utility to solve a separable variables differential equation. Use a symbolic integration utility to solve the differential equation

$$y' = \frac{x}{y^2 + 1}.$$

Essentially, the technique of separation of variables is just what its name implies. For a differential equation involving x and y, you separate the x variables to one side and the y variables to the other. After separating variables, integrate each side to obtain the general solution. Here is an example.

EXAMPLE 1   **Solving a Differential Equation**

Find the general solution of

$$\frac{dy}{dx} = \frac{x}{y^2 + 1}.$$

SOLUTION   Begin by separating variables, then integrate each side.

$$\frac{dy}{dx} = \frac{x}{y^2 + 1} \qquad \text{Differential equation}$$

$$(y^2 + 1)\, dy = x\, dx \qquad \text{Separate variables.}$$

$$\int (y^2 + 1)\, dy = \int x\, dx \qquad \text{Integrate each side.}$$

$$\frac{y^3}{3} + y = \frac{x^2}{2} + C \qquad \text{General solution}$$

Applications

EXAMPLE 6   **Modeling National Income**

Let y represent the national income, let a represent the income spent on necessities, and let b represent the percent of the remaining income spent on luxuries. A commonly used economic model that relates these three quantities is

$$\frac{dy}{dt} = k(1 - b)(y - a)$$

where t is the time in years. Assume that b is 75%, and solve the resulting differential equation.

Corporate profits in the United States are closely monitored by New York City's Wall Street executives. Corporate profits, however, represent only about 10.5% of the national income. In 2003, the national income was more than $9.5 trillion. Of this, about 65% was employee compensation.

SOLUTION   Because b is 75%, it follows that $(1 - b)$ is 0.25. So, you can solve the differential equation as shown.

$$\frac{dy}{dt} = k(0.25)(y - a)$$ Differential equation

$$\frac{1}{y - a}\, dy = 0.25k\, dt$$ Separate variables.

$$\int \frac{1}{y - a}\, dy = \int 0.25k\, dt$$ Integrate each side.

$$\ln(y - a) = 0.25kt + C_1$$ Find antiderivatives, given $y - a > 0$.

$$y - a = Ce^{0.25kt}$$ Exponentiate each side.

$$y = a + Ce^{0.25kt}$$ Add a to each side.

The graph of this solution is shown in Figure A.11. In the figure, note that the national income is spent in three ways.

$$(\text{National income}) = (\text{necessities}) + (\text{luxuries}) + (\text{capital investment})$$

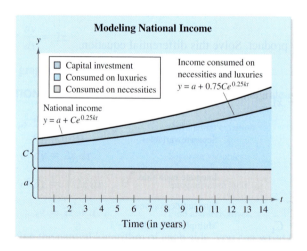

Modeling National Income

□ Capital investment
□ Consumed on luxuries
□ Consumed on necessities

Income consumed on necessities and luxuries
$y = a + 0.75Ce^{0.25kt}$

National income
$y = a + Ce^{0.25kt}$

Time (in years)

FIGURE A.11

EXAMPLE 7 **Using Graphical Information**

Find the equation of the graph that has the characteristics listed below.

1. At each point (x, y) on the graph, the slope is $-x/2y$.

2. The graph passes through the point $(2, 1)$.

SOLUTION Using the information about the slope of the graph, you can write the differential equation

$$\frac{dy}{dx} = -\frac{x}{2y}.$$

Using the point on the graph, you can determine the initial condition $y = 1$ when $x = 2$.

$$\frac{dy}{dx} = -\frac{x}{2y} \qquad \text{Differential equation}$$

$$2y\,dy = -x\,dx \qquad \text{Separate variables.}$$

$$\int 2y\,dy = \int -x\,dx \qquad \text{Integrate each side.}$$

$$y^2 = -\frac{x^2}{2} + C_1 \qquad \text{Find antiderivatives.}$$

$$2y^2 = -x^2 + C \qquad \text{Multiply each side by 2.}$$

$$x^2 + 2y^2 = C \qquad \text{Simplify.}$$

Applying the initial condition yields

$$(2)^2 + 2(1)^2 = C$$

which implies that $C = 6$. So, the equation that satisfies the two given conditions is

$$x^2 + 2y^2 = 6. \qquad \text{Particular solution}$$

As shown in Figure A.12, the graph of this equation is an ellipse.

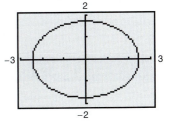

FIGURE A.12

TAKE ANOTHER LOOK

Classifying Differential Equations

In which of the differential equations can the variables be separated?

a. $\dfrac{dy}{dx} = \dfrac{3x}{y}$

b. $\dfrac{dy}{dx} = \dfrac{3x}{y} + 1$

c. $x^2\dfrac{dy}{dx} = \dfrac{3x}{y}$

d. $\dfrac{dy}{dx} = \dfrac{3x + y}{y}$

The following warm-up exercises involve skills that were covered in earlier sections. You will use these skills in the exercise set for this section.

In Exercises 1–6, find the indefinite integral and check your result by differentiating.

1. $\displaystyle\int x^{3/2}\, dx$

2. $\displaystyle\int (t^3 - t^{1/3})\, dt$

3. $\displaystyle\int \frac{2}{x-5}\, dx$

4. $\displaystyle\int \frac{y}{2y^2 + 1}\, dy$

5. $\displaystyle\int e^{2y}\, dy$

6. $\displaystyle\int xe^{1-x^2}\, dx$

In Exercises 7–10, solve the equation for C or k.

7. $(3)^2 - 6(3) = 1 + C$

8. $(-1)^2 + (-2)^2 = C$

9. $10 = 2e^{2k}$

10. $(6)^2 - 3(6) = e^{-k}$

EXERCISES C.2

In Exercises 1–6, decide whether the variables in the differential equation can be separated.

1. $\dfrac{dy}{dx} = \dfrac{x}{y+3}$

2. $\dfrac{dy}{dx} = \dfrac{x+1}{x}$

3. $\dfrac{dy}{dx} = \dfrac{1}{x} + 1$

4. $\dfrac{dy}{dx} = \dfrac{x}{x+y}$

5. $\dfrac{dy}{dx} = x - y$

6. $x\dfrac{dy}{dx} = \dfrac{1}{y}$

In Exercises 7–26, use separation of variables to find the general solution of the differential equation.

7. $\dfrac{dy}{dx} = 2x$

8. $\dfrac{dy}{dx} = \dfrac{1}{x}$

9. $3y^2\dfrac{dy}{dx} = 1$

10. $\dfrac{dy}{dx} = x^2 y$

11. $(y+1)\dfrac{dy}{dx} = 2x$

12. $(1+y)\dfrac{dy}{dx} - 4x = 0$

13. $y' - xy = 0$

14. $y' - y = 5$

15. $\dfrac{dy}{dt} = \dfrac{e^t}{4y}$

16. $e^y\dfrac{dy}{dt} = 3t^2 + 1$

17. $\dfrac{dy}{dx} = \sqrt{1-y}$

18. $\dfrac{dy}{dx} = \sqrt{\dfrac{x}{y}}$

19. $(2+x)y' = 2y$

20. $y' = (2x-1)(y+3)$

21. $xy' = y$

22. $y' - y(x+1) = 0$

23. $y' = \dfrac{x}{y} - \dfrac{x}{1+y}$

24. $\dfrac{dy}{dx} = \dfrac{x^2+2}{3y^2}$

25. $e^x(y'+1) = 1$

26. $yy' - 2xe^x = 0$

In Exercises 27–32, use the initial condition to find the particular solution of the differential equation.

Differential Equation	Initial Condition
27. $yy' - e^x = 0$	$y = 4$ when $x = 0$
28. $\sqrt{x} + \sqrt{y}\,y' = 0$	$y = 4$ when $x = 1$
29. $x(y+4) + y' = 0$	$y = -5$ when $x = 0$
30. $\dfrac{dy}{dx} = x^2(1+y)$	$y = 3$ when $x = 0$
31. $dP - 6P\, dt = 0$	$P = 5$ when $t = 0$
32. $dT + k(T - 70)\, dt = 0$	$T = 140$ when $t = 0$

In Exercises 33 and 34, find an equation for the graph that passes through the point and has the specified slope. Then graph the equation.

33. Point: $(-1, 1)$

Slope: $y' = \dfrac{6x}{5y}$

34. Point: $(8, 2)$

Slope: $y' = \dfrac{2y}{3x}$

Velocity In Exercises 35 and 36, solve the differential equation to find velocity v as a function of time t if $v = 0$ when $t = 0$. The differential equation models the motion of two people on a toboggan after consideration of the force of gravity, friction, and air resistance.

35. $12.5\dfrac{dv}{dt} = 43.2 - 1.25v$

36. $12.5\dfrac{dv}{dt} = 43.2 - 1.75v$

Chemistry: Newton's Law of Cooling In Exercises 37–39, use Newton's Law of Cooling, which states that the rate of change in the temperature T of an object is proportional to the difference between the temperature T of the object and the temperature T_0 of the surrounding environment. This is described by the differential equation $dT/dt = k(T - T_0)$.

37. A steel ingot whose temperature is 1500°F is placed in a room whose temperature is a constant 90°F. One hour later, the temperature of the ingot is 1120°F. What is the ingot's temperature 5 hours after it is placed in the room?

38. A room is kept at a constant temperature of 70°F. An object placed in the room cools from 350°F to 150°F in 45 minutes. How long will it take for the object to cool to a temperature of 80°F?

39. Food at a temperature of 70°F is placed in a freezer that is set at 0°F. After 1 hour, the temperature of the food is 48°F.

(a) Find the temperature of the food after it has been in the freezer 6 hours.

(b) How long will it take the food to cool to a temperature of 10°F?

40. ***Biology: Cell Growth*** The rate of growth of a spherical cell with volume V is proportional to its surface area S. For a sphere, the surface area and volume are related by $S = kV^{2/3}$. So, a model for the cell's growth is

$$\frac{dV}{dt} = kV^{2/3}.$$

Solve this differential equation.

41. ***Learning Theory*** The management of a factory has found that a worker can produce at most 30 units per day. The number of units N per day produced by a new employee will increase at a rate proportional to the difference between 30 and N. This is described by the differential equation

$$\frac{dN}{dt} = k(30 - N)$$

where t is the time in days. Solve this differential equation.

42. ***Sales*** The rate of increase in sales S (in thousands of units) of a product is proportional to the current level of sales and inversely proportional to the square of the time t. This is described by the differential equation

$$\frac{dS}{dt} = \frac{kS}{t^2}$$

where t is the time in years. The saturation point for the market is 50,000 units. That is, the limit of S as $t \to \infty$ is 50. After 1 year, 10,000 units have been sold. Find S as a function of the time t.

43. ***Economics: Pareto's Law*** According to the economist Vilfredo Pareto (1848–1923), the rate of decrease of the number of people y in a stable economy having an income of at least x dollars is directly proportional to the number of such people and inversely proportional to their income x. This is modeled by the differential equation

$$\frac{dy}{dx} = -k\frac{y}{x}.$$

Solve this differential equation.

44. ***Economics: Pareto's Law*** In 2001, 15.2 million people in the United States earned more than $75,000 and 90.9 million people earned more than $25,000 (see figure). Assume that Pareto's Law holds and use the result of Exercise 43 to determine the number of people (in millions) who earned (a) more than $20,000 and (b) more than $100,000.   *(Source: U.S. Census Bureau)*

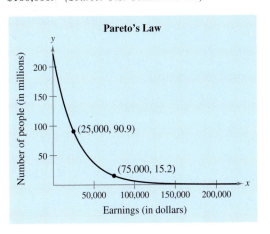

Pareto's Law

$(25,000, 90.9)$

$(75,000, 15.2)$

Number of people (in millions)

Earnings (in dollars)

C.3 FIRST–ORDER LINEAR DIFFERENTIAL EQUATIONS

Solve first-order linear differential equations. • Use first-order linear differential equations to model and solve real-life problems.

First-Order Linear Differential Equations

> **Definition of a First-Order Linear Differential Equation**
>
> A **first-order linear differential equation** is an equation of the form
>
> $$y' + P(x)y = Q(x)$$
>
> where P and Q are functions of x. An equation that is written in this form is said to be in **standard form.**

STUDY TIP

The term "first-order" refers to the fact that the highest-order derivative of y in the equation is the first derivative.

To solve a linear differential equation, write it in standard form to identify the functions $P(x)$ and $Q(x)$. Then integrate $P(x)$ and form the expression

$$u(x) = e^{\int P(x)\,dx} \qquad \text{Integrating factor}$$

which is called an **integrating factor.** The general solution of the equation is

$$y = \frac{1}{u(x)} \int Q(x)u(x)\,dx. \qquad \text{General solution}$$

EXAMPLE 1 **Solving a Linear Differential Equation**

Find the general solution of

$$y' + y = e^x.$$

SOLUTION For this equation, $P(x) = 1$ and $Q(x) = e^x$. So, the integrating factor is

$$u(x) = e^{\int dx} \qquad \text{Integrating factor}$$
$$= e^x.$$

This implies that the general solution is

$$y = \frac{1}{e^x} \int e^x(e^x)\,dx$$
$$= e^{-x}\left(\frac{1}{2}e^{2x} + C\right)$$
$$= \frac{1}{2}e^x + Ce^{-x}. \qquad \text{General solution}$$

In Example 1, the differential equation was given in standard form. For equations that are not written in standard form, you should first convert to standard form so that you can identify the functions $P(x)$ and $Q(x)$.

EXAMPLE 2 **Solving a Linear Differential Equation**

Find the general solution of

$$xy' - 2y = x^2.$$

Assume $x > 0$.

SOLUTION Begin by writing the equation in standard form.

$$y' - \left(\frac{2}{x}\right)y = x \qquad \text{Standard form, } y' + P(x)y = Q(x)$$

In this form, you can see that $P(x) = -2/x$ and $Q(x) = x$. So,

$$\int P(x)\,dx = -\int \frac{2}{x}\,dx$$
$$= -2\ln x$$
$$= -\ln x^2$$

which implies that the integrating factor is

$$u(x) = e^{\int P(x)\,dx}$$
$$= e^{-\ln x^2}$$
$$= \frac{1}{x^2}. \qquad \text{Integrating factor}$$

This implies that the general solution is

$$y = \frac{1}{u(x)}\int Q(x)u(x)\,dx \qquad \text{Form of general solution}$$
$$= \frac{1}{1/x^2}\int x\left(\frac{1}{x^2}\right)dx \qquad \text{Substitute.}$$
$$= x^2\int \frac{1}{x}\,dx \qquad \text{Simplify.}$$
$$= x^2(\ln x + C). \qquad \text{General solution}$$

Guidelines for Solving a Linear Differential Equation

1. Write the equation in standard form

$$y' + P(x)y = Q(x).$$

2. Find the integrating factor

$$u(x) = e^{\int P(x)\,dx}.$$

3. Evaluate the integral below to find the general solution.

$$y = \frac{1}{u(x)}\int Q(x)u(x)\,dx$$

Application

> **EXAMPLE 3** **Finding a Balance**

You are setting up a "continuous annuity" trust fund. For 20 years, money is continuously transferred from your checking account to the trust fund at the rate of $1000 per year (about $2.74 per day). The account earns 8% interest, compounded continuously. What is the balance in the account after 20 years?

SOLUTION Let A represent the balance after t years. The balance increases in two ways: with interest *and* with additional deposits. The rate at which the balance is changing can be modeled by

$$\frac{dA}{dt} = \underbrace{0.08A}_{\text{Interest}} + \underbrace{1000}_{\text{Deposits}}.$$

In standard form, this linear differential equation is

$$\frac{dA}{dt} - 0.08A = 1000 \qquad \text{Standard form}$$

which implies that $P(t) = -0.08$ and $Q(t) = 1000$. The general solution is

$$A = -12{,}500 + Ce^{0.08t}. \qquad \text{General solution}$$

Because $A = 0$ when $t = 0$, you can determine that $C = 12{,}500$. So, the revenue after 20 years is

$$A = -12{,}500 + 12{,}500e^{0.08(20)}$$
$$\approx -12{,}500 + 61{,}912.91$$
$$= \$49{,}412.91.$$

TAKE ANOTHER LOOK

Why an Integrating Factor Works

When both sides of the first-order linear differential equation

$$y' + P(x)y = Q(x)$$

are multiplied by the integrating factor $e^{\int P(x)\,dx}$, you obtain

$$y'e^{\int P(x)\,dx} + P(x)e^{\int P(x)\,dx}y = Q(x)e^{\int P(x)\,dx}.$$

Show that the left side is the derivative of $ye^{\int P(x)\,dx}$, which implies that the general solution is given by

$$ye^{\int P(x)\,dx} = \int Q(x)e^{\int P(x)\,dx}\,dx.$$

PREREQUISITE REVIEW C.3

The following warm-up exercises involve skills that were covered in earlier sections. You will use these skills in the exercise set for this section.

In Exercises 1–4, simplify the expression.

1. $e^{-x}(e^{2x} + e^x)$

2. $\dfrac{1}{e^{-x}}(e^{-x} + e^{2x})$

3. $e^{-\ln x^3}$

4. $e^{2\ln x + x}$

In Exercises 5–10, find the indefinite integral.

5. $\displaystyle\int e^x(2 + e^{-2x})\, dx$

6. $\displaystyle\int e^{2x}(xe^x + 1)\, dx$

7. $\displaystyle\int \frac{1}{2x + 5}\, dx$

8. $\displaystyle\int \frac{x + 1}{x^2 + 2x + 3}\, dx$

9. $\displaystyle\int (4x - 3)^2\, dx$

10. $\displaystyle\int x(1 - x^2)^2\, dx$

EXERCISES C.3

In Exercises 1–6, write the linear differential equation in standard form.

1. $x^3 - 2x^2y' + 3y = 0$

2. $y' - 5(2x - y) = 0$

3. $xy' + y = xe^x$

4. $xy' + y = x^3y$

5. $y + 1 = (x - 1)y'$

6. $x = x^2(y' + y)$

In Exercises 7–18, solve the differential equation.

7. $\dfrac{dy}{dx} + 3y = 6$

8. $\dfrac{dy}{dx} + 5y = 15$

9. $\dfrac{dy}{dx} + y = e^{-x}$

10. $\dfrac{dy}{dx} + 3y = e^{-3x}$

11. $\dfrac{dy}{dx} + \dfrac{y}{x} = 3x + 4$

12. $\dfrac{dy}{dx} + \dfrac{2y}{x} = 3x + 1$

13. $y' + 5xy = x$

14. $y' + 5y = e^{5x}$

15. $(x - 1)y' + y = x^2 - 1$

16. $xy' + y = x^2 + 1$

17. $x^3y' + 2y = e^{1/x^2}$

18. $xy' + y = x^2 \ln x$

In Exercises 19–22, solve for y in two ways.

19. $y' + y = 4$

20. $y' + 10y = 5$

21. $y' - 2xy = 2x$

22. $y' + 4xy = x$

In Exercises 23–26, match the differential equation with its solution.

Differential Equation	Solution
23. $y' - 2x = 0$	(a) $y = Ce^{x^2}$
24. $y' - 2y = 0$	(b) $y = -\frac{1}{2} + Ce^{x^2}$
25. $y' - 2xy = 0$	(c) $y = x^2 + C$
26. $y' - 2xy = x$	(d) $y = Ce^{2x}$

In Exercises 27–34, find the particular solution that satisfies the initial condition.

Differential Equation	Initial Condition
27. $y' + y = 6e^x$	$y = 3$ when $x = 0$
28. $y' + 2y = e^{-2x}$	$y = 4$ when $x = 1$
29. $xy' + y = 0$	$y = 2$ when $x = 2$
30. $y' + y = x$	$y = 4$ when $x = 0$
31. $y' + 3x^2y = 3x^2$	$y = 6$ when $x = 0$
32. $y' + (2x - 1)y = 0$	$y = 2$ when $x = 1$
33. $xy' - 2y = -x^2$	$y = 5$ when $x = 1$
34. $x^2y' - 4xy = 10$	$y = 10$ when $x = 1$

35. Sales The rate of change (in thousands of units) in sales S is modeled by

$$\frac{dS}{dt} = 0.2(100 - S) + 0.2t$$

where t is the time in years. Solve this differential equation and use the result to complete the table.

t	0	1	2	3	4	5	6	7	8	9	10
S	0										

36. Sales The rate of change in sales S is modeled by

$$\frac{dS}{dt} = k_1(L - S) + k_2 t$$

where t is the time in years and $S = 0$ when $t = 0$. Solve this differential equation for S as a function of t.

Elasticity of Demand In Exercises 37 and 38, find the demand function $p = f(x)$. Recall from Section 3.5 that the price elasticity of demand was defined as $\eta = (p/x)/(dp/dx)$.

37. $\eta = 1 - \dfrac{400}{3x}$, $\quad p = 340$ when $x = 20$

38. $\eta = 1 - \dfrac{500}{3x}$, $\quad p = 2$ when $x = 100$

Supply and Demand In Exercises 39 and 40, use the demand and supply functions to find the price p as a function of time t. Begin by setting $D(t)$ equal to $S(t)$ and solving the resulting differential equation. Find the general solution, and then use the initial condition to find the particular solution.

39. $D(t) = 480 + 5p(t) - 2p'(t)$ Demand function

    $S(t) = 300 + 8p(t) + p'(t)$ Supply function

    $p(0) = \$75.00$ Initial condition

40. $D(t) = 4000 + 5p(t) - 4p'(t)$ Demand function

    $S(t) = 2800 + 7p(t) + 2p'(t)$ Supply function

    $p(0) = \$1000.00$ Initial condition

41. Investment A brokerage firm opens a new real estate investment plan for which the earnings are equivalent to continuous compounding at the rate of r. The firm estimates that deposits from investors will create a net cash flow of Pt dollars, where t is the time in years. The rate of change in the total investment A is modeled by

$$\frac{dA}{dt} = rA + Pt.$$

(a) Solve the differential equation and find the total investment A as a function of t. Assume that $A = 0$ when $t = 0$.

(b) Find the total investment A after 10 years given that $P = \$500,000$ and $r = 9\%$.

42. Investment Let $A(t)$ be the amount in a fund earning interest at the annual rate of r, compounded continuously. If a continuous cash flow of P dollars per year is withdrawn from the fund, then the rate of decrease of A is given by the differential equation

$$\frac{dA}{dt} = rA - P$$

where $A = A_0$ when $t = 0$.

(a) Solve this equation for A as a function of t.

(b) Use the result of part (a) to find A when $A_0 = \$2,000,000$, $r = 7\%$, $P = \$250,000$, and $t = 5$ years.

(c) Find A_0 if a retired person wants a continuous cash flow of $\$40,000$ per year for 20 years. Assume that the person's investment will earn 8%, compounded continuously.

43. Velocity A booster rocket carrying an observation satellite is launched into space. The rocket and satellite have mass m and are subject to air resistance proportional to the velocity v at any time t. A differential equation that models the velocity of the rocket and satellite is

$$m\frac{dv}{dt} = -mg - kv$$

where g is the acceleration due to gravity. Solve the differential equation for v as a function of t.

44. Health An infectious disease spreads through a large population according to the model

$$\frac{dy}{dt} = \frac{1 - y}{4}$$

where y is the percent of the population exposed to the disease, and t is the time in years.

(a) Solve this differential equation, assuming $y(0) = 0$.

(b) Find the number of years it takes for half of the population to have been exposed to the disease.

(c) Find the percentage of the population that has been exposed to the disease after 4 years.

45. Research Project Use your school's library, the Internet, or some other reference source to find an article in a scientific or business journal that uses a differential equation to model a real-life situation. Write a short paper describing the situation. If possible, describe the solution of the differential equation.

C.4 APPLICATIONS OF DIFFERENTIAL EQUATIONS

Use differential equations to model and solve real-life problems.

EXAMPLE 1 **Modeling Advertising Awareness**

The new cereal product from Example 3 in Section C.1 is introduced through an advertising campaign to a population of 1 million potential customers. The rate at which the population hears about the product is assumed to be proportional to the number of people who are not yet aware of the product. By the end of 1 year, half of the population has heard of the product. How many will have heard of it by the end of 2 years?

SOLUTION   Let y be the number (in millions) of people at time t who have heard of the product. This means that $(1 - y)$ is the number of people who have not heard of it, and dy/dt is the rate at which the population hears about the product. From the given assumption, you can write the differential equation as shown.

$$\frac{dy}{dt} = k(1 - y)$$

Rate of change of y | is proportional to | the difference between 1 and y.

Using separation of variables *or* a symbolic integration utility, you can find the general solution to be

$$y = 1 - Ce^{-kt}. \qquad \text{General solution}$$

To solve for the constants C and k, use the initial conditions. That is, because $y = 0$ when $t = 0$, you can determine that $C = 1$. Similarly, because $y = 0.5$ when $t = 1$, it follows that $0.5 = 1 - e^{-k}$, which implies that

$$k = \ln 2 \approx 0.693.$$

So, the particular solution is

$$y = 1 - e^{-0.693t}. \qquad \text{Particular solution}$$

This model is shown graphically in Figure A.13. Using the model, you can determine that the number of people who have heard of the product after 2 years is

$$y = 1 - e^{-0.693(2)}$$

$$\approx 0.75 \text{ or } 750,000 \text{ people.}$$

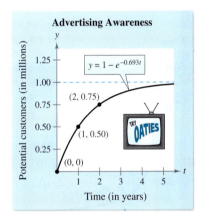

Advertising Awareness

$y = 1 - e^{-0.693t}$

FIGURE A.13

In genetics, a commonly used hybrid selection model is based on the differential equation

$$\frac{dy}{dt} = ky(1 - y)(a - by).$$

In this model, y represents the portion of the population that has a certain characteristic and t represents the time (measured in generations). The numbers a, b, and k are constants that depend on the genetic characteristic that is being studied.

EXAMPLE 4   **Modeling Hybrid Selection**

You are studying a population of beetles to determine how quickly characteristic D will pass from one generation to the next. At the beginning of your study ($t = 0$), you find that half the population has characteristic D. After four generations ($t = 4$), you find that 80% of the population has characteristic D. Use the hybrid selection model above with $a = 2$ and $b = 1$ to find the percent of the population that will have characteristic D after 10 generations.

SOLUTION   Using $a = 2$ and $b = 1$, the differential equation for the hybrid selection model is

$$\frac{dy}{dt} = ky(1 - y)(2 - y).$$

Using separation of variables *or* a symbolic integration utility, you can find the general solution to be

$$\frac{y(2 - y)}{(1 - y)^2} = Ce^{2kt}. \qquad \text{General solution}$$

To solve for the constants C and k, use the initial conditions. That is, because $y = 0.5$ when $t = 0$, you can determine that $C = 3$. Similarly, because $y = 0.8$ when $t = 4$, it follows that

$$\frac{0.8(1.2)}{(0.2)^2} = 3e^{8k}$$

which implies that

$$k = \frac{1}{8} \ln 8 \approx 0.2599.$$

So, the particular solution is

$$\frac{y(2 - y)}{(1 - y)^2} = 3e^{0.5199t}. \qquad \text{Particular solution}$$

Using the model, you can estimate the percent of the population that will have characteristic D after 10 generations to be given by

$$\frac{y(2 - y)}{(1 - y)^2} = 3e^{0.5199(10)}.$$

Using a symbolic algebra utility, you can solve this equation for y to obtain $y \approx 0.96$. The graph of the model is shown in Figure A.16.

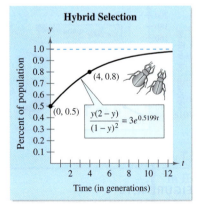

FIGURE A.16

EXAMPLE 5 **Modeling a Chemical Mixture**

A tank contains 40 gallons of a solution composed of 90% water and 10% alcohol. A second solution containing half water and half alcohol is added to the tank at the rate of 4 gallons per minute. At the same time, the tank is being drained at the rate of 4 gallons per minute, as shown in Figure A.17. Assuming that the solution is stirred constantly, how much alcohol will be in the tank after 10 minutes?

SOLUTION Let y be the number of gallons of alcohol in the tank at any time t. The *percent* of alcohol in the 40-gallon tank at any time is $y/40$. Moreover, because 4 gallons of solution is being drained each minute, the rate of change of y is

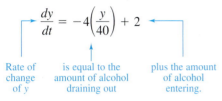

4 gal/min

4 gal/min

FIGURE A.17

where 2 represents the number of gallons of alcohol entering each minute in the 50% solution. In standard form, this linear differential equation is

$$y' + \frac{1}{10}y = 2. \qquad \text{Standard form}$$

Using an integrating factor *or* a symbolic integration utility, you can find the general solution to be

$$y = 20 + Ce^{-t/10}. \qquad \text{General solution}$$

Because $y = 4$ when $t = 0$, you can conclude that $C = -16$. So, the particular solution is

$$y = 20 - 16e^{-t/10}. \qquad \text{Particular solution}$$

Using this model, you can determine that the amount of alcohol in the tank when $t = 10$ is

$$y = 20 - 16e^{-(10)/10}$$
$$\approx 14.1 \text{ gallons.}$$

TAKE ANOTHER LOOK

Chemical Mixture

Sketch the particular solution obtained in Example 5. Describe the rate of change of the amount of alcohol in the tank. Does the amount approach 0 as t increases? Explain your reasoning.

PREREQUISITE REVIEW C.4

The following warm-up exercises involve skills that were covered in earlier sections. You will use these skills in the exercise set for this section.

In Exercises 1–4, use separation of variables to find the general solution of the differential equation.

1. $\dfrac{dy}{dx} = 3x$

2. $2y\dfrac{dy}{dx} = 3$

3. $\dfrac{dy}{dx} = 2xy$

4. $\dfrac{dy}{dx} = \dfrac{x - 4}{4y^3}$

In Exercises 5–8, use an integrating factor to solve the first-order linear differential equation.

5. $y' + 2y = 4$

6. $y' + 2y = e^{-2x}$

7. $y' + xy = x$

8. $xy' + 2y = x^2$

In Exercises 9 and 10, write the equation that models the statement.

9. The rate of change of y with respect to x is proportional to the square of x.

10. The rate of change of x with respect to t is proportional to the difference of x and t.

EXERCISES C.4

In Exercises 1–6, assume that the rate of change of y is proportional to y. Solve the resulting differential equation $dy/dx = ky$ and find the particular solution that passes through the points.

1. $(0, 1), (3, 2)$

2. $(0, 4), (1, 6)$

3. $(0, 4), (4, 1)$

4. $(0, 60), (5, 30)$

5. $(2, 2), (3, 4)$

6. $(1, 4), (2, 1)$

7. *Investment*   The rate of growth of an investment is proportional to the amount A of the investment at any time t. An investment of \$2000 increases to a value of \$2983.65 in 5 years. Find its value after 10 years.

8. *Population Growth*   The rate of change of the population of a city is proportional to the population P at any time t. In 1998, the population was 400,000, and the constant of proportionality was 0.015. Estimate the population of the city in the year 2005.

9. *Sales Growth*   The rate of change in sales S (in thousands of units) of a new product is proportional to the difference between L and S (in thousands of units) at any time t. When $t = 0$, $S = 0$. Write and solve the differential equation for this sales model.

10. *Sales Growth*   Use the result of Exercise 9 to write S as a function of t if (a) $L = 100$, $S = 25$ when $t = 2$, and (b) $L = 500$, $S = 50$ when $t = 1$.

In Exercises 11–14, the rate of change of y is proportional to the product of y and the difference of L and y. Solve the resulting differential equation $dy/dx = ky(L - y)$ and find the particular solution that passes through the points for the given value of L.

11. $L = 20$; $(0, 1), (5, 10)$

12. $L = 100$; $(0, 10), (5, 30)$

13. $L = 5000$; $(0, 250), (25, 2000)$

14. $L = 1000$; $(0, 100), (4, 750)$

15. Biology At any time t, the rate of growth of the population N of deer in a state park is proportional to the product of N and $L - N$, where $L = 500$ is the maximum number of deer the park can maintain. When $t = 0$, $N = 100$, and when $t = 4$, $N = 200$. Write N as a function of t.

16. Sales Growth The rate of change in sales S (in thousands of units) of a new product is proportional to the product of S and $L - S$. L (in thousands of units) is the estimated maximum level of sales, and $S = 10$ when $t = 0$. Write and solve the differential equation for this sales model.

Learning Theory In Exercises 17 and 18, assume that the rate of change in the proportion P of correct responses after n trials is proportional to the product of P and $L - P$, where L is the limiting proportion of correct responses.

17. Write and solve the differential equation for this learning theory model.

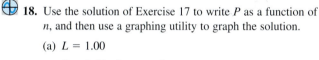 **18.** Use the solution of Exercise 17 to write P as a function of n, and then use a graphing utility to graph the solution.

(a) $L = 1.00$

   $P = 0.50$ when $n = 0$

   $P = 0.85$ when $n = 4$

(b) $L = 0.80$

   $P = 0.25$ when $n = 0$

   $P = 0.60$ when $n = 10$

Chemical Reaction In Exercises 19 and 20, use the chemical reaction model in Example 2 to find the amount y as a function of t, and use a graphing utility to graph the function.

19. $y = 45$ grams when $t = 0$; $y = 4$ grams when $t = 2$

20. $y = 75$ grams when $t = 0$; $y = 12$ grams when $t = 1$

In Exercises 21 and 22, use the Gompertz growth model described in Example 3 to find the growth function, and sketch its graph.

21. $L = 500$; $y = 100$ when $t = 0$; $y = 150$ when $t = 2$

22. $L = 5000$; $y = 500$ when $t = 0$; $y = 625$ when $t = 1$

23. Biology A population of eight beavers has been introduced into a new wetlands area. Biologists estimate that the maximum population the wetlands can sustain is 60 beavers. After 3 years, the population is 15 beavers. If the population follows a Gompertz growth model, how many beavers will be in the wetlands after 10 years?

24. Biology A population of 30 rabbits has been introduced into a new region. It is estimated that the maximum population the region can sustain is 400 rabbits. After 1 year, the population is estimated to be 90 rabbits. If the population follows a Gompertz growth model, how many rabbits will be present after 3 years?

Biology In Exercises 25 and 26, use the hybrid selection model in Example 4 to find the percent of the population that has the indicated characteristic.

25. You are studying a population of mayflies to determine how quickly characteristic A will pass from one generation to the next. At the start of the study, half the population has characteristic A. After four generations, 75% of the population has characteristic A. Find the percent of the population that will have characteristic A after 10 generations. (Assume $a = 2$ and $b = 1$.)

26. A research team is studying a population of snails to determine how quickly characteristic B will pass from one generation to the next. At the start of the study, 40% of the snails have characteristic B. After five generations, 80% of the population has characteristic B. Find the percent of the population that will have characteristic B after eight generations. (Assume $a = 2$ and $b = 1$.)

27. Chemical Reaction In a chemical reaction, a compound changes into another compound at a rate proportional to the unchanged amount, according to the model

$$\frac{dy}{dt} = ky.$$

(a) Solve the differential equation.

(b) If the initial amount of the original compound is 20 grams, and the amount remaining after 1 hour is 16 grams, when will 75% of the compound have been changed?

28. Chemical Mixture A 100-gallon tank is full of a solution containing 25 pounds of a concentrate. Starting at time $t = 0$, distilled water is admitted to the tank at the rate of 5 gallons per minute, and the well-stirred solution is withdrawn at the same rate.

(a) Find the amount Q of the concentrate in the solution as a function of t. (*Hint:* $Q' + Q/20 = 0$)

(b) Find the time when the amount of concentrate in the tank reaches 15 pounds.

29. Chemical Mixture A 200-gallon tank is half full of distilled water. At time $t = 0$, a solution containing 0.5 pound of concentrate per gallon enters the tank at the rate of 5 gallons per minute, and the well-stirred mixture is withdrawn at the same rate. Find the amount Q of concentrate in the tank after 30 minutes. (*Hint:* $Q' + Q/20 = \frac{5}{2}$)

30. Safety Assume that the rate of change in the number of miles s of road cleared per hour by a snowplow is inversely proportional to the depth h of snow. That is,

$$\frac{ds}{dh} = \frac{k}{h}.$$

Find s as a function of h if $s = 25$ miles when $h = 2$ inches and $s = 12$ miles when $h = 6$ inches ($2 \leq h \leq 15$).

31. *Chemistry* A wet towel hung from a clothesline to dry loses moisture through evaporation at a rate proportional to its moisture content. If after 1 hour the towel has lost 40% of its original moisture content, after how long will it have lost 80%?

32. *Biology* Let x and y be the sizes of two internal organs of a particular mammal at time t. Empirical data indicate that the relative growth rates of these two organs are equal, and can be modeled by

$$\frac{1}{x}\frac{dx}{dt} = \frac{1}{y}\frac{dy}{dt}.$$

Use this differential equation to write y as a function of x.

33. *Population Growth* When predicting population growth, demographers must consider birth and death rates as well as the net change caused by the difference between the rates of immigration and emigration. Let P be the population at time t and let N be the net increase per unit time due to the difference between immigration and emigration. So, the rate of growth of the population is given by

$$\frac{dP}{dt} = kP + N, \quad N \text{ is constant.}$$

Solve this differential equation to find P as a function of time.

34. *Meteorology* The barometric pressure y (in inches of mercury) at an altitude of x miles above sea level decreases at a rate proportional to the current pressure according to the model

$$\frac{dy}{dx} = -0.2y$$

where $y = 29.92$ inches when $x = 0$. Find the barometric pressure (a) at the top of Mt. St. Helens (8364 feet) and (b) at the top of Mt. McKinley (20,320 feet).

35. *Investment* A large corporation starts at time $t = 0$ to invest part of its receipts at a rate of P dollars per year in a fund for future corporate expansion. Assume that the fund earns r percent interest per year compounded continuously. So, the rate of growth of the amount A in the fund is given by

$$\frac{dA}{dt} = rA + P$$

where $A = 0$ when $t = 0$. Solve this differential equation for A as a function of t.

Investment In Exercises 36–38, use the result of Exercise 35.

36. Find A for each situation.

(a) $P = \$100,000$, $r = 12\%$, and $t = 5$ years

(b) $P = \$250,000$, $r = 15\%$, and $t = 10$ years

37. Find P if the corporation needs $\$120,000,000$ in 8 years and the fund earns $16\frac{1}{4}\%$ interest compounded continuously.

38. Find t if the corporation needs $\$800,000$ and it can invest $\$75,000$ per year in a fund earning 13% interest compounded continuously.

Medical Science In Exercises 39–41, a medical researcher wants to determine the concentration C (in moles per liter) of a tracer drug injected into a moving fluid. Solve this problem by considering a single-compartment dilution model (see figure). Assume that the fluid is continuously mixed and that the volume of fluid in the compartment is constant.

Figure for 39–41

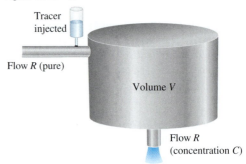

Tracer injected

Flow R (pure)

Volume V

Flow R (concentration C)

39. If the tracer is injected instantaneously at time $t = 0$, then the concentration of the fluid in the compartment begins diluting according to the differential equation

$$\frac{dC}{dt} = \left(-\frac{R}{V}\right)C, \quad C = C_0 \text{ when } t = 0.$$

(a) Solve this differential equation to find the concentration as a function of time.

(b) Find the limit of C as $t \to \infty$.

40. Use the solution of the differential equation in Exercise 39 to find the concentration as a function of time, and use a graphing utility to graph the function.

(a) $V = 2$ liters, $R = 0.5$ L/min, and $C_0 = 0.6$ mol/L

(b) $V = 2$ liters, $R = 1.5$ L/min, and $C_0 = 0.6$ mol/L

41. In Exercises 39 and 40, it was assumed that there was a single initial injection of the tracer drug into the compartment. Now consider the case in which the tracer is continuously injected (beginning at $t = 0$) at the rate of Q mol/min. Considering Q to be negligible compared with R, use the differential equation

$$\frac{dC}{dt} = \frac{Q}{V} - \left(\frac{R}{V}\right)C, \quad C = 0 \text{ when } t = 0.$$

(a) Solve this differential equation to find the concentration as a function of time.

(b) Find the limit of C as $t \to \infty$.

Answers to Selected Exercises

CHAPTER 0

SECTION 0.1   *(page 0-7)*

1. Rational   **3.** Irrational   **5.** Rational

7. Rational   **9.** Irrational

11. (a) Yes   (b) No   (c) Yes   (d) No

13. (a) Yes   (b) No   (c) No   (d) Yes

15. $x \geq 12$ **17.** $x < -\frac{1}{2}$

19. $x > 1$ **21.** $-\frac{1}{2} < x < \frac{7}{2}$

23. $-\frac{3}{4} < x < -\frac{1}{4}$ **25.** $x > 6$

27. $-\frac{3}{2} < x < 2$ **29.**

Hydrochloric acid   Pure water   Oven cleaner
Lemon juice   Black coffee   Baking soda

31. $x \geq 36$ units   **33.** $285.71 < x < 428.57$ miles

35. (a) False   (b) True   (c) True   (d) False

SECTION 0.2   *(page 0-12)*

1. (a) -51 (b) 51 (c) 51

3. (a) -14.99 (b) 14.99 (c) 14.99

5. (a) $-\frac{128}{75}$ (b) $\frac{128}{75}$ (c) $\frac{128}{75}$ **7.** $|x| \leq 2$

9. $|x| > 2$ **11.** $|x - 4| \leq 2$ **13.** $|x - 2| > 2$

15. $|x - 4| < 2$ **17.** $|y - a| \leq 2$

19. $-5 < x < 5$ **21.** $x < -6$ or $x > 6$

23. $-7 < x < 3$ **25.** $x \leq -7$ or $x \geq 13$

27. $x < 6$ or $x > 14$ **29.** $4 < x < 5$

31. $a - b \leq x \leq a + b$ **33.** $\dfrac{a - 8b}{3} < x < \dfrac{a + 8b}{3}$

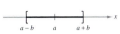 

35. 14   **37.** 1.25   **39.** $\frac{1}{8}$ **41.** $|M - 1083.4| < 0.2$

43. $65.8 \leq h \leq 71.2$ **45.** $175,000 \leq x \leq 225,000$

47. (a) $|4750 - E| \leq 500$, $|4750 - E| \leq 237.50$

    (b) At variance

49. (a) $|20,000 - E| \leq 500$, $|20,000 - E| \leq 1000$

    (b) At variance

SECTION 0.3   *(page 0-18)*

1. -24 **3.** $\frac{1}{2}$ **5.** 3   **7.** 44   **9.** 5   **11.** 9

13. $\frac{1}{2}$ **15.** $\frac{1}{4}$ **17.** 908.3483   **19.** -5.3601

21. $\dfrac{3}{4y^{14}}$ **23.** $10x^4$ **25.** $7x^5$

27. $\frac{4}{3}(x + y)^5$, $x \neq -y$ **29.** $3x$, $x > 0$

31. (a) $2\sqrt{2}$ (b) $3\sqrt{2}$

33. (a) $2x\sqrt[3]{2x^2}$ (b) $2|x|z\sqrt[4]{2z}$

35. (a) $\dfrac{2x^3z}{y}\sqrt[3]{\dfrac{18z^2}{y}}$ (b) $2(3x + 5)^3\sqrt{3(3x + 5)}$

37. $2x(2x^2 - 3)$ **39.** $(2x^3 + 1)/x^{1/2}$

41. $3(x + 1)^{1/2}(x + 2)(x - 1)$

43. $\dfrac{2(x - 1)^2}{(x + 1)^2}$ **45.** $\dfrac{(x^2 + 1)(3x^2 - 2x + 1)}{\sqrt{x - 1}}$

47. $x \geq 1$ **49.** $(-\infty, \infty)$ **51.** $(-\infty, 1) \cup (1, \infty)$

53. $x \neq 4$, $x \geq -2$ **55.** $1 \leq x \leq 5$ **57.** \$19,121.84

59. \$11,345.46   **61.** $\dfrac{\sqrt{2}}{2}\pi$ seconds or about 2.22 seconds

SECTION 0.4   *(page 0-24)*

1. $\frac{1}{2}, -\frac{1}{3}$ **3.** $\frac{3}{2}$ **5.** $-2 \pm \sqrt{3}$ **7.** $\dfrac{-3 \pm \sqrt{41}}{4}$

9. $(x - 2)^2$ **11.** $(2x + 1)^2$ **13.** $(x + 2)(x - 1)$

15. $(3x - 2)(x - 1)$ **17.** $(x - 2y)^2$

19. $(3 + y)(3 - y)(9 + y^2)$ **21.** $(x - 2)(x^2 + 2x + 4)$

23. $(y + 4)(y^2 - 4y + 16)$ **25.** $(x - 3)(x^2 + 3x + 9)$

27. $(x - 4)(x - 1)(x + 1)$ **29.** $(2x - 3)(x^2 + 2)$

31. $(x - 2)(2x^2 - 1)$ **33.** $(x + 4)(x - 4)(x^2 + 1)$

35. $0, 5$ **37.** ± 3 **39.** $\pm\sqrt{3}$ **41.** $0, 6$

43. $-2, 1$ **45.** $2, 3$ **47.** -4 **49.** ± 2 **51.** $1, \pm 2$

53. $(-\infty, -2] \cup [2, \infty)$ **55.** $(-\infty, 3] \cup [4, \infty)$

57. $(x + 1)(x^2 - 4x - 2)$ **59.** $(x - 1)(2x^2 + x - 1)$

61. $-2, -1, 4$ **63.** $1, 2, 3$ **65.** $-\frac{2}{3}, -\frac{1}{2}, 3$

67. 4 **69.** 2000 units   **71.** 3.4×10^{-5}

SECTION 0.5   *(page 0-32)*

1. $\dfrac{x + 5}{x - 1}$ **3.** $\dfrac{5x - 1}{x^2 + 2}$ **5.** $-\dfrac{x}{x^2 - 4}$ **7.** $\dfrac{2}{x - 3}$

9. $\dfrac{(A + C)x^2 - (A - B - 2C)x - (2A + 2B - C)}{(x + 1)^2(x - 2)}$

11. $\dfrac{(A + B)x^2 - (6B - C)x + 3(A - 2C)}{(x - 6)(x^2 + 3)}$

13. $-\dfrac{(x - 1)^2}{x(x^2 + 1)}$ **15.** $-\dfrac{x^2 + 3}{(x + 1)(x - 2)(x - 3)}$

17. $\dfrac{x + 2}{(x + 1)^{3/2}}$ **19.** $-\dfrac{3t}{2\sqrt{1 + t}}$ **21.** $\dfrac{x(x^2 + 2)}{(x^2 + 1)^{3/2}}$

23. $\dfrac{2}{x^2\sqrt{x^2 + 2}}$ **25.** $\dfrac{1}{2\sqrt{x}(x + 1)^{3/2}}$ **27.** $\dfrac{3x + 4}{2(x + 2)^{1/2}}$

29. $\dfrac{3x(x + 2)}{(2x + 3)^{3/2}}$ **31.** $\dfrac{\sqrt{6}}{2}$ **33.** $\dfrac{x\sqrt{x - 4}}{x - 4}$

35. $\dfrac{49\sqrt{x^2 - 9}}{x + 3}$ **37.** $\dfrac{\sqrt{14} + 2}{2}$ **39.** $\dfrac{x(5 + \sqrt{3})}{11}$

41. $\sqrt{6} - \sqrt{5}$ **43.** $\sqrt{x} - \sqrt{x - 2}$

45. $\dfrac{4 - 3x^2}{x^4(4 - x^2)^{3/2}}$ **47.** $232.68

CHAPTER 1

SECTION 1.1   *(page 8)*

Prerequisite Review

1. $3\sqrt{5}$ **2.** $2\sqrt{5}$ **3.** $\frac{1}{2}$ **4.** -2 **5.** $5\sqrt{3}$

6. $-\sqrt{2}$ **7.** $x = -3, x = 9$

8. $y = -8, y = 4$ **9.** $x = 19$ **10.** $y = 1$

1. (a) $a = 4, b = 3, c = 5$

(b) $4^2 + 3^2 = 5^2$

3. (a) $a = 10, b = 3, c = \sqrt{109}$

(b) $10^2 + 3^2 = \left(\sqrt{109}\right)^2$

5. (a) $a = 4, b = 5, c = \sqrt{41}$

(b) $4^2 + 5^2 = \left(\sqrt{41}\right)^2$

7. (a)

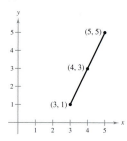

(b) $d = 2\sqrt{5}$

(c) Midpoint: $(4, 3)$

9. (a)

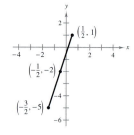

(b) $d = 2\sqrt{10}$

(c) Midpoint: $\left(-\frac{1}{2}, -2\right)$

11. (a)

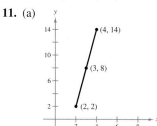

(b) $d = 2\sqrt{37}$ (c) Midpoint: $(3, 8)$

13. (a)

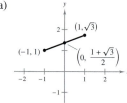

(b) $d = \sqrt{8 - 2\sqrt{3}}$ (c) Midpoint: $\left(0, \dfrac{1 + \sqrt{3}}{2}\right)$

15. $d_1 = \sqrt{45}, d_2 = \sqrt{20},$ **17.** $d_1 = d_2 = d_3 = d_4$

$d_3 = \sqrt{65}$ $= \sqrt{5}$

$d_1^2 + d_2^2 = d_3^2$

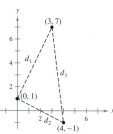

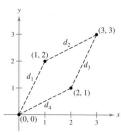

19. Collinear, because

$$d_1 + d_2 = d_3$$
$$2\sqrt{5} + \sqrt{5} = 3\sqrt{5}$$

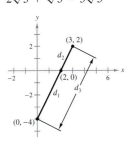

21. Not collinear, because

$$d_1 + d_2 \neq d_3$$
$$d_1 = \sqrt{18}, d_2 = \sqrt{41}$$
$$d_3 = \sqrt{113}$$

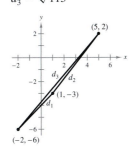

23. $x = 4, -2$ **25.** $y = \pm\sqrt{55}$

27. $\left(\dfrac{3x_1 + x_2}{4}, \dfrac{3y_1 + y_2}{4}\right), \left(\dfrac{x_1 + x_2}{2}, \dfrac{y_1 + y_2}{2}\right),$

$\left(\dfrac{x_1 + 3x_2}{4}, \dfrac{y_1 + 3y_2}{4}\right)$

29. (a) $\left(\tfrac{7}{4}, -\tfrac{7}{4}\right), \left(\tfrac{5}{2}, -\tfrac{5}{2}\right), \left(\tfrac{13}{4}, -\tfrac{5}{4}\right)$

(b) $\left(-\tfrac{3}{2}, -\tfrac{9}{4}\right), \left(-1, -\tfrac{3}{2}\right), \left(-\tfrac{1}{2}, -\tfrac{3}{4}\right)$

31. (a) 16.76 feet   (b) 1341.04 square feet

33. Answers will vary. Sample answer:

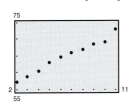

The number of subscribers appears to be increasing linearly.

35. (a) 10,400   (b) 8900   (c) 8500   (d) 10,500

37. (a) $85 thousand   (b) $100 thousand

(c) $122 thousand   (d) $159 thousand

39. (a) Revenue: $25,172 million

Profit: $890.7 million

(b) Actual 2001 revenue: $24,623 million

Actual 2001 profit: $885.6 million

(c) Yes, the increase in revenue per year is ≈ $3667 million.

Yes, the increase in profit per year is ≈ $133 million.

(d) Expenses for 1999: $17,214.9 million

Expenses for 2001: $23,737.4 million

Expenses for 2003: $31,347.7 million

(e) Answers will vary.

41. (a) $(-1, 2), (1, 1), (2, 3)$

(b)

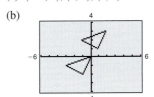

43. (a)

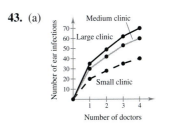

(b) The larger the clinic, the more patients a doctor can treat.

SECTION 1.2   *(page 21)*

Prerequisite Review

1. $y = \tfrac{1}{5}(x + 12)$ **2.** $y = x - 15$

3. $y = \dfrac{1}{x^3 + 2}$

4. $y = \pm\sqrt{x^2 + x - 6} = \pm\sqrt{(x + 3)(x - 2)}$

5. $y = -1 \pm \sqrt{9 - (x - 2)^2}$

6. $y = 5 \pm \sqrt{81 - (x + 6)^2}$ **7.** $x^2 - 4x + 4$

8. $x^2 + 6x + 9$ **9.** $x^2 - 5x + \tfrac{25}{4}$

10. $x^2 + 3x + \tfrac{9}{4}$ **11.** $(x - 2)(x - 1)$

12. $(x + 3)(x + 2)$ **13.** $\left(y - \tfrac{3}{2}\right)^2$ **14.** $\left(y - \tfrac{7}{2}\right)^2$

1. (a) Not a solution point   (b) Solution point

(c) Solution point

3. (a) Solution point   (b) Not a solution point

(c) Not a solution point

5. (a) Not a solution point   (b) Solution point

(c) Solution point

7. e   **8.** b   **9.** c   **10.** f   **11.** a   **12.** d

13. $(0, -3), \left(\tfrac{3}{2}, 0\right)$ **15.** $(0, -2), (-2, 0), (1, 0)$

17. $(0, 0), (-3, 0), (3, 0)$ **19.** $(-2, 0), (0, 2)$

21. $(0, 0)$

23.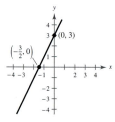

25.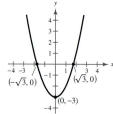

27.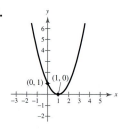

29.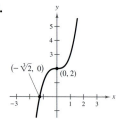

31.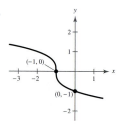

33.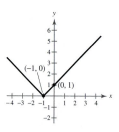

35.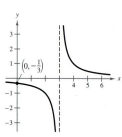

37.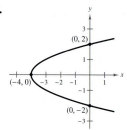

39. $x^2 + y^2 - 9 = 0$ **41.** $x^2 + y^2 - 4x + 2y - 11 = 0$

43. $x^2 + y^2 + 2x - 4y = 0$ **45.** $x^2 + y^2 - 6y = 0$

47. $(x - 1)^2 + (y + 3)^2 = 4$

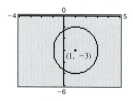

49. $(x + 2)^2 + (y + 3)^2 = 16$

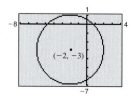

51. $\left(x - \frac{1}{2}\right)^2 + \left(y - \frac{1}{2}\right)^2 = 2$

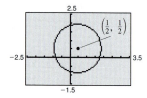

53. $\left(x + \frac{1}{2}\right)^2 + \left(y + \frac{5}{4}\right)^2 = \frac{9}{4}$

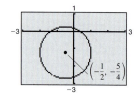

55. $(1, 1)$ **57.** $(3, 4), (5, 0)$

59. $(0, 0), \left(\sqrt{2}, 2\sqrt{2}\right), \left(-\sqrt{2}, -2\sqrt{2}\right)$

61. $(-1, 0), (0, 1), (1, 0)$

63. (a) $C = 11.8x + 15,000$; $R = 19.3x$ (b) 2000 units

(c) 2134 units

65. 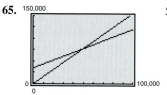   50,000 units

67. 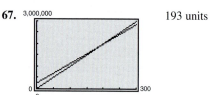   193 units

69. $(15, 120)$

71. (a) The model is good. Explanations will vary.

(b) $11,547 million

73. (a)

Year	1997	1998	1999
Salary	488.18	512.28	535.56

Year	2000	2001	2004
Salary	558.07	579.84	641.06

(b) Answers will vary.

(c) $678.79; answers will vary.

75.

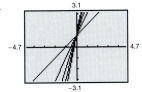

The greater the value of c, the steeper the line.

77.

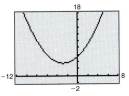

$(0, 5.36)$

79.

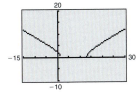

$(1.4780, 0), (12.8553, 0), (0, 2.3875)$

81.

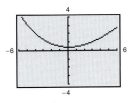

$(0, 0.4167)$

SECTION 1.3   (page 33)

Prerequisite Review

1. -1 **2.** $-\frac{7}{3}$ **3.** $\frac{1}{3}$ **4.** $-\frac{7}{6}$

5. $y = 4x + 7$ **6.** $y = 3x - 7$

7. $y = 3x - 10$ **8.** $y = -x - 7$

9. $y = 7x - 17$ **10.** $y = \frac{2}{3}x + \frac{5}{3}$

1. 1   **3.** 0

5.

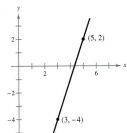

$m = 3$

7.

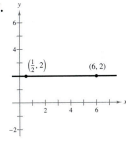

$m = 0$

9.

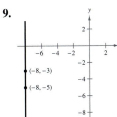

m is undefined.

11.

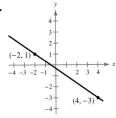

$m = -\frac{2}{3}$

13.

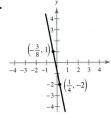

$m = -\frac{24}{5}$

15.

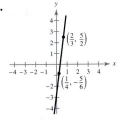

$m = 8$

17. $(0, 1), (1, 1), (3, 1)$ **19.** $(3, -6), (9, -2), (12, 0)$

21. $(0, 10), (2, 4), (3, 1)$ **23.** $(-8, 0), (-8, 2), (-8, 3)$

25. $m = -\frac{1}{5}, (0, 4)$ **27.** $m = \frac{7}{5}, (0, -3)$

29. $m = 3, (0, -15)$ **31.** m is undefined; no y-intercept.

33. $m = 0, (0, 4)$

35. $y = 2x - 5$ **37.** $3x + y = 0$

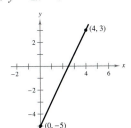

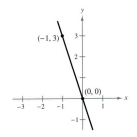

39. $x - 2 = 0$

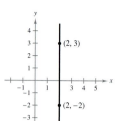

41. $y + 1 = 0$

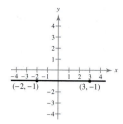

63. (a) $x + y + 1 = 0$ (b) $x - y + 5 = 0$

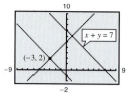

43. $3x - 6y + 7 = 0$

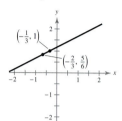

45. $4x - y + 6 = 0$

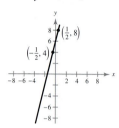

65. (a) $6x + 8y - 3 = 0$ (b) $96x - 72y + 127 = 0$

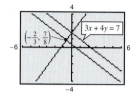

47. $3x - 4y + 12 = 0$

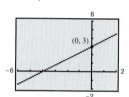

49. $x + 1 = 0$

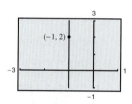

67. (a) $y = 0$ (b) $x + 1 = 0$

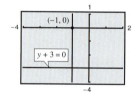

51. $y - 7 = 0$

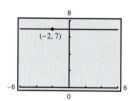

53. $y = -4x - 2$

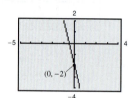

69. (a) $x - 1 = 0$ (b) $y - 1 = 0$

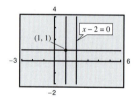

55. $9x - 12y + 8 = 0$

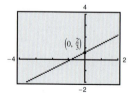

57. The points are not collinear.
Explanations of methods will vary.

59. $x - 3 = 0$ **61.** $y + 10 = 0$

71.

73.

75.

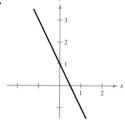

77.

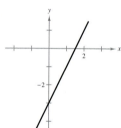

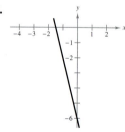

79. (a) $y = 49.4t + 3514.2$; The slope $m = 49.4$ tells you that the population increases 49.4 thousand each year.

(b) 3958.8 thousand (3,958,800)

(c) 4057.6 thousand (4,057,600)

(d) 1999: 3975 thousand (3,975,000);

2001: 4062 thousand (4,062,000)

The estimates were close to the actual populations.

(e) The model could possibly be used to predict the population in 2006 if the population continues to grow at the same linear rate.

81. $F = \frac{9}{5}C + 32$ or $C = \frac{5}{9}F - \frac{160}{9}$

83. $C = 0.34x + 150$

85. (a) $C = \frac{8}{5}A - 45$ or $A = \frac{5}{8}C + \frac{225}{8}$

(b) $F = \frac{72}{25}A - 49$ or $A = \frac{25}{72}F + \frac{1225}{72}$

(c) 75° (d) $C = 92.6°; F = 198.68°$ (e) $A = 56.25°$

87. (a) $y = 1025 - 205t, 0 \le t \le 5$

(b)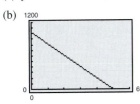

(c) \$410 (d) $t = 2.07$ years

89. (a) $x = -\frac{1}{15}p + \frac{226}{3}$ (b) 45 units (c) 49 units

91. (a) $Y = 437t + 3878$ (b) \$7811 billion

(c) \$9122 billion

(d) 1999: \$7786.5 billion

2002: \$8929.1 billion

93.

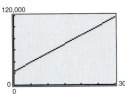

$x \le 24$ units

95.

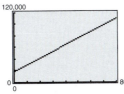

$x \le 70$ units

97.

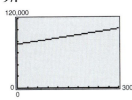

$x \le 275$ units

99.

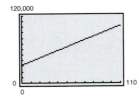

$x \le 104$ units

101.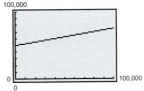

$x \le 200,000$ units

SECTION 1.4   *(page 45)*

Prerequisite Review

1. 20   **2.** 10   **3.** $x^2 + x - 6$

4. $x^3 + 9x^2 + 26x + 30$ **5.** $\frac{1}{x}$ **6.** $\frac{2x - 1}{x}$

7. $y = -2x + 17$ **8.** $y = \frac{6}{5}x^2 + \frac{1}{5}$

9. $y = 3 \pm \sqrt{5 + (x + 1)^2}$ **10.** $y = \pm \sqrt{4x^2 + 2}$

11. $y = 2x + \frac{1}{2}$ **12.** $y = \frac{x^3}{2} + \frac{1}{2}$

1. y is not a function of x.   **3.** y is a function of x.

5. y is a function of x.   **7.** y is not a function of x.

9. 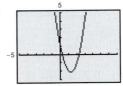   **11.**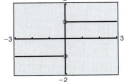

Domain: $(-\infty, \infty)$ Domain: $(-\infty, 0) \cup (0, \infty)$

Range: $[-2.125, \infty)$ Range: $y = -1$ or $y = 1$

13.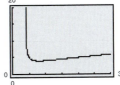

Domain: $(4, \infty)$

Range: $[4, \infty)$

15.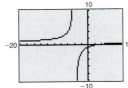

Domain: $(-\infty, -4) \cup (-4, \infty)$

Range: $(-\infty, 1) \cup (1, \infty)$

17. Domain: $(-\infty, \infty)$ **19.** Domain: $(-\infty, \infty)$

Range: $(-\infty, \infty)$ Range: $(-\infty, 4]$

21. (a) -3 (b) -9 (c) $2x - 5$ (d) $2x + 2\Delta x - 3$

23. (a) $\dfrac{1}{2}$ (b) 4 (c) $\dfrac{1}{x + 4}$ (d) $-\dfrac{\Delta x}{x(x + \Delta x)}$

25. $2x + \Delta x - 4$, $\Delta x \neq 0$

27. $\dfrac{1}{\sqrt{x + \Delta x + 3} + \sqrt{x + 3}}$, $\Delta x \neq 0$

29. $-\dfrac{1}{(x + \Delta x - 2)(x - 2)}$, $\Delta x \neq 0$

31. y is not a function of x.   **33.** y is a function of x.

35. (a) $2x$ (b) $10x - 25$ (c) $\dfrac{2x - 5}{5}$ (d) 5 (e) 5

37. (a) $x^2 + x$ (b) $(x^2 + 1)(x - 1) = x^3 - x^2 + x - 1$

(c) $\dfrac{x^2 + 1}{x - 1}$ (d) $x^2 - 2x + 2$ (e) x^2

39. (a) $\dfrac{x + 1}{x^2}$ (b) $\dfrac{1}{x^3}$ (c) $x, x \neq 0$

(d) $x^2, x \neq 0$ (e) $x^2, x \neq 0$

41. (a) 0 (b) 0 (c) -1 (d) $\sqrt{15}$

(e) $\sqrt{x^2 - 1}$ (f) $x - 1, x \geq 0$

43. The data fit the function (b), $g(x) = cx^2$, with $c = -2$.

45. The data fit the function (d), $r(x) = c/x$, with $c = 32$.

47. $f(g(x)) = 5\left(\dfrac{x - 1}{5}\right) + 1 = x$

$g(f(x)) = \dfrac{5x + 1 - 1}{5} = x$

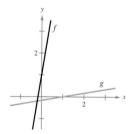

49. $f(g(x)) = 9 - \left(\sqrt{9 - x}\right)^2 = 9 - (9 - x) = x$

$g(f(x)) = \sqrt{9 - (9 - x^2)} = \sqrt{x^2} = x$

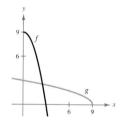

51. $f(x) = 2x - 3,\ f^{-1}(x) = \dfrac{x + 3}{2}$

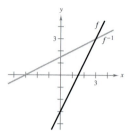

53. $f(x) = x^5,\ f^{-1}(x) = \sqrt[5]{x}$

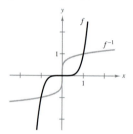

55. $f(x) = \sqrt{9 - x^2}, 0 \leq x \leq 3$

$f^{-1}(x) = \sqrt{9 - x^2}, 0 \leq x \leq 3$

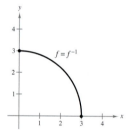

57. $f(x) = x^{2/3}, x \geq 0$

$f^{-1}(x) = x^{3/2}, x \geq 0$

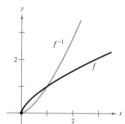

59.

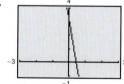

$f(x)$ is one-to-one. $f^{-1}(x) = \dfrac{3 - x}{7}$

61.

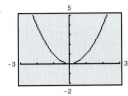

$f(x)$ is not one-to-one.

63.

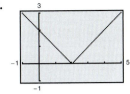

$f(x)$ is not one-to-one.

65. (a)

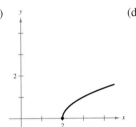

(b)

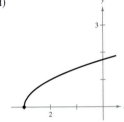

(c)

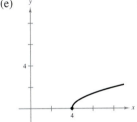

(d)

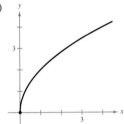

(e)

(f)

67. (a) $y = (x + 3)^2$ (b) $y = x^2 + 3$

(c) $y = -(x - 3)^2 + 6$ (d) $y = -(x + 6)^2 - 3$

69. $R_T = R_1 + R_2 = -0.8t^2 - 7.22t + 734$,
$t = 0, 1, \ldots, 6$

71. (a) $x = \dfrac{1475}{p} - 100$ (b) 47.5 units

73. $C(x(t)) = 2800t + 375$

C is the weekly cost in terms of t hours of manufacturing.

75. (a) $p = \begin{cases} 90, & 0 \le x \le 100 \\ 91 - 0.01x, & 100 < x \le 1600 \\ 75, & x > 1600 \end{cases}$

(b) $P = \begin{cases} 30x, & 0 \le x \le 100 \\ 31x - 0.01x^2, & 100 < x \le 1600 \\ 15x, & x > 1600 \end{cases}$

77. (a) $R = rn = [8 - 0.05(n - 80)]n$

(b)

n	90	100	110	120
R	675	700	715	720

n	130	140	150
R	715	700	675

(c) Answers will vary.

79.

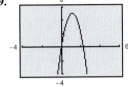

Zeros: $x = 0, \frac{9}{4}$
$f(x)$ is not one-to-one.

81.

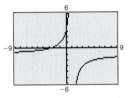

Zero: $t = -3$
$g(t)$ is one-to-one.

83.

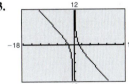

Zeros: $x = \pm 2$
$f(x)$ is not one-to-one.

85.

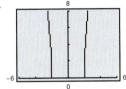

Zero: ± 2
$g(t)$ is not one-to-one.

87. Answers will vary.

SECTION 1.5   *(page 58)*

Prerequisite Review

1. (a) 7   (b) $c^2 - 3c + 3$

(c) $x^2 + 2xh + h^2 - 3x - 3h + 3$

2. (a) -4 (b) 10   (c) $3t^2 + 4$

3. h **4.** 4

5. Domain: $(-\infty, 0) \cup (0, \infty)$
   Range: $(-\infty, 0) \cup (0, \infty)$

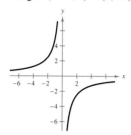

6. Domain: $[-5, 5]$ **7.** Domain: $(-\infty, \infty)$
   Range: $[0, 5]$ Range: $[0, \infty)$

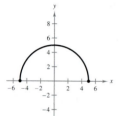

8. Domain: $(-\infty, 0) \cup (0, \infty)$
   Range: $-1, 1$

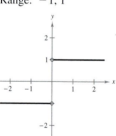

9. y is not a function of x.

10. y is a function of x.

1.

x	1.9	1.99	1.999	2
$f(x)$	13.5	13.95	13.995	?

x	2.001	2.01	2.1
$f(x)$	14.005	14.05	14.5

$\lim\limits_{x \to 2} (5x + 4) = 14$

3.

x	1.9	1.99	1.999	2
$f(x)$	0.2564	0.2506	0.2501	?

x	2.001	2.01	2.1
$f(x)$	0.2499	0.2494	0.2439

$\lim\limits_{x \to 2} \dfrac{x - 2}{x^2 - 4} = \dfrac{1}{4}$

5.

x	-0.1	-0.01	-0.001	0
$f(x)$	0.2911	0.2889	0.2887	?

x	0.001	0.01	0.1
$f(x)$	0.2887	0.2884	0.2863

$\lim\limits_{x \to 0} \dfrac{\sqrt{x + 3} - \sqrt{3}}{x} \approx 0.2887$

$\left(\text{The actual limit is } \dfrac{1}{2\sqrt{3}}. \right)$

7.

x	-0.5	-0.1	-0.01	-0.001	0
$f(x)$	-0.0714	-0.0641	-0.0627	-0.0625	?

$\lim\limits_{x \to 0^-} \dfrac{\dfrac{1}{x + 4} - \dfrac{1}{4}}{x} = -\dfrac{1}{16}$

9. (a) 1   (b) 3

11. (a) 1   (b) 3

13. (a) 12   (b) 27   (c) $\frac{1}{3}$

15. (a) 4   (b) 48   (c) 256

17. (a) 1   (b) 1   (c) 1

19. (a) 0   (b) 0   (c) 0

21. (a) 3   (b) -3 (c) Limit does not exist.

23. 16   **25.** -7 **27.** 0   **29.** 2   **31.** -2

33. $-\frac{3}{4}$ **35.** $\frac{35}{9}$ **37.** $\frac{1}{3}$ **39.** $-\frac{1}{20}$ **41.** -2

43. Limit does not exist.   **45.** $\frac{1}{10}$ **47.** 12

49. Limit does not exist.

51. -1 **53.** 2   **55.** $\dfrac{1}{2\sqrt{x + 2}}$ **57.** $2t - 5$

59.

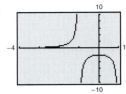

x	0	0.5	0.9	0.99
$f(x)$	-2	-2.67	-10.53	-100.5

x	0.999	0.9999	1
$f(x)$	-1000.5	$-10,000.5$	Undefined

$-\infty$

61.

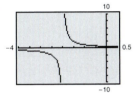

x	-3	-2.5	-2.1	-2.01
$f(x)$	-1	-2	-10	-100

x	-2.001	-2.0001	-2
$f(x)$	-1000	$-10,000$	Undefined

$-\infty$

63.

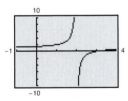

Limit does not exist.

65.

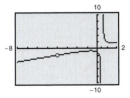

$-\frac{17}{9} \approx -1.8889$

67. (a)

x	-0.01	-0.001	-0.0001	0
$f(x)$	2.732	2.720	2.718	Undefined

x	0.0001	0.001	0.01
$f(x)$	2.718	2.717	2.705

$\lim_{x \to 0} (1 + x)^{1/x} \approx 2.718$

(b)

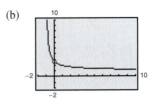

(c) Domain: $(-1, 0) \cup (0, \infty)$

Range: $(1, e) \cup (e, \infty)$

69. (a) $25,000   (b) 80%

(c) ∞; The cost function increases without bound as x approaches 100 from the left. Therefore, according to the model, it is not possible to remove 100% of the pollutants.

71. (a)

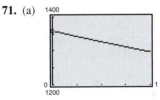

(b) For $x = 0.25$, $A \approx 1342.53$.

For $x = \frac{1}{365}$, $A \approx 1358.95$.

(c) $\lim_{x \to 0^+} 500(1 + 0.1x)^{10/x} = 500e \approx 1359.14$; continuous compounding

SECTION 1.6   *(page 69)*

Prerequisite Review

1. $\dfrac{x + 4}{x - 8}$ **2.** $\dfrac{x + 1}{x - 3}$ **3.** $\dfrac{x + 2}{2(x - 3)}$ **4.** $\dfrac{x - 4}{x - 2}$

5. $x = 0, -7$ **6.** $x = -5, 1$ **7.** $x = -\frac{2}{3}, -2$

8. $x = 0, 3, -8$ **9.** 13   **10.** -1

1. Continuous; The function is a polynomial.

3. Not continuous $(x \neq \pm 2)$

5. Continuous; The rational function's domain is the set of real numbers.

7. Not continuous ($x \neq 3$ and $x \neq 5$)

9. Not continuous ($x \neq \pm 2$)

11. $(-\infty, 0)$ and $(0, \infty)$ **13.** $(-\infty, -1)$ and $(-1, \infty)$

15. $(-\infty, \infty)$ **17.** $(-\infty, -1)$ and $(-1, 1)$ and $(1, \infty)$

19. $(-\infty, \infty)$ **21.** $(-\infty, 4)$ and $(4, 5)$ and $(5, \infty)$

23. Continuous on all intervals $\left(\dfrac{c}{2}, \dfrac{c}{2} + \dfrac{1}{2}\right)$ where c is an integer.

25. $(-\infty, \infty)$ **27.** $(-\infty, 2]$ and $(2, \infty)$

29. $(-\infty, -1)$ and $(-1, \infty)$

31. Continuous on all intervals $(c, c + 1)$, where c is an integer.

33. $(1, \infty)$ **35.** Continuous

37. Nonremovable discontinuity at $x = 2$

39.

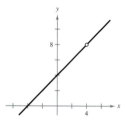

Continuous on $(-\infty, 4)$ and $(4, \infty)$

41.

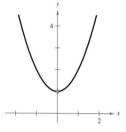

Continuous on $(-\infty, 0)$ and $(0, \infty)$

43.

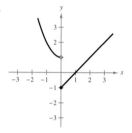

Continuous on $(-\infty, 0)$ and $(0, \infty)$

45. $a = 2$

47.

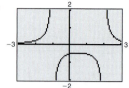

Not continuous at $x = 2$ and $x = -1$

49.

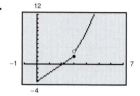

Not continuous at $x = 3$

51.

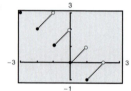

Not continuous at all integers c

53. $(-\infty, \infty)$

55. Continuous on all intervals $\left(\dfrac{c}{2}, \dfrac{c + 1}{2}\right)$, where c is an integer.

57.

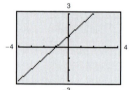

The graph of $f(x) = \dfrac{x^2 + x}{x}$ appears to be continuous on $[-4, 4]$, but f is not continuous at $x = 0$.

59. (a)

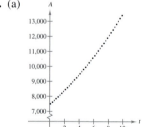

The graph has nonremovable discontinuities at $t = \frac{1}{4}, \frac{1}{2}, \frac{3}{4}, 1, \frac{5}{4}, \ldots$

(b) $11,379.17

61. $C = 9.80 - 2.50[\![1 - x]\!]$

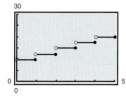

C is not continuous at $x = 1, 2, 3, \ldots$

63. (a)

$$C(t) = \begin{cases} 1.04, & 0 < t \le 2 \\ 1.04 + 0.36[\![t - 1]\!], & t > 2, t \text{ is not an integer.} \\ 1.04 + 0.36(t - 2), & t > 2, t \text{ is an integer.} \end{cases}$$

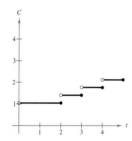

C is not continuous at $t = 2, 3, 4, \ldots$

(b) $3.56

65. (a)

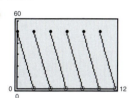

Nonremovable discontinuities at $t = 2, 4, 6, 8, \ldots$

(b) Every 2 months

67.

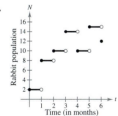

Nonremovable discontinuities at $t = 1, 2, 3, 4, 5,$ and 6

REVIEW EXERCISES FOR CHAPTER 1
(page 76)

1. a   **2.** c   **3.** b   **4.** d   **5.** $\sqrt{29}$ **7.** $3\sqrt{2}$

9. $(7, 4)$ **11.** $(-8, 6)$

13. The tallest bars in the graph represent revenues. The middle bars represent costs. The bars on the left in each group represent profits, because $P = R - C$.

15. $(4, 7), (5, 8), (8, 10)$

17.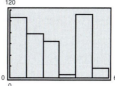

19. 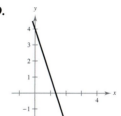   **21.**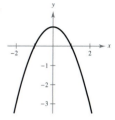

23. 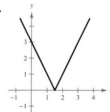   **25.**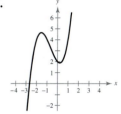

27.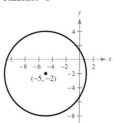

29. $(0, 1), (1, 0), (-1, 0)$ **31.** $(x - 2)^2 + (y + 1)^2 = 73$

33. $(x + 5)^2 + (y + 2)^2 = 36$

Center: $(-5, -2)$

Radius: 6

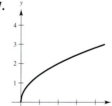

35. $(2, 1), (-1, -2)$ **37.** $(0, 0), (1, 1)$

39. (a) $C = 6000 + 6.50x$

$R = 13.90x$

(b) ≈ 811 units

41. Slope: -3

y-intercept: $(0, -2)$

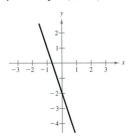

43. Slope: 0 (horizontal line)   **45.** Slope: $-\frac{2}{5}$

y-intercept: $\left(0, -\frac{5}{3}\right)$ y-intercept: $(0, -1)$

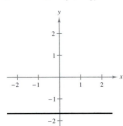   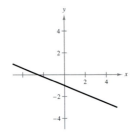

47. $\frac{6}{7}$ **49.** $\frac{20}{21}$

51. $y = -2x + 5$

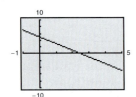

53. (a) $7x - 8y + 69 = 0$ (b) $2x + y = 0$

(c) $2x + y = 0$ (d) $2x + 3y - 12 = 0$

55. (a) $x = -10p + 1070$ (b) 725 units   (c) 650 units

57. y is a function of x.   **59.** y is not a function of x.

61. (a) 7   (b) $3x + 7$ (c) $10 + 3\Delta x$

63. 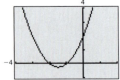   Domain: $(-\infty, \infty)$

Range: $\left[-\frac{1}{4}, \infty\right)$

65. 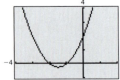   Domain: $[-1, \infty)$

Range: $[0, \infty)$

67.   Domain: $(-\infty, \infty)$

Range: $(-\infty, 3]$

69. (a) $x^2 + 2x$ (b) $x^2 - 2x + 2$

(c) $2x^3 - x^2 + 2x - 1$ (d) $\dfrac{1 + x^2}{2x - 1}$

(e) $4x^2 - 4x + 2$ (f) $2x^2 + 1$

71. $f^{-1}(x) = \frac{2}{3}x$

73. $f(x)$ does not have an inverse function.

75. 7   **77.** 49   **79.** $\frac{10}{3}$ **81.** -2

83. $-\frac{1}{4}$ **85.** $-\infty$ **87.** Limit does not exist.

89. $-\frac{1}{16}$ **91.** $3x^2 - 1$ **93.** 0.5774

95. False, limit does not exist.

97. False, limit does not exist.

99. False, limit does not exist.

101. $(-\infty, -4)$ and $(-4, \infty)$ **103.** $(-\infty, -1)$ and $(-1, \infty)$

105. Continuous on all intervals $(c, c + 1)$, where c is an integer

107. $(-\infty, 0)$ and $(0, \infty)$ **109.** $a = 2$

111. (a)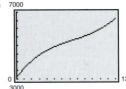

(b)

t	0	1	2	3	4
Debt	3206.3	3598.2	4001.8	4351.0	4643.3
Model	3103.6	3644.1	4078.9	4424.7	4697.8

t	5	6	7	8	9
Debt	4920.6	5181.5	5369.2	5478.2	5605.5
Model	4914.8	5092.2	5246.5	5394.1	5551.7

t	10	11	12	13
Debt	5628.7	5769.9	6198.4	6752.0
Model	5735.6	5962.4	6248.6	6610.8

(c) $10,137.2 billion

113.

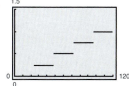

(a) Discontinuous at $x = 24n$, where n is a positive integer

(b) $15.50

SAMPLE POST-GRAD EXAM QUESTIONS
(page 80)

1. d   **2.** d   **3.** b   **4.** a   **5.** e

6. c   **7.** e   **8.** b   **9.** b

CHAPTER 2

SECTION 2.1   *(page 90)*

Prerequisite Review

1. $x = 2$ **2.** $y = 2$ **3.** $2x$ **4.** $3x^2$

5. $\dfrac{1}{x^2}$ **6.** $2x$ **7.** $(-\infty, 1) \cup (1, \infty)$

8. $(-\infty, \infty)$ **9.** $(-\infty, 0) \cup (0, \infty)$

10. $(-\infty, -4) \cup (-4, 3) \cup (3, \infty)$

1.

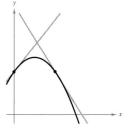

3.

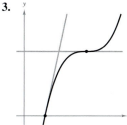

5. $m = 1$ **7.** $m = 0$ **9.** $m = -\frac{1}{3}$

11. 1997: $m \approx 250$ **13.** $t = 1$: $m \approx 65$

2000: $m \approx 200$ $t = 8$: $m \approx 0$

2002: $m \approx 0$ $t = 12$: $m \approx -1000$

15. $f(x) = 3$

$f(x + \Delta x) = 3$

$f(x + \Delta x) - f(x) = 0$

$\dfrac{f(x + \Delta x) - f(x)}{\Delta x} = 0$

$\lim\limits_{\Delta x \to 0} \dfrac{f(x + \Delta x) - f(x)}{\Delta x} = 0$

17. $f(x) = -5x + 3$

$f(x + \Delta x) = -5x - 5\Delta x + 3$

$f(x + \Delta x) - f(x) = -5\Delta x$

$\dfrac{f(x + \Delta x) - f(x)}{\Delta x} = -5$

$\lim\limits_{\Delta x \to 0} \dfrac{f(x + \Delta x) - f(x)}{\Delta x} = -5$

19. $f(x) = x^2 - 4$

$f(x + \Delta x) = x^2 + 2x\Delta x + (\Delta x)^2 - 4$

$f(x + \Delta x) - f(x) = 2x\Delta x + (\Delta x)^2$

$\dfrac{f(x + \Delta x) - f(x)}{\Delta x} = 2x + \Delta x$

$\lim\limits_{\Delta x \to 0} \dfrac{f(x + \Delta x) - f(x)}{\Delta x} = 2x$

21. $h(t) = \sqrt{t - 1}$

$h(t + \Delta t) = \sqrt{t + \Delta t - 1}$

$h(t + \Delta t) - h(t) = \sqrt{t + \Delta t - 1} - \sqrt{t - 1}$

$\dfrac{h(t + \Delta t) - h(t)}{\Delta t} = \dfrac{1}{\sqrt{t + \Delta t - 1} + \sqrt{t - 1}}$

$\lim\limits_{\Delta t \to 0} \dfrac{h(t + \Delta t) - h(t)}{\Delta t} = \dfrac{1}{2\sqrt{t - 1}}$

23. $f(t) = t^3 - 12t$

$f(t + \Delta t) = t^3 + 3t^2\Delta t + 3t(\Delta t)^2$
$\qquad\qquad + (\Delta t)^3 - 12t - 12\Delta t$

$f(t + \Delta t) - f(t) = 3t^2\Delta t + 3t(\Delta t)^2 + (\Delta t)^3 - 12\Delta t$

$\dfrac{f(t + \Delta t) - f(t)}{\Delta t} = 3t^2 + 3t\Delta t + (\Delta t)^2 - 12$

$\lim\limits_{\Delta t \to 0} \dfrac{f(t + \Delta t) - f(t)}{\Delta t} = 3t^2 - 12$

25. $f(x) = \dfrac{1}{x + 2}$

$f(x + \Delta x) = \dfrac{1}{x + \Delta x + 2}$

$f(x + \Delta x) - f(x) = \dfrac{-\Delta x}{(x + \Delta x + 2)(x + 2)}$

$\dfrac{f(x + \Delta x) - f(x)}{\Delta x} = \dfrac{-1}{(x + \Delta x + 2)(x + 2)}$

$\lim\limits_{\Delta x \to 0} \dfrac{f(x + \Delta x) - f(x)}{\Delta x} = -\dfrac{1}{(x + 2)^2}$

27. $f'(x) = -2$ **29.** $f'(x) = 0$

$f'(2) = -2$ $f'(0) = 0$

31. $f'(x) = 2x$ **33.** $f'(x) = 3x^2 - 1$

$f'(2) = 4$ $f'(2) = 11$

35. $f'(x) = -\dfrac{1}{\sqrt{1-2x}}$

$f'(-4) = -\dfrac{1}{3}$

37. $y = 2x - 2$

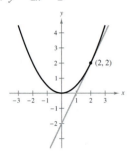

39. $y = -6x - 3$

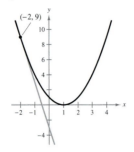

41. $y = \dfrac{x}{4} + 2$

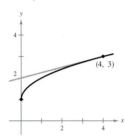

43. $y = -x + 2$

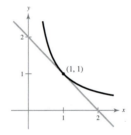

45. $y = -x + 1$

47. $y = -6x + 8$

$y = -6x - 8$

49. $x \neq -3$ (node)   **51.** $x \neq 3$ (cusp)   **53.** $x > 1$

55. $x \neq 0$ (nonremovable discontinuity)

57.

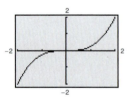

$f'(x) = \frac{3}{4}x^2$

x	-2	$-\frac{3}{2}$	-1	$-\frac{1}{2}$
$f(x)$	-2	-0.8438	-0.25	-0.0313
$f'(x)$	3	1.6875	0.75	0.1875

x	0	$\frac{1}{2}$	1	$\frac{3}{2}$	2
$f(x)$	0	0.0313	0.25	0.8438	2
$f'(x)$	0	0.1875	0.75	1.6875	3

59.

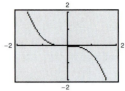

$f'(x) = -\frac{3}{2}x^2$

x	-2	$-\frac{3}{2}$	-1	$-\frac{1}{2}$
$f(x)$	4	1.6875	0.5	0.0625
$f'(x)$	-6	-3.375	-1.5	-0.375

x	0	$\frac{1}{2}$	1	$\frac{3}{2}$	2
$f(x)$	0	-0.0625	-0.5	-1.6875	-4
$f'(x)$	0	-0.375	-1.5	-3.375	-6

61. $f'(x) = 2x - 4$

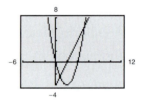

The x-intercept of the derivative indicates a point of horizontal tangency for f.

63. $f'(x) = 3x^2 - 3$

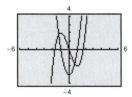

The x-intercepts of the derivative indicate points of horizontal tangency for f.

65. Answers will vary. Sample answer: $f(x) = -x$

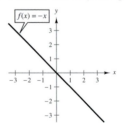

$f'(x) = -1$

67.

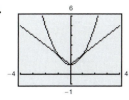

The graph of f is smooth at $(0, 1)$, but the graph of g has a sharp point at $(0, 1)$. The function g is not differentiable at $x = 0$.

69. False. $f(x) = |x|$ is continuous, but not differentiable, at $x = 0$.

71. True

SECTION 2.2   *(page 102)*

Prerequisite Review

1. (a) 8   (b) 16   (c) $\frac{1}{2}$

2. (a) $\frac{1}{36}$ (b) $\frac{1}{32}$ (c) $\frac{1}{64}$

3. $4x(3x^2 + 1)$ **4.** $\frac{3}{2}x^{1/2}(x^{3/2} - 1)$ **5.** $\frac{1}{4x^{3/4}}$

6. $x^2 - \frac{1}{x^{1/2}} + \frac{1}{3x^{2/3}}$ **7.** $0, -\frac{2}{3}$

8. $0, \pm 1$ **9.** $-10, 2$ **10.** $-2, 12$

---

1. (a) 2   (b) $\frac{1}{2}$ **3.** (a) -1 (b) $-\frac{1}{3}$ **5.** 0

7. 4   **9.** $2x + 4$ **11.** $-6t + 2$ **13.** $3t^2 - 2$

15. $\frac{16}{3}t^{1/3}$ **17.** $\frac{2}{\sqrt{x}}$ **19.** $-\frac{8}{x^3} + 4x$

21. Function: $y = \frac{1}{4x^3}$

Rewrite: $y = \frac{1}{4}x^{-3}$

Differentiate: $y' = \frac{-3}{4}x^{-4}$

Simplify: $y' = -\frac{3}{4x^4}$

23. Function: $y = \frac{1}{(4x)^3}$

Rewrite: $y = \frac{1}{64}x^{-3}$

Differentiate: $y' = -\frac{3}{64}x^{-4}$

Simplify: $y' = -\frac{3}{64x^4}$

25. Function: $y = \frac{\sqrt{x}}{x}$

Rewrite: $y = x^{-1/2}$

Differentiate: $y' = -\frac{1}{2}x^{-3/2}$

Simplify: $y' = -\frac{1}{2x^{3/2}}$

27. -1 **29.** -2 **31.** 4   **33.** $2x + \frac{4}{x^2} + \frac{6}{x^3}$

35. $2x - 2 + \frac{8}{x^5}$ **37.** $3x^2 + 1$ **39.** $6x^2 + 16x - 1$

41. $\frac{2x^3 - 6}{x^3}$ **43.** $\frac{4x^3 - 2x - 10}{x^3}$ **45.** $\frac{4}{5x^{1/5}} + 1$

47. $y = 2x - 2$ **49.** $y = \frac{8}{15}x + \frac{22}{15}$

51. $(0, -1), \left(-\frac{\sqrt{6}}{2}, \frac{5}{4}\right), \left(\frac{\sqrt{6}}{2}, \frac{5}{4}\right)$ **53.** $(-5, -12.5)$

55. (a)

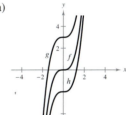

(b) $f'(1) = g'(1)$
    $= h'(1)$
    $= 3$

(c)

57. (a) 3   (b) 6   (c) -3 (d) 6

59. (a) 1998: 401.1

2001: -79.1

2003: 516.2

(b) The results are similar.

(c) Millions of dollars per year per year

61. $p = 0.40x - 250$

$p' = 0.40$

63. 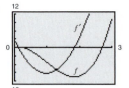 $(0.11, 0.14), (1.84, -10.49)$

65. False. Let $f(x) = x$ and $g(x) = x + 1$.

SECTION 2.3   *(page 116)*

Prerequisite Review

1. 3   **2.** -7 **3.** $y' = 8x - 2$

4. $y' = -9t^2 + 4t$ **5.** $s' = -32t + 24$

6. $y' = -32x + 54$ **7.** $A' = -\frac{3}{5}r^2 + \frac{3}{5}r + \frac{1}{2}$

8. $y' = 2x^2 - 4x + 7$ **9.** $y' = 12 - \dfrac{x}{2500}$

10. $y' = 74 - \dfrac{3x^2}{10,000}$

---

1. (a) \$11 billion per year   (b) \$7 billion per year

(c) \$6 billion per year   (d) \$16 billion per year

(e) \$9.5 billion per year   (f) \$10.4 billion per year

3.

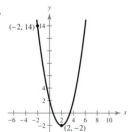

Average rate: 2

Instantaneous rates: $f'(1) = 2, f'(2) = 2$

5.

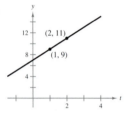

Average rate: -4

Instantaneous rates: $h'(-2) = -8, h'(2) = 0$

7.

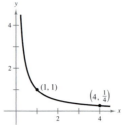

Average rate: $-\frac{1}{4}$

Instantaneous rates: $f'(1) = -1, f'(4) = -\frac{1}{16}$

9.

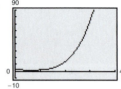

Average rate: 36

Instantaneous rates: $g'(1) = 2, g'(3) = 102$

11. (a) -500

The number of visitors to the park is decreasing at an average rate of 500 hundred thousand people per month from September to December.

(b) Answers will vary. The instantaneous rate of change at $t = 8$ is approximately 0.

13. (a) Average rate: $\frac{11}{27}$

Instantaneous rates: $E'(0) = \frac{1}{3}, E'(1) = \frac{4}{9}$

(b) Average rate: $\frac{11}{27}$

Instantaneous rates: $E'(1) = \frac{4}{9}, E'(2) = \frac{1}{3}$

(c) Average rate: $\frac{5}{27}$

Instantaneous rates: $E'(2) = \frac{1}{3}, E'(3) = 0$

(d) Average rate: $-\frac{7}{27}$

Instantaneous rates: $E'(3) = 0, E'(4) = -\frac{5}{9}$

15. (a) -80 feet per second

(b) $s'(2) = -64$ feet per second,
$s'(3) = -96$ feet per second

(c) $\dfrac{\sqrt{555}}{4} \approx 5.89$ seconds

(d) $-8\sqrt{555} \approx -188.5$ feet per second

17. 1.47 dollars   **19.** $470 - 0.5x$ dollars, $0 \le x \le 940$

21. $50 - x$ dollars   **23.** $-18x^2 + 16x + 200$ dollars

25. $-4x + 72$ dollars   **27.** $-0.0005x + 12.2$ dollars

29. (a) \$0.58   (b) \$0.60

(c) The results are nearly the same.

31. (a) $4.95   (b) $5.00

(c) The results are nearly the same.

33. (a)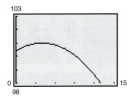

(b) For $t < 4$, increasing; for $t > 4$, decreasing; shows when fever is going up and down.

(c) $T(0) = 100.4$

$T(4) = 101$

$T(8) = 100.4$

$T(12) = 98.6$

(d) $T'(t) = -0.075t + 0.3$

The rate of change of temperature

(e) $T'(0) = 0.3$

$T'(4) = 0$

$T'(8) = -0.3$

$T'(12) = -0.6$

35. (a) $R = 5x - 0.001x^2$

(b) $P = -0.001x^2 + 3.5x - 35$

(c)

x	600	1200	1800	2400	3000
dR/dx	3.8	2.6	1.4	0.2	-1
dP/dx	2.3	1.1	-0.1	-1.3	-2.5
P	1705	2725	3025	2605	1465

37. (a) $P = -0.0025x^2 + 1.7x - 20$

(b)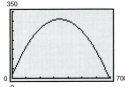

When $x = 200$, slope is positive.

When $x = 400$, slope is negative.

(c) $P'(200) = 0.7$

$P'(400) = -0.3$

39. (a) $P = -\dfrac{1}{6000}x^2 + 11.8x - 85,000$

(b)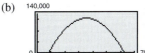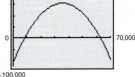

When $x = 18,000$, slope is positive.

When $x = 36,000$, slope is negative.

(c) $P'(18,000) = 5.8$ dollars

$P'(36,000) = -0.2$ dollars

41. (a) $0.33 per unit   (b) $0.13 per unit

(c) $0 per unit   (d) $$-0.08$ per unit

$p'(2500) = 0$ indicates that $x = 2500$ is the optimal value of x. So, $p = \dfrac{50}{\sqrt{x}} = \dfrac{50}{\sqrt{2500}} = \$1.00$.

43. (a) ≈ 4.7 miles per year

(b) ≈ 3.5 miles per year

(c) ≈ 3.85 miles per year

45. (a) The rate of change of the number of gallons of gasoline sold when the price is $1.479 per gallon.

(b) In general, the rate of change when $p = 1.479$ should be negative.

47.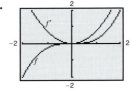

f has a horizontal tangent at $x = 0$.

49. The population in each phase is increasing. During the acceleration phase the population's growth is the greatest, so the slopes of the tangent lines are greater than the slopes of the tangent lines during the lag phase and the deceleration phase. Possible reasons for the changing rates could be seasonal growth and food supplies.

SECTION 2.4   *(page 128)*

Prerequisite Review

1. $2(3x^2 + 7x + 1)$ **2.** $4x^2(6 - 5x^2)$

3. $8x^2(x^2 + 2)^3 + (x^2 + 4)$

4. $(2x)(2x + 1)[2x + (2x + 1)^3]$

5. $\dfrac{23}{(2x + 7)^2}$ **6.** $-\dfrac{x^2 + 8x + 4}{(x^2 - 4)^2}$

7. $-\dfrac{2(x^2 + x - 1)}{(x^2 + 1)^2}$ **8.** $\dfrac{4(3x^4 - x^3 + 1)}{(1 - x^4)^2}$

9. $\dfrac{4x^3 - 3x^2 + 3}{x^2}$ **10.** $\dfrac{x^2 - 2x + 4}{(x - 1)^2}$

11. 11   **12.** 0   **13.** $-\dfrac{1}{4}$ **14.** $\dfrac{17}{4}$

1. $f'(1) = 13$ **3.** $f'(0) = 0$ **5.** $g'(4) = 11$

7. $h'(6) = -5$ **9.** $f'(3) = \dfrac{3}{4}$ **11.** $g'(6) = -11$

13. $f'(1) = \dfrac{2}{5}$

15. Function: $y = \dfrac{x^2 + 2x}{x}$

Rewrite: $y = x + 2,\ x \neq 0$

Differentiate: $y' = 1,\ x \neq 0$

Simplify: $y' = 1,\ x \neq 0$

17. Function: $y = \dfrac{7}{3x^3}$

Rewrite: $y = \dfrac{7}{3}x^{-3}$

Differentiate: $y' = -7x^{-4}$

Simplify: $y' = -\dfrac{7}{x^4}$

19. Function: $y = \dfrac{4x^2 - 3x}{8\sqrt{x}}$

Rewrite: $y = \dfrac{1}{2}x^{3/2} - \dfrac{3}{8}x^{1/2}$

Differentiate: $y' = \dfrac{3}{4}x^{1/2} - \dfrac{3}{16}x^{-1/2}$

Simplify: $y' = \dfrac{3}{4}\sqrt{x} - \dfrac{3}{16\sqrt{x}}$

21. Function: $y = \dfrac{x^2 - 4x + 3}{x - 1}$

Rewrite: $y = x - 3,\ x \neq 1$

Differentiate: $y' = 1,\ x \neq 1$

Simplify: $y' = 1,\ x \neq 1$

23. $10x^4 + 12x^3 - 3x^2 - 18x - 15$

25. $12t^2(2t^3 - 1)$

27. $\dfrac{5}{6x^{1/6}} + \dfrac{1}{x^{2/3}}$ **29.** $-\dfrac{5}{(2x - 3)^2}$

31. $\dfrac{2}{(x + 1)^2},\ x \neq 1$ **33.** $\dfrac{x^2 + 2x - 1}{(x + 1)^2}$

35. $\dfrac{3s^2 - 2s - 5}{2s^{3/2}}$ **37.** $\dfrac{2x^3 + 11x^2 - 8x - 17}{(x + 4)^2}$

39. $y = 5x - 2$ **41.** $y = \dfrac{3}{4}x - \dfrac{5}{4}$

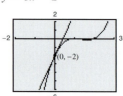   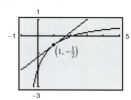

43. $y = -16x - 5$

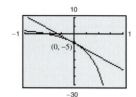

45. $(0, 0),\ (2, 4)$ **47.** $(0, 0),\ (\sqrt[3]{-4},\ -2.117)$

49.   **51.**

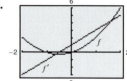   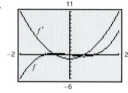

53. -1.87 **55.** (a) -0.480 (b) 0.120 (c) 0.015

57. 31.55 bacteria per hour

59. (a) $p = \dfrac{4000}{\sqrt{x}}$ (b) $C = 250x + 10{,}000$

(c) $P = 4000\sqrt{x} - 250x - 10{,}000$

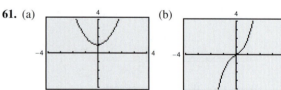

$500 per unit

61. (a)   (b)

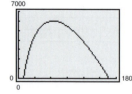

(c)

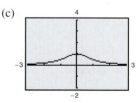

The graph of (c) would most likely represent a demand function. As the number of units increases, demand is likely to decrease, not increase as in (a) and (b).

63. (a) -38.125

(b) -10.37

(c) -3.80

Increasing the order size reduces the cost per item.

65. $\dfrac{dP}{dt} = -\dfrac{1.25(40{,}900 - 8360t + 403t^2)}{(2500 - 515t + 28t^2)^2}$

$P'(5) = -0.0294$

$P'(7) = -0.0373$

$P'(9) = 0.1199$

$P'(11) = 0.0577$

The rate of change in price at year t

SECTION 2.5   *(page 138)*

Prerequisite Review

1. $(1 - 5x)^{2/5}$ **2.** $(2x - 1)^{3/4}$

3. $(4x^2 + 1)^{-1/2}$ **4.** $(x - 6)^{-1/3}$

5. $x^{1/2}(1 - 2x)^{-1/3}$ **6.** $(2x)^{-1}(3 - 7x)^{3/2}$

7. $(x - 2)(3x^2 + 5)$ **8.** $(x - 1)(5\sqrt{x} - 1)$

9. $(x^2 + 1)^2(4 - x - x^3)$

10. $(3 - x^2)(x - 1)(x^2 + x + 1)$

$y = f(g(x))$	$u = g(x)$	$y = f(u)$
1. $y = (6x - 5)^4$	$u = 6x - 5$	$y = u^4$
3. $y = (4 - x^2)^{-1}$	$u = 4 - x^2$	$y = u^{-1}$
5. $y = \sqrt{5x - 2}$	$u = 5x - 2$	$y = \sqrt{u}$
7. $y = \dfrac{1}{3x + 1}$	$u = 3x + 1$	$y = u^{-1}$

9. c   **11.** b   **13.** a   **15.** c   **17.** $6(2x - 7)^2$

19. $-6(4 - 2x)^2$ **21.** $6x(6 - x^2)(2 - x^2)$

23. $\dfrac{4x}{3(x^2 - 9)^{1/3}}$ **25.** $\dfrac{1}{2\sqrt{t + 1}}$ **27.** $\dfrac{4t + 5}{2\sqrt{2t^2 + 5t + 2}}$

29. $\dfrac{6x}{(9x^2 + 4)^{2/3}}$ **31.** $\dfrac{27}{4(2 - 9x)^{3/4}}$ **33.** $\dfrac{4x^2}{(4 - x^3)^{7/3}}$

35. $y = 216x - 378$ **37.** $y = \frac{8}{3}x - \frac{7}{3}$

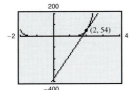

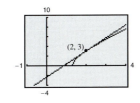

39. $y = x - 1$

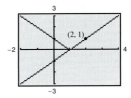

41. $f'(x) = \dfrac{1 - 3x^2 - 4x^{3/2}}{2\sqrt{x}(x^2 + 1)^2}$

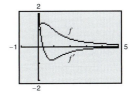

The zero of $f'(x)$ corresponds to the point on the graph of $f(x)$ where the tangent line is horizontal.

43. $f'(x) = -\dfrac{\sqrt{(x + 1)/x}}{2x(x + 1)}$

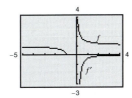

$f'(x)$ has no zeros.

45. $-\dfrac{1}{(x - 2)^2}$ **47.** $\dfrac{8}{(t + 2)^3}$ **49.** $-\dfrac{2(2x - 3)}{(x^2 - 3x)^3}$

51. $-\dfrac{2t}{(t^2 - 2)^2}$ **53.** $27(x - 3)^2(4x - 3)$

55. $\dfrac{3(x + 1)}{\sqrt{2x + 3}}$ **57.** $\dfrac{t(5t - 8)}{2\sqrt{t - 2}}$

59. $-\dfrac{3}{4x^{3/2}\sqrt{3 - 2x}}$ **61.** $\dfrac{x}{\sqrt{x^2 + 1}} - \dfrac{x}{\sqrt{x^2 - 1}}$

63. $\dfrac{2(6 - 5x)(5x^2 - 12x + 5)}{(x^2 - 1)^3}$

65. $y = \frac{8}{3}t + 4$ **67.** $y = -6t - 14$

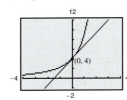

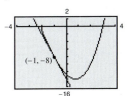

69. $y = -2x + 7$

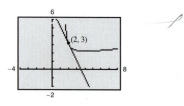

71. (a) $74.00 per 1\%$ (b) $81.59 per 1\%$

(c) $89.94 per 1\%$

73.

t	0	1	2	3	4
$\dfrac{dN}{dt}$	0	177.78	44.44	10.82	3.29

The rate of growth of N is decreasing.

75. (a) $V = \dfrac{10{,}000}{\sqrt[3]{t+1}}$

(b) $-\$1322.83$ per year

(c) $-\$524.97$ per year

77. False. $y' = \frac{1}{2}(1-x)^{-1/2}(-1) = -\frac{1}{2}(1-x)^{-1/2}$

SECTION 2.6   *(page 145)*

> ### Prerequisite Review
>
> **1.** $t = 0, \frac{3}{2}$ **2.** $t = -2, 7$ **3.** $t = -2, 10$
>
> **4.** $t = \dfrac{9 \pm 3\sqrt{10{,}249}}{32}$ **5.** $\dfrac{dy}{dx} = 6x^2 + 14x$
>
> **6.** $\dfrac{dy}{dx} = 8x^3 + 18x^2 - 10x - 15$
>
> **7.** $\dfrac{dy}{dx} = \dfrac{2x(x+7)}{(2x+7)^2}$ **8.** $\dfrac{dy}{dx} = -\dfrac{6x^2 + 10x + 15}{(2x^2 - 5)^2}$
>
> **9.** Domain: $(-\infty, \infty)$ **10.** Domain: $[7, \infty)$
>
> Range: $[-4, \infty)$ Range: $[0, \infty)$

1. 0   **3.** 2   **5.** $2t - 8$ **7.** $\dfrac{9}{2t^4}$

9. $18(2 - x^2)(5x^2 - 2)$ **11.** $\dfrac{4}{(x-1)^3}$

13. $12x^2 + 24x + 16$ **15.** $60x^2 - 72x$

17. $120x + 360$ **19.** $-\dfrac{9}{2x^5}$ **21.** 260 **23.** $-\dfrac{1}{648}$

25. -126 **27.** $4x$ **29.** $\dfrac{2}{x^2}$ **31.** 2

33. $f''(x) = 6(x - 3) = 0$ when $x = 3$.

35. $f''(x) = 2(3x + 4) = 0$ when $x = -\frac{4}{3}$.

37. $f''(x) = \dfrac{x(2x^2 - 3)}{(x^2 - 1)^{3/2}} = 0$ when $x = \pm\dfrac{\sqrt{6}}{2}$

39. $f''(x) = \dfrac{2x(x+3)(x-3)}{(x^2 + 3)^3}$

$= 0$ when $x = 0$ or $x = \pm 3$.

41. (a) $s(t) = -16t^2 + 144t$

(b) $v(t) = -32t + 144$

$a(t) = -32$

(c) 4.5 seconds; 324 feet

(d) $v(9) = -144$ feet per second, which is the same speed as the initial velocity

43.

t	0	10	20	30	40	50	60
$\dfrac{ds}{dt}$	0	45	60	67.5	72	75	77.1
$\dfrac{d^2s}{dt^2}$	9	2.25	1	0.56	0.36	0.25	0.18

As time increases, velocity increases and acceleration decreases.

45. $f(x) = x^2 - 6x + 6$

$f'(x) = 2x - 6$

$f''(x) = 2$

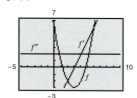

The degrees of the successive derivatives decrease by 1.

47.

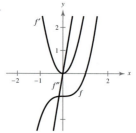

We know the degrees of the successive derivatives decrease by 1.

49. (a)
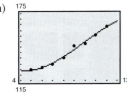

(b) $y' = -0.2484t^2 + 4.886t - 17.06$

$y'' = -0.4968t + 4.886$

(c) $y' > 0$ on $[5, 12]$ (d) $1999\ (t = 9.835)$

(e) The first derivative is used to show that the price of homes is increasing in (c), and the second derivative is used to show the maximum increase in (d).

51. False. The product rule is

$[f(x)g(x)]' = f'(x)g(x) + f(x)g'(x).$

53. True   **55.** True   **57.** $xf^{(n)}(x) + nf^{(n-1)}(x)$

SECTION 2.7   *(page 152)*

Prerequisite Review

1. $y = x^2 - 2x$ **2.** $y = \dfrac{x-3}{4}$

3. $y = 1, x \neq -6$ **4.** $y = -4, x \neq \pm\sqrt{3}$

5. $y = \pm\sqrt{5 - x^2}$ **6.** $y = \pm\sqrt{6 - x^2}$ **7.** $\frac{8}{3}$

8. $-\frac{1}{2}$ **9.** $\frac{5}{7}$ **10.** 1

1. $-\dfrac{y}{x}$ **3.** $-\dfrac{x}{y}$ **5.** $\dfrac{xy^2}{2 - x^2y}$ **7.** $\dfrac{y}{8y - x}$

9. $-\dfrac{1}{10y - 2}$ **11.** $\dfrac{1}{2}$ **13.** $-\dfrac{x}{y}, 0$

15. $-\dfrac{y}{x+1}, -\dfrac{1}{4}$ **17.** $\dfrac{y - 3x^2}{2y - x}, \dfrac{1}{2}$ **19.** $\dfrac{1 - 3x^2y^3}{3x^3y^2 - 1}, -1$

21. $-\sqrt{\dfrac{y}{x}}, -\dfrac{5}{4}$ **23.** $-\sqrt[3]{\dfrac{y}{x}}, -\dfrac{1}{2}$ **25.** 3

27. 0   **29.** $-\dfrac{\sqrt{5}}{3}$ **31.** $-\dfrac{x}{y}, \dfrac{4}{3}$ **33.** $\dfrac{1}{2y}, -\dfrac{1}{2}$

35. At $(5, 12)$: $5x + 12y - 169 = 0$

At $(-12, 5)$: $12x - 5y + 169 = 0$

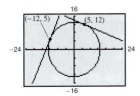

37. At $\left(1, \sqrt{5}\right)$: $15x - 2\sqrt{5}y - 5 = 0$

At $\left(1, -\sqrt{5}\right)$: $15x + 2\sqrt{5}y - 5 = 0$

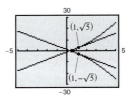

39. At $(0, 2)$: $y = 2$

At $(2, 0)$: $x = 2$

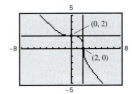

41. $\dfrac{1}{0.024x^3 + 0.04x}$ **43.** $-\dfrac{4xp}{2p^2 + 1}$

45. (a) -2

(b)
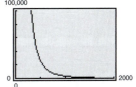

As more labor is used, less capital is available.

As more capital is used, less labor is available.

SECTION 2.8   *(page 160)*

Prerequisite Review

1. $A = \pi r^2$ **2.** $V = \frac{4}{3}\pi r^3$ **3.** $S = 6s^2$

4. $V = s^3$ **5.** $V = \frac{1}{3}\pi r^2 h$ **6.** $A = \frac{1}{2}bh$

7. $-\dfrac{x}{y}$ **8.** $\dfrac{2x - 3y}{3x}$ **9.** $-\dfrac{2x + y}{x + 2}$

10. $-\dfrac{y^2 - y + 1}{2xy - 2y - x}$

1. (a) 62   (b) $\frac{32}{85}$ **3.** (a) $-\frac{5}{8}$ (b) $\frac{3}{2}$

5. (a) 24π square inches per minute

(b) 96π square inches per minute

7. If $\dfrac{dr}{dt}$ is constant, $\dfrac{dA}{dt} = 2\pi r \dfrac{dr}{dt}$ and so is proportional to r.

9. (a) $\dfrac{5}{\pi}$ feet per minute   (b) $\dfrac{5}{4\pi}$ feet per minute

11. (a) 112.5 dollars per week

(b) 7500 dollars per week

(c) 7387.5 dollars per week

13. (a) 9 cubic centimeters per second

(b) 900 cubic centimeters per second

15. (a) -12 centimeters per minute

(b) 0 centimeters per minute

(c) 4 centimeters per minute

(d) 12 centimeters per minute

17. (a) $-\dfrac{7}{12}$ foot per second   (b) $-\dfrac{3}{2}$ feet per second

(c) $-\dfrac{48}{7}$ feet per second

19. (a) -750 miles per hour   (b) 20 minutes

21. -8.33 feet per second   **23.** ≈ 188.5 ft^3 per minute

25. 5 units per week

REVIEW EXERCISES FOR CHAPTER 2
(page 166)

1. -2 **3.** 0

5. $t = 7$: slope $\approx \$5500$ million per year per year
(sales are increasing)

$t = 10$: slope $\approx \$7500$ million per year per year
(sales are increasing)

$t = 12$: slope $\approx \$3500$ million per year per year
(sales are increasing)

7. $t = 0$: slope ≈ 180
$t = 4$: slope ≈ -70
$t = 6$: slope ≈ -900

9. -3; -3 **11.** $2x - 4$; -2

13. $\dfrac{1}{2\sqrt{x+9}}$; $\dfrac{1}{4}$ **15.** $-\dfrac{1}{(x-5)^2}$; -1

17. -5 **19.** 0   **21.** $\frac{1}{6}$ **23.** -5 **25.** 1   **27.** 0

29. $y = -\dfrac{4}{3}t + 2$ **31.** $y = 2x + 2$

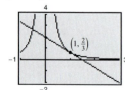   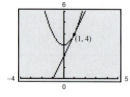

33. $y = -34x - 27$ **35.** $y = x - 1$

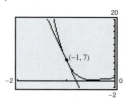   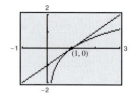

37. $y = -2x + 6$

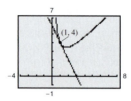

39. Average rate of change: 4
Instantaneous rate of change when $x = 0$: 3
Instantaneous rate of change when $x = 1$: 5

41. (a) 6976 million per year

(b) 1998: $7317 million per year
2002: $3876 million per year

(c) Sales were increasing in 1998 and 2002, and grew at a rate of $6976 million over the period 1998–2002.

43. (a) $P'(t) = -0.004236t^3 + 0.09045t^2 - 0.57t + 1.007$

(b) $-\$0.004$ per pound in 1997
$\$0.116$ per pound in 2000
$-\$0.128$ per pound in 2002

(c)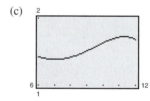

The price is increasing from 1997 to 2001, and decreasing from 1996 to 1997 and from 2001 to 2002.

(d) Negative slope: $6 < t < 7$ and $11 < t < 12$
Positive slope: $7 < t < 11$

(e) When the price increases, the slope is positive.
When the price decreases, the slope is negative.

45. (a) $s(t) = -16t^2 + 276$ (b) -32 feet per second

(c) $t = 2$: -64 feet per second
$t = 3$: -96 feet per second

(d) 4.15 second   (e) 132.8 feet per second

47. $R = 27.50x$

$C = 15x + 2500$

$P = 12.50x - 2500$

49. $\dfrac{dC}{dx} = 320$ **51.** $\dfrac{dC}{dx} = \dfrac{1.275}{\sqrt{x}}$

53. $\dfrac{dR}{dx} = 200 - \dfrac{2}{5}x$ **55.** $\dfrac{dR}{dx} = \dfrac{35(x-4)}{2(x-2)^{3/2}}$

57. $\dfrac{dP}{dx} = -0.0006x^2 + 12x - 1$ **59.** $15x^2(1-x^2)$

61. $16x^3 - 33x^2 + 12x$ **63.** $\dfrac{2(3 + 5x - 3x^2)}{(x^2+1)^2}$

65. $30x(5x^2 + 2)^2$ **67.** $-\dfrac{1}{(x+1)^{3/2}}$ **69.** $\dfrac{2x^2 + 1}{\sqrt{x^2+1}}$

71. $80x^4 - 24x^2 + 1$ **73.** $18x^5(x+1)(2x+3)^2$

75. $x(x-1)^4(7x-2)$ **77.** $\dfrac{3(9t+5)}{2\sqrt{3t+1}(1-3t)^3}$

79. (a) $t = 1$: -6.63 $t = 3$: -6.5

   $t = 5$: -4.33 $t = 10$: -1.36

   (b)    The rate of decrease is
approaching zero.

81. 6   **83.** $-\dfrac{120}{x^6}$ **85.** $\dfrac{35x^{3/2}}{2}$ **87.** $\dfrac{2}{x^{2/3}}$

89. (a) $s(t) = -16t^2 + 5t + 30$ (b) 1.534 seconds

   (c) -44.09 feet per second

   (d) -32 feet per second squared

91. $-\dfrac{2x + 3y}{3(x+y^2)}$ **93.** $\dfrac{2x-8}{2y-9}$ **95.** $y = \dfrac{1}{3}x + \dfrac{1}{3}$

97. $y = \dfrac{4}{3}x + \dfrac{2}{3}$ **99.** $\dfrac{1}{64}$ feet per minute

SAMPLE POST-GRAD EXAM QUESTIONS

(page 170)

   1. (c)   **2.** (e)   **3.** (e)   **4.** (c)   **5.** (c)   **6.** (a)

CHAPTER 3

SECTION 3.1   *(page 179)*

Prerequisite Review

   1. $x = 0, x = 8$ **2.** $x = 0, x = 24$ **3.** $x = \pm5$

   4. $x = 0$ **5.** $(-\infty, 3) \cup (3, \infty)$ **6.** $(-\infty, 1)$

   7. $(-\infty, -2) \cup (-2, 5) \cup (5, \infty)$ **8.** $\left(-\sqrt{3}, \sqrt{3}\right)$

   9. $x = -2$: -6 **10.** $x = -2$: 60

   $x = 0$: 2 $x = 0$: -4

   $x = 2$: -6 $x = 2$: 60

   11. $x = -2$: $-\dfrac{1}{3}$ **12.** $x = -2$: $\dfrac{1}{18}$

   $x = 0$: 1 $x = 0$: $-\dfrac{1}{8}$

   $x = 2$: 5 $x = 2$: $-\dfrac{3}{2}$

1. $f'(-1) = -\dfrac{8}{25}$ **3.** $f'(-3) = -\dfrac{2}{3}$

   $f'(0) = 0$ $f'(-2)$ is undefined.

   $f'(1) = \dfrac{8}{25}$ $f'(-1) = \dfrac{2}{3}$

5. Increasing on $(-\infty, -1)$

   Decreasing on $(-1, \infty)$

7. Increasing on $(-1, 0)$ and $(1, \infty)$

   Decreasing on $(-\infty, -1)$ and $(0, 1)$

9. No critical numbers   **11.** Critical number: $x = 1$

   Increasing on $(-\infty, \infty)$ Increasing on $(-\infty, 1)$

   Decreasing on $(1, \infty)$

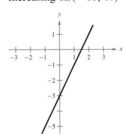   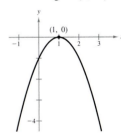

13. Critical number: $x = \dfrac{5}{2}$

   Decreasing on $\left(-\infty, \dfrac{5}{2}\right)$

   Increasing on $\left(\dfrac{5}{2}, \infty\right)$

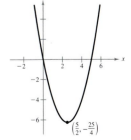

27. $(3, 0)$ **29.** $(1, 0), (3, -16)$

31. No inflection points   **33.** $\left(\frac{3}{2}, -\frac{1}{16}\right), (2, 0)$

35.
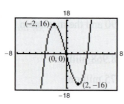
Relative maximum: $(-2, 16)$

Relative minimum: $(2, -16)$

Point of inflection: $(0, 0)$

37.
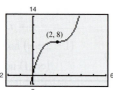
No relative extrema

Point of inflection: $(2, 8)$

39.
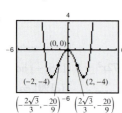
Relative maximum: $(0, 0)$

Relative minima: $(\pm 2, -4)$

Points of inflection:

$$\left(\pm \frac{2\sqrt{3}}{3}, -\frac{20}{9}\right)$$

41.
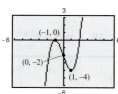
Relative maximum: $(-1, 0)$

Relative minimum: $(1, -4)$

Point of inflection: $(0, -2)$

43.
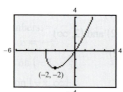
Relative minimum: $(-2, -2)$

No inflection points

45.
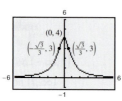
Relative maximum: $(0, 4)$

Points of inflection:

$$\left(\pm \frac{\sqrt{3}}{3}, 3\right)$$

47.
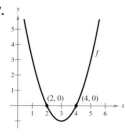

49.
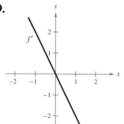

(a) f': Positive on $(-\infty, 0)$

f: Increasing on $(-\infty, 0)$

(b) f': Negative on $(0, \infty)$

f: Decreasing on $(0, \infty)$

(c) f': Not increasing

f: Not concave upward

(d) f': Decreasing on $(-\infty, \infty)$

f: Concave downward on $(-\infty, \infty)$

51. $(200, 320)$ **53.** 100 units   **55.** 8:30 P.M.

57. $\sqrt{3} \approx 1.732$ years

59.
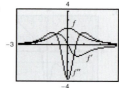
Relative minimum: $(0, -5)$

Relative maximum: $(3, 8.5)$

Point of inflection:

$$\left(\frac{2}{3}, -3.2963\right)$$

61.
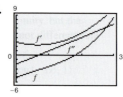
Relative maximum: $(0, 2)$

Points of inflection:

$(0.58, 1.5), (-0.58, 1.5)$

63. (a) Relative maximum: $(2, 2150)$

Relative minimum: $(1, 2050)$

Absolute maximum: $(0, 2250)$

Absolute minimum: $(7.5, 1740)$

The market opened at the maximum for the day and closed at the minimum. At approximately 9:30 A.M., the market started to recover after falling, and at approximately 10:30 A.M., the market started to fall again.

(b) $(4, 2010)$;

At approximately 12:30 P.M., the market began to fall at a greater rate.

65. (a) At $t = 8$, 256 people will be infected.

(b) At $t = 4$, the virus will be spreading most rapidly.

(c)
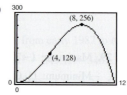

SECTION 3.4   *(page 207)*

Prerequisite Review

1. $x + \frac{1}{2}y = 12$ **2.** $2xy = 24$ **3.** $xy = 24$

4. $\sqrt{(x_2 - x_1)^2 + (y_2 - y_1)^2} = 10$

5. $x = -3$ **6.** $x = -\frac{2}{3}, 1$ **7.** $x = \pm 5$

8. $x = 4$ **9.** $x = \pm 1$ **10.** $x = \pm 3$

1. 55, 55   **3.** 18, 9   **5.** $\sqrt{192}, \sqrt{192}$ **7.** 1

9. $l = w = 25$ meters   **11.** $l = w = 8$ feet

13. $x = 25$ feet, $y = \frac{100}{3}$ feet

15. (a) Proof

    (b) $V_1 = 99$ cubic inches

        $V_2 = 125$ cubic inches

        $V_3 = 117$ cubic inches

    (c) 5 inches × 5 inches × 5 inches

17. Rectangular portion: $\dfrac{16}{\pi + 4} \times \dfrac{32}{\pi + 4}$ feet

19. 1.056 cubic feet

21. Plant 18 trees to yield 1296 apples.

23. $2\sqrt{15} + 4$ inches × $\sqrt{15} + 2$ inches

25. (a) $L = \sqrt{x^2 + 4 + \dfrac{8}{x - 1} + \dfrac{4}{(x - 1)^2}}, \quad x > 1$

    (b)
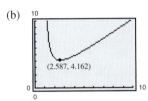
(2.587, 4.162)

    Minimum when $x \approx 2.587$

    (c) $(0, 0), (2, 0), (0, 4)$

27. $\sqrt{2}r \times \dfrac{\sqrt{2}}{2}r$ **29.** $\dfrac{4\pi r^3}{3\sqrt{3}}$ **31.** $\left(\pm\sqrt{\dfrac{5}{2}}, \dfrac{7}{2}\right)$

33. 18 inches × 18 inches × 36 inches

35. Radius of circle: $\dfrac{8}{\pi + 4}$

    Side of square: $\dfrac{16}{\pi + 4}$

37. $x = 1$ mile

39. Harvesting in the 6th week will yield 140 bushels for a maximum value of $490.

41. Length $= 10\sqrt{2} \approx 14.14$
Width $= 5\sqrt{2} \approx 7.07$

SECTION 3.5   *(page 217)*

Prerequisite Review

1. 1   **2.** $\frac{6}{5}$ **3.** 2   **4.** $\frac{1}{2}$

5. $\dfrac{dC}{dx} = 1.2 + 0.006x$ **6.** $\dfrac{dP}{dx} = 0.02x + 11$

7. $\dfrac{dR}{dx} = 14 - \dfrac{x}{1000}$ **8.** $\dfrac{dR}{dx} = 3.4 - \dfrac{x}{750}$

9. $\dfrac{dP}{dx} = -1.4x + 7$ **10.** $\dfrac{dC}{dx} = 4.2 + 0.003x^2$

1. 2000 units   **3.** 200 units   **5.** 80 units

7. 50 units   **9.** $60   **11.** $69.68

13. 3 units

$\overline{C}(3) = 17; \dfrac{dC}{dx} = 4x + 5$; when $x = 3$, $\dfrac{dC}{dx} = 17$

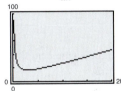

15. (a) $80   (b) $45.93

17. The maximum profit occurs when $s = 10$ (or $10,000).
The point of diminishing returns occurs at $s = \frac{35}{6}$ (or $5833.33).

19. 200 radios   **21.** $50

23. $C =$ cost under water + cost on land
$= 8\sqrt{0.25 + x^2} + 6(6 - x)$

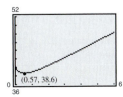
(0.57, 38.6)

The line should run from the power station to a point across the river approximately 0.57 mile downstream.

$\left(\text{Exact: } 3/\left(2\sqrt{7}\right) \text{ mile}\right)$

25. 77.46 miles per hour

27. $-\frac{17}{3}$, elastic

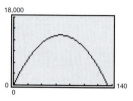

Elastic: $\left(0, \frac{200}{3}\right) \approx (0, 66.7)$

Inelastic: $\left(\frac{200}{3}, \frac{400}{3}\right) \approx (66.7, 133.3)$

29. $-\frac{9997}{3}$, elastic

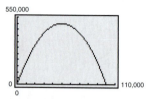

Elastic: $(0, 50{,}000)$

Inelastic: $(50{,}000, 100{,}000)$

31. $-\frac{3}{2}$, elastic

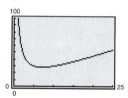

Elastic: $\left(5\sqrt{2}, \infty\right) \approx (7.1, \infty)$

Inelastic: $\left(0, 5\sqrt{2}\right) \approx (0, 7.1)$

33. (a) -2.48% (b) -0.496

(c) -0.5; The results are approximately the same.

(d) $R = p^3 - 20p^2 + 100p$, $x = \frac{400}{9}$ units, $p = \$3.33$

35. (a) $-\frac{11}{14}$ (b) $x = 500$ units, $p = \$10$

(c) Answers will vary.

37. No; when $p = 5$, $x = 350$ and $\eta = -\frac{5}{7}$.

Because $|\eta| = \frac{5}{7} < 1$, demand is inelastic.

39. Proof

41. (a) Revenue was greatest in 2001 and least in 1994.

(b) Revenue was increasing at the greatest rate in 1998 and decreasing at the greatest rate in 2003.

(c)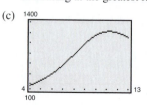

43. Answers will vary.

SECTION 3.6   *(page 228)*

Prerequisite Review

1. 3   **2.** 1   **3.** -11 **4.** 4   **5.** $-\frac{1}{4}$

6. -2 **7.** 0   **8.** 1

9. $\overline{C} = \dfrac{150}{x} + 3$ **10.** $\overline{C} = \dfrac{1900}{x} + 1.7 + 0.002x$

$\dfrac{dC}{dx} = 3$ $\dfrac{dC}{dx} = 1.7 + 0.004x$

11. $\overline{C} = 0.005x + 0.5 + \dfrac{1375}{x}$ **12.** $\overline{C} = \dfrac{760}{x} + 0.05$

$\dfrac{dC}{dx} = 0.01x + 0.5$ $\dfrac{dC}{dx} = 0.05$

1. Vertical asymptote: $x = 0$

Horizontal asymptote: $y = 1$

3. Vertical asymptotes: $x = -1$, $x = 2$

Horizontal asymptote: $y = 1$

5. Vertical asymptote: none

Horizontal asymptote: $y = \frac{3}{2}$

7. Vertical asymptotes: $x = \pm 2$

Horizontal asymptote: $y = \frac{1}{2}$

9. f   **10.** b   **11.** c   **12.** a   **13.** e   **14.** d

15. ∞ **17.** $-\infty$ **19.** $-\infty$ **21.** $-\infty$ **23.** $\frac{2}{3}$

25. 0   **27.** $-\infty$ **29.** ∞ **31.** 5

33.

x	10^0	10^1	10^2	10^3
$f(x)$	2.000	0.348	0.101	0.032

x	10^4	10^5	10^6
$f(x)$	0.010	0.003	0.001

$$\lim_{x \to \infty} \frac{x + 1}{x\sqrt{x}} = 0$$

35.

x	10^0	10^1	10^2	10^3
$f(x)$	0	49.5	49.995	49.99995

x	10^4	10^5	10^6
$f(x)$	50.0	50.0	50.0

$$\lim_{x\to\infty} \frac{x^2-1}{0.02x^2} = 50$$

37.

x	-10^6	-10^4	-10^2	10^0
$f(x)$	-2	-2	-1.9996	0.8944

x	10^2	10^4	10^6
$f(x)$	1.9996	2	2

$$\lim_{x\to-\infty} \frac{2x}{\sqrt{x^2+4}} = -2, \quad \lim_{x\to\infty} \frac{2x}{\sqrt{x^2+4}} = 2$$

39.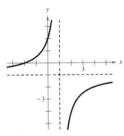

41.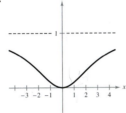

43.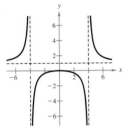

45.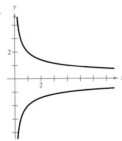

47.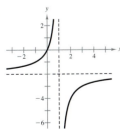

49.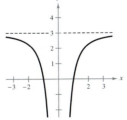

51.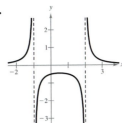

53.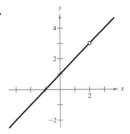

55.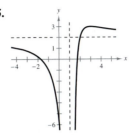

57. (a) $\overline{C} = 1.35 + \dfrac{4570}{x}$ (b) \$47.05, \$5.92   (c) \$1.35

59. (a) 25%; \$176 million; 50%; \$528 million;

  75%; \$1584 million

  (b) ∞

  (c)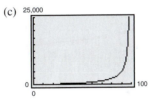

61. a

63. (a) 5 years: 153 elk   (b) 400 elk

   10 years: 215 elk

   25 years: 294 elk

65. (a) $\overline{P} = 50 - \dfrac{1000}{x}$ (b) \$40, \$48, \$49   (c) \$50

SECTION 3.7   *(page 238)*

Prerequisite Review

1. Vertical asymptote: $x = 0$

  Horizontal asymptote: $y = 0$

2. Vertical asymptote: $x = 2$

  Horizontal asymptote: $y = 0$

3. Vertical asymptote: $x = -3$

  Horizontal asymptote: $y = 40$

4. Vertical asymptotes: $x = 1, x = 3$

  Horizontal asymptote: $y = 1$

5. Decreasing on $(-\infty, -2)$
   Increasing on $(-2, \infty)$

6. Increasing on $(-\infty, -4)$
   Decreasing on $(-4, \infty)$

7. Increasing on $(-\infty, -1)$ and $(1, \infty)$
   Decreasing on $(-1, 1)$

8. Decreasing on $(-\infty, 0)$ and $\left(\sqrt[3]{2}, \infty\right)$
   Increasing on $\left(0, \sqrt[3]{2}\right)$

9. Increasing on $(-\infty, 1)$ and $(1, \infty)$

10. Decreasing on $(-\infty, -3)$ and $\left(\frac{1}{3}, \infty\right)$
   Increasing on $\left(-3, \frac{1}{3}\right)$

1.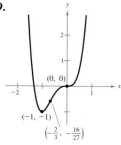

3.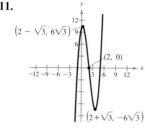

5.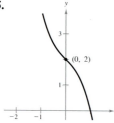

7.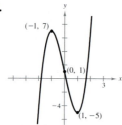

9.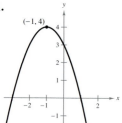

11.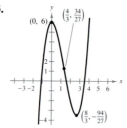

13.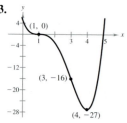

15.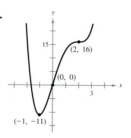

17.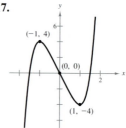

19.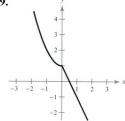

21.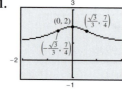

23.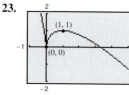

25.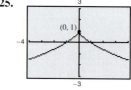

27.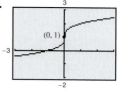

29.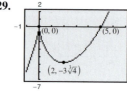

31.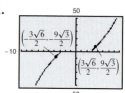

33.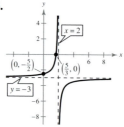

Domain: $(-\infty, 2) \cup (2, \infty)$

35.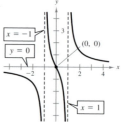

Domain: $(-\infty, -1) \cup (-1, 1) \cup (1, \infty)$

37.

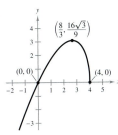

$\left(\frac{8}{3}, \frac{16\sqrt{3}}{9}\right)$

$(0, 0)$ $(4, 0)$

Domain: $(-\infty, 4]$

39.

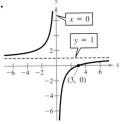

$x = 0$

$y = 1$

$(3, 0)$

Domain: $(-\infty, 0) \cup (0, \infty)$

41.

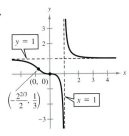

$y = 1$

$(0, 0)$

$\left(-\frac{2^{2/3}}{2}, \frac{1}{3}\right)$ $x = 1$

Domain: $(-\infty, 1) \cup (1, \infty)$

43. Answers will vary.

Sample answer: $f(x) = -x^3 + x^2 + x + 1$

45. Answers will vary. Sample answer: $f(x) = x^3 + 1$

47. Answers will vary.

Sample answer:

49. Answers will vary.

Sample answer:

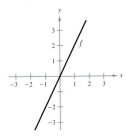

f

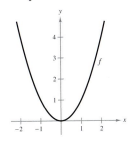

f

51. Answers will vary. Sample answer:

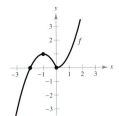

f

53. (a) $C = 0.36s + \dfrac{900}{s}, \quad 40 \le s \le 65$

(b)

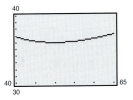

Most economical speed ≈ 50 miles per hour

55.

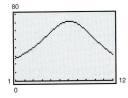

Absolute minimum: $(1, 27.23)$

Absolute maximum: $(7.13, 71.23)$

The monthly normal temperature reaches a minimum of $27.23°F$ in January and a maximum of $71.23°F$ at the beginning of July.

57.

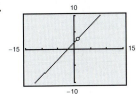

The rational function has the common factor of $x - 1$ in the numerator and denominator. At $x = 1$, there is a hole in the graph, not a vertical asymptote.

SECTION 3.8 *(page 246)*

Prerequisite Review

1. $\dfrac{dC}{dx} = 0.18x$ **2.** $\dfrac{dC}{dx} = 0.15$

3. $\dfrac{dR}{dx} = 1.25 + 0.03\sqrt{x}$ **4.** $\dfrac{dR}{dx} = 15.5 - 3.1x$

5. $\dfrac{dP}{dx} = -\dfrac{0.01}{\sqrt[3]{x^2}} + 1.4$ **6.** $\dfrac{dP}{dx} = -0.04x + 25$

7. $\dfrac{dA}{dx} = \dfrac{\sqrt{3}}{2}x$ **8.** $\dfrac{dA}{dx} = 12x$ **9.** $\dfrac{dC}{dr} = 2\pi$

10. $\dfrac{dP}{dw} = 4$ **11.** $\dfrac{dS}{dr} = 8\pi r$ **12.** $\dfrac{dP}{dx} = 2 + \sqrt{2}$

13. $A = \pi r^2$ **14.** $A = x^2$

15. $V = x^3$ **16.** $V = \frac{4}{3}\pi r^3$

1. $dy = 6x\,dx$ **3.** $dy = 12(4x - 1)^2\,dx$

5. $dy = \dfrac{x}{\sqrt{x^2 + 1}}\,dx$ **7.** 0.1005   **9.** -0.013245

11. $dy = 0.3$ **13.** $dy = -0.04$

$\Delta y = 0.331$ $\Delta y \approx -0.0394$

15.

$dx = \Delta x$	dy	Δy	$\Delta y - dy$	$\dfrac{dy}{\Delta y}$
1.000	4.000	5.0000	1.0000	0.8000
0.500	2.000	2.2500	0.2500	0.8889
0.100	0.400	0.4100	0.0100	0.9756
0.010	0.040	0.0401	0.0001	0.9975
0.001	0.004	0.0040	0.0000	1.0000

17.

$dx = \Delta x$	dy	Δy	$\Delta y - dy$	$\dfrac{dy}{\Delta y}$
1.000	-0.25000	-0.13889	0.11111	1.79999
0.500	-0.12500	-0.09000	0.03500	1.38889
0.100	-0.02500	-0.02324	0.00176	1.07573
0.010	-0.00250	-0.00248	0.00002	1.00806
0.001	-0.00025	-0.00025	0.00000	1.00000

19.

$dx = \Delta x$	dy	Δy	$\Delta y - dy$	$\dfrac{dy}{\Delta y}$
1.000	0.14865	0.12687	-0.02178	1.17167
0.500	0.07433	0.06823	-0.00610	1.08940
0.100	0.01487	0.01459	-0.00028	1.01919
0.010	0.00149	0.00148	-0.00001	1.00676
0.001	0.00015	0.00015	0.00000	1.00000

21. $y = 28x + 37$

For $\Delta x = -0.01$, $f(x + \Delta x) = -19.281302$ and
$y(x + \Delta x) = -19.28$
For $\Delta x = 0.01$, $f(x + \Delta x) = -18.721298$ and
$y(x + \Delta x) = -18.72$

23. $y = x$

For $\Delta x = -0.01$, $f(x + \Delta x) = -0.009999$ and
$y(x + \Delta x) = -0.01$
For $\Delta x = 0.01$, $f(x + \Delta x) = 0.009999$ and
$y(x + \Delta x) = 0.01$

25. A79(a) $\Delta p = -0.25 = dp$ (b) $\Delta p = -0.25 = dp$

27. \$5.20   **29.** \$7.50

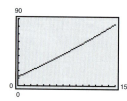
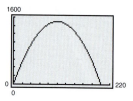

31. $-\$1250$ **33.** $R = -\dfrac{1}{3}x^2 + 100x$; \$6

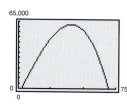
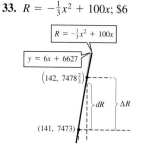

35. $P = -\dfrac{1}{2000}x^2 + 23x - 275{,}000$; $-\$5$

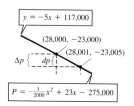

37. (a) $dA = 2x\,\Delta x$, $\Delta A = 2x\,\Delta x + (\Delta x)^2$
    (b) and (c)

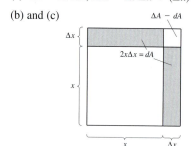

39. $\pm\dfrac{5}{2}\pi$ square inches, $\pm\dfrac{1}{40}$

41. $\pm 2.88\pi$ cubic inches, ± 0.01 **43.** True

REVIEW EXERCISES FOR CHAPTER 3
(page 252)

1. $x = 1$ **3.** $x = 0, x = 1$

5. Increasing on $\left(-\dfrac{1}{2}, \infty\right)$
    Decreasing on $\left(-\infty, -\dfrac{1}{2}\right)$

7. Increasing on $(-\infty, 3)$ and $(3, \infty)$

9. (a) $(1.48, 7.28)$ (b) $(1, 1.48), (7.28, 12)$

(c) Normal monthly temperature is rising from early January to July.

Normal monthly temperature is decreasing in early January and from early July to December.

(d)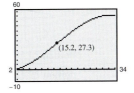

11. Relative maximum: $(0, -2)$

Relative minimum: $(1, -4)$

13. Relative minimum: $(8, -52)$

15. Relative maxima: $(-1, 1), (1, 1)$

Relative minimum: $(0, 0)$

17. Relative maximum: $(0, 6)$

19. Relative maximum: $(0, 0)$

Relative minimum: $(4, 8)$

21. Maximum: $(0, 6)$ **23.** Maxima: $(-2, 17), (4, 17)$

Minimum: $\left(-\frac{5}{2}, -\frac{1}{4}\right)$ Minima: $(-4, -15), (2, -15)$

25. Maximum: $(1, 3)$

Minimum: $\left(3, 4\sqrt{3} - 9\right)$

27. Maximum: $(2, 26)$ **29.** Maximum: $(1, 1)$

Minimum: $(1, -1)$ Minimum: $(-1, -1)$

31.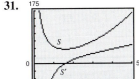

$r \approx 1.58$ inches

33. Concave upward on $(2, \infty)$

Concave downward on $(-\infty, 2)$

35. Concave upward on $\left(-\frac{2}{\sqrt{3}}, \frac{2}{\sqrt{3}}\right)$

Concave downward on $\left(-\infty, -\frac{2}{\sqrt{3}}\right)$ and $\left(\frac{2}{\sqrt{3}}, \infty\right)$

37. $(0, 0), (4, -128)$

39. $(0, 0), (1.0652, 4.5244), (2.5348, 3.5246)$

41. Relative maximum: $\left(-\sqrt{3}, 6\sqrt{3}\right)$

Relative minimum: $\left(\sqrt{3}, -6\sqrt{3}\right)$

43. Relative maximum: $(-4, 0)$

Relative minimum: $(-2, -108)$

45. $\left(50, 166\frac{2}{3}\right)$

47. 13, 13

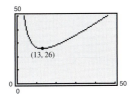

49. (a) Absolute maximum: $(4.23, 1764.29)$

Absolute minimum: $(30, 1485)$

(b) 1989

(c) The maximum number of daily newspapers in circulation was 1764.29 million in 1974 and the minimum number was 1485 million in 2000.

The year 1989 was when circulation was changing at the greatest rate.

51. $x = \frac{137}{9} \approx 15.2$ years

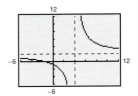

53. $s'(r) = -2cr$

$-2cr = 0 \implies r = 0$

$s''(r) = -2c < 0$ for all r

Therefore, $r = 0$ yields a maximum value of s.

55. $N = 85$ (maximizes revenue)   **57.** 125 units

59. Elastic: $(0, 75)$

Inelastic: $(75, 150)$

Demand is of unit elasticity when $x = 75$.

61. Elastic: $(0, 200)$

Inelastic: $(200, 300)$

Demand is of unit elasticity when $x = 200$.

63. Vertical asymptote: $x = 4$

Horizontal asymptote: $y = 2$

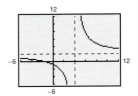

65.

SECTION 4.2   *(page 270)*

67.

Prerequisite Review

1. Continuous on $(-\infty, \infty)$
2. Discontinuous for $x = \pm 2$
3. Discontinuous for $x = \pm\sqrt{3}$
4. Removable discontinuity at $x = 4$
5. 0   6. 0   7. 4   8. $\frac{1}{2}$ 9. $\frac{3}{2}$
10. 6   11. 0   12. 0

69.
77.

1. (a) e^7 (b) e^{12} (c) $\dfrac{1}{e^6}$ (d) 1

3. (a) e^5 (b) $e^{5/2}$ (c) e^6 (d) e^7

5. $-\frac{1}{3}$ 7. 9   9. $\pm e$ 11. $\sqrt[3]{3}e$ 13. f

14. e   15. d   16. b   17. c   18. a

19.
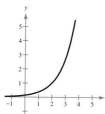

21.
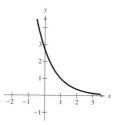

79.

23.
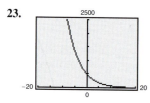

25.
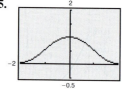

27.
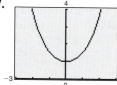

No horizontal asymptotes

Continuous on the entire
real number line

29.
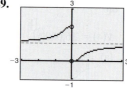

Horizontal asymptote: $y = 1$

Discontinuous at $x = 0$

83.

31.

n	1	2	4	12
A	1343.92	1346.86	1348.35	1349.35

n	365	Continuous compounding
A	1349.84	1349.86

33.

n	1	2	4	12
A	3262.04	3290.66	3305.28	3315.15

n	365	Continuous compounding
A	3319.95	3320.12

35.

t	1	10	20
P	96,078.94	67,032.00	44,932.90

t	30	40	50
P	30,119.42	20,189.65	13,533.53

37.

t	1	10	20
P	95,132.82	60,716.10	36,864.45

t	30	40	50
P	22,382.66	13,589.88	8251.24

39. (a) 9%   (b) 9.2%   (c) 9.31%   (d) 9.38%

41. $12,500   43. $8751.92

45. (a) $849.53   (b) $421.12

$\lim\limits_{x \to \infty} p = 0$

47. (a) 0.1535   (b) 0.4866   (c) 0.8111

49. (a) 1995: $4510.69 million; 2000: $5719.90 million

2003: $6595.92 million

(b) Yes. There is a positive correlation between sales and
time in years.

(c) 2011

51. (a)
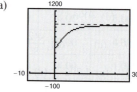

(b) Yes, $\lim\limits_{t \to \infty} \dfrac{925}{1 + e^{-0.3t}} = 925$

(c) $\lim\limits_{t \to \infty} \dfrac{1000}{1 + e^{-0.3t}} = 1000$

Models similar to this logistic growth model where
$y = \dfrac{a}{1 + be^{-ct}}$ have a limit of a as $t \to \infty$.

53. (a) 0.731   (b) 11   (c) Yes, $\lim\limits_{n\to\infty} \dfrac{0.83}{1 + e^{-0.2n}} = 0.83$

SECTION 4.3   *(page 279)*

Prerequisite Review

1. $\dfrac{1}{2}e^x(2x^2 - 1)$ **2.** $\dfrac{e^x(x + 1)}{x}$ **3.** $e^x(x - e^x)$

4. $e^{-x}(e^{2x} - x)$ **5.** $-\dfrac{6}{7x^3}$ **6.** $6x - \dfrac{1}{6}$

7. $6(2x^2 - x + 6)$ **8.** $\dfrac{t + 2}{2t^{3/2}}$

9. Relative maximum: $\left(-\dfrac{4\sqrt{3}}{3}, \dfrac{16\sqrt{3}}{9}\right)$

   Relative minimum: $\left(\dfrac{4\sqrt{3}}{3}, -\dfrac{16\sqrt{3}}{9}\right)$

10. Relative maximum: $(0, 5)$

   Relative minima: $(-1, 4), (1, 4)$

---

1. 3   **3.** -1 **5.** $4e^{4x}$ **7.** $-2xe^{-x^2}$

9. $\dfrac{2}{x^3}e^{-1/x^2}$ **11.** $e^{4x}(4x^2 + 2x + 4)$

13. $-\dfrac{6(e^x - e^{-x})}{(e^x + e^{-x})^4}$ **15.** $xe^x + e^x + 4e^{-x}$

17. $y = -2x + 1$ **19.** $y = \dfrac{4}{e^2}$ **21.** $y = 24x + 8$

23. $\dfrac{dy}{dx} = \dfrac{1}{2}(-x - 1 - 2y)\left(\text{Equivalently, } \dfrac{dy}{dx} = -\dfrac{1}{2}\right)$

25. $\dfrac{dy}{dx} = \dfrac{e^{-x}(x^2 - 2x) + y}{4y - x}$

27. $6(3e^{3x} + 2e^{-2x})$ **29.** $5e^{-x} - 50e^{-5x}$

31.

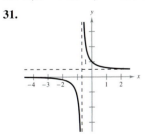

No relative extrema

No points of inflection

Horizontal asymptote to the right: $y = \dfrac{1}{2}$

Horizontal asymptote to the left: $y = 0$

Vertical asymptote: $x \approx -0.693$

33.

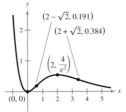

Relative minimum: $(0, 0)$

Relative maximum: $(2, 4e^{-2})$

Points of inflection: $\left(2 - \sqrt{2}, 0.191\right), \left(2 + \sqrt{2}, 0.384\right)$

35.

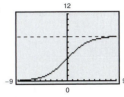

Asymptotes: $y = 0, y = 8$

37. (a)

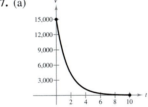

(b) $-\$5028.84/\text{year}$ (c) $-\$406.89/\text{year}$

(d) $v = -1497.2t + 15,000$

(e) In the exponential function, the initial rate of depreciation is greater than in a linear model. The linear model has a constant rate of depreciation.

39. (a)

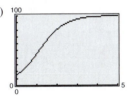

(b) 80.3%

(c) $x \approx 1.1$ or approximately 1100 egg masses

41. (a) $\$433.31/\text{year}$ (b) $\$890.22/\text{year}$

(c) $\$21,839.26/\text{year}$

43. $t = 5$: 14.44

   $t = 10$: 3.63

   $t = 25$: 0.58

45. (a) $f(x) = \dfrac{1}{12.5\sqrt{2\pi}} e^{-(x-650)^2/2(12.5)^2}$

$= \dfrac{1}{12.5\sqrt{2\pi}} e^{-(x-650)^2/312.5}$

(b)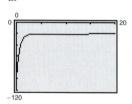

(c) $f'(x) = \dfrac{-4\sqrt{2}(x-650)e^{-2(x-650)^2/624}}{15625\sqrt{\pi}}$

(d) Answers will vary.

47.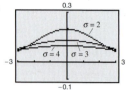

As σ increases, the graph becomes flatter.

49. (a) $\dfrac{dh}{dt} = -80e^{-1.6t} - 20$

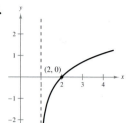

(b) $-100, -36.15, -20.03, -20.00, -20.00$

(c) The values in (b) are rates of descent in feet per second. As time increases, the rate is approximately constant at -20 feet/second.

SECTION 4.4   (page 287)

Prerequisite Review

1. $\frac{1}{4}$ **2.** 64   **3.** 3^6 **4.** $\left(\frac{2}{3}\right)^3$ **5.** 1

6. $81e^4$ **7.** $\dfrac{e^3}{2}$ **8.** $\dfrac{125}{8e^3}$ **9.** $x > -4$

10. Any real number x **11.** $x < -1$ or $x > 1$

12. $x > 5$ **13.** \$3462.03   **14.** \$3374.65

1. $e^{0.6931\ldots} = 2$ **3.** $e^{-1.6094\ldots} = 0.2$ **5.** $\ln 1 = 0$

7. $\ln(0.0498\ldots) = -3$ **9.** c   **10.** d

11. b   **12.** a

13. 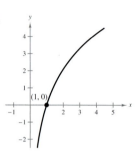   **15.**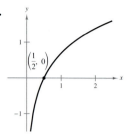

17.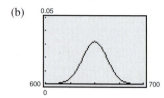

19. Answers will vary.   **21.** Answers will vary.

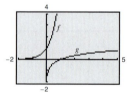   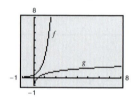

23. x^2 **25.** $5x + 2$ **27.** $2x - 1$

29. (a) 1.7917   (b) 0.4055   (c) 4.3944   (d) 0.5493

31. $\ln 2 - \ln 3$ **33.** $\ln x + \ln y + \ln z$

35. $\frac{1}{2}\ln(x^2 + 1)$ **37.** $\ln z + 2\ln(z - 1)$

39. $\ln 3 + \ln x + \ln(x + 1) - 2\ln(2x + 1)$

41. $\ln \dfrac{x-2}{x+2}$ **43.** $\ln \dfrac{x^3 y^2}{z^4}$ **45.** $\ln \left[\dfrac{x(x+3)}{x+4}\right]^3$

47. $\ln \left[\dfrac{x(x^2+1)}{x+1}\right]^{3/2}$ **49.** $\ln \dfrac{(x+1)^{1/3}}{(x-1)^{2/3}}$

51. $x = 4$ **53.** $x = 1$ **55.** $x = \ln 4 - 1 \approx 0.3863$

57. $t = \dfrac{\ln 7 - \ln 3}{-0.2} \approx -4.2365$

59. $x = \frac{1}{2}\left(1 + \ln\frac{3}{2}\right) \approx 0.7027$

61. $x = -100\ln\frac{3}{4} \approx 28.7682$

63. $x = \dfrac{\ln 15}{2\ln 5} \approx 0.8413$ **65.** $t = \dfrac{\ln 2}{\ln 1.07} \approx 10.2448$

67. $t = \dfrac{\ln 3}{12\ln[1 + (0.07/12)]} \approx 15.740$

69. (a) 14.21 years   (b) 13.89 years

(c) 13.86 years   (d) 13.86 years

71. $t \approx -12,194$ years

73. (a) $P(20) \approx 1,681,900$ (b) 2011

75. 9395 years   **77.** 12,484 years

79. (a) 80   (b) 57.5   (c) 10 months

81.

x	y	$\dfrac{\ln x}{\ln y}$	$\ln \dfrac{x}{y}$	$\ln x - \ln y$
1	2	0	-0.6931	-0.6931
3	4	0.7925	-0.2877	-0.2877
10	5	1.4307	0.6931	0.6931
4	0.5	-2.0000	2.0794	2.0794

83.

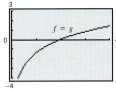

85. False. $f(x) = \ln x$ is undefined for $x \le 0$.

87. False. $f\left(\dfrac{x}{2}\right) = f(x) - f(2)$ **89.** False. $u = v^2$

SECTION 4.5   (page 296)

Prerequisite Review

1. $2 \ln(x + 1)$ **2.** $\ln x + \ln(x + 1)$

3. $\ln x - \ln(x + 1)$ **4.** $3[\ln x - \ln(x - 3)]$

5. $\ln 4 + \ln x + \ln(x - 7) - 2 \ln x$

6. $3 \ln x + \ln(x + 1)$

7. $-\dfrac{y}{x + 2y}$ **8.** $\dfrac{3 - 2xy + y^2}{x(x - 2y)}$

9. $-12x + 2$ **10.** $-\dfrac{6}{x^4}$

---

1. 3   **3.** 2   **5.** $\dfrac{2}{x}$ **7.** $\dfrac{2x}{x^2 + 3}$ **9.** $\dfrac{2(x^3 - 1)}{x(x^3 - 4)}$

11. $\dfrac{3}{x}(\ln x)^5$ **13.** $x(1 + \ln x^2)$ **15.** $\dfrac{2x^2 - 1}{x(x^2 - 1)}$

17. $\dfrac{1}{x(x + 1)}$ **19.** $\dfrac{2}{3(x^2 - 1)}$ **21.** $-\dfrac{4}{x(4 + x^2)}$

23. $e^{-x}\left(\dfrac{1}{x} - \ln x\right)$ **25.** $\dfrac{e^x - e^{-x}}{e^x + e^{-x}}$ **27.** $e^{x(\ln 2)}$

29. $\dfrac{1}{\ln 4} \ln x$ **31.** 5.585   **33.** -0.631

35. -2.134 **37.** $(\ln 3)3^x$ **39.** $\dfrac{1}{x \ln 2}$

41. $(2 \ln 4)4^{2x-3}$ **43.** $\dfrac{2x + 6}{(x^2 + 6x) \ln 10}$

45. $2^x(1 + x \ln 2)$ **47.** $y = x - 1$

49. $y = \dfrac{1}{3 \ln 3}x - \dfrac{2}{9 \ln 3} + 2$ or $y = 0.303x + 1.798$

51. $\dfrac{2xy}{3 - 2y^2}$ **53.** $\dfrac{y(1 - 6x^2)}{1 + y}$ **55.** $\dfrac{1}{2x}$ **57.** $(\ln 5)^2\, 5^x$

59. $\dfrac{d\beta}{dI} = \dfrac{10}{(\ln 10)I}$; for $I = 10^{-4}$, the rate of change is about 43,429.4 decibels per watt per square centimeters.

61. $2, y = 2x - 1$ **63.** $-\dfrac{8}{5}, y = -\dfrac{8}{5}x - 4$

65. $\dfrac{1}{\ln 2}, y = \dfrac{1}{\ln 2}x - \dfrac{1}{\ln 2}$

67.

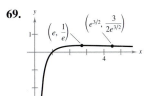

Relative minimum: $(1, 1)$

69.

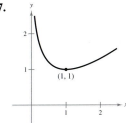

Relative maximum: $\left(e, \dfrac{1}{e}\right)$

Point of inflection: $\left(e^{3/2}, \dfrac{3}{2e^{3/2}}\right)$

71.

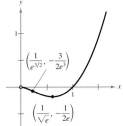

Relative minimum: $\left(\dfrac{1}{\sqrt{e}}, -\dfrac{1}{2e}\right)$

Point of inflection: $\left(\dfrac{1}{e^{3/2}}, -\dfrac{3}{2e^3}\right)$

73. $-\dfrac{1}{p}, -\dfrac{1}{10}$　　**75.** $-\dfrac{1000p}{(p^2+1)[\ln(p^2+1)]^2}, -4.65$

77. $p = 1000e^{-x}$

$\dfrac{dp}{dx} = -1000e^{-x}$

At $p = 10$, rate of change $= -10$.

$\dfrac{dp}{dx}$ and $\dfrac{dx}{dp}$ are reciprocals of each other.

79. (a) $\overline{C} = \dfrac{500 + 300x - 300\ln x}{x}$

(b) Minimum of 279.15 at $e^{8/3}$

(c)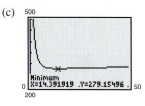

81. (a)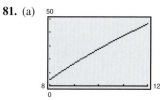

(b) $27.56 billion　　(c) 10.33

83. Answers will vary.

SECTION 4.6 *(page 305)*

Prerequisite Review

1. $-\dfrac{1}{4}\ln 2$　**2.** $\dfrac{1}{5}\ln\dfrac{10}{3}$　**3.** $-\dfrac{\ln(25/16)}{0.01}$

4. $-\dfrac{\ln(11/16)}{0.02}$　**5.** $7.36e^{0.23t}$　**6.** $1.296e^{0.072t}$

7. $-33.6e^{-1.4t}$　**8.** $-0.025e^{-0.001t}$　**9.** 4

10. 12　　**11.** $2x + 1$　　**12.** $x^2 + 1$

1. $y = 2e^{0.1014t}$　　**3.** $y = 4e^{-0.4159t}$

5. $y = 0.6687e^{0.4024t}$　　**7.** $y = 10e^{2t}$, exponential growth

9. $y = 30e^{-4t}$, exponential decay

11. *Amount after 1000 years:* 6.48 grams

Amount after 10,000 years: 0.13 gram

13. *Initial quantity:* 6.73 grams

Amount after 1000 years: 5.96 grams

15. *Initial quantity:* 2.16 grams

Amount after 10,000 years: 1.62 grams

17. 68%　　**19.** 15,642 years

21. $k_1 = \dfrac{\ln 4}{12} \approx 0.1155$, so $y_1 = 5e^{0.1155t}$.

$k_2 = \dfrac{1}{6}$, so $y_2 = 5(2)^{t/6}$

Explanations will vary.

23. (a) 1350　　(b) $\dfrac{5\ln 2}{\ln 3} \approx 3.15$ hours

(c) No. Answers will vary.

25. *Time to double:* 5.78 years

Amount after 10 years: $3320.12

Amount after 25 years: $20,085.54

27. *Annual rate:* 8.94%

Amount after 10 years: $1833.67

Amount after 25 years: $7009.86

29. *Annual rate:* 9.50%

Time to double: 7.30 years

Amount after 25 years: $5375.51

31. (a) Answers will vary.　　(b) 6.17%

33.

Number of compoundings/yr	4	12
Effective yield	5.095%	5.116%

Number of compoundings/yr	365	Continuous
Effective yield	5.127%	5.127%

35. Answers will vary.

37. (a) $1034.08 million　　(b) $628.25 million

(c)

$t = 0$ corresponds to 1993.

Answers will vary.

39. (a) $C = 30$

$k = \ln\left(\dfrac{1}{6}\right) \approx -1.7918$

(b) $30e^{-0.35836} = 20.9646$ or 20,965 units

(c)
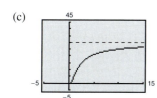

41. About 36 days   **43.** \$496,806

45. (a) $C = \frac{625}{64}$

$k = \frac{1}{100} \ln \frac{4}{5}$

(b) $x = 448$ units; $P = \$3.59$

47. 2046

REVIEW EXERCISES FOR CHAPTER 4
(page 312)

1. 8   **3.** 125   **5.** 1   **7.** $\frac{1}{6}$ **9.** 4

11. 16   **13.** $\frac{1}{2}$ **15.** $e\sqrt[3]{3}$

17. (a) 1995: $R(5) \approx \$524.04$ million

1998: $R(8) \approx \$636.58$ million

2001: $R(11) \approx \$773.30$ million

(b) Answers will vary.

19.
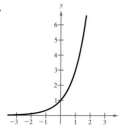

21.
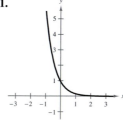

23.
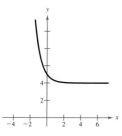

25.
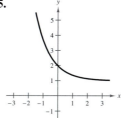

27.
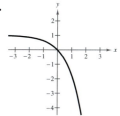

29. \$7500
Explanations will vary.

31. (a) $2e \approx 5.4366$ (b) $2e^{-1/2} \approx 1.2131$

(c) $2e^9 \approx 16,206.168$

33. (a) $12e^{-3.4} \approx 0.4005$ (b) $12e^{-10} \approx 0.0005$

(c) $12e^{-20} \approx 2.4734 \times 10^{-8}$

35. (a)
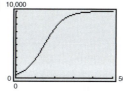

(b) $P \approx 1049$ fish

(c) Yes, P approaches 10,000 fish as t approaches ∞.

(d) The population is increasing most rapidly at the inflection point, which occurs around $t = 15$ months.

37.

n	1	2	4	12
A	\$1216.65	\$1218.99	\$1220.19	\$1221.00

n	365	Continuous compounding
A	\$1221.39	\$1221.40

39. b   **41.** (a) 6.14%   (b) 6.17%   **43.** \$9889.50

45. 1980: $P(0) = 24.196$ million

1995: $P(15) \approx 31.675$ million

2000: $P(20) = 32.700$ million

47. $8xe^{x^2}$ **49.** $\dfrac{1 - 2x}{e^{2x}}$ **51.** $4e^{2x}$ **53.** $\dfrac{-10e^{2x}}{(1 + e^{2x})^2}$

55.
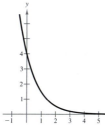
No relative extrema
No points of inflection
Horizontal asymptote: $y = 0$

57.
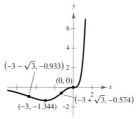

Relative minimum: $(-3, -1.344)$

Inflection points: $(0, 0)$, $(-3 + \sqrt{3}, -0.574)$, and $(-3 - \sqrt{3}, -0.933)$

Horizontal asymptote: $y = 0$

59.

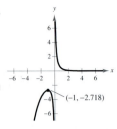

Relative maximum: $(-1, -2.718)$

Horizontal asymptote: $y = 0$

Vertical asymptote: $x = 0$

61.

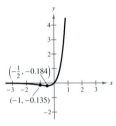

Relative minimum: $\left(-\dfrac{1}{2}, -\dfrac{1}{2e}\right)$

Inflection point: $\left(-1, -\dfrac{1}{e^2}\right)$

Horizontal asymptote: $y = 0$

63. $e^{2.4849} \approx 12$ **65.** $\ln 4.4817 \approx 1.5$

67.

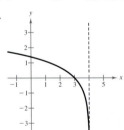

69.

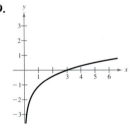

71. $\ln x + \dfrac{1}{2}\ln(x - 1)$ **73.** $2\ln x - 3\ln(x + 1)$

75. $3[\ln(1 - x) - \ln 3 - \ln x]$ **77.** 3

79. $e^{3e^{-1}} \approx 3.0151$ **81.** 1

83. $\dfrac{1}{2}(\ln 6 + 1) \approx 1.3959$

85. $\dfrac{3 + \sqrt{13}}{2} \approx 3.3028$ **87.** $-\dfrac{\ln(0.25)}{1.386} \approx 1.0002$

89. $\dfrac{\ln 1.1}{\ln 1.21} = 0.5$ **91.** $100\ln\left(\dfrac{25}{4}\right) \approx 183.2581$

93. (a)

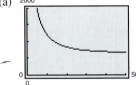

(b) A 30-year term has a smaller monthly payment, but it takes more time to pay off than a 20-year term.

95. $\dfrac{2}{x}$ **97.** $\dfrac{1}{x} + \dfrac{1}{x - 1} - \dfrac{1}{x - 2} = \dfrac{x^2 - 4x + 2}{x(x - 2)(x - 1)}$

99. 2   **101.** $\dfrac{1 - 3\ln x}{x^4}$ **103.** $\dfrac{4x}{3(x^2 - 2)}$

105. $\dfrac{2}{x} + \dfrac{1}{2(x + 1)}$ **107.** $\dfrac{1}{1 + e^x}$

109.

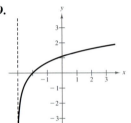

111.

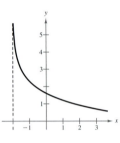

No relative extrema   No relative extrema

No points of inflection   No points of inflection

113. 2   **115.** 0   **117.** 1.431   **119.** 1.500

121. $\dfrac{2}{(2x - 1)\ln 3}$ **123.** $-\dfrac{2}{x \ln 2}$

125. (a)

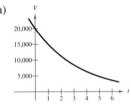

$t = 2$: \$11,250

(b) $t = 1$: -4315.23 dollars/year

$t = 4$: -1820.49 dollars/year

(c) $t \approx 4.8$ years

127. $A = 500e^{-0.01277t}$ **129.** 27.9 years

131. \$896.10 million

SAMPLE POST-GRAD EXAM QUESTIONS
(page 316)

1. d   **2.** b   **3.** c   **4.** a   **5.** c

6. d   **7.** a   **8.** b   **9.** d

CHAPTER 5

SECTION 5.1 (page 326)

Prerequisite Review

1. $x^{-1/2}$ **2.** $(2x)^{4/3}$ **3.** $5^{1/2}x^{3/2} + x^{5/2}$

4. $x^{-1/2} + x^{-2/3}$ **5.** $(x+1)^{5/2}$ **6.** $x^{1/6}$

7. -12 **8.** -10 **9.** 14 **10.** 14

1–7. Answers will vary.

9. $6x + C$

$$\frac{d}{dx}[6x + C] = 6$$

11. $\frac{5}{3}t^3 + C$ **13.** $-\frac{5}{2x^2} + C$

$$\frac{d}{dt}\left[\frac{5}{3}t^3 + C\right] = 5t^2 \qquad \frac{d}{dx}\left[-\frac{5}{2x^2} + C\right] = 5x^{-3}$$

15. $u + C$ **17.** $et + C$

$$\frac{d}{du}[u + C] = 1 \qquad \frac{d}{dt}[et + C] = e$$

19. $\frac{2}{5}y^{5/2} + C$

$$\frac{d}{dy}\left[\frac{2}{5}y^{5/2} + C\right] = y^{3/2}$$

Rewrite	Integrate	Simplify
21. $\int x^{1/3}\, dx$	$\dfrac{x^{4/3}}{4/3} + C$	$\dfrac{3}{4}x^{4/3} + C$
23. $\int x^{-3/2}\, dx$	$\dfrac{x^{-1/2}}{-1/2} + C$	$-\dfrac{2}{\sqrt{x}} + C$
25. $\int (x^3 + 3x)\, dx$	$\dfrac{x^4}{4} + \dfrac{3x^2}{2} + C$	$\dfrac{x^2}{4}(x^2 + 6) + C$
27. $\dfrac{1}{2}\int x^{-3}\, dx$	$\dfrac{1}{2}\left(\dfrac{x^{-2}}{-2}\right) + C$	$-\dfrac{1}{4x^2} + C$

29. **31.**

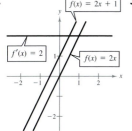

33. $\frac{1}{4}x^4 + 2x + C$ **35.** $\frac{3}{4}x^{4/3} - \frac{3}{4}x^{2/3} + C$

37. $\frac{3}{5}x^{5/3} + x + C$

39. $-\dfrac{1}{9x^3} + C$ **41.** $2x - \dfrac{1}{2x^2} + C$

43. $\frac{3}{4}u^4 + \frac{1}{2}u^2 + C$ **45.** $2x^3 - \frac{11}{2}x^2 + 5x + C$

47. $\frac{2}{7}y^{7/2} + C$ **49.** $f(x) = 2x^{3/2} + 3x - 1$

51. $f(x) = 2x^3 - 3x^2$ **53.** $f(x) = -\dfrac{1}{x^2} + \dfrac{1}{x} + \dfrac{1}{2}$

55. $y = -\frac{5}{2}x^2 - 2x + 2$ **57.** $f(x) = 4x^{3/2} - 10x + 10$

59. $f(x) = x^2 + x + 4$ **61.** $f(x) = \frac{9}{4}x^{4/3}$

63. $C = 85x + 5500$ **65.** $C = \frac{1}{10}\sqrt{x} + 4x + 750$

67. $R = 225x - \frac{3}{2}x^2,\ p = 225 - \frac{3}{2}x$

69. $R = 225x + x^2 - \frac{1}{3}x^3, p = 225 + x - \frac{1}{3}x^2$

71. $P = -9x^2 + 1650x$ **73.** $P = -12x^2 + 805x + 68$

75. 56.25 feet **77.** $v_0 = 40\sqrt{22} \approx 187.617$ feet/second

79. (a) $C = x^2 - 12x + 125$ (b) \$2025

$$\overline{C} = x - 12 + \frac{125}{x}$$

(c) \$125 is fixed.

\$1900 is variable.

Examples will vary.

81. (a) $M = 0.212t^3 - 14.24t^2 + 632.7t + 44{,}608$
(in thousands)

(b) $60{,}700{,}000$; Yes, this seems reasonable.

83. Answers will vary.

SECTION 5.2 (page 335)

Prerequisite Review

1. $\frac{1}{2}x^4 + x + C$ **2.** $\frac{3}{2}x^2 + \frac{2}{3}x^{3/2} - 4x + C$

3. $-\dfrac{1}{x} + C$ **4.** $-\dfrac{1}{6t^2} + C$

5. $\frac{4}{7}t^{7/2} + \frac{2}{5}t^{5/2} + C$ **6.** $\frac{4}{5}x^{5/2} - \frac{2}{3}x^{3/2} + C$

7. $\dfrac{5x^3 - 4}{2x} + C$ **8.** $\dfrac{-6x^2 + 5}{3x^3} + C$

9. $\frac{1}{5}x^5 + \frac{2}{3}x^3 + x + C$

10. $\frac{1}{7}x^7 - \frac{4}{5}x^5 + \frac{1}{2}x^4 + \frac{4}{3}x^3 - 2x^2 + x + C$

11. $-\dfrac{5(x-2)^4}{16}$ **12.** $-\dfrac{1}{12(x-1)^2}$

13. $9(x^2 + 3)^{2/3}$ **14.** $-\dfrac{5}{(1 - x^3)^{1/2}}$

$\displaystyle\int u^n \frac{du}{dx}\,dx$	u	$\dfrac{du}{dx}$
1. $\displaystyle\int (5x^2 + 1)^2(10x)\,dx$	$5x^2 + 1$	$10x$
3. $\displaystyle\int \sqrt{1 - x^2}\,(-2x)\,dx$	$1 - x^2$	$-2x$
5. $\displaystyle\int \left(4 + \frac{1}{x^2}\right)^5\left(\frac{-2}{x^3}\right) dx$	$4 + \dfrac{1}{x^2}$	$-\dfrac{2}{x^3}$
7. $\displaystyle\int (1 + \sqrt{x})^3\left(\frac{1}{2\sqrt{x}}\right) dx$	$1 + \sqrt{x}$	$\dfrac{1}{2\sqrt{x}}$

9. $\frac{1}{5}(1 + 2x)^5 + C$ **11.** $\frac{2}{3}(5x^2 - 4)^{3/2} + C$

13. $\frac{1}{5}(x - 1)^5 + C$ **15.** $\frac{1}{16}(x^2 - 1)^8 + C$

17. $-\dfrac{1}{3(1 + x^3)} + C$ **19.** $-\dfrac{1}{2(x^2 + 2x - 3)} + C$

21. $\sqrt{x^2 - 4x + 3} + C$ **23.** $-\frac{15}{8}(1 - u^2)^{4/3} + C$

25. $4\sqrt{1 + y^2} + C$ **27.** $-3\sqrt{2t + 3} + C$

29. $-\frac{1}{2}\sqrt{1 - x^4} + C$ **31.** $-\dfrac{1}{24}\left(1 + \dfrac{4}{t^2}\right)^3 + C$

33. $\frac{1}{6}(x^3 + 3x)^2 + C$ **35.** $\frac{1}{48}(6x^2 - 1)^4 + C$

37. $-\frac{2}{45}(2 - 3x^3)^{5/2} + C$ **39.** $\sqrt{x^2 + 25} + C$

41. $\frac{2}{3}\sqrt{x^3 + 3x + 4} + C$

43. (a) $\frac{1}{6}(2x - 1)^3 + C_1 = \frac{4}{3}x^3 - 2x^2 + x + C_2$

 (b) Answers differ by a constant: $C_2 = C_1 - \frac{1}{6}$

 (c) Answers will vary.

45. (a) $\dfrac{(x^2 - 1)^3}{6} + C_1 = \frac{1}{6}x^6 - \frac{1}{2}x^4 + \frac{1}{2}x^2 + C_2$

 (b) Answers differ by a constant: $C_2 = C_1 - \dfrac{1}{6}$

 (c) Answers will vary.

47. $f(x) = \frac{1}{3}[5 - (1 - x^2)^{3/2}]$

49. (a) $C = 8\sqrt{x + 1} + 18$

 (b)

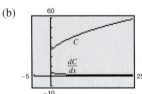

51. $x = \frac{1}{3}(p^2 - 25)^{3/2} + 24$ **53.** $x = \dfrac{6000}{\sqrt{p^2 - 16}} + 3000$

55. (a) $h = \sqrt{17.6t^2 + 1} + 5$ (b) 26 inches

57. (a) $Q = (x - 19,999)^{0.95} + 19,999$

(b)

x	20,000	50,000	100,000	150,000
Q	20,000	37,916.56	65,491.59	92,151.16
$x - Q$	0	12,083.44	34,508.41	57,848.84

(c)

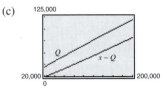

59. $-\frac{2}{3}x^{3/2} + \frac{2}{3}(x + 1)^{3/2} + C$

SECTION 5.3   *(page 342)*

Prerequisite Review

1. $\left(\frac{5}{2}, \infty\right)$ **2.** $(-\infty, 2) \cup (3, \infty)$

3. $x + 2 - \dfrac{2}{x + 2}$ **4.** $x - 2 + \dfrac{1}{x - 4}$

5. $x + 8 + \dfrac{2x - 4}{x^2 - 4x}$ **6.** $x^2 - x - 4 + \dfrac{20x + 22}{x^2 + 5}$

7. $\frac{1}{4}x^4 - \dfrac{1}{x} + C$ **8.** $\frac{1}{2}x^2 + 2x + C$

9. $\frac{1}{2}x^2 - \dfrac{4}{x} + C$ **10.** $-\dfrac{1}{x} - \dfrac{3}{2x^2} + C$

1. $e^{2x} + C$ **3.** $\frac{1}{4}e^{4x} + C$ **5.** $-\frac{9}{2}e^{-x^2} + C$

7. $\frac{5}{3}e^{x^3} + C$ **9.** $\frac{1}{3}e^{x^3 + 3x^2 - 1} + C$ **11.** $-5e^{2-x} + C$

13. $\ln|x + 1| + C$ **15.** $-\frac{1}{2}\ln|3 - 2x| + C$

17. $\frac{2}{3}\ln|3x + 5| + C$

19. $\ln\sqrt{x^2 + 1} + C$ **21.** $\frac{1}{3}\ln|x^3 + 1| + C$

23. $\frac{1}{2}\ln|x^2 + 6x + 7| + C$ **25.** $\ln|\ln x| + C$

27. $\ln|1 - e^{-x}| + C$ **29.** $-\frac{1}{2}e^{2/x} + C$ **31.** $2e^{\sqrt{x}} + C$

33. $\frac{1}{2}e^{2x} - 4e^x + 4x + C$ **35.** $-\ln(1 + e^{-x}) + C$

37. $-2\ln|5 - e^{2x}| + C$ **39.** $e^x + 2x - e^{-x} + C$

41. $-\dfrac{2}{3}(1 - e^x)^{3/2} + C$ **43.** $-\dfrac{1}{x - 1} + C$

45. $2e^{2x-1} + C$ **47.** $\frac{1}{4}x^2 - 4\ln|x| + C$

49. $2\ln(e^x + 1) + C$ **51.** $\frac{1}{2}x^2 + 3x + 8\ln|x - 1| + C$

53. $\ln|e^x + x| + C$

55. $f(x) = \frac{1}{2}x^2 + 5x + 8\ln|x - 1| - 8$

57. (a) $P(t) = 1000[1 + \ln(1 + 0.25t)^{12}]$

(b) $P(3) \approx 7715$ (c) $t \approx 6$ days

59. (a) $p = -50e^{-x/500} + 45.06$

(b)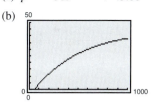

The price increases as the demand increases.

(c) 387

61. (a) $S = 37{,}452.86e^{0.07t} + 37{,}606.58$ (in dollars)

(b) \$107,928.47

63. False. $\ln x^{1/2} = \frac{1}{2}\ln x$

SECTION 5.4   (page 353)

Prerequisite Review

1. $\frac{3}{2}x^2 + 7x + C$ **2.** $\frac{2}{5}x^{5/2} + \frac{4}{3}x^{3/2} + C$

3. $\frac{1}{5}\ln|x| + C$ **4.** $-\frac{1}{6e^{6x}} + C$ **5.** $-\frac{8}{5}$

6. $-\frac{62}{3}$ **7.** $C = 0.008x^{5/2} + 29{,}500x + C$

8. $R = x^2 + 9000x + C$

9. $P = 25{,}000x - 0.005x^2 + C$

10. $C = 0.01x^3 + 4600x + C$

1.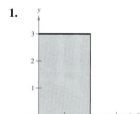

Area = 6

3.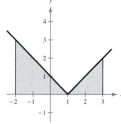

Area = $\frac{35}{2}$

5.

Area = $\frac{13}{2}$

7.

Area = $\frac{9\pi}{2}$

9. (a) 11   (b) 5   (c) -32 (d) -1

11. $\frac{1}{6}$ **13.** $\frac{1}{2}$ **15.** $6\left(1 - \frac{1}{e^2}\right)$ **17.** $8\ln 2 + \frac{15}{2}$

19. 1   **21.** 0   **23.** $\frac{14}{3}$ **25.** $-\frac{15}{4}$ **27.** -4

29. $\frac{22}{3}$ **31.** $-\frac{27}{20}$ **33.** 2   **35.** $\frac{1}{2}(1 - e^{-2}) \approx 0.432$

37. $\frac{e^3 - e}{3} \approx 5.789$ **39.** $\frac{1}{3}\left[(e^2 + 1)^{3/2} - 2\sqrt{2}\right] \approx 7.157$

41. $\frac{1}{8}\ln 17 \approx 0.354$ **43.** 4   **45.** 4

47. $\frac{1}{2}\ln 5 - \frac{1}{2}\ln 8 \approx -0.235$

49. $2\ln(2 + e^3) - 2\ln 3 \approx 3.993$

51. Area = 10   **53.** Area = $\frac{1}{4}$

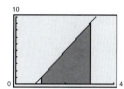   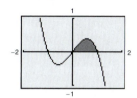

55. Area = $\ln 9$

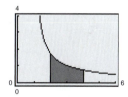

57. 10   **59.** $4 + 5\ln 5 \approx 12.047$

61.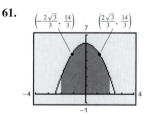

Average = $\frac{14}{3}$

$x = \pm\frac{2\sqrt{3}}{3} \approx \pm 1.155$

63.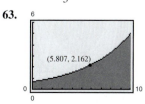

Average = $\frac{5}{2} - \frac{5}{2}e^{-2}$

$x = 5\ln\left(\frac{e^2 - 1}{2}\right) \approx 5.807$

65.

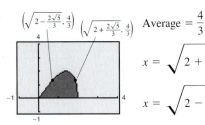

$\left(\sqrt{2 - \frac{2\sqrt{5}}{3}}, \frac{4}{3}\right)$ $\left(\sqrt{2 + \frac{2\sqrt{5}}{3}}, \frac{4}{3}\right)$

Average $= \dfrac{4}{3}$

$x = \sqrt{2 + \dfrac{2\sqrt{5}}{3}} \approx 1.868$

$x = \sqrt{2 - \dfrac{2\sqrt{5}}{3}} \approx 0.714$

67.

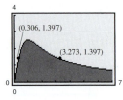

(0.306, 1.397)
(3.273, 1.397)

Average $= \frac{5}{14}\ln 50$

$x \approx 0.306$

$x \approx 3.273$

69. Even   **71.** Neither odd nor even

73. (a) $\frac{8}{3}$ (b) $\frac{16}{3}$ (c) $-\frac{8}{3}$

Explanations will vary.

75. $6.75   **77.** $22.50   **79.** $3.97   **81.** $1925.23

83. $16,605.21   **85.** $2500   **87.** $4565.65

89. (a) $137,000   (b) $214,720.93   (c) $338,393.53

91. $3082.95   **93.** $\dfrac{2kR^2}{3}$ **95.** $\dfrac{2}{3}\sqrt{7} - \dfrac{1}{3}$ **97.** $\dfrac{39}{200}$

SECTION 5.5   *(page 362)*

Prerequisite Review

1. $-x^2 + 3x + 2$ **2.** $-2x^2 + 4x + 4$

3. $-x^3 + 2x^2 + 4x - 5$ **4.** $x^3 - 6x - 1$

5. $(0, 4), (4, 4)$ **6.** $(1, -3), (2, -12)$

7. $(-3, 9), (2, 4)$ **8.** $(-2, -4), (0, 0), (2, 4)$

9. $(1, -2), (5, 10)$ **10.** $(1, e)$

1. 36   **3.** 9   **5.** $\frac{3}{2}$ **7.** $e - 2$

9.

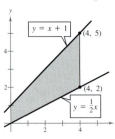

$y = x + 1$
(4, 5)
(4, 2)
$y = \frac{1}{2}x$

11.

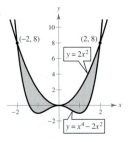

(-2, 8)
(2, 8)
$y = 2x^2$
$y = x^4 - 2x^2$

13.

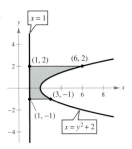

$x = 1$
(1, 2)   (6, 2)
(3, -1)
(1, -1)
$x = y^2 + 2$

15.

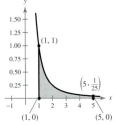

(1, 1)
$\left(5, \frac{1}{25}\right)$
(1, 0)   (5, 0)

Area $= \frac{4}{5}$

17.

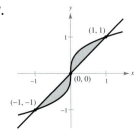

(1, 1)
(0, 0)
(-1, -1)

Area $= \frac{1}{2}$

19.

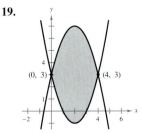

(0, 3)   (4, 3)

Area $= \frac{64}{3}$

21.

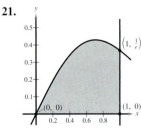

$\left(1, \frac{1}{e}\right)$
(0, 0)   (1, 0)

Area $= -\frac{1}{2}e^{-1} + \frac{1}{2}$

23.

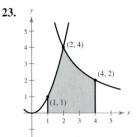

(2, 4)
(4, 2)
(1, 1)

Area $= \frac{7}{3} + 8\ln 2$

25.

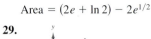

(2, e)
$(1, e^{0.5})$
$\left(2, -\frac{1}{2}\right)$
(1, -1)

Area $= (2e + \ln 2) - 2e^{1/2}$

27.

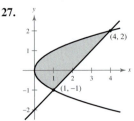

(4, 2)
(1, -1)

Area $= \frac{9}{2}$

29.

(0, 9)   (3, 9)
(0, 0)

Area $= 18$

31.

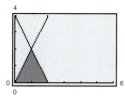

$$\text{Area} = \int_0^1 2x\,dx + \int_1^2 (4 - 2x)\,dx$$

33.

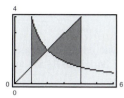

$$\text{Area} = \int_1^2 \left(\frac{4}{x} - x\right) dx + \int_2^4 \left(x - \frac{4}{x}\right) dx$$

35.

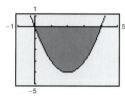

$\text{Area} = \frac{32}{3}$

37.

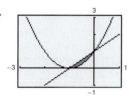

$\text{Area} = \frac{1}{6}$

39. 8

41. Consumer surplus = 1600

Producer surplus = 400

43. Consumer surplus ≈ 1666.67

Producer surplus = 1250

45. Consumer surplus = 50,000

Producer surplus ≈ 25,497

47. A typical demand function is decreasing, whereas a typical supply function is increasing.

49. R_1, $3.12 billion

51. $287.64 million; Explanations will vary.

53. (a)

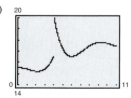

(b) 1412.47 pounds more

55. Consumer surplus = $625,000

Producer surplus = $1,375,000

57. $337.33 million

59.

Quintile	Lowest	2nd	3rd	4th	Highest
Percent	2.92	7.09	14.58	26.74	44.89

SECTION 5.6   *(page 369)*

Prerequisite Review

1. $\frac{1}{6}$ **2.** $\frac{3}{20}$ **3.** $\frac{7}{40}$ **4.** $\frac{13}{12}$ **5.** $\frac{61}{30}$ **6.** $\frac{53}{18}$

7. $\frac{2}{3}$ **8.** $\frac{4}{7}$ **9.** 0   **10.** 5

1. Midpoint Rule: 2

Exact area: 2

3. Midpoint Rule: 0.6730

Exact area: $\frac{2}{3} \approx 0.6667$

5. Midpoint Rule: 4.6250

Exact area: $\frac{14}{3} = 4.\overline{6}$

7. Midpoint Rule: 17.2500

Exact area: $\frac{52}{3} = 17.\overline{3}$

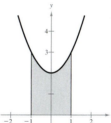

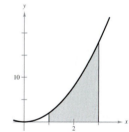

9. Midpoint Rule: 0.0859

Exact area: $\frac{1}{12} = 0.08\overline{3}$

11. Midpoint Rule: 0.0859

Exact area: $\frac{1}{12} = 0.08\overline{3}$

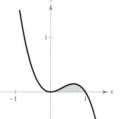

13. Area ≈ 54.6667,

$n = 31$

15. Area ≈ 4.16,

$n = 5$

17. Midpoint Rule: 1.5

Exact area: 1.5

19. Midpoint Rule: 25

Exact area: $\frac{76}{3} = 25.\overline{3}$

21. Exact: 4

Trapezoidal Rule: 4.0625

Midpoint Rule: 3.9688

Midpoint Rule is better in this example.

23. 1.1167   **25.** 1.55

27.

n	Midpoint Rule	Trapezoidal Rule
4	15.3965	15.6055
8	15.4480	15.5010
12	15.4578	15.4814
16	15.4613	15.4745
20	15.4628	15.4713

29. 4.8103   **31.** 916.25 feet

33. Midpoint Rule: $\pi \approx 3.146801$

Trapezoidal Rule: $\pi \approx 3.131176$

Graphing utility: $\pi \approx 3.141593$

SECTION 5.7   *(page 376)*

Prerequisite Review

1. 0, 2   **2.** 0, 2   **3.** 0, 2, -2 **4.** $-1, 2$

5. 2, 4   **6.** 1, 5   **7.** $e^4 - 1$ **8.** $\ln 7$

9. $\dfrac{5\sqrt{5}}{3} - \dfrac{1}{3}$ **10.** $\dfrac{(\ln 5)^3}{3}$

1. $\dfrac{16\pi}{3}$ **3.** $\dfrac{15\pi}{2}$ **5.** $\dfrac{512\pi}{15}$ **7.** $\dfrac{32\pi}{15}$ **9.** $\dfrac{\pi}{3}$

11. $\dfrac{171\pi}{2}$ **13.** $\dfrac{128\pi}{5}$ **15.** $\dfrac{\pi}{2}(e^2 - 1)$ **17.** 8π

19. $\dfrac{2\pi}{3}$ **21.** $\dfrac{\pi}{4}$ **23.** $\dfrac{256\pi}{15}$ **25.** 18π

27. $V = \pi \displaystyle\int_0^h \left(\dfrac{r}{h}x\right)^2 dx = \dfrac{1}{3}\pi r^2 h$ **29.** 100π **31.** $\dfrac{\pi}{30}$

33. (a) 1,256,637 cubic feet   (b) 2513 fish

35. 58.434

REVIEW EXERCISES FOR CHAPTER 5
(page 382)

1. $16x + C$ **3.** $\frac{2}{3}x^3 + \frac{5}{2}x^2 + C$ **5.** $x^{2/3} + C$

7. $\frac{3}{7}x^{7/3} + \frac{3}{2}x^2 + C$ **9.** $\frac{4}{9}x^{9/2} - 2\sqrt{x} + C$

11. $f(x) = \frac{3}{2}x^2 + x - 2$ **13.** $f(x) = \frac{1}{6}x^4 - 8x + \frac{33}{2}$

15. (a) 2.5 seconds   (b) 100 feet

(c) 1.25 seconds   (d) 75 feet

17. $x + 5x^2 + \frac{25}{3}x^3 + C$ or $\frac{1}{15}(1 + 5x)^3 + C_1$

19. $\frac{2}{5}\sqrt{5x - 1} + C$ **21.** $\frac{1}{2}x^2 - x^4 + C$

23. $\frac{1}{4}(x^4 - 2x)^2 + C$

25. (a) 30.5 board-feet   (b) 125.2 board-feet

27. $-e^{-3x} + C$ **29.** $\frac{1}{2}e^{x^2 - 2x} + C$

31. $-\frac{1}{3}\ln|1 - x^3| + C$ **33.** $\frac{2}{3}x^{3/2} + 2x + 2x^{1/2} + C$

35. $A = 4$ **37.** $A = \frac{8}{3}$ **39.** $A = 2\ln 2$

41. 16   **43.** 0   **45.** 2   **47.** $\frac{1}{8}$ **49.** 3.899

51. 0   **53.** Increases $700.25

55. Average value: $\frac{8}{5}$, $x = \frac{29}{4}$

57. Average value: $\frac{1}{3}(-1 + e^3) \approx 6.362$, $x \approx 3.150$

59. $520.54; Explanations will vary.

61. (a) $B = -0.0243t^2 + 0.564t - 1.58$

(b) The price of beef per pound does not surpass $2.00. The highest price is $1.69 during 2001, and after that the prices are decreasing.

63. $17,492.94

65. $\displaystyle\int_{-2}^2 6x^5\, dx = 0$ **67.** $\displaystyle\int_{-2}^{-1} \dfrac{4}{x^2}\, dx = \int_1^2 \dfrac{4}{x^2}\, dx = 2$

(Odd function)   (Symmetric about y-axis)

69.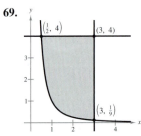

Area $= \frac{25}{3}$

71.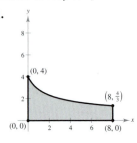

Area $= 16$

73.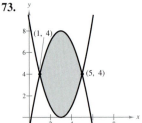

Area $= \frac{64}{3}$

75.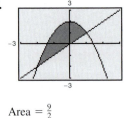

Area $= \frac{9}{2}$

77. Consumer surplus: 11,250   **79.** $5511 million less

Producer surplus: 14,062.5

81. (a)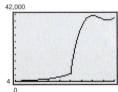

(b) Decreased

(c) $85,834 million less

83. $n = 4$: 13.3203
$\quad n = 20$: 13.7167

85. $n = 4$: 0.7867
$\quad n = 20$: 0.7855

87. $\pi \ln 4 \approx 4.355$

89. $\frac{\pi}{2}(e^2 - e^{-2}) \approx 11.394$

91. $\frac{56\pi}{3}$

93. $\frac{2\pi}{35}$

95. $\frac{5\pi}{16}\sqrt{15}$

SAMPLE POST-GRAD EXAM QUESTIONS
(page 386)

1. d **2.** b **3.** c **4.** b **5.** a
6. b **7.** d **8.** d **9.** a

CHAPTER 6

SECTION 6.1   *(page 394)*

Prerequisite Review

1. $5x + C$ **2.** $\frac{1}{3}x + C$ **3.** $\frac{2}{5}x^{5/2} + C$

4. $\frac{3}{5}x^{5/3} + C$ **5.** $\frac{1}{4}(x^2 + 1)^4 + C$

6. $\frac{(x^3 - 1)^3}{3} + C$ **7.** $e^{6x} + C$

8. $\ln|2x + 1| + C$ **9.** $x(x - 1)(2x - 1)$

10. $3x(x + 4)^2(x + 8)$

11. $(x + 21)(x + 7)^{-1/2}$ **12.** $x(x + 5)^{-2/3}$

1. $\frac{1}{5}(x - 2)^5 + C$ **3.** $-\frac{2}{9 - t} + C$

5. $\ln|t^2 - t + 2| + C$ **7.** $\frac{2}{3}(1 + x)^{3/2} + C$

9. $\ln(3x^2 + x)^2 + C$ **11.** $-\frac{1}{10(5x + 1)^2} + C$

13. $2\sqrt{x + 1} + C$ **15.** $-\frac{1}{3}\ln|1 - e^{3x}| + C$

17. $-\frac{1}{3}e^{-3x^2} + C$ **19.** $\frac{1}{2}x^2 + x + \ln|x - 1| + C$

21. $\frac{1}{3}(x^2 + 4)^{3/2} + C$ **23.** $\frac{1}{5}e^{5x} + C$

25. $-\ln|e^{-x} + 2| + C$ **27.** $\frac{-1}{2(x + 1)^2} + \frac{1}{3(x + 1)^3} + C$

29. $\frac{1}{9}\left(\ln|3x - 1| - \frac{1}{3x - 1}\right) + C$

31. $2(\sqrt{t} - 1) + 2\ln|\sqrt{t} - 1| + C$

33. $4\sqrt{t} + \ln|t| + C$ **35.** $\frac{1}{3}(x - 1)\sqrt{2x + 1} + C$

37. $\left\{-\frac{2}{105}(1 - t)^{3/2}[35 - 42(1 - t) + 15(1 - t)^2]\right\} + C = -\frac{2}{105}(15t^2 + 12t + 8)(1 - t)^{3/2} + C$

39. $\frac{26}{3}$ **41.** $\frac{3}{2}(e - 1) \approx 2.577$

43. $\ln 2 - \frac{1}{2} \approx 0.193$ **45.** $\frac{13}{320}$

47. Area $= \frac{144}{5}$

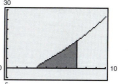

49. Area $= \frac{1696}{105}$

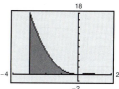

51. Area $= \frac{224}{15}$

53. Area $= \frac{1209}{28}$

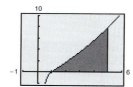

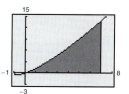

55. $\frac{16}{15}\sqrt{2}$ **57.** $\frac{4}{3}$ **59.** $\frac{4\pi}{15} \approx 0.838$ **61.** $\frac{1}{2}$

63. (a) 0.547 (b) 0.586

65. (a)

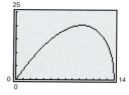

(b) About 13.97 inches (c) About 195.56 inches

67. 5.885

SECTION 6.2   *(page 403)*

Prerequisite Review

1. $\frac{1}{x + 1}$ **2.** $\frac{2x}{x^2 - 1}$ **3.** $3x^2e^{x^3}$

4. $-2xe^{-x^2}$ **5.** $e^x(x^2 + 2x)$ **6.** $e^{-2x}(1 - 2x)$

7. $\frac{64}{3}$ **8.** $\frac{4}{3}$ **9.** 36 **10.** 8

1. $\frac{1}{3}xe^{3x} - \frac{1}{9}e^{3x} + C$ **3.** $-x^2e^{-x} - 2xe^{-x} - 2e^{-x} + C$

5. $x \ln 2x - x + C$ **7.** $\frac{1}{4}e^{4x} + C$

9. $\frac{e^{4x}}{16}(4x - 1) + C$ **11.** $\frac{1}{2}e^{x^2} + C$

13. $x^2e^x - 2e^x x + 2e^x + C$

15. $\frac{1}{2}t^2 \ln|t + 1| - \frac{1}{2}\ln|t + 1| - \frac{1}{4}(t - 1)^2 + C$

17. $-e^{1/t} + C$ **19.** $\frac{x^2}{2}(\ln x)^2 - \frac{x^2}{2}\ln x + \frac{x^2}{4} + C$

21. $\frac{1}{3}(\ln x)^3 + C$ **23.** $\frac{2}{15}(x - 1)^{3/2}(3x + 2) + C$

25. $\frac{1}{4}x^4 + \frac{2}{3}x^3 + \frac{1}{2}x^2 + C$ **27.** $\frac{e^{2x}}{4(2x+1)} + C$

29. $e - 2 \approx 0.718$ **31.** $\frac{5}{36}e^6 + \frac{1}{36} \approx 56.060$

33. $2 \ln 2 - 1 \approx 0.386$

35. Area $= 2e^2 + 6$ **37.** Area $= \frac{1}{9}(2e^3 + 1)$

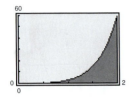

39. $\frac{2}{5}(2x - 3)^{3/2}(x + 1) + C$

41. $\frac{2}{75}\sqrt{4 + 5x}(5x - 8) + C$ **43.** Proof

45. $\frac{e^{5x}}{125}(25x^2 - 10x + 2) + C$ **47.** $-\frac{1}{x} - \frac{\ln x}{x} + C$

49. $1 - 5e^{-4} \approx 0.908$ **51.** $\frac{1}{4}(e^2 + 1) \approx 2.097$

53. (a) 2   (b) $4\pi(e - 2) \approx 9.026$

55. $\frac{3}{128} - \frac{379}{128}e^{-8} \approx 0.022$

57. $\frac{1,171,875}{256}\pi \approx 14,381.070$

59.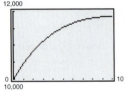

  (a) Increase

  (b) 113,212 units   (c) 11,321 units per year

61. (a) $3.2 \ln 2 - 0.2 \approx 2.018$

  (b) $12.8 \ln 4 - 7.2 \ln 3 - 1.8 \approx 8.035$

63. \$18,126.92   **65.** \$1,332,474.72   **67.** \$4103.07

69. (a) \$1,200,000   (b) \$1,094,142.26

71. \$45,957.78   **73.** (a) \$17,378.62   (b) \$3681.26

75. 4.254

SECTION 6.3   *(page 413)*

Prerequisite Review

1. $(x - 4)(x + 4)$ **2.** $(x - 5)(x + 5)$

3. $(x - 4)(x + 3)$ **4.** $(x - 2)(x + 3)$

5. $x(x - 2)(x + 1)$ **6.** $x(x - 2)^2$

7. $(x - 2)(x - 1)^2$ **8.** $(x - 3)(x - 1)^2$

9. $\frac{1}{x - 2} + x$ **10.** $-\frac{1}{1 - x} + 2x - 2$

11. $-\frac{2}{x - 2} + x^2 - x - 2$

12. $-\frac{4}{x + 1} + x^2 - x + 3$

13. $\frac{6}{x - 1} + x + 4, \quad x \neq -1$

14. $\frac{1}{x + 1} + x + 3 \quad x \neq 1$

1. $\frac{5}{x - 5} - \frac{3}{x + 5}$ **3.** $\frac{9}{x - 3} - \frac{1}{x}$ **5.** $\frac{1}{x - 5} + \frac{3}{x + 2}$

7. $\frac{3}{x} - \frac{5}{x^2}$ **9.** $\frac{1}{3(x - 2)} + \frac{1}{(x - 2)^2}$

11. $\frac{8}{x + 1} - \frac{1}{(x + 1)^2} + \frac{2}{(x + 1)^3}$ **13.** $\frac{1}{2}\ln\left|\frac{x - 1}{x + 1}\right| + C$

15. $\frac{1}{4}\ln\left|\frac{x + 4}{x - 4}\right| + C$ **17.** $\ln\left|\frac{3x - 1}{x}\right| + C$

19. $\ln\left|\frac{x}{2x + 1}\right| + C$ **21.** $\ln\left|\frac{x - 1}{x + 2}\right| + C$

23. $\frac{3}{2}\ln|2x - 1| - 2\ln|x + 1| + C$

25. $5 \ln|x - 2| - \ln|x + 2| - 3 \ln|x| + C$

27. $\frac{1}{2}(3 \ln|x - 4| - \ln|x|) + C$

29. $-3 \ln|x - 1| - \frac{1}{x - 1} + C$

31. $\ln|x| + 2 \ln|x + 1| + \frac{1}{x + 1} + C$

33. $\frac{1}{6}\ln\frac{4}{7} \approx -0.093$ **35.** $-\frac{4}{5} + 2 \ln\frac{5}{3} \approx 0.222$

37. $\frac{1}{2} - \ln 2 \approx -0.193$ **39.** $4 \ln 2 + \frac{1}{2} \approx 3.273$

41. $12 - \frac{7}{2}\ln 7 \approx 5.189$ **43.** $5 \ln 2 - \ln 5 \approx 1.856$

45. $\frac{1}{2a}\left(\frac{1}{a + x} + \frac{1}{a - x}\right)$ **47.** $\frac{1}{a}\left(\frac{1}{x} + \frac{1}{a - x}\right)$

49.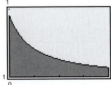

$\frac{\pi}{165}\left[136 - 33 \ln\frac{11}{3}\right] \approx 1.7731$

51.

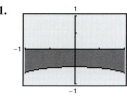

$$\pi\left(\tfrac{1}{3} + \tfrac{1}{4}\ln 3\right) \approx 1.9100$$

53. $y = \dfrac{1000}{1 + 9e^{-0.1656t}}$

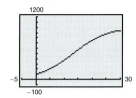

55. \$1.077 thousand   **57.** \$6188.4 million; \$773.6 million

59. (a) 103   (b) 200   **61.** Answers will vary.

SECTION 6.4   *(page 424)*

Prerequisite Review

1. $x^2 + 8x + 16$ **2.** $x^2 - 2x + 1$

3. $x^2 + x + \tfrac{1}{4}$ **4.** $x^2 - \tfrac{2}{3}x + \tfrac{1}{9}$

5. $\dfrac{2}{x} - \dfrac{2}{x + 2}$ **6.** $-\dfrac{3}{4x} + \dfrac{3}{4(x - 4)}$

7. $\dfrac{3}{2(x - 2)} - \dfrac{2}{x^2} - \dfrac{3}{2x}$ **8.** $-\dfrac{3}{x + 1} + \dfrac{2}{x - 2} + \dfrac{4}{x}$

9. $2e^x(x - 1) + C$ **10.** $x^3 \ln x - \dfrac{x^3}{3} + C$

---

1. $\dfrac{1}{9}\left(\dfrac{2}{2 + 3x} + \ln|2 + 3x|\right) + C$

3. $\dfrac{2(3x - 4)}{27}\sqrt{2 + 3x} + C$ **5.** $\ln\left(x^2 + \sqrt{x^4 - 9}\right) + C$

7. $\dfrac{1}{2}(x^2 - 1)e^{x^2} + C$ **9.** $\ln\left|\dfrac{x}{1 + x}\right| + C$

11. $-\dfrac{1}{3}\ln\left|\dfrac{3 + \sqrt{x^2 + 9}}{x}\right| + C$

13. $-\dfrac{1}{2}\ln\left|\dfrac{2 + \sqrt{4 - x^2}}{x}\right| + C$

15. $\tfrac{1}{4}x^2(-1 + 2\ln x) + C$ **17.** $3x^2 - \ln(1 + e^{3x^2}) + C$

19. $\tfrac{1}{4}\left(x^2\sqrt{x^4 - 4} - 4\ln|x^2 + \sqrt{x^4 - 4}|\right) + C$

21. $\dfrac{1}{27}\left[\dfrac{4}{2 + 3t} - \dfrac{2}{(2 + 3t)^2} + \ln|2 + 3t|\right] + C$

23. $\dfrac{1}{\sqrt{3}}\ln\left|\dfrac{\sqrt{3 + s} - \sqrt{3}}{\sqrt{3 + s} + \sqrt{3}}\right| + C$

25. $\dfrac{1}{8}\left[\dfrac{-1}{2(3 + 2x)^2} + \dfrac{2}{(3 + 2x)^3} - \dfrac{9}{4(3 + 2x)^4}\right] + C$

27. $-\dfrac{\sqrt{1 - x^2}}{x} + C$ **29.** $\dfrac{1}{9}x^3(-1 + 3\ln x) + C$

31. $\dfrac{1}{27}\left(3x - \dfrac{25}{3x - 5} + 10\ln|3x - 5|\right) + C$

33. $\tfrac{1}{9}(3\ln x - 4\ln|4 + 3\ln x|) + C$

35. Area $= \dfrac{40}{3}$

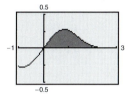

Area $= 13.\overline{3}$

37. Area $= \dfrac{1}{2}\left[4 + \ln\left(\dfrac{2}{1 + e^4}\right)\right]$

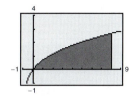

Area ≈ 0.3375

39. Area $= \tfrac{1}{4}\left[21\sqrt{5} - 8\ln(\sqrt{5} + 3) + 8\ln 2\right]$

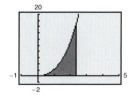

Area ≈ 9.8145

41. $\dfrac{5\sqrt{5}}{3}$ **43.** $12\left(2 + \ln\left|\dfrac{2}{1 + e^2}\right|\right) \approx 6.7946$

45. $(x^2 - 2x + 2)e^x + C$ **47.** $-\left(\dfrac{1}{x} + \ln\left|\dfrac{x}{x + 1}\right|\right) + C$

49. (a) $(x + 3)^2 - 9$ (b) $(x - 4)^2 - 7$
   (c) $(x^2 + 1)^2 - 6$ (d) $4 - (x + 1)^2$

51. (a) $4\left(x + \tfrac{3}{2}\right)^2 + 6$ (b) $3(x - 2)^2 - 21$
   (c) $(x - 1)^2 - 1$ (d) $25 - (x - 4)^2$

53. $\dfrac{1}{2\sqrt{17}}\ln\left|\dfrac{x + 3 - \sqrt{17}}{x + 3 + \sqrt{17}}\right| + C$

55. $-\ln\left|\dfrac{1 + \sqrt{x^2 - 2x + 2}}{x - 1}\right| + C$

57. $\dfrac{1}{8}\ln\left|\dfrac{x - 3}{x + 1}\right| + C$

59. $\tfrac{1}{2}\ln|x^2 + 1 + \sqrt{x^4 + 2x^2 + 2}| + C$

61.

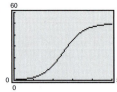

Average value: 42.58

63. $1138.43   **65.** $0.40 billion per year

SECTION 6.5   *(page 433)*

Prerequisite Review

1. $\dfrac{2}{x^3}$ **2.** $-\dfrac{96}{(2x+1)^4}$ **3.** $-\dfrac{12}{x^4}$ **4.** $6x - 4$

5. $16e^{2x}$ **6.** $e^{x^2}(4x^2 + 2)$ **7.** $(3, 18)$

8. $(1, 8)$ **9.** $n < -5\sqrt{10},\ n > 5\sqrt{10}$

10. $n < -5, n > 5$

	Exact value	*Trapezoidal Rule*	*Simpson's Rule*
1.	2.6667	2.7500	2.6667
3.	8.4000	9.0625	8.4167
5.	4.0000	4.0625	4.0000
7.	0.6931	0.6941	0.6932
9.	5.3333	5.2650	5.3046
11.	0.6931	0.6970	0.6933

13. (a) 0.783   (b) 0.785

15. (a) 0.749   (b) 0.771

17. (a) 0.877   (b) 0.830

19. (a) 1.880   (b) 1.890

21. $21,831.20   **23.** $678.36   **25.** 0.3413 = 34.13%

27. 0.4999 = 49.99%   **29.** 89,500 square feet

31. (a) 2   (b) $\dfrac{2^5}{180(4^4)}(24) \approx 0.017$

33. (a) $\dfrac{5e}{64} \approx 0.212$ (b) $\dfrac{13e}{1024} \approx 0.035$

35. (a) $n = 101$ (b) $n = 8$

37. (a) $n = 3280$ (b) $n = 60$

39. 19.5215   **41.** 3.6558

43. Exact value: $\displaystyle\int_0^1 x^3\,dx = \dfrac{x^4}{4}\Big]_0^1 = \dfrac{1}{4}$

Simpson's Rule: $\displaystyle\int_0^1 x^3\,dx = \dfrac{1}{6}\left[0^3 + 4\left(\dfrac{1}{2}\right)^3 + 1^3\right] = \dfrac{1}{4}$

45. 416.1 feet

47. 58.876 milligrams (Simpson's Rule with $n = 100$)

49. 1876 subscribers (Simpson's Rule with $n = 100$)

SECTION 6.6   *(page 444)*

Prerequisite Review

1. 9   **2.** 3   **3.** $-\dfrac{1}{8}$ **4.** Limit does not exist.

5. Limit does not exist.   **6.** -4

7. (a) $\dfrac{32}{3}b^3 - 16b^2 + 8b - \dfrac{4}{3}$ (b) $-\dfrac{4}{3}$

8. (a) $\dfrac{b^2 - b - 11}{(b-2)^2(b-5)}$ (b) $\dfrac{11}{20}$

9. (a) $\ln\left(\dfrac{5 - 3b^2}{b+1}\right)$ (b) $\ln 5 \approx 1.609$

10. (a) $e^{-3b^2}(e^{6b^2} + 1)$ (b) 2

1. 1   **3.** 1   **5.** Diverges   **7.** Diverges

9. Diverges   **11.** Diverges   **13.** 0   **15.** 4

17. 6   **19.** Diverges   **21.** 6   **23.** Diverges

25. 0   **27.** $\ln(4 + \sqrt{7}) - \ln 3 \approx 0.7954$

29. (a) 1   (b) $\dfrac{\pi}{3}$

31.

x	1	10	25	50
xe^{-x}	0.3679	0.0005	0.0000	0.0000

33.

x	1	10	25	50
$x^2e^{-(1/2)x}$	0.6065	0.6738	0.0023	0.0000

35. 2   **37.** $\dfrac{1}{4}$ **39.** (a) $4,637,228   (b) $5,555,556

41. (a) $748,367.34   (b) $808,030.14   (c) $900,000.00

43. (a) 0.9687   (b) 0.0724   (c) 0.0009

REVIEW EXERCISES FOR CHAPTER 6
(page 450)

1. $t + C$ **3.** $\dfrac{(x+5)^4}{4} + C$ **5.** $\dfrac{1}{10}e^{10x} + C$

7. $\dfrac{1}{5}\ln|x| + C$ **9.** $\dfrac{1}{3}(x^2 + 4)^{3/2} + C$

11. $2\ln(3 + e^x) + C$ **13.** $\dfrac{(x-2)^5}{5} + \dfrac{(x-2)^4}{2} + C$

15. $\dfrac{2}{15}(x+1)^{3/2}(3x - 2) + C$

17. $\dfrac{4}{5}(x-3)^{3/2}(x+2) + C$

19. $-\dfrac{2}{15}(1-x)^{3/2}(3x + 7) + C$

21. $\dfrac{26}{15}$ **23.** $\dfrac{412}{15}$ **25.** (a) 0.696   (b) 0.693

27. (a) \$2661.667 million   (b) \$15,970.002 million

29. $2\sqrt{x}\ln x - 4\sqrt{x} + C$ **31.** $xe^x - 2e^x + C$

33. $x^2 e^{2x} - xe^{2x} + \frac{1}{2}e^{2x} + C$ **35.** \$45,317.31

37. \$432,979.25

39. (a) \$4423.98, \$3934.69, \$3517.56   (b) \$997,629.35

41. \$45,118.84   **43.** $\frac{1}{5}\ln\left|\dfrac{x}{x+5}\right| + C$

45. $\ln|x - 5| + 3\ln|x + 2| + C$

47. $x - \frac{25}{8}\ln|x + 5| + \frac{9}{8}\ln|x - 3| + C$

49. (a) $y = \dfrac{10,000}{1 + 7e^{-0.106873t}}$

(b)

Time, t	0	3	6	12	24
Sales, y	1250	1645	2134	3400	6500

(c) $t \approx 28$ weeks

51. $\sqrt{x^2 + 25} - 5\ln\left|\dfrac{5 + \sqrt{x^2 + 25}}{x}\right| + C$

53. $\frac{1}{4}\ln\left|\dfrac{x - 2}{x + 2}\right| + C$ **55.** $\frac{8}{3}$

57. $2\sqrt{1 + x} + \ln\left|\dfrac{\sqrt{1 + x} - 1}{\sqrt{1 + x} + 1}\right| + C$

59. $(x - 5)^3 e^{x-5} - 3(x - 5)^2 e^{x-5} + 6(x - 6)e^{x-5} + C$

61. $\frac{1}{10}\ln\left|\dfrac{x - 3}{x + 7}\right| + C$

63. $\frac{1}{2}\big[(x - 5)\sqrt{(x - 5)^2 - 25}$
$\qquad - 25\ln|(x - 5) + \sqrt{(x - 5)^2 - 25}|\big] + C$

65. 0.705   **67.** 0.741   **69.** 0.376   **71.** 0.289

73. 9.0997   **75.** 0.017   **77.** 1   **79.** Diverges

81. 2   **83.** 2   **85.** (a) \$494,525.28   (b) \$833,333.33

87. (a) 0.431   (b) 0.108   (c) 0.013

SAMPLE POST-GRAD EXAM QUESTIONS
(page 454)

1. a   **2.** d   **3.** a   **4.** b   **5.** c

6. d   **7.** b   **8.** a   **9.** b

CHAPTER 7
SECTION 7.1   *(page 462)*

Prerequisite Review

1. $2\sqrt{5}$ **2.** 5   **3.** 8   **4.** 8   **5.** $(4, 7)$

6. $(1, 0)$ **7.** $(0, 3)$ **8.** $(-1, 1)$

9. $(x - 2)^2 + (y - 3)^2 = 4$

10. $(x - 1)^2 + (y - 4)^2 = 25$

1.   **3.**

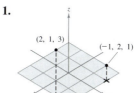   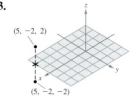

5. $3\sqrt{2}$ **7.** $\sqrt{206}$ **9.** $(2, -5, 3)$

11. $\left(\frac{1}{2}, \frac{1}{2}, -1\right)$ **13.** $(6, -3, 5)$ **15.** $(1, 2, 1)$

17. $3, 3\sqrt{5}, 6$; right triangle

19. $2, 2\sqrt{5}, 2\sqrt{2}$; neither right nor isosceles

21. $x^2 + (y - 2)^2 + (z - 2)^2 = 4$

23. $\left(x - \frac{3}{2}\right)^2 + (y - 2)^2 + (z - 1)^2 = \frac{21}{4}$

25. $(x - 1)^2 + (y - 1)^2 + (z - 5)^2 = 9$

27. $(x - 1)^2 + (y - 3)^2 + z^2 = 10$

29. $(x + 2)^2 + (y - 1)^2 + (z - 1)^2 = 1$

31. Center: $\left(\frac{5}{2}, 0, 0\right)$ **33.** Center: $(1, -3, -4)$
Radius: $\frac{5}{2}$ Radius: 5

35. Center: $(1, 3, 2)$
Radius: $\dfrac{5\sqrt{2}}{2}$

37.   **39.**

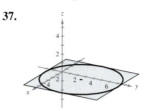   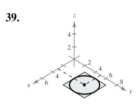

41.

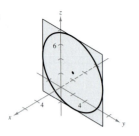

43. (a)

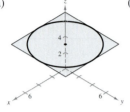

(b)

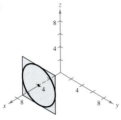

45. (a)

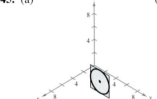

(b)

47. $(3, 3, 3)$

SECTION 7.2   *(page 472)*

Prerequisite Review

1. $(4, 0), (0, 3)$ **2.** $\left(-\frac{4}{3}, 0\right), (0, -8)$

3. $(1, 0), (0, -2)$ **4.** $(-5, 0), (0, -5)$

5. $(x - 1)^2 + (y - 2)^2 + (z - 3)^2 + 1 = 0$

6. $(x - 4)^2 + (y + 2)^2 - (z + 3)^2 = 0$

7. $(x + 1)^2 + (y - 1)^2 - z = 0$

8. $(x - 3)^2 + (y + 5)^2 + (z + 13)^2 = 1$

9. $x^2 - y^2 + z^2 = \frac{1}{4}$ **10.** $x^2 - y^2 + z^2 = 4$

1.

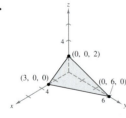

3.

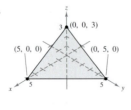

5.

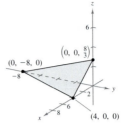

7.

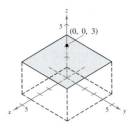

9.

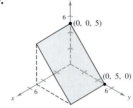

11.

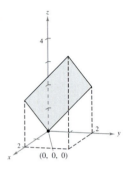

13. Perpendicular   **15.** Parallel   **17.** Parallel

19. Neither parallel nor perpendicular   **21.** Perpendicular

23. $\dfrac{6\sqrt{14}}{7}$ **25.** $\dfrac{8\sqrt{14}}{7}$ **27.** $\dfrac{13\sqrt{29}}{29}$ **29.** $\dfrac{28\sqrt{29}}{29}$

31. c   **32.** e   **33.** f   **34.** g

35. d   **36.** b   **37.** a   **38.** h

39. Trace in xy-plane $(z = 0)$: $y = x^2$ (parabola)

    Trace in plane $y = 1$: $x^2 - z^2 = 1$ (hyperbola)

    Trace in yz-plane $(x = 0)$: $y = -z^2$ (parabola)

41. Trace in xy-plane $(z = 0)$: $\dfrac{x^2}{4} + y^2 = 1$ (ellipse)

    Trace in xz-plane $(y = 0)$: $\dfrac{x^2}{4} + z^2 = 1$ (ellipse)

    Trace in yz-plane $(x = 0)$: $y^2 + z^2 = 1$ (circle)

43. Ellipsoid   **45.** Hyperboloid of one sheet

47. Elliptic paraboloid   **49.** Hyperbolic paraboloid

51. Hyperboloid of two sheets   **53.** Elliptic cone

55. Hyperbolic paraboloid

57.

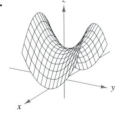

59.

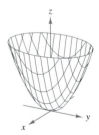

61. $\dfrac{x^2}{3963^2} + \dfrac{y^2}{3963^2} + \dfrac{z^2}{3950^2} = 1$

SECTION 7.3   *(page 480)*

Prerequisite Review

1. 11   **2.** -16 **3.** 7

4. 4   **5.** $(-\infty, \infty)$

6. $(-\infty, -3) \cup (-3, 0) \cup (0, \infty)$

7. $[5, \infty)$ **8.** $\left(-\infty, -\sqrt{5}\,\right] \cup \left[\sqrt{5}, \infty\right)$

9. 55.0104   **10.** 6.9165

1. (a) $\dfrac{3}{2}$ (b) $-\dfrac{1}{4}$ (c) 6   (d) $\dfrac{5}{y}$ (e) $\dfrac{x}{2}$ (f) $\dfrac{5}{t}$

3. (a) 5   (b) $3e^2$ (c) $2e^{-1}$

  (d) $5e^y$ (e) xe^2 (f) te^t

5. (a) $\frac{2}{3}$ (b) 0   **7.** (a) 90π (b) 50π

9. (a) \$20,655   (b) \$1,397,673   **11.** (a) 0   (b) 6

13. (a) $x^2 + 2x\,\Delta x + (\Delta x)^2 - 2y$ (b) $-2, \Delta y \neq 0$

15. Domain: all points (x, y) inside and on the circle

  $x^2 + y^2 = 16$

  Range: $[0, 4]$

17. Domain: all points (x, y) such that $y \neq 0$

  Range: $(0, \infty)$

19. All points inside and on the ellipse $9x^2 + y^2 = 9$

21. All points (x, y) such that $y \neq 0$

23. All points (x, y) such that $x \neq 0$ nor $y \neq 0$

25. All points (x, y) such that $y \geq 0$

27. The half-plane below the line $y = -x + 4$

29. b   **30.** d   **31.** a   **32.** c

33. The level curves are   **35.** The level curves are circles.
  parallel lines.

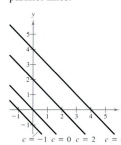

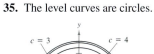

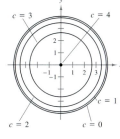

37. The level curves are hyperbolas.

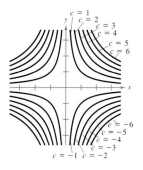

39. The level curves are circles.

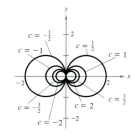

41. 135,540 units   **43.** \$21,960

45. (a) \$13,250.00   (b) \$15,925.00

47.

R \\ I	0	0.03	0.05
0	\$2593.74	\$1929.99	\$1592.33
0.28	\$2004.23	\$1491.34	\$1230.42
0.35	\$1877.14	\$1396.77	\$1152.40

49. (a) The different colors represent various amplitudes.

  (b) No, the level curves are uneven and sporadically spaced.

SECTION 7.4   *(page 491)*

Prerequisite Review

1. $\dfrac{x}{\sqrt{x^2 + 3}}$ **2.** $-6x(3 - x^2)^2$ **3.** $e^{2t+1}(2t + 1)$

4. $\dfrac{e^{2x}(2 - 3e^{2x})}{\sqrt{1 - e^{2x}}}$ **5.** $-\dfrac{2}{3 - 2x}$ **6.** $\dfrac{3(t^2 - 2)}{2t(t^2 - 6)}$

7. $-\dfrac{10x}{(4x - 1)^3}$ **8.** $-\dfrac{(x + 2)^2(x^2 + 8x + 27)}{(x^2 - 9)^3}$

9. $f'(2) = 8$ **10.** $g'(2) = \frac{7}{2}$

1. $f_x(x, y) = 2$ **3.** $f_x(x, y) = \dfrac{5}{2\sqrt{x}}$

  $f_y(x, y) = -3$ $f_y(x, y) = -12y$

5. $f_x(x, y) = \dfrac{1}{y}$

$f_y(x, y) = -\dfrac{x}{y^2}$

7. $f_x(x, y) = \dfrac{x}{\sqrt{x^2 + y^2}}$

$f_y(x, y) = \dfrac{y}{\sqrt{x^2 + y^2}}$

9. $\dfrac{\partial z}{\partial x} = 2xe^{2y}$

$\dfrac{\partial z}{\partial y} = 2x^2 e^{2y}$

11. $h_x(x, y) = -2xe^{-(x^2+y^2)}$

$h_y(x, y) = -2ye^{-(x^2+y^2)}$

13. $\dfrac{\partial z}{\partial x} = \dfrac{3y - x}{x^2 - y^2}$

$\dfrac{\partial z}{\partial y} = \dfrac{y - 3x}{x^2 - y^2}$

15. $f_x(x, y) = 3xye^{x-y}(2 + x)$

17. $g_x(x, y) = 3y^2 e^{y-x}(1 - x)$ **19.** 9

21. $f_x(x, y) = 6x + y$, 13; $f_y(x, y) = x - 2y$, 0

23. $f_x(x, y) = 3ye^{3xy}$, 12; $f_y(x, y) = 3xe^{3xy}$, 0

25. $f_x(x, y) = -\dfrac{y^2}{(x - y)^2}, -\dfrac{1}{4}$

$f_y(x, y) = \dfrac{x^2}{(x - y)^2}, \dfrac{1}{4}$

27. $f_x(x, y) = \dfrac{2x}{x^2 + y^2}, 2$

$f_y(x, y) = \dfrac{2y}{x^2 + y^2}, 0$

29. $w_x = 6xy - 5yz$

$w_y = 3x^2 - 5xz + 10z^2$

$w_z = -5xy + 20yz$

31. $w_x = \dfrac{y(y + z)}{(x + y + z)^2}$

$w_y = \dfrac{x(x + z)}{(x + y + z)^2}$

$w_z = -\dfrac{xy}{(x + y + z)^2}$

33. $w_x = \dfrac{x}{\sqrt{x^2 + y^2 + z^2}}, \dfrac{2}{3}$

$w_y = \dfrac{y}{\sqrt{x^2 + y^2 + z^2}}, -\dfrac{1}{3}$

$w_z = \dfrac{z}{\sqrt{x^2 + y^2 + z^2}}, \dfrac{2}{3}$

35. $w_x = \dfrac{x}{x^2 + y^2 + z^2}, \dfrac{3}{25}$

$w_y = \dfrac{y}{x^2 + y^2 + z^2}, 0$

$w_z = \dfrac{z}{x^2 + y^2 + z^2}, \dfrac{4}{25}$

37. $w_x = 2z^2 + 3yz$, 2

$w_y = 3xz - 12yz$, 30

$w_z = 4xz + 3xy - 6y^2$, -1

39. $(-6, 4)$ **41.** $(1, 1)$

43. (a) 2   (b) -3 **45.** (a) 6   (b) -18

47. (a) $-\dfrac{3}{4}$ (b) 0   **49.** (a) -2 (b) -2

51. $\dfrac{\partial^2 z}{\partial x \partial y} = \dfrac{\partial^2 z}{\partial y \partial x} = -2$ **53.** $\dfrac{\partial^2 z}{\partial x \partial y} = \dfrac{\partial^2 z}{\partial y \partial x} = ye^{2xy}$

55. $\dfrac{\partial^2 z}{\partial x^2} = 6x$

$\dfrac{\partial^2 z}{\partial y^2} = -8$

$\dfrac{\partial^2 z}{\partial y \partial x} = \dfrac{\partial^2 z}{\partial x \partial y} = 0$

57. $\dfrac{\partial^2 z}{\partial x^2} = 24x$

$\dfrac{\partial^2 z}{\partial y^2} = 6x - 24y$

$\dfrac{\partial^2 z}{\partial y \partial x} = \dfrac{\partial^2 z}{\partial x \partial y} = 6y$

59. $\dfrac{\partial^2 z}{\partial x^2} = \dfrac{2y^2}{(x - y)^3}$

$\dfrac{\partial^2 z}{\partial y^2} = \dfrac{2x^2}{(x - y)^3}$

$\dfrac{\partial^2 z}{\partial y \partial x} = \dfrac{\partial^2 z}{\partial x \partial y}$

$\qquad = -\dfrac{2xy}{(x - y)^3}$

61. $\dfrac{\partial^2 z}{\partial x^2} = 0$

$\dfrac{\partial^2 z}{\partial y^2} = 2xe^{-y^2}(2y^2 - 1)$

$\dfrac{\partial^2 z}{\partial y \partial x} = \dfrac{\partial^2 z}{\partial x \partial y} = -2ye^{-y^2}$

63. $f_{xx}(x, y) = 12x^2 - 6y^2$, 12

$f_{xy}(x, y) = -12xy$, 0

$f_{yy}(x, y) = -6x^2 + 2$, -4

$f_{yx}(x, y) = -12xy$, 0

65. $f_{xx}(x, y) = -\dfrac{1}{(x - y)^2}, -1$

$f_{xy}(x, y) = \dfrac{1}{(x - y)^2}, 1$

$f_{yy}(x, y) = -\dfrac{1}{(x - y)^2}, -1$

$f_{yx}(x, y) = \dfrac{1}{(x - y)^2}, 1$

67. At $(120, 160)$, $\dfrac{\partial C}{\partial x} \approx 154.77$

At $(120, 160)$, $\dfrac{\partial C}{\partial y} \approx 193.33$

69. (a) $f_x(x, y) = 60\left(\dfrac{y}{x}\right)^{0.4}$, $f_x(1000, 500) \approx 45.47$

(b) $f_y(x, y) = 40\left(\dfrac{x}{y}\right)^{0.6}$, $f_x(1000, 500) \approx 60.63$

71. (a) Complementary   (b) Substitute

(c) Complementary

73. An increase in either price will cause a decrease in the number of applicants.

75. (a) At $t = 90°$ and $h = 0.80$, $\dfrac{\partial A}{\partial t} = 1.845$.

At $t = 90°$ and $h = 0.80$, $\dfrac{\partial A}{\partial h} = 29.3$.

(b) The humidity has a greater effect since the coefficient of h is greater.

77. Answers will vary.

SECTION 7.5   *(page 501)*

Prerequisite Review

1. $(3, 2)$ **2.** $(11, 6)$ **3.** $(1, 4)$ **4.** $(4, 4)$

5. $(5, 2)$ **6.** $(3, -2)$ **7.** $(0, 0), (-1, 0)$

8. $(-2, 0), (2, -2)$

9. $\dfrac{\partial z}{\partial x} = 12x^2 \qquad \dfrac{\partial^2 z}{\partial y^2} = -6$

$\dfrac{\partial z}{\partial y} = -6y \qquad \dfrac{\partial^2 z}{\partial x \partial y} = 0$

$\dfrac{\partial^2 z}{\partial x^2} = 24x \qquad \dfrac{\partial^2 z}{\partial y \partial x} = 0$

10. $\dfrac{\partial z}{\partial x} = 10x^4 \qquad \dfrac{\partial^2 z}{\partial y^2} = -6y$

$\dfrac{\partial z}{\partial y} = -3y^2 \qquad \dfrac{\partial^2 z}{\partial x \partial y} = 0$

$\dfrac{\partial^2 z}{\partial x^2} = 40x^3 \qquad \dfrac{\partial^2 z}{\partial y \partial x} = 0$

11. $\dfrac{\partial z}{\partial x} = 4x^3 - \dfrac{\sqrt{xy}}{2x} \qquad \dfrac{\partial^2 z}{\partial y^2} = \dfrac{\sqrt{xy}}{4y^2}$

$\dfrac{\partial z}{\partial y} = -\dfrac{\sqrt{xy}}{2y} + 2 \qquad \dfrac{\partial^2 z}{\partial x \partial y} = -\dfrac{\sqrt{xy}}{4xy}$

$\dfrac{\partial^2 z}{\partial x^2} = 12x^2 + \dfrac{\sqrt{xy}}{4x^2} \qquad \dfrac{\partial^2 z}{\partial y \partial x} = -\dfrac{\sqrt{xy}}{4xy}$

12. $\dfrac{\partial z}{\partial x} = 4x - 3y \qquad \dfrac{\partial^2 z}{\partial y^2} = 2$

$\dfrac{\partial z}{\partial y} = 2y - 3x \qquad \dfrac{\partial^2 z}{\partial x \partial y} = -3$

$\dfrac{\partial^2 z}{\partial x^2} = 4 \qquad \dfrac{\partial^2 z}{\partial y \partial x} = -3$

13. $\dfrac{\partial z}{\partial x} = y^3 e^{xy^2} \qquad \dfrac{\partial^2 z}{\partial y^2} = 4x^2 y^3 e^{xy^2} + 6xy e^{xy^2}$

$\dfrac{\partial z}{\partial y} = 2xy^2 e^{xy^2} + e^{xy^2} \qquad \dfrac{\partial^2 z}{\partial x \partial y} = 2xy^4 e^{xy^2} + 3y^2 e^{xy^2}$

$\dfrac{\partial^2 z}{\partial x^2} = y^5 e^{xy^2} \qquad \dfrac{\partial^2 z}{\partial y \partial x} = 2xy^4 e^{xy^2} + 3y^2 e^{xy^2}$

14. $\dfrac{\partial z}{\partial x} = e^{xy}(xy + 1) \qquad \dfrac{\partial^2 z}{\partial y^2} = x^3 e^{xy}$

$\dfrac{\partial z}{\partial y} = x^2 e^{xy} \qquad \dfrac{\partial^2 z}{\partial x \partial y} = x e^{xy}(xy + 2)$

$\dfrac{\partial^2 z}{\partial x^2} = y e^{xy}(xy + 2) \qquad \dfrac{\partial^2 z}{\partial y \partial x} = x e^{xy}(xy + 2)$

1. Critical point: $(-2, -4)$

No relative extrema

$(-2, -4, 1)$ is a saddle point.

3. Critical point: $(0, 0)$

Relative minimum: $(0, 0, 1)$

5. Relative minimum: $(1, 3, 0)$

7. Relative minimum: $(-1, 1, -4)$

9. Relative maximum: $(8, 16, 74)$

11. Relative minimum: $(2, 1, -7)$

13. Saddle point: $(-2, -2, -8)$

15. Saddle point: $(0, 0, 0)$

17. Relative maxima: $(0, \pm 1, 4)$

Relative minimum: $(0, 0, 0)$

Saddle points: $(\pm 1, 0, 1)$

19. Saddle point: $(0, 0, 1)$

21. Insufficient information

23. $f(x_0, y_0)$ is a saddle point.

25. Relative minima: $(a, 0, 0), (0, b, 0)$

Second-Partials Test fails at $(a, 0)$ and $(0, b)$.

27. Saddle point: $(0, 0, 0)$

Second-Partials Test fails at $(0, 0)$.

29. Relative minimum: $(0, 0, 0)$

Second-Partials Test fails at $(0, 0)$.

31. Relative minimum: $(1, -3, 0)$

33. 10, 10, 10   **35.** 10, 10, 10

37. $x_1 = 3, x_2 = 6$ **39.** $p_1 = 2500, p_2 = 3000$

41. $x_1 \approx 94, x_2 \approx 157$

43. 48 inches $\times$ 24 inches $\times$ 24 inches

45. Proof

47. $D_x(x, y) = 2x - 18 + 2y$

$D_y(x, y) = 4y - 24 + 2x$

To minimize the duration of the infection, 600 mg of the first drug and 300 mg of the second drug are necessary.

49. True

51. False. The origin is a minimum.

SECTION 7.6   *(page 511)*

Prerequisite Review

1. $\left(\frac{7}{8}, \frac{1}{12}\right)$ 2. $\left(-\frac{1}{24}, -\frac{7}{8}\right)$ 3. $\left(\frac{55}{12}, -\frac{25}{12}\right)$

4. $\left(\frac{22}{23}, -\frac{3}{23}\right)$ 5. $\left(\frac{5}{3}, \frac{1}{3}, 0\right)$ 6. $\left(\frac{14}{19}, -\frac{10}{19}, -\frac{32}{57}\right)$

7. $f_x = 2xy + y^2$ 8. $f_x = 50y^2(x + y)$

   $f_y = x^2 + 2xy$ $f_y = 50y(x + y)(x + 2y)$

9. $f_x = 3x^2 - 4xy + yz$ 10. $f_x = yz + z^2$

   $f_y = -2x^2 + xz$ $f_y = xz + z^2$

   $f_z = xy$ $f_z = xy + 2xz + 2yz$

1. $f(5, 5) = 25$ 3. $f(2, 2) = 8$

5. $f\left(\frac{\sqrt{2}}{2}, \frac{1}{2}\right) = \frac{1}{4}$ 7. $f\left(\frac{25}{2}, \frac{25}{2}\right) = 231.25$

9. $f(1, 1) = 2$ 11. $f(2, 2) = e^4$ 13. $f(9, 6, 9) = 432$

15. $f\left(\frac{1}{3}, \frac{1}{3}, \frac{1}{3}\right) = \frac{1}{3}$ 17. $f\left(\frac{1}{\sqrt{3}}, \frac{1}{\sqrt{3}}, \frac{1}{\sqrt{3}}\right) = \sqrt{3}$

19. $f\left(\frac{4}{15}, \frac{8}{15}, \frac{4}{15}, \frac{2}{15}\right) = \frac{8}{15}$ 21. $f(9, 6, 9) = 486$

23. $f\left(\sqrt{\frac{10}{3}}, \frac{1}{2}\sqrt{\frac{10}{3}}, \sqrt{\frac{5}{3}}\right) = \frac{5\sqrt{15}}{9}$

25. $x = 4, y = \frac{2}{3}, z = 2$ 27. 40, 40, 40   29. $\frac{S}{3}, \frac{S}{3}, \frac{S}{3}$

31. $\sqrt{5}$ 33. $\sqrt{3}$

35. 36 inches × 18 inches × 18 inches

37. Length = width = $\sqrt[3]{360} \approx 7.1$ feet

   Height = $\dfrac{480}{360^{2/3}} \approx 9.5$ feet

39. $x_1 = 752.5, x_2 = 1247.5$

   To minimize cost, let $x_1 = 753$ units and $x_2 = 1247$ units.

41. (a) $x = 50\sqrt{2} \approx 71$ (b) Answers will vary.

   $y = 200\sqrt{2} \approx 283$

43. (a) $f\left(\frac{3125}{6}, \frac{6250}{3}\right) \approx 147{,}314$ (b) 1.473   (c) 184,142

45. $x = \sqrt[3]{0.065} \approx 0.402$ liter

   $y = \frac{1}{2}\sqrt[3]{0.065} \approx 0.201$ liter

   $z = \frac{1}{3}\sqrt[3]{0.065} \approx 0.134$ liter

47. (a) 50 feet × 120 feet   (b) \$2400

49. Stock G: \$138,333.33

   Stock P: \$7000.00

   Stock S: \$154,666.67

51. Answers will vary.

SECTION 7.7   *(page 521)*

Prerequisite Review

1. 5.0225   2. 0.0189

3. $S_a = 2a - 4 - 4b$ 4. $S_a = 8a - 6 - 2b$

   $S_b = 12b - 8 - 4a$ $S_b = 18b - 4 - 2a$

5. 15   6. 42   7. $\frac{25}{12}$

8. 14   9. 31   10. 95

1. (a) $y = \frac{3}{4}x + \frac{4}{3}$ (b) $\frac{1}{6}$

3. (a) $y = -2x + 4$ (b) 2

5. $y = x + \frac{2}{3}$ 7. $y = -2.3x - 0.9$ 9. $y = \frac{7}{10}x + \frac{7}{5}$

11. $y = x + 4$ 13. $y = -\frac{13}{20}x + \frac{7}{4}$

15. $y = \frac{37}{43}x + \frac{7}{43}$ 17. $y = -\frac{175}{148}x + \frac{945}{148}$

19. $y = \frac{3}{7}x^2 + \frac{6}{5}x + \frac{26}{35}$ 21. $y = 1.25x^2 - 1.75x + 0.25$

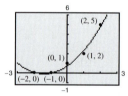 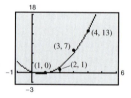

23. Linear: $y = 1.4x + 6$

   Quadratic: $y = 0.12x^2 + 1.7x + 6$

   The quadratic model is a better fit.

25. Linear: $y = -68.9x + 754$

   Quadratic: $y = 2.82x^2 - 83.0x + 763$

   The quadratic model is a better fit.

27. (a) $y = -240x + 685$ (b) 349   (c) \$0.77

29. (a) $y = 13.8x + 22.1$ (b) 44.18 bushels/acre

31. (a) $y = -0.48t + 19.74$; In 2010, $y \approx 0.54$ deaths

   (b) $y = 0.0027t^2 - 0.51t + 19.03$

     In 2010, $y \approx 2.95$ deaths

33. (a) $y = -\frac{25}{112}x^2 + \frac{541}{56}x - \frac{25}{14}$ (b) 40.9 miles per hour

35. Linear: $y = 3.757x + 9.03$

   Quadratic: $y = 0.006x^2 + 3.625x + 9.43$

   Either model is a good fit for the data.

37. Quadratic: $y = -0.087x^2 + 2.82x + 0.4$

39.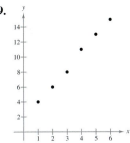

Positive correlation,
$r = 0.9981$

41.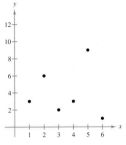

No correlation, $r = 0$

43. $y = -49.95t^2 + 4442.6t - 41,941$, where $t = 20$ represents 20-year-olds; $\approx \$43,291$

45. True   **47.** True

SECTION 7.8   (page 530)

Prerequisite Review

1. 1   **2.** 6   **3.** 42   **4.** $\frac{1}{2}$ **5.** $\frac{19}{4}$

6. $\frac{16}{3}$ **7.** $\frac{1}{7}$ **8.** 4   **9.** $\ln 5$ **10.** $\ln|e - 1|$

11. $\frac{e}{2}(e^4 - 1) \approx 72.8474$ **12.** $\frac{1}{2}\left(1 - \frac{1}{e^2}\right)$

13. 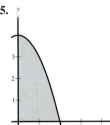   **14.**

15.   **16.**

1. $\dfrac{3x^2}{2}$ **3.** $y \ln|2y|$ **5.** $\dfrac{x^2}{2}(9 - x^2)$

7. $\frac{1}{2}y[(\ln y)^2 - y^2]$ **9.** $x^2(1 - e^{-x^2} - x^2e^{-x^2})$ **11.** 1

13. 36   **15.** $\frac{2}{3}$ **17.** $\frac{20}{3}$ **19.** 5   **21.** $\frac{16}{3}$ **23.** 4

25.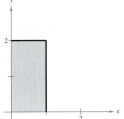

$$\int_0^1 \int_0^2 dy\,dx = \int_0^2 \int_0^1 dx\,dy = 2$$

27.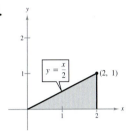

$$\int_0^1 \int_{2y}^2 dx\,dy = \int_0^2 \int_0^{x/2} dy\,dx = 1$$

29.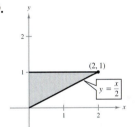

$$\int_0^2 \int_{x/2}^1 dy\,dx = \int_0^1 \int_0^{2y} dx\,dy = 1$$

31.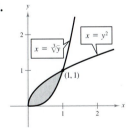

$$\int_0^1 \int_{y^2}^{\sqrt[3]{y}} dx\,dy = \int_0^1 \int_{x^3}^{\sqrt{x}} dy\,dx = \frac{5}{12}$$

33. $\frac{1}{2}(e^9 - 1) \approx 4051.042$ **35.** 24   **37.** $\frac{16}{3}$

39. $\frac{8}{3}$ **41.** $\frac{500}{3}$ **43.** $\frac{45}{14}$ **45.** 2   **47.** 0.6588

49. 8.1747   **51.** 0.4521   **53.** 1.1190

55. False, because $dA = dy\,dx = dx\,dy$, it doesn't matter in what order the integration is performed.

SECTION 7.9   (page 538)

Prerequisite Review

1.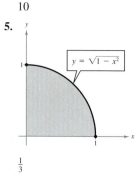

2.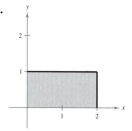

3.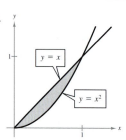

4.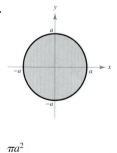

5. 1   **6.** 6   **7.** $\frac{1}{3}$ **8.** $\frac{40}{3}$

9. $\frac{28}{3}$ **10.** $\frac{7}{6}$

1.

10

3.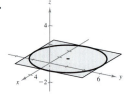

$\frac{1}{54}$

5.

$\frac{1}{3}$

7.

πa^2

9. $\int_0^3 \int_0^5 xy\, dy\, dx = \int_0^5 \int_0^3 xy\, dx\, dy = \dfrac{225}{4}$

11. $\displaystyle\int_0^2 \int_x^{2x} \frac{y}{x^2 + y^2}\, dy\, dx = \int_0^2 \int_{y/2}^{y} \frac{y}{x^2 + y^2}\, dx\, dy$

$\displaystyle + \int_2^4 \int_{y/2}^{2} \frac{y}{x^2 + y^2}\, dx\, dy = \ln\frac{5}{2}$

13. $\displaystyle\int_0^{1/2} \int_0^{2x} e^{-x^2}\, dy\, dx = 0.2212$ **15.** 4

17. 22.5   **19.** 12   **21.** $\frac{3}{8}$ **23.** $\frac{40}{3}$

25. $\frac{1}{3}$ **27.** 4   **29.** $\frac{32}{3}$ **31.** 10,000

33. 2   **35.** $\frac{8}{3}$ **37.** $75,125

REVIEW EXERCISES FOR CHAPTER 7
(page 544)

1. 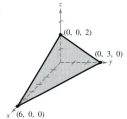   **3.** $\sqrt{110}$ **5.** $(-1, 4, 6)$

7. $x^2 + (y - 1)^2 + z^2 = 25$

9. $(x - 4)^2 + (y - 6)^2 + (z - 1)^2 = 6$

11. Center: $(-2, 1, 4)$; radius: 4

13. 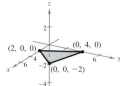   **15.**

17.

19. Sphere   **21.** Ellipsoid   **23.** Elliptic Paraboloid

25. Top half of a circular cone

27. (a) 18   (b) 0   (c) -245 (d) -32

29. The domain is the set of all points inside or on the circle $x^2 + y^2 = 1$, and the range is $[0, 1]$.

31. The level curves are lines of slope $-\frac{2}{5}$.

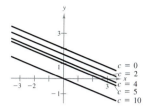

33. The level curves are hyperbolas.

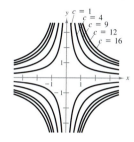

35. (a) As the color darkens from light green to dark green, the average yearly precipitation increases.

(b) The small eastern portion containing Davenport

(c) The northwestern portion containing Sioux City

37. Southwest   **39.** $2.50   **41.** $f_x = 2xy + 3y + 2$

$f_y = x^2 + 3x - 5$

43. $z_x = 12x\sqrt{y} + \frac{3}{2}\sqrt{\frac{y}{x}} - 7y$ **45.** $f_x = \frac{2}{2x + 3y}$

$z_y = \frac{3x^2}{\sqrt{y}} + \frac{3}{2}\sqrt{\frac{x}{y}} - 7x$ $f_y = \frac{3}{2x + 3y}$

47. $f_x = 2xe^y - y^2e^x$ **49.** $w_x = yz^2$

$f_y = x^2e^y - 2ye^x$ $w_y = xz^2$

$w_z = 2xyz$

51. (a) $z_x = 3$ (b) $z_y = -4$

53. (a) $z_x = -2x$ (b) $z_y = -2y$

At $(1, 2, 3)$, $z_x = -2$.   At $(1, 2, 3)$, $z_y = -4$.

55. $f_{xx} = 6x$

$f_{yy} = -8x + 6y$

$f_{xy} = f_{yx} = -8y$

57. $f_{xx} = \dfrac{y^2 - 64}{(64 - x^2 - y^2)^{3/2}}$

$f_{yy} = \dfrac{x^2 - 64}{(64 - x^2 - y^2)^{3/2}}$

$f_{xy} = f_{yx} = \dfrac{-xy}{(64 - x^2 - y^2)^{3/2}}$

59. $C_x(250, 175) \approx 99.70$

$C_y(250, 175) \approx 140.01$

61. (a) $A_w = 43.095w^{-0.575}h^{0.725}$

$A_h = 73.515w^{0.425}h^{-0.275}$

(b) ≈ 47.35;

The surface area of an average human body increases approximately 47.35 square centimeters per pound for a human who weighs 180 pounds and is 70 inches tall.

63. Relative minimum: $(x, -x, 0)$

65. Saddle point: $\left(\frac{3}{2}, -\frac{3}{2}, \frac{25}{2}\right)$

67. Relative minimum: $\left(\frac{1}{6}, \frac{1}{12}, -\frac{1}{432}\right)$

Saddle point: $(0, 0, 0)$

69. Relative minimum: $(1, 1, -2)$

Relative maximum: $(-1, -1, 6)$

Saddle points: $(1, -1, 2), (-1, 1, 2)$

71. (a) $R = -x_1^2 - \frac{1}{2}x_2^2 + 100x_1 + 200x_2$

(b) $x_1 = 50, x_2 = 200$ (c) $22,500.00

73. At $\left(\frac{4}{3}, \frac{1}{3}\right)$, the relative maximum is $\frac{16}{27}$.

At $(0, 1)$, the relative minimum is 0.

75. At $\left(\frac{4}{3}, \frac{2}{3}, \frac{4}{3}\right)$, the relative maximum is $\frac{32}{27}$.

77. At $\left(\frac{4}{3}, \frac{10}{3}, \frac{14}{3}\right)$, the relative minimum is $34\frac{2}{3}$.

79. $x = 2\sqrt{2}, y = 2\sqrt{2}, z = \sqrt{2}$

81. $f(49.4, 253) \approx 13,202$

83. (a) $y = \frac{60}{59}x - \frac{15}{59}$ (b) 2.746

85. (a) $y = 14x + 19$ (b) 21.8 bushels/acre

87. $y = 1.71x^2 - 2.57x + 5.56$

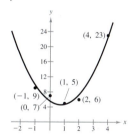

89. $\frac{29}{6}$ **91.** $\frac{7}{4}$

93. $\displaystyle\int_{-2}^{2}\int_{5}^{9-x^2} dy\, dx = \int_{5}^{9}\int_{-\sqrt{9-y}}^{\sqrt{9-y}} dx\, dy = \frac{32}{3}$

95. $\displaystyle\int_{-3}^{6}\int_{1/3(x+3)}^{\sqrt{x+3}} dy\, dx = \int_{0}^{3}\int_{3y-3}^{y^2-3} dx\, dy = \frac{9}{2}$

97. $\frac{4096}{9}$ **99.** 0.0833 mile

SAMPLE POST-GRAD EXAM QUESTIONS
(page 548)

1. b   **2.** b   **3.** c   **4.** d

5. a   **6.** c   **7.** b

CHAPTER 8

SECTION 8.1   *(page 555)*

> **Prerequisite Review**
>
> **1.** 35 square centimeters   **2.** 12 square inches
>
> **3.** $c = 13$ **4.** $b = 4$ **5.** $b = 15$
>
> **6.** $a = 6$ **7.** Equilateral triangle
>
> **8.** Isosceles triangle   **9.** Right triangle
>
> **10.** Isosceles triangle and right triangle

1. (a) $405°, -315°$ (b) $319°, -401°$

3. (a) $660°, -60°$ (b) $20°, -340°$

5. (a) $\dfrac{19\pi}{9}, -\dfrac{17\pi}{9}$ (b) $\dfrac{8\pi}{3}, -\dfrac{4\pi}{3}$

7. (a) $\dfrac{7\pi}{4}, -\dfrac{\pi}{4}$ (b) $\dfrac{28\pi}{15}, -\dfrac{32\pi}{15}$

9. $\dfrac{\pi}{6}$ **11.** $\dfrac{5\pi}{4}$ **13.** $\dfrac{7\pi}{4}$ **15.** $-\dfrac{\pi}{6}$ **17.** $-\dfrac{3\pi}{2}$

19. $\dfrac{13\pi}{6}$ **21.** $270°$ **23.** $330°$ **25.** $-300°$

27. $405°$ **29.** $570°$ **31.** $-\dfrac{3}{2}\pi$ **33.** $\dfrac{4}{5}\pi$

35. $c = 10, \theta = 60°$ **37.** $a = 4\sqrt{3}, \theta = 30°$

39. $\theta = 40°$ **41.** $s = \sqrt{3}, \theta = 60°$

43. $9\sqrt{3}$ square inches   **45.** $\dfrac{25\sqrt{3}}{4}$ square feet

47. 18 feet

49.

r	8 ft	15 in.	85 cm	24 in.	$\dfrac{12{,}963}{\pi}$ mi
s	12 ft	24 in.	200.28 cm	96 in.	8642 mi
θ	1.5	1.6	$\dfrac{3\pi}{4}$	4	$\dfrac{2\pi}{3}$

51. (a) $\dfrac{5\pi}{12}$ (b) $\dfrac{125\pi}{16}$ inches

53. (a) ≈ 500 revolutions per minute   (b) 20 minutes

55. False. An obtuse angle is between $90°$ and $180°$.

57. True

SECTION 8.2   *(page 566)*

> **Prerequisite Review**
>
> **1.** $\dfrac{3\pi}{4}$ **2.** $\dfrac{7\pi}{4}$ **3.** $-\dfrac{7\pi}{6}$ **4.** $-\dfrac{5\pi}{3}$
>
> **5.** $-\dfrac{2\pi}{3}$ **6.** $-\dfrac{5\pi}{4}$ **7.** 3π **8.** $\dfrac{13\pi}{6}$
>
> **9.** $x = 0, 1$ **10.** $x = 0, -\dfrac{1}{2}$
>
> **11.** $x = -\dfrac{1}{2}, 1$ **12.** $x = -1, 3$ **13.** $x = 1$
>
> **14.** $x = -1, \dfrac{1}{2}$ **15.** $x = 2, 3$ **16.** $x = -2, 1$
>
> **17.** $t = 10$ **18.** $t = \dfrac{7}{2}$ **19.** $t = \dfrac{445}{8}$ **20.** $t = 7$

1. $\sin\theta = \dfrac{4}{5},$ $\csc\theta = \dfrac{5}{4}$
$\cos\theta = \dfrac{3}{5},$ $\sec\theta = \dfrac{5}{3}$
$\tan\theta = \dfrac{4}{3},$ $\cot\theta = \dfrac{3}{4}$

3. $\sin\theta = -\dfrac{5}{13},$ $\csc\theta = -\dfrac{13}{5}$
$\cos\theta = -\dfrac{12}{13},$ $\sec\theta = -\dfrac{13}{12}$
$\tan\theta = \dfrac{5}{12},$ $\cot\theta = \dfrac{12}{5}$

5. $\sin\theta = \dfrac{1}{2},$ $\csc\theta = 2$
$\cos\theta = -\dfrac{\sqrt{3}}{2},$ $\sec\theta = -\dfrac{2\sqrt{3}}{3}$
$\tan\theta = -\dfrac{\sqrt{3}}{3},$ $\cot\theta = -\sqrt{3}$

7. $\csc\theta = 2$ **9.** $\cot\theta = \dfrac{4}{3}$ **11.** $\sec\theta = \dfrac{17}{15}$

13.

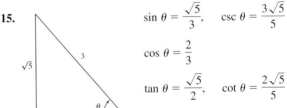

$\cos\theta = \dfrac{2\sqrt{2}}{3},$ $\csc\theta = 3$

$\tan\theta = \dfrac{\sqrt{2}}{4},$ $\sec\theta = \dfrac{3\sqrt{2}}{4}$

$\cot\theta = 2\sqrt{2}$

15.

$\sin\theta = \dfrac{\sqrt{5}}{3},$ $\csc\theta = \dfrac{3\sqrt{5}}{5}$

$\cos\theta = \dfrac{2}{3}$

$\tan\theta = \dfrac{\sqrt{5}}{2},$ $\cot\theta = \dfrac{2\sqrt{5}}{5}$

17.

$$\sin \theta = \frac{7\sqrt{53}}{53}, \qquad \csc \theta = \frac{\sqrt{53}}{7}$$

$$\cos \theta = \frac{2\sqrt{53}}{53}, \qquad \sec \theta = \frac{\sqrt{53}}{2}$$

$$\cot \theta = \frac{2}{7}$$

19. Quadrant IV   **21.** Quadrant I   **23.** Quadrant II

25. (a) $\sin 60° = \dfrac{\sqrt{3}}{2}$ (b) $\sin\left(-\dfrac{2\pi}{3}\right) = -\dfrac{\sqrt{3}}{2}$

$\cos 60° = \dfrac{1}{2}$ $\cos\left(-\dfrac{2\pi}{3}\right) = -\dfrac{1}{2}$

$\tan 60° = \sqrt{3}$ $\tan\left(-\dfrac{2\pi}{3}\right) = \sqrt{3}$

27. (a) $\sin\left(-\dfrac{\pi}{6}\right) = -\dfrac{1}{2}$ (b) $\sin 150° = \dfrac{1}{2}$

$\cos\left(-\dfrac{\pi}{6}\right) = \dfrac{\sqrt{3}}{2}$ $\cos 150° = -\dfrac{\sqrt{3}}{2}$

$\tan\left(-\dfrac{\pi}{6}\right) = -\dfrac{\sqrt{3}}{3}$ $\tan 150° = -\dfrac{\sqrt{3}}{3}$

29. (a) $\sin 225° = -\dfrac{\sqrt{2}}{2}$ (b) $\sin(-225°) = \dfrac{\sqrt{2}}{2}$

$\cos 225° = -\dfrac{\sqrt{2}}{2}$ $\cos(-225°) = -\dfrac{\sqrt{2}}{2}$

$\tan 225° = 1$ $\tan(-225°) = -1$

31. (a) $\sin 750° = \dfrac{1}{2}$ (b) $\sin 510° = \dfrac{1}{2}$

$\cos 750° = \dfrac{\sqrt{3}}{2}$ $\cos 510° = -\dfrac{\sqrt{3}}{2}$

$\tan 750° = \dfrac{\sqrt{3}}{3}$ $\tan 510° = -\dfrac{\sqrt{3}}{3}$

33. (a) 0.2079   (b) 4.8097

35. (a) 0.3640   (b) 0.3640

37. (a) -0.3420 (b) -0.3420

39. (a) 2.0070   (b) 2.0000

41. (a) $\dfrac{\pi}{6}, \dfrac{5\pi}{6}$ (b) $\dfrac{7\pi}{6}, \dfrac{11\pi}{6}$

43. (a) $\dfrac{\pi}{3}, \dfrac{2\pi}{3}$ (b) $\dfrac{3\pi}{4}, \dfrac{7\pi}{4}$

45. (a) $\dfrac{3\pi}{4}, \dfrac{7\pi}{4}$ (b) $\dfrac{5\pi}{6}, \dfrac{11\pi}{6}$ **47.** $\dfrac{\pi}{4}, \dfrac{3\pi}{4}, \dfrac{5\pi}{4}, \dfrac{7\pi}{4}$

49. $0, \dfrac{\pi}{4}, \pi, \dfrac{5\pi}{4}, 2\pi$ **51.** $\dfrac{\pi}{6}, \dfrac{\pi}{2}, \dfrac{5\pi}{6}, \dfrac{3\pi}{2}$ **53.** $\dfrac{\pi}{4}, \dfrac{5\pi}{4}$

55. $0, \dfrac{\pi}{2}, \pi, 2\pi$ **57.** $\dfrac{100\sqrt{3}}{3}$ **59.** $\dfrac{25\sqrt{3}}{3}$

61. 15.5572   **63.** $20\sin 75° \approx 19.32$ feet

65. $150\cot 3° \approx 2862.2$ feet

67. (a) 102.6°F   (b) 102.1°F   (c) 100.6°F

At 4 P.M. the following afternoon, the patient's temperature should return to normal. This is determined by setting the function equal to 98.6 and solving for t.

69.

x	0	2	4	6	8	10
$f(x)$	0	2.7021	2.7756	1.2244	1.2979	4

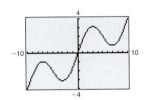

SECTION 8.3   *(page 575)*

Prerequisite Review

1. 14   **2.** 10   **3.** 0   **4.** 0   **5.** 1

6. $-\dfrac{\sqrt{3}}{3}$ **7.** $-\dfrac{1}{2}$ **8.** $-\dfrac{\sqrt{3}}{2}$ **9.** $\dfrac{1}{2}$

10. $-\dfrac{\sqrt{3}}{2}$ **11.** 0.9659   **12.** -0.6428

13. -0.9962 **14.** 0.6428   **15.** 0.9744

16. 0.3090   **17.** -0.6494 **18.** -0.8391

1. Period: π **3.** Period: 4π
Amplitude: 2   Amplitude: $\dfrac{3}{2}$

5. Period: 2   **7.** Period: 2π
Amplitude: $\dfrac{1}{2}$ Amplitude: 2

9. Period: $\dfrac{\pi}{5}$ **11.** Period: 3π
Amplitude: 2   Amplitude: $\dfrac{1}{2}$

13. Period: $\dfrac{1}{2}$
Amplitude: 3

15. $\dfrac{\pi}{2}$ **17.** $\dfrac{2\pi}{5}$ **19.** 6   **21.** c; π **22.** e; π

23. f; 2   **24.** a; 2π **25.** b; 4π **26.** d; 2π

27.

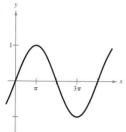

29.

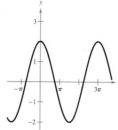

31.

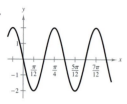

33.

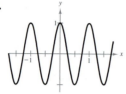

35.

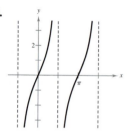

37.

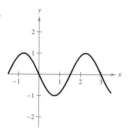

39.

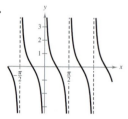

41.

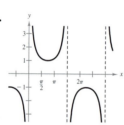

43.

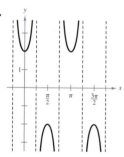

45.

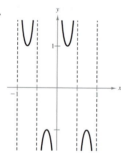

47.

x	-0.1	-0.01	-0.001
$f(x)$	-0.0997	-0.01	-0.001

x	0.001	0.01	0.1
$f(x)$	0.001	0.01	0.0997

$$\lim_{x \to 0} \frac{1 - \cos 2x}{2x} = 0$$

49.

x	-0.1	-0.01	-0.001
$f(x)$	0.1997	0.2000	0.2000

x	0.001	0.01	0.1
$f(x)$	0.2000	0.2000	0.1997

$$\lim_{x \to 0} \frac{\sin x}{5x} = \frac{1}{5}$$

51.

x	-0.1	-0.01	-0.001
$f(x)$	-0.1499	-0.0150	-0.0015

x	0.001	0.01	0.1
$f(x)$	0.0015	0.0150	0.1499

$$\lim_{x \to 0} \frac{3(1 - \cos x)}{x} = 0$$

53.

x	-0.1	-0.01	-0.001
$f(x)$	2.027	2.000	2.000

x	0.001	0.01	0.1
$f(x)$	2.000	2.000	2.027

$$\lim_{x \to 0} \frac{\tan 2x}{x} = 2$$

55.

x	-0.1	-0.01	-0.001
$f(x)$	-1.516	-0.160	-0.016

x	0.001	0.01	0.1
$f(x)$	0.016	0.160	1.516

$$\lim_{x \to 0} \frac{1 - \cos^2 4x}{x} = 0$$

57. $a = -1, b = 2, c = 0, d = 1$

$y = -\sin(2x) + 1$

59. (a) 6 sec   (b) 10

(c)

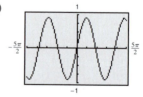

61. (a) $\frac{1}{440}$ (b) 440

(c)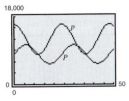

63. (a)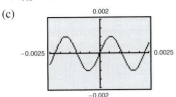

(b) As the population of the prey increases, the population of the predator increases as well. At some point, the predator eliminates the prey faster than the prey can reproduce, and the prey population decreases rapidly. As the prey becomes scarce, the predator population decreases, releasing the prey from predator pressure, and the cycle begins again.

65.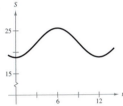

67. $P(8930) \approx 0.9977$

$E(8930) \approx -0.4339$

$I(8930) \approx -0.6182$

69.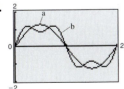

71.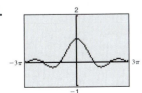

$$\lim_{x \to 0} \frac{\sin x}{x} = 1$$

73.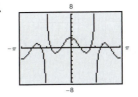

$$\lim_{x \to 0} \frac{\sin 5x}{\sin 2x} = \frac{5}{2}$$

75. (a)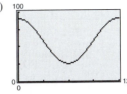

(b) January, October, November, December

77. Answers will vary.

79. Whenever the population of hares increases, the population of lynx can increase as well. At some point, the lynx devour the hares faster than the hares can reproduce, and the hare population decreases rapidly. As food becomes scarce, the lynx population plummets, releasing the hares from predation pressure, and the cycle begins again.

81. False. The period is $\frac{\pi}{4/3} = \frac{3\pi}{4}$.

83. False. $\tan\left(\frac{5\pi/4}{2}\right) \neq 1$

SECTION 8.4   (page 585)

Prerequisite Review

1. $f'(x) = 9x^2 - 4x + 4$ **2.** $g'(x) = 12x^2(x^3 + 4)^3$

3. $f'(x) = 3x^2 + 2x + 1$ **4.** $g'(x) = \dfrac{2(5 - x^2)}{(x^2 + 5)^2}$

5. Relative minimum: $(-2, -3)$

6. Relative maximum: $\left(-2, \frac{22}{3}\right)$

Relative minimum: $\left(2, -\frac{10}{3}\right)$

7. $x = \dfrac{\pi}{3}, x = \dfrac{2\pi}{3}$ **8.** $x = \dfrac{2\pi}{3}, x = \dfrac{4\pi}{3}$

9. $x = \pi$ **10.** No solution

---

1. $-3 \cos x$ **3.** $2x + \sin x$ **5.** $\dfrac{2}{\sqrt{x}} - 3 \sin x$

7. $-t^2 \sin t + 2t \cos t$ **9.** $-\dfrac{t \sin t + \cos t}{t^2}$

11. $\sec^2 x + 2x$ **13.** $e^{x^2}\sec x(\tan x + 2x)$

15. $-3\sin 3x + 2\sin x \cos x$ **17.** $\pi \cos \pi x$

19. $\sin\dfrac{1}{x} - \dfrac{1}{x}\cos\dfrac{1}{x}$ **21.** $12\sec^2 4x$

23. $16\tan 4x + 16\tan^3 4x$ **25.** $2e^{2x}(\cos 2x + \sin 2x)$

27. $-2\cos x \sin x = -\sin 2x$

29. $-4\cos x \sin x = -2\sin 2x$

31. $\dfrac{1}{\sin x}\cos x = \cot x$ **33.** $2x\csc x^2$

35. $\sec^2 x - 1 = \tan^2 x$ **37.** $\dfrac{2\cos x}{\sin x} = 2\cot x$

39. $y = 2x + \dfrac{\pi}{2} - 1$ **41.** $y = 4x - 4\pi$

43. $y = -2x + \dfrac{3}{2}\pi - 1$ **45.** $y = -2x + \dfrac{\pi}{2}$

47. $\dfrac{\cos x}{2\sin 2y}, 0$

49. $y'' + y = (-2\sin x - 3\cos x) + (2\sin x + 3\cos x) = 0$

51. $y'' + 4y = (-4\cos 2x - 4\sin 2x)$
$\qquad\qquad + 4(\cos 2x + \sin 2x) = 0$

53. $\frac{5}{4}$; one complete cycle   **55.** 2; two complete cycles

57. 1; one complete cycle

59. Relative maximum: $\left(\dfrac{\pi}{3}, \dfrac{3\sqrt{3}}{2}\right)$

Relative minimum: $\left(\dfrac{5\pi}{3}, -\dfrac{3\sqrt{3}}{2}\right)$

61. Relative maximum: $\left(\dfrac{5\pi}{3}, \dfrac{5\pi}{3} + \sqrt{3}\right)$

Relative minimum: $\left(\dfrac{\pi}{3}, \dfrac{\pi}{3} - \sqrt{3}\right)$

63. Relative maximum: $\left(\dfrac{7\pi}{4}, 0.0029\right)$

Relative minimum: $\left(\dfrac{3\pi}{4}, -0.0670\right)$

65. (a) $h'(t)$ is a maximum when $t = 0$, or at midnight.

(b) $h'(t)$ is a minimum when $t = \frac{1}{2}$, or at noon.

67. (a)

(b) $\dfrac{dh}{dx} = \dfrac{3}{10}\pi \cos\dfrac{\pi x}{150}; 0.15\pi, -0.3\pi, -0.15\pi, 0.15\pi$;

dh/dx is the rate at which the ride is rising or falling as it moves horizontally from its starting point.

(c) Maximum height: 95 feet

Minimum height: 5 feet

(d) 150 feet

69. Proof

71. (a) 0.2 radian   (b) ≈ 1.449 radians per second

73. (a)

(b) 0, 2.2889, 5.0870

(c) $f' > 0$ on $(0, 2.2889)$, $(5.0870, 2\pi)$

$f' < 0$ on $(2.2889, 5.0870)$

75. (a)

(b) 0.5236, $\pi/2$, 2.6180

(c) $f' > 0$ on $(0, 0.5236)$, $(0.5236, \pi/2)$

$f' < 0$ on $(\pi/2, 2.6180)$, $(2.6180, \pi)$

77. (a)

(b) 1.8366, 4.8158

(c) $f' > 0$ on $(0, 1.8366)$, $(4.8158, 2\pi)$

$f' < 0$ on $(1.8366, 4.8158)$

79. Relative maximum: $(4.49, -4.60)$

81. Relative maximum: $(1.27, 0.07)$

Relative minimum: $(3.38, -1.18)$

83. Relative maximum: $(3.96, 1)$

85. False. $y' = -\frac{1}{2}(1 - x)^{-1/2}$

87. False. You must apply the Product Rule.

SECTION 8.5   *(page 594)*

Prerequisite Review

1. $-\dfrac{\sqrt{2}}{2}$ 2. $-\dfrac{1}{2}$ 3. $-\dfrac{\sqrt{3}}{2}$ 4. $\dfrac{\sqrt{3}}{2}$

5. $-\dfrac{\sqrt{3}}{3}$ 6. $-\dfrac{\sqrt{3}}{3}$ 7. -1 8. 0

9. $\tan x$ 10. $\cot x$ 11. $\sin^2 x$ 12. $\cos^2 x$

13. 1 14. $\cos x$ 15. $\csc x$ 16. $\cos x \sin x$

17. $\frac{88}{3}$ 18. $\frac{4}{3}$ 19. 4 20. $\frac{17}{4}$

1. $-2\cos x + 3\sin x + C$ 3. $t + \csc t + C$

5. $-\cot \theta - \sin \theta + C$ 7. $-\frac{1}{2}\cos 2x + C$

9. $\frac{1}{2}\sin x^2 + C$ 11. $2\tan \dfrac{x}{2} + C$

13. $-\frac{1}{3}\ln|\cos 3x| + C$ 15. $\frac{1}{4}\tan^4 x + C$

17. $\dfrac{1}{\pi}\ln|\sin \pi x| + C$

19. $\frac{1}{2}\ln|\csc 2x - \cot 2x| + C$ 21. $\frac{1}{2}\ln|\tan 2x| + C$

23. $\ln|\sec x - 1| + C$ 25. $-\ln|1 + \cos x| + C$

27. $\frac{1}{2}\tan^2 x + C$ 29. $-\cos e^x + C$

31. $e^{\sin x} + C$ 33. $x - \frac{1}{4}\cos 4x + C$

35. $x\sin x + \cos x + C$ 37. $x\tan x + \ln|\cos x| + C$

39. $\dfrac{3\sqrt{3}}{8} \approx 0.6495$ 41. $2(\sqrt{3} - 1) \approx 1.4641$

43. $\frac{1}{2}$ 45. $\ln(\cos 0) - \ln(\cos 1) \approx 0.6156$

47. 4 49. $\dfrac{\pi^2}{2} + 2 \approx 6.9348$ 51. 2 53. π

55. Trapezoidal Rule: 1.3655

   Simpson's Rule: 1.3708

   Graphing utility: 1.3708

57. (a) 225.28 million barrels

   (b) 225.28 million barrels

   (c) 217 million barrels

59. 18.54 inches

61. (a) $C \approx \$9.17$ (b) $C \approx \$3.14$, savings $\approx \$6.03$

63. 0.5093 liter   65. 0.9777   67. 3.8202

69. True   71. False. $\displaystyle\int \sin^2 2x \cos 2x\, dx = \frac{1}{6}\sin^3 2x + C$

SECTION 8.6   *(page 603)*

Prerequisite Review

1. ∞ 2. ∞ 3. 0 4. 0 5. $\frac{2}{3}$ 6. 1

7. ∞ 8. ∞ 9. $-2x\sin x^2$

10. $5\cos(5x - 1)$ 11. $4\sec(4x)\tan(4x)$

12. $2x\sec^2(x^2 - 2)$ 13. $-4\sin(2x + 3)$

14. $-\dfrac{\cos(x/2)}{4}$ 15. $2\sec^2 x \tan x$

16. $2\csc^2 x \cot x$

1. Yes   3. No   5. Yes

7.

x	-0.1	-0.01	-0.001	0
$f(x)$	-0.35	-0.335	-0.3335	?

x	0.001	0.01	0.1
$f(x)$	-0.3332	-0.332	-0.32

$\displaystyle\lim_{x\to 0} \frac{e^{-x} - 1}{3x} = -\frac{1}{3}$

9.

x	-0.1	-0.01	-0.001	0
$f(x)$	0.1997	0.2	0.2	?

x	0.001	0.01	0.1
$f(x)$	0.2	0.2	0.1997

$\displaystyle\lim_{x\to 0} \frac{\sin x}{5x} = \frac{1}{5}$

11. 0 13. -1 15. -3 17. -1 19. $\frac{1}{5}$

21. 0 23. -3 25. 1 27. 0 29. $\frac{4}{3}$ 31. 0

33. $\frac{1}{2}$ 35. $\frac{2}{5}$ 37. $\frac{2}{3}$ 39. 1 41. $-\frac{1}{2}$ 43. ∞

45. 0 47. ∞ 49. 1 51. 0 53. 0 55. 0

57. 0, so e^{4x} grows more rapidly than x^2.

59. 0, so x grows more rapidly than $(\ln x)^4$.

61. 0, so x^m grows more rapidly than $(\ln x)^n$.

63.

x	10	10^2	10^3
$\dfrac{(\ln x)^5}{x}$	6.47	20.71	15.73

x	10^4	10^5	10^6
$\dfrac{(\ln x)^5}{x}$	6.63	2.02	0.503

$$\lim_{x \to \infty} \frac{(\ln x)^5}{x} = 0$$

65. The limit of the denominator is not 0.

67. The limit of the numerator is not 0.

69. (a) 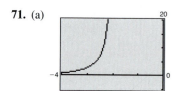   (b) $\frac{1}{3}$

71. (a) 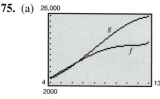   (b) Limit does not exist.

73. Proof

75. (a)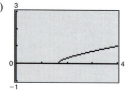

(b) CVS appears to have had a greater rate of growth for $4 \le t \le 13$.

(c) Because Rite Aid's model includes an exponential term, it ultimately will have a greater rate of growth than CVS's polynomial model.

(d) 2004

77. Answers will vary.   **79.** False. $\displaystyle\lim_{x \to \infty} \frac{x}{1-x} = -1$

81. False. For example, $\displaystyle\lim_{x \to \infty} \frac{x+1}{x} = 1$, but $x+1 \ne x$.

REVIEW EXERCISES FOR CHAPTER 8
(page 610)

1. 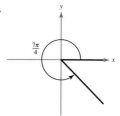   **3.**

$\dfrac{15\pi}{4}, -\dfrac{\pi}{4}$ $\dfrac{7\pi}{2}, -\dfrac{\pi}{2}$

5. 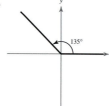   **7.**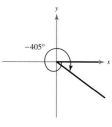

$495°, -225°$ $315°, -45°$

9. $\dfrac{7\pi}{6}$ **11.** $-\dfrac{\pi}{3}$ **13.** $-\dfrac{8\pi}{3}$ **15.** $\dfrac{3\pi}{4}$ **17.** $240°$

19. $-120°$ **21.** $b = 4\sqrt{3}, \theta = 60°$

23. $a = \dfrac{5\sqrt{3}}{2}, c = 5, \theta = 60°$ **25.** 15.38 feet

27. $\dfrac{\pi}{3}$ **29.** $\dfrac{\pi}{6}$ **31.** $60°$ **33.** $60°$ **35.** $\dfrac{\sqrt{2}}{2}$

37. $\sqrt{3}$ **39.** $-\dfrac{\sqrt{3}}{2}$ **41.** $\sqrt{3}$ **43.** -1

45. $-\dfrac{1}{2}$ **47.** 0.6494 **49.** 3.2361 **51.** -0.3420

53. -0.2588 **55.** $r \approx 146.19$ **57.** $x \approx 68.69$

59. $\dfrac{2\pi}{3} + 2k\pi, \dfrac{4\pi}{3} + 2k\pi$

61. $\dfrac{7\pi}{6} + 2k\pi, \dfrac{3\pi}{2} + 2k\pi, \dfrac{11\pi}{6} + 2k\pi$

63. $\dfrac{\pi}{3} + 2k\pi, \pi + 2k\pi, \dfrac{5\pi}{3} + 2k\pi$ **65.** 81.18 feet

67. 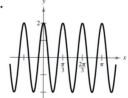   **69.**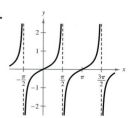

71.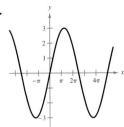

73.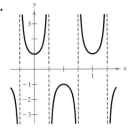

75.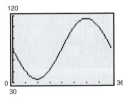

77. $5\pi \cos 5\pi x$

79. $-x \sec^2 x - \tan x$ **81.** $\dfrac{-x \sin x - 2 \cos x}{x^3}$

83. $24 \sin 4x \cos 4x + 1 = 12 \sin 8x + 1$

85. $-6 \csc^3 x \cot x$ **87.** $e^x(\cot x - \csc^2 x)$

89. $y = 8x - 8\pi$ **91.** $y = \frac{1}{4}$ **93.** $y = 2x$

95. Relative maximum: $(0.523, 1.128)$

Relative minimum: $(2.616, 0.443)$

97. Relative maxima: $\left(\dfrac{\pi}{2}, 2\right), \left(\dfrac{3\pi}{2}, 0\right)$

Relative minima: $\left(\dfrac{7\pi}{6}, -\dfrac{1}{4}\right), \left(\dfrac{11\pi}{6}, -\dfrac{1}{4}\right)$

99. (a) 116.25 thousand, day 274

(b) 34.75 thousand, day 90

101. $-3 \cos x - 2 \sin x + C$ **103.** $\frac{1}{4} \sin^4 x + C$

105. $\pi + 2$ **107.** $\dfrac{2\sqrt{3}}{3}$ **109.** 0   **111.** 2

113. 1   **115.** $\frac{5}{3}$ **117.** 6.9 quadrillion Btu

119. 21.11 inches   **121.** $\frac{1}{5}$ **123.** $\frac{19}{12}$ **125.** $-\frac{5}{4}$

127. $\frac{1}{2}$ **129.** 0   **131.** 0   **133.** ∞

135. $g(x)$ grows faster than $f(x)$.

SAMPLE POST-GRAD EXAM QUESTIONS
(page 614)

1. d   **2.** d   **3.** d   **4.** c   **5.** b

CHAPTER 9

SECTION 9.1   *(page 622)*

Prerequisite Review

1. 1   **2.** 1   **3.** 2   **4.** 2   **5.** $\frac{1}{2}$

6. 1   **7.** 37.50%   **8.** $81\frac{9}{11}\%$

9. $54\frac{1}{6}\%$ **10.** 43.75%

1. (a) $S = \{HHH, HHT, HTH, HTT, THH,$

$THT, TTH, TTT\}$

(b) $A = \{HHH, HHT, HTH, THH\}$

(c) $B = \{HTT, THT, TTH, TTT\}$

3. (a) $S = \{3, 6, 9, 12, 15, 18, 21, 24, 27, 30, 33,$

$36, 39, 42, 45, 48\}$

(b) $A = \{12, 24, 36, 48\}$

(c) $B = \{9, 36\}$

5. 0.24   **7.** 0.0145

9.

Random variable	0	1	2
Frequency	1	2	1

11.

Random variable	0	1	2	3
Frequency	1	3	3	1

13.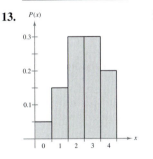

15.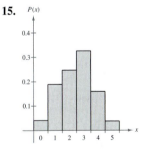

(a) $\frac{3}{4}$ (b) $\frac{4}{5}$ (a) 0.803   (b) 0.197

17. (a) $S = \{gggg, gggb, ggbg, gbgg, bggg, ggbb,$

$gbgb, gbbg, bgbg, bbgg, bggb, gbbb,$

$bgbb, bbgb, bbbg, bbbb\}$

(b)

x	0	1	2	3	4
$P(x)$	$\frac{1}{16}$	$\frac{4}{16}$	$\frac{6}{16}$	$\frac{4}{16}$	$\frac{1}{16}$

(c) 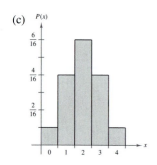   (d) $\frac{15}{16}$

19. There are 16 possibilities, as indicated in the following chart.

	RY	Ry	rY	ry
RY	RRYY	RRYy	RrYY	RrYy
Ry	RRYy	RRyy	RrYy	Rryy
rY	RrYY	RrYy	rrYY	rrYy
ry	RrYy	Rryy	rrYy	rryy

(a) $\frac{9}{16}$ (b) $\frac{3}{16}$ (c) $\frac{3}{16}$ (d) $\frac{1}{16}$

21. $E(x) = 3$ **23.** $E(x) = 0.8$
$V(x) = 0.875$ $V(x) = 8.16$
$\sigma = 0.9354$ $\sigma = 2.8566$

25. (a) Mean: 2.5   **27.** (a) $E(x) = 18.5$
Variance: 1.25 $\sigma = 8.0777$
(b) Mean: 5   (b) \$54,575
Variance: 2.5

29. \$201   **31.** $-\$0.0526$ **33.** City 1

35. (a)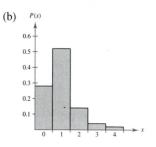
(b) 0.5488 or 54.88%
(c) 0.9567 or 95.67%
(d) 0.0054 or 0.54%

37. (a)

x	0	1	2	3	4
$P(x)$	$\frac{14}{50}$	$\frac{26}{50}$	$\frac{7}{50}$	$\frac{2}{50}$	$\frac{1}{50}$

(b)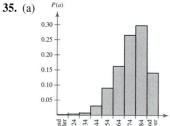

(c) $\frac{35}{50}$ (d) $E(x) = 1$, $V(x) = 0.76$, $\sigma \approx 0.87$
Answers will vary.

39. Mean: 4.4
Standard deviation: 2.816

SECTION 9.2   *(page 631)*

Prerequisite Review

1. Yes   **2.** No   **3.** No   **4.** Yes   **5.** 1
6. $\frac{1}{2}$ **7.** 1   **8.** $\frac{1}{4}$ **9.** 1   **10.** 1

1.

$$\int_0^8 \frac{1}{8}\, dx = \left[\frac{1}{8}x\right]_0^8 = 1$$

3.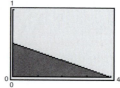

$$\int_0^4 \frac{4-x}{8}\, dx = \left[\frac{1}{2}x - \frac{1}{16}x^2\right]_0^4 = 1$$

5.

$$\int_0^1 6x(1-x)\, dx = \left[3x^2 - 2x^3\right]_0^1 = 1$$

7.

$$\int_0^\infty \frac{1}{5}e^{-x/5}\, dx = \lim_{b\to\infty}\left[-e^{-x/5}\right]_0^b = 1$$

9.

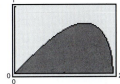

$$\int_0^2 \frac{3}{8}x\sqrt{4-x^2}\,dx = \left[-\frac{(4-x^2)^{3/2}}{8}\right]_0^2 = 1$$

11.

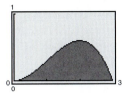

$$\int_0^3 \frac{4}{27}x^2(3-x)\,dx = \frac{4}{27}\left[x^3 - \frac{x^4}{4}\right]_0^3 = 1$$

13.

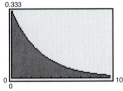

$$\int_0^\infty \frac{1}{3}e^{-x/3}\,dx = \lim_{b\to\infty}\left[-e^{-x/3}\right]_0^b = 1$$

15. $\frac{2}{15}$ **17.** $\frac{3}{32}$ **19.** $\frac{1}{2}$

21.

(a) $\frac{3}{5}$ (b) $\frac{1}{5}$ (c) $\frac{1}{5}$ (d) $\frac{4}{5}$

23.

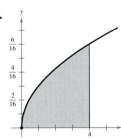

(a) $\frac{\sqrt{2}}{4} \approx 0.354$

(b) $1 - \frac{\sqrt{2}}{4} \approx 0.646$

(c) $\frac{1}{8}(3\sqrt{3} - 1) \approx 0.525$

(d) $\frac{3\sqrt{3}}{8} \approx 0.650$

25.

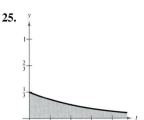

(a) $e^{-0/3} - e^{-2/3} \approx 0.4866$

(b) $e^{-2/3} - 0 \approx 0.5134$

(c) $e^{-1/3} - e^{-4/3} \approx 0.4529$

(d) 0

27. (a) $\frac{1}{6}$ (b) $\frac{2}{5}$

29. (a) $1 - e^{-2/3} \approx 0.487$ (b) $e^{-2/3} - e^{-4/3} \approx 0.250$

   (c) $e^{-2/3} \approx 0.513$

31. $1 - e^{-4/3} \approx 0.736$

33. (a) $-\frac{1}{6}(12e^{-1} - 6) \approx 0.264$ (b) $2e^{-1} - 3e^{-2} \approx 0.330$

   (c) $1 - (1 - 3e^{-2}) \approx 0.406$

35. (a) 0.75. There is a 75% probability that the community will receive up to 10 inches of rain.

   (b) 0.25. There is a 25% probability that the community will receive 10 to 15 inches of rain.

   (c) 0.25. There is a 25% probability that the community will receive up to 5 inches of rain.

   (d) ≈ 0.095. There is a 9.5% probability that the community will receive 12 to 15 inches of rain.

37. (a) $0.987 = 98.7\%$ (b) $0.366 = 36.6\%$

SECTION 9.3   *(page 641)*

Prerequisite Review

1. 5   **2.** 8   **3.** 3 ln 2   **4.** 9 ln 2

5. $\frac{4}{3}$ **6.** $\frac{4}{3}$ **7.** $\frac{3}{4}$ **8.** $\frac{2}{9}$

9. (a) $\frac{1}{4}$ (b) $\frac{1}{2}$ **10.** (a) $\frac{1}{2}$ (b) $\frac{11}{16}$

1. (a) 4   (b) $\frac{16}{3}$ (c) $\frac{4}{\sqrt{3}} = \frac{4\sqrt{3}}{3}$

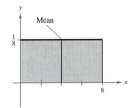

3. (a) 4   (b) 2   (c) $\sqrt{2}$

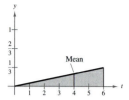

5. (a) $\frac{5}{7}$ (b) $\frac{20}{441}$ (c) $\frac{2\sqrt{5}}{21}$

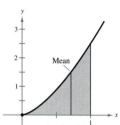

7.

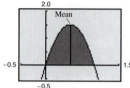

Mean: $\frac{1}{2}$

9.

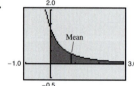

Mean ≈ 0.848

11. $-9 \ln \frac{1}{2} \approx 6.238$

13. Uniform density function

Mean: 5

Variance: $\frac{25}{3}$

Standard deviation: $\frac{5\sqrt{12}}{6} \approx 2.887$

15. Exponential density function

Mean: 8

Variance: 64

Standard deviation: 8

17. Normal density function

Mean: 100

Variance: 121

Standard deviation: 11

19. Mean: 0

Standard deviation: 1

$P(0 \le x \le 0.85) \approx 0.3023$

21. Mean: 6

Standard deviation: 6

$P(x \ge 2.23) \approx 0.6896$

23. Mean: 8

Standard deviation: 2

$P(3 \le x \le 13) \approx 0.9876$

25. (a) $P(x > 64) \approx 0.3694$ (b) $P(x > 70) \approx 0.2023$

(c) $P(x < 70) \approx 0.7977$ (d) $P(33 < x < 65) \approx 0.6493$

27. (a) Mean: 10:05 A.M.   (b) $\frac{3}{10}$

Standard deviation: $\frac{5\sqrt{3}}{3} \approx 2.9$ minutes

29. (a) $f(t) = \frac{1}{2}e^{-t/2}$

(b) $P(0 < t < 1) = 1 - e^{-1/2} \approx 0.3935$

31. (a) $f(t) = \frac{1}{5}e^{-t/5}$ (b) $0.865 = 86.5\%$

33. (a) 1.5 standard deviations   (b) 93.32%

35. (a) $\mu = 3$, $\sigma = \frac{3\sqrt{5}}{5} \approx 1.342$

(b) 3   (c) $0.626 = 62.6\%$

37. $\mu = \frac{4}{7}$

$V(x) = \frac{8}{147}$

39. (a) 10   (b) $P(x \le 4) \approx 0.1912$

41. Mean: $\frac{11}{2}$ = median

43. Mean: $\frac{1}{6}$ **45.** Mean: 5

Median: 0.1465   Median: $5 \ln 2 \approx 3.4657$

47. $\frac{1}{-0.28} \ln 0.5 \approx 2.4755$

49. (a) Expected value: 6

Standard deviation: $3\sqrt{2} \approx 4.243$

(b) 0.615

51. $\mu \approx 12.25$ **53.** 40.68%

55. (a)   (b) $0.252 = 25.2\%$

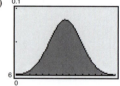

CHAPTER 9 REVIEW EXERCISES   *(page 647)*

1. $S = \{$January, February, March, April, May, June, July, August, September, October, November, December$\}$

3. If the essays are numbered 1, 2, 3, and 4,

$S = \{123, 124, 134, 234\}$.

5. $S = \{0, 1, 2, 3\}$

7.

x	0	1	2	3
$n(x)$	1	3	3	1

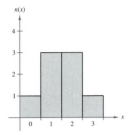

9.

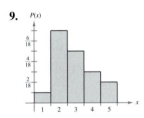

(a) $\frac{5}{6}$ (b) $\frac{5}{9}$

11. (a) $\frac{5}{36}$ (b) $\frac{5}{6}$ (c) $\frac{1}{6}$ (d) $\frac{1}{36}$

13. 19.5 **15.** (a) 20.5 (b) \$15,375

17. $V(x) = 218{,}243.7500$ **19.** $V(x) \approx 1.1611$

$\sigma \approx 467.1657$ $\sigma \approx 1.0775$

21.

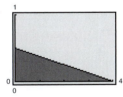

$\int_0^4 \frac{1}{8}(4 - x)\,dx = \left[\frac{1}{2}x - \frac{x^2}{16}\right]_0^4 = (2 - 1) = 1$

23.

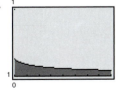

$\int_1^9 \frac{1}{4\sqrt{x}}\,dx = \frac{1}{4}\left[2\sqrt{x}\right]_1^9 = 1$

25. $\frac{9}{25}$ **27.** $\frac{2}{3}$ **29.** (a) $\frac{1}{2}$ (b) $\frac{1}{4}$

31. $\frac{1}{2}$ **33.** 2.5 **35.** 6

37. Variance: $\frac{9}{20}$

Standard deviation: $\dfrac{3}{2\sqrt{5}}$

39. Variance: 4

Standard deviation: 2

41. $\frac{1}{2}$ **43.** 2.7726 **45.** (a) 0.4866 (b) 0.2498

47. 0.00383 **49.** 0.3829

51.

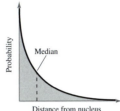

SAMPLE POST-GRAD EXAM QUESTIONS
(page 650)

1. c **2.** b **3.** c **4.** e **5.** a **6.** c

CHAPTER 10
SECTION 10.1   *(page 658)*

Prerequisite Review

1. 0 **2.** 0 **3.** 2 **4.** ∞ **5.** 0 **6.** 0

7. $\dfrac{n - 2}{n}$ **8.** $\dfrac{n - 3}{n - 4}$ **9.** $\dfrac{3n^2 + 1}{n^3}$

10. $\dfrac{2n + 1}{(n - 1)(n + 2)}$

1. 2, 4, 8, 16, 32 **3.** $\frac{1}{2}, \frac{2}{3}, \frac{3}{4}, \frac{4}{5}, \frac{5}{6}$ **5.** $3, \frac{9}{2}, \frac{27}{6}, \frac{81}{24}, \frac{243}{120}$

7. $-1, \frac{1}{4}, -\frac{1}{9}, \frac{1}{16}, -\frac{1}{25}$ **9.** Converges to 0

11. Converges to 1 **13.** Converges to $\frac{1}{2}$ **15.** Diverges

17. Converges to 0 **19.** Diverges **21.** Converges to 3

23. Converges to 0 **25.** Diverges **27.** Diverges

29. Diverges **31.** $3n - 2$ **33.** $5n - 6$

35. $\dfrac{n + 1}{n + 2}$ **37.** $\dfrac{(-1)^{n-1}}{2^{n-2}}$ **39.** $\dfrac{n + 1}{n}$ **41.** $2(-1)^n$

43. $\dfrac{(-1)^n x^n}{n}$ **45.** 2, 5, 8, 11, 14, 17, . . .

47. $1, \frac{5}{3}, \frac{7}{3}, 3, \frac{11}{3}, \frac{13}{3}, . . .$ **49.** $3, -\frac{3}{2}, \frac{3}{4}, -\frac{3}{8}, \frac{3}{16}, -\frac{3}{32}, . . .$

51. 2, 6, 18, 54, 162, 486, . . . **53.** Geometric, $20\left(\frac{1}{2}\right)^{n-1}$

55. Arithmetic, $\frac{2}{3}n + 2$ **57.** $\frac{3n + 1}{4n}$

59. $9045.00, $9090.23, $9135.68, $9181.35, $9227.26,
$9273.40, $9319.76, $9366.36, $9413.20, $9460.26

61. (a)

Year	1	2	3
Balance	$2200	$4620	$7282

Year	4	5	6
Balance	$10,210.20	$13,431.22	$16,974.34

(b) $126,005.00   (c) $973,703.62

63. $S_6 = 240$, $S_7 = 440$, $S_8 = 810$, $S_9 = 1490$, $S_{10} = 2740$

65. (a) 2.40, 2.50, 2.77, 3.16, 3.58, 4.00, 4.38, 4.71, 4.98, 5.22,
5.41, 5.57, 5.71, 5.82, 5.92, 6.00

(b)

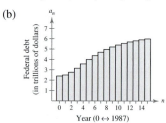

67. (a) $S_1 = 1$ (b) $S_{20} = 2870$
$S_2 = 5$
$S_3 = 14$
$S_4 = 30$
$S_5 = 55$

69. (a) $1.3(0.85)^n$ billion dollars

(b)

Year	1	2
Budget amount	$1.105 billion	$0.939 billion

Year	3	4
Budget amount	$0.798 billion	$0.679 billion

(c) Converges to 0

71. $2095

73. (a) $a_n = -0.265625n^3 + 5.32271n^2 - 9.7470n + 90.192$

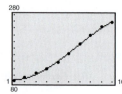

(b) $157.7 billion

75. $a_1 = 2$
$a_{10} = 2.5937$
$a_{100} = 2.7048$
$a_{1000} = 2.7169$
$a_{10,000} = 2.7181$

SECTION 10.2   *(page 670)*

Prerequisite Review

1. $\frac{77}{60}$ **2.** $\frac{73}{24}$ **3.** $\frac{31}{16}$ **4.** $\frac{40}{9}$ **5.** $\frac{21}{8}$

6. $\frac{31}{32}$ **7.** $\frac{3}{4}$ **8.** 0   **9.** 1   **10.** $\frac{1}{2}$

1. $S_1 = 1$ **3.** $S_1 = 3$
$S_2 = \frac{5}{4} = 1.25$ $S_2 = \frac{9}{2} = 4.5$
$S_3 = \frac{49}{36} \approx 1.361$ $S_3 = \frac{21}{4} = 5.25$
$S_4 = \frac{205}{144} \approx 1.424$ $S_4 = \frac{45}{8} = 5.625$
$S_5 = \frac{5269}{3600} \approx 1.464$ $S_5 = \frac{93}{16} = 5.8125$

5. nth-Term Test: $\lim\limits_{n \to \infty} \dfrac{n}{n + 1} = 1 \neq 0$

7. nth-Term Test: $\lim\limits_{n \to \infty} \dfrac{n^2}{n^2 + 1} = 1 \neq 0$

9. Geometric series: $r = \frac{3}{2} > 1$

11. Geometric series: $r = 1.055 > 1$

13. $r = \frac{3}{4} < 1$ **15.** $r = 0.9 < 1$ **17.** 2   **19.** $\frac{2}{3}$

21. $4 + 2\sqrt{2} \approx 6.828$ **23.** $\frac{10}{9}$ **25.** $\frac{3}{2}$ **27.** $\frac{1}{2}$

29. $\dfrac{17}{6}$ **31.** $\lim\limits_{n \to \infty} \dfrac{n + 10}{10n + 1} = \dfrac{1}{10} \neq 0$; diverges

33. $\lim\limits_{n \to \infty} \dfrac{n! + 1}{n!} = 1 \neq 0$; diverges

35. $\lim\limits_{n \to \infty} \dfrac{3n - 1}{2n + 1} = \dfrac{3}{2} \neq 0$; diverges

37. Geometric series: $r = 1.075 > 1$; diverges

39. Geometric series: $r = \frac{1}{4} < 1$; converges   **41.** $\frac{2}{3}$

43. $\frac{9}{11}$ **45.** (a) $80,000(1 - 0.9^n)$ (b) 80,000

47. ≈ 72.89 feet   **49.** $7808.24

51. $\sum\limits_{n=0}^{\infty} 100(0.75)^n = 400 million   **53.** $10,485.75

55. 2   **57.** ≈ 71.12 ppm

59. (a) $\sum\limits_{i=1}^{10} 880i = 8800

(b) $\sum\limits_{i=1}^{168} 880i - 100,000 = 147,840 - 100,000$

$= $47,840$ more

61. 6   **63.** ≈ 0.5431 **65.** $\dfrac{e^2}{e-1} \approx 4.3003$

67. False. $\lim\limits_{n\to 0}\dfrac{1}{n} = 0$, but $\sum\limits_{n=1}^{\infty}\dfrac{1}{n}$ diverges.

SECTION 10.3   *(page 679)*

Prerequisite Review

1. $\dfrac{1}{n+1}$ **2.** $n+1$ **3.** $\dfrac{3n}{n+1}$ **4.** $\dfrac{n+1}{n^2}$

5. 1   **6.** 5   **7.** 1   **8.** $\frac{1}{3}$

9. Geometric series   **10.** Not a geometric series

1. p-series   **3.** Not a p-series   **5.** Not a p-series

7. Converges   **9.** Diverges   **11.** Converges

13. Diverges   **15.** Converges   **17.** Converges

19. Diverges   **21.** Converges   **23.** Diverges

25. Converges   **27.** Diverges   **29.** Converges

31. ≈ 1.1777; maximum error $\leq \frac{1}{32}$.

33. ≈ 1.9953; maximum error $\leq \dfrac{2}{\sqrt{10}} \approx 0.6325$.

35. $\lim\limits_{n\to\infty}\left|\dfrac{a_{n+1}}{a_n}\right| = \lim\limits_{n\to\infty}\dfrac{1/[(n+1)^{3/2}]}{1/(n^{3/2})}$

$\qquad = \lim\limits_{n\to\infty}\left(\dfrac{n}{n+1}\right)^{3/2} = 1$

37. a; diverges: $p = \frac{3}{4} < 1$

38. d; diverges: $p = 1$, harmonic series

39. b; converges: $p = \frac{3}{2} > 1$

40. c; converges: $p = 2 > 1$

41. Diverges; nth-Term Test

43. Converges; p-Series Test; ≈ 3.6009

45. Converges; Geometric Series Test; $\frac{3}{5}$

47. Converges; p-Series Test; ≈ 0.4429

49. Diverges; Geometric Series Test

51. Diverges; Ratio Test

53. Diverges; Ratio Test   **55.** Converges; Ratio Test; $\frac{10}{3}$

57. $\sum\limits_{n=1}^{100}\dfrac{1}{n^2} \approx 1.635,\ \dfrac{\pi^2}{6} \approx 1.644934$

59. (a) $\sum\limits_{t=1}^{\infty}(0.1396n^2 + 0.309n + 12.32)$

$\qquad$ (b) No, the Ratio Test yields a limit equal to 1.

SECTION 10.4   *(page 689)*

Prerequisite Review

1. $f(g(x)) = (x-1)^2$ **2.** $f(g(x)) = 6x + 3$

$\quad g(f(x)) = x^2 - 1$ $\qquad g(f(x)) = 6x + 1$

3. $f(g(x)) = \sqrt{x^2+4}$

$\quad g(f(x)) = x + 4,\ x \geq -4$

4. $f(g(x)) = e^{x^2}$ **5.** $f'(x) = 5e^x$

$\quad g(f(x)) = e^{2x}$ $\qquad f''(x) = 5e^x$

$\qquad\qquad\qquad\qquad f'''(x) = 5e^x$

$\qquad\qquad\qquad\qquad f^{(4)}(x) = 5e^x$

6. $f'(x) = \dfrac{1}{x}$ **7.** $f'(x) = 6e^{2x}$

$\quad f''(x) = -\dfrac{1}{x^2}$ $\qquad f''(x) = 12e^{2x}$

$\qquad\qquad\qquad\qquad f'''(x) = 24e^{2x}$

$\quad f'''(x) = \dfrac{2}{x^3}$ $\qquad f^{(4)}(x) = 48e^{2x}$

$\quad f^{(4)}(x) = -\dfrac{6}{x^4}$

8. $f'(x) = \dfrac{1}{x}$ **9.** $\dfrac{n+1}{3}$ **10.** $\dfrac{n+3}{n+1}$

$\quad f''(x) = -\dfrac{1}{x^2}$

$\quad f'''(x) = \dfrac{2}{x^3}$

$\quad f^{(4)}(x) = -\dfrac{6}{x^4}$

1. $1, \dfrac{x}{4}, \left(\dfrac{x}{4}\right)^2, \left(\dfrac{x}{4}\right)^3, \left(\dfrac{x}{4}\right)^4$

3. $-1, (x+1), -\dfrac{(x+1)^2}{2}, \dfrac{(x+1)^3}{6}, -\dfrac{(x+1)^4}{24}$

5. 2   **7.** 1   **9.** ∞ **11.** 0   **13.** 4

15. 5   **17.** 1   **19.** c **21.** ∞ **23.** ∞

25. $\sum\limits_{n=0}^{\infty}\dfrac{x^n}{n!},\ \ R = \infty$ **27.** $\sum\limits_{n=0}^{\infty}\dfrac{(2x)^n}{n!},\ \ R = \infty$

29. $\sum\limits_{n=0}^{\infty}(-1)^n x^n,\ \ R = 1$

31. $1 + \dfrac{1}{2}(x-1)\sum\limits_{n=2}^{\infty}\dfrac{(-1)^{n+1}\,1\cdot 3\cdot 5\cdots(2n-3)(x-1)^n}{2^n\cdot n!},$

$\quad R = 1$

33. $\sum\limits_{n=0}^{\infty}(-1)^n\dfrac{(n+2)(n+1)}{2}x^n,\ \ R = 1$

35. $1 + \sum_{n=1}^{\infty} \frac{(-1)^n 1 \cdot 3 \cdot 5 \cdots (2n-1)}{2^n n!} x^n$, $R = 1$

37. $R = 2$ (all parts)   **39.** $R = 1$ (all parts)

41. $\sum_{n=0}^{\infty} \frac{x^{3n}}{n!}$ **43.** $3 \sum_{n=0}^{\infty} \frac{x^{3n+2}}{n!}$ **45.** $\sum_{n=0}^{\infty} (-1)^n x^{4n}$

47. $\sum_{n=0}^{\infty} \frac{(-1)^n x^{2n+2}}{n+1}$ **49.** $\sum_{n=0}^{\infty} \frac{(-1)^n (x-1)^{n+1}}{n+1}$

51. $\sum_{n=1}^{\infty} (-1)^{n+1} n(x-1)^{n-1}$

53. 1.6487   **55.** -0.6931 **57.** -2.3018

SECTION 10.5   *(page 698)*

Prerequisite Review

1. $\sum_{n=0}^{\infty} \frac{3^n x^n}{n!}$ **2.** $\sum_{n=0}^{\infty} \frac{(-1)^n 3^n x^n}{n!}$

3. $4 \sum_{n=0}^{\infty} (-1)^n (x-1)^n$

4. $\ln 5 + \sum_{n=1}^{\infty} \frac{(-1)^{n-1}(x-1)^n}{n}$

5. $1 + \frac{x}{4} - \frac{3x^2}{4^2 2!} + \frac{3 \cdot 7 x^3}{4^3 3!} - \frac{3 \cdot 7 \cdot 11 x^4}{4^4 4!} + \cdots$

6. $1 + \frac{x}{2} - \frac{x^2}{2^2 2!} + \frac{1 \cdot 3 x^3}{2^3 3!} - \frac{1 \cdot 3 \cdot 5 x^4}{2^4 4!} + \cdots$

7. $\frac{47}{60}$ **8.** $\frac{311}{576}$ **9.** $\frac{5}{12}$ **10.** $\frac{77}{192}$

---

1. (a) $S_1(x) = 1 + x$ (b) $S_2(x) = 1 + x + \frac{x^2}{2}$

   (c) $S_3(x) = 1 + x + \frac{x^2}{2} + \frac{x^3}{6}$

   (d) $S_4(x) = 1 + x + \frac{x^2}{2} + \frac{x^3}{6} + \frac{x^4}{24}$

3. (a) $S_1(x) = 1 + \frac{x}{2}$ (b) $S_2(x) = 1 + \frac{x}{2} - \frac{x^2}{8}$

   (c) $S_3(x) = 1 + \frac{x}{2} - \frac{x^2}{8} + \frac{x^3}{16}$

   (d) $S_4(x) = 1 + \frac{x}{2} - \frac{x^2}{8} + \frac{x^3}{16} - \frac{5x^4}{128}$

5. (a) $S_1(x) = x$ (b) $S_2(x) = x - x^2$

   (c) $S_3(x) = x - x^2 + x^3$

   (d) $S_4(x) = x - x^2 + x^3 - x^4$

7.

x	0	0.25	0.50	0.75	1.00
$f(x)$	1.0000	1.1331	1.2840	1.4550	1.6487
$S_1(x)$	1.0000	1.1250	1.2500	1.3750	1.5000
$S_2(x)$	1.0000	1.1328	1.2813	1.4453	1.6250
$S_3(x)$	1.0000	1.1331	1.2839	1.4541	1.6458
$S_4(x)$	1.0000	1.1331	1.2840	1.4549	1.6484

9. (a) $S_2(x) = 1 - x^2$ (b) $S_4(x) = 1 - x^2 + x^4$

   (c) $S_6(x) = 1 - x^2 + x^4 - x^6$

   (d) $S_8(x) = 1 - x^2 + x^4 - x^6 + x^8$

11. $S_4(x) = 1 - x^2 + x^4$ **13.** d   **14.** c   **15.** a

16. b   **17.** 0.607   **19.** 0.4055   **21.** 0.74286

23. 0.481   **25.** 7   **27.** $\frac{1}{6!} \approx 0.00139$

29. (a) Answers will vary.   (b) 1   (c) \$10

SECTION 10.6   *(page 705)*

Prerequisite Review

1. $f(2.4) = -0.04$ **2.** $f(-0.6) = 0.064$
        $f'(2.4) = 2.8$ $f'(-0.6) = 3.48$

3. $f(0.35) = 0.01$ **4.** $f(1.4) = 0.30$
        $f'(0.35) = 4.03$ $f'(1.4) = 12.88$

5. $4.9 \le x \le 5.1$ **6.** $0.798 \le x \le 0.802$

7. $5.97 \le x \le 6.03$ **8.** $-3.505 \le x \le -3.495$

9. $\left(\frac{\sqrt{13}+3}{2}, \sqrt{13}+2 \right), \left(\frac{3-\sqrt{13}}{2}, 2 - \sqrt{13} \right)$

10. $\left(\frac{1-\sqrt{5}}{2}, \frac{3-\sqrt{5}}{2} \right), \left(\frac{1+\sqrt{5}}{2}, \frac{3+\sqrt{5}}{2} \right)$

---

1. 2.2364   **3.** 0.682   **5.** 1.25   **7.** 0.567   **9.** ± 0.753

11. $-4.596, -1.042, 5.638$ **13.** 2.926   **15.** 2.893

17. 11.8033   **19.** $\pm 1.9021, \pm 1.1756$

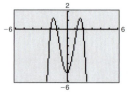

21. 0.9, 1.1, 1.9

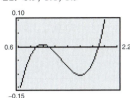

23. 1.1459, 7.8541

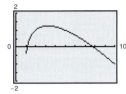

25. 0.2359, 1.3385

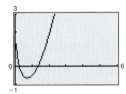

27. 0.8655

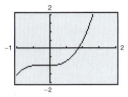

29. Newton's Method fails because $f'(x_1) = 0$.

31. Newton's Method fails because $1 = x_1 = x_3 = \ldots$;
$0 = x_2 = x_4 = \ldots$. Therefore, the limit does not exist.

33. $x_{n+1} = \dfrac{x_n^2 + a}{2x_n}$ **35.** 2.646   **37.** 1.565

39. $f(x) = \dfrac{1}{x} - a$

$f'(x) = -\dfrac{1}{x^2}$

Newton's Method: $x_{n+1} = x_n - \dfrac{f(x_n)}{f'(x_n)}$

$x_{n+1} = x_n - \dfrac{\dfrac{1}{x_n} - a}{-\dfrac{1}{x_n^2}}$

$= x_n(2 - ax_n)$

41. $x \approx 1.563$ miles down the coast   **43.** $t \approx 4.486$ hours

45. 15.9 years $\approx$ 2016

47. False. Let $f(x) = \dfrac{x^2 - 1}{x - 1}$.   **49.** True

CHAPTER 10 REVIEW EXERCISES   (page 712)

1. $-\frac{1}{3}, \frac{1}{9}, -\frac{1}{27}, \frac{1}{81}, -\frac{1}{243}$ **3.** $4, 8, 10\frac{2}{3}, 10\frac{2}{3}, \frac{128}{15}$

5. Converges to 0   **7.** Diverges   **9.** Converges to 5

11. Converges to 0   **13.** $\dfrac{n}{3n}$ or $\dfrac{1}{3}$, $n \neq 0$

15. $(-1)^n \dfrac{2^n}{3^{n+1}}$, $n = 0, 1, 2, \ldots$

17. (a) $15{,}000 + 10{,}000(n - 1)$ (b) \$175,000

19. \$1.07, \$1.14, \$1.23, \$1.31, \$1.40, \$1.50, \$1.61, \$1.72, \$1.84, \$1.97

21. $S_0 = 1$

$S_1 = \frac{5}{2} = 2.5$

$S_2 = \frac{19}{4} = 4.75$

$S_3 = \frac{65}{8} = 8.125$

$S_4 = \frac{211}{16} = 13.1875$

23. $S_1 = \frac{1}{2} = 0.5$

$S_2 = \frac{11}{24} \approx 0.4583$

$S_3 = \frac{331}{720} \approx 0.4597$

$S_4 = \frac{18{,}535}{40{,}320} \approx 0.4597$

$S_5 = \frac{1{,}668{,}151}{3{,}628{,}800} \approx 0.4597$

25. Diverges   **27.** Converges

29. $\displaystyle\lim_{n\to\infty} \dfrac{2n}{n+5} = 2 \neq 0$ **31.** $\displaystyle\lim_{n\to\infty} \left(\dfrac{5}{4}\right)^n = \infty \neq 0$

33. $\frac{5}{4}\left[1 - \left(\frac{1}{5}\right)^{N+1}\right]$ **35.** $2\left[1 - \left(\frac{1}{2}\right)^{N+1}\right] + \frac{4}{3}\left[1 - \left(\frac{1}{4}\right)^{N+1}\right]$

37. Diverges   **39.** Converges to $\frac{13}{4}$

41. (a) $D = -8 + 16 + 16(0.7) + 16(0.7)^2 + \cdots$

(b) $\frac{136}{3}$ feet

43. $\approx 2.782\%$ **45.** Converges   **47.** Converges

49. a   **50.** c   **51.** d   **52.** b

53. 1.0172; error $\leq \dfrac{1}{(5)4^5} \approx 1.9531 \times 10^{-4}$

55. 2.09074; error $\leq \dfrac{1}{(1/4)(6)^{1/4}} < 2.5558$

57. Converges   **59.** Diverges   **61.** Converges

63. $R = 1$ **65.** $R = 0$ **67.** $\displaystyle\sum_{n=0}^{\infty} \left(-\dfrac{1}{2}\right)^n \dfrac{x^n}{n!}$

69. $-\displaystyle\sum_{n=0}^{\infty} (x + 1)^n$ **71.** $\ln 2 + \displaystyle\sum_{n=1}^{\infty} (-1)^{n+1} \dfrac{(x/2)^n}{n}$

73. $1 + 2x^2 + x^4 + \cdots$ **75.** $x^2 \displaystyle\sum_{n=0}^{\infty} \dfrac{x^n}{n!} = \displaystyle\sum_{n=0}^{\infty} \dfrac{x^{n+2}}{n!}$

77. $x^2 \displaystyle\sum_{n=0}^{\infty} (-1)^n x^n = \displaystyle\sum_{n=0}^{\infty} (-1)^n x^{n+2}$

79. $\frac{1}{9} - \frac{2}{27}x + \frac{1}{27}x^2 - \frac{4}{243}x^3 + \frac{5}{729}x^4 - \frac{2}{729}x^5 + \frac{7}{6561}x^6$

81. $\ln 3 + \frac{1}{3}(x - 1) - \frac{1}{18}(x - 1)^2 + \frac{1}{81}(x - 1)^3$
$- \frac{1}{324}(x - 1)^4 + \frac{1}{1215}(x - 1)^5 - \frac{1}{4374}(x - 1)^6$

83. 4.7705   **85.** 0.9163   **87.** $\frac{1}{32}$ **89.** 0.301

91. 0.1233   **93.** 0.5, \$11.50   **95.** 0.313

97. 0.258   **99.** 1.341   **101.** 0.773

SAMPLE POST-GRAD EXAM QUESTIONS
(page 715)

1. b   **2.** b   **3.** b   **4.** b   **5.** c   **6.** d

APPENDIX A   *(page A10)*

1. Left Riemann sum: 0.518

Right Riemann sum: 0.768

3. Left Riemann sum: 0.746

Right Riemann sum: 0.646

5. Left Riemann sum: 0.859

Right Riemann sum: 0.659

7. Midpoint Rule: 0.673

9. (a) 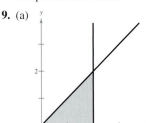   (b) Answers will vary.

(c) Answers will vary.

(d) Answers will vary.

(e)

n	5	10	50	100
Left sum, S_L	1.6	1.8	1.96	1.98
Right sum, S_R	2.4	2.2	2.04	2.02

(f) Answers will vary.

11. $\displaystyle\int_0^5 3\,dx$

13. $\displaystyle\int_{-4}^4 (4 - |x|)\,dx = \int_{-4}^0 (4 + x)\,dx + \int_0^4 (4 - x)\,dx$

15. $\displaystyle\int_{-2}^2 (4 - x^2)\,dx$ **17.** $\displaystyle\int_0^2 \sqrt{x + 1}\,dx$

19.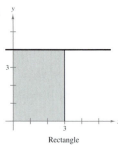

Rectangle

$A = 12$

21.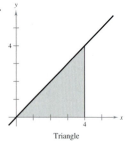

Triangle

$A = 8$

23.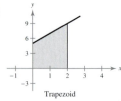

Trapezoid

$A = 14$

25.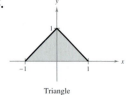

Triangle

$A = 1$

27.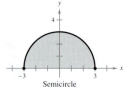

Semicircle

$A = \dfrac{9\pi}{2}$

29. Answers will vary.   **31.** >

APPENDIX C

Section C.1   *(page A24)*

Prerequisite Review

1. $y' = 6x + 2$ **2.** $y' = -6x^2 - 8$

$y'' = 6$ $y'' = -12x$

3. $y' = -6e^{2x}$ **4.** $y' = -6xe^{x^2}$

$y'' = -12e^{2x}$ $y'' = -6e^{x^2}(2x^2 + 1)$

5. $\dfrac{1 - x}{y}$ **6.** $\dfrac{2}{3y^2 + 4}$ **7.** $-\dfrac{y}{2x}$

8. $-\dfrac{y}{x}$ **9.** $k = 2 \ln 3 - \ln \dfrac{17}{2} \approx 0.0572$

10. $k = \ln 10 - \dfrac{\ln 41}{2} \approx 0.4458$

1. $y = 3x^2$

3. $y' = -2e^{-2x}$ and $y' + 2y = -2e^{-2x} + 2(e^{-2x}) = 0$

5. $y' = 6x^2$ and $y' - \dfrac{3}{x}y = 6x^2 - \dfrac{3}{x}(2x^3) = 0$

7. $y'' = 2$ and $x^2y'' - 2y = x^2(2) - 2(x^2) = 0$

9. $y' = 4e^{2x}$, $y'' = 8e^{2x}$, and

$y'' - y' - 2y = 8e^{2x} - 4e^{2x} - 2(2e^{2x}) = 0$

11. $\dfrac{dy}{dx} = -\dfrac{1}{x^2}$ **13.** $\dfrac{dy}{dx} = 4Ce^{4x} = 4y$

15. $\dfrac{dy}{dt} = -\dfrac{1}{3}Ce^{-t/3}$ and

$3\dfrac{dy}{dt} + y - 7 = 3\left(-\dfrac{1}{3}Ce^{-t/3}\right) + (Ce^{-t/3} + 7) - 7 = 0$

17. $xy' - 3x - 2y = x(2Cx - 3) - 3x - 2(Cx^2 - 3x) = 0$

19. $xy' + y = x\left(2x + 2 - \dfrac{C}{x^2}\right) + \left(x^2 + 2x + \dfrac{C}{x}\right)$

$= x(3x + 4)$

21. $2y'' + 3y' - 2y = 2\left(\frac{1}{4}C_1 e^{x/2} + 4C_2 e^{-2x}\right)$

$\qquad\qquad\qquad + 3\left(\frac{1}{2}C_1 e^{x/2} - 2C_2 e^{-2x}\right)$

$\qquad\qquad\qquad - 2(C_1 e^{x/2} + C_2 e^{-2x}) = 0$

23. $y' - \dfrac{ay}{x} = \left(\dfrac{4bx^3}{4-a} + aCx^{a-1}\right) - \dfrac{a}{x}\left(\dfrac{bx^4}{4-a} + Cx^a\right)$

$\qquad\quad = bx^3$

25. $y' + 2xy = -\dfrac{4Cxe^{x^2}}{(1 - Ce^{x^2})^2} + 2x\left(\dfrac{2}{1 + Ce^{x^2}}\right) = xy^2$

27. $y' = \ln x + 1 + C$ and

$\quad x(y' - 1) - (y - 4) = x(\ln x + 1 + C - 1)$

$\qquad\qquad\qquad\qquad\qquad - (x \ln x + Cx + 4 - 4) = 0$

29. $2x + 2yy' = Cy'$

$\quad y' = \dfrac{2x}{C - 2y} = \dfrac{2xy}{Cy - 2y^2}$

$\quad\;\; = \dfrac{2xy}{(x^2 + y^2) - 2y^2} = \dfrac{2xy}{x^2 - y^2}$

31. $x + y = \dfrac{C}{x}$

$\quad y'' = \dfrac{2C}{x^3}$

$\quad x^2 y'' - 2(x + y) = \dfrac{2C}{x} - \dfrac{2C}{x} = 0$

33. Solution   **35.** Not a solution   **37.** Not a solution

39. Solution   **41.** $y = 3e^{-2x}$ **43.** $y = 5 + \ln\sqrt{|x|}$

45. $y = 3e^{4x} + 2e^{-3x}$ **47.** $y = \frac{4}{3}(3 - x)e^{2x/3}$

49. 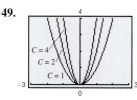   **51.**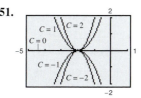

53. $y = x^3 + C$ **55.** $y = x + 3\ln|x| + C$

57. $y = \dfrac{1}{2}\ln\left(\dfrac{x-1}{x+1}\right) + C$

59. $y = \frac{2}{5}(x - 3)^{3/2}(x + 2) + C$

61. $y^2 = \frac{1}{4}x^3$ **63.** $y = 3e^x$

65. (a) $N = 750 - 650e^{-0.0484t}$

(b)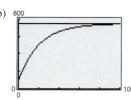

(c) $N \approx 214$

67.

Year, t	2	4	6	8	10
Units, x	3867	7235	10,169	12,725	14,951

69. Because

$$\frac{ds}{dh} = -\frac{13}{\ln 3}\left(\frac{1/2}{h/2}\right) = -\frac{13}{\ln 3}\frac{1}{h}, \text{ and } -\frac{13}{\ln 3}$$

is a constant, we can conclude that the equation is a solution of $ds/dh = k/h$ where $k = -13/(\ln 3)$.

71. $k = 0.07$

73. False. From Example 1, $y = e^x$ is a solution of $y'' - y = 0$, but $y = e^x + 1$ is not.

Section C.2   (page A32)

Prerequisite Review

1. $\frac{2}{5}x^{5/2} + C$ **2.** $\frac{1}{4}t^4 - \frac{3}{4}t^{4/3} + C$

3. $2\ln|x - 5| + C$ **4.** $\frac{1}{4}\ln|2y^2 + 1| + C$

5. $\frac{1}{2}e^{2y} + C$ **6.** $-\frac{1}{2}e^{1-x^2} + C$ **7.** $C = -10$

8. $C = 5$ **9.** $k = \dfrac{\ln 5}{2} \approx 0.8047$

10. $k = -2\ln 3 - \ln 2 \approx -2.8904$

1. Yes   **3.** Yes

$\quad (y + 3)\,dy = x\,dx$ $\quad dy = \left(\dfrac{1}{x} + 1\right)dx$

5. No. The variables cannot be separated.

7. $y = x^2 + C$ **9.** $y = \sqrt[3]{x + C}$

11. $C = 2x^2 - (y + 1)^2$ **13.** $y = Ce^{x^2/2}$

15. $y^2 = \frac{1}{2}e^t + C$

17. $y = 1 - \left(C - \dfrac{x}{2}\right)^2$

19. $y = C(2 + x)^2$ **21.** $y = Cx$

23. $3y^2 + 2y^3 = 3x^2 + C$ **25.** $y = -e^{-x} - x + C$

27. $y^2 = 2e^x + 14$ **29.** $y = -4 - e^{-x^2/2}$

31. $P = 5e^{6t}$

33. $5y^2 = 6x^2 - 1$ or $6x^2 - 5y^2 = 1$

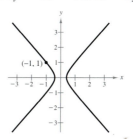

35. $v = 34.56(1 - e^{-0.1t})$ **37.** $T \approx 383.298°\text{F}$

39. (a) $T \approx 7.277°\text{F}$ (b) $t \approx 5.158$ hours

41. $N = 30 + Ce^{-kt}$ **43.** $y = Cx^{-k}$

Section C.3   *(page A37)*

Prerequisite Review

1. $e^x + 1$ **2.** $e^{3x} + 1$ **3.** $\dfrac{1}{x^3}$ **4.** $x^2 e^x$

5. $2e^x - e^{-x} + C$ **6.** $e^{3x}\left(\dfrac{x}{3} - \dfrac{1}{9}\right) + \dfrac{1}{2}e^{2x} + C$

7. $\dfrac{1}{2}\ln|2x + 5| + C$ **8.** $\dfrac{1}{2}\ln|x^2 + 2x + 3| + C$

9. $\dfrac{1}{12}(4x - 3)^3 + C$ **10.** $\dfrac{1}{6}(x^2 - 1)^3 + C$

1. $y' + \dfrac{-3}{2x^2}y = \dfrac{x}{2}$ **3.** $y' + \dfrac{1}{x}y = e^x$

5. $y' + \dfrac{1}{1 - x}y = \dfrac{1}{x - 1}$ **7.** $y = 2 + Ce^{-3x}$

9. $y = e^{-x}(x + C)$ **11.** $y = x^2 + 2x + \dfrac{C}{x}$

13. $y = \dfrac{1}{5} + Ce^{-(5/2)x^2}$ **15.** $y = \dfrac{x^3 - 3x + C}{3(x - 1)}$

17. $y = e^{1/x^2}\left(-\dfrac{1}{2x^2} + C\right)$ **19.** $y = Ce^{-x} + 4$

21. $y = Ce^{x^2} - 1$

23. c   **24.** d   **25.** a   **26.** b

27. $y = 3e^x$ **29.** $xy = 4$

31. $y = 1 + 5e^{-x^3}$ **33.** $y = x^2(5 - \ln|x|)$

35. $S = t + 95(1 - e^{-t/5})$

t	0	1	2	3	4	5
S	0	18.22	33.32	45.86	56.31	65.05

t	6	7	8	9	10
S	72.39	78.57	83.82	88.30	92.14

37. $p = 400 - 3x$ **39.** $p = 15(4 + e^{-t})$

41. (a) $A = \dfrac{P}{r^2}(rt - 1 + e^{-rt})$ (b) $A \approx \$18,924,053.07$

43. $v = -\dfrac{gm}{k} + Ce^{-kt/m}$ **45.** Answers will vary.

Section C.4   *(page A44)*

Prerequisite Review

1. $y = \dfrac{3}{2}x^2 + C$ **2.** $y^2 = 3x + C$

3. $y = Ce^{x^2}$ **4.** $y^4 = \dfrac{1}{2}(x - 4)^2 + C$

5. $y = 2 + Ce^{-2x}$ **6.** $y = xe^{-2x} + Ce^{-2x}$

7. $y = 1 + Ce^{-x^2/2}$ **8.** $y = \dfrac{1}{4}x^2 + Cx^{-2}$

9. $\dfrac{dy}{dx} = Cx^2$ **10.** $\dfrac{dx}{dt} = C(x - t)$

1. $y = e^{(x \ln 2)/3} \approx e^{0.2310x}$ **3.** $y = 4e^{-(x \ln 4)/4}$
$\approx 4e^{-0.3466x}$

5. $y = \dfrac{1}{2}e^{(\ln 2)x} \approx \dfrac{1}{2}e^{0.6931x}$ **7.** \$4451.08

9. $S = L(1 - e^{-kt})$ **11.** $y = \dfrac{20}{1 + 19e^{-0.5889x}}$

13. $y = \dfrac{5000}{1 + 19e^{-0.10156x}}$ **15.** $N = \dfrac{500}{1 + 4e^{-0.2452t}}$

17. $\dfrac{dP}{dn} = kP(L - P), \; P = \dfrac{CL}{e^{-Lkn} + C}$

19. $y = \dfrac{360}{8 + 41t}$ **21.** $y = 500e^{-1.6094e^{-0.1451t}}$

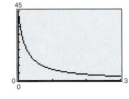

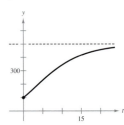

23. 34 beavers   **25.** 92%

27. (a) $y = Ce^{kt}$ (b) ≈ 6.2 hours

29. 38.843 pounds per gallon   **31.** ≈ 3.15 hours

33. $P = Ce^{kt} - \dfrac{N}{k}$ **35.** $A = \dfrac{P}{r}(e^{rt} - 1)$

37. \$7,305,295.15   **39.** (a) $C = C_0 e^{-Rt/V}$ (b) 0

41. (a) $C(t) = \dfrac{Q}{R}(1 - e^{-Rt/V})$ (b) $\dfrac{Q}{R}$

Answers to Try Its

CHAPTER 0

SECTION 0.1

Try It 1 $x < 5$ or $(-\infty, 5)$

Try It 2 $x < -2$ or $x > 5$; $(-\infty, -2) \cup (5, \infty)$

Try It 3 $200 \le x \le 400$; so the daily production levels during the month varied between a low of 200 units and a high of 400 units.

SECTION 0.2

Try It 1 $8; 8; -8$

Try It 2 $2 \le x \le 10$

Try It 3 $\$4027.50 \le C \le \$11,635$

SECTION 0.3

Try It 1 $\frac{4}{9}$

Try It 2 8

Try It 3   (a) $3x^6$ (b) $8x^{7/2}$ (c) $4x^{4/3}$

Try It 4   (a) $x(x^2 - 2)$ (b) $2x^{1/2}(1 + 4x)$

Try It 5 $\dfrac{(3x - 1)^{3/2}(13x - 2)}{(x + 2)^{1/2}}$

Try It 6 $\dfrac{10(x - 4)}{(x + 1)^3}$

Try It 7   (a) $[2, \infty)$ (b) $(2, \infty)$ (c) $(-\infty, \infty)$

SECTION 0.4

Try It 1   (a) $\dfrac{-2 \pm \sqrt{2}}{2}$ (b) 4 (c) No real zeros

Try It 2   (a) $x = -3$ and $x = 5$ (b) $x = -1$
(c) $x = \frac{3}{2}$ and $x = 2$

Try It 3 $(-\infty, -2] \cup [1, \infty)$

Try It 4 $-1, \frac{1}{2}, 2$

SECTION 0.5

Try It 1   (a) $\dfrac{x^2 + 2}{x}$ (b) $\dfrac{3x + 1}{(x + 1)(2x + 1)}$

Try It 2   (a) $\dfrac{3x + 4}{(x + 2)(x - 2)}$ (b) $-\dfrac{x + 1}{3x(x + 2)}$

Try It 3

(a) $\dfrac{(A + B + C)x^2 + (A + 3B)x + (-2A + 2B - C)}{(x + 1)(x - 1)(x + 2)}$

(b) $\dfrac{(A + C)x^2 + (-A + B + 2C)x + (-2A - 2B + C)}{(x + 1)^2(x - 2)}$

Try It 4   (a) $\dfrac{3x + 8}{4(x + 2)^{3/2}}$ (b) $\dfrac{1}{\sqrt{x^2 + 4}}$

Try It 5 $\dfrac{\sqrt{x^2 + 4}}{x^2}$

Try It 6   (a) $\dfrac{5\sqrt{2}}{4}$ (b) $\dfrac{x + 2}{4\sqrt{x + 2}}$ (c) $\dfrac{\sqrt{6} + \sqrt{3}}{3}$
(d) $\dfrac{\sqrt{x + 2} - \sqrt{x}}{2}$

CHAPTER 1

SECTION 1.1

Try It 1

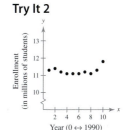

Try It 2

Try It 3 5

Try It 4 $d_1 = \sqrt{20}, d_2 = \sqrt{45}, d_3 = \sqrt{65}$
$d_1^2 + d_2^2 = 20 + 45 = 65 = d_3^2$

Try It 5   25 yards

Try It 6 $(-2, 5)$

Try It 7 $\$4.56$ billion

Try It 8 $(-1, -4), (1, -2), (1, 2), (-1, 0)$

SECTION 1.2

Try It 1

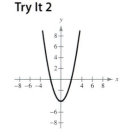

Try It 2

Try It 3   (a) *x*-intercepts: $(3, 0), (-1, 0)$

y-intercept: $(0, -3)$

(b) *x*-intercept: $(-4, 0)$

y-intercepts: $(0, 2), (0, -2)$

Try It 4 $(x + 2)^2 + (y - 1)^2 = 25$

Try It 5 $(x - 2)^2 + (y + 1)^2 = 4$

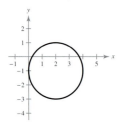

Try It 6   12,500 units

Try It 7   4 million units at \$122/unit

Try It 8   The projection obtained from the model is \$7962.02 million, which is close to the *Value Line* projection.

SECTION 1.3

Try It 1

(a)   (b)

Try It 2   Yes, $\frac{27}{312} \approx 0.08654 > \frac{1}{12} = 0.08\overline{3}$.

Try It 3   The *y*-intercept $(0, 875)$ tells you that the original value of the copier is \$875. The slope of $m = -175$ tells you that the value decreases \$175/year.

Try It 4   (a) 2   (b) $-\frac{1}{2}$

Try It 5 $y = 2x + 4$

Try It 6 $S = 1.72t + 16.49$; \$19.93

Try It 7   (a) $y = \frac{1}{2}x$ (b) $y = -2x + 5$

Try It 8 $V = -1375t + 12,000$

SECTION 1.4

Try It 1   (a) Yes, $y = x - 1$.   (b) No, $y = \pm\sqrt{4 - x^2}$.

(c) No, $y = \pm\sqrt{2 - x}$.   (d) Yes, $y = x^2$.

Try It 2   (a) Domain: $[-1, \infty)$ (b) Domain: $(-\infty, \infty)$

Range: $[0, \infty)$ Range: $[0, \infty)$

Try It 3 $f(0) = 1, f(1) = -3, f(4) = -3$

No, *f* is not one-to-one.

Try It 4   (a) $x^2 + 2x\,\Delta x + (\Delta x)^2 - 2x - 2\,\Delta x + 3$

(b) $2x + \Delta x - 2, \Delta x \neq 0$

Try It 5   (a) $2x^2 + 5$ (b) $4x^2 + 4x + 3$

Try It 6   (a) $f^{-1}(x) = 5x$ (b) $f^{-1}(x) = \frac{1}{3}(x - 2)$

Try It 7 $f^{-1}(x) = \sqrt{x - 2}$

Try It 8 $$f(x) = x^2 + 4$$
$$y = x^2 + 4$$
$$x = y^2 + 4$$
$$x - 4 = y^2$$
$$\pm\sqrt{x - 4} = y$$

SECTION 1.5

Try It 1   6

Try It 2   (a) 4   (b) Does not exist   (c) 4

Try It 3   5

Try It 4   12

Try It 5   7

Try It 6 $\frac{1}{4}$

Try It 7   (a) -1 (b) 1

Try It 8   1

Try It 9 $\lim\limits_{x \to 1^-} f(x) = 8$ and $\lim\limits_{x \to 1^+} f(x) = 10$

$\lim\limits_{x \to 1^-} f(x) \neq \lim\limits_{x \to 1^+} f(x)$

Try It 10   Does not exist

SECTION 1.6

Try It 1   (a) *f* is continuous on the entire real line.

(b) *f* is continuous on the entire real line.

Try It 2   (a) *f* is continuous on $(-\infty, 1)$ and $(1, \infty)$.

(b) *f* is continuous on $(-\infty, 2)$ and $(2, \infty)$.

(c) *f* is continuous on the entire real line.

Try It 3   *f* is continuous on $[2, \infty)$.

Try It 4   *f* is continuous on $[-1, 5]$.

Try It 5

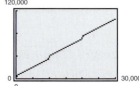

Try It 6 $A = 10,000(1 + 0.02)^{[4t]}$

CHAPTER 2

SECTION 2.1

Try It 1   3

Try It 2   For the months on the graph to the left of July, the tangent lines have positive slopes. For the months to the right of July, the tangent lines have negative slopes. The average daily temperature is increasing prior to July and decreasing after July.

Try It 3   4

Try It 4   2

Try It 5 $m = 8x$

At $(0, 1)$, $m = 0$.

At $(1, 5)$, $m = 8$.

Try It 6 $2x - 5$

Try It 7 $-\dfrac{4}{t^2}$

SECTION 2.2

Try It 1   (a) 0   (b) 0   (c) 0   (d) 0

Try It 2   (a) $4x^3$ (b) $-\dfrac{3}{x^4}$ (c) $2w$ (d) $-\dfrac{1}{t^2}$

Try It 3 $f'(x) = 3x^2$

$m = f'(-1) = 3$;

$m = f^{-1}(0) = 0$;

$m = f^{-1}(1) = 3$

Try It 4   (a) $8x$ (b) $\dfrac{8}{\sqrt{x}}$

Try It 5   (a) $\frac{1}{4}$ (b) $-\frac{2}{5}$

Try It 6   (a) $-\dfrac{9}{2x^3}$ (b) $-\dfrac{9}{8x^3}$

Try It 7   (a) $\dfrac{\sqrt{5}}{2\sqrt{x}}$ (b) $\dfrac{1}{3x^{2/3}}$

Try It 8 -1

Try It 9 $y = -x + 2$

Try It 10 $R'(8) \approx \$0.76/\text{year}$

SECTION 2.3

Try It 1   (a) $0.5\overline{6}$ milligrams per milliliter/minute

(b) 0 milligrams per milliliter/minute

(c) -1.5 milligrams per milliliter/minute

Try It 2   (a) -16 feet/second   (b) -48 feet/second

(c) -80 feet/second

Try It 3   When $t = 1.75$, $h'(1.75) = -56$ feet/second.

When $t = 2$, $h'(2) = -64$ feet/second.

Try It 4 $h = -16t^2 + 16t + 12$

$v = h' = -32t + 16$

Try It 5   When $x = 100$, $\dfrac{dP}{dx} = \$16/\text{unit}$.

Actual gain $= \$16.06$

Try It 6 $p = 11 - \dfrac{x}{2000}$

Try It 7   Revenue: $R = 2000x - 4x^2$

Marginal revenue: $\dfrac{dR}{dx} = 2000 - 8x$

Try It 8 $\dfrac{dP}{dx} = \$1.44/\text{unit}$

Actual increase in profit $\approx \$1.44$

SECTION 2.4

Try It 1 $-27x^2 + 12x + 24$

Try It 2 $\dfrac{2x^2 - 1}{x^2}$

Try It 3   (a) $18x^2 + 30x$ (b) $12x + 15$

Try It 4 $-\dfrac{22}{(5x - 2)^2}$

Try It 5 $y = \frac{8}{25}x - \frac{4}{5}$;

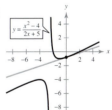

Try It 6 $\dfrac{-3x^2 + 4x + 8}{x^2(x + 4)^2}$

Try It 7   (a) $\frac{2}{5}x + \frac{4}{5}$ (b) $3x^3$

Try It 8 $\dfrac{2x^2 - 4x}{(x - 1)^2}$

Try It 9

t	0	1	2	3	4	5	6	7
$\dfrac{dP}{dt}$	0	-50	-16	-6	-2.77	-1.48	-0.88	-0.56

As t increases, the rate at which the blood pressure drops decreases.

SECTION 2.5

Try It 1   (a) $u = g(x) = x + 1$

$$y = f(u) = \frac{1}{\sqrt{u}}$$

(b) $u = g(x) = x^2 + 2x + 5$

$$y = f(u) = u^3$$

Try It 2 $6x^2(x^3 + 1)$

Try It 3 $4(2x + 3)(x^2 + 3x)^3$

Try It 4 $y = \frac{1}{3}x + \frac{8}{3}$

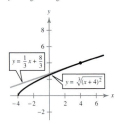

Try It 5   (a) $-\dfrac{8}{(2x + 1)^2}$ (b) $-\dfrac{6}{(x - 1)^4}$

Try It 6 $\dfrac{x(3x^2 + 2)}{\sqrt{x^2 + 1}}$

Try It 7 $-\dfrac{12(x + 1)}{(x - 5)^3}$

Try It 8   About $2.65/year

SECTION 2.6

Try It 1 $f'(x) = 18x^2 - 4x,\ f''(x) = 36x - 4,$
$f'''(x) = 36,\ f^{(4)}(x) = 0$

Try It 2   18

Try It 3 $\dfrac{120}{x^6}$

Try It 4 $s(t) = -16t^2 + 64t + 80$
$v(t) = s'(t) = -32t + 64$
$a(t) = v'(t) = s''(t) = -32$

Try It 5 -9.8 meters/seconds squared

Try It 6

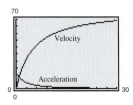

Acceleration approaches zero.

SECTION 2.7

Try It 1 $-\dfrac{2}{x^3}$

Try It 2   (a) $12x^2$ (b) $6y\dfrac{dy}{dx}$ (c) $1 + 5\dfrac{dy}{dx}$

(d) $y^3 + 3xy^2\dfrac{dy}{dx}$

Try It 3 $\frac{3}{4}$

Try It 4 $\dfrac{dy}{dx} = \dfrac{x + 2}{-y + 1}$

Try It 5 $\frac{5}{9}$

Try It 6 $\dfrac{dx}{dp} = -\dfrac{1000}{xp^2}$

SECTION 2.8

Try It 1   9

Try It 2 $12\pi \approx 37.7$ squared feet/second

Try It 3 $72\pi \approx 226.2$ squared inches/minute

Try It 4 $1500/day

Try It 5 $28,400/week

CHAPTER 3

SECTION 3.1

Try It 1 $f'(x) = 4x^3$

$f'(x) < 0$ if $x < 0$; therefore, f is decreasing on $(-\infty, 0)$.

$f'(x) > 0$ if $x > 0$; therefore, f is increasing on $(0, \infty)$.

Try It 2 $\dfrac{dW}{dt} = 0.198t + 0.17 > 0$ if $t > 0$, which implies that the consumption of bottled water was increasing from 1990 through 1999.

Try It 3   Increasing on $(-\infty, -2)$ and $(2, \infty)$
Decreasing on $(-2, 2)$

Try It 4   Increasing on $(0, \infty)$
Decreasing on $(-\infty, 0)$

Try It 5   Because $f'(x) = -3x^2 = 0$ when $x = 0$ and because f is decreasing on $(-\infty, 0) \cup (0, \infty)$, f is decreasing on $(-\infty, \infty)$.

Try It 6 $(0, 3000)$

SECTION 3.2

Try It 1   Relative maximum at $(-1, 5)$

Relative minimum at $(1, -3)$

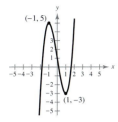

Try It 2   Relative minimum at $(3, -27)$

Try It 3   Relative maximum at $(1, 1)$

Relative minimum at $(0, 0)$

Try It 4   Absolute maximum at $(0, 10)$

Absolute minimum at $(4, -6)$

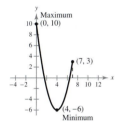

Try It 5

x (units)	24,000	24,200	24,300	24,400
P (profit)	$24,760	$24,766	$24,767.50	$24,768

x (units)	24,500	24,600	24,800	25,000
P (profit)	$24,767.50	$24,766	$24,760	$24,750

SECTION 3.3

Try It 1   (a) $f'' = -4$; because $f''(x) < 0$ for all x, f is concave downward for all x.

(b) $f''(x) = \dfrac{1}{2x^{3/2}}$; because $f''(x) > 0$ for all $x > 0$, f is concave upward for all $x > 0$.

Try It 2   Because $f''(x) > 0$ for $x < -2$ and $x > 2$, f is concave upward on $(-\infty, -2)$ and $(2, \infty)$.

Because $f''(x) < 0$ for $-2 < x < 2$, f is concave downward on $(-2, 2)$.

Try It 3 f is concave upward on $(-\infty, 0)$ and $(1, \infty)$.

f is concave downward on $(0, 1)$.

Points of inflection: $(0, 1)$, $(1, 0)$

Try It 4   Relative minimum: $(3, -26)$

Try It 5   Point of diminishing returns: $x = \$150$ thousand

SECTION 3.4

Try It 1 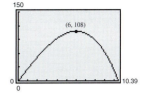

Maximum volume $= 108$ cubic inches

Try It 2 $x = 6, y = 12$

Try It 3 $\left(\sqrt{\tfrac{1}{2}}, \tfrac{7}{2} \right)$ and $\left(-\sqrt{\tfrac{1}{2}}, \tfrac{7}{2} \right)$

Try It 4   8 inches by 12 inches

SECTION 3.5

Try It 1   125 units yield a maximum revenue of $1,562,500.

Try It 2   400 units

Try It 3 $6.25/unit

Try It 4 $4.00

Try It 5   Demand is elastic when $0 < x < 1156/9$.

Demand is inelastic when $1156/9 < x < 289$.

Demand is of unit elasticity when $x = 1156/9$.

SECTION 3.6

Try It 1   (a) $\displaystyle \lim_{x \to 2^-} \frac{1}{x-2} = -\infty; \ \lim_{x \to 2^+} \frac{1}{x-2} = \infty$

(b) $\displaystyle \lim_{x \to -3^-} \frac{-1}{x+3} = \infty; \ \lim_{x \to -3^+} \frac{-1}{x+3} = -\infty$

Try It 2 $x = 0, x = 4$

Try It 3 $x = 3$

Try It 4 $\displaystyle \lim_{x \to 2^-} \frac{x^2 - 4x}{x-2} = \infty; \ \lim_{x \to 2^+} \frac{x^2 - 4x}{x-2} = -\infty$

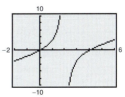

Try It 5   2

Try It 6   (a) $y = 0$ (b) $y = \tfrac{1}{2}$

(c) No horizontal asymptote

Try It 7 $C = 0.75x + 25,000$

$$\overline{C} = 0.75 + \frac{25,000}{x}$$

$$\lim_{x \to \infty} \overline{C} = \$0.75/\text{unit}$$

Try It 8 No, the cost function is not defined at $p = 100$, which implies that it is not possible to remove 100% of the pollutants.

SECTION 3.7

Try It 1

	$f(x)$	$f'(x)$	$f''(x)$	Shape of graph
x in $(-\infty, -1)$		$-$	$+$	Decreasing, concave upward
$x = -1$	-32	0	$+$	Relative minimum
x in $(-1, 1)$		$+$	$+$	Increasing, concave upward
$x = 1$	-16	$+$	0	Point of inflection
x in $(1, 3)$		$+$	$-$	Increasing, concave downward
$x = 3$	0	0	$-$	Relative maximum
x in $(3, \infty)$		$-$	$-$	Decreasing, concave downward

Try It 2

	$f(x)$	$f'(x)$	$f''(x)$	Shape of graph
x in $(-\infty, 0)$		$-$	$+$	Decreasing, concave upward
$x = 0$	5	0	0	Point of inflection
x in $(0, 2)$		$-$	$-$	Decreasing, concave downward
$x = 2$	-11	$-$	0	Point of inflection
x in $(2, 3)$		$-$	$+$	Decreasing, concave upward
$x = 3$	-22	0	$+$	Relative minimum
x in $(3, \infty)$		$+$	$+$	Increasing, concave upward

Try It 3

	$f(x)$	$f'(x)$	$f''(x)$	Shape of graph
x in $(-\infty, 0)$		$+$	$-$	Increasing, concave downward
$x = 0$	0	0	$-$	Relative maximum
x in $(0, 1)$		$-$	$-$	Decreasing, concave downward
$x = 1$	Undef.	Undef.	Undef.	Vertical asymptote
x in $(1, 2)$		$-$	$+$	Decreasing, concave upward
$x = 2$	4	0	$+$	Relative minimum
x in $(2, \infty)$		$+$	$+$	Increasing, concave upward

Try It 4

	$f(x)$	$f'(x)$	$f''(x)$	Shape of graph
x in $(-\infty, -1)$		$+$	$+$	Increasing, concave upward
$x = -1$	Undef.	Undef.	Undef.	Vertical asymptote
x in $(-1, 0)$		$+$	$-$	Increasing, concave downward
$x = 0$	-1	0	$-$	Relative maximum
x in $(0, 1)$		$-$	$-$	Decreasing, concave downward
$x = 1$	Undef.	Undef.	Undef.	Vertical asymptote
x in $(1, \infty)$		$-$	$+$	Decreasing, concave upward

Try It 5

	$f(x)$	$f'(x)$	$f''(x)$	Shape of graph
x in $(0, 1)$		$-$	$+$	Decreasing, concave upward
$x = 1$	-4	0	$+$	Relative minimum
x in $(1, \infty)$		$+$	$+$	Increasing, concave upward

SECTION 3.8

Try It 1 $dy = 0.32$; $\Delta y = 0.32240801$

Try It 2 $dR = \$5$; $\Delta R = \$4.94$

Try It 3 $dP = \$10.96$; $\Delta P = \$10.98$

Try It 4   (a) $dy = 12x^2\, dx$ (b) $dy = \frac{2}{3}\, dx$

(c) $dy = (6x - 2)\, dx$ (d) $dy = -\dfrac{2}{x^3}\, dx$

Try It 5 $S = 1.96\pi$ squared inches ≈ 6.1575 squared inches

$dS = \pm 0.056\pi$ squared inches $\approx \pm 0.1759$ squared inches

CHAPTER 4

SECTION 4.1

Try It 1   (a) 243   (b) 3   (c) 64

(d) 8   (e) $\frac{1}{2}$ (f) $\sqrt{10}$

Try It 2   (a) 5.453×10^{-13} (b) 1.621×10^{-13}

(c) 2.629×10^{-14}

Try It 3

x	-3	-2	-1	0	1	2	3
$f(x)$	$\frac{1}{125}$	$\frac{1}{25}$	$\frac{1}{5}$	1	5	25	125

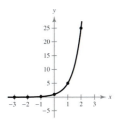

Try It 4

x	-3	-2	-1	0	1	2	3
$f(x)$	9	5	3	2	$\frac{3}{2}$	$\frac{5}{4}$	$\frac{9}{8}$

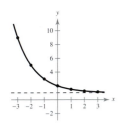

Horizontal asymptote: $y = 1$

SECTION 4.2

Try It 1

x	-2	-1	0	1	2
$f(x)$	$e^2 \approx 7.389$	$e \approx 2.718$	1	$\dfrac{1}{e} \approx 0.368$	$\dfrac{1}{e^2} \approx 0.135$

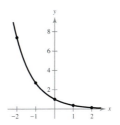

Try It 2   After 0 hours, $y = 1.25$ grams.

After 1 hour, $y \approx 1.338$ grams.

After 10 hours, $y \approx 1.498$ grams.

$\displaystyle \lim_{t \to \infty} \frac{1.50}{1 + 0.2e^{-0.5t}} = 1.50$ grams.

Try It 3   (a) \$4870.38   (b) \$4902.71

(c) \$4918.66   (d) \$4919.21

All else being equal, the more often interest is compounded, the greater the balance.

Try It 4   (a) 7.12%   (b) 7.25%

Try It 5   \$16,712.90

SECTION 4.3

Try It 1   At $(0, 1)$, $y = x + 1$.

At $(1, e)$, $y = ex$.

Try It 2   (a) $3e^{3x}$ (b) $-\dfrac{6x^2}{e^{2x^3}}$ (c) $8xe^{x^2}$ (d) $-\dfrac{2}{e^{2x}}$

Try It 3   (a) $xe^x(x + 2)$ (b) $\frac{1}{2}(e^x - e^{-x})$

(c) $\dfrac{e^x(x - 2)}{x^3}$ (d) $e^x(x^2 + 2x - 1)$

Try It 4

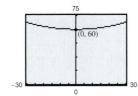

Try It 5   \$18.39/unit (80,000 units)

Try It 6

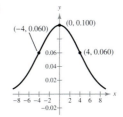

Points of inflection: $(-4, 0.060)$, $(4, 0.060)$

SECTION 4.4

Try It 1

x	-1.5	-1	-0.5	0	0.5	1
$f(x)$	-0.693	0	0.405	0.693	0.916	1.099

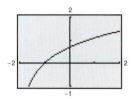

Try It 2 (a) 3 (b) $x + 1$

Try It 3 (a) $\ln 2 - \ln 5$ (b) $\frac{1}{3}\ln(x + 2)$

(c) $\ln x - \ln 5 - \ln y$ (d) $\ln x + 2\ln(x + 1)$

Try It 4 (a) $\ln x^4 y^3$ (b) $\ln\dfrac{x + 1}{(x + 3)^2}$

Try It 5 (a) $\ln 6$ (b) $5\ln 5$

Try It 6 (a) e^4 (b) e^3

Try It 7 7.9 years

SECTION 4.5

Try It 1 $\dfrac{1}{x}$

Try It 2 (a) $\dfrac{2x}{x^2 - 4}$ (b) $x(1 + 2\ln x)$ (c) $\dfrac{2\ln x - 1}{x^3}$

Try It 3 $\dfrac{1}{3(x + 1)}$

Try It 4 $\dfrac{2}{x} + \dfrac{x}{x^2 + 1}$

Try It 5 Relative minimum: $(2, 2 - 2\ln 2) \approx (2, 0.6137)$

Try It 6 $\dfrac{dp}{dt} = -1.3\%$/month

The average score would decrease at a greater rate than the model in Example 6.

Try It 7 (a) 4 (b) -2 (c) -5 (d) 3

Try It 8 (a) 2.322 (b) 2.631 (c) 3.161 (d) -0.5

Try It 9

As time increases, the derivative approaches 0. The rate of change of the amount of carbon isotopes is proportional to the amount present.

SECTION 4.6

Try It 1 About 2113.7 years

Try It 2 $y = 25e^{0.6931t}$

Try It 3 $r = \frac{1}{8}\ln 2 \approx 0.0866$ or 8.66%

Try It 4 About 12.42 months

CHAPTER 5

SECTION 5.1

Try It 1 (a) $\displaystyle\int 3\, dx = 3x + C$ (b) $\displaystyle\int 2x\, dx = x^2 + C$

(c) $\displaystyle\int 9t^2\, dt = 3t^3 + C$

Try It 2 (a) $5x + C$ (b) $-r + C$ (c) $2t + C$

Try It 3 $\frac{5}{2}x^2 + C$

Try It 4 (a) $-\dfrac{1}{x} + C$ (b) $\dfrac{3}{4}x^{4/3} + C$

Try It 5 (a) $\frac{1}{2}x^2 + 4x + C$ (b) $x^4 - \frac{5}{2}x^2 + 2x + C$

Try It 6 $\frac{2}{3}x^{3/2} + 4x^{1/2} + C$

Try It 7 General solution: $F(x) = 2x^2 + 2x + C$

Particular solution: $F(x) = 2x^2 + 2x + 4$

Try It 8 $s(t) = -16t^2 + 32t + 48$. The ball hits the ground 3 seconds after it is thrown, with a velocity of -64 feet per second.

Try It 9 $C = -0.01x^2 + 28x + 12.01$

$C(200) = \$5212.01$

SECTION 5.2

Try It 1 (a) $\dfrac{(x^3 + 6x)^3}{3} + C$ (b) $\frac{2}{3}(x^2 - 2)^{3/2} + C$

Try It 2 $\frac{1}{36}(3x^4 + 1)^3 + C$

Try It 3 $2x^9 + \frac{12}{5}x^5 + 2x + C$

Try It 4 $\frac{5}{3}(x^2 + 1)^{3/2} + C$

Try It 5 $-\frac{1}{3}(1 - 2x)^{3/2} + C$

Try It 6 Approximately $30,045

SECTION 5.3

Try It 1 (a) $3e^x + C$ (b) $e^{5x} + C$ (c) $e^x - \dfrac{x^2}{2} + C$

Try It 2 $\frac{1}{2}e^{2x+3} + C$

Try It 3 $2e^{x^2} + C$

Try It 4 (a) $2 \ln|x| + C$ (b) $\ln|x^3| + C$

(c) $\ln|2x + 1| + C$

Try It 5 $\frac{1}{4} \ln|4x + 1| + C$

Try It 6 $\frac{3}{2} \ln(x^2 + 4) + C$

Try It 7 (a) $4x - 3 \ln|x| - \dfrac{2}{x} + C$

(b) $2 \ln(1 + e^x) + C \, dx$

(c) $\dfrac{x^2}{2} + x + 3 \ln|x + 1| + C$

SECTION 5.4

Try It 1 $\frac{1}{2}(3)(12) = 18$

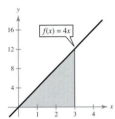

Try It 2 $\frac{22}{3}$ square units

Try It 3 68

Try It 4 (a) $\frac{1}{4}(e^4 - 1) \approx 13.3995$

(b) $-\ln 5 + \ln 2 \approx -0.9163$

Try It 5 $\frac{13}{2}$

Try It 6 (a) About $14.18   (b) $141.79

Try It 7 $13.70

Try It 8 (a) $\frac{2}{5}$ (b) 0

Try It 9 About $15,319.26

SECTION 5.5

Try It 1 $\frac{8}{3}$ square units

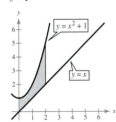

Try It 2 $\frac{32}{3}$ square units

Try It 3 $\frac{9}{2}$ square units

Try It 4 $\frac{253}{12}$ square units

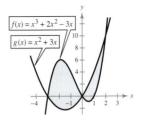

Try It 5 Consumer surplus: 40

Producer surplus: 20

Try It 6 The company can save $80.04 million.

SECTION 5.6

Try It 1 $\frac{37}{8}$ square units

Try It 2 0.436 square unit

Try It 3 5.642 square units

Try It 4 ≈ 1.463

SECTION 5.7

Try It 1 $\dfrac{512\pi}{15}$ cubic units

Try It 2 $\dfrac{832\pi}{15}$ cubic units

Try It 3 About 45 cubic inches

CHAPTER 6
SECTION 6.1

Try It 1 $\ln|x - 2| - \dfrac{2}{x - 2} + C$

Try It 2 $\frac{1}{3}(x^2 + 4)^{3/2} + C$

Try It 3 $\frac{1}{2} \ln|1 + e^{2x}| + C$

Try It 4 $\frac{2}{15}(x + 2)^{3/2}(3x - 4) + C$

Try It 5 $\frac{149}{30}$

Try It 6 50.4%

SECTION 6.2

Try It 1 $\frac{1}{2}xe^{2x} - \frac{1}{4}e^{2x} + C$

Try It 2 $\dfrac{x^2}{2} \ln x - \dfrac{1}{4}x^2 + C$

Try It 3 $\dfrac{d}{dx}[x \ln x - x + C] = x\left(\dfrac{1}{x}\right) + \ln x - 1$

$$= \ln x$$

Try It 4 $e^x(x^3 - 3x^2 + 6x - 6) + C$

Try It 5 $e - 2$

Try It 6 $538,145$

Try It 7 $528,482$

SECTION 6.3

Try It 1 $\dfrac{5}{x + 3} - \dfrac{4}{x + 4}$

Try It 2 $\ln|x(x + 2)^2| + \dfrac{1}{x + 2} + C$

Try It 3 $\dfrac{1}{2}x^2 - 2x - \dfrac{1}{x} + 4\ln|x + 1| + C$

Try It 4 $ky(1 - y) = \dfrac{kbe - kt}{(1 + be^{-kt})^2}$

$y = (1 + be^{-kt})^{-1}$

$\dfrac{dy}{dt} = \dfrac{kbe^{-kt}}{(1 + be^{-kt})^2}$

Therefore, $\dfrac{dy}{dt} = ky(1 - y)$

Try It 5 $y = 4$

Try It 6 $y = \dfrac{4000}{1 + 39e^{-0.31045t}}$

SECTION 6.4

Try It 1 $\frac{2}{3}(x - 4)\sqrt{2 + x} + C$ (Formula 19)

Try It 2 $\sqrt{x^2 + 16} - 4\ln\left|\dfrac{4 + \sqrt{x^2 + 16}}{x}\right| + C$

(Formula 23)

Try It 3 $\dfrac{1}{4}\ln\left|\dfrac{x - 2}{x + 2}\right| + C$ (Formula 29)

Try It 4 $\frac{1}{3}[1 - \ln(1 + e) + \ln 2] \approx 0.12663$ (Formula 37)

Try It 5 $x(\ln x)^2 + 2x - 2x\ln x + C$ (Formula 42)

Try It 6 $\dfrac{\sqrt{10}}{20}\ln\left|\dfrac{x - 3 - \sqrt{10}}{x - 3 + \sqrt{10}}\right| + C$ (Formula 29)

SECTION 6.5

Try It 1 3.2608

Try It 2 3.1956

Try It 3 1.154

SECTION 6.6

Try It 1 (a) Converges; $\frac{1}{2}$ (b) Diverges

Try It 2 1

Try It 3 $\frac{1}{2}$

Try It 4 2

Try It 5 Diverges

Try It 6 Diverges

Try It 7 0.0038 or $\approx 0.4\%$

Try It 8 (c) Converges, $\displaystyle\int_1^{\infty} \dfrac{1}{x^p}\,dx$ diverges for $0 < p \le 1$ and converges for $p > 1$.

CHAPTER 7

SECTION 7.1

Try It 1

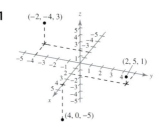

Try It 2 $2\sqrt{6}$

Try It 3 $\left(-\frac{5}{2}, 2, -2\right)$

Try It 4 $(x - 4)^2 + (y - 3)^2 + (z - 2)^2 = 25$

Try It 5 $(x - 1)^2 + (y - 3)^2 + (z - 2)^2 = 38$

Try It 6 Center: $(-3, 4, -1)$; radius: 6

Try It 7 $(x + 1)^2 + (y - 2)^2 = 16$

SECTION 7.2

Try It 1 x-intercept: $(4, 0, 0)$;
y-intercept: $(0, 2, 0)$;
z-intercept: $(0, 0, 8)$

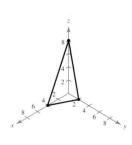

Try It 2 Hyperboloid of one sheet
xy-trace: circle, $x^2 + y^2 = 1$; yz-trace: hyperbola, $y^2 - z^2 = 1$; xz-trace: hyperbola, $x^2 - z^2 = 1$; $z = 3$ trace: circle, $x^2 + y^2 = 10$

Try It 3 (a) $\dfrac{x^2}{9} + \dfrac{y^2}{4} = z$; elliptic paraboloid

(b) $\dfrac{x^2}{4} + \dfrac{y^2}{9} - z^2 = 0$; elliptic cone

SECTION 7.3

Try It 1 (a) 0   (b) $\frac{9}{4}$

Try It 2 Domain: $x^2 + y^2 \leq 9$

Range: $0 \leq z \leq 3$

Try It 3 Steep; nearly level

Try It 4 Alaska is mainly used for forest land. Alaska does not contain any manufacturing centers, but it does contain a mineral deposit of petroleum.

Try It 5 $f(1500, 1000) \approx 127{,}542$ units

$f(1000, 1500) \approx 117{,}608$ units

x, person-hours, has a greater effect on production.

Try It 6 (a) $M = \$733.76$/month

(b) Total paid $= (30 \times 12) \times 733.76 = \$264{,}153.60$

SECTION 7.4

Try It 1 $\dfrac{\partial z}{\partial x} = 4x - 8xy^3$

$\dfrac{\partial z}{\partial y} = -12x^2y^2 + 4y^3$

Try It 2 $f_x(x, y) = 2xy^3;\ f_x(1, 2) = 16$

$f_y(x, y) = 3x^2y^2;\ f_y(1, 2) = 12$

Try It 3 In the x-direction: $f_x(1, -1, 49) = 8$

In the y-direction: $f_y(1, -1, 49) = -18$

Try It 4 Substitute product relationship

Try It 5 $\dfrac{\partial w}{\partial x} = xy + 2xy \ln(xz)$

$\dfrac{\partial w}{\partial y} = x^2 \ln xz$

$\dfrac{\partial w}{\partial z} = \dfrac{x^2 y}{z}$

Try It 6 $f_{xx} = 8y^2$

$f_{yy} = 8x^2 + 8$

$f_{xy} = 16xy$

$f_{yx} = 16xy$

Try It 7 $f_{xx} = 0 \qquad f_{xy} = e^y \qquad f_{xz} = 2$

$f_{yx} = e^y \qquad f_{yy} = xe^y + 2 \qquad f_{yz} = 0$

$f_{zx} = 2 \qquad f_{zy} = 0 \qquad f_{zz} = 0$

SECTION 7.5

Try It 1 $f(-8, 2) = -64$: relative minimum

Try It 2 $f(0, 0) = 1$: relative maximum

Try It 3 $f(0, 0) = 0$: saddle point

Try It 4 $P(3.11, 3.81) = \$744.81$ maximum profit

Try It 5 $V\left(\frac{4}{3}, \frac{2}{3}, \frac{8}{3}\right) = \frac{64}{27}$ cubic units

SECTION 7.6

Try It 1 $V\left(\frac{4}{3}, \frac{2}{3}, \frac{8}{3}\right) = \frac{64}{27}$ cubic units

Try It 2 $f(187.5, 50) \approx 13{,}474$ units

Try It 3 $\approx 26{,}740$ units

Try It 4 $f(x, y, z, w) = \frac{15}{2}$

Try It 5 $P(3.35, 4.26) = \$758.08$ maximum profit

Try It 6 $f(2, 0, 2) = 8$

SECTION 7.7

Try It 1 For $f(x)$, $S \approx 9.1$.

For $g(x)$, $S \approx 0.47515$.

The quadratic model is a better fit.

Try It 2 $f(x) = \frac{6}{5}x + \frac{23}{10}$

Try It 3 $y = 18{,}544.2t - 78{,}667$

In 2006, $y \approx 218{,}040{,}000$ subscribers

Try It 4 $y = 19.49t^2 - 106.1t + 3638$

In 2006, $y \approx 6930$ dollars.

SECTION 7.8

Try It 1 (a) $\frac{1}{4}x^4 + 2x^3 - 2x - \frac{1}{4}$ (b) $\ln|y^2 + y| - \ln|2y|$

Try It 2 $\frac{25}{2}$

Try It 3 $\displaystyle\int_2^4 \int_1^5 dx\, dy = 8$

Try It 4 $\frac{4}{3}$

Try It 5 (a)

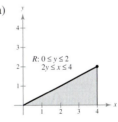

$R: 0 \leq y \leq 2$
$2y \leq x \leq 4$

(b) $\displaystyle\int_0^4 \int_0^{x/2} dy\, dx$ (c) $\displaystyle\int_0^2 \int_{2y}^4 dx\, dy = 4 = \int_0^4 \int_0^{x/2} dy\, dx$

Try It 6 $\displaystyle\int_{-1}^3 \int_{x^2}^{2x+3} dy\, dx = \frac{32}{3}$

SECTION 7.9

Try It 1 $\frac{16}{3}$

Try It 2 $e - 1$

Try It 3 $\frac{176}{15}$

Try It 4 Integration by parts

Try It 5 3

Try It 5 $\mu = 1$; $\sigma \approx 0.577$

Try It 6   0.61 or 61%

Try It 7   0.159

CHAPTER 10

SECTION 10.1

Try It 1   (a) $a_1 = 2, a_2 = 5, a_3 = 8, a_4 = 11$

(b) $a_1 = \frac{1}{2}, a_2 = \frac{2}{5}, a_3 = \frac{3}{10}, a_4 = \frac{4}{17}$

Try It 2   (a) 0   (b) 2

Try It 3   0

Try It 4

n	1	2	3	4	$\cdots$	n
$f^{(n-1)}(x)$	e^{2x}	$2e^{2x}$	$4e^{2x}$	$8e^{2x}$	$\cdots$	$2^{n-1}e^{2x}$
$f^{(n-1)}(0)$	1	2^1	2^2	2^3	$\cdots$	2^{n-1}

Diverges

Try It 5 $\dfrac{(-1)^{n+1}n^2}{(n+1)!}$

Try It 6 $A_n = 1000(1.015)^n$

SECTION 10.2

Try It 1 $\displaystyle\sum_{i=1}^{4} 4\left(-\frac{1}{2}\right)^n$

Try It 2 $\frac{1}{3}$

Try It 3   (a) 4   (b) $\frac{1}{2}$

Try It 4   (a) Diverges   (b) Diverges

Try It 5 $S_5 \approx 5.556, S_{50} = 5.\overline{5}, S_{500} = 5.\overline{5}$

Try It 6 $1021.17

Try It 7   (a) Converges to $\frac{5}{30}$ (b) Diverges

(c) Converges to $\frac{20}{3}$

Try It 8   About 40,000 units

Try It 9   140 feet

SECTION 10.3

Try It 1   (a) p-series with $p = \pi$ (b) p-series with $p = \frac{3}{2}$

(c) Geometric series

Try It 2   (a) Converges   (b) Converges   (c) Diverges

Try It 3   Converges to approximately 20.086

Try It 4   Diverges

Try It 5   Diverges

SECTION 10.4

Try It 1   (a) 2   (b) -3 (c) 0

Try It 2   Radius of convergence is infinite.

Try It 3 $R = 3$

Try It 4 $\displaystyle\sum_{n=0}^{\infty} \frac{(-x)^n}{n!}$; radius of convergence is infinite.

Try It 5 $\displaystyle\sum_{n=1}^{\infty} \frac{(-1)^{n+1}(x-1)^n}{n}$

Try It 6   (a) $\displaystyle\sum_{n=0}^{\infty} \frac{(2x)^n}{n!}$ (b) $\displaystyle\sum_{n=0}^{\infty} \frac{(-1)^n(2x)^n}{n!}$

Try It 7

$$(1+x)^{1/2} = 1 + \frac{x}{2} - \frac{x^2}{2^2 2!} + \frac{3x^3}{2^3 3!} - \frac{3 \cdot 5x^4}{2^4 4!} + \cdots$$

Try It 8   (a) $1 - \displaystyle\sum_{n=0}^{\infty} \frac{(-x)^n}{n!}$

(b) $\rho^2 \displaystyle\sum_{n=0}^{\infty} \frac{(-x)^n}{n!}$

(c) $\displaystyle\sum_{n=1}^{\infty} \frac{(-1)^{n-1}(4x-1)^n}{n}$

SECTION 10.5

Try It 1 $S_{12}(x) = 1 + 2x^3 + 2x^6 + \dfrac{4x^9}{3} + \dfrac{2x^{12}}{3}$

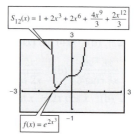

Try It 2 $e^{-0.5} \approx 0.607$ with a maximum error of 0.0003.

Try It 3   1.970

Try It 4   0.0005; 8 times

Try It 5   1

SECTION 10.6

Try It 1   1.735811

Try It 2 -0.453398

Try It 3   1.319074

Try It 4   0.567143

Index

APPLICATIONS *(continued from front endsheets)*